IB	IIB	IIIB	IVB	VB	VIB	VIIB	O
								2
			B 5 10.81 $2s^22p^1$ ~ 2103 —	C 6 12.011 $2s^22p^2$ (3836) —	N 7 14.0067 $2s^22p^3$ −210.0 —	15.9994 $2s^22p^4$ −218.8 —	18.9984 $2s^22p^5$ −219.7 —	20.179 $2s^22p^6$ −248.6 —
			Al 13 26.9815 $3s^23p^1$ 660.4 fcc	Si 14 28.086 $3s^23p^2$ 1414 dc	P 15 30.9738 $3s^23p^3$ 44.1 —	S 16 32.06 $3s^23p^4$ 112.8 —	Cl 17 35.453 $3s^23p^5$ −101.0 —	Ar 18 39.948 $3s^23p^6$ −189.3 —
28 8.71 d^84s^2 455 cc	Cu 29 63.546 $3d^{10}4s^1$ 1084.5 fcc	Zn 30 65.37 $4s^2$ 419.6 hcp	Ga 31 69.72 $4s^24p^1$ 29.8 —	Ge 32 72.59 $4s^24p^2$ 938.3 dc	As 33 74.9216 $4s^24p^3$ (603) —	Se 34 78.96 $4s^24p^4$ 221 —	Br 35 79.904 $4s^24p^5$ −7.2 —	Kr 36 83.80 $4s^24p^6$ −157.4 —
46 06.4 $d^{10}5s^0$ 554 cc	Ag 47 107.868 $4d^{10}5s^1$ 961.9 fcc	Cd 48 112.40 $5s^2$ 321.1 hcp	In 49 114.82 $5s^25p^1$ 156.6 —	Sn 50 118.69 $5s^25p^2$ 232.0 —	Sb 51 121.75 $5s^25p^3$ 630.7 —	Te 52 127.60 $5s^25p^4$ 449.6 —	I 53 126.9045 $5s^25p^5$ 113.6 —	Xe 54 131.30 $5s^25p^6$ −111.8 —
78 95.09 d^96s^1 772 cc	Au 79 196.9665 $5d^{10}6s^1$ 1064.4 fcc	Hg 80 200.59 $6s^2$ −38.9 —	Tl 81 204.37 $6s^26p^1$ 304 hcp (‡)	Pb 82 207.19 $6s^26p^2$ 327.5 fcc	Bi 83 208.9806 $6s^26p^3$ 271.4 —	Po 84 (210) $6s^26p^4$ 254 —	At 85 (210) $6s^26p^5$ — —	Rn 86 (222) $6s^26p^6$ −71 —

64 7.25 $^75d^16s^2$ 314 p (‡)	Tb 65 158.9254 $4f^96s^2$ 1359 hcp (‡)	Dy 66 160.50 $4f^{10}6s^2$ 1411 hcp (‡)	Ho 67 164.9303 $4f^{11}6s^2$ 1472 hcp (‡)	Er 68 167.26 $4f^{12}6s^2$ 1524 hcp (‡)	Tm 69 168.9342 $4f^{13}6s^2$ 1547 hcp	Yb 70 173.04 $4f^{14}6s^2$ 825 fcc (‡)	Lu 71 174.97 $4f^{14}5d^16s^2$ 1665 hcp (‡)
96 47) $^76d^17s^2$ 42 uble p (‡)	Bk 97 (247) $5f^97s^2$ 987 double hcp (‡)	Cf 98 (251) $5f^{10}7s^2$ — —	Es 99 (254) $5f^{11}7s^2$ — —	Fm 100 (257) $5f^{12}7s^2$ — —	Md 101 (256) $5f^{13}7s^2$ — —	No 102 (254) $5f^{14}7s^2$ — —	Lr 103 (257) $5f^{14}6d^17s^2$ — —

INTRODUCTION TO

MATERIALS SCIENCE AND ENGINEERING

Board of Advisors, Engineering

INTRODUCTION TO MATERIALS SCIENCE AND ENGINEERING

Kenneth M. Ralls
Department of Mechanical Engineering
The University of Texas at Austin

Thomas H. Courtney
Department of Metallurgical Engineering
Michigan Technological University

John Wulff
Department of Metallurgy and Materials Science
Massachusetts Institute of Technology

JOHN WILEY & SONS

New York • Chichester • Brisbane • Toronto

Library of Congress Cataloging in Publication Data:

Ralls, Kenneth M 1938-
Introduction to materials science and engineering.

1. Materials. I. Courtney Thomas H., joint author.
I. Courtney Thomas H., joint author. II. Wulff, John,
1903- joint author. III. Title.
TA403.R354 620.1'12 76-10813
ISBN 0-471-70665-5

Printed in the United States of America

10 9 8 7

PREFACE

This book is intended for students who want to learn about the nature of solid substances and, especially, for beginning engineering students who are making their first serious contact with the structure and properties of real solids. It presents, clearly and logically, the chemical and physical principles on which the properties of materials depend. The basic relationships introduced in general chemistry and physics courses are reviewed and extended in order to permit the student to relate the properties of ceramic, metallic, and polymeric solids to their internal structure and external environment.

To make this book more useful, we assumed that beginning students have only a limited knowledge of the elementary aspects of the chemistry and physics of solids. Therefore, each topic is developed so that it is almost complete in itself. Because of this, a science elective subject, based on this textbook, is appropriate for first-year students in science and engineering. Within a single semester we have found that the first two thirds of the book can be adequately covered, but a condensed version of the last third may be included for first-year students who have completed one semester (or quarter) of chemistry, physics, and calculus and who are studying electricity in a second-semester physics course. In second-year subjects, such as introductory materials science or materials engineering, where students have completed two semesters (or quarters) of college chemistry, physics, and calculus, the treatment of some topics in the first half of the book can be shortened in order to allow more detailed coverage of the second half of the book. Instructors may wish to emphasize certain topics at the expense of others because of the specific needs of their students. For instance, Chapters 19 to 21, which deal with mechanical behavior, will be covered in classes composed mainly of mechanical or civil engineering students. However, Chapters 22 to 27, which present electronic properties, will be of greater interest to classes that consist mainly of electrical engineering students. To facilitate this, we repeated basic ideas and relationships throughout and employed frequent cross-references.

We think that an early introduction to materials science (as a scientific rather than as a descriptive subject) in an engineering student's program will be of great educational value, since it will help him to bridge the intellectual gap between introductory chemistry and physics courses, and later engineering courses. With this in mind (and because of space limitations), we have not included detailed discussions of the processing, utilization, or selection of materials. Although we frequently discuss these topics in the subjects that we teach, which are based on this book, the inclusion of them here would interrupt the continuity and clarity of the presentation. Subjects dealing with materials processing, utilization, and selection are valuable in their own right, but the many factors involved in the proper selection of a material and its processing schedule only will be clear after the student has obtained a firm grasp of the basic principles underlying the behavior and properties of materials.

With respect to units, we express each quantity, when first introduced in the text, in terms of the recommended International System of Units (SI). This was done for pedagogic reasons. However, we also mention the equivalent traditional units. In many of the tables and graphs, traditional units are used in parallel with SI units and, for the discussion of mechanical properties in Chapters 19 to 21, we use English units alongside SI units for better communication. Conversion factors for particular units are listed on the end cover of the book.

We are grateful to the publications, individuals, and organizations from whom we borrowed data and illustrations. Appropriate sources are given in the text. We also thank our students and fellow instructors who gave us suggestions. Specifically, we thank Professor Kenneth C. Russell and James Coke of the Massachusetts Institute of Technology, Professor Robert W. Bene of the University of Texas at Austin, Warren G. Wonka, formerly of Stanford University, who read the manuscript and criticized it. Nevertheless, any errors that remain are ours.

Finally, we must thank Mary Lou Hibbs, Patricia Adams, and Patricia Teinert, for the many hours spent in typing the manuscript, and Robert J. Farrell, for preparing many photomicrographs.

K. M. Ralls
T. H. Courtney
J. Wulff

CONTENTS

16 PHASE DIAGRAMS 312

17 PHASE TRANSFORMATIONS 346

18 STRUCTURAL AND PROPERTY CHANGES 372

Matter and Energy

This chapter introduces the general concepts and ideas concerning matter and energy which will be employed throughout the text.

1.1 THE CONSTITUENTS OF MATTER

The ancient Greek idea, that the matter of the universe is all composed of atoms, has been firmly established by experimental evidence over the last century. The 90 naturally occurring elements that make up the earth are present in greatly

TABLE 1.1 Components of the Earth's Atmosphere

Component	*Atomic or Molecular Weight*	*Percent by Volume*
N_2	28.01	78.0
O_2	32.00	21.0
Ar	39.95	0.93
CO_2	44.01	0.03
Ne	20.18	1.8×10^{-3}
CH_4	16.04	7.4×10^{-4}
He	4.00	5.2×10^{-4}
CO	28.01	1×10^{-4}
H_2	2.02	5×10^{-5}
H_2O	18.02	Variable

differing amounts and are unevenly distributed with respect to the atmosphere, the hydrosphere, and the lithosphere. Only the inert gas elements, which constitute a very small fraction of the atmosphere (Table 1.1), are found to exist in the atomic state. The majority of elements in the atmosphere exist as molecules consisting of two or more like or unlike atoms (e.g., O_2, N_2 or CO_2). The hydrosphere or ocean is mainly water containing dissolved substances in varying amounts (Table 1.2). Most of these substances exist in the ionic or charged state, rather than the atomic or neutral state. The rocks, sands, and clays of the lithosphere or earth's crust are principally solid aggregates of compounds involving the elements listed in Table 1.3. All reasonably sized samples of the lithosphere are heterogeneous mixtures of various elements, ions, and compounds from which more uniform or homogeneous substances can be developed only by physical and chemical processing.

TABLE 1.2 Approximate Composition of Sea Water

Element	*Atomic Percent*
H	66.4
O	33
Cl	0.33
Na	0.28
Mg	0.034
S	0.017
Ca	6×10^{-3}
K	6×10^{-3}
C	1.4×10^{-3}
Br	5×10^{-4}

TABLE 1.3 Approximate Composition of the Earth's Crust

Element	*Atomic Percent*
O	60.4
Si	20.5
Al	6.2
Na	2.5
H	~2
Fe	1.9
Ca	1.9
Mg	1.8
K	1.4
Ti	0.3

1.2 THE STATES OF MATTER

The term phase is used to denote a structurally homogeneous part of matter. Although a pure gas or a mixture of gases possesses no internally regular atomic or molecular arrangement because of the random motion and large separation of its constituents, it is considered to be a single phase system identical to the gaseous state. Liquids, in one sense, resemble highly compressed gases. However, the interatomic and intermolecular forces responsible for the stability of a liquid between the melting and boiling temperatures promote a *local ordering* in the liquid state. Here the spatial relationship between an atom or molecule and its neighbors is regular at any instant, but continuously changes with time, and particle motion is vigorous enough to prevent *long-range ordering*. Unlike gases, liquids may exist as distinct phases that do not mix. A combination of oil and water or of mercury and water produces two phases separated by a boundary, whereas alcohol and water form a single phase because of complete mutual solubility; that is, they are intimately "mixed" on the scale of atomic dimensions.

Some solids such as crystalline quartz (SiO_2), galena (PbS), ice (H_2O), and a noncrystalline solid like window glass are single-phase substances. Others like granite, wood, and steel are multiphase substances; that is, they are aggregates of two or more phases that differ in chemical composition as well as in structure. Multiphase solids can be recognized by optical examination of their *microstructure* at sufficiently high magnifications or by their X-ray diffraction patterns. Figures 1.1 and 1.2 show the microstructures of a single-phase metal and a two-phase metal, respectively.

Although the atoms or molecules in a solid are not much more closely packed than in a liquid, they do have relatively fixed geometrical positions with respect to one another. Indeed, they often are arranged in an orderly fashion. If a geometrical array of ordered atoms or molecules persists in three dimensions over a spatial region large in comparison with the atom or molecule itself, the solid is considered crystalline. Alternatively, if the order is only short-range or local (comparable to

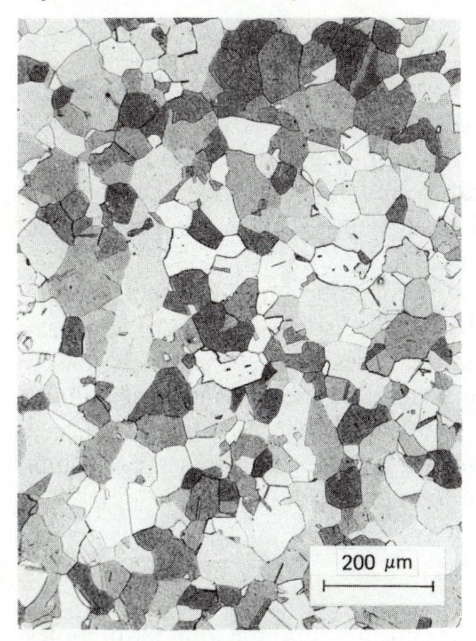

FIGURE 1.1 Microstructure of single-phase polycrystalline nickel. Grain boundaries that separate individual crystallites or grains have been revealed by etching. (Courtesy of D. J. Tillack, International Nickel Co.)

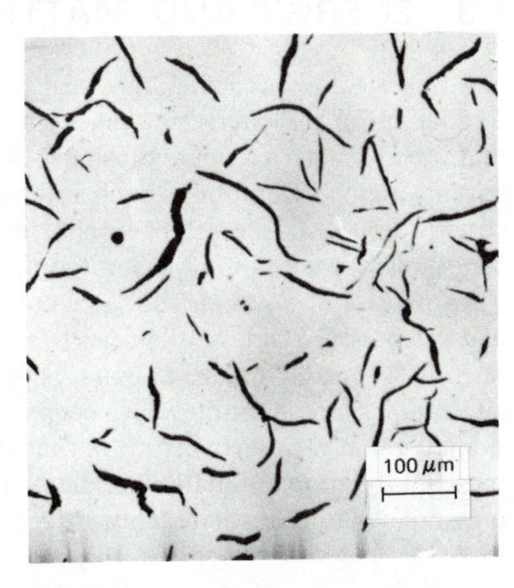

FIGURE 1.2 Microstructure of ferritic gray cast iron, showing two solid phases. The dark particles are graphite flakes and the white matrix is Fe containing a small amount of carbon in solid solution. (Reproduced by permission from *Metals Handbook*, Vol. 7, 8th Edition, American Society for Metals, 1972.)

that in a liquid), the solid is considered amorphous or noncrystalline. Metals are crystalline, whereas glasses and many polymeric substances are noncrystalline.

Single-phase crystalline substances may exist as single crystals or as polycrystalline aggregates. Most engineering materials are polycrystalline. Single crystals, however, are used extensively in electronic devices and for scientific research. In single crystals the order present extends from one external surface to the other. In polycrystalline materials, on the other hand, many small single crystals called *grains* are separated from one another by internal surfaces called grain boundaries. Crystalline order exists within each grain, but the orientation of the regular array of atoms or molecules changes when a grain boundary is crossed. Thus, each grain has a different orientation than its neighboring grains, and grain boundaries represent one type of structural discontinuity. An example of grain boundaries in a single-phase solid is shown in Fig. 1.1.

The short-range or long-range regularity with which atoms or molecules are packed in solids arises from geometrical conditions imposed by bonding and packing. Both of these influences are examined in greater detail in subsequent chapters.

1.3 ENERGY AND MATTER

All physical phenomena are revealed by the interaction of energy with matter. The nature of the interaction gives rise to the observed properties and behavior that are characteristic of a substance. In turn, properties are determined by externally imposed conditions, such as temperature and pressure, and by the internal structure of a substance. The behavior of a substance also is related to these conditions as well as to its macroscopic shape.

In the case of a gas at a suitable combination of sufficiently high temperature and low pressure, almost all properties can be specified solely in terms of temperature, pressure, and molecular mass. This situation arises because the widely spaced and rapidly moving molecules occupy such a small fraction of the total volume available that they virtually do not interact with each other. As a consequence, a particularly simple equation of state, termed the ideal gas law, serves to link pressure, volume, absolute temperature, and the number of moles:

$$PV = nRT \tag{1.1}$$

Here, $R(=8.314 \text{ J/K} \cdot \text{mol})$ is the universal gas constant, and the number of moles is given either by $n = m/W$ (for m grams of gas having molecular weight W) or by $n = N/N_0$ (where N is the number of molecules and $N_0 = 6.022 \times 10^{23}$ is Avogadro's number). According to Eq. 1.1, equal volumes of ideal gases at the same pressure and temperature must consist of the same number of moles. This remarkable generality, independent of the chemical makeup of the molecules, results because no significant interaction occurs between molecules.

A gas in contact with a solid or liquid exerts a pressure on the surface. The pressure is related directly to the average momentum change per unit time suffered by the gas particles upon collision with a unit area of solid or liquid surface. In a mixture of gases, the pressure due to one component is called the partial pressure of that component. When different ideal gases are mixed, the partial pressure of the *i*th component is given by

$$P_i = n_i \frac{RT}{V} \tag{1.2}$$

where n_i is the number of moles of the *i*th component. (The partial pressure is the pressure that a single component would exert in a fixed volume and at a given temperature if all other components were removed.) Necessarily, the total pressure of an ideal gas mixture is the sum of individual partial pressures, and such a mixture obeys Eq. 1.1 because the total number of moles is summed over the components.

For a monatomic gas, the average kinetic energy of all atoms is termed the internal energy, E. According to the kinetic theory of gases, this is given by

$$E = \tfrac{3}{2}RT \tag{1.3}$$

for a mole of ideal monatomic gas or by

$$E = \tfrac{3}{2}kT \tag{1.4}$$

per atom. In Eq. 1.4, $k(\equiv R/N_0 = 1.381 \times 10^{-23}$ J/K) is the Boltzmann constant. (The internal energy of a molecular gas also involves the average kinetic energy associated with rotation of molecules and, at high temperatures, that associated with vibrations within molecules.) The significance of Eqs. 1.3 and 1.4 is that temperature is related, on a microscopic basis, to the average kinetic energy of the gas particles. But temperature is a macroscopic phenomenon, and not all gas particles possess the same kinetic energy. Rather, there is a distribution of velocities and, hence, of kinetic energies at any given temperature, and these distributions change as temperature is changed (Fig. 1.3).

When heat is transferred to a substance constrained to constant volume, the internal energy increases by an amount exactly equal to the heat input. For example, the heat added to a monatomic gas increases the average kinetic energy of the atoms. The amount of heat necessary to raise the temperature of *any* substance at constant volume by one kelvin is a property called the heat capacity at constant volume (C_v). This is the same as the slope of the internal energy *versus*

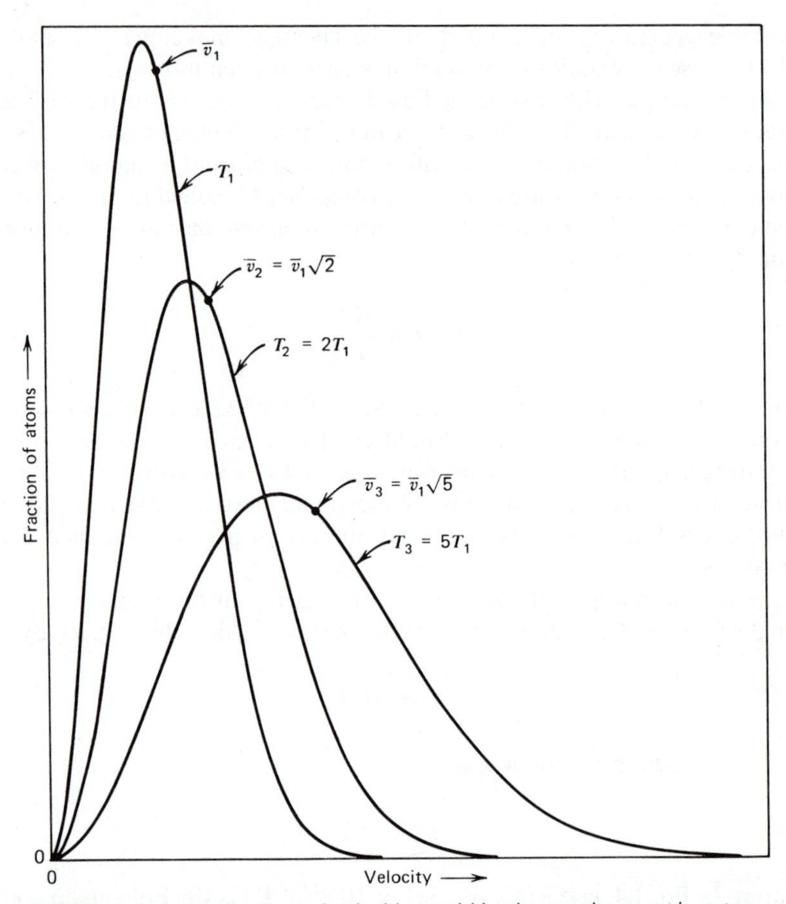

FIGURE 1.3a Distributions of velocities and kinetic energies at various temperatures for a monatomic ideal gas. The fraction of atoms within an incremental velocity range is plotted as a function of atom velocity, and the average velocity ($\overline{v}$) is indicated at each temperature. As temperature rises, the distribution broadens, the average velocity increases, and the fraction of high-velocity atoms increases greatly.

temperature curve when volume is maintained constant. Thus, $C_v = \frac{3}{2}R$ on a molar basis for an ideal monatomic gas (cf. Eq. 1.3). If any substance is subjected to constant pressure while heat is added, an energy quantity called enthalpy (H) increases by an amount exactly equal to the heat input. Enthalpy is defined by

$$H = E + PV \tag{1.5}$$

An enthalpy increase at constant pressure involves the effect of thermal expansion as well as the increase in internal energy. The amount of heat required to raise the

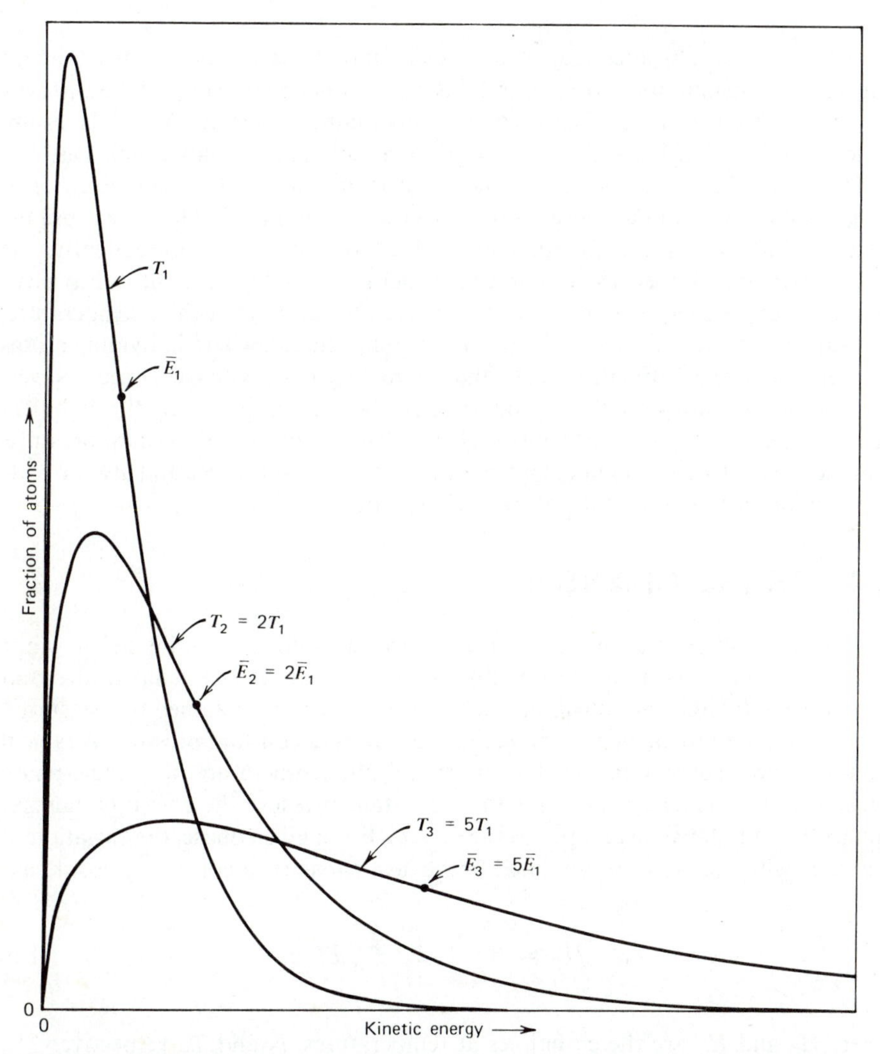

FIGURE 1.3b Distributions of velocities and kinetic energies at various temperatures for a monatomic ideal gas. The fraction of atoms within an incremental energy range is plotted as a function of atom energy. As temperature rises, the distribution broadens, the average energy ($\bar{E}$) increases, and the fraction of high-energy atoms increases even more. (The average kinetic energy for a monatomic gas, designated as $\bar{E}$ in this figure, is identical to the internal energy, designated simply as $\bar{E}$ in the text.)

temperature of a substance subjected to constant pressure by one kelvin is the heat capacity at constant pressure (C_p), and this corresponds to the slope of the enthalpy *versus* temperature curve when pressure is maintained constant. As can be established with Eqs. 1.1, 1.3, and 1.5, $C_p = \frac{5}{2}R$ for a mole of ideal monatomic gas.

When gases are compressed or cooled sufficiently, the ideal gas law (Eq. 1.1) no longer holds and internal energy is no longer given by Eq. 1.3. This is because the particles interact with each other as a result of collisions. Consequently, the thermodynamic properties (i.e., internal energy, enthalpy, and heat capacity), defined above, become more complex functions of variables such as temperature, pressure, or volume. This also is true for a solid or liquid, where individual atoms or molecules are in intimate contact and interact strongly with each other. Nevertheless, for any substance these properties can be measured directly through heat effects, depending upon constraints such as constant volume or constant pressure, and they relate to certain important characteristics of a substance and also provide useful information about the nature of a substance.

1.4 PHASE CHANGES

Practically all pure elements can exist in the crystalline solid, liquid, or vapor state, depending upon the externally imposed conditions of temperature and pressure. Useful information about such substances can be obtained by measuring the heat required to melt or vaporize the substances at constant pressure. A typical heating curve is shown in Fig. 1.4. In general, the temperature of a single-phase substance rises as heat is added, and at constant pressure the enthalpy increase equals the cumulative heat input to the system. For a given phase, the variation of enthalpy with temperature is related to the heat capacity at constant pressure as

$$H_2 = H_1 + \int_{T_1}^{T_2} C_p \, dT \tag{1.6}$$

where H_2 and H_1 are the enthalpies at temperatures T_2 and T_1, respectively. In Eq. 1.6, C_p is the reciprocal of the slope of either line A, C, or E in Fig. 1.4, depending on the phase of interest.

During a phase change of a pure substance (e.g., solid to liquid or liquid to vapor in Fig. 1.4), the heat added does not cause a temperature change. At the fusion or melting temperature T_m, a discrete amount of heat (ΔH_m) must be transferred to the substance before melting is complete. The quantity ΔH_m is called the enthalpy of fusion. Similarly, a discrete amount of heat (ΔH_v), called the enthalpy of vaporization, must be added at the vaporization or boiling temperature T_v before the liquid completely transforms to vapor. As shown in Table 1.4, the enthalpy of vaporization always is greater than the corresponding enthalpy of

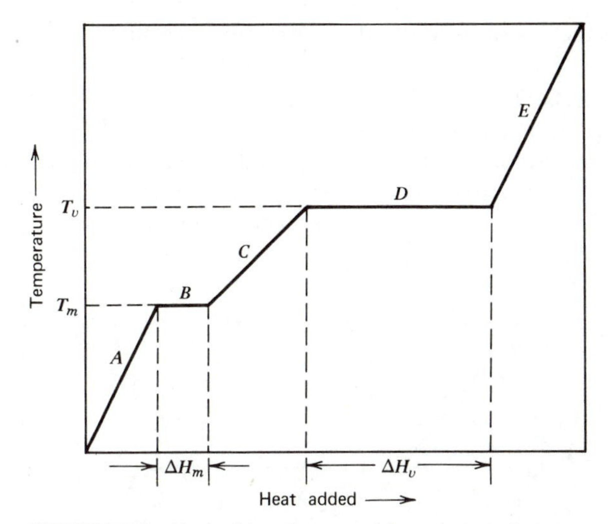

FIGURE 1.4 Typical heating curve for a pure substance at constant pressure. Lines labeled *A*, *C*, and *E* represent the temperature change with heat input for the crystalline solid, liquid, and vapor, respectively; lines labeled *B* and *D* show the *thermal arrests* associated with melting and vaporization.

fusion. This indicates that conversion of a solid to a liquid, with the resulting change in internal structure, requires less energy than does the conversion of a liquid to vapor.

The situation of cooling vapor to liquid and then to a crystalline solid involves the reverse of the processes described above. The enthalpy of each phase decreases as temperature is lowered, and condensation and crystallization are accompanied by a release of energy equal to $-\Delta H_v$ and $-\Delta H_m$, respectively. In other words,

TABLE 1.4 Some Molar Enthalpies of Fusion and Vaporization

Substance	ΔH_m *(kJ/mol)*	ΔH_v *(kJ/mol)*	*Crystal Type*
Ar	1.2	6.5	Atomic
H_2O	6.0	40.7	Molecular
C_6H_6 (benzene)	10.0	30.8	Molecular
Li	3.0	148	Metallic
Fe	15.2	350	Metallic
NaCl	28.7	171	Ionic
Ge	34.0	334	Covalent

the energy of the system is lowered successively by condensation and crystallization.

Not all materials show a marked and definite evolution of heat on solidification. Glass-forming elements like Se, glass-forming compounds like SiO_2 and many organic polymers, when cooled from the liquid, provide examples where the liquid portion of the cooling curve continues without discontinuity down to temperatures where such materials are solids. This indicates that the liquid structure has been

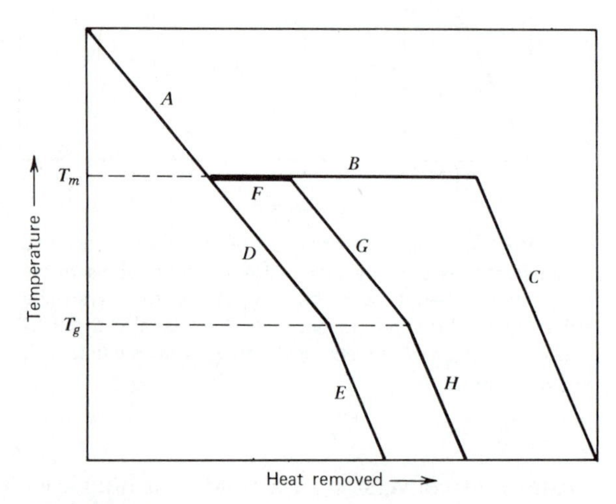

FIGURE 1.5 Schematic cooling curves for a substance that does not readily crystallize. The effect of cooling rate is indicated: (1) for very slow cooling, *A*, *B*, and *C* represent liquid, complete crystallization process, and solid, respectively; (2) for relatively rapid cooling, *A*, *D*, and *E* represent liquid, supercooled liquid, and glass, respectively; (3) for an intermediate cooling rate, *A*, *F*, *G*, and *H* represent liquid, partial crystallization process, crystallites in supercooled liquid matrix, and crystallites in glass matrix, respectively.

frozen in. Whether or not the liquid structure is retained in the solid state depends on chemical composition and cooling rate. At very slow cooling rates, atomic or molecular rearrangements necessary for the introduction of the long-range order of the crystalline state may have time to occur, and a definite evolution of heat is observed. At intermediate cooling rates, a two-phase noncrystalline and crystalline solid aggregate may result, and at high cooling rates a completely noncrystalline solid may occur. Cooling curves representing these cases are shown in Fig. 1.5.

1.5 STABILITY: ENTROPY AND FREE ENERGY

In materials, as in mechanical, systems it is universally true that the lower the energy the greater the stability of the systems. All systems, therefore, exhibit a tendency to lower their energy to achieve a more stable configuration. This statement alone is not sufficient to explain many well-known physical and chemical reactions in which an additional tendency for a system to disorder itself exists. At high temperatures, the more disordered yet higher energy phase is stable. At lower temperatures, the more ordered but lower energy phase is stable. Which of these two natural tendencies dominates, in a given phase change or chemical reaction, determines the direction in which a particular reaction takes place. To adequately discuss this matter it is necessary to extend our concepts of energy.

Entropy (S) is a measure of the randomness or degree of disorder of the internal arrangements of atoms or molecules in a single-phase materials system. To quantify S, we assume that the entropy of a pure perfect crystal is zero at 0 K, corresponding to a state of complete order. Entropy is a definite property of a system and, like enthalpy, the entropy or disorder of a system increases with temperature. The product of entropy and absolute temperature, TS, is called the entropy factor and has the units of energy. The difference between the enthalpy H of a system and its entropy factor is called the Gibbs free energy G:

$$G = H - TS \tag{1.7}$$

For a phase change or chemical reaction at constant temperature and pressure, the change in Gibbs free energy (ΔG) due to the reaction is given by

$$\Delta G = \Delta H - T\,\Delta S \tag{1.8}$$

If a reaction is to occur spontaneously, there must be a net decrease in the Gibbs free energy; that is, ΔG must be negative. Figure 1.6 shows the variation in G with temperature for the liquid and crystalline phases of water. Even though the enthalpy change is positive when ice transforms to water, the increase in the entropy factor, at sufficiently high temperatures (0°C), results in a phase change.

The negative of the free energy change, $-\Delta G$, often is called the driving force for a reaction. As shown in Fig. 1.6 ice will melt spontaneously above 0°C and supercooled water will freeze spontaneously below 0°C. In the former case the phase change results in higher enthalpy and entropy, while in the latter the phase change results in a lower enthalpy and entropy. The equilibrium melting temperature, T_m, is defined as that temperature at which $\Delta G = 0$, where the two phases coexist in equilibrium. At this temperature, the enthalpy change on melting ΔH_m is related uniquely to the entropy change ΔS_m by

$$\Delta H_m = T_m\,\Delta S_m \tag{1.9}$$

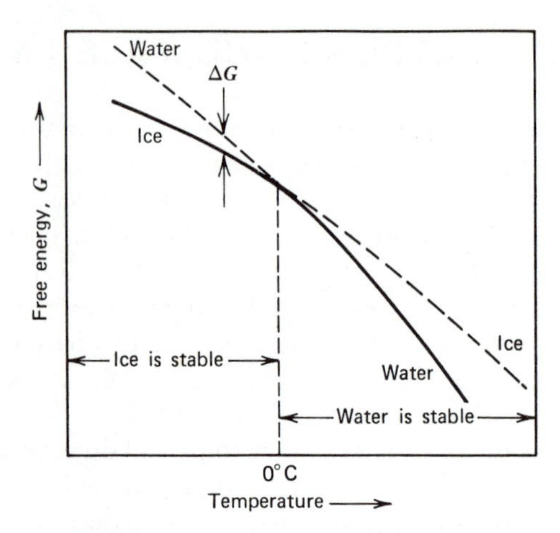

FIGURE 1.6 Free energy as a function of temperature for crystalline (ice) and liquid (water) forms of H_2O. Below 0°C, the equilibrium melting temperature, and at atmospheric pressure, the free energy of ice is lower than that of water; therefore, ice is the stable phase. Above 0°C water is the stable phase. At any temperature the difference in free energy, ΔG, is represented by the vertical separation between the curves; at the equilibrium melting point the free energies are equal and both ice and water are stable.

1.6 RATES OF CHANGE

All phase changes and chemical reactions require time to occur, and their rates are governed by factors other than a decrease in free energy. Some changes, which involve only minor geometrical or bonding rearrangements, may proceed at a relatively rapid rate. Others, such as the crystallization of a glass, require considerable time. In the latter case, fairly complex groupings of atoms must move around or diffuse before they can arrange themselves in a crystalline array, but such motion is restricted by dense packing. A higher temperature usually increases the rate of a reaction by increasing the rate of diffusion of atoms or molecules as well as by increasing the rate of breaking old bonds and forming new ones. Glass, for example, will crystallize over a period of time if it is heated to a temperature somewhat below its equilibrium melting point or crystallization temperature whereas, at lower temperatures, diffusion is so sluggish that the rate of crystallization is negligible.

Reaction rates are related to the state of a substance because this determines both atomic mobility and how free atomic motion is. Thus, under suitable conditions, a chemical reaction involving gases proceeds more rapidly than one involving liquids, and the latter is faster than a chemical reaction involving solids. An increase in temperature serves to increase atomic mobility by causing a larger fraction of atoms to have high energies, a feature illustrated in Fig. 1.3 for a monatomic ideal gas. With a gas, only the high-energy atoms or molecules possess sufficient kinetic energies to break and form bonds when collisions occur.

All processes that take a system from an initial state to a final state of lower free energy require that an energy barrier be surmounted. For example, when an atom is attached to the surface of a solid, it preferentially occupies sites of lowest energy, and for it to move from one favored site to another involves passing through a higher energy position. The simple mechanical analogue shown in Fig. 1.7 serves to illustrate an energy barrier. Shown is a rectangular block of square cross section that is initially standing on its square end. This position represents

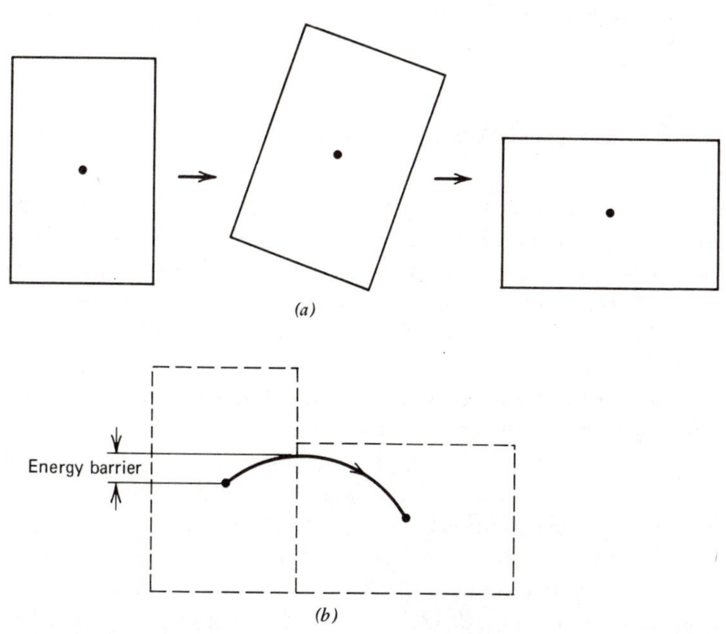

FIGURE 1.7 (*a*) When a rectangular block having a square cross-section is tilted about one edge, the center of mass (indicated by the dots) rises initially before decreasing to a lower value. (*b*) Since the height of the center of mass is directly proportional to the potential energy of the block, the tilting process used to reduce the potential energy involves passing through a maximum energy position (energy barrier).

a higher potential energy than if the block were lying on one of its rectangular faces. The smallest amount of work required to lower the potential energy of the block is that involved in tilting it about one of the bottom edges. Upon initial tilting, the potential energy is raised until its maximum is reached. This maximum is the energy barrier that must be surmounted, because further tilting results in a continuous decrease in potential energy.

The rate of a given process in a materials system depends on what fraction of atoms have sufficient energies to overcome the energy barrier associated with the process, and this is determined in part by the height of the energy barrier, commonly referred to as the activation energy, and other factors such as temperature. The usefulness of many materials comes about as a result of negligibly low rates of change. Thus, aluminum can persist indefinitely at room temperature in the metallic form, even though aluminum oxide is much more stable, because after a thin layer of oxide forms on its surface, the rate of further oxidation approaches zero.

Considerations of both equilibrium and rates of change are important for all substances, and they play significant roles in the preparation, alteration, and utilization of real materials.

1.7 THE IMPORTANCE OF REAL MATERIALS

Since mankind relies upon materials for many purposes, it is important to understand their properties, behavior, and limitations. Some materials are chosen for specific applications because they possess requisite characteristics. For example, copper is used commonly to conduct electricity because it has a high electrical conductivity. In addition to specific properties, cost, ease of fabrication, and mechanical strength are of almost universal importance. Furthermore, the availability of a natural resource and how difficult it is to convert this into a useful material are significant considerations.

Real materials are much more complex than ideal gases, because their characteristics depend upon many factors. To a large extent, the properties and behavior of a real solid are related directly to its structure, including bonding, atomic arrangements, phases, defects, and flaws. In turn, structure is determined largely by the chemical composition of a solid and by the processing it receives.

One can measure the strength, chemical durability, and other characteristics of a specific material through testing, but it is preferable to develop an understanding of the behavior of materials based on a combination of fundamental principles and empirical experience. Since energy, the structure of matter, and structural changes are of primary importance, the principles relating to these topics are presented in the remainder of this book.

REFERENCES

1. Chemical Bond Approach Project. *Chemical Systems*, McGraw-Hill, New York, 1964.
2. Mahan, B. H. *College Chemistry*, Addison-Wesley, Reading, Mass., 1969.
3. Hägg, G. *General and Inorganic Chemistry*, Wiley, New York, 1969.

QUESTIONS AND PROBLEMS

1.1 (a) Define a mole of a substance in terms of atomic or molecular weight.
(b) Define a mole in terms of Avogadro's number.
(c) How many moles are there in 100 g of solid CO_2?

1.2 (a) What is the volume of one mole of an ideal gas at 0°C and one atmosphere pressure?
(b) What effect would a temperature increase at constant pressure have on your answer to part (a)?
(c) What effect would a pressure increase at constant temperature have on your answer to part (a)?

1.3 (a) Show that the univeral gas constant R equals 0.0821 liter atm K^{-1} mol^{-1}.
(b) By using conversion factors for energy (or for pressure and volume), show that $R = 1.987$ cal K^{-1} mol^{-1} $= 8.314 \times 10^7$ erg K^{-1} mol^{-1} $= 8.314$ J K^{-1} mol^{-1}.

1.4 If minor components are ignored, the atmosphere consists essentially of 78% N_2, 21% O_2, and 1% Ar, all by volume. Assume that this is an ideal gas.
(a) What are the mole fractions of N_2, O_2, and Ar in the atmosphere?
(b) In a sample of atmosphere at $P = 1$ atm, what are the partial pressures of N_2, O_2, and Ar?

1.5 Magnesia (MgO) has a density of 3.58 g/cm^3. How many total atoms (Mg and O) are in 1 cm^3 of MgO?

1.6 Heat is extracted from a vessel containing one mole of water vapor at a rate of 1 cal/sec, pressure being maintained at 1 atm. Plot temperature versus time between 150°C and −10°C. Given: $C_p(g) = 8.6$ cal °C^{-1} mol^{-1}, $C_p(l) = 18.0$ cal °C^{-1} mol^{-1}, $C_p(s) = 8.9$ cal °C^{-1} mol^{-1}, $\Delta H_m = 1.44$ kcal/mol, and $\Delta H_v = 9.77$ kcal/mol. (1 cal = 4.184 J.)

1.7 The enthalpy of a substance is given by $H = E + PV$, where E is the internal energy and PV is the product of pressure and volume.
(a) Show that PV has the dimensions of energy.
(b) For a single-phase substance at constant pressure, $C_p = dH/dT$. If H_1 is the enthalpy of the substance at temperature T_1, what is the enthalpy H_2 at $T_2 > T_1$?

1.8 Differentiate the expression $G = H - TS$ with respect to temperature (at constant pressure). Substitute into your equation $dH/dT = C_p$ and $dS/dT = C_p/T$ (for constant pressure), and show that $dG/dT = -S$ at constant pressure.

1.9 (a) Is a combination of ice and water a single-phase or multiphase system?
(b) Is water, containing a small amount of salt (NaCl), a single-phase or multiphase system?
(c) In the microstructure of steel, iron carbide (Fe_3C) particles are found in a matrix of almost pure iron. Is this a single-phase or multiphase system?
(d) Distinguish between homogeneous and heterogeneous systems.

1.10 Using the appropriate equations in the text, show that $C_p = \frac{5}{2}R$ for an ideal monatomic gas.

2

Atomic Structure

This chapter introduces the electronic structure of atoms, ionization potential, and electron affinity.

2.1 ATOMS

The familiar model of an atom is that of a tiny nucleus composed of protons and neutrons, surrounded by rapidly moving electrons. Typically, the atomic diameter is on the order of 10^{-10} m and the nucleus is on the order of 10^{-15} m in diameter, yet the nucleus contains most of the mass. A proton and a neutron have about the same mass, and each is about 1800 times as massive as an electron. A neutron is electrically neutral, but a proton has a definite positive charge ($+1.602 \times 10^{-19}$ coulomb) that is exactly the opposite of the negative charge of an electron. In a neutral atom the number of electrons equals the number of protons.

The number of protons in the nucleus Z, called the atomic number, characterizes a chemical element. Atoms of an element all have the same number of protons, yet may have different masses. The atomic mass number of an atom A is given by $A = Z + N$, where N is the number of neutrons in the nucleus. Since an element is characterized solely by Z, it follows that atoms of a given chemical element may have a varying number of neutrons. Subspecies of chemical elements, with the same Z but differing N and A, are called isotopes. The atomic weight of an element is the weighted average of the atomic masses of the various naturally occurring isotopes of the element, and the atomic weight scale is based on a value of exactly 12 for the carbon isotope that has an atomic mass number of 12.

2.2 NUCLEI

The nucleus of an atom weighs less than the sum of the weights of its isolated component particles. The difference between the actual mass and that of the components is called the mass defect. The mass defect, Δm, is related to the binding energy of the nucleus, ΔE, through Einstein's equation

$$\Delta E = \Delta mc^2 \tag{2.1}$$

where c is the velocity of light. The nuclear forces that bind protons and neutrons together are of great strength, and the binding energy amounts to about 8.5 million electron volts (MeV) per nucleus particle (nucleon). [One electron volt is the amount of energy an electron acquires on being accelerated through a voltage drop of one volt. This is equal to 1.602×10^{-19} J (3.83×10^{-20} cal).] The greatest binding energy per nucleon is found in nuclei of medium atomic number such as Fe. In these nuclei, N is approximately equal to Z. For nuclei of larger atomic number, such as uranium, N is about equal to $1.5Z$ and the binding energy per nucleon is less. As a consequence of this lesser nuclear stability, some isotopes of uranium are unstable with respect to fission. That is, if the uranium isotope ($Z = 92$, $A = 235$) is bombarded with a neutron, the following reaction can take place:

$$^{235}\mathrm{U} + n \longrightarrow {}^{94}\mathrm{Y} + {}^{140}\mathrm{I} + 2n \tag{2.2}$$

Release of considerable energy accompanies fission, and the reaction products are smaller nuclei and neutrons. Each superscript represents the mass number and the atomic number is indicated by the chemical symbol.

Some heavy nuclei or even light nuclei that have an imbalance between the number of protons and neutrons can decay spontaneously by the emission of α particles (helium ions), β particles (electrons), or γ rays (short wavelength electromagnetic radiation). The length of time over which decay occurs for such unstable nuclei varies greatly. A measure of this time is the half-life of the material. In one half-life period, half of the unstable nuclei will emit some type of radiation and thus change their character. In two half-life periods, only $\frac{1}{4}$ of the nuclei will not have decayed; in three half lives, only $\frac{1}{8}$ of the original nuclei remain; etc. The half life of gamma-emitting ^{60}Co, for example, is 5.3 years, whereas that of radioactive ^{14}C is 5700 years.

2.3 ELECTRONS

According to the early quantum theory, electrons move about the nucleus of an atom in definite orbits, each of which is designated by numbers called quantum numbers. An electron moving in an orbit close to the nucleus has a lower energy than one in an outer orbit. Work is therefore necessary to move an electron from an inner to an outer orbit. Conversely, energy is released in the form of electromagnetic radiation if an electron is able to move from an outer to an inner orbit. The electromagnetic radiation consists of discrete entities or quanta, which are called photons. The energy of an emitted photon is the difference in energy between the two orbits or energy levels and is related to the frequency ν, or the wavelength λ, of the electromagnetic radiation by

$$E_2 - E_1 = h\nu = h\frac{c}{\lambda} \tag{2.3}$$

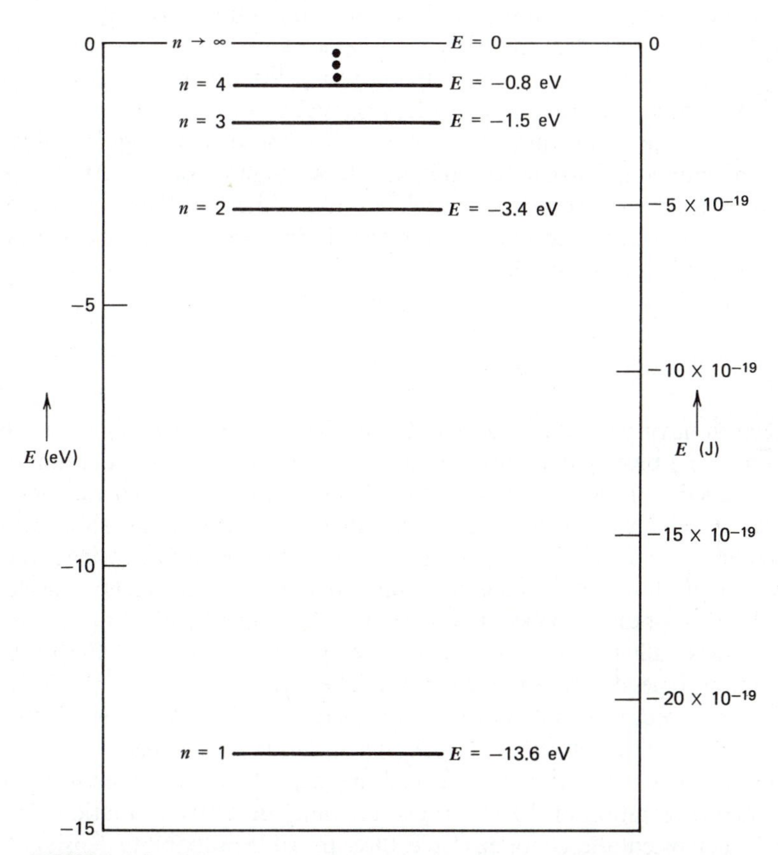

FIGURE 2.1 Some of the possible discrete atomic energy levels in the hydrogen atom. By convention, the zero of energy corresponds to infinite separation between electron and nucleus; hence, the ground state energy of the hydrogen atom is −13.6 eV.

where h is Planck's constant, E_2 and E_1 are the energies of the outer and inner orbits, respectively, and c is the velocity of light. Through spectroscopic observations of the frequencies or wavelengths of radiation emitted by an excited atom, the relative electron energy levels can be determined. Figure 2.1 represents such energy levels obtained from measurements of radiation wavelengths emitted by atomic hydrogen excited electrically in a gas discharge tube.

2.4 QUANTUM NUMBERS

To represent the spatial location and energy of an electron in an atom requires the use of four quantum numbers. The first or principal quantum number, n, corresponds to the orbit number in the original quantum theory. It is restricted to integral values, $n = 1, 2, 3$, and so forth. A value of $n = 1$ signifies that the electron exists in the lowest energy level or the *orbit* as close to the nucleus as possible. Successively higher energy levels are represented by $n = 2, 3$, etc.

The second quantum number, l, often called the orbital angular momentum quantum number, is restricted to integral values, ranging from $l = 0$ to $l = (n - 1)$. When $n = 1$, only $l = 0$ is possible; when $n = 2$, either $l = 0$ or $l = 1$ is possible; etc. For convenience, letters are used to specify the *electronic state* corresponding to the second quantum number:

$$\begin{array}{rccccc} l\text{:} & 0 & 1 & 2 & 3 & 4 \\ \text{state:} & s & p & d & f & g. \end{array}$$

An electron having $l = 0$ is in an s state; one having $l = 1$ is in a p state. Thus, an electron with a principal quantum number $n = 3$ and a second quantum number $l = 1$ is called a $3p$ electron. Unlike the old quantum theory that was based on a planetary model, wave mechanics asserts that an electron in an atom cannot be considered as a particle having an orbit with a definite radius. Instead, there is a *probability* of an electron being at certain spatial positions. Hence, the location of an electron is best described in terms of its *probability density distribution*, which is sometimes called an *electron cloud*. The spatial symmetry of the probability distribution depends upon the value of l. The electron cloud is spherically symmetric for s electrons, but more complicated for other electrons. Examples of these distributions are shown in Fig. 2.2 for a $1s$ and a $2p$ electron.

The third quantum number, m_l, called the magnetic quantum number, governs the spatial orientation of the electron probability density distribution. There are $(2l + 1)$ such orientations; for example, three possible probability density distributions for $l = 1$ are shown in Fig. 2.3. The possible characteristic values of m_l are $-l, -(l - 1), \ldots -1, 0, +1, \ldots +(l - 1), +l$; that is, there are $(2l + 1)$ values of m_l. For the example shown in Fig. 2.3, the three possible orientations do not

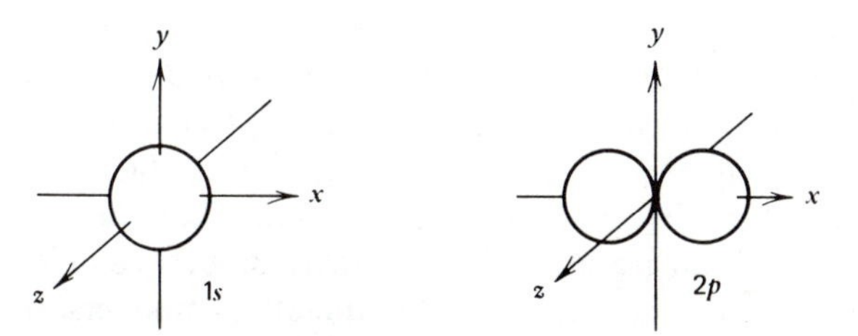

FIGURE 2.2 A schematic representation of the probability density distributions for a 1*s* and a 2*p* orbital. These diagrams indicate the angular and radial dependence of the electron distribution functions. The combination of angular and radial density distributions defines the spatial extent and shape of an orbital.

correspond to $m_l = -1, 0, +1$; rather $m_l = \pm 1$ corresponds to linear combinations of the p_x and p_y distributions while $m_l = 0$ corresponds to the p_z distribution. Alternatively, the p_x and p_y distributions can be considered to be linear combinations of the electron distributions with $m_l = \pm 1$. Since the bonding characteristics of the electrons are better described with reference to the p_x, p_y, and p_z density distributions (Fig. 2.3), these are most commonly used. Indeed, linear combinations of atomic *orbitals* are frequently made to more clearly reflect observed electron bonding characteristics. *Hybridization* of atomic orbitals in carbon, which will be discussed later, is a most important example of such a procedure.

The first three quantum numbers, *n*, *l*, and m_l, define an atomic electron *orbital.* They are related, respectively, to the size, shape, and spatial orientation of the

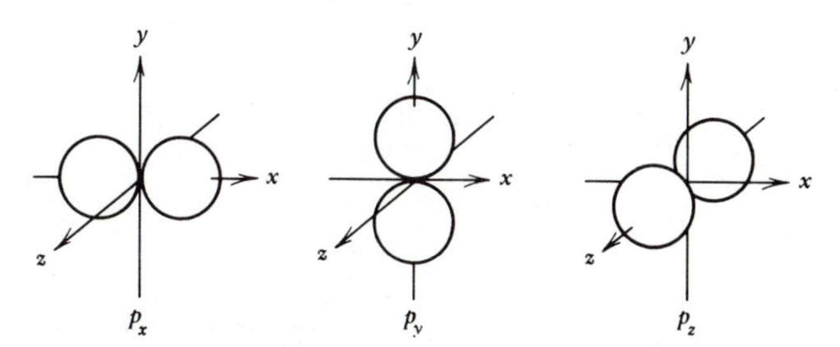

FIGURE 2.3 Spatial arrangement of three possible 2*p* (p_x, p_y, and p_z) orbitals. The number of possible orientations is $(2l + 1)$ or 3 for $l = 1$. If each of these orbitals is filled with electrons, then the resulting 2*p* electron cloud distribution is symmetric about the nucleus. Asymmetry and directionality result if any orbital is only partially filled.

electron orbital. A fourth quantum number, m_s, called the electron spin quantum number, can have values of $+\frac{1}{2}$ and $-\frac{1}{2}$. For notational purposes a positive m_s is often represented as ↑ (*spin up*); a negative m_s, as ↓ (*spin down*). According to the Pauli exclusion principle, each electron in an atom must have a unique set of the four quantum numbers; hence, an atomic orbital may contain a maximum of two electrons that have opposite spins. The interactions between electrons in atoms are discussed in the next section using the quantum numbers and the notation introduced in this section.

2.5 ENERGY LEVELS

The energy levels associated with atomic orbitals are shown in Fig. 2.4. These energy levels and their occupancy by electrons are intimately related to the quantum numbers discussed previously. The lower the principal quantum number (shell) is, the lower the energy level. Similarly, the lower the orbital angular momentum quantum number (subshell) is, the lower the energy level. An energy level for a free atom not subjected to an external magnetic or electric field depends

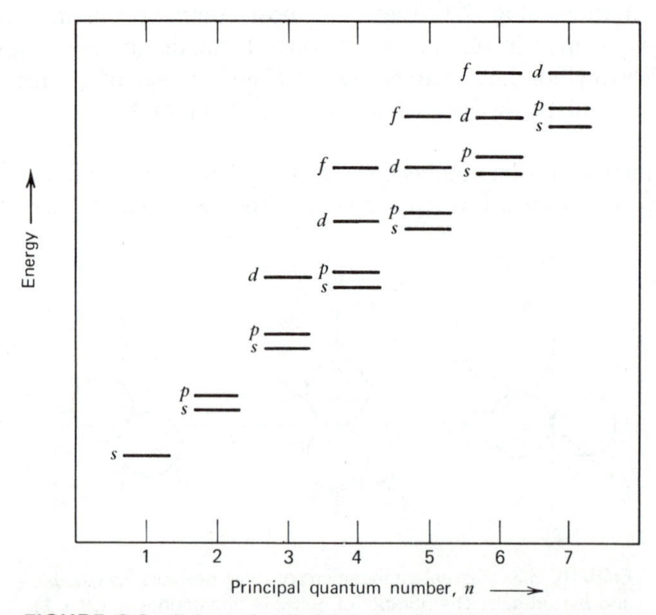

FIGURE 2.4 Schematic diagram showing the relative energy levels of bound electrons in different atomic orbitals. This diagram indicates the order in which energy levels are filled as the atomic number increases.

only on n and l, although in the case of a one-electron atom such as hydrogen (Fig. 2.1), the energy depends solely on n. The maximum number of electrons in each energy level is thus dictated by the Pauli exclusion principle, since an orbital can hold only two electrons. Consequently, the subshells, characterized by different l values, can accommodate different numbers of electrons, as shown in Table 2.1. The maximum number of possible electrons in a given subshell is given by the progression formula, number $= 2(2l + 1)$, and the corresponding number of states in a subshell equals $(2l + 1)$.

TABLE 2.1 Electron Populations

Subshell	*Number of Orbitals*	*Maximum Number of Electrons*	*Quantum Numbers*
s ($l = 0$)	1	2	$l = 0, m_l = 0, m_s = +\frac{1}{2}$
			$l = 0, m_l = 0, m_s = -\frac{1}{2}$
p ($l = 1$)	3	6	$l = 1, m_l = -1, m_s = +\frac{1}{2}$
			$l = 1, m_l = -1, m_s = -\frac{1}{2}$
			$l = 1, m_l = 0, m_s = +\frac{1}{2}$
			$l = 1, m_l = 0, m_s = -\frac{1}{2}$
			$l = 1, m_l = +1, m_s = +\frac{1}{2}$
			$l = 1, m_l = +1, m_s = -\frac{1}{2}$
d ($l = 2$)	5	10	$l = 2 \ldots$

The maximum number of electrons for a given principal quantum number, n (i.e., the maximum number of electrons in a given shell) is the sum of the maximum number of electrons in the various subshells or $\sum_{l=0}^{(n-1)} 2(2l + 1)$. Thus, the maximum occupancy for the shell with $n = 1$ is identical to the maximum number of electrons in the $1s$ subshell (i.e., 2); the maximum occupancy for the shell with $n = 2$ is the sum of the possible number of electrons in the $2s$ (i.e., 2) and $2p$ (i.e., 6) subshells or 8; for the shell with $n = 3$, the maximum occupancy is 18. Similar computations can be carried out readily for larger principal quantum numbers.

2.6 ELECTRONIC CONFIGURATIONS

From experimental observations of excited atoms, as mentioned in Sect. 2.3, the occupancy of the various electronic energy levels has been found to progress as shown in Table 2.2. For an atom in the unexcited or *ground state*, electrons successively fill up the lowest possible energy levels, provided that the Pauli exclusion principle is not violated.

TABLE 2.2 Occupancy of Electronic Energy Levels for Free Atoms in Their Ground States

Element		$(n = 1)$	$(n = 2)$		$(n = 3)$			$(n = 4)$			
Symbol	*Number*	$1s$	$2s$	$2p$	$3s$	$3p$	$3d$	$4s$	$4p$	$4d$	$4f$
H	1	1									
He	2	2									
Li	3	2	1								
Be	4	2	2								
B	5	2	2	1							
C	6	2	2	2							
N	7	2	2	3							
O	8	2	2	4							
F	9	2	2	5							
Ne	10	2	2	6							
Na	11	2	2	6	1						
Mg	12	2	2	6	2						
Al	13	2	2	6	2	1					
Si	14	2	2	6	2	2					
P	15	2	2	6	2	3					
S	16	2	2	6	2	4					
Cl	17	2	2	6	2	5					
Ar	18	2	2	6	2	6					
K	19	2	2	6	2	6		1			
Ca	20	2	2	6	2	6		2			
Sc	21	2	2	6	2	6	1	2			
Ti	22	2	2	6	2	6	2	2			
V	23	2	2	6	2	6	3	2			
Cr	24	2	2	6	2	6	5	1			
Mn	25	2	2	6	2	6	5	2			
Fe	26	2	2	6	2	6	6	2			
Co	27	2	2	6	2	6	7	2			
Ni	28	2	2	6	2	6	8	2			
Cu	29	2	2	6	2	6	10	1			
Zn	30	2	2	6	2	6	10	2			
Ga	31	2	2	6	2	6	10	2	1		
Ge	32	2	2	6	2	6	10	2	2		
As	33	2	2	6	2	6	10	2	3		
Se	34	2	2	6	2	6	10	2	4		
Br	35	2	2	6	2	6	10	2	5		
Kr	36	2	2	6	2	6	10	2	6		

Ground state electronic configurations are represented by a conventional notation. For hydrogen with a single electron, $1s^1$ is used. The notation for helium with two electrons in the $1s$ level is $1s^2$. For lithium ($Z = 3$) the notation is $1s^2 2s^1$; for beryllium ($Z = 4$), $1s^2 2s^2$; for boron ($Z = 5$) $1s^2 2s^2 2p^1$. Carbon, with six electrons, has the configuration $1s^2 2s^2 2p^2$. Since the $2p$ level consists of three orbitals, we can either put the final two electrons into the same orbital (same values of m_l) with opposite spins or put them into different $2p$ orbitals (different values of m_l) with the same spins. According to *Hund's rule*, the most energetically favorable configuration is the latter. The resulting electronic configuration ($1s^2 2s^2 2p^2$) can be represented diagramatically so as to show the parallel spins in the $2p$ level (Fig. 2.5).

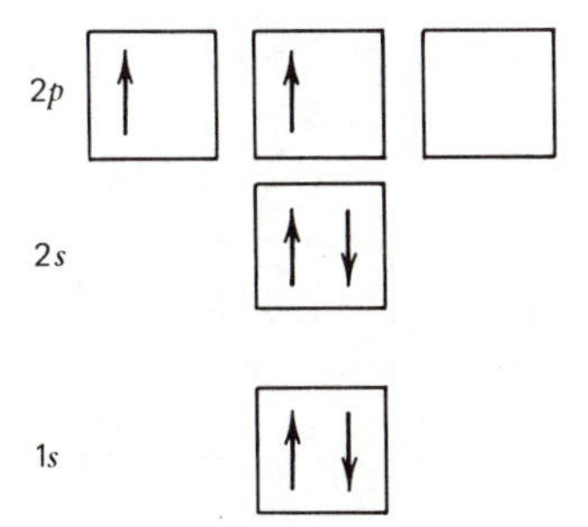

FIGURE 2.5 Ground state electronic structure of a free carbon atom. Each box represents an orbital; arrows indicate electrons with spin up or spin down.

The filling of electronic energy levels for the various elements results in a periodic outer electronic structure beyond the last filled shell. This periodicity corresponds directly to the periodicity in chemical behavior observed for the elements in the periodic table (Fig. 2.6). Thus, the outer electronic structure of an element determines its chemical behavior.

2.7 IONIZATION ENERGY

Electrons may be removed from isolated atoms by bombardment with other electrons. The energy required to remove the most weakly bound electron from an isolated atom is known as the ionization energy (or, more precisely, the first

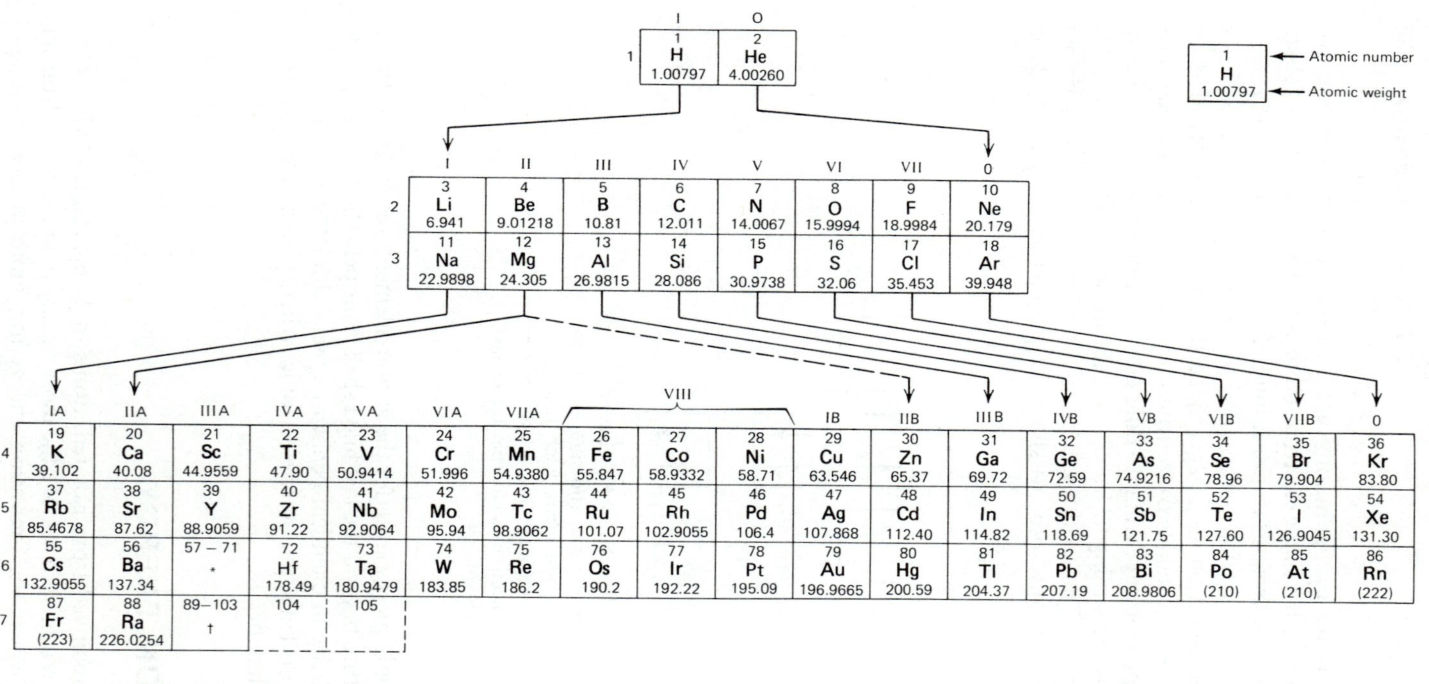

FIGURE 2.6 The periodic table of the elements. Atomic numbers and atomic weights (based on 12 for ^{12}C) are indicated. Similar outer electronic structures exist for elements in a given group.

ionization energy) and often is called the ionization potential. Ionization potentials usually are expressed in units of electron volts (eV). Some values are given in Table 2.3.

It is important to note in Table 2.3 that inert or noble gases have some of the highest ionization energies. This is because these elements have electrons that completely fill a shell or a subshell and are thus in a very stable configuration.

TABLE 2.3 First Ionization Energies (eV per atom[a])

H 13.60							He 24.48
Li 5.39	Be 9.32	B 8.30	C 11.26	N 14.53	O 13.61	F 17.42	Ne 21.56
Na 5.14	Mg 7.64	Al 5.98	Si 8.15	P 10.48	S 10.36	Cl 13.01	Ar 15.76
K 4.34						Br 11.84	Kr 14.00
Rb 4.18						I 10.45	Xe 12.13
Cs 3.89							

[a]1 eV/atom is equivalent to 96,490 J/mol or 23,060 cal/mol.

The ionization energies of alkali metal atoms (Li, Na, etc.) are low; the lowest, that of Cs, is only 3.89 eV. The reason is that alkali metal atoms have one outer *s* electron beyond the stable electronic structure of an inert gas atom. Consequently, this electron can be removed relatively easily. Figure 2.7 shows the variation of ionization energy among the elements of the periodic table. Note that the second and successive ionization energies of an atom are increasingly greater.

From a chemical and physical viewpoint, it is found that whether an atom is neutral or ionized as well as isolated or combined leads to distinctly different characteristics. Since it is useful to visualize the electronic structures of ions, the same notation as used previously for neutral atoms can be employed. The electronic configurations of an iron atom and iron ions are shown below:

$$\text{Fe}: \quad 1s^2 2s^2 2p^6 3s^2 3p^6 3d^6 4s^2 \quad \text{(neutral atom, ground state)}$$

$$\text{Fe}^{2+}: \quad 1s^2 2s^2 2p^6 3s^2 3p^6 3d^6 \quad \text{(ferrous ion)}$$

$$\text{Fe}^{3+}: \quad 1s^2 2s^2 2p^6 3s^2 3p^6 3d^5 \quad \text{(ferric ion)}$$

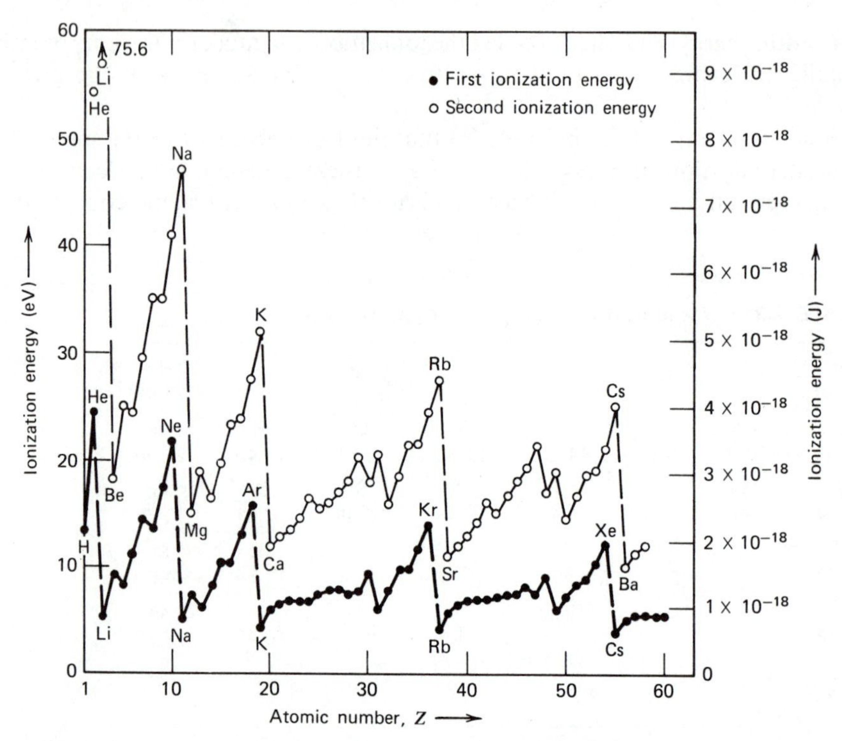

FIGURE 2.7 Variation of first and second ionization energies of free atoms with atomic number. Note the periodicity.

2.8 ELECTRON AFFINITY

As we shall see in the following chapter, the tendency for some atoms to accept, as well as lose, electrons is important in determining how atoms combine with each other. Whereas removal of an electron always requires the expenditure of energy, acceptance of one extra electron by an atom generally is accompanied by the release of energy. The amount of energy released is called the electron affinity (or first electron affinity). Table 2.4 lists the few electron affinities that have been measured as well as those that have been calculated. The elements on the left of the periodic table (metals) and the inert gases have low electron affinities, whereas the nonmetals have higher ones. The particularly high electron affinities of halogen atoms (F, Cl, etc.) may be attributed to the fact that they lack one electron in order to have the stable electronic structure of an inert gas. By accepting an electron they increase their stability. In this case the additional electron goes into the orbital lacking one electron and, accordingly, is firmly held.

TABLE 2.4 First Electron Affinities (eV per atom[a])

H 0.77 (0.75)						
Li — (0.58)	Be — (—)	B — (0.30)	C 1.25 (1.17)	N — (−0.27)	O 1.47 (1.22)	F 3.45 (3.37)
Na — (0.78)	Mg — (—)	Al — (0.49)	Si — (1.39)	P — (0.78)	S 2.07 (2.12)	Cl 3.61 (3.56)
						Br 3.36 —
						I 3.06 (—)

[a]1 eV/atom is equivalent to 96,490 J/mol or 23,060 cal/mol. (Theoretical values are enclosed in parentheses; all other values are experimental.)

2.9 ATOMIC SIZE

The size or volume of an isolated atom is difficult to define explicitly since, in the electron cloud model of the atom, the probability density distribution theoretically reaches zero only at infinity. Nevertheless, the electron density falls off so rapidly at a short distance from the nucleus that some approximation of size can be made. In the hydrogen atom, for example, the electron density is very nearly zero at a distance of 0.12 nm (1.2 Å) from the nucleus. The problem of defining atomic sizes is simplified in molecules and solids because rather precise radii can be determined from interatomic distances, which can be measured by X-ray diffraction techniques. Thus, in the H_2 molecule the atoms are only 0.074 nm apart, as determined by the distance between nuclei. In this case the radius of the hydrogen atom is taken to be 0.037 nm, even though the *size* of the H_2 molecule is considerably more than four times this value.

From the above examples, it is apparent that atomic radius depends upon whether an atom is isolated or combined with other atoms. The radius of an isolated atom is called the *van der Waals* radius; that of a bound atom in a molecule, the *covalent radius*; and that of a bound atom in a metal, the *metallic radius.* Van der Waals, covalent, and metallic radii for some elements are listed in Table 2.5.

TABLE 2.5 Some Atomic Radii

H							He
—							—
0.028[a]							0.093
0.12							0.177
Li	Be	B	C	N	O	F	Ne
0.152	0.111	—	—	—	—	—	—
0.134	0.090	0.082	0.077	0.075	0.073	0.072	0.131
—	—	—	—	0.15	0.140	0.135	0.160
Na	Mg	Al	Si	P	S	Cl	Ar
0.186	0.160	0.143	—	—	—	—	—
0.154	0.130	0.118	0.118	0.109	0.102	0.101	0.174
0.186	—	—	—	0.194	0.185	0.180	0.193
K	Ca	Ga	Ge	As	Se	Br	Kr
0.227	0.197	0.122	—	—	—	—	—
0.196	0.174	0.126	0.123	0.125	0.116	0.114	0.189
0.231	—	—	—	0.200	0.200	0.195	0.200

Sc	Ti	V	Cr	Mn	Fe	Co	Ni	Cu	Zn
0.163	0.145	0.131	0.125	0.112	0.124	0.125	0.125	0.128	0.133
0.144	0.136	—	—	—	—	—	—	0.138	0.131
—	—	—	—	—	—	—	—	—	—

[a] In the elemental state, the covalent radius of H is 0.037 nm; in covalent compounds, its radius is 0.028 nm. (*Note*: 1 nm = 10 Å.)

Key:

Na
0.186 ⟵ Metallic radius (nm)
0.154 ⟵ Single bond covalent radius (nm)
0.186 ⟵ Van der Waals radius (nm)

The radii of positive ions or cations and those of negative ions or anions also differ from the van der Waals radii. Ionic radii of various elements are listed in Table 2.6.

Atomic volume is another measure of atomic size. In its most useful form, atomic volume is obtained directly from the total volume occupied by an atom in the elemental solid or in the elemental liquid, if the latter is more dense. Figure 2.8 shows the variation of atomic volume for some solid elements.

2.10 INTERATOMIC INTERACTIONS

All similar atoms on approaching one another experience some attraction. Even inert gas atoms are weakly attracted by neighboring atoms, and this attraction is sufficiently large at very low temperatures to cause condensation and

TABLE 2.6 Some Ionic Radii

H 0.208 (1−)							He —
Li 0.060 (1+)	Be 0.031 (2+)	B 0.020 (3+)	C 0.015 (4+) 0.260 (4−)	N 0.011 (5+) 0.171 (3−)	O 0.009 (6+) 0.140 (2−)	F 0.007 (7+) 0.136 (1−)	Ne —
Na 0.095 (1+)	Mg 0.065 (2+)	Al 0.050 (3+)	Si 0.041 (4+) 0.271 (4−)	P 0.034 (5+) 0.212 (3−)	S 0.029 (6+) 0.184 (2−)	Cl 0.026 (7+) 0.181 (1−)	Ar —
K 0.133 (1+)	Ca 0.099 (2+)	Ga 0.062 (3+) 0.148 (1+)	Ge 0.053 (4+) 0.093 (2+)	As 0.047 (5+) 0.222 (3−)	Se 0.042 (6+) 0.198 (2−)	Br 0.039 (7+) 0.195 (1−)	Kr —

Sc	Ti	V	Cr	Mn	Fe	Co	Ni	Cu	Zn
0.081 (3+)	0.068 (4+) 0.090 (2+)	0.059 (5+) 0.074 (3+)	0.052 (6+) 0.069 (3+)	0.046 (7+) 0.080 (2+)	0.064 (3+) 0.076 (2+)	0.062 (3+) 0.078 (2+)	0.062 (3+) 0.078 (2+)	0.069 (2+) 0.096 (1+)	0.074 (2+)

Note: 1 nm = 10 Å.

Key:

Si
0.041 (4+) ⟵ Ionic radius (nm) for Si^{4+}
0.271 (4−) ⟵ Ionic radius (nm) for Si^{4-}

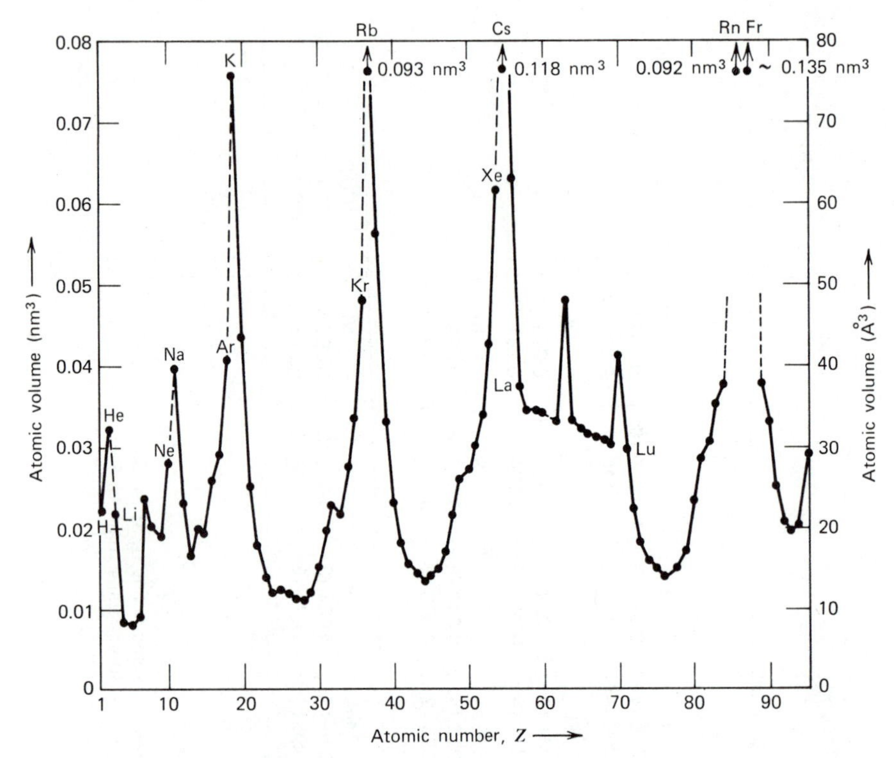

FIGURE 2.8 Variation of atomic volume (as determined from the elemental condensed state) with atomic number. Note the periodicity.

subsequent solidification. The forces involved here are of electrostatic origin, even though the average electric field strength near an inert gas atom is zero. At any given instant, however, the electric field strength is not zero if the number of electrons is instantaneously greater on one side of the nucleus than the other. This situation is shown in Fig. 2.9, and the combination of positive–negative charge separation by the distance a is called an electric dipole. The strength of the dipole is defined by the magnitude of the charge times the separation distance a. These fluctuating dipoles interact with other fluctuating dipoles that are nearby. At very low temperatures, where thermal agitation is minimal, the number and attraction of such dipoles is sufficient to cause the cohesion necessary for formation of the liquid or solid states. The weakness of such physical or van der Waals bonding, however, is reflected in the low boiling and melting points of the inert gases (Table 2.7).

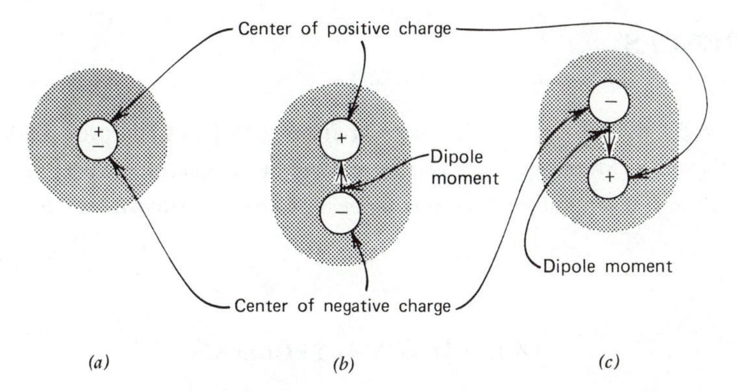

FIGURE 2.9 Schematic representation of an atom at different instances of time. The total electron cloud of an atom is not necessarily symmetric about the nucleus at any instance of time as shown in (*b*) and (*c*), although the time average is symmetrical as shown in (*a*). The instantaneous separation of centers of positive and negative charge gives rise to a dipole moment.

Fluctuating dipoles occur in all atoms and molecules, and some molecules have permanent electric dipoles that interact somewhat more strongly. It is, however, much stronger electrostatic or electronic interactions of various other kinds (chemical bonding) that generally account for the cohesion of most solid materials. The different types of primary or strong chemical bonds (ionic, covalent, and metallic) and of secondary or weak physical bonds, are discussed in the next chapter.

TABLE 2.7 Melting and Boiling Points of the Inert Gas Elements at Atmospheric Pressure

Element	T_m (°C)	T_v (°C)
He	−272[a]	−269
Ne	−249	−246
Ar	−189	−186
Kr	−157	−153
Xe	−112	−108
Rn	−71	−62

[a] At 26 atm.

REFERENCES

1. Chemical Bond Approach Project. *Chemical Systems*, McGraw-Hill, New York, 1964.
2. Hochstrasser, R. M. *Behavior of Electrons in Atoms*, Benjamin, New York, 1964.
3. Pimental, G. E. and Spratley, R. D. *Understanding Chemistry*, Holden-Day, San Francisco, 1971.

QUESTIONS AND PROBLEMS

2.1 (a) Calculate the change in mass (Δm) that would result if one nucleus of ^{20}Ne were separated into individual protons and individual neutrons. (Use a binding energy of 8.5 MeV per nucleon.)
(b) Compare Δm from part (a) with the mass of a ^{20}Ne atom.

2.2 Radiocarbon dating is a method of estimating the age of carbon-containing materials (such as wood in a growing tree). It is based on the assumption that the proportion of radioactive ^{14}C in the carbon reservoir of the earth is constant. Cosmic rays cause the formation of ^{14}C in the upper atmosphere; the ^{14}C subsequently enters the carbon cycle whereby plants take up CO_2 from the atmosphere and form carbon compounds that are incorporated into the structure of the plants or animals. Describe how radiocarbon dating could be used to establish the age of an archeological specimen.

2.3 Calculate the energy change (in kJ/mol) associated with the electronic transition in Hg atoms responsible for the emission of photons of wavelength 253.7 nm.

2.4 How many electrons are there in the 3*d* subshell of Fe, Co, and Ni, and how are their spins aligned? In the 4*s* subshell, for these atoms, how are the electron spins aligned? (Use a box diagram to indicate your answer.)

2.5 Show that a maximum number of 14 electrons can be accommodated in a $f(l = 3)$ subshell.

2.6 Write out the electronic configurations (e.g., $1s^2$ . . .) for Be ($Z = 4$), F ($Z = 9$), Na ($Z = 11$), S ($Z = 16$), Ar ($Z = 18$), F^- ($Z = 9$), and Na^+ ($Z = 11$).

2.7 Show schematically the electronic configurations and electron spin alignment in (a) an oxygen atom, (b) an oxygen ion O^{2-}, (c) a neon atom, and (d) an argon atom.

2.8 If 1*s* electrons in the atoms of a piece of solid copper are excited to higher energy levels that are empty, radiation of various wavelengths is emitted as the electrons reestablish the ground-state configuration. In particular, an electronic transition from the 2*p* level to the 1*s* level results in radiation of wavelength 0.154 nm. What is the energy difference between these two levels?

2.9 The atomic numbers of the inert gases are 2, 10, 18, 36, 54, and 86, respectively. Nuclei having either a number of protons or a number of neutrons equal to 2, 8, 20, 28, 50, 82, and 126 are stable in a radioactive sense. What does this observation concerning nuclei suggest to you?

2.10 Compare the first ionization energies of (a) elements that are good electrical conductors in the solid state (e.g., Li, Na, Al, Cu) and (b) elements that are poor electrical conductors in the solid state (e.g., P, S, Cl).

2.11 Compare the ionic radii of cations and anions to the radii of the parent atoms. Comment on the pattern noted.

2.12 In Figure 2.7, the first ionization energies of the inert gases are relative maxima and those of the alkali metal atoms (e.g., Li and Na) are relative minima. Explain why the *second ionization energies* show relative maxima for atoms such as Li and Na and relative minima for atoms such as Be, Mg, and Zn.

2.13 (a) Elements 21 to 29, 39 to 47 and 72 to 79 are known as transition elements. Look up their electronic configurations and state the characteristic common to each of these series.

(b) Elements 57 to 71 (the rare earths or lanthanum series) and 89 on (the actinium series) also have a common characteristic. What is it?

2.14 (a) Using the atomic volume (i.e., 0.0166 nm^3), calculate the number of atoms per cm^3 in solid aluminum.

(b) The density of solid aluminum is 2.70 g/cm^3. What is the mass of a single aluminum atom?

3

Chemical Bonding

This chapter describes the ways in which atoms combine to form molecules and solids as a consequence of their electronic configurations.

3.1 TYPES OF CHEMICAL BONDS

The chemical reactivity of atoms of different elements depends mainly on the number and spatial distribution of their outer electrons. As mentioned previously, inert gas atoms such as He, Ne, and Ar are very unreactive chemically because they have filled outer electron shells or subshells. As a result, they attract each other only slightly and thus condense only at very low temperatures. Many molecules, for example, methane (CH_4), are strongly bonded internally but are weakly attracted to one another. Consequently, molecular substances such as these also liquify and solidify at very low temperatures. Other substances, like sodium chloride, silicon, and copper, have high melting points, indicative of strong bonding in the solid state. These three materials exemplify the three types of primary chemical bonding: ionic (NaCl), covalent (Si), and metallic (Cu).

The atoms or ions in a crystalline solid usually are arrayed geometrically in such a manner to minimize the free energy of the solid. In crystalline NaCl the arrangement is a three-dimensional array of Na^+ and Cl^- ions as represented in Fig. 3.1. The ionic bonding in this material arises from electrostatic attraction between the two types of ions that have been formed by the transfer of one valence electron from each Na atom to each Cl atom. In contrast, covalent bonding involves electron sharing between adjacent atoms. Each silicon atom in a silicon crystal

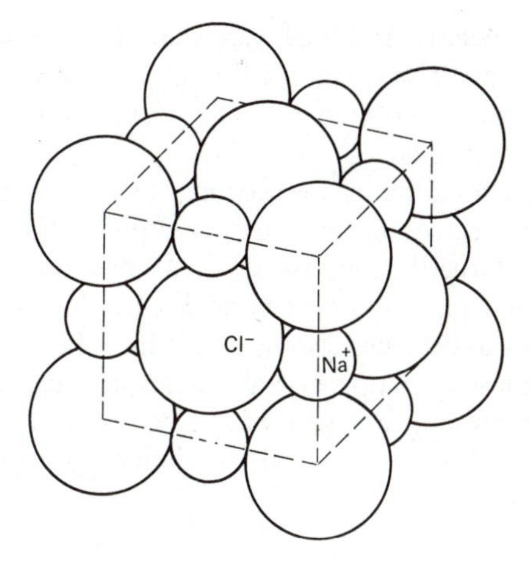

FIGURE 3.1 Ionic arrangement in a NaCl crystal. The dotted lines outline the basic repeating unit or unit cell for this crystal structure. Each cation (Na^+) is coordinated by six anions (Cl^-).

is bonded discretely to four nearest neighbors by pairs of electrons having opposite spins, and the three-dimensional array of atoms (Fig. 3.2) is determined by the network of directed covalent bonds. Because of the charateristic electron sharing, the number of adjacent atoms bonding to any given atom usually is restricted to four or less for covalent bonding. Bonding in metals bears some resemblance to both ionic and covalent bonding but differs from either in that the valence electrons

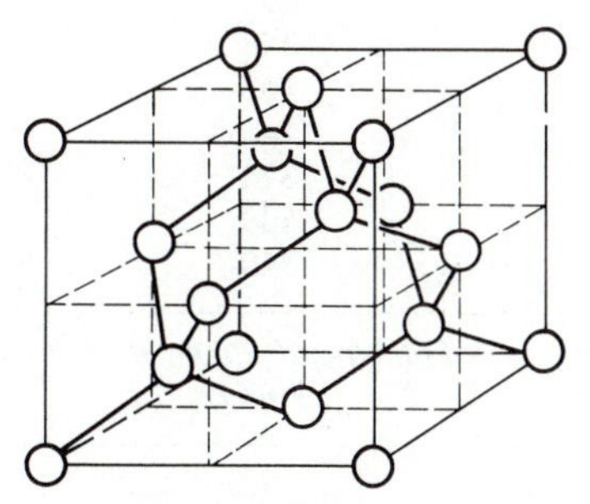

FIGURE 3.2 Atomic arrangement in the unit cell of a silicon crystal. Each Si atom forms four covalent bonds with neighboring Si atoms as denoted by the heavy lines. Diamond, one form of carbon, has the same structure.

which lead to bonding behave as if they were shared by the solid as a whole rather than by one or two atoms. Thus, metallic bonding consists of collective sharing of *free* electrons, such that an ordered array of positive ion cores (Fig. 3.3) is held together by the free electron gas.

In general, chemical bonding arises because almost all atoms have a tendency to attain more stable electronic configurations (often those of the inert gas elements). Electrons are directly responsible for all chemical bonds, and the more electrons per atom that participate in bonding, the higher the bonding energy. The energies necessary to dissociate some solids into the atomic gaseous state are listed in Table 3.1; these enthalpies of atomization are approximately equal to the total interatomic binding energies of the solids. For purposes of comparison, binding energies of some physical or secondary bonds also are given in Table 3.1.

TABLE 3.1 Binding Energies for Selected Solids

Substance	*Chemical Bond Type*	*Enthalpy of Atomization (kJ/mol) at 25°C*
C (diamond)	Covalent	713
Si	Covalent	450
SiC	Covalent	1230
InSb	Covalent	523
LiF	Ionic	849
NaCl	Ionic	640
MgO	Ionic	1000
CaF_2	Ionic	1548
Na	Metallic	108
Mg	Metallic	149
Al	Metallic	324
Cu	Metallic	339
W	Metallic	849

Substance	*Physical Bond Type*	*Enthalpy of Sublimation (kJ/mol)*[a]
Ar	van der Waals	7.5
O_2	van der Waals	7.5
CO_2	van der Waals	25
CH_4	van der Waals	18
H_2O	Hydrogen	51
NH_3	Hydrogen	35

[a]At the melting temperature of each substance.

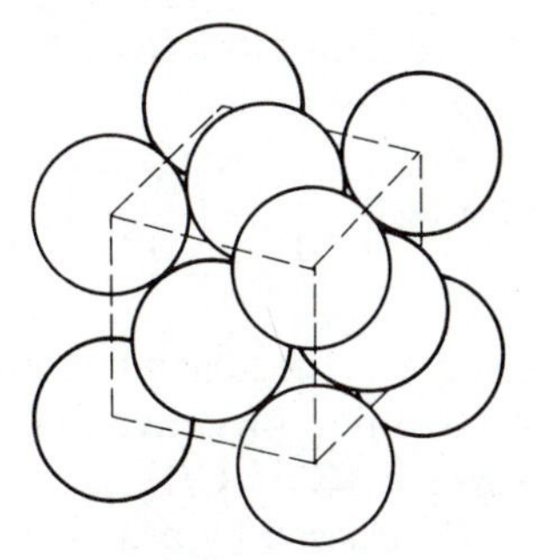

FIGURE 3.3 Atomic arrangement in a copper crystal. The atoms (or positive ion cores) shown are held together by a free electron gas. Each Cu atom is coordinated by 12 other Cu atoms. This structure is called face-centered cubic.

3.2 IONIC BONDING

Ionic or heteropolar bonding is the simplest type of chemical bonding to visualize since it is almost totally electrostatic in nature. It occurs between electropositive elements (metals, i.e., those elements on the left side of the periodic table) and electronegative elements (nonmetals, i.e., those elements on the right side of the periodic table). NaCl and MgO are examples of solids in which ionic bonding dominates. The formation of NaCl involves a transfer of valence electrons from Na to Cl atoms. To remove the valence electrons from a mole of free sodium atoms requires an expenditure of $I = 496$ kJ. (This quantity is the first ionization energy per mole of Na atoms.) This results in a mole of singly charged Na^+ cations whose electronic configuration is that of the inert gas neon. Acquisition of the electrons by a mole of free chlorine atoms results in a release of $E_a = 349$ kJ (the sum of the first electron affinities for a mole of Cl atoms) and the formation of a mole of singly charged Cl^- anions having the same electronic configuration as the inert gas argon. Thus, to initiate a reaction between a mole of free neutral sodium atoms and a mole of free neutral chlorine atoms requires a net expenditure of $I - E_a = 496 - 349 = 147$ kJ (=35.3 kcal). Once started, the reaction proceeds vigorously with the evolution of light and heat because of the sizable energy decrease resulting from ionic bonding. The final product usually is a large number of salt crystallites.

The electrostatic energy of an ionic bond in a *diatomic* molecule can be calculated using Coulomb's law, which gives the force between two particles having opposite charges of $+ze$ and $-ze$. This attractive force is

$$F = -\frac{1}{4\pi\varepsilon_0}\left(\frac{z^2e^2}{r^2}\right) \tag{3.1}$$

where r is the interionic spacing, z is the absolute number of unit charges on the cation (positive ion) and on the anion (negative ion), e is the magnitude of the electron charge, and ε_0 $(=8.854 \times 10^{-12}\ \mathrm{C/N \cdot m^2})$ is a conversion factor. The electrostatic potential energy, E, released by bringing the cation and anion from infinite separation to their equilibrium interionic spacing, r_0, is

$$E = -\int_{\infty}^{r_0} F\,dr = -\frac{1}{4\pi\varepsilon_0}\left(\frac{z^2e^2}{r_0}\right) \tag{3.2}$$

The value of r_0 is finite since the two ions are *not* point charges that would tend to approach a zero spacing. When the electron clouds of the two ions begin to overlap, there is a strong repulsion, almost as if between two impenetrable spheres. The resulting repulsive force and its corresponding positive potential energy term are proportional to $e^{-r/\rho}$, where ρ is an empirical parameter on the order of 0.03 nm. Thus, the repulsive force is very short range in extent and rises rapidly with decreasing r. At the equilibrium interionic spacing the attractive and repulsive forces balance each other, and the total potential energy is a minimum given by

$$E_0 = -\frac{1}{4\pi\varepsilon_0}\left(\frac{z^2e^2}{r_0}\right) + \lambda e^{-r_0/\rho} \tag{3.3a}$$

where λ and ρ are constants of the substance that can be determined experimentally. The total bond energy U_0 is simply the negative of the total potential energy. The force and associated potential energy for the ion pair as a function of interionic separation are shown schematically in Fig. 3.4.

The binding energy, *relative to neutral free atoms*, is

$$U_0 - (I - E_a) = \frac{1}{4\pi\varepsilon_0}\left(\frac{z^2e^2}{r_0}\right) - \lambda e^{-r_0/\rho} - I + E_a \tag{3.3b}$$

where I is the ionization potential to form the cation and E_a is the electron affinity to form the anion. For a mole of NaCl molecules in the gaseous state, the value of U_0 is 556 kJ and the binding energy is about 410 kJ.

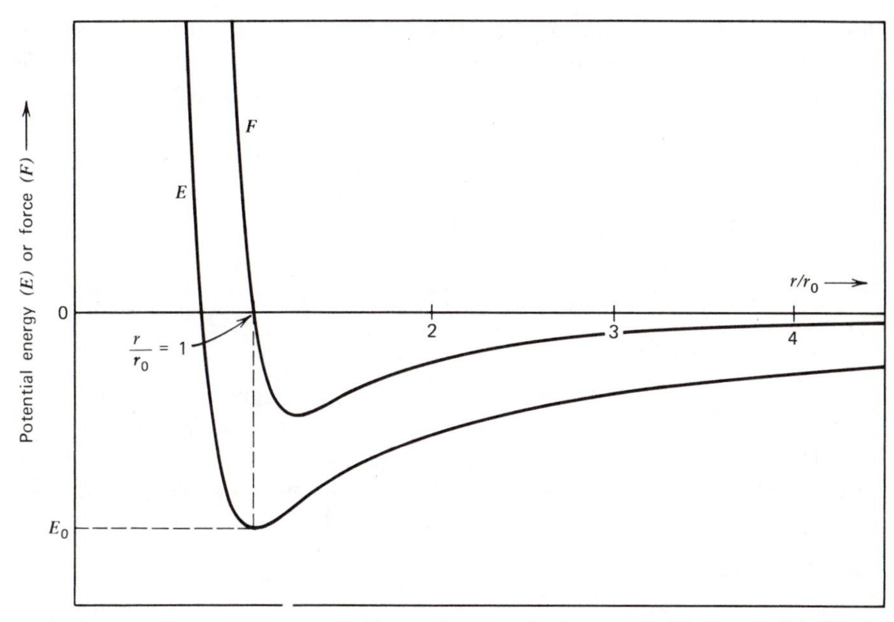

FIGURE 3.4 Variation with interionic separation of interionic force and potential energy of a cation–anion pair. The potential energy is a minimum (E_0) and the repulsive and attractive forces balance at the equilibrium separation r_0.

3.3 LATTICE ENERGY CALCULATIONS

The lattice energy of an ionic solid will differ from the bond energy of diatomic ionic molecules since, in the former case, there will be interactions between more than two ions. To illustrate how the lattice energy calculation is performed, consider the imaginary one-dimensional ionic crystal of Fig. 3.5. The two Cl^- ion neighbors of the Na^+ ion at position 0 contribute $-2e^2/4\pi\varepsilon_0 r_0$ to the electrostatic energy, whereas the two nearest Na^+ ions to either side contribute $+2e^2/8\pi\varepsilon_0 r_0$.

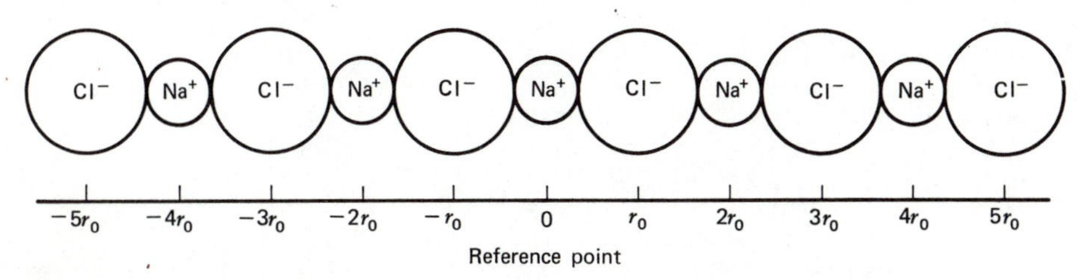

FIGURE 3.5 A hypothetical one-dimensional NaCl ionic crystal.

Thus, there are both negative (attractive) and positive (repulsive) electrostatic energy terms. Continuing this procedure by considering all electrostatic interactions of the reference Na^+ ion at 0, we obtain an infinite number of terms that may be written as

$$E = \frac{1}{4\pi\varepsilon_0}\left(-\frac{2e^2}{r_0} + \frac{2e^2}{2r_0} - \frac{2e^2}{3r_0} + \cdots\right)$$

$$= -\frac{1}{2\pi\varepsilon_0}\frac{e^2}{r_0}\left[\left(1 - \frac{1}{2}\right) + \left(\frac{1}{3} - \frac{1}{4}\right) + \left(\frac{1}{5} - \frac{1}{6}\right) + \cdots\right] \tag{3.4}$$

Since each term enclosed in parentheses is a positive number, the value of the whole expression in brackets must be greater than $\frac{1}{2}$, the magnitude of the first term. Therefore, E must be algebraically less than $-e^2/4\pi\varepsilon_0 r_0$ and the electrostatic potential energy associated with a Na^+ ion in this one-dimensional crystal is lower (i.e., of greater magnitude) than it is for a diatomic NaCl molecule.

TABLE 3.2 Comparison of Measured and Calculated Lattice Energies of Some Ionic Solids

Substance	*U_0, Experimental*[a] *(kJ/mol)*	*U_0, Calculated*[b] *(kJ/mol)*	*Percent Deviation*
LiF	1038	1013	−2.4
LiCl	862	807	−6.4
LiBr	803	757	−5.7
LiI	732	695	−5.1
NaF	923	900	−2.5
NaCl	788	747	−5.2
NaBr	736	708	−3.8
NaI	673	655	−2.7
KF	820	791	−3.5
KCl	717	676	−5.7
KBr	673	646	−4.0
KI	617	605	−1.9
RbCl	687	650	−5.4
CsCl	659	613	−7.0
AgCl	915	837	−8.5
AgI	859	782	−9.0
CaF_2	2624	2601	−0.9
BaF_2	2342	2317	−1.1
MgO	3891	3753	−3.5

[a] Born–Haber cycle.
[b] Eq. 3.5.

A similar calculation can be made for a three-dimensional ionic crystal. The resulting total potential energy, including the ion-core repulsion term, for one mole of formula units is given in general by

$$E_0 = -\frac{N_0 M z^2 e^2}{4\pi\varepsilon_0 r_0} + N_0 n\lambda e^{-r_0/\rho} = -U_0 \tag{3.5}$$

where N_0 is Avogadro's number, M is the Madelung constant (which represents the effect of a specific geometrical array of ions on the electrostatic potential energy), n is the number of anions that are nearest neighbors to a cation, λ and ρ are materials constants, and U_0 is the lattice energy per mole. For NaCl-type crystals the value of M is approximately 1.75; the remaining parameters (r_0, n, λ, and ρ) in Eq. 3.5 can be determined by suitable experiments. Calculated lattice energies often are within a few percent of the values determined experimentally (Table 3.2).

3.4 THE BORN–HABER CYCLE

As was shown in Sect. 3.2, the formation of a NaCl molecule may be visualized as occurring in different steps, each of which results in energy absorption or evolution. Likewise, the complete chemical reaction

$$\text{Na}(s) + \tfrac{1}{2}\text{Cl}_2(g) \longrightarrow \text{NaCl}(s) \tag{3.6}$$

can be analyzed in terms of a number of separate steps as shown in the Born–Haber cycle (Fig. 3.6). The Born–Haber cycle calculation is based on the fact that the enthalpy difference between any two states of a system is independent of the path taken between the states. Thus in Fig. 3.6, the left-hand side of Eq. 3.6 represents state [1] and the right-hand side, state [4]. The corresponding enthalpy change. ΔH_{1-4}, represents the enthalpy of formation of solid NaCl from the standard, or common, states of Na and Cl, which are solid and gaseous, respectively, at room temperature and pressure. The complete Eq. 3.6 can be considered to be the sum of three separate reactions: (1) solid sodium and gaseous diatomic chlorine going to isolated (i.e., gaseous) atoms of each; (2) ionization of the isolated atoms; and (3) the *condensation* of the ions to form the solid sodium chloride lattice. The enthalpy change for the first of these is simply the enthalpy of atomization of solid sodium and $\frac{1}{2}$ the bond energy of diatomic chlorine. The second and third reactions have been discussed previously and their corresponding enthalpy changes are thus $N_0(I_{\text{Na}} - E_{a-\text{Cl}})$ and $-U_0$, the sodium chloride lattice energy. The enthalpy

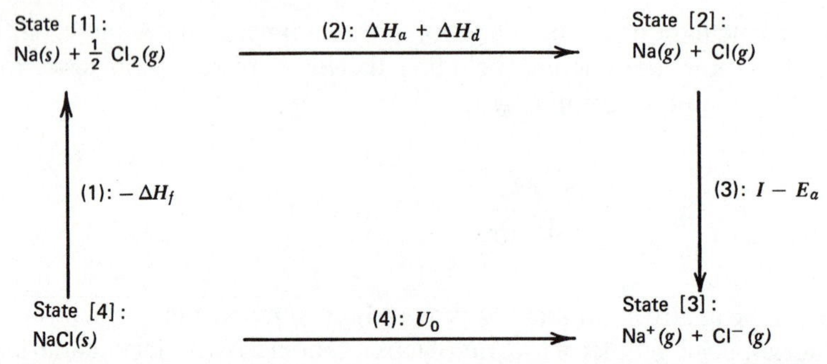

FIGURE 3.6a The Born–Haber cycle. Going from state [1] to state [4] represents the reaction $Na(s) + \frac{1}{2}Cl_2(g) \rightarrow NaCl(s)$. According to thermodynamics, the enthalpy of this reaction is the same as the sum of the enthalpies for the steps from state [1] to state [2], state [2] to state [3], and state [3] to state [4].

Key: $(\Delta H_a)_{Na}$ = enthalpy of atomization for one mole of solid sodium
$(\Delta H_d)_{Cl_2}$ = molar dissociation energy for Cl_2
I_{Na} = first ionization energy for sodium
E_{a-Cl} = electron affinity for chlorine
ΔH_f = enthalpy of formation for one mole of solid NaCl
U_0 = lattice energy

change resulting from these three reactions must be equivalent to ΔH_{1-4} (Fig. 3.6*b*).

(1) Negative of enthalpy of formation,
$NaCl(s) \rightarrow Na(s) + \frac{1}{2}Cl_2(g)$: $-\Delta H_f = +411$ kJ/mol

(2) Enthalpy of atomization,
$Na(s) \rightarrow Na(g)$: $\Delta H_a = +108$ kJ/mol

Enthalpy of molecular dissociation,
$\frac{1}{2}Cl_2(g) \rightarrow Cl(g)$: $\Delta H_d = +121$ kJ/mol
(molecular bond energy)

(3) Ionization potential,
$Na(g) \rightarrow Na^+(g) + e^-$: $I = +496$ kJ/mol

Electron affinity,
$Cl(g) + e^- \rightarrow Cl^-(g)$: $E_a = -349$ kJ/mol

---

(4) Summation = lattice energy,
$NaCl(s) \rightarrow Na^+(g) + Cl^-(g)$: $U_0 = +787$ kJ/mol

(negative of ionic bond energy)

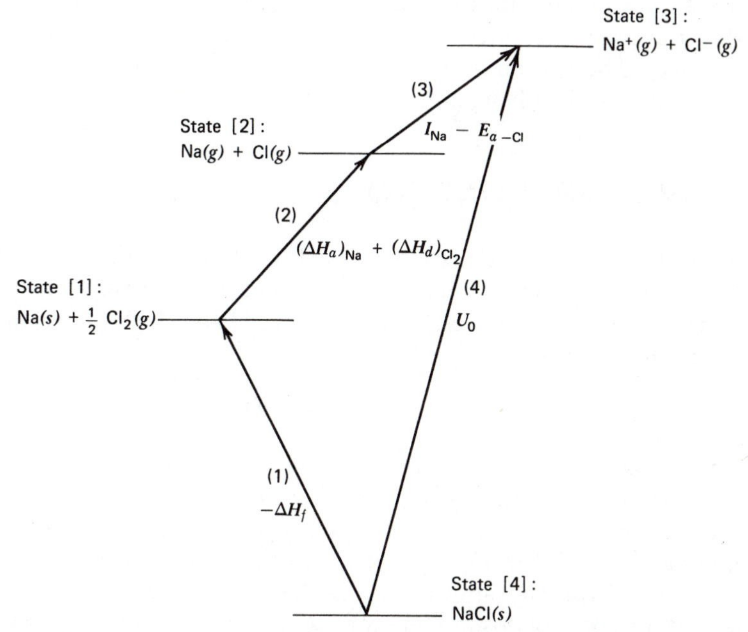

FIGURE 3.6b Energy level diagram representation of the Born–Haber cycle. The relative enthalpy of each state is shown schematically.
Key: $(\Delta H_a)_{Na}$ = enthalpy of atomization for one mole of solid sodium
$(\Delta H_d)_{Cl_2}$ = molar dissociation energy for Cl_2
I_{Na} = first ionization energy for sodium
E_{a-Cl} = electron affinity for chlorine
ΔH_f = enthalpy of formation for one mole of solid NaCl
U_0 = lattice energy

In its most convenient form, the Born–Haber cycle can be used to compare theoretical and experimental values of lattice energy; for example, if ΔH_{1-4}, ΔH_{1-2}, and ΔH_{2-3} are measured, U_0 $(= -\Delta H_{3-4})$ is thus obtained and can be compared with the theoretical value (Table 3.2). Utilization of the Born–Haber cycle for such purposes is limited, however, because electron affinities are imperfectly known for most atoms. The Born–Haber cycle is, however, a convenient device for remembering the various energy terms representing changes from one state to another. Thus the enthalpy of formation of a solid in a chemical reaction always represents the change in enthalpy upon going from the standard state of the elements to the combined state. Similarly the *binding energy* represents the decrease in enthalpy upon going from the gaseous state of isolated atoms to the combined state, whereas the lattice energy of NaCl represents the decrease in enthalpy upon going from isolated ions in the gaseous state to the combined state. The Born–Haber cycle, as presented here, has been restricted to ionic type materials; however, it should be mentioned that similar types of cycles can be used for thermodynamic calculations involving nonionic solids.

3.5 COVALENT BONDING

The simplest case of a single covalent bond occurs in the hydrogen molecule. When two hydrogen atoms combine to form a molecule, neither atom gets complete possession of the bonding electrons to form a closed shell; rather, they share the electrons in a valence bond:

$$\text{H}^{\cdot} + \text{H}^{\cdot} \longrightarrow \text{H:H}$$

Although the simple dot notation employed above is useful in representing many covalent bonds, consideration of valence electron density distributions is more fundamental. Thus, as two widely separated hydrogen atoms in their ground states approach one another (Fig. 3.7*a*), their electron density distributions interact and overlap (Fig. 3.7*b*). The latter represents the valence bond orbital, containing a pair of electrons with opposite spins, formed from atomic orbitals. As shown in Fig. 3.7*b*, the hydrogen molecule has a high electron density (i.e., a concentration of negative charge) between the nuclei (protons), which is reasonable from electrostatic considerations alone.

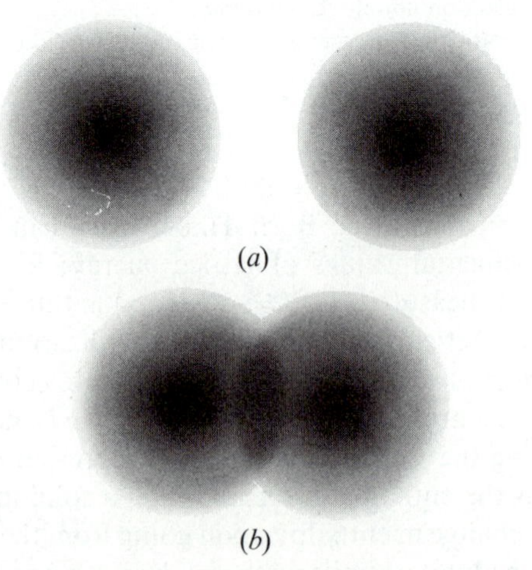

FIGURE 3.7 The electron charge distributions (*a*) for two separate hydrogen atoms and (*b*) for a hydrogen molecule.

Electron sharing is characteristic of covalent or homopolar bonding in other elemental molecules such as N_2, O_2, and F_2, as well. In these cases, p electrons are shared. The F_2 molecule has a single bond (one electron pair) since a fluorine atom ($1s^2 2s^2 2p^5$) is one electron short of having the neon configuration. Similarly, O_2 and N_2 have double and triple bonds, respectively, because of the electronic structures of an oxygen atom ($1s^2 2s^2 2p^4$) and a nitrogen atom ($1s^2 2s^2 2p^3$). Multiple bonding is reflected in the binding energies: 154 kJ/mol for F_2, 494 kJ/mol for O_2, and 942 kJ/mol for N_2.

Dissimilar atoms also can combine to form covalently bonded molecules:

$$\mathrm{H}:\underset{\cdot\cdot}{\overset{\cdot\cdot}{\mathrm{F}}}: \qquad \begin{array}{c} \\ \mathrm{H}:\overset{\cdot\cdot}{\mathrm{O}}: \\ \cdot\cdot \\ \mathrm{H} \end{array} \qquad \begin{array}{c} \mathrm{H} \\ \cdot\cdot \\ \mathrm{H}:\mathrm{N}: \\ \cdot\cdot \\ \mathrm{H} \end{array} \qquad \begin{array}{c} \mathrm{H} \\ \cdot\cdot \\ \mathrm{H}:\mathrm{C}:\mathrm{H} \\ \cdot\cdot \\ \mathrm{H} \end{array}$$

hydrogen fluoride **water** **ammonia** **methane**

In these cases each atom contributes one electron to form an electron-pair bond. Another type of covalent bonding, called coordinate covalent bonding, occurs when both of the shared electrons are supplied by but one of the bonded atoms. Thus, the combination of H^+ with NH_3 leads to an ammonium ion:

$$\begin{array}{c} \mathrm{H} \\ \cdot\cdot \\ \mathrm{H}:\mathrm{N}: \\ \cdot\cdot \\ \mathrm{H} \end{array} + \mathrm{H}^+ \longrightarrow \left[\begin{array}{c} \mathrm{H} \\ \cdot\cdot \\ \mathrm{H}:\mathrm{N}:\mathrm{H} \\ \cdot\cdot \\ \mathrm{H} \end{array}\right]^+$$

Another example of coordinate covalent bonding is provided by the compound $POCl_3$ where both of the electrons in the P—O bond are donated by the phosphorus atom whereas in the P—Cl bonds one electron is donated by Cl and one by P:

$$\begin{array}{c} \cdot\cdot \\ :\mathrm{Cl}: \\ \cdot\cdot\ \ \cdot\cdot\ \ \cdot\cdot \\ :\mathrm{Cl}:\mathrm{P}:\mathrm{O}: \\ \cdot\cdot\ \ \cdot\cdot\ \ \cdot\cdot \\ :\mathrm{Cl}: \\ \cdot\cdot \end{array}$$

Binding energies and a list of bond lengths associated with a number of covalent bonds are given in Table 3.3. It should be noted that multiple carbon–carbon bonds have shorter lengths and higher energies, although the double bond is not twice as strong nor the triple bond three times as strong as a single carbon–carbon bond. This apparent anomaly is related to the molecular orbitals responsible for multiple bonding, a topic that will be deferred until Chapter 4. At this point it is

TABLE 3.3 Covalent Bond Lengths and Associated Binding Energies

Bond	Bond Length (nm)	Binding Energy (kJ/mol of bonds)
H—F	0.092	561
H—O	0.096	464
H—N	0.101	389
H—C	0.109	414
O=O	0.121	494
N≡N	0.110	942
C—C	0.154	347
C=C	0.134	611
C≡C	0.120	837
H—H	0.074	436
F—F	0.144	154
Cl—Cl	0.202	239
Br—Br	0.228	190
I—I	0.267	149
H—Cl	0.128	428
H—Br	0.142	362
H—I	0.161	295

important to realize that the utilization of the electron dot notation to represent covalent bonding is insufficient because it affords very little information about the nature of the bonds and no information about bonding geometry.

3.6 POLAR COVALENT BONDING

Whereas the bonding in molecules like H_2, F_2, or O_2 is considered to be purely covalent, that in HF, H_2O, NH_3, or CH_4 has a considerable degree of ionic character because of an asymmetrical electron density distribution. This in combination with a structural asymmetry leads to polar molecules in many cases. As shown in Fig. 3.8, the HF molecule is a polar molecule having a permanent electric dipole because the centers of positive and negative charge do not coincide. The water molecule is also polar but nonlinear, with an average angle of 104.5° between the two O—H bonds (Fig. 3.9*a*). That the bond angle is 104.5° rather than the 90° angle between the atomic 2*p* orbitals of oxygen may be attributed to repulsion between the electron clouds constituting the two O—H bonds. The ammonia molecule, also polar, has a pyramidal structure with average angles of 107.3° between the N—H bonds (Fig. 3.9*b*). Although the atomic 2*p* orbitals of nitrogen

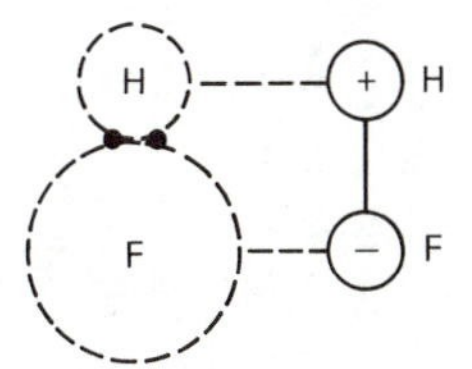

FIGURE 3.8 The electron charge distribution for an HF molecule. The partial ionic character of the HF molecule results in separate centers of negative and positive charge as shown. This gives rise to a permanent electric dipole.

that contribute to the bonding are at right angles to one another, the actual bond angles are spread as a result of electron repulsions. The electric dipole moment of H_2O is less than that of HF but greater than that of NH_3 (Table 3.4). On the other hand, methane has no permanent dipole moment because it possesses a symmetrical structure with average bond angles of 109.5° between the four C—H bonds (Fig. 3.9*c*).

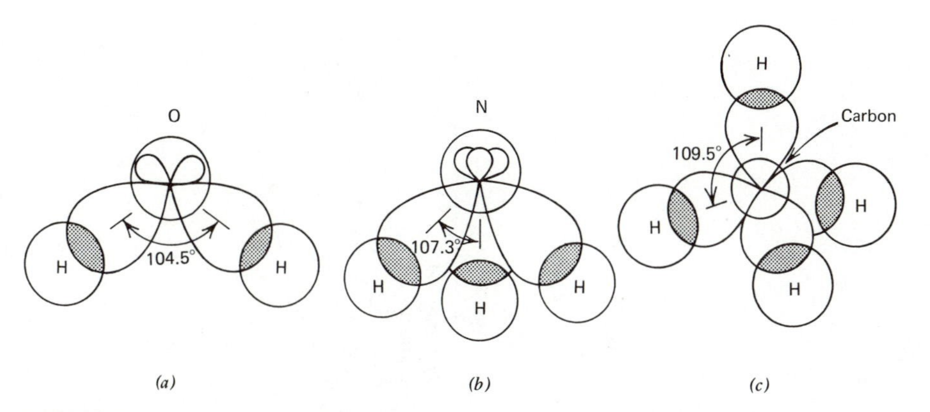

FIGURE 3.9 Schematic representation of (*a*) the H_2O molecule, (*b*) the NH_3 molecule, and (*c*) the CH_4 molecule. The average bond angles and geometrical arrangement of the bonding orbitals are indicated.

TABLE 3.4 Properties of Neon and of Compounds Having the Same Number of Electrons as Neon

Substance	T_m (°C)	ΔH_m (kJ/mol)	T_v (°C)	ΔH_v (kJ/mol)	*Molecular Dipole Moment* $\times 10^{30}$ (C · m)	*Type of Bonding Between Molecules in Solid*
Methane, CH_4	−182.5	0.96	−161.5	8.16	0	Van der Waals
Ammonia, NH_3	−77.7	5.65	−33.4	23.35	4.9	Hydrogen
Water, H_2O	0	6.01	100	40.88	6.2	Hydrogen
Hydrogen fluoride, HF	−83	4.56	20	7.53	6.6	Hydrogen
Neon, Ne	−248.7	0.34	−245.9	1.76	0	Van der Waals

3.7 HYBRIDIZATION

The tetrahedral structure of methane, with average bond angles of 109.5° (Fig. 3.9*c*), is somewhat more difficult to rationalize than those of HF, H_2O, or NH_3 since carbon in the ground state has a $1s^22s^22p^2$ electron configuration with only two $2p$ orbitals capable of bonding. However, if one $2s$ electron is promoted to the empty $2p$ orbital, a higher energy electron configuration ($1s^22s^12p^3$) results. This gives four unpaired electrons, all having parallel spins, and makes possible the bonding with the four hydrogen atoms in methane. In reality, the four molecular bonding orbitals become equivalent as a result of hybridization (suitable combination of the atomic bonding orbitals). The resulting molecular bonding orbitals in this case are termed sp^3 hybrid orbitals and are directed symmetrically toward the corners of a tetrahedron (Fig. 3.10*a*).

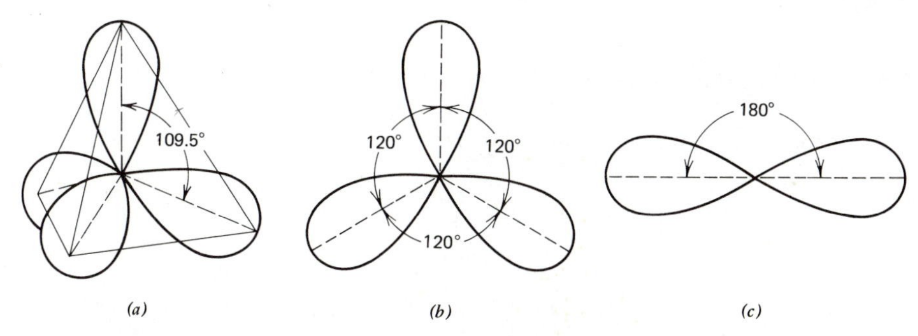

FIGURE 3.10 (*a*) sp^3 hybrid orbitals. These are directed symmetrically toward the corners of a rectangular tetrahedron. (*b*) sp^2 hybrid orbitals. These are directed at angles of 120° within a plane. (*c*) *sp* hybrid orbitals. These are directed at 180° with respect to each other.

Although the sequence of promotion and then hybridization

$$1s^22s^22p^2 \longrightarrow 1s^22s^12p^3 \longrightarrow 1s^22(sp^3)$$

requires an increase in energy, the formation of four covalent bonds results in a lower final energy state than would the formation of two bonds, and CH_4 exists as a stable molecule but CH_2 does not. The definite directionality of hybrid orbitals is accompanied by greater spatial extension of these orbitals, which facilitates more electron cloud overlap during bonding. It is important to point out that the geometry of both H_2O and NH_3 can also be discussed in terms of sp^3 hybrid orbitals, although in these cases each of the four orbitals is not equivalent since some are occupied by electron pairs not involved in bonding.

Hybridization is a common phenomenon that can be used to explain the definite directionality found in many cases of covalent bonding. Thus, sp^2 hybrid orbitals, which are directed at 120° angles within a plane (Fig. 3.10*b*), account partially for the bonding of carbon to three neighboring atoms in ethylene, C_2H_4 or

```
H       H
 \     /
  C=C
 /     \
H       H
```

In addition, *sp* hybrid orbitals, which are directed at 180° with respect to each other (Fig. 3.10*c*), play an important role in leading to the linear structure of acetylene, H—C≡C—H. (The nature of the double and triple bonds in molecules such as ethylene and acetylene will be elucidated in Chapter 4.) More involved hybridization schemes involving *s*, *p*, and *d* atomic orbitals can be used to rationalize the structures of many complex compounds.

3.8 RESONANCE

The bonding in benzene is commonly represented as

```
        H
        |
  H   C   H
   \ // \ /
    C     C
    |     ||
    C     C
   / \\ / \
  H   C    H
      |
      H
```

however, all carbon–carbon bond lengths in this molecule are equal (0.139 nm) and longer than a C=C double bond (0.134 nm) but shorter than a C—C single bond (0.154 nm). A better representation is provided by

```
        H                          H
        |                          |
  H   C   H                  H   C   H
   \ // \ /                   \ / \\ /
    C     C                    C     C
    |     ||       <-->        ||    |
    C     C                    C     C
   / \\ / \                   / \ // \
  H   C    H                 H   C    H
      |                          |
      H                          H
```

(*a*) (*b*)

where the molecule may be considered as resonating or rapidly alternating between structures (*a*) and (*b*). In reality, neither of these structures ever exists, and the actual bonding structure, called a resonance hybrid, is intermediate between them. The necessity of invoking the concept of resonance arises because of our rather simplified approach to covalent bonding in terms of valence bonds. Nevertheless, it is an empirical fact that when resonance structures must be used to represent an actual bonding structure, the actual bonding is more stable. In other words, resonance confers additional stability on a covalent bond. For example, if the binding energy of the benzene molecule is calculated on the basis of six C—H, three C—C, and three C═C bonds (Table 3.3), it amounts to 5358 kJ/mol; however, the binding energy found experimentally is 5515 kJ/mol, or an excess of 157 kJ/mol.

Resonance is a general concept that also plays a role in polar molecules like HF and HCl, where resonance between purely covalent and purely ionic forms can be conceived. Indeed, an important consequence of the resonance concept is that there is no absolute distinction between ionic and covalent bonding, and a continuous transition from one to the other is possible.

3.9 ELECTRONEGATIVITY

The possible resonance structures for the HF molecule are

$$\underset{\substack{\textbf{ionic}\\(a)}}{H^+F^-} \longleftrightarrow \underset{\substack{\textbf{covalent}\\(b)}}{H\text{—}F} \longleftrightarrow \underset{\substack{\textbf{ionic}\\(c)}}{H^-F^+}$$

If (*a*) and (*c*) have an equal weight in contributing to the actual resonance hybrid, then we would expect HF to be a nonpolar molecule. Actually, HF is a polar molecule with structure (*a*) predominating over structure (*c*). That is, the fluorine atom attracts electrons preferentially in comparison to the hydrogen atom. This feature can be quantified through the introduction of the concept of electronegativity. Electronegativity is the ability of an atom in a molecule or in a solid to attract bonding electrons to itself. This, of course, is not the same as electron affinity (see Sect 2.8). The polar nature of the HF molecule can be described by saying that the fluorine atom has a larger electronegativity than does the hydrogen atom.

If purely covalent bonding existed in the HF molecule, it would be plausible to expect that the binding energy would be given by

$$E_{\text{H—F}} = \tfrac{1}{2}(E_{\text{H—H}} + E_{\text{F—F}}) \tag{3.7}$$

that is, the arithmetic mean of the normal covalent binding energies for a single H—H bond and a single F—F bond. This amounts to about 295 kJ/mol; however, the actual HF binding energy is significantly larger (561 kJ/mol). This difference is due to the extra stability afforded by resonance or inequality of electron sharing:

$$\Delta_{\mathrm{H-F}} = E_{\mathrm{H-F}} - \tfrac{1}{2}(E_{\mathrm{H-H}} + E_{\mathrm{F-F}}) \tag{3.8}$$

On the other hand, the binding energy of HI (295 kJ/mol) is just slightly larger than the arithmetic mean of the H—H and I—I binding energies (292 kJ/mol), and $\Delta_{\mathrm{H-I}}$ is small. Extensive empirical studies have shown that Δ is greater than or equal to zero for any particular bond and increases as the inequality of electron sharing becomes greater. The electronegativity difference between two atoms provides a measure of the inequality of electron sharing, and Pauling has suggested the relationship

$$\Delta_{A-B} = 96.5(x_A - x_B)^2, \tag{3.9}$$

where Δ_{A-B} is the excess binding energy (in kJ/mol of bonds), and x_A and x_B are the electronegativities of A and B. By using Eq. 3.9 and relations such as Eq. 3.8, Pauling and others have developed an empirical scale of relative electronegativities (Table 3.5). With such a table and a knowledge of certain normal covalent bond binding energies, one can predict the binding energy of a compound:

$$E_{A-B} = \tfrac{1}{2}(E_{A-A} + E_{B-B}) + 96.5(x_A - x_B)^2 \tag{3.10}$$

The result often is sufficiently accurate so as to be very useful; however, calculations involving electronegativity values are necessarily approximate because the electronegativity of an atom depends upon its bonding situation.

A perusal of Table 3.5 should lead to the important conclusion that metallic elements possess relatively low electronegativities and nonmetallic elements possess relatively high electronegativities, with $x = 2$ approximately dividing the two classes of elements. When two elements having a small electronegativity difference combine to form a compound, the bonding will be predominately nonpolar covalent or metallic, depending upon the nature of the elements involved. With an increasing electronegativity difference, the bonding becomes more polar, tending toward an ionic nature. Thus, the sequence

Si AlP MgS NaCl

goes from purely covalent bonding (Si) to largely ionic bonding (NaCl), and the sequence

HF HCl HBr HI

TABLE 3.5 Electronegativities of the Elements

H 2.1																
Li 1.0	Be 1.5											B 2.0	C 2.5	N 3.0	O 3.5	F 4.0
Na 0.9	Mg 1.2											Al 1.5	Si 1.8	P 2.1	S 2.5	Cl 3.0
K 0.8	Ca 1.0	Sc 1.3	Ti 1.5	V 1.6	Cr 1.6	Mn 1.5	Fe 1.8	Co 1.8	Ni 1.8	Cu 1.9	Zn 1.5	Ga 1.6	Ge 1.8	As 2.0	Se 2.4	Br 2.8
Rb 0.8	Sr 1.0	Y 1.2	Zr 1.4	Nb 1.6	Mo 1.8	Tc 1.9	Ru 2.2	Rh 2.2	Pd 2.2	Ag 1.9	Cd 1.7	In 1.7	Sn 1.8	Sb 1.9	Te 2.1	I 2.5
Cs 0.7	Ba 0.9	La 1.1	Hf 1.3	Ta 1.5	W 1.7	Re 1.9	Os 2.2	Ir 2.2	Pt 2.2	Au 2.4	Hg 1.9	Tl 1.8	Pb 1.8	Bi 1.9	Po 2.0	At 2.2

goes from a highly polar character (HF) to a weakly polar character (HI). Purely covalent and purely ionic bonding represent extremes on a continuous bonding scale. Indeed, if the four extreme bond types (ionic, covalent, metallic, and van der Waals) are pictorially positioned at the corners of a tetrahedron (Fig. 3.11), then any actual bond can be located on the surface of or within the tetrahedron.

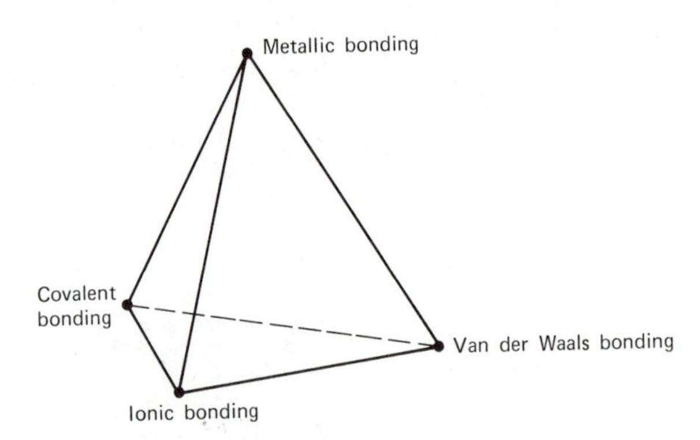

FIGURE 3.11 Tetrahedron of bond types. The tips of the tetrahedron represent pure ionic, covalent, metallic, and van der Waals bonding. The bonding in most actual materials is represented by a point on the surface of or within the tetrahedron.

Electronegativity can be approached in another manner that provides physical insight into the concept. Since electronegativity represents the attractive power that an atom has for bonding electrons, it is reasonable to propose that it is proportional to the average of the ionization energy and the electron affinity:

$$x \propto \frac{I + E_a}{2} \tag{3.11}$$

The first ionization energy is the energy of attraction between an electron and a monovalent cation, and the electron affinity is the energy of attraction between an electron and a neutral atom. Computation of electronegativities with Eq. 3.11, using values of I and E_a (Tables 2.3 and 2.4), gives values of x that are generally consistent with those in Table 3.5 when a suitable constant scaling factor is employed. Furthermore, defining electronegativity in terms of Eq. 3.12 naturally leads to the dependence of x upon the valence state of an ion, for example, because both I and E_a depend upon the valence state. This provides a rationalization for the observation that electronegativity has different values for an atom in different bonding states.

3.10 METALLIC BONDING

The stability of both covalent and metallic bonds arises from the potential energy lowering experienced by valence electrons under the influence of more than one nucleus. In metals, where valence electrons are not tightly bound to the ion cores, we cannot expect the formation of strong electron-pair bonds. The bond energies of known diatomic molecules of metallic elements are, in fact, smaller than those for molecules consisting of nonmetallic elements: 112 kJ/mol for Li_2, 77 kJ/mol for Na_2, and 17 kJ/mol for Hg_2. Only diatomic molecules of the semimetals have relatively high binding energies (384 kJ/mol for As_2, 299 kJ/mol for Sb_2, and 200 kJ/mol for Bi_2), and these values reflect multiple bonding. Much greater stability is possible in larger aggregates of atoms such as bulk metals (see Table 3.1). Indeed, it is only in large aggregates that metallic bonding occurs; the diatomic molecules mentioned above are covalently bonded.

The known properties of metals, such as high electrical conductivities, support the conceptual view that the valence electrons never remain near any particular atom very long but drift in a random manner throughout the whole metal. We may therefore visualize the ordered array of ion cores in a metal as being held together by a gas of free electrons. This situation arises because the number of valence electrons is too small to provide complete shells for all the atoms. Consequently, the bonding structure of a metal can be regarded as an extreme form of a resonance hybrid structure in which electron-pair bonds occur transiently between various pairs of atoms; however, this oversimplified view offers little in the way of physical understanding, and a more useful approach to metallic bonding will be discussed in Chapter 4.

3.11 THE HYDROGEN BOND

The hydrogen bond is the name given to a unique kind of physical bond that is significantly stronger than van der Waals or permanent dipole bonds but appreciably weaker than chemical bonds (see Table 3.1). Hydrogen bonding is particularly important in many polymers, including proteins, and plays a dominant role in determining the nature of H_2O. It occurs between polar molecules that contain hydrogen covalently bonded to a highly electronegative atom (N, O, or F). The bond polarity arises because the electronegative atom attracts the bonding electrons more strongly than does hydrogen so that the center of negative charge is much closer to the electronegative atom than is the center of positive charge. On this account and because the sole electron of hydrogen is involved in covalent bonding, a rather strong electrostatic attraction exists between a hydrogen atom of one molecule and the electronegative atom of another molecule. In effect, the

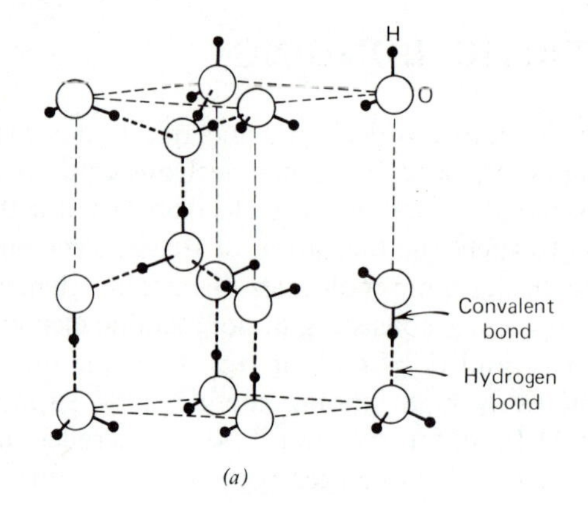

(a)

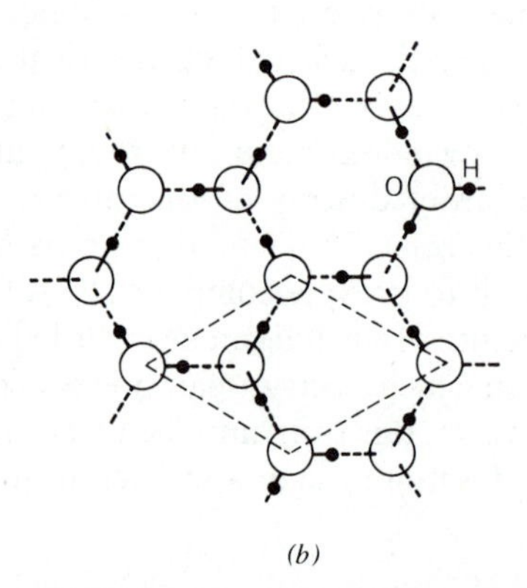

(b)

FIGURE 3.12 Model of an ice crystal. As shown in the perspective representation (*a*), each oxygen atom is coupled to two hydrogen atoms in other H_2O molecules by hydrogen bonds. In (*b*), the hexagonal symmetry of ice is shown. Hydrogen bonds are indicated as (----) and covalent bonds, as (——).

hydrogen atom forms a bridge between two electronegative atoms as shown in Fig. 3.12 for ice. Hydrogen bonding is directional because of the definite geometry of the molecules. Thus, the structure of an ice crystal is more open and less dense than that of water. Hydrogen bonding between H_2O molecules in water is still effective but decreases with increasing temperature. At room temperature, water is composed of small groups of H_2O molecules associated by hydrogen bonding. The size of these associated groups decreases with increasing temperature.

It is worth noting that HF, H_2O, and NH_3 have melting and boiling points that are considerably higher than those of CH_4 and Ne (Table 3.4). Hydrogen bonding predominates between molecules in the former group, whereas only van der Waals bonding occurs between CH_4 molecules or Ne atoms. Another important characteristic, revealed in Table 3.4, is the relatively wide temperature range of liquid stability that exists for the hydrogen bonded substances as compared to the narrow temperature range of liquid stability that exists for the van der Waals bonded substances.

3.12 OTHER TYPES OF PHYSICAL BONDS

The weakest type of interatomic or intermolecular bonding is van der Waals bonding, which occurs between inert gas atoms (see Sects. 2.10 and 3.1) and also between nonpolar molecules in the liquid or solid state. Van der Waals forces are ubiquitous, occurring in all solids and liquids as a result of fluctuating electric dipole interactions; however, for a substance in which chemical bonding or hydrogen bonding exists, the contribution of van der Waals forces is almost negligible. Hence, they are truly important only with inert gas elements and many molecular substances.

Polar covalent molecules that do not form hydrogen bonds (e.g., HCl, H_2S, and ClF) interact with each other somewhat more strongly than do nonpolar molecules, although the attraction is still much weaker than that due to hydrogen bonding. In such a case both permanent dipole interactions and fluctuating dipole interactions contribute to the attraction. Even intermolecular bonding between a polar molecule and a nonpolar molecule can be slightly stronger than van der Waals bonding alone because a polar molecule, due to its associated charge distribution, induces a polarity in the nonpolar molecule.

REFERENCES

1. Chemical Bond Approach Project. *Chemical Systems*, McGraw-Hill, New York, 1964.
2. Mahan, B. H. *College Chemistry*, Addison-Wesley, Reading, Mass., 1969.
3. Hägg, G. *General and Inorganic Chemistry*, Wiley, New York, 1969.

QUESTIONS AND PROBLEMS

3.1 How does chemical or primary bonding differ from physical or secondary bonding?

3.2 The radius of a Na^+ cation is 0.098 nm and that of a Cl^- anion is 0.181 nm.

(a) What is the attractive force between these two ions, assuming an equilibrium separation of 0.279 nm? (Equation 3.1 gives the force in newtons if r is in meters and $e = 1.602 \times 10^{-19}$ C.)

(b) What is the repulsive force at the same separation?

3.3 For an ionic crystal the potential energy (in J/mol) can be written as

$$E = -\frac{1}{4\pi\varepsilon_0}\left(\frac{N_0 M z^2 e^2}{r}\right) + N_0 n \lambda e^{-r/\rho}$$

At the equilibrium cation–anion separation (r_0) the interionic force is given by

$$F = -\left(\frac{dE}{dr}\right)_{r_0} = 0$$

(a) Differentiate the above equation with respect to r and solve for $n\lambda$ in terms of $Mz^2e^2/4\pi\varepsilon_0$, ρ, and r_0.

(b) Substitute your answer to part (a) into the equation and obtain the lattice energy U_0 (for $r = r_0$) in terms of $(1/4\pi\varepsilon_0)(N_0 M z^2 e^2)$, ρ, and r_0.

3.4 (a) Using the answer to part (b) of problem 3.3, calculate the lattice energy for NaCl. (For NaCl, $M = 1.748$, $\rho = 0.033$ nm, and $r_0 = 0.282$ nm; $e = 1.602 \times 10^{-19}$ C.)

(b) What is the lattice energy for MgO? (MgO has the same crystal structure as NaCl; $\rho = 0.039$ nm and $r_0 = 0.210$ nm.)

(c) The melting point of MgO is 2800°C and that of NaCl is 801°C. Account for this difference.

3.5 Instead of the equation given in problem 3.3, the potential energy of an ionic crystal is sometimes written as

$$E = -\frac{1}{4\pi\varepsilon_0}\left(\frac{N_0 M z^2 e^2}{r}\right) + \frac{N_0 B}{r^n}$$

where B and n are materials constants. (Here, n is a number between about 5 and 12, and is not the same as n in problem 3.3.)

(a) Differentiate the above equation with respect to r and solve for B.

(b) Substitute your answer to part (a) into the equation and obtain the lattice energy U_0 (for $r = r_0$) in terms of $(1/4\pi\varepsilon_0)(N_0 M z^2 e^2)$, n, and r_0.

(c) Use the data for NaCl given in part (a) of problem 3.4 and $n = 8$ to calculate the lattice energy for NaCl. How does this value compare to the value obtained in problem 3.4?

3.6 Distinguish between binding energy and bond energy (lattice energy) for an ionic solid, using sketches and words.

3.7 Compare covalent bond lengths (Table 3.3) with covalent radii of the elements (Table 2.5). Comment on the accuracy of the comparison.

3.8 Sketch a bonding tetrahedron similar to Fig. 3.11. Locate on it the bonding characteristics of CH_4, solid NaCl, solid Na, solid argon, HF, and solid chromium.

3.9 A somewhat inaccurate, but geometrically convenient way of visualizing carbon bonding is to consider the carbon nucleus at the center of a tetrahedron with four valence electron clouds extending to corners of the tetrahedron. In this scheme, a carbon–carbon single bond represents tetrahedra joined tip-to-tip, a double bond represents tetrahedra joined edge-to-edge, and a triple bond represents tetrahedra joined face-to-face. Calculate the expected ratio of single, double, and triple bond lengths according to this geometrical interpretation and compare with the measured bond lengths (Table 3.3). Comment on your results.

3.10 The Lewis octet rule states that an atom other than hydrogen tends to form covalent bonds until it is surrounded by eight electrons.

(a) Show that this is consistent with the behavior of C, N, O, and F in their compounds with H.

(b) Show that this is consistent with the existence of an S_8 ring molecule.

3.11 Can you form a lithium molecule that satisfies the octet rule? An aluminum molecule? Explain.

3.12 NaCl dissolves in water. Indicate, with a sketch, how you would expect water molecules to be arrayed around a Na^+ ion and around a Cl^- ion.

3.13 (a) Classify ionic, covalent, and metallic bonding as directional or nondirectional. Provide a brief justification for each.

(b) Classify van der Waals and hydrogen bonding as directional or nondirectional. Provide a brief justification for each.

3.14 List the number of atoms bonded to a carbon atom that exhibits sp^3, sp^2, and sp hybridization. Also state the resulting atomic configuration (i.e., arrangement).

Molecular Orbitals

Molecular orbitals, which are particularly useful in describing bonding in covalent molecules, covalent solids, and metals, are discussed in this chapter.

4.1 THE CONCEPT OF MOLECULAR ORBITALS

There are two chief types of approximation that are commonly used to describe chemical bonding: the valence bond approach and the molecular orbital approach. In the valence bond theory, which was employed to discuss covalent bonding in Chapter 3, a molecule is regarded as being composed of atoms that to some extent preserve their distinct character even when chemically combined. Bonding occurs by electron-pair sharing or, more precisely, by overlap of valence electron density distributions.

It is conceptually simpler to treat a molecule from the same point of view as that used for atoms. In the molecular orbital theory, a molecule is considered to have unique molecular orbitals and energy states that can be occupied by valence electrons, and in the molecular ground state the valence electrons fill up the energy levels sequentially in a manner consistent with the Pauli exclusion principle and with Hund's rule (see Sect. 2.6). Since both the valence bond theory and the molecular orbital theory are used to describe chemical bonding, they should give consistent results. However, because each is an approximation, one or the other often provides a more useful means of visualizing bonding.

4.2 CLASSIFICATION OF MOLECULAR ORBITALS

In the quantum theory of covalent bonding, it is found that the electrostatic interaction between nuclei and valence electrons results in a simultaneous attraction of the nuclei for the bonding electrons. This corresponds to the overlapping of valence electron density distributions, with the bond strength increasing as the amount of overlap increases, and may also be viewed as the formation of a molecular orbital from atomic orbitals.

Just as atomic orbitals have definite spatial symmetries, so do molecular orbitals. The simplest molecular orbitals are designated σ (sigma) and π (pi); the former is axially symmetric and located directly between two bonding atoms, whereas the latter is not axially symmetric. These are the molecular analogs of atomic *s* and *p* orbitals. The *s–s* overlap in H_2, *s–p* overlap in HF, *p–p* overlap in F_2, and s–sp^3 overlap in CH_4 all give rise to σ molecular orbitals (Fig. 4.1). That is, endwise overlap of atomic or hybrid orbitals results in a σ orbital that can be occupied by two electrons having opposite spins. Sidewise or lateral overlap of atomic *p* orbitals yields a π molecular orbital (Fig. 4.2*a*), with the electron density distribution concentrated above and below a line drawn between the atomic nuclei. Because atomic *p* orbitals are mutually orthogonal, it is possible to have simultaneous sidewise overlap of two *p* orbital pairs, which gives two π orbitals (Fig. 4.2*b*). When this situation exists and both of the π orbitals are filled with electrons, the resulting electron density distribution is cylindrically symmetric about an axis between the two nuclei (Fig. 4.2*c*).

Up to this point we have ignored an important feature that derives from the quantum mechanical treatment of electrons in orbitals. That is, when orbitals overlap they may either reinforce and add together or they may cancel and subtract from one another. To take this into account, we introduce a mathematical artificiality in which an entire atomic orbital is labeled + or −, or discrete parts of a single atomic orbital are labeled + and −. This labeling has no physical significance with respect to the nature of an orbital or its occupancy by electrons, but it permits combination of atomic orbitals to form appropriate molecular orbitals. For example, when two *s* orbitals are brought together, they may form either a σ orbital or a σ^* orbital (Fig. 4.3*a*). The former is called a bonding orbital since it is at a lower energy than the atomic *s* orbitals and, when occupied, will concentrate electrons between the nuclei. The latter is called an antibonding oribtal since it is at a higher energy than the atomic *s* orbitals and, when occupied, will concentrate electrons away from the region between the nuclei. Similarly, endwise overlap of *p* orbitals may give rise to either σ or σ^* orbitals (Fig. 4.3*b*). Lateral overlap of *p* orbitals will result in either π or π^* orbitals (Fig. 4.3*c*). A significant consequence of the above is that the total number of molecular orbitals

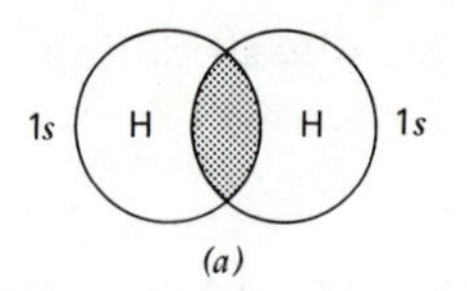

(a)

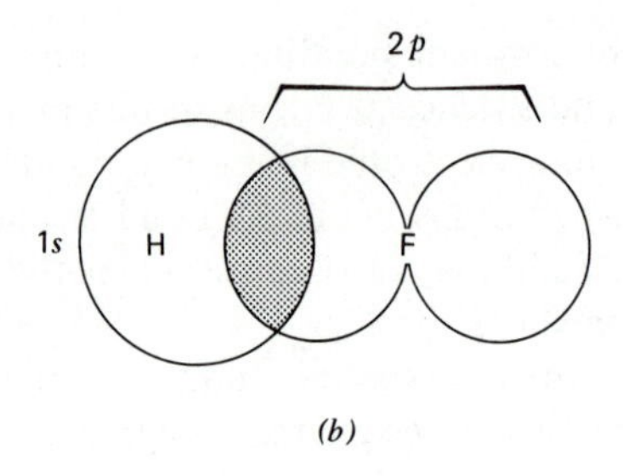

(b)

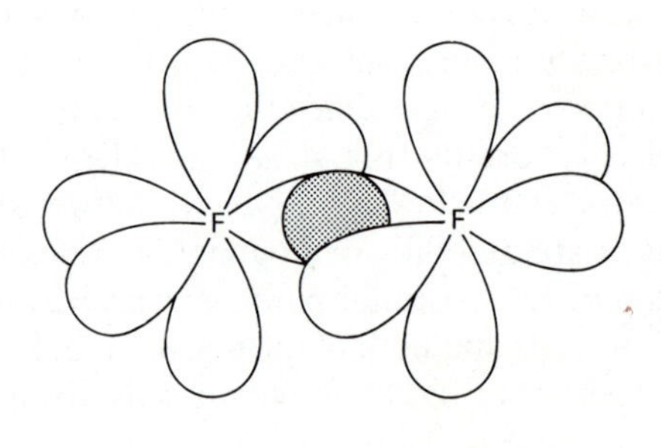

(c)

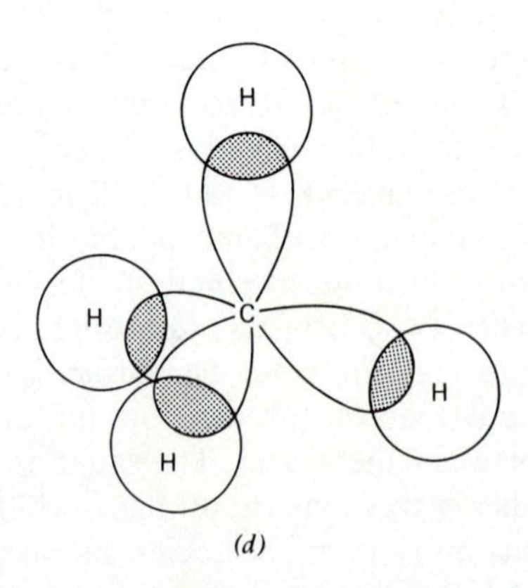

(d)

FIGURE 4.1 σ molecular orbitals arising from (*a*) *s* atomic orbital overlap in H_2, (*b*) *s-p* orbital overlap in HF, (*c*) *p-p* overlap in F_2, and (*d*) $s\text{-}sp^3$ overlap in CH_4.

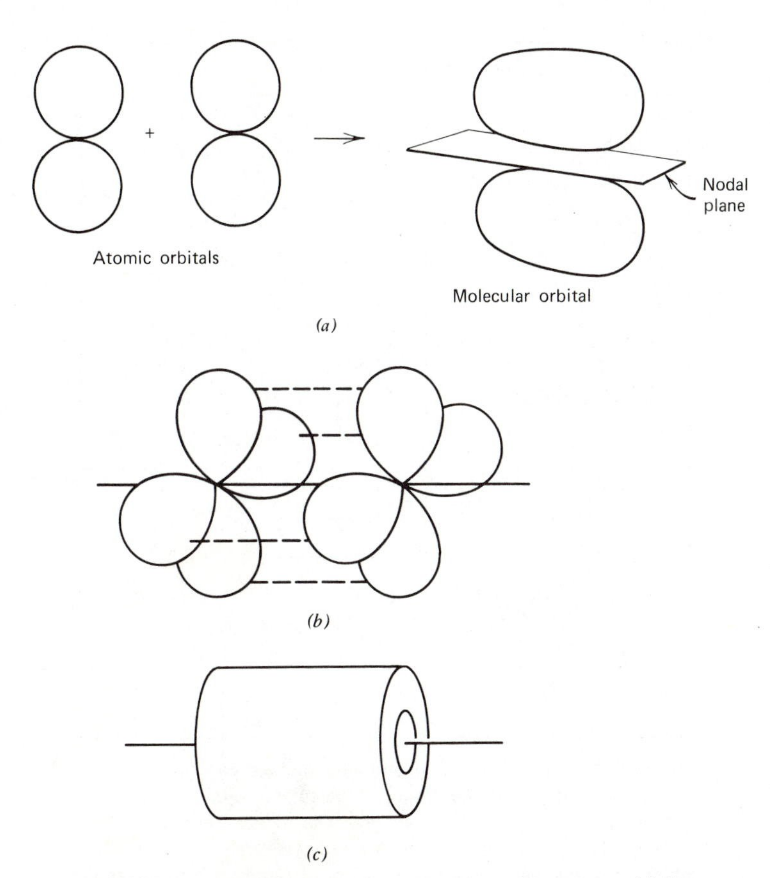

FIGURE 4.2 (*a*) Sidewise or lateral overlap of atomic *p* orbitals to produce a π molecular orbital. (*b*) Simultaneous sidewise overlap of two atomic *p* orbital pairs; the resulting electron density distribution (*c*) is cylindrically symmetric about an axis between the two nuclei.

equals the summation of all atomic orbitals that overlap when atoms are brought together to form a molecule. Half of the molecular orbitals are bonding in nature while the other half are antibonding.

4.3 ELECTRONIC ENERGY LEVELS IN SIMPLE MOLECULES

The occupancy of molecular orbitals by electrons depends upon their energy levels. A simple example is provided by H_2. Here, the pair of bonding electrons occupy a σ orbital. This molecular configuration is designated $(\sigma 1s)^2$, where $\sigma 1s$

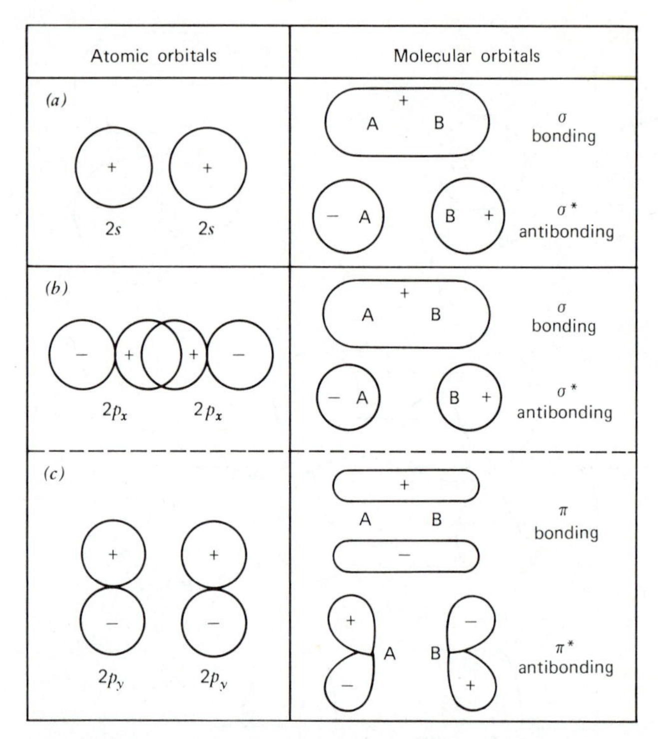

FIGURE 4.3 (*a*) Overlap of atomic *s* orbitals can give rise to a σ (bonding) or σ^* (antibonding) molecular orbital. (*b*) Endwise overlap of atomic *p* orbitals can give rise to a σ (bonding) or σ^* (antibonding) molecular orbital. (*c*) Sidewise overlap of atomic *p* orbitals can give rise to a π (bonding) or π^* (antibonding) molecular orbital.

indicates that the occupied molecular orbital is formed from atomic 1*s* orbitals and the superscript indicates that two electrons are in this orbital. Figure 4.4*a* shows an energy level diagram for H_2. Consideration of a hypothetical He_2 molecule leads to the designation $(\sigma 1s)^2(\sigma^* 1s)^2$ and the energy level diagram shown in Fig. 4.4*b*. The He_2 molecule is unstable because complete occupancy of an antibonding orbital and its corresponding bonding orbital usually leads either to no bonding or to a small degree of antibonding, which means that an He_2 molecule has either the same or a higher energy than two separate He atoms. On the other hand, a He_2^+ ion with $(\sigma 1s)^2(\sigma^* 1s)^1$ does exhibit binding.

For most homonuclear diatomic molecules in the second period of the periodic table, the sequence of energy levels is

$$\sigma 1s < \sigma^* 1s < \sigma 2s < \sigma^* 2s < \sigma 2p < \pi 2p < \pi^* 2p < \sigma^* 2p$$

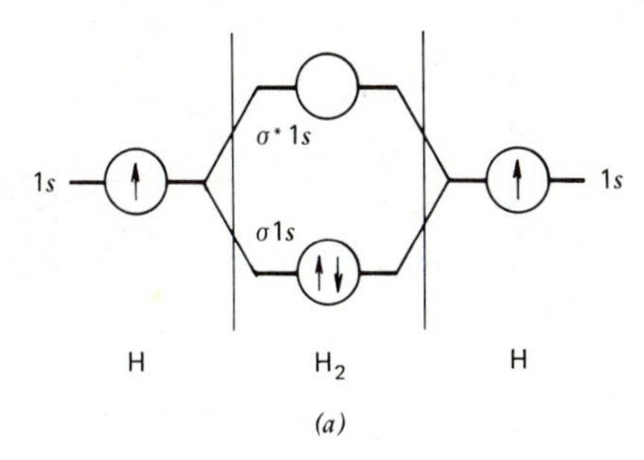

(a)

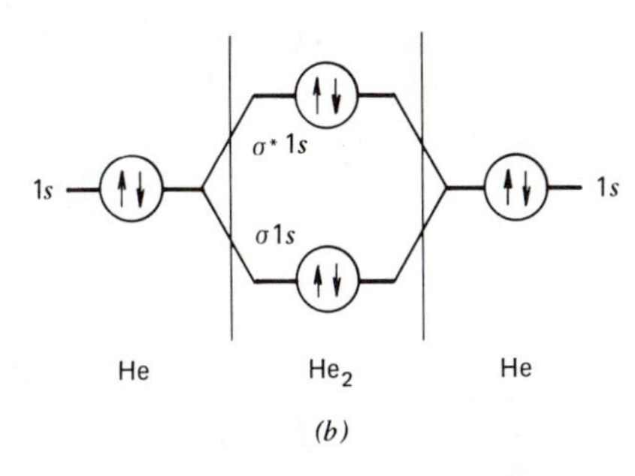

(b)

FIGURE 4.4 Energy level diagrams for (*a*) the H_2 and (*b*) He_2 molecules. In H_2, the two electrons occupy a $\sigma 1s$ bonding orbital and the energy of the H_2 molecule is less than that of isolated atoms. In He_2, the four electrons occupy both the $\sigma 1s$ bonding and the $\sigma^* 1s$ antibonding orbitals with a resultant total energy that is *not* less than the isolated He atoms. As a consequence, He is a monatomic substance.

Inclusion of $\sigma 1s$ and $\sigma^* 1s$ in this list is not really necessary, however, because inner shells usually play no effective part in molecular bonding. In other words, inner-shell electrons usually stay in atomic orbitals that remain part of individual ion cores. Energy level diagrams for Li_2 and Be_2 are shown in Fig. 4.5. Li_2 is a stable molecule, but Be_2 is only marginally stable since there are two electrons in a bonding orbital and two in an antibonding orbital. In the cases of N_2, O_2, and F_2, all of the valence shell electrons go into molecular orbitals (Fig. 4.6), and bonding results because of the net excess of occupied bonding orbitals as compared to occupied antibonding orbitals. This net excess of occupied bonding orbitals is consistent with the number of bonds expected from the valence bond approach.

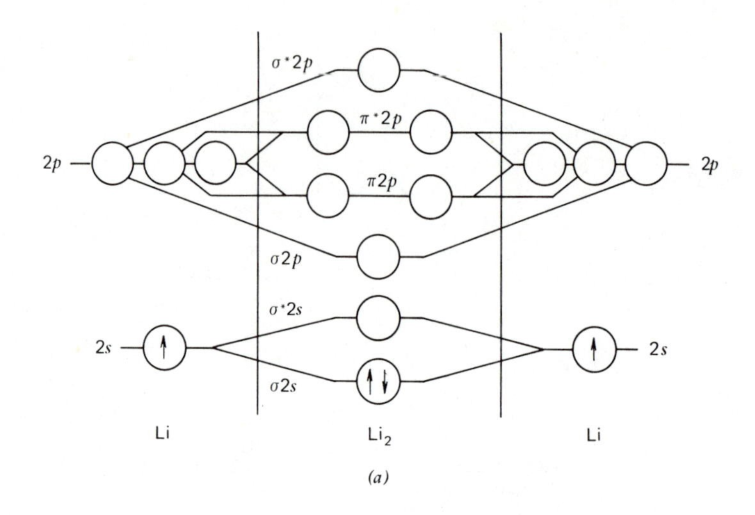

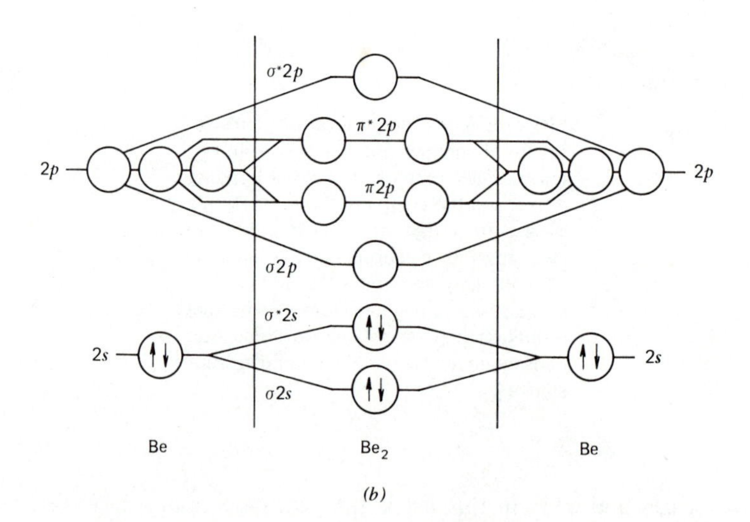

FIGURE 4.5 Molecular and atomic energy level diagrams for (*a*) Li and Li_2 and (*b*) Be and Be_2. This situation is somewhat analogous to that between H_2 and He_2 (Fig. 4.4); Li_2 is a stable molecule with respect to Li atoms; by contrast Be_2 is marginal with respect to the stability of isolated Be atoms because the σ^*2s orbital as well as the bonding $\sigma 2s$ orbital are occupied.

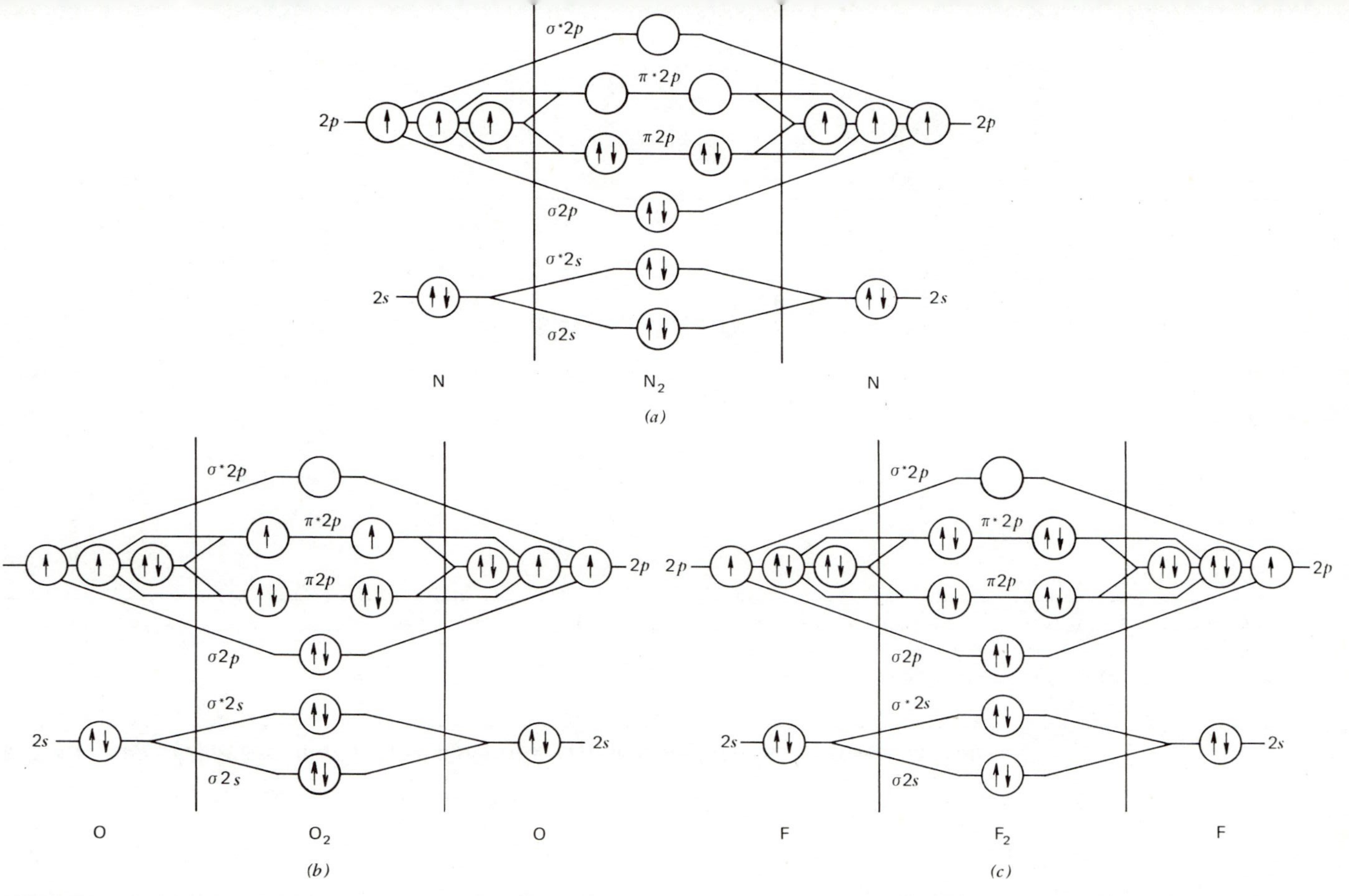

FIGURE 4.6 Atomic and molecular energy level diagrams for (*a*) nitrogen, (*b*) oxygen, and (*c*) fluorine. All of these form stable diatomic molecules because the number of bonding electrons exceeds the number of antibonding electrons. Although the positioning of the various molecular energy levels is schematic, the bonding energy is greatest for N_2, which has an excess of six bonding electrons, and lowest in F_2, which has an excess of two bonding electrons (Table 4.1).

TABLE 4.1 Electronic Configurations and Bond Dissociation Energies of Some Diatomic Molecules

	$\sigma 1s$	$\sigma^* 1s$	$\sigma 2s$	$\sigma^* 2s$	$\sigma 2p$	$\pi 2p$	$\pi^* 2p$	$\sigma^* 2p$	*Total Number of Bonding Electrons*	*Total Number of Antibonding Electrons*	*Net Number of Bonding Electrons*	*Dissociation Energy (eV)*
H_2^+	1								1		1	2.65
H_2	2								2		2	4.48
He_2^+	2	1							2	1	1	~2.6
He_2	2	2							2	2	0	0
Li_2	—	—	2						2		2	1.10
Be_2	—	—	2	2					2	2	0	0.7
B_2	—	—	2	2		2			4	2	2	~2.9
C_2	—	—	2	2	1	3			6	2	4	6.2
N_2^+	—	—	2	2	1	4			7	2	5	8.71
N_2	—	—	2	2	2	4			8	2	6	9.76
O_2^+	—	—	2	2	2	4	1		8	3	5	6.66
O_2	—	—	2	2	2	4	2		8	4	4	5.11
F_2^+	—	—	2	2	2	4	3		8	5	3	3.3
F_2	—	—	2	2	2	4	4		8	6	2	1.59
Ne_2	—	—	2	2	2	4	4	2	8	8	0	0

Table 4.1 provides a summary of molecular orbital occupancy for some homonuclear diatomic molecules.

Comments concerning N_2 and O_2 are worthwhile. The triple bond that exists in N_2 consists of one $\sigma 2p$ orbital and two $\pi 2p$ orbitals, each containing paired electrons. The amount of overlap of the atomic orbitals that give rise to the σ and π orbitals differs, and this leads to different energy levels (Fig. 4.6). That is, the triple bond does not consist of three equivalent bonds that contribute equally to the total binding energy. This situation is common in many organic compounds. In the case of O_2, the molecular configuration is

$$(\sigma 2s)^2(\sigma^* 2s)^2(\sigma 2p)^2(\pi 2p)^4(\pi^* 2p)^2$$

with each of the two $\pi^* 2p$ orbitals containing a single electron (a consequence of Hund's rule). These unpaired electrons have parallel spins which, in the presence of an applied magnetic field, tend to be aligned so as to reinforce the magnetic field weakly. The existence of such paramagnetism (see Chapter 25) due to unpaired electrons would not be expected from simple valence bond considerations. In addition, on the basis of the valence bond approach, it would seem reasonable to expect that the binding energy for O_2^+ is less than that for O_2; however, that the opposite is true (see Table 4.1) follows directly from the molecular orbital approach.

Finally, the molecular orbital treatment does not require pairing of electrons for bond formation. For example, a single electron in a π orbital (or a single electron in each of two π orbitals) will contribute to bonding. Nevertheless, it is conventional to refer to two electrons in a σ orbital as a σ bond and to two electrons in a π orbital as a π bond.

4.4 CARBON COMPOUNDS

Carbon, which is the basis for organic chemistry, combines with itself and other elements to form a wide variety of covalently bonded molecules. The bonding in many of these molecules can best be described in terms of molecular orbitals. For molecules other than those discussed in Sect. 4.3, the sequence of molecular orbital energy levels no longer follows a simple scheme. This is similar to the changing pattern of atomic energy levels with increasing atomic number. In addition, the designation of molecular orbitals so as to indicate the parent atomic orbitals becomes difficult. Consequently, we shall utilize only the designations introduced in Sect. 4.2 to describe the bonding in more complex molecules, and the emphasis will be placed on the bonding orbitals that actually contribute to the binding.

In methane (CH_4; see Fig. 3.9*c*), the sp^3 hybrid orbitals of the C atom overlap with the s orbitals of the four H atoms to form four separate σ bonds. In ethylene (C_2H_4; see Fig. 3.10*b*), σ bonds from sp^2–s overlap also occur between a C atom and

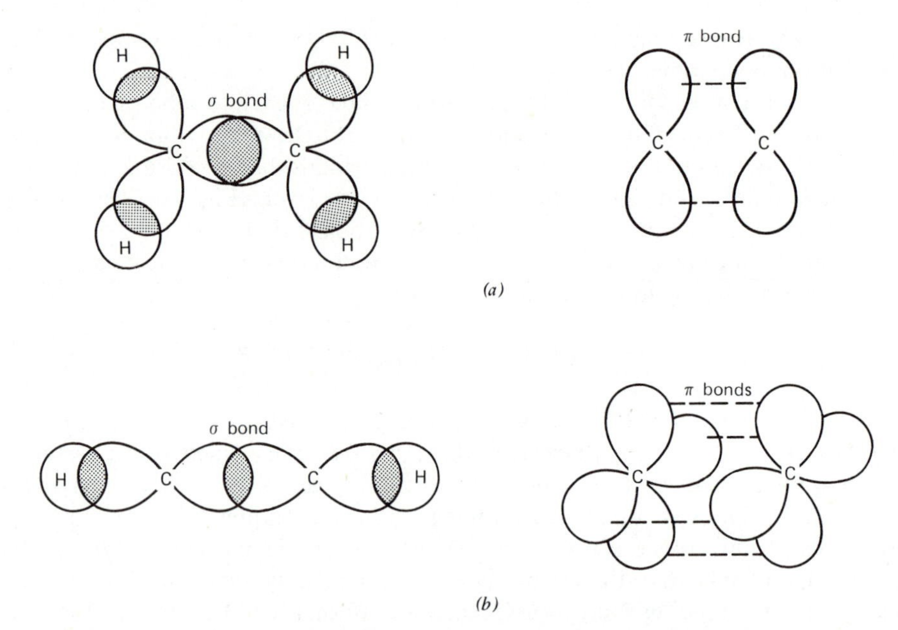

FIGURE 4.7 A molecular orbital view of the bonding in (*a*) ethylene, C_2H_4 and (*b*) acetylene, C_2H_2. The double bond in ethylene is in part a σ bond resulting from sp^2–sp^2 overlap and in part a π bond because of the sidewise overlap of the remaining atomic *p* orbitals. The triple bond in acetylene is due to a σ bond resulting from *sp*–*sp* overlap and two π bonds arising from sidewise overlap of two pairs of atomic *p* orbitals. Compare with Fig. 3.10.

two H atoms; however, a double bond joins the carbon atoms. Part of this double bond is a σ bond resulting from sp^2–sp^2 overlap between the two C atoms, and part is a π bond due to sidewise overlap of the remaining atomic *p* orbitals (Fig. 4.7*a*). Thus, the C═C double bond consists of two unequal molecular orbitals. The fact that ethylene is a planar molecule is a direct consequence of the atomic and hybrid orbital directionality and how overlap must occur. In other words, for the π bond to form, there cannot be free rotation of the two CH_2 groups with respect to each other. The triple bond in acetylene (C_2H_2; see Fig. 3.10*c*) also consists of unequal molecular orbitals. A σ bond arises from *sp*–*sp* overlap, and two π bonds form from sidewise overlap of two pairs of atomic *p* orbitals (Fig. 4.7*b*). Recalling the relative strengths of single, double and triple carbon–carbon bonds (Table 3.3), we must conclude that the amount of overlap in the π bonds is less than that in the σ bond, such that the π bonds are weaker than the σ bond. This is reflected in the greater chemical reactivity exhibited by molecules in which double or triple carbon–carbon bonds exist (e.g., $H_2C{=}CH_2$) as compared to molecules with only single carbon–carbon bonds (e.g., $H_3C{-}CH_3$). At this point, it should be apparent that it is the hybrid orbitals developed for the valence bond treatment (Sect. 3.7) that play a dominant role in determining molecular geometry.

The molecular orbital treatment of the benzene molecule provides an attractive physical picture of the resonance hybrid structure that was used in conjunction with valence bond considerations (Sect. 3.8). In benzene (C_6H_6) each carbon atom of the hexagonal ring forms a σ bond with each neighboring carbon atom (sp^2–sp^2 overlap) and with one hydrogen atom (sp^2–s overlap). The atomic p orbitals, which project perpendicular to the planar ring, overlap sidewise with each other to form π bonds (Fig. 4.8*a*). The fused-ring representation of the π orbitals shown in Fig. 4.8*b* is more realistic because simultaneous p–p overlap occurs to both sides

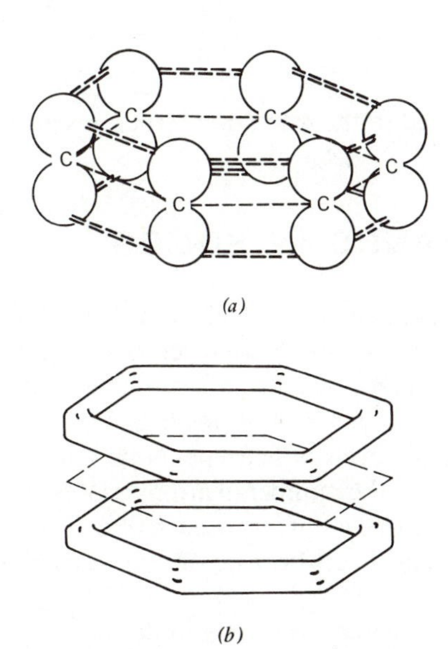

FIGURE 4.8 (*a*) The atomic *p* orbitals of carbon in the benzene ring project perpendicular to the plane of the benzene ring and overlap sidewise with each other to form π bonds. (*b*) The fused ring representation of π orbitals in benzene is more realistic because the *p* orbital of one carbon atom overlaps with similar *p* orbitals of both neighboring carbon atoms. This overlapping gives rise to a delocalization of the π bonded electrons. A *fused ring* lies both above and below the plane of the carbon nuclei.

of a given carbon atom. That is, there are really three π orbitals in benzene, and the six electrons in these *delocalized* orbitals belong to the whole molecule rather than being localized in a bond between only two atoms. An appropriate two-dimensional structural formula for benzene is

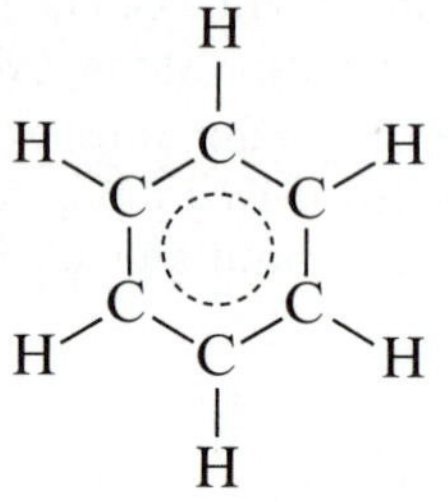

```
         H
         |
  H     C     H
   \  /   \  /
    C       C
    |  ( )  |
    C       C
   /  \   /  \
  H     C     H
        |
        H
```

which should be compared to the resonance structures given in Sect. 3.8.

4.5 ENERGY BANDS IN SOLIDS

Just as electronic states in an atom usually correspond to different energy levels (e.g., 1*s*, 2*s*, 2*p*, . . . , in Fig. 2.4), there are discrete molecular energy levels as was discussed in Sect. 4.3. Furthermore, a certain number of states may have the same energy, a situation that also exists in atoms (e.g., the three 2*p* states). Figure 4.9 shows the π and π^* energy levels of benzene. Note that the number of molecular states (i.e., six) is identical to the number of atomic states from which these molecular states were derived (i.e., one *p* state from each of six C atoms). This conservation of the number of states has direct bearing on the discussion of solids in terms of molecular orbitals.

As atoms are brought together to form a molecule, the atomic orbitals interact with each other and with the various nuclei. This causes each atomic state to split into multiple molecular states, which often have different energies. In the formation of a diatomic H_2 molecule, for example, the 1*s* level splits into two molecular levels whose energies depend on the interatomic separation (Fig. 4.10). This coincides with the formation of σ and σ^* molecular orbitals because of the overlap of atomic 1*s* orbitals. A similar situation occurs if N hydrogen atoms are brought together to form a hypothetical crystalline solid (metallic hydrogen), in that the atomic 1*s* states will give rise to N states in the solid. In this case, the 1*s* orbital of a given H atom will overlap with those of all neighboring H atoms, corresponding to the formation of delocalized molecular orbitals that extend throughout the crystal. The different energy levels that result will be spaced so closely that it becomes useful to speak of a band of energy levels or an energy band. In actuality,

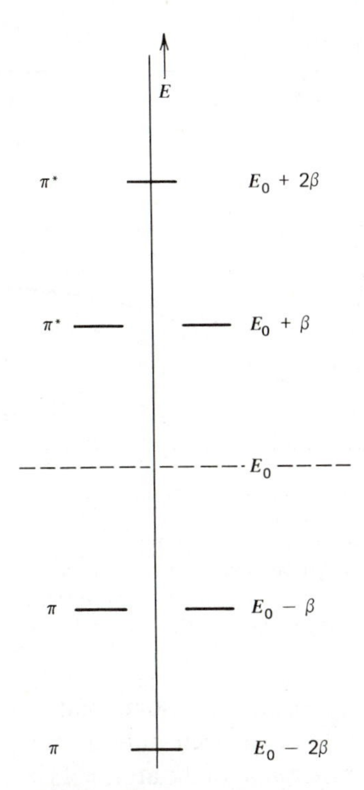

FIGURE 4.9 A schematic energy level diagram of bonding (π) and antibonding (π^*) levels in benzene. The number of possible molecular states is exactly equal to the number of participating atomic orbitals, which in this case is one *p* state from each of six different carbon atoms. Only the bonding orbitals are occupied.

several energy bands are formed (Fig. 4.11). As the hydrogen atoms approach each other, the outermost atomic states (i.e., those having the highest principal quantum numbers) split into bands first, while the lower lying atomic states remain associated with individual atoms. With closer interatomic approach, more and more atomic states give rise to energy bands. Because the individual energy levels within a band are so close to one another, only the envelope curves of the various energy bands for hypothetical metallic hydrogen are shown in Fig. 4.11.

States in an energy band are filled sequentially by electrons. The only restriction, imposed by the Pauli exclusion principle, is that a maximum of two electrons,

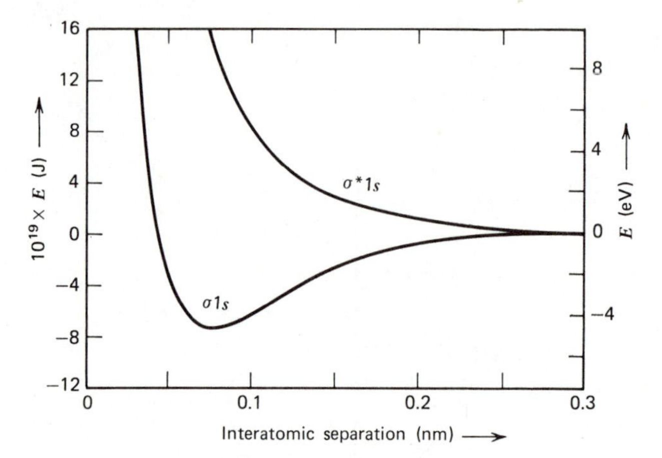

FIGURE 4.10 The energy of the $\sigma 1s$ and $\sigma^* 1s$ molecular orbital states in hydrogen as a function of the distance between the hydrogen nuclei. (Adapted from *Introduction to Solid State Physics*, Third Edition, by C. Kittel, Wiley, New York, 1966.)

having opposite spins, may occupy the same state. Accordingly, cohesion will occur in hypothetical metallic hydrogen only if more electrons are in bonding states (i.e., at energies lower than that of the atomic 1*s* level) than are in antibonding states (i.e., at energies higher than that of the atomic 1*s* level). Since this would require sufficiently close mutual interatomic approach to form the 1*s* band, metallic hydrogen should exist only at extremely high pressures. Under normal conditions only discrete H_2 molecules form.

The above designation of hypothetical solid hydrogen as metallic and the statement that such a crystal would have delocalized molecular orbitals extending throughout the solid require justification. In general, a molecular orbital will be delocalized if the atomic orbital of any one atom overlaps significantly with those of more than one of it neighboring atoms. This is a necessary but not a sufficient condition for a metallic character. To pursue this point, let us consider an actual metal like sodium. A calculated energy diagram showing the bands and the equilibrium interatomic separation for sodium metal is shown in Fig. 4.12. Since a free Na atom possesses a single 3*s* electron, the 3*s* band (also called the valence band) is only half full, and calculations show that the spacing between energy levels in this band is on the order of 10^{-9} eV. Consequently, it requires very little energy to promote (or excite) electrons from the highest filled states to the lowest empty states. This can be accomplished readily through the application of an electric field (or voltage), which also causes a shift in the occupancy of states so as to favor those excited states that have electrons moving in a direction opposite to the applied electric field. Thus, if a piece of sodium is connected to the terminals of a

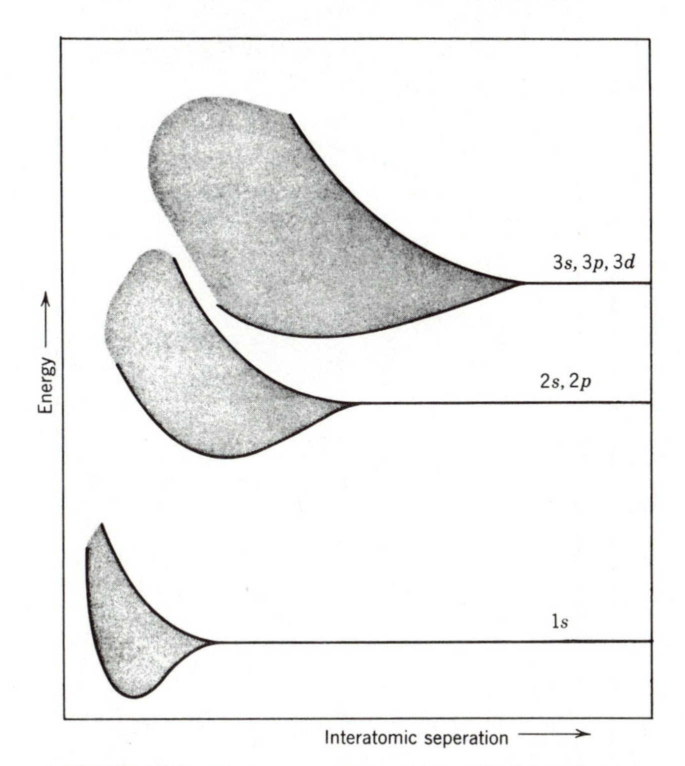

FIGURE 4.11 The energy states of "metallic" hydrogen as a function of distance of separation between the hydrogen nuclei. The number of separate energy levels formed within an energy band is N times the number of energy levels within an atomic orbital where N is the number of hydrogen nuclei present. Thus the 1*s* and 2*s* energy bands contain N energy levels and the 2*p* band, $3N$ energy levels. Each level can accommodate two electrons.

battery, an electric current will flow, and sodium is classified as a metal because of its good electrical conductivity. The nature of electrical conduction will be elaborated upon in Chapter 22. For our present purposes, it is sufficient to realize that a solid is metallic if it has a partially filled valence band.

4.6 ELECTRICAL INSULATORS AND CONDUCTORS

Of the four types of crystalline solids, molecular crystals, ionic crystals, and covalent crystals are considered as electrical insulators, and metallic crystals as electrical conductors. To be sure, molecular solids and, for the most part, ionic

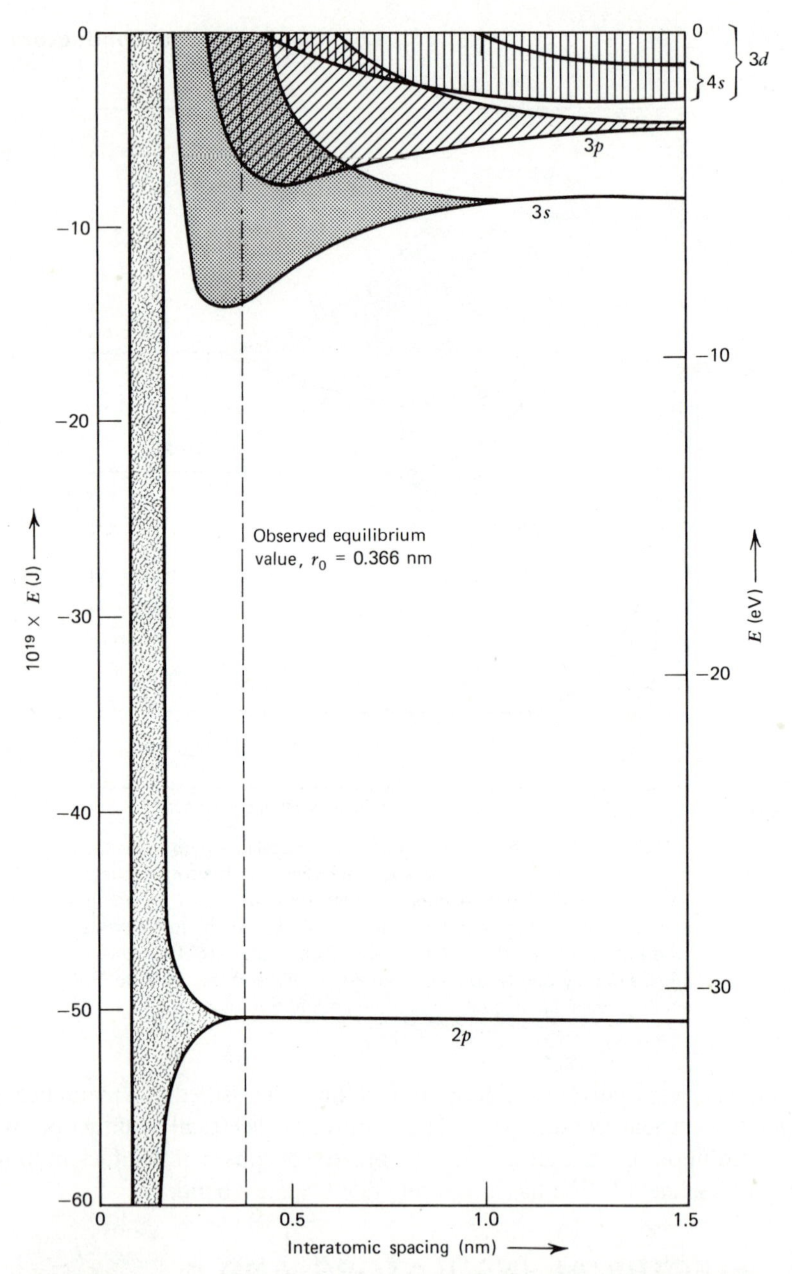

FIGURE 4.12 An energy band diagram for metallic sodium as a function of interatomic spacing. The overall pattern is similar to that for metallic hydrogen (Fig. 4.16), but the core electron states (i.e., 1*s* and 2*s*, which are not shown, and 2*p*) have not split at the equilibrium spacing. This indicates that these core electrons are still localized on their respective nuclei. Energy band overlap is also noted; however, in sodium all of the valence electrons lie at energies in the 3*s* band below the energy at which the 3*p* band overlaps the 3*s* band. [After J. C. Slater, *Phys. Rev.*, **45**, 794 (1934).]

crystals are insulators; however, some covalent crystals are capable of electrical conduction in spite of their nonmetallic nature. The reasons for this relate to the molecular orbitals in covalent solids.

A covalent crystal such as diamond has a three-dimensional network of sp^3 bonds (see Fig. 3.2 and Sect. 3.7). In terms of valence bond theory, all valence electrons in diamond occupy bonding orbitals and, accordingly, are not free to move through the crystal. Molecular orbital theory provides an alternative view of diamond in which the valence electrons are not actually localized in the bonds but exist in orbitals that extend throughout the crystal. The sp^3 hybrid atomic orbitals give rise to two energy bands, one of which is completely full (the valence band) and one of which is completely empty (the conduction band) at the absolute zero of temperature (Fig. 4.13). These two bands are separated by an energy gap or range of forbidden energies that electrons cannot have in the solid (Fig. 4.13). The energy gap is analogous to the forbidden energies for electrons in an atom (see Sect. 2.5) or in a molecule (see Sect. 4.3).

Diamond, silicon, and germanium share the same crystal structure and the same type of bonding. Nevertheless, diamond is an electrical insulator because of its

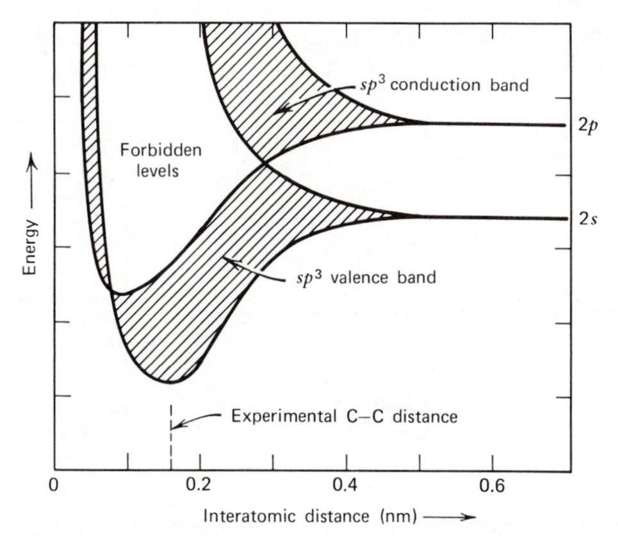

FIGURE 4.13 Schematic of the valence and conduction bands arising from sp^3 hybrid atomic orbitals in diamond. At the absolute zero of temperature all electrons are in the completely filled valence band and none in the conduction band. The region between the bands (the energy gap) represents a forbidden region of energy. [After G. E. Kimball, *J. Chem. Phys.*, **3**, 560 (1935).]

large energy gap (~5.4 eV), and silicon and germanium are classified as semiconductors because of their smaller energy gaps (~1.2 eV and ~0.7 eV, respectively). That is, some electrons in Si or Ge are thermally excited and *jump* across the energy gap at normal temperatures, and the resulting occupancy of some states in the conduction band permits electrical conduction when a voltage is applied. At absolute zero, Si and Ge become insulators like diamond because all valence electrons reside in a full valence band and an applied electrical field cannot cause a shift in occupancy of allowed states in a full band.

The origin of energy gaps in the band structures of nonmetallic crystals is beyond the scope of our treatment. Basically, however, we can expect one band to form from each atomic orbital. If the individual atomic levels are well separated in energy or if the distance between adjacent atoms in a crystal is so large that the amount of atomic orbital overlap is small, then the bands will be separated by energy gaps. This allows the existence of completely full and completely empty bands, a feature that is characteristic of covalent crystals at $T = 0$ K. Even ionic crystals may be considered from this point of view, although the *valence band* in a highly ionic solid essentially corresponds to a single energy level because the valence electrons occupy orbitals that are very localized on anions. Covalent and ionic crystals may be either electrical insulators or semiconductors, depending on the magnitude of the energy gap between the valence and conduction bands.

If the original atomic levels are close together in energy or if neighboring atoms in a crystal are sufficiently near one another so that atomic orbital overlap is

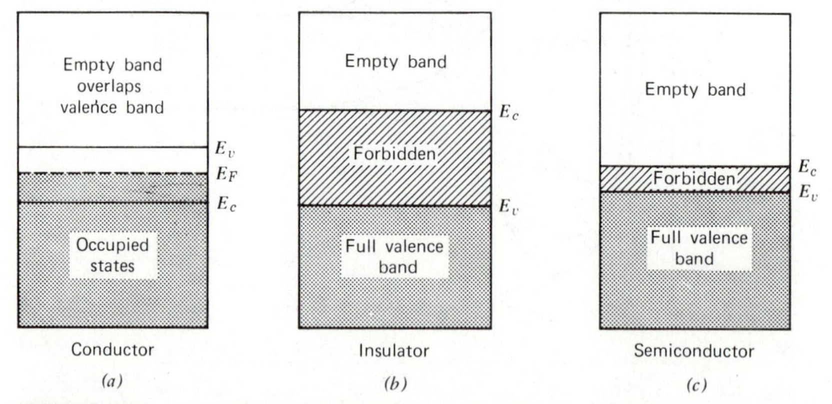

FIGURE 4.14 A diagrammatic viewpoint of the differences between metals, semiconductors, and electrical insulators in terms of energy bands. (*a*) In metals, unoccupied energy levels lie infinitestinally above occupied levels because of either incomplete filling of an energy band or band overlap; (*b*) insulators and (*c*) semiconductors, at 0 K, are characterized by completely filled valence and completely empty conduction bands separated by an energy gap. In the latter, the band gap is sufficiently small, however, that finite temperatures can excite electrons from the valence to the conduction band.

large, then the bands will overlap or merge into each other without energy gaps. This situation has already been illustrated for sodium (Fig. 4.12). The molecular orbitals having energies in a range where band overlap occurs derive some of their characteristics from each type of parent atomic orbital and, in effect, can be considered as the solid-state equivalent of hybrid atomic orbitals. If it were not for band overlapping, a characteristic of metals, we might expect the divalent alkaline earth elements to be nonmetals having full *s* bands. Actually, they are metals, and magnesium, for example, has a band structure similar to that of sodium. The difference between the band structure of metals, semiconductors, and insulators is summarized diagrammatically in Fig. 4.14.

4.7 COHESION IN CRYSTALS AND THE DENSITY OF STATES

Complete specification of the band structure of a solid involves not only the positions of the band boundaries but also the distribution of electronic states within the bands. Knowledge of band structures facilitates computation of the cohesive or binding energies of solids through a summation of the contributions of all occupied electronic states. The problem, although more complex, is analogous to summing the occupied electronic energy levels in a molecule (Sect. 4.3) and comparing this sum to the individual atomic levels.

The distribution of electronic states within a band is treated best in terms of the density of states, $N(E)$, which is the number of electronic states per unit energy range. That is, for an incremental energy range between E and $E + dE$, there are $N(E)\,dE$ states. The quantity $N(E)$ depends upon both the spacing between individual energy levels and the number of electronic states at each energy level. Theoretical calculation of the density of states as a function of energy can be quite complex, particularly when there is band overlap, and its experimental determination is difficult also. For the case of a simple metal such as sodium, however, the $N(E)$ vs. E curve for the occupied 3*s* band has been calculated and experimentally verified (Fig. 4.15). At $T = 0$ K, electrons fill states from the bottom of the valence band to a maximum energy level called the Fermi energy, E_F. That is, the Fermi level by definition is the highest occupied level in a metal at absolute zero. The total number of occupied states [i.e., the area under the $N(E)$ curve between the bottom of the 3*s* band and E_F] equals 0.5 per Na atom since each state contains two electrons.

The cohesive energy of a solid is the difference between the energy of a free atom and the crystal energy per atom. With knowledge of the density of states, one can trace the average electron energy in sodium as a function of interatomic separation (Fig. 4.16). This curve is similar to the potential energy curve for ionic bonding (see Fig. 3.4 and Sect. 3.3), and the minimum energy value occurs at the

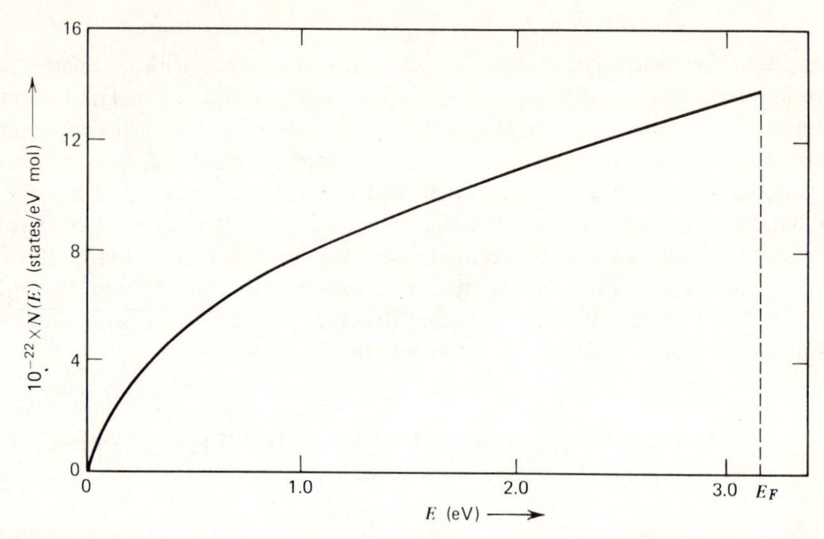

FIGURE 4.15 The density of states curve [$N(E)$ vs. E] for the occupied portion of the 3*s* band of sodium at 0 K. All energy levels are occupied from the lowest level to a maximum energy, E_F, the Fermi energy.

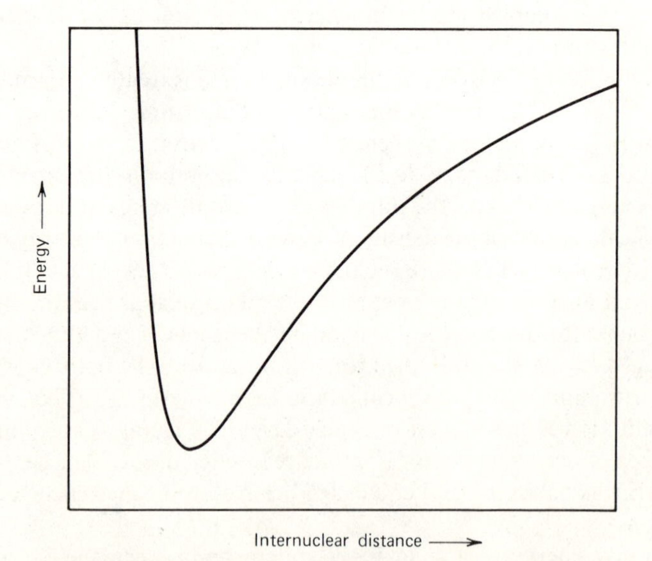

FIGURE 4.16 The average energy of a 3*s* electron in sodium as a function of internuclear separation distance. The energy is at a minimum at the equilibrium spacing r_0 and lies at a level $(\frac{3}{5})E_F$ above the lowest energy in the 3*s* band.

equilibrium interatomic spacing. In the case of sodium, the Fermi level has approximately the same energy as the atomic 3*s* level, and it can be shown that the average electron energy in the solid equals $\frac{3}{5}E_F$. Thus, the cohesive energy is given by $(E_F - \frac{3}{5}E_F) = \frac{2}{5}E_F$ or 1.30 eV per Na atom. Cohesive energies obtained in this manner for the alkali metals are listed in Table 4.2, which also contains experimental values for comparison.

TABLE 4.2 Fermi Energies and Cohesive Energies of the Alkali Metals at 0 K

		Approximate Theoretical Cohesive Energy, $\frac{2}{5}E_F$		*Experimental Cohesive Energy,*[a] ΔH_a	
Element	E_F *(eV)*	*kcal/mol*	*kJ/mol*	*kcal/mol*	*kJ/mol*
Li	4.79	44.1	185	38.1	159
Na	3.24	29.9	125	26.0	109
K	2.12	19.6	82	21.7	91
Rb	1.86	17.1	72	19.8	83
Cs	1.59	14.7	62	19.1	80

[a] Enthalpy of atomization.

Evaluation of cohesive energies for most metals and covalent crystals is hampered by the complexity of their density of states curves. Nevertheless, we can qualitatively illustrate the variation of cohesive energy through the transition metals in one period of the periodic table. Transition metals have overlapping *s* and *d* bands (Fig. 4.17), with the energy of a free atom falling near the middle of the bands. Starting with Ba and proceeding through W, the bonding levels in the *s–d* band fill up, and the melting and vaporization temperatures and cohesive energies increase sequentially (Table 4.3). Proceeding from W through Au, whose bands are filled above the top of the *d* band, the opposite trend occurs (Table 4.3) as the antibonding levels fill up. It is worth noting that the noble metals (i.e., Cu, Ag, Au) exhibit greater cohesive energies than would be expected from one net bonding electron per atom. This is because the filled *d* band contributes to binding in Cu, Ag, or Au, unlike the analogous situation in the case of molecules (see Sect. 4.3).

Although the concept of density of states has been introduced in order to discuss cohesion in crystals, the density of states at the Fermi level is also important with respect to many physical properties of metals. In particular, it essentially gives the number of electrons in a metal that can participate in electrical conduction (Chapter 22) and that can be thermally excited to higher energy states (Chapter 24).

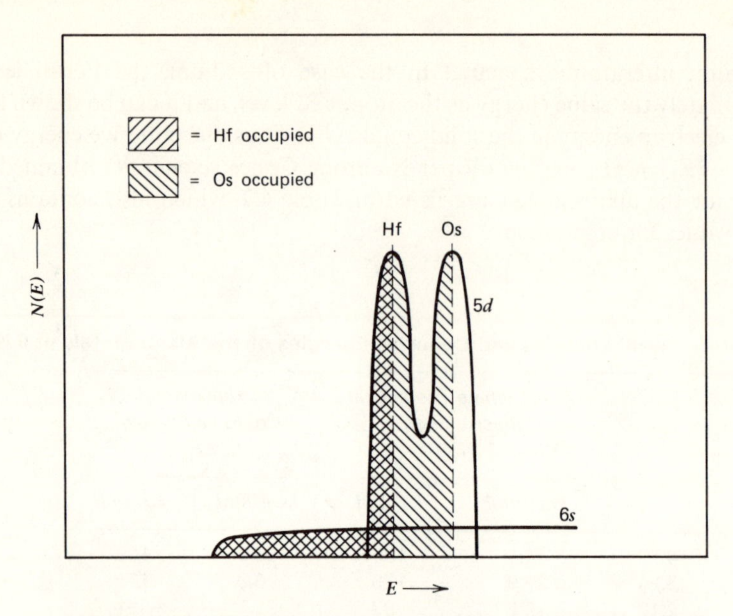

FIGURE 4.17 Schematic of the density of states curve for transition metals in the sixth period of the periodic table. This shows that *s* and *d* bands overlap and that electrons occupy levels in both bands. Bonding levels occur at the lower energies and antibonding levels at the higher energies.

TABLE 4.3 Some Properties of Metallic Elements in the Third Transition Period

Element	*Valence Electrons per Atom*	T_m *(K)*	T_v *(K)*	*Experimental Cohesive Energy*[a] *kcal/mol*	*kJ/mol*
Ba	2	1002	1910	43	178
La	3	1194	3742	103	431
Hf	4	2504	4875	148	619
Ta	5	3293	5698	187	782
W	6	3660	6000	203	849
Re	7	3459	5900	187	782
Os	8	3306	5300	188	787
Ir	9	2720	4820	160	669
Pt	10	2045	4100	135	565
Au	11	1338	3239	88	368

[a] Enthalpy of atomization at 25°C.

REFERENCES

1. Mahan, B. H. *College Chemistry*, Addison-Wesley, Reading, Mass., 1969.
2. Hägg, G. *General and Inorganic Chemistry*, Wiley, New York, 1969.
3. Mott, N. The Solid State, *Scientific American*, Vol. 217, No. 3, September, 1967, pp. 80–89.

QUESTIONS AND PROBLEMS

4.1 Distinguish between a bonding molecular orbital and its conjugate antibonding molecular orbital in terms of (i) the location of the electron clouds (i.e., the electron density distribution functions) when the orbital is full and (ii) the energy levels relative to that of the parent atomic states.

4.2 Using energy level diagrams, explain why the nitrogen molecule ion N_2^+ has a smaller binding energy than N_2, whereas the oxygen molecule ion O_2^+ has a larger binding energy than O_2.

4.3 (a) Write the electronic configuration for a C_2 molecule and draw the corresponding energy diagram.
(b) How many bonds exist between the carbon atoms?
(c) Should a C_2^+ molecule ion have a larger or smaller binding energy than a C_2 molecule? Explain.
(d) Should a C_2^- molecule ion have a larger or smaller binding energy than a C_2 molecule? Explain.

4.4 List the orbital interactions that produce σ bonds and those that produce π bonds.

4.5 The dissociation energy for an O_2 molecule is 5.11 eV, and that for an O_2^+ molecule ion is 6.66 eV.
(a) Does this mean that O_2^+ is more stable than O_2? (*Hint.* What are the respective dissociation products?)
(b) Use the following data to verify the dissociation energies for H_2^+, N_2^+, O_2^+, and F_2^+ given in Table 4.1.

Substance	*First Ionization Potential (eV)*	*Dissociation (or Binding) Energy (eV)*
H	13.60	—
N	14.53	—
O	13.61	—
F	17.42	—
H_2	15.43	4.48
N_2	15.58	9.76
O_2	12.06	5.11
F_2	15.7	1.59

4.6 For the halogens the following data have been determined:

Substance	*First Ionization Potential of a Free Atom (eV)*	*First Ionization Potential of a Diatomic Molecule (eV)*	*Dissociation Energy of a Diatomic Molecule (eV)*
F	17.42	15.7	1.59
Cl	13.01	11.48	2.52
Br	11.84	10.5	2.01
I	10.45	9.3	1.54
At	9.5	—	0.82

(a) Explain why it is reasonable that the ionization potential for a diatomic halogen molecule is less than that for the corresponding free atom.
(b) Neglecting F_2, justify the decrease in dissociation energy with increasing atomic number.

4.7 The shape of a NH_3 molecule is pyramidal with angles of 107° between N—H bonds. Diagram the overlapping of p and s orbitals that form σ bonds in this compound.

4.8 In the formaldehyde molecule H_2CO, a double bond exists between the carbon and oxygen atoms.
(a) What type of hybridization is involved?
(b) The molecule is found to be planar; one bond between the C and O atoms is a σ bond and the other is a π bond. With a simple sketch show the atomic orbital overlap that is responsible for the π bond.

4.9 Bonding orbitals (σ type) are concentrated directly between atoms in crystalline silicon (refer to Fig. 3.2).
(a) What is the likely location of the antibonding orbitals (σ^* type) in Si?
(b) Sketch a schematic energy band diagram for silicon and label the bonding and antibonding levels.

4.10 With the aid of simple diagrams compare the electron population in the energy bands of Mg and Si crystals.

4.11 Graphite, the stable form of carbon under normal conditions, consists of planar layers of C atoms arrayed in a hexagonal manner:

```
       \   /       \   /
        C—C         C—C
       /   \       /   \
   —C       C—C         C—
       \   /       \   /
        C—C         C—C
       /   \       /   \
   —C       C—C         C—
       \   /       \   /
        C—C         C—C
       /   \       /   \
```

Intralayer bonding is strong (C—C distance = 0.142 nm), but interlayer bonding is weaker (interlayer distance = 0.335 nm).

(a) Account for the geometry of a layer (i.e., the 120° angles between bonds).

(b) What is the nature of the molecular orbital that contains the fourth valence electron of each carbon atom?

(c) On the basis of your answer to part (b), do you expect graphite to be an electrical insulator or conductor? Explain.

4.12 The following data are for the alkali metal elements:

Element	*Enthalpy of Atomization of the Metal (eV/atom)*	*Interatomic Distance in the Metal (nm)*	*Dissociation Energy for the Diatomic Molecule (eV/atom)*	*Interatomic Distance in the Diatomic Molecule (nm)*
Li	1.67	0.304	0.58	0.267
Na	1.12	0.372	0.40	0.308
K	0.93	0.461	0.28	0.392
Rb	0.85	0.494	0.26_5	—
Cs	0.81	0.535	0.24_5	—

Why does the metallic form of each element have both a larger binding energy (per atom) and a larger interatomic distance?

4.13 The density of states for an alkali metal is a simple function of energy over the range of occupied levels:

$$N(E) = 3.41 \times 10^{21} VE^{1/2}$$

where V is the molar volume in cm^3/mol, E is in eV, and $N(E)$ is in states/eV · mol.

(a) Using the above expression show that

$$E_F = 3.64 \times 10^{-15}(N_0/V)^{2/3}$$

where E_F is in eV and N_0 is Avogadro's number.

(*Hint.* The area under the density of states curve from the bottom of the band ($E = 0$) to E_F equals the number of occupied states per mole.)

(b) What is the density of states at the Fermi level for each alkali metal? (The molar volumes near $T = 0$ K are Li, 12.66 cm^3/mol; Na, 22.71 cm^3/mol; K, 42.95 cm^3/mol; Rb, 52.45 cm^3/mol; and Cs, 66.52 cm^3/mol.)

4.14 In a semimetal (e.g., As, Sb, and Bi) one or two bands are slightly filled or slightly empty even at the absolute zero of temperature.

(a) Schematically compare the band structures of a semimetal, a metal, and a semiconductor at (i) $T = 0$ K and (ii) room temperature.

(b) Given that $N(E)$ goes to zero at the edge of an energy band, justify why semimetals have much lower electrical conductivities than metals.

4.15 (a) The number of electrons per unit volume thermally excited across the energy gap of pure substances such as diamond, silicon, and germanium is proportional to $e^{-E_g/2kT}$ Calculate the ratio of the number of conduction electrons in Si and in Ge to that in diamond at 100 K, 300 K, and 1000 K. The room temperature energy gaps are diamond, 5.33 eV; Si, 1.14 eV; and Ge, 0.67 eV.

(b) The electrical conductivity of a pure semiconductor is proportional to the number of conduction electrons per unit volume. What are the approximate relative conductivities of Si and Ge as compared to diamond at 100 K, 300 K and 1000 K?

5

Organic Compounds

This chapter provides a brief introduction to organic chemistry and some of the important substances that serve as raw materials for synthetic polymers.

5.1 HYDROCARBONS

Few elements are as amenable to different bonding schemes, or as capable of forming molecules so varied in geometry, as carbon. It can form the backbone of chain molecules either branched or unbranched and it can form a variety of cyclic or ring molecules (Fig. 5.1). The propensity of carbon atoms to bond with other carbon atoms is relatively unique, and this in combination with different types of hybridization and of molecular orbitals leads to the existence of numerous carbon compounds. Although metal carbonates and carbides are regarded as inorganic substances, the vast majority of molecular carbon compounds are termed organic since they generally are derived from fossil fuels, wood, and other plant or animal sources. Hydrocarbons, consisting solely of carbon and hydrogen and their derivatives, comprise the most important groups of organic compounds. Indeed, more than a million different hydrocarbons are known to exist.

Hydrocarbon compounds are classified conveniently into homologous series, each series possessing similar structural and bonding arrangements. The alkane or paraffin series, in which there are straight or branched chains of carbon atoms with single carbon–carbon bonds only, has the general formula C_nH_{2n+2}. Methane (CH_4) is the smallest member of this series. Table 5.1 lists the first 10 members of the normal alkanes (i.e., those alkanes consisting of straight-chain molecules) as

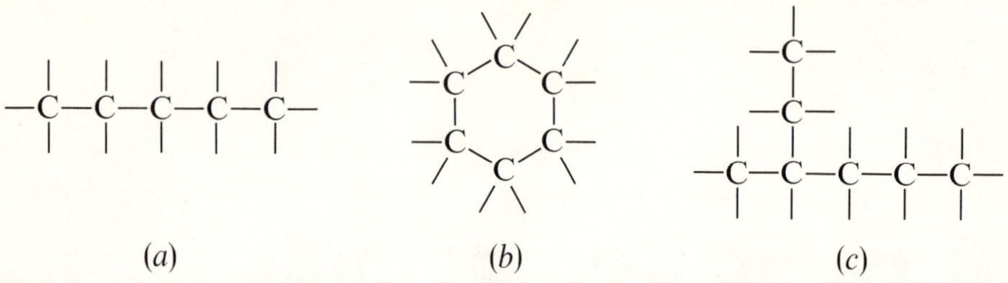

(a) (b) (c)

FIGURE 5.1 Schematic of the carbon "backbone" in (*a*) chain, (*b*) ring, and (*c*) branched structures of organic compounds.

well as their melting and boiling points. The tendency of the latter to increase with increasing molecular weight is a consequence of greater intermolecular binding as molecular size increases. With the exception of methane, ethane, and propane, all normal alkanes have structural isomers that possess the same chemical formula but a different molecular architecture. For example, the various isomers of butane (C_4H_{10}) and pentane (C_5H_{12}) are represented schematically in Fig. 5.2. Isomers become more numerous as molecular size increases; thus, decane ($C_{10}H_{22}$) has 75 isomers that differ in geometry as well as in properties.

Straight- or branched-chain hydrocarbons containing one double carbon–carbon bond are called alkenes or olefins and have the general formula C_nH_{2n} (where $n \geq 2$). Although the double bond is shorter and stronger than a single bond, the alkenes are more reactive chemically than the alkanes. A multiple bond can be broken more easily than a single bond, and a hydrocarbon containing a double (or triple) carbon–carbon bond combines more readily with other atoms than does

TABLE 5.1 The First Ten Members of the Normal Alkane Series and Some of Their Properties

Substance	*Molecular Weight*	*T_m (°C)*	*T_v (°C)*
Methane, CH_4	16.04	−182.5	−164
Ethane, C_2H_6	30.07	−183.3	−88.6
Propane, C_3H_8	44.10	−189.7	−42.1
Butane, C_4H_{10}	58.13	−138.4	−0.5
Pentane, C_5H_{12}	72.15	−129.7	36.1
Hexane, C_6H_{14}	86.18	−94.3	69.0
Heptane, C_7H_{16}	100.21	−90.6	98.4
Octane, C_8H_{18}	114.23	−56.8	125.7
Nonane, C_9H_{20}	128.26	−53.7	150.8
Decane, $C_{10}H_{22}$	142.29	−29.7	174.1

n-butane, C_4H_{10}

$$\begin{array}{ccccccccccc}
 & & H & & H & & H & & H & & \\
 & & \vert & & \vert & & \vert & & \vert & & \\
H & - & C & - & C & - & C & - & C & - & H \\
 & & \vert & & \vert & & \vert & & \vert & & \\
 & & H & & H & & H & & H & &
\end{array}$$

isobutane, C_4H_{10}

$$\begin{array}{ccccccccc}
 & & & & H & & & & \\
 & & & & \vert & & & & \\
 & & H & - & C & - & H & & \\
 & & & & \vert & & & & \\
 & & H & & \vert & & H & & \\
 & & \vert & & \vert & & \vert & & \\
H & - & C & - & C & - & C & - & H \\
 & & \vert & & \vert & & \vert & & \\
 & & H & & H & & H & &
\end{array}$$

n-pentane, C_5H_{12}

$$\begin{array}{ccccccccccccc}
 & & H & & H & & H & & H & & H & & \\
 & & \vert & & \vert & & \vert & & \vert & & \vert & & \\
H & - & C & - & C & - & C & - & C & - & C & - & H \\
 & & \vert & & \vert & & \vert & & \vert & & \vert & & \\
 & & H & & H & & H & & H & & H & &
\end{array}$$

isopentane, C_5H_{12}

$$\begin{array}{ccccccccccc}
 & & & & H & & & & & & \\
 & & & & \vert & & & & & & \\
 & & H & - & C & - & H & & & & \\
 & & & & \vert & & & & & & \\
 & & H & & \vert & & H & & H & & \\
 & & \vert & & \vert & & \vert & & \vert & & \\
H & - & C & - & C & - & C & - & C & - & H \\
 & & \vert & & \vert & & \vert & & \vert & & \\
 & & H & & H & & H & & H & &
\end{array}$$

neopentane, C_5H_{12}

$$\begin{array}{ccccccccc}
 & & & & H & & & & \\
 & & & & \vert & & & & \\
 & & H & - & C & - & H & & \\
 & & & & \vert & & & & \\
 & & H & & \vert & & H & & \\
 & & \vert & & \vert & & \vert & & \\
H & - & C & - & C & - & C & - & H \\
 & & \vert & & \vert & & \vert & & \\
 & & H & & \vert & & H & & \\
 & & & & \vert & & & & \\
 & & H & - & C & - & H & & \\
 & & & & \vert & & & & \\
 & & & & H & & & &
\end{array}$$

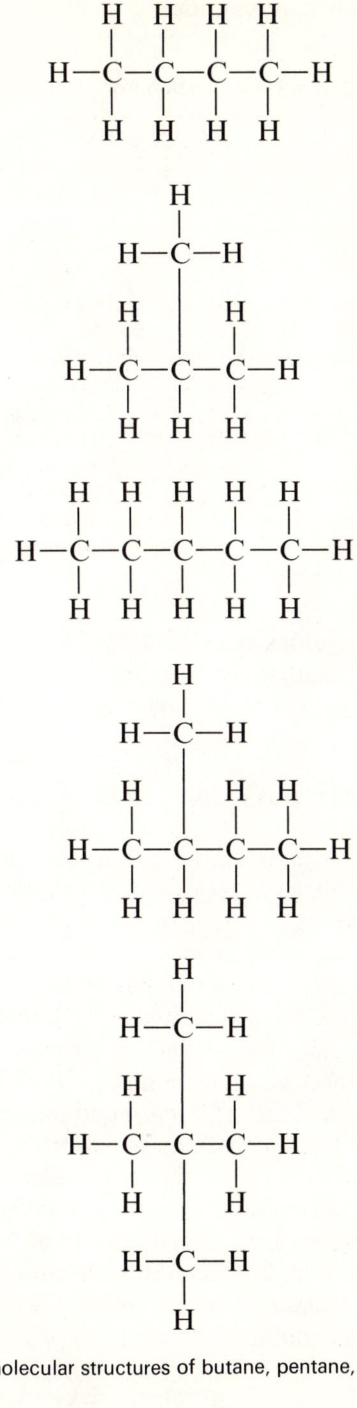

FIGURE 5.2 Schematic molecular structures of butane, pentane, and their isomers.

TABLE 5.2 Some Hydrocarbon Series and Corresponding General Formulas

Chain Hydrocarbons	*Formula*	*Characteristics*
Alkane series	C_nH_{2n+2}	All single C—C bonds
Alkene series	C_nH_{2n}	One double C=C bond
Alkadiene series	C_nH_{2n-2}	Two double C=C bonds
Alkyne series	C_nH_{2n-2}	One triple C≡C bond
Alkadiyne series	C_nH_{2n-6}	Two triple C≡C bonds
Cyclic Hydrocarbons	*Formula*	*Characteristics*
Cycloalkane series	C_nH_{2n}	All single C—C bonds, cyclic
Cycloalkene series	C_nH_{2n-2}	One double C=C bond, cyclic
Aromatic	Various	Ring structures, based on the benzene ring, in which single and double carbon bonds alternate

an alkane. Hydrocarbons with only single bonds are designated saturated, and those with multiple bonds, unsaturated. Selected homologous hydrocarbon series for chain and cyclic molecules are listed in Table 5.2.

5.2 BONDING IN HYDROCARBONS

The covalent bonding in hydrocarbon molecules can be described best with molecular orbitals. In the case of the alkanes each carbon atom has four σ bonds with neighboring atoms as a result of sp^3 hybrid orbitals. Methane has been discussed previously (Sects. 3.6, 3.7, and 4.4), and the same tetrahedral bond angle of 109.5° occurs in all compounds of the alkane series. Thus, each carbon atom in ethane (C_2H_6) is bonded tetrahedrally to three H atoms and one C atom. Likewise, all carbon atoms in an alkene series molecule, except those which have a double bond, also bond with four tetrahedrally directed σ orbitals. The double carbon–carbon bond consists of one σ bond and one π bond. The π bond results from sidewise overlap of unhybridized p orbitals and the σ bond, from endwise overlap of sp^2 hybrid orbitals, as shown for ethylene (Fig. 4.7*a*). Similar considerations apply to the alkyne series compounds as well.

In methane, ethylene, acetylene, and other compounds of various hydrocarbon series, σ bonds are fixed between ion cores and correspond to localized electrons. Certain molecules such as benzene (see Sect. 4.4) and butadiyne (C_4H_2) also have nonlocalized bonding electrons. In the case of butadiyne,

$$\mathrm{H{-}C{\equiv}C{-}C{\equiv}C{-}H}$$

there are four occupied π orbitals, and each pair of π orbitals actually extends beyond the two carbon atoms joined by a triple bond. Evidence for this is that the C—C bond length in C_4H_2 (~0.137 nm) is considerably shorter than the single bond in ethane (~0.154 nm) but longer than the double bond in ethylene (~0.134 nm), and each C≡C bond length in C_4H_2 (~0.121 nm) is slightly longer that the triple bond in acetylene (~0.120 nm). Thus, the central pair of carbon atoms are bonded by more than a single bond because of the nonlocalized π molecular orbitals. A few bond lengths pertaining to hydrocarbons are given in Table 5.3.

TABLE 5.3 Selected Carbon–Carbon Bond Lengths in Hydrocarbons

	Bond Length (nm)
Single bond:	
In alkanes	0.154
Partial double bond:	
In propyne	0.146
(H_3C—C≡C—H; arrow at the C—C single bond)	
In aromatic compounds	0.140
In butadiyne	0.137
(H—C≡C—C≡C—H; arrow at the central C—C bond)	
Double bond:	0.134
Partial triple bond:	
In propadiene	0.131
(H_2C=C=CH_2; arrows at both C=C bonds)	
In pentadiyne	0.121
(H_3C—C≡C—C≡C—H; arrows at both C≡C bonds)	
Triple bond:	0.120

5.3 RADICALS AND FUNCTIONAL GROUPS

All organic compounds in addition to hydrocarbons may be considered as made up of hydrocarbon molecule fragments or radicals to which functional groups are attached. A simple example of a radical is the methyl radical ($\cdot CH_3$), which has one half-full orbital. That is, $\cdot CH_3$ is just methane less one hydrogen atom. This leaves one electron available for bonding. Radicals are important units. They maintain their integrity during many chemical reactions but, as a rule, they exist only transiently. Studies of their nature are valuable for they can lead to an understanding of organic chemical reactions and of the compounds that result.

A functional group is a particular group of atoms that occurs in many molecules and that imparts characteristic chemical properties to the molecule regardless of

TABLE 5.4 Some Common Functional Groups in Organic Compounds

Group	*General Formula*[a]	*Compound Class*	*Example*
—Cl	*R*—Cl	Chlorides	Methyl chloride, CH_3Cl
—OH	*R*—OH	Alcohols	Ethyl alcohol, CH_3CH_2OH
—C(=O)—OH	*R*—C(=O)—OH	Acids (or carboxylic acids)	Acetic acid, CH_3COOH
>C=O	*R*(*R*)C=O	Ketones	Acetone, CH_3COCH_3
H(—)C=O	H(*R*)C=O	Aldehydes	Propionaldehyde, CH_3CH_2CHO
—O—	*R*—O—*R*	Ethers	Ethyl ether, $CH_3CH_2OCH_2CH_3$
$—NH_2$	$R—NH_2$	Amines	Propylamine, $CH_3(CH_2)_2NH_2$
$—C(=O)—NH_2$	$R—C(=O)—NH_2$	Amides	Acetamide, CH_3CONH_2
—C(=O)—O—*R′*	*R*—C(=O)—O—*R′*	Esters	Ethyl acetate, $CH_3COOC_2H_5$

[a] *R* represents hydrocarbon radicals; *R′* represents a hydrocarbon radical that is part of a functional group.

the form of the carbon skeleton. When viewed separately, a functional group has at least one unfilled bonding orbital and, thus, can combine with a hydrocarbon radical to form a molecule. Alternatively, a functional group may replace one hydrogen atom in a hydrocarbon molecule to form another molecule.

Because of the characteristic chemical properties that are imparted, organic compounds are classified in accordance with their functional group (Table 5.4). Thus, a methyl radical ($\cdot CH_3$) combined with a hydroxyl group is methanol or methyl alcohol, and a methyl radical combined with a carboxyl group is acetic acid. In many cases, the placement of functional groups leads to geometrically distinct molecules which possess different physical properties.

5.4 ISOMERISM

As mentioned previously, molecules of the same chemical composition that have different arrangements of their constituent atoms are called isomers. Structural isomers have different pairs of atoms bonded together, and stereoisomers have the same pairs of atoms linked together by the same set of bonds. These two categories, in turn, contain distinct classes.

In positional isomers, one class of structural isomers, the same functional groups are located at different positions in the molecule. An example is provided by propyl alcohol and isopropyl alcohol:

$$\begin{array}{ccccccccc} & & H & & H & & H & & \\ & & \mid & & \mid & & \mid & & \\ H & — & C & — & C & — & C & — & OH \\ & & \mid & & \mid & & \mid & & \\ & & H & & H & & H & & \end{array} \qquad \begin{array}{ccccccccc} & & H & & OH & & H & & \\ & & \mid & & \mid & & \mid & & \\ H & — & C & — & C & — & C & — & H \\ & & \mid & & \mid & & \mid & & \\ & & H & & H & & H & & \end{array}$$

propyl alcohol **isopropyl alcohol**

The hydrocarbon structural isomers considered in Sect. 5.1 are strictly positional isomers (see Fig. 5.2). Positional isomers have different chemical and physical properties, yet differ very little in total covalent binding energy.

When atoms are so arranged that the functional groups in molecules are different, then the molecules are termed functional isomers, another class of structural isomers. This is the case for methyl ether and ethanol, for example:

$$\begin{array}{ccccccc} & & H & & H & & \\ & & \mid & & \mid & & \\ H & — & C & —O— & C & — & H \\ & & \mid & & \mid & & \\ & & H & & H & & \end{array} \qquad \begin{array}{ccccccc} & & H & & H & & \\ & & \mid & & \mid & & \\ H & — & C & — & C & — & OH \\ & & \mid & & \mid & & \\ & & H & & H & & \end{array}$$

methyl ether **ethanol**

The chemical and physical properties of such isomers differ radically, because the type of functional group largely determines the nature of organic compounds.

If the functional groups (or side groups) in a molecule are the same but can occupy different relative geometric positions, the molecules are called geometrical isomers, a class of sterioisomers. There are, for example, *cis* and *trans* forms of 2-butene:

```
CH3       CH3          CH3        H
   \     /                \      /
    C=C                    C=C
   /     \                /      \
 H         H            H        CH3
```

cis-**2-butene** *trans*-**2-butene**

These are distinct because no rotation can occur about the carbon–carbon double bond axis. Physical properties of these isomers differ somewhat, but not greatly, and the total covalent binding energies differ only slightly. If chlorine atoms were substituted for the methyl groups, the resulting *cis* and *trans* isomers would have different dipole moments.

Optical isomers, another class of stereoisomers, consist of molecules that have all atoms or functional groups in the same positions relative to one another, but the molecules themselves are distinct mirror images. Necessarily such isomers cannot be superimposed. The chemical and most physical properties of such isomers, with the important exception of optical properties, usually are identical.

Various chemical reactions can convert one isomer to another. Reactions that involve positional or geometrical isomers are accompanied by small changes in free energy and enthalpy, whereas those involving functional isomers often are accompanied by sizable changes in free energy and enthalpy. In addition to having different chemical and physical properties, isomers have a different propensity for crystallizing. From geometrical considerations alone, it is apparent that some molecules can pack together better than others.

5.5 CRYSTALLIZATION OF ORGANIC COMPOUNDS

Although interatomic bonding is primarily covalent within organic molecules, intermolecular bonding is due to either van der Waals bonds, which are weak, or hydrogen bonds. Strong intramolecular bonding is manifested by the fact that methane, for example, will decompose to carbon and hydrogen only at very high temperatures, yet the attraction *between* methane molecules is so slight that it condenses at $-164°C$ and freezes at $-182.5°C$. For larger molecules the van der

Waals bonding per molecule becomes greater. Thus, octane [$CH_3(CH_2)_6CH_3$] condenses at 125.7°C and freezes at −56.8°C. The variation of freezing and boiling points with molecular size for some normal alkanes is shown in Fig. 5.3. Both T_m and T_v provide a qualitative measure of intermolecular bonding strength, and the systematic trend of T_m and T_v with molecular size is evident, except for the smallest alkanes. Crystalline polymers, which are made up of very large molecules, have correspondingly higher melting points (e.g., T_m = 137°C for linear polyethylene) and most actually decompose before vaporizing.

Molecular arrangements in crystals of simple organic compounds are similar to atomic arrangements found in crystalline inorganic solids. For example, the

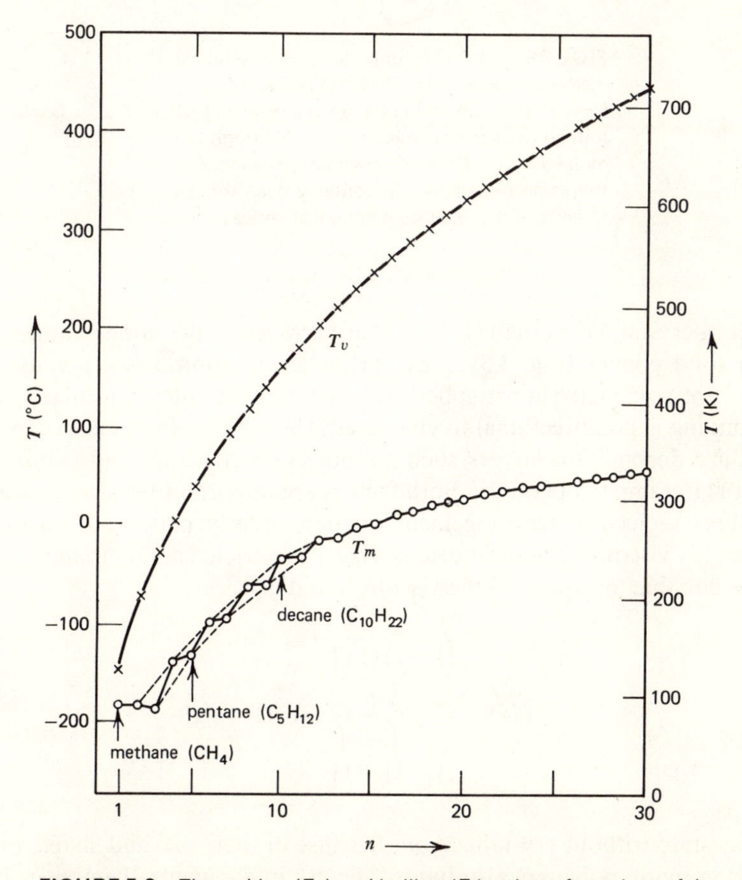

FIGURE 5.3 The melting (T_m) and boiling (T_v) points of members of the normal alkane series, (C_nH_{2n+2}), as a function of molecular size (n). The systematic increase of T_m and T_v with n is indicative of greater intermolecular physical bonding as the size of the molecule increases.

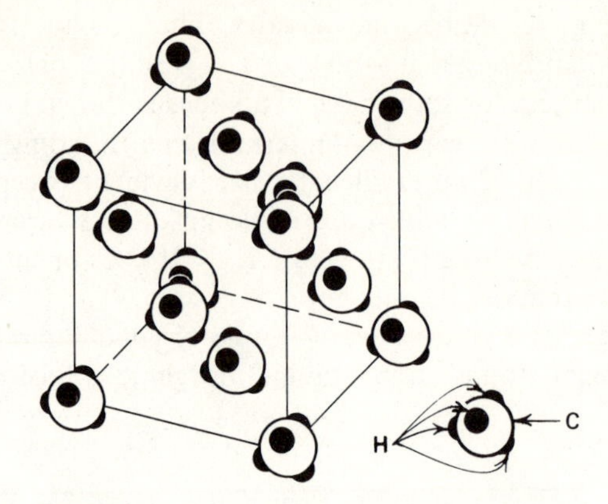

FIGURE 5.4 The unit cell of crystalline methane above 20 K. The molecules of CH_4 form a face-centered cubic array similar to that formed by atoms in solid copper. Although the molecules are depicted schematically as having the same orientation, in actuality each rotates freely and thus acts as a spherical entity.

molecules in crystalline methane (Fig. 5.4) are arrayed in the same manner as the atoms in solid copper (Fig. 3.3). This particular situation arises because CH_4 molecules are compact, almost spherical units and the intermolecular van der Waals bonding is nondirectional in character. The ability of organic compounds to crystallize depends on factors such as molecular structure and cooling rate through the freezing temperature. Both factors are important because crystallization requires molecular rearrangement into an orderly pattern. If an organic liquid is highly viscous, then the molecules have a restricted ability to move about. Thus, it is possible to supercool the viscous liquid glycerin,

$$\begin{array}{c} \quad\;\; \text{H} \quad \text{OH} \; \text{H} \\ \quad\;\; | \quad\;\; | \quad\; | \\ \text{HO}-\text{C}-\text{C}-\text{C}-\text{OH} \\ \quad\;\; | \quad\;\; | \quad\; | \\ \quad\;\; \text{H} \quad \text{H} \quad \text{H} \end{array}$$

to a glassy state without crystallization. Because of their size and shape, macromolecules in many polymers also have difficulty in arranging themselves into a completely ordered crystalline pattern, and a supercooled liquid or *amorphous* arrangement often persists even when the polymer is solid. Frequently, however, a solid polymer consists of a mixture of noncrystalline and crystalline phases of the same composition, with the degree of crystallinity controlled by the factors mentioned above as well as by subsequent thermal treatments given to the solid polymer.

5.6 ORGANIC REACTIONS

Any hydrocarbon can participate in chemical reactions, but unsaturated hydrocarbons in general are more reactive chemically than saturated hydrocarbons. Unsaturated hydrocarbons have a greater reactivity because electrons in π molecular orbitals are more exposed than those in σ orbitals. Perhaps the most familiar organic reaction is the combustion of methane (the main constituent of natural gas) in air. Here, C—H bonds are broken at elevated temperatures, and carbon–oxygen and hydrogen–oxygen bonds are formed. In some cases the reaction may proceed in steps. For example, when methane is combusted the initial reaction forms methanol:

$$CH_4 + \tfrac{1}{2}O_2 \longrightarrow CH_3OH \tag{5.1}$$

In turn, formaldehyde and water form from methanol and oxygen:

$$CH_3OH + \tfrac{1}{2}O_2 \longrightarrow HCHO + H_2O \tag{5.2}$$

Next, formaldehyde combines with oxygen, giving formic acid, carbon monoxide, and water:

$$HCHO + \tfrac{1}{2}O_2 \longrightarrow \tfrac{1}{2}HCOOH + \tfrac{1}{2}CO + \tfrac{1}{2}H_2O \tag{5.3}$$

Finally, formic acid and carbon monoxide combine with oxygen to form carbon dioxide and water:

$$\tfrac{1}{2}HCOOH + \tfrac{1}{2}CO + \tfrac{1}{2}O_2 \longrightarrow CO_2 + \tfrac{1}{2}H_2O \tag{5.4}$$

Even the above sequence of reactions is incomplete, for it does not indicate fully all of the intermediate species that may occur. For example, in the first reaction (Eq. 5.1) free methyl radicals ($\cdot CH_3$) and free hydroxyl groups ($\cdot OH$) will form from CH_4 and O_2 before methanol molecules can form. The net result of combusting methane is the sum of the above reactions (Eqs. 5.1 to 5.4):

$$CH_4 + 2O_2 \longrightarrow CO_2 + 2H_2O \tag{5.5}$$

As a rule combustion reactions are so violent that hydrocarbon chains are broken in the process. Other types of reactions involving organic compounds can be grouped into certain classes, and the most important of these are discussed below.

It is possible for some functional groups to be oxidized, commonly in the presence of an oxidizing agent. Although oxidation in a general chemical sense denotes the raising of an oxidation state through removal of electrons, oxidation of an organic compound usually corresponds to the removal of a hydrogen atom from a functional group, the addition of an oxygen atom to a functional group, or both. An example of such an oxidation reaction is provided by Eq. 5.3. In this case the functional group is oxidized: —CHO → —COOH. Reduction reactions, which usually are carried out at elevated temperatures or high pressures in the presence of a reducing agent or a catalyst, are the opposite of oxidation reactions. An example is the conversion of acetone to isopropyl alcohol, utilizing Pt as a catalyst

$$(CH_3)_2C{=}O + H_2 \xrightarrow{\text{Pt catalyst}} H{-}\overset{\displaystyle CH_3}{\underset{\displaystyle CH_3}{C}}{-}OH \tag{5.6}$$

Oxidation-reduction reactions comprise one important class of organic reactions.

Another class is substitution reactions, where one functional group replaces another or a hydrogen atom. A simple example involves the reaction of methane with chlorine

$$CH_4 + Cl_2 \longrightarrow CH_3Cl + HCl \tag{5.7}$$

which proceeds at room temperature only when the starting gas mixture is exposed to ultraviolet radiation. This reaction may be broken down into steps:

$$Cl_2 \xrightarrow{\text{UV}} 2Cl \tag{5.8a}$$

$$CH_4 + Cl \longrightarrow \cdot CH_3 + HCl \tag{5.8b}$$

$$\cdot CH_3 + Cl \longrightarrow CH_3Cl \tag{5.8c}$$

As with other series reactions, the slowest step determines the overall reaction rate.

Elimination reactions involve the formation of double or triple bonds, either in a functional group or in the carbon chain of a molecule, because of the removal of side atoms or groups. If X and X' represent side atoms or groups, then an example of an elimination reaction is

$$\begin{array}{c} \mathrm{H}\ \ \mathrm{H} \\ |\ \ \ | \\ R\text{—}\mathrm{C}\text{—}\mathrm{C}\text{—}R' \\ |\ \ \ | \\ X\ \ X' \end{array} \longrightarrow \begin{array}{c} \mathrm{H}\ \ \mathrm{H} \\ |\ \ \ | \\ R\text{—}\mathrm{C}{=}\mathrm{C}\text{—}R' + X\text{—}X' \end{array} \tag{5.9}$$

where R and R' are hydrocarbon radicals. Addition reactions correspond to the addition of side atoms or groups as a result of breaking double or triple bonds in a carbon chain. This is just the opposite of an elimination reaction, and the reverse of Eq. 5.9 serves as an example. Necessarily at least one of the reactants must have a multiple bond; hence, unsaturated hydrocarbons are susceptible to addition reactions. These are particularly important with ethylene and its derivatives in that large molecules can be formed from the small molecules. That is, a suitable addition or chain polymerization reaction of ethylene leads to the giant molecules that are found in the synthetic polymer polyethylene. Such molecules often are composed of tens of thousands of basic repeating units, each derived from one of the initial ethylene molecules or monomers. The resulting substance possesses distinctly different properties than the monomeric substance from which it was formed.

To start chain polymerization a free radical is required, and this can be generated by the decomposition of an organic peroxide [e.g., $(CH_3)_3COOC(CH_3)_3$]. The first stage, initiation, involves formation of free radicals and their combination with ethylene molecules:

$$\begin{array}{l} R\text{—O—O—}R \longrightarrow R\text{—O}\cdot + \cdot\text{O—}R \\ R\text{—O}\cdot + \begin{array}{c} \mathrm{H}\ \ \mathrm{H} \\ |\ \ \ | \\ \mathrm{C}{=}\mathrm{C} \\ |\ \ \ | \\ \mathrm{H}\ \ \mathrm{H} \end{array} \longrightarrow R\text{—O—}\begin{array}{c} \mathrm{H}\ \ \mathrm{H} \\ |\ \ \ | \\ \mathrm{C}\text{—}\mathrm{C}\cdot \\ |\ \ \ | \\ \mathrm{H}\ \ \mathrm{H} \end{array} \end{array} \tag{5.10}$$

The second stage, propagation, is the addition of more monomer to the growing chain end:

$$R\text{—O—}\begin{array}{c} \mathrm{H}\ \ \mathrm{H} \\ |\ \ \ | \\ \mathrm{C}\text{—}\mathrm{C}\cdot \\ |\ \ \ | \\ \mathrm{H}\ \ \mathrm{H} \end{array} + \begin{array}{c} \mathrm{H}\ \ \mathrm{H} \\ |\ \ \ | \\ \mathrm{C}{=}\mathrm{C} \\ |\ \ \ | \\ \mathrm{H}\ \ \mathrm{H} \end{array} \longrightarrow R\text{—O—}\begin{array}{c} \mathrm{H}\ \ \mathrm{H}\ \ \mathrm{H}\ \ \mathrm{H} \\ |\ \ \ |\ \ \ |\ \ \ | \\ \mathrm{C}\text{—}\mathrm{C}\text{—}\mathrm{C}\text{—}\mathrm{C}\cdot \\ |\ \ \ |\ \ \ |\ \ \ | \\ \mathrm{H}\ \ \mathrm{H}\ \ \mathrm{H}\ \ \mathrm{H} \end{array} \tag{5.11}$$

This process proceeds rapidly until a final size is reached when the supply of ethylene monomers is reduced significantly. The third stage, termination, corresponds either to the coupling of chain radicals with one another:

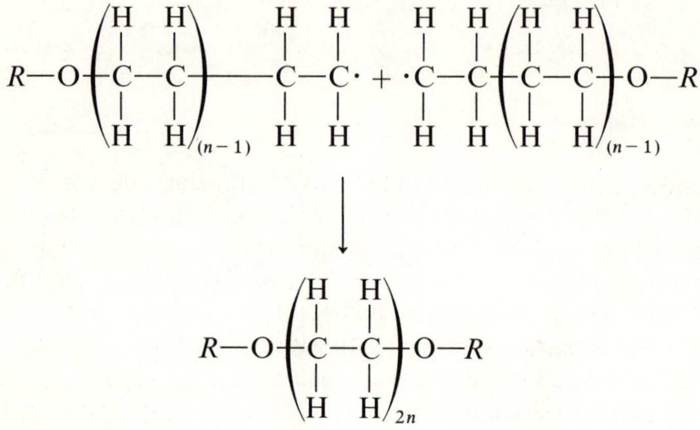

$$R\text{—O}\left(\begin{matrix}\text{H} & \text{H}\\ | & |\\ \text{C}— & \text{C}\\ | & |\\ \text{H} & \text{H}\end{matrix}\right)_{(n-1)}\!\!—\begin{matrix}\text{H} & \text{H}\\ | & |\\ \text{C}— & \text{C}\cdot\\ | & |\\ \text{H} & \text{H}\end{matrix} + \begin{matrix}\text{H} & \text{H}\\ | & |\\ \cdot\text{C}— & \text{C}\\ | & |\\ \text{H} & \text{H}\end{matrix}\!\left(\begin{matrix}\text{H} & \text{H}\\ | & |\\ \text{C}— & \text{C}\\ | & |\\ \text{H} & \text{H}\end{matrix}\right)_{(n-1)}\!\!\text{O—}R$$

$$\downarrow \qquad (5.12a)$$

$$R\text{—O}\left(\begin{matrix}\text{H} & \text{H}\\ | & |\\ \text{C}— & \text{C}\\ | & |\\ \text{H} & \text{H}\end{matrix}\right)_{2n}\!\!\text{O—}R$$

or to disproportionation:

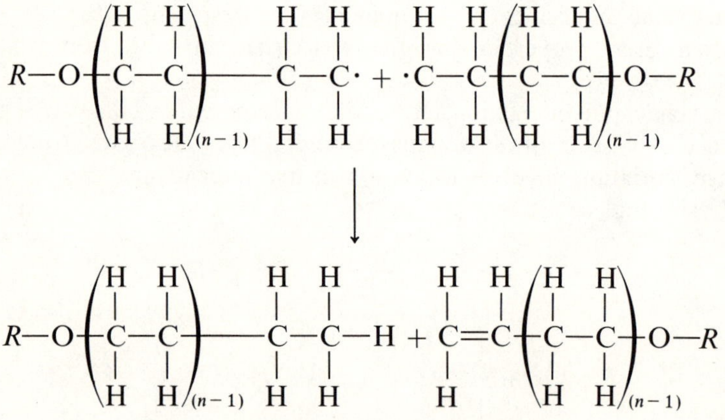

$$R\text{—O}\left(\begin{matrix}\text{H} & \text{H}\\ | & |\\ \text{C}— & \text{C}\\ | & |\\ \text{H} & \text{H}\end{matrix}\right)_{(n-1)}\!\!—\begin{matrix}\text{H} & \text{H}\\ | & |\\ \text{C}— & \text{C}\cdot\\ | & |\\ \text{H} & \text{H}\end{matrix} + \begin{matrix}\text{H} & \text{H}\\ | & |\\ \cdot\text{C}— & \text{C}\\ | & |\\ \text{H} & \text{H}\end{matrix}\!\left(\begin{matrix}\text{H} & \text{H}\\ | & |\\ \text{C}— & \text{C}\\ | & |\\ \text{H} & \text{H}\end{matrix}\right)_{(n-1)}\!\!\text{O—}R$$

$$\downarrow \qquad (5.12b)$$

$$R\text{—O}\left(\begin{matrix}\text{H} & \text{H}\\ | & |\\ \text{C}— & \text{C}\\ | & |\\ \text{H} & \text{H}\end{matrix}\right)_{(n-1)}\!\!—\begin{matrix}\text{H} & \text{H}\\ | & |\\ \text{C}— & \text{C—H}\\ | & |\\ \text{H} & \text{H}\end{matrix} + \begin{matrix}\text{H} & \text{H}\\ | & |\\ \text{C}= & \text{C}\\ | & \\ \text{H} & \end{matrix}\!\left(\begin{matrix}\text{H} & \text{H}\\ | & |\\ \text{C}— & \text{C}\\ | & |\\ \text{H} & \text{H}\end{matrix}\right)_{(n-1)}\!\!\text{O—}R$$

The overall reaction can be represented symbolically as

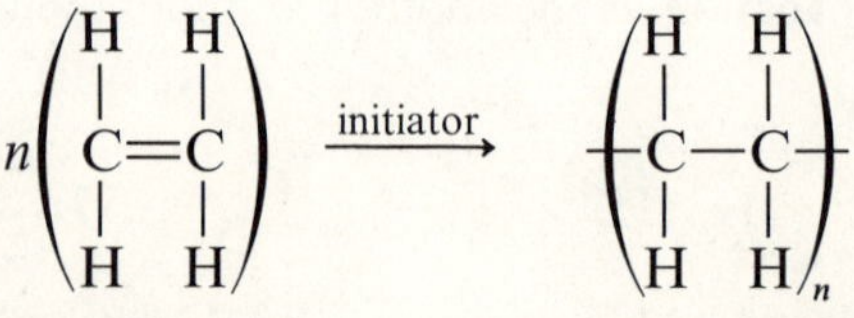

$$n\left(\begin{matrix}\text{H} & \text{H}\\ | & |\\ \text{C}= & \text{C}\\ | & |\\ \text{H} & \text{H}\end{matrix}\right) \xrightarrow{\text{initiator}} \left(\begin{matrix}\text{H} & \text{H}\\ | & |\\ \text{C}— & \text{C}\\ | & |\\ \text{H} & \text{H}\end{matrix}\right)_{n} \qquad (5.13)$$

ethylene **polyethylene**

and the final molecular weight depends upon the type of termination.

The principle of polymerization is based on the capability of a micromolecule or monomer to form at least two new bonds. The number of bonds that a given monomer can form in a given reaction is called the functionality of the monomer. Ethylene and its vinyl analogs, which have the general formula

$$\begin{array}{ccc} H & & H \\ | & & | \\ C & = & C \\ | & & | \\ H & & R \end{array}$$

have a functionality of two. Accordingly, they can be converted into long-chain polymers. For a functionality greater than two, three-dimensional network polymers can form.

Whereas in chain polymerization the final size of each molecule is reached rapidly and further growth of the macromolecules does not occur, in stepwise polymerization each macromolecule that is formed can react further with monomers or with other macromolecules of various sizes. That is, further reaction in stepwise polymerization will lead to increasingly larger macromolecules. Substances that undergo stepwise reactions lead to polymer chains or networks that contain atoms other than carbon with two single bonds.

Stepwise polymerization invariably involves the reaction of two substances, and in many cases one of the products is a small molecule. For example, a monomer like ethylene glycol ($HO—CH_2—CH_2—OH$), with a functionality of two, can react with adipic acid, which is also bifunctional, to form one type of polyester:

$$HO-CH_2-CH_2-O\boxed{H} + \boxed{HO}-C(=O)-CH_2-CH_2-CH_2-CH_2-C(=O)-OH \longrightarrow$$

ethylene glycol **adipic acid**

$$HO-CH_2-CH_2-O-C(=O)-CH_2-CH_2-CH_2-CH_2-C(=O)-OH + H_2O \quad (5.14)$$

ester linkage

This stepwise polymerization proceeds by a condensation reaction, so-called because a small molecule (e.g., H_2O) is eliminated in the process. It should be apparent that other ethylene glycol and adipic acid molecules can join to the unit formed in Eq. 5.14 and that the repeating units in a resulting macromolecule will be joined by ester linkages. This type of polymerization continues even after one of the reactants is used up because the ends on macromolecules can react with each other. Some common polymers formed by chain polymerization and by stepwise polymerization are listed in Table 5.5.

TABLE 5.5 Some Polymers Formed by Chain and by Stepwise Polymerization

	Chain	
Monomer	*Polymer Unit*	*Polymer*
Ethylene $CH_2{=}CH_2$	$-CH_2-CH_2-$	Polyethylene
Vinyl chloride $CH_2{=}CHCl$	$-CH_2-CHCl-$	Polyvinyl chloride
Styrene $CH_2{=}CHC_6H_5$	$-CH_2-CHC_6H_5-$	Polystyrene
Isobutylene $CH_2{=}C(CH_3)_2$	$-CH_2-C(CH_3)_2-$	Polyisobutylene
Butadiene $CH_2{=}CH_2-CH_2{=}CH_2$	$-CH_2-CH_2{=}CH_2-CH_2-$	Polybutadiene
	Stepwise	
Monomers	*Polymer Linkage*	*Polymer*
$HOOC-R-COOH$, $HO-R-OH$	$-\overset{\displaystyle O}{\overset{\|\|}{C}}-O-$	Polyester
CH_2O, phenol (benzene ring–OH)	$-CH_2-$ $-CH_2OCH_2-$	Bakelite
$H_2N-R-NH_2$, $HOOC-R-COOH$, or $H_2N-R-COOH$	$-\overset{\displaystyle O}{\overset{\|\|}{C}}-\overset{\displaystyle H}{\overset{\|}{N}}-$	Polyamide (nylon)
$Cl-(CH_2)_n-Cl$, Na_2S	$-S-$ $-S-S-$	Polysulfide

5.7 MOLECULAR WEIGHTS

Studies of polymers in solution have shown that they contain molecules of many different sizes such that a polymer cannot be characterized by a single molecular weight. The nature of the distribution can be determined experimentally by different techniques such as fractionating the polymer from dilute solution and then weighing and determining the molecular sizes of a series of narrow fractions. The size distribution in many cases follows the type of curve shown in Fig. 5.5.

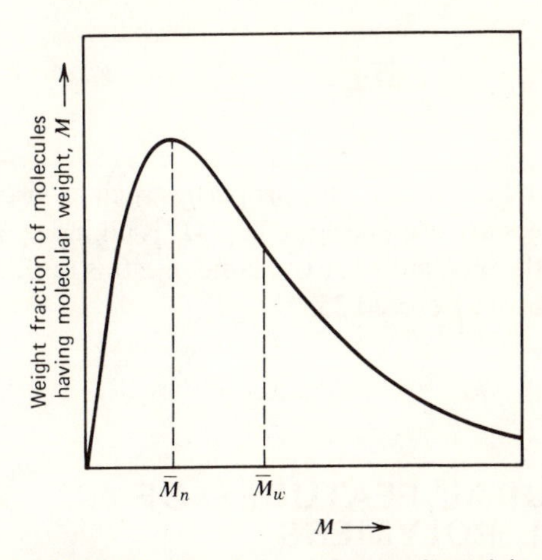

FIGURE 5.5 Distribution of molecular weights in a typical polymer.

Since the physical properties of a given polymer depend on the molecular size distribution as well as the average degree of polymerization ($\bar{x}$), which is defined as the average number of repeating units in the polymer molecules present, it is necessary to characterize both. One value used is the number-average molecular weight $\overline{M}_n$. If n_i is the number fraction (or mole fraction) of molecules with molecular weight M_i, then

$$\overline{M}_n = \sum_i n_i M_i \tag{5.15a}$$

If end groups are neglected, this equation can be written as

$$\overline{M}_n = M_0 \bar{x} \tag{5.15b}$$

where M_0 is the molecular weight of a repeating unit. Another value is the weight-average molecular weight $\overline{M}_w$. If w_i is the weight fraction of molecules having molecular weight M_i, then

$$\overline{M}_w = \sum_i w_i M_i = \frac{\sum_i n_i M_i^2}{\sum_i n_i M_i} \tag{5.16a}$$

which can be rewritten as

$$\overline{M}_w = \frac{M_0}{\bar{x}} \sum_i n_i x_i^2 \tag{5.16b}$$

The quantity $\overline{M}_w$ correlates better with properties such as viscosity than does $\overline{M}_n$; however, both averages are used because $\overline{M}_w/\overline{M}_n$ is an index of the breadth of the distribution. If all of the molecules are the same size, this ratio is unity. Usually it is between 1.5 and 3, but may exceed 25.

5.8 STRUCTURAL FEATURES OF NATURAL POLYMERS

Although synthetic polymers are produced in large quantities from petrochemicals, several natural polymers also are used for engineering purposes, and many are of great importance in living organisms. One of the most common natural polymers is cellulose, the principal constituent of the cell walls in higher plants. Cellulose is a polymer of β-glucose,

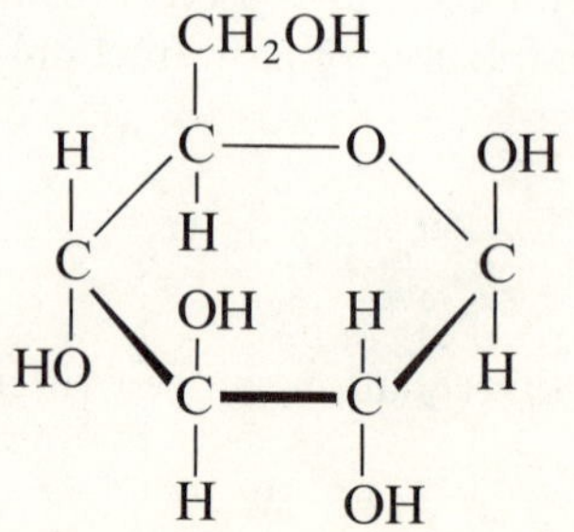

CH_2OH
H C—O OH
H
C C
OH H
HO C—C H
H OH

and cellulose has the following repeating unit:

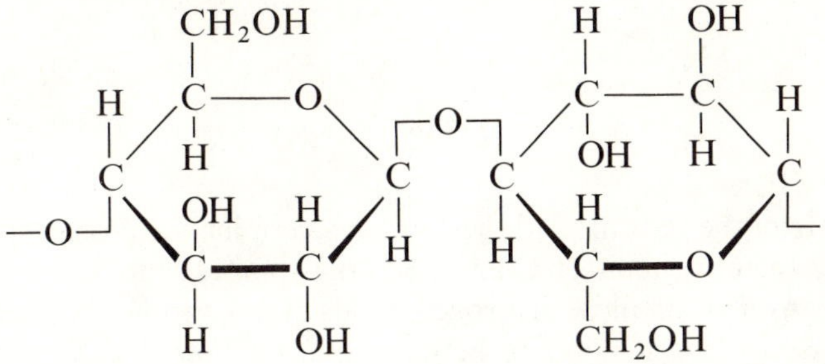

The side groups effectively extend above and below the *nonplanar* carbon–oxygen ring as do the oxygen atom linkages. This facilitates hydrogen bonding between parallel chains, and the combination of relatively strong intermolecular bonding with the regular molecular structure leads to a high degree of crystallinity.

Wood and cotton are two of the most useful forms of native cellulose. Cotton consists of about 90% cellulose, but wood is only about 50% cellulose and 30% lignin, a more complex polymer. In both of these materials, structures on a much larger scale than the molecules themselves are important in determining their natures. On a gross scale, wood has growth rings that correspond to its visible *grain*. Typically, the spring growth layers have a lighter color and lower density than the summer growth layers. These layers are made up predominantly of bundles of thin-walled, tubular biological cells whose axes are parallel to the trunk or a limb of a tree, and it is the variation of cross-sectional cell size with seasonal growth that leads to the growth rings. The cell walls, containing cellulose, lignin, and other substances, provide structural support. Since the cells and the grain of wood are oriented, wood has anisotropic properties.

A fairly sizable amount of regenerated cellulose is produced annually in the form of rayon fibers and cellophane film, and there are a number of chemical derivatives of cellulose. Cotton, rayon, cellophane and other polysaccharides remain important materials even though many synthetic polymers have been developed.

Proteins are another important type of natural polymer found in all living cells. The repeating units are derived from amino acids having the form

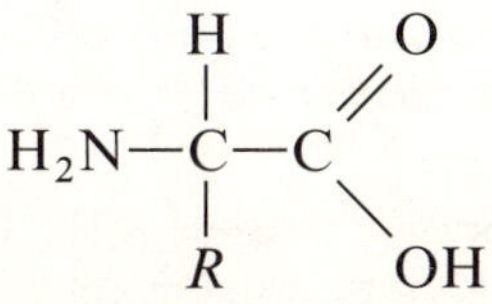

$$H_2N-\underset{R}{\overset{H}{C}}-C\begin{matrix}\diagup\!\!\diagup O \\ \diagdown OH\end{matrix}$$

Each protein macromolecule possesses a definite order of amino acid groups

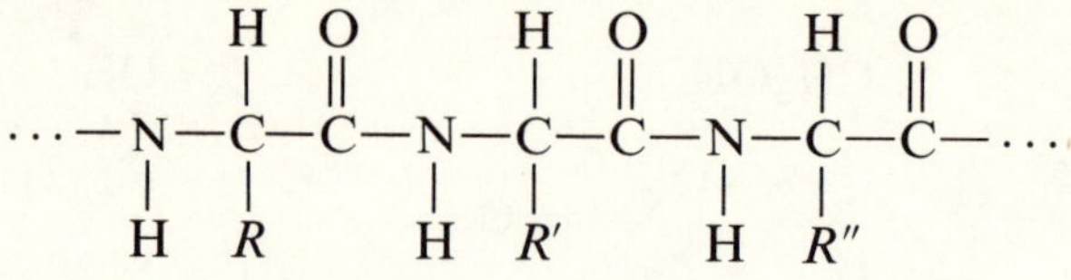

$$\cdots-\underset{\mathrm{H}}{\underset{|}{\mathrm{N}}}-\overset{\mathrm{H}}{\overset{|}{\underset{R}{\underset{|}{\mathrm{C}}}}}-\overset{\mathrm{O}}{\overset{\|}{\mathrm{C}}}-\underset{\mathrm{H}}{\underset{|}{\mathrm{N}}}-\overset{\mathrm{H}}{\overset{|}{\underset{R'}{\underset{|}{\mathrm{C}}}}}-\overset{\mathrm{O}}{\overset{\|}{\mathrm{C}}}-\underset{\mathrm{H}}{\underset{|}{\mathrm{N}}}-\overset{\mathrm{H}}{\overset{|}{\underset{R''}{\underset{|}{\mathrm{C}}}}}-\overset{\mathrm{O}}{\overset{\|}{\mathrm{C}}}-\cdots$$

and the nature of the radicals and their order determine the characteristics of the protein. Since there are about 30 amino acids, the number of possible combinations is quite large. As with cellulose, hydrogen bonds play a significant role in proteins. In many cases hydrogen bonding between different parts of the same protein molecule leads to a helical molecular configuration. Hydrogen bonding between molecules is responsible for the fibrous structure of hair and the sheet structure of hides. Although silk, wool, and hides are utilized widely, most proteins are of biological interest in that they serve structural purposes, transmit information, and regulate life functions in animals and humans. For engineering applications the synthetic polyamides or nylons, which are related to proteins by having amide linkages $\left(-\overset{\mathrm{O}}{\overset{\|}{\mathrm{C}}}-\underset{\mathrm{H}}{\underset{|}{\mathrm{N}}}-\right)$, are more important.

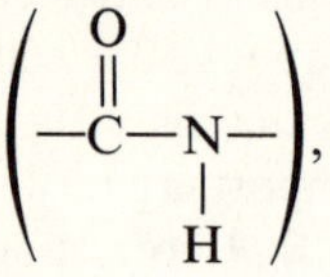

REFERENCES

1. Pimentel, G. E. and Spratley, R. D. *Understanding Chemistry*, Holden-Day, San Francisco, 1971.
2. Allinger, N. and Allinger, J. *Structures of Organic Molecules*, Prentice-Hall, Englewood Cliffs, N.J., 1965.

QUESTIONS AND PROBLEMS

5.1 Account for the greater strength of the hydrogen bond as compared to the van der Waals intermolecular bond. Does the small size of hydrogen have anything to do with it?

5.2 The term degree of polymerization ($\bar{x}$) refers to the number of repeating units in a polymer molecule. What is the molecular weight of a polyethylene molecule having $\bar{x} =$ 10,000.

5.3 Consider Eq. 5.10. If the organic peroxide $(CH_3)_3COOC(CH_3)_3$ is used to initiate chain polymerization, what is the radical R in Eq. 5.10?

5.4 The monomer of the polyester polymer discussed in Sect. 5.6 is shown in Eq. 5.14. Continue the reaction, Eq. 5.14, one more step to show how additional ethylene glycol and adipic acid molecules can join the unit shown in Eq. 5.14.

5.5 Hexamethylene diamine [$NH_2—(CH_2)_6—NH_2$] and adipic acid react to form the mer of nylon 66. The reaction is a stepwise condensation reaction with liberation of water. Write the analog of Eq. 5.14 for the condensation reaction leading to the formation of nylon 66.

5.6 Show how two molecules of β-glucose can react to form the repeating unit of cellulose.

5.7 Sketch schematic structural formulas of the alkene series and the cycloalkane series that have the same chemical formula C_nH_{2n} (Table 5.2). Do this through $n = 6$. What types of isomers are these?

5.8 Repeat problem 5.7 for the alkadiene series and the cycloalkene series (C_nH_{2n-2}).

5.9 Consider Eq. 5.16a. Show that $\sum_i w_i M_i = \sum_i n_i M_i^2 / \sum n_i M_i$. Show that Eq. 5.16b follows from Eqs. 5.16a and 5.15.

5.10 The text states that $\overline{M}_w/\overline{M}_n$ is an index of the breadth of the distribution in molecular weights. Physically explain why this is so.

5.11 Draw structural formulas comparing starch with cellulose.

5.12 Cotton consists of cellulose molecules containing some 9000 glucose units arranged in strands cross-linked by hydrogen bonds. Diagram such hydrogen bonding between portions of two glucose chains.

5.13 Carbon is the basic "building block" of a vast number of molecules and compounds. What other element would possibly be the basis for a large number of molecules and compounds? Explain your choice.

5.14 (a) The carbon–carbon double bond in hexene (C_6H_{12}), which has straight-chain molecules, may be located in three distinct positions along the carbon chain. Sketch these three molecules.

(b) When the double bond is between the first and second carbon atoms, the compound is called 1-hexene; when it is between the second and third carbon atoms, the compound is 2-hexene; etc. Explain why the designations 4-hexene and 5-hexene are never used.

5.15 Consider the substitution of two side groups (or functional groups) for two hydrogens in benzene. Give the structural formulas of all possible isomers.

5.16 Show the centers of positive and negative charge in (i) CCl_4, (ii) $C_2H_2Cl_2$, and (iii) CH_3Cl. Which of these molecules are polar molecules? Which can have two forms?

Polymers

This chapter is concerned with the structure of some common synthetic polymers.

6.1 CLASSIFICATION

Cellulose, starches, proteins, and enzymes are all naturally occurring substances that are polymeric in nature. Some natural polymeric materials are used directly (e.g., wood, wool, silk, and cotton), but others are reconstituted into other useful forms (e.g., cellulose into rayon and cellophane). It is, however, the manmade or synthetic polymers, with repeating units derived from small organic molecules found in fossil fuels, that are increasingly important since these constitute many of the useful plastics, rubbers, and fibers commonly encountered. Designations such as Bakelite, Dacron, Orlon, and nylon, some of which are trademarks, may be familiar. It is usual to associate the word polymer with a high molecular weight molecule (high polymer); however, almost a continuum on the molecular weight scale is possible, and lower molecular weight polymers are used as oils, greases, waxes, paints, lacquers, and glues.

Strictly speaking, the word polymer refers to a large molecule or macromolecule made up of many repeating units. It will be used here also to designate the class of solids constituted by macromolecules having molecular weights between about 10,000 and 1,000,000. Synthetic organic polymers may be classified in a number of ways. For example, polymers whose backbones contain only carbon atoms with single bonds are called carbon chain polymers, and those whose backbones contain other atoms in addition to carbon or carbon with multiple bonds are called heterochain polymers. The former class is made typically by chain polymerization reactions, and the latter class, by stepwise polymerization reactions (see Sect. 5.6). Formability provides another classification scheme for many polymers. Thermoplastics (e.g., polyethylene), within a certain temperature range, can be repeatedly shaped and reshaped because they soften appreciably as temperature increases and also eventually liquify. Thermosets (e.g., Bakelite), once shaped, cannot be reshaped at any temperature. In fact, decomposition of thermosets occurs before any appreciable softening. Similarly when elastomers (e.g., vulcanized rubber) are formed and suitably treated, they cannot be reshaped significantly. The reasons for such formability behavior are structural in origin and may be explained in terms of molecular architecture and bonding in polymeric solids. During the final thermosetting reaction of phenol-formaldehyde (Bakelite), for example, covalent cross-linking develops a three-dimensional network structure. Such bonds are not broken readily and, as a consequence, Bakelite remains a hard, rigid material up to its decomposition temperature. On the other hand, thermoplastics consist of long-chain molecules with little or no covalent cross-linking, and the intermolecular bonding is of a physical type. Long-chain polymer molecules tend to bend and kink, such that their end-to-end distance is only a small fraction of the total length of the polymer chain. Polymers with this molecular architecture usually are shaped and reshaped readily within a definite temperature range where the molecules slide by one another without breaking covalent bonds. Elastomers are rather intermediate in structure between thermosets and thermoplastics. They consist of coiled long-chain molecules with occasional covalent cross-links that prevent irreversible sliding of molecules past one another but allow the sizable reversible extensibility characteristic of rubbers. Below definite temperatures, both thermoplastics and elastomers become glasslike.

In view of structural differences and corresponding differences in properties, polymers may also therefore be classified as network polymers, long-chain polymers, and elastomers. This classification scheme proves valuable for the various applications of polymeric materials. Nevertheless, grouping of polymers according to the type of molecular backbone will be employed because this relates to the type of polymerization reaction.

6.2 CARBON CHAIN POLYMERS

A number of compounds having the formula

```
H   H
|   |
C = C
|   |
H   R
```

TABLE 6.1 Some Carbon-Chain Polymers and Their Uses

Name	*Monomer*	*Repeating Unit Structure*	*Typical Uses*
Polyethylene	$CH_2{=}CH_2$	$-CH_2-CH_2-$	Film for packaging applications (branched PE); bottles and other containers (linear PE).
Polypropylene	$CH_2{=}CH(CH_3)$	$-CH_2-CH(CH_3)-$	Rope and filaments.
Polystyrene	$CH_2{=}CH(C_6H_5)$	$-CH_2-CH(C_6H_5)-$	Molded packaging containers, foams, plastic optical components.
Polymethyl methacrylate (PMMA; Lucite; Plexiglas)	$CH_2{=}C(CH_3)(C{=}O-O-CH_3)$	$-CH_2-C(CH_3)(C{=}O-O-CH_3)-$	Plastic windows, fixtures, paints.

Polyacrylonitrile (Orlon)	$H_2C{=}CH{-}C{\equiv}N$	$-CH_2{-}CH(C{\equiv}N)-$	Acrylic fibers.
Polyvinyl acetate	$H_2C{=}CH{-}O{-}C({=}O){-}CH_3$	$-CH_2{-}CH(O{-}C({=}O){-}CH_3)-$	Water-based emulsion paints, adhesives.
Polyvinyl alcohol	$H_2C{=}CH{-}OH$	$-CH_2{-}CH(OH)-$	Textile fiber, wet-strength adhesives.
Polyvinyl chloride (PVC)	$H_2C{=}CHCl$	$-CH_2{-}CHCl-$	Insulation of electrical wire and cable, pipe, phonograph records.
Polyvinylidene chloride (saran, if copolymerized with < 20% PVC)	$H_2C{=}CCl_2$	$-CH_2{-}CCl_2-$	Transparent film, fiber.
Polytetrafluoroethylene (Tefon, TFE)	$F_2C{=}CF_2$	$-CF_2{-}CF_2-$	Chemically resistant coatings, laboratory ware, coatings for nonstick food processing equipment.

are called vinyl compounds and have a functionality of two. Vinyl monomers and other monomers based on ethylene with more than one side group can undergo chain polymerization (or addition) reactions. The sequential process of initiation, propagation, and termination was described in Sect. 5.6, and the monomer concentration decreases continuously as the reaction proceeds. Either branched or unbranched carbon chain molecules result. A list of some carbon chain polymers along with their monomers and repeating units is given in Table 6.1.

Since the introduction of special catalysts, it has become possible not only to carry out chain polymerization reactions at much lower temperatures and pressures than previously, but also to control the structure to a greater extent. For example, the amount of branching in polyethylene can be altered, and virtually unbranched or linear polyethylene can be produced. In addition, a wide range of

TABLE 6.2 Some Heterochain Polymers and Their Uses

Name	*Monomers*
Polyhexamethylene adipamide (a polyamide) (nylon 66)	$NH_2(CH_2)_6NH_2$ **(hexamethylene-diamine)** + $COOH(CH_2)_4COOH$ **(adipic acid)**
Polyethylene terephthalate (a polyester) (Dacron; Mylar)	$C_6H_4(COOCH_3)_2$ **(terephthalic acid, dimethyl ester)** + CH_2OHCH_2OH **(glycol)**
Phenol-formaldehyde (Bakelite)	C_6H_5OH + $H-CHO$

copolymers, which, unlike *homopolymers*, contain more than one kind of repeating unit, can be formed from a mixture of monomeric species.

6.3 HETEROCHAIN POLYMERS

Stepwise polymerization is carried out with monomers that contain more than one reactive functional group, and the reaction can proceed continuously (Sect. 5.6). The monomer is used up early, and further polymerization involves reaction of molecules larger than the monomer. The process usually is carried out at high temperatures and pressures, and any condensation product (e.g., H_2O) can be removed by vacuum pumping. Often, the final product has a greater molecular size distribution than polymers formed by chain polymerization. A list of some heterochain polymers, their monomers and repeating units is given in Table 6.2.

Repeating Unit Structure	*Typical Uses*
O H H H H O H H H H H H —C—C—C—C—C—C—N—C—C—C—C—C—C—N— H H H H H H H H H H H H	Fibers, rollers, bearings, jacket over primary electrical insulation.
O O H H —C—(benzene ring)—C—O—C—C—O— H H	Textile fibers, films, recording tape.
H H C OH C H H (benzene ring) H—C—H	Dielectric parts, heat-resistant appliance parts, laminates.

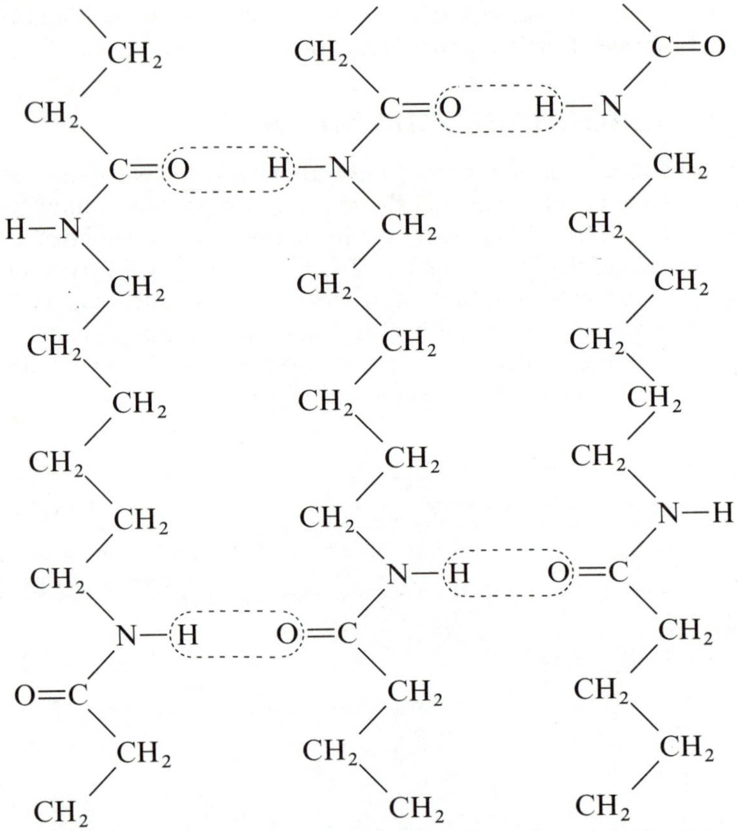

```
     CH2          CH2                       C=O
CH2                  C=O ········ H—N
     C=O ········ H—N                          CH2
H—N                  CH2                 CH2
     CH2          CH2                          CH2
CH2                  CH2                 CH2
     CH2          CH2                          CH2
CH2                  CH2                 CH2
     CH2          CH2                          N—H
CH2                  N—H ········ O=C
     N—H ········ O=C                          CH2
O=C                  CH2                 CH2
     CH2          CH2                          CH2
CH2                  CH2                 CH2
```

FIGURE 6.1 The arrangement of molecular chains in the crystalline form of the polymer nylon 66. The hydrogen bonding between chains (noted by the dotted lines) provides relatively strong interchain bonding and a corresponding more rigid structure on a macroscopic level.

When the monomers have an average functionality of two, long-chain molecules result. Functionalities greater than two lead to highly branched or network structures.

Many heterochain polymers such as polyesters and polyamides (nylons) have polar side groups on the linking units within the molecule chain:

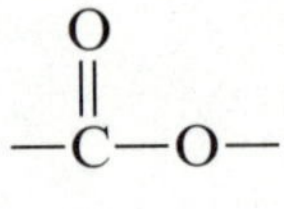

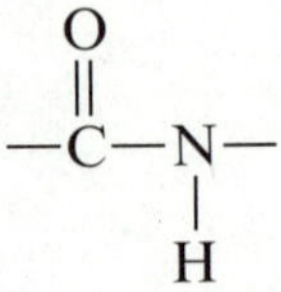

```
   O              O
   ||             ||
 —C—O—          —C—N—
                    |
                    H
```

ester linkage **amide linkage**

The negative polarity of each oxygen to the side of the chain in these polymers facilitates hydrogen bonding between chains, and this is stronger than the van der Waals bonding between molecules in most carbon chain polymers (Fig. 6.1). Polyamides, therefore, resist softening more than polyethylene as temperature rises. Somewhat surprisingly, however, polyesters soften more readily than polyethylene. This is primarily a consequence of the greater chain flexibility resulting from the oxygen atoms in the polyester chain.

Network polymers like Bakelite fit into the heterochain classification and are formed by stepwise polymerization reactions. In this case, partially polymerized powder is produced initially. Such powder can be densified by pressing in a mold and curing (heating under pressure) to form a fully polymerized and shaped solid. The resulting network of covalent bonds, completed during curing, gives the solid a rigid, noncrystalline structure comparable to that of an inorganic glass (Chapter 7).

6.4 CRYSTALLIZATION AND GLASS TRANSITION

Different polymers exhibit different tendencies toward crystallinity, depending upon their composition and molecular structure. Network polymers and elastomers are noncrystalline due to essentially random three-dimensional covalent bonding that prevents any rearrangements necessary for long-range ordering.

Whether or not a long-chain polymer crystallizes depends on the character of the side groups, the amount of chain branching and, to a lesser extent, the molecular chain length. Polymers made up of regular and identical repeating units can sometimes fit into an orderly crystalline arrangement. Those with bulky side groups or extensive chain branching are more noncrystalline than those with smaller side groups or no branching. The geometric arrangement of side groups also influences the tendency toward crystallinity. If the side groups are located randomly along the chain as in Fig. 6.2*a*, the configuration is referred to as atactic. Such polymers are difficult to crystallize. If, on the other hand, the configuration of side groups is such that all are on one side (isotactic, Fig. 6.2*b*) or they are regularly alternating (syndiotactic, Fig. 6.2*c*), the polymer can crystallize more easily, even when the side group is bulky. Actual polymers are not purely atactic, isotactic, or syndiotactic, but have different degrees of tacticity.

Copolymers, which have more than one repeating unit, crystallize less readily than homopolymers, which have only one repeating unit. Again, the tendency toward crystallinity relates to whether or not the different repeating units making up the copolymer chain are spaced in a regular or random manner (Fig. 6.3). Lack of regularity favors noncrystallinity. Copolymers formed from three or more different monomers usually are completely noncrystalline.

Crystalline bulk polymers seldom are completely crystalline. They generally consist of a two-phase crystalline and noncrystalline aggregate. This relates

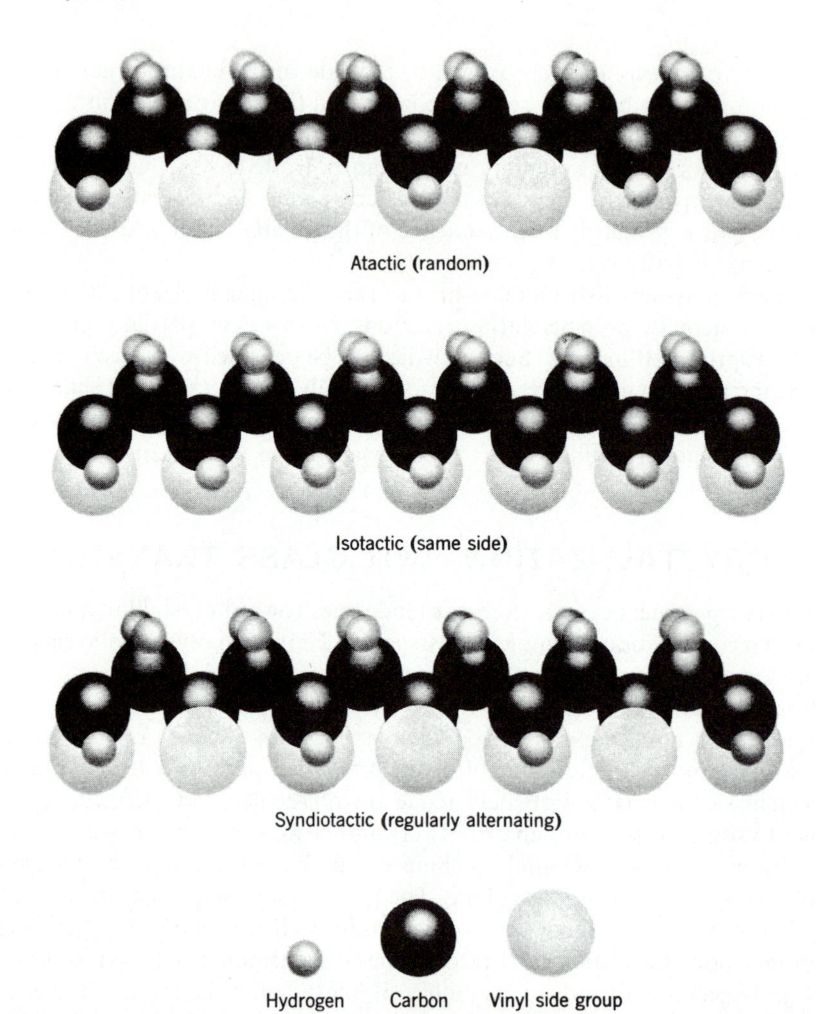

FIGURE 6.2 The possible arrangements of side groups along a simple vinyl polymer chain: (*a*) atactic or random arrangement; (*b*) isotactic or all on one side; and (*c*) syndiotactic or regularly alternating. The side group may be a single atom, such as chloride in polyvinyl chloride, or it may be a group of atoms, such as a benzene ring in polystyrene. (From *The Structure and Properties of Materials,* Vol. I, by W. G. Moffatt, G. W. Pearsall, and J. Wulff, Wiley, New York, 1964.)

primarily to structural considerations and secondarily to cooling rate from the melt. Figure 6.4 shows schematically the effect of cooling from the liquid region to room temperature on the volume of a polymer. At temperatures well above the equilibrium melting point (region *A*) the polymer is a rather viscous liquid. As temperature decreases (region *B*), the liquid becomes more viscous. If the polymer is inherently noncrystalline (i.e., atactic or an irregular copolymer), then crystalliza-

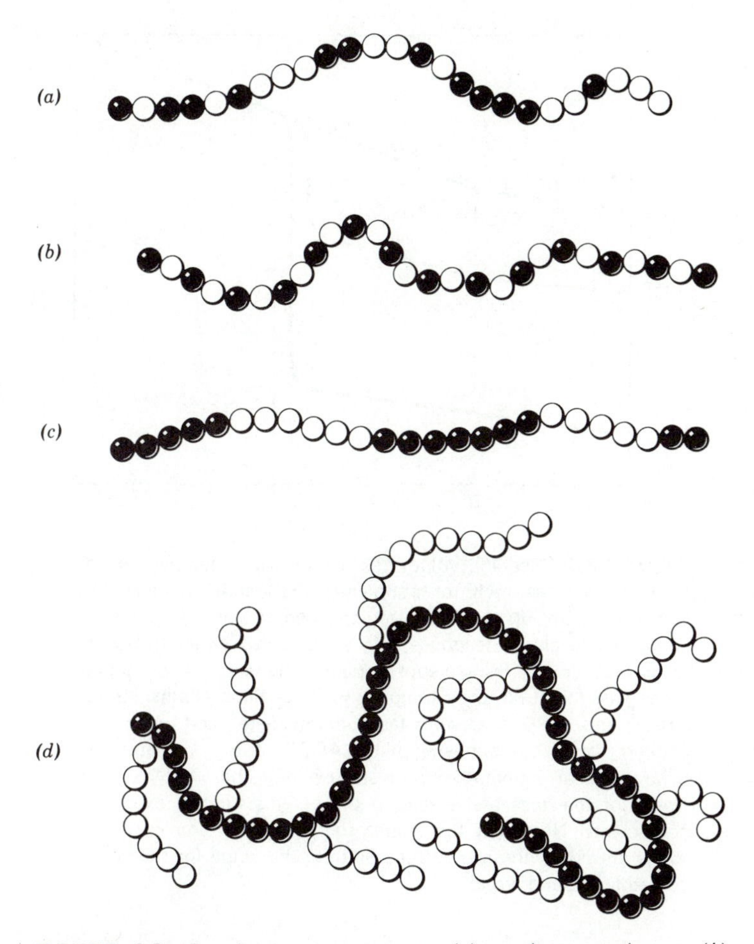

FIGURE 6.3 Copolymer arrangements: (*a*) random copolymer; (*b*) regular copolymer; (*c*) block copolymer; and (*d*) graft copolymer. Each dark sphere represents one type of repeating unit, and each light sphere represents another type of repeating unit. The regular arrangement is most readily crystallized.

tion will not occur (path *ABC*) and the liquid structure will be retained as a pliable and rubbery supercooled liquid (region *C*). (This type of behavior can be expected for other long-chain polymers that are cooled extremely rapidly.) A polymer that tends to be highly crystalline (e.g., linear polyethylene) follows path *ABG*, and crystallization is accompanied by a fairly sharp decrease in volume (ideally, a discontinuous decrease in volume) since the molecules pack more efficiently in the long-range ordered crystallites than in the liquid. A crystalline polymer is less

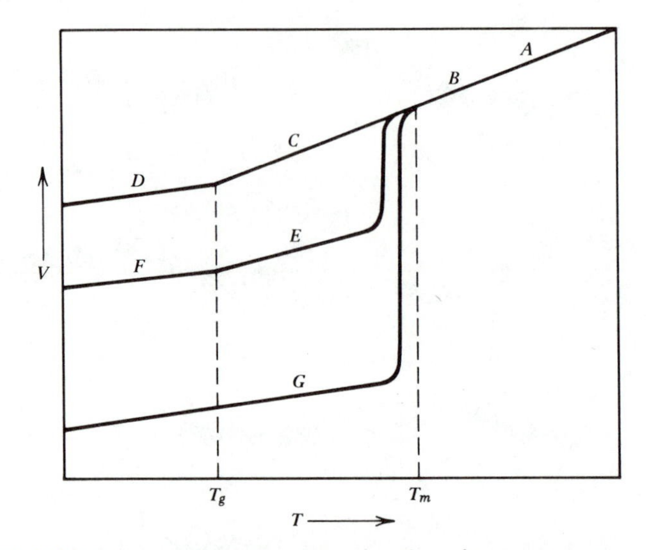

FIGURE 6.4 Specific volume as a function of temperature on cooling from the melt for a polymer that tends to crystallize. (Compare with Fig. 1.5.) The regions denoted are: *A*, liquid; *B*, liquid (some elastic response); *C*, supercooled liquid (rubbery); *D*, glass; *E*, crystallites in a supercooled liquid matrix; *F*, crystallites in a glassy matrix; *G*, "completely" crystalline. Paths *ABCD*, *ABEF*, and *ABG* represent fast, intermediate, and very slow cooling rates, respectively; path *ABCD* also represents the behavior of an inherently noncrystalline long-chain polymer. The slopes of the lines are related to the volume thermal expansion coefficients. Note that the slopes for the glassy and crystalline states are about the same, but less than the slope for the supercooled liquid and liquid.

pliable than a noncrystalline polymer. Because complete long-range order never occurs in a bulk polymer, some noncrystalline material exists even in region *G*. In the case intermediate between a totally noncrystalline polymer and a highly crystalline polymer (e.g., path *ABE*), less material crystallizes and the sharp decrease in volume is less pronounced.

If the polymer is cooled to the glass transition temperatures T_g, the viscosity of the noncrystalline portions increases to such an extent that they become glasslike and brittle. A totally noncrystalline polymer becomes completely glassy (region *D*), and a partially crystalline polymer consists of crystallites in a glassy matrix (region *F*). The results of cooling can be reversed by heating since no fundamental changes in the basic molecular structure accompany changes in state.

The crystalline melting temperature depends upon the strength of physical bonding between molecules. Because both T_m and T_g determine the usefulness of a

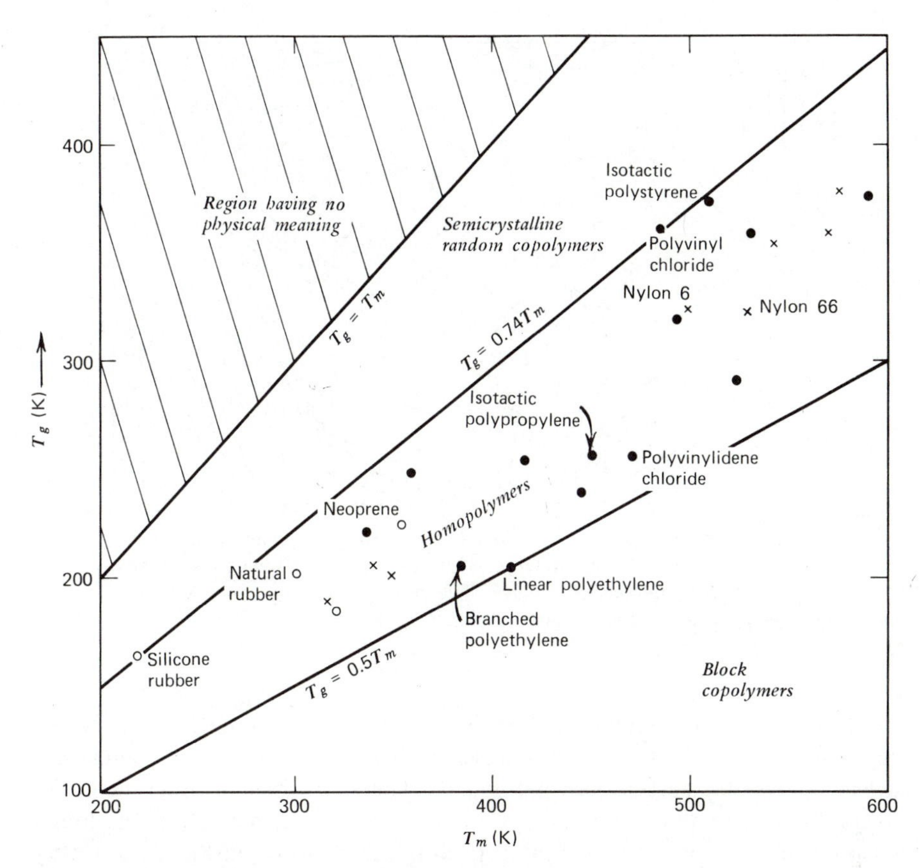

FIGURE 6.5 Glass transition temperature (T_g) as a function of crystalline melting temperature (T_m) for homopolymers. *Key*: ● addition homopolymers; ○ elastomers; × condensation homopolymers. Symmetrical homopolymers tend to be close to $T_g = 0.5T_m$, whereas unsymmetrical homopolymers are closer to $T_g = 0.74T_m$. Random copolymers occupy the region where $T_g > 0.74T_m$; block copolymers occupy the region where $T_g < 0.5T_m$.

polymeric material, it is important to note that polymer structure affects T_m and T_g similarly for homopolymers and regular copolymers (Fig. 6.5). For nonregular copolymers no definite relationship exists between T_m and T_g.

6.5 POLYMER CRYSTALS

The morphology or spatial arrangement of long-chain molecules in crystallites as yet has not been resolved fully. The molecules themselves are known in many cases to be substantially larger than individual crystallites. Thus, one molecule may

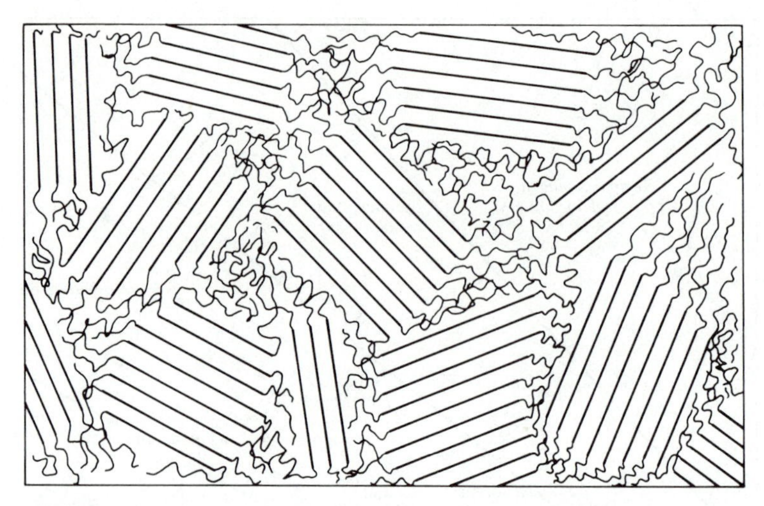

FIGURE 6.6 Fringed micelle model of a mixed crystalline-noncrystalline polymer. Regions of crystallinity (crystallites), where molecules are aligned, are separated by regions of noncrystalline material (either supercooled liquid or glassy), where molecules are not aligned.

extend through both crystalline and noncrystalline regions within a bulk polymer. For many years the model used to represent polymers prepared from the melt was the fringed micelle structure shown in Fig. 6.6. Here individual polymer molecules traverse many crystallites and noncrystalline regions. More recently, it was found that polyethylene could be condensed from dilute solutions in the form of very small single crystals. In this case, the individual molecular chain is folded back and forth many times within the crystal (Fig. 6.7). Consequently, it is now felt that

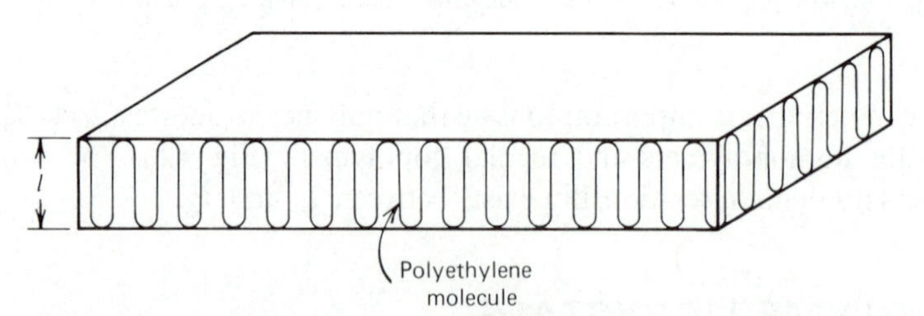

FIGURE 6.7 Model of the folded-chain structure of single crystal polyethylene. The length of a molecular chain is much greater than the thickness of the single crystal (l); thus, each molecule extends back and forth many times through the crystal, with chain folds at the top and bottom faces.

the fringed micelle model is not entirely correct even for bulk polymers crystallized from the melt. The structure of crystalline polymers is perhaps best described as being intermediate between the fringed micelle and the folded chain models.

Crystals in polymers prepared from the melt often take the form of spherulites (Fig. 6.8). This type of crystallization phenomenon is observed in polyethylene, nylon, and other crystalline polymers. Each spherulite is a spherical region, containing crystalline material, which grew radially at the expense of noncrystalline

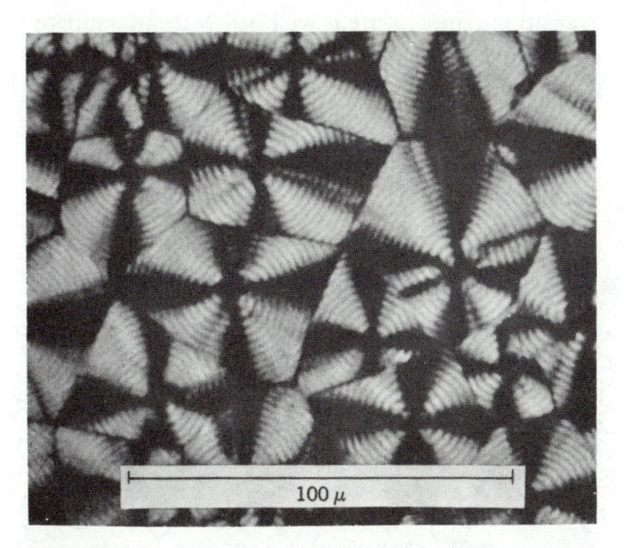

FIGURE 6.8 Spherulites in polyethylene crystallized from the melt. This transmitted polarized light photomicrograph shows polygonal regions that are contacting spherulites. (Courtesy of F. P. Price, General Electric Co.)

material about it. Within a given spherulite, crystallites have a radial orientation such that the polymer chains are oriented tangentially. A spherulite should be regarded as a grouping of crystallites with a small amount of noncrystalline material between them rather than as a highly faulted single crystal. Totally spherulitic polymers are not completely crystalline because the regions of misorientation between crystallites represent a significant deviation from long-range ordering. In many polymers, noncrystalline material separates spherulites as well, although neighboring spherulites may partially penetrate one another through the noncrystalline material, and this leads to greater stiffness for the polymer than if the spherulites were not connected.

6.6 ELASTOMERS

The structure of network and long-chain polymers is developed mainly during polymerization. In elastomers, however, the cross-linked network is developed in a separate postpolymerization step. Elastomers are polymers that exhibit a large and reversible extensibility at room temperature. They can be stretched to between 100 and 1000% of their original length and will snap back to their original dimensions when the load is released. Necessary but not sufficient conditions for a polymer to be an elastomer are that it is noncrystalline at room temperature and that its glass transition temperature is well below room temperature. In addition, the polymer chains must be (1) very long with many bends, (2) in constant motion at room temperature, and (3) cross-linked or connected every few hundred atoms.

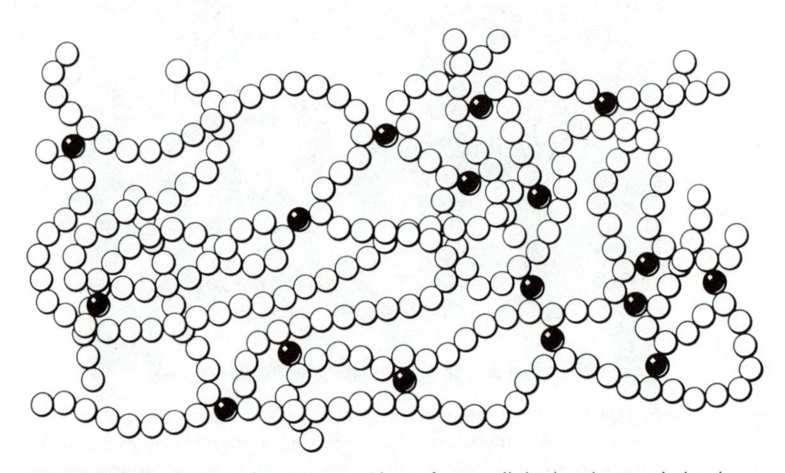

FIGURE 6.9 Schematic representation of cross-linked polymer chains in an elastomer. The black atoms, for example, represent sulfur-bridge cross-links between *cis*-polyisoprene chains in vulcanized natural rubber.

Such a structure might be compared with a bowl of kinked, wriggling spaghetti (the motion arising from thermal energy), with the strands of spaghetti glued together at occasional points (Fig. 6.9). On stretching, the chains are straightened out. When the load is released, the *pinning points* and the tendency for kinking enable the elastomer to return to its original shape. Without the presence of such pinning points, permanent deformation by intermolecular sliding could occur.

The best known elastomer is derived from *cis*-polyisoprene (natural rubber), which is a gummy liquid obtained from certain tropical trees. The repeating unit is

```
            H
            |
  H    H—C—H    H   H
  |         |       |   |
—C————C═══C—C—
  |                     |
  H                     H
```

with the H and CH_3 side groups that are joined to double-bonded carbon atoms being on the same side of the polymer chain. The geometrical hindrance between these side groups causes the chain to bend with the result that it is highly coiled. The isomer *trans*-polyisoprene (gutta-percha), with the H and CH_3 side groups on opposite sides of the chain, has considerably less coiled molecules and therefore cannot be converted into an elastomer.

The elastomeric condition is achieved through a process that causes cross-linking between neighboring chains. One example is vulcanization, which involves heating natural rubber with small amounts of sulfur. Periodic breaks in the carbon–carbon double bond may occur, allowing sulfur-bridge cross-links to form. These may tie the chains together as indicated:

```
 H  CH3  H  H                    H  CH3  H  H
 |   |   |  |                    |   |   |  |
—C—C═══C—C—                     —C—C————C—C—
 |          |                    |   |   |  |
 H          H                    H   |   |  H
                 + 2S  ——→           S   S
 H          H                    H   |   |  H
 |          |                    |   |   |  |
—C—C═══C—C—                     —C—C————C—C—
 |   |   |  |                    |   |   |  |
 H  CH3  H  H                    H  CH3  H  H
```

or polysulfide cross-links may develop:

```
 H  CH3  H  H                   H  CH3  H  H
 |   |   |  |                   |   |   |  |
—C—C═══C—C—                    —C—C═══C—C—
 |          |                   |          |
 H          H                   S          H
                + 3S ——→        |                + H2S
 H          H                   S          H
 |          |                   |          |
—C—C═══C—H—                    —C—C═══C—C—
 |   |   |  |                   |   |   |  |
 H  CH3  H  H                   H  CH3  H  H
```

Cross-linking also can be effected by other thermochemical reactions.

The relative stiffness of vulcanized rubber depends upon the frequency of cross-links. For example, stiffness can be controlled by the amount of sulfur present. Indeed, if an excess of sulfur is present, hard rubber (ebonite) results. Ebonite contains so many cross-links that it is considered a network polymer and is hard and rather stiff.

6.7 SEMIORGANIC POLYMERS

Silicon, because its outer electronic structure is similar to carbon's, might be expected to be the basis for a series of polymers. It differs from carbon, however, in that long Si chains are unstable, and multiple bonds do not occur between Si atoms. Nevertheless, a combination of silicon and oxygen serves as the basis for silicone polymers that have chains or networks made up of alternating silicon and oxygen atoms (siloxane links). These are formed by stepwise polymerization following hydrolysis of CH_3Cl:

$$2CH_3Cl + Si \xrightarrow{\text{catalyst}} Cl{-}\underset{\displaystyle CH_3}{\overset{\displaystyle CH_3}{\underset{|}{\overset{|}{Si}}}}{-}Cl \xrightarrow{\text{hydrolysis}} HO{-}\underset{\displaystyle CH_3}{\overset{\displaystyle CH_3}{\underset{|}{\overset{|}{Si}}}}{-}OH + 2HCl;$$

$$(n+1)HO{-}\underset{\displaystyle CH_3}{\overset{\displaystyle CH_3}{\underset{|}{\overset{|}{Si}}}}{-}OH \xrightarrow[\text{polymerization}]{\text{stepwise}}$$

$$HO{-}\underset{\displaystyle CH_3}{\overset{\displaystyle CH_3}{\underset{|}{\overset{|}{Si}}}}{-}O{-}\underset{\displaystyle CH_3}{\overset{\displaystyle CH_3}{\underset{|}{\overset{|}{Si}}}}{-}O{-}\cdots{-}\underset{\displaystyle CH_3}{\overset{\displaystyle CH_3}{\underset{|}{\overset{|}{Si}}}}{-}OH + nH_2O$$

In practice, suitable mixtures of chlorosilanes, such as monofunctional $(CH_3)_3SiCl$ and bifunctional $(CH_3)_2SiCl_2$, are hydrolyzed together, and the polymerization reaction yields

$$CH_3{-}\underset{\displaystyle CH_3}{\overset{\displaystyle CH_3}{\underset{|}{\overset{|}{Si}}}}{-}O{-}\underset{\displaystyle CH_3}{\overset{\displaystyle CH_3}{\underset{|}{\overset{|}{Si}}}}{-}O{-}\cdots{-}\underset{\displaystyle CH_3}{\overset{\displaystyle CH_3}{\underset{|}{\overset{|}{Si}}}}{-}CH_3$$

Depending upon the functionality of the chlorosilanes, either long-chain or network polymers result. Low molecular weight silicones are oily liquids used as lubricants or incorporated with fillers like lithium stearate to use as greases. Those with high molecular weights are made into silicone rubbers, which have excellent flexibility to temperatures as low as −90°C.

6.8 ADDITIVE AGENTS

To facilitate polymerization, control structure, or modify properties, various agents may be added to a monomer or polymer mix. Some of these, like peroxide initiators and cross-linking agents, have been mentioned previously. Others are used as inhibitors or accelerators of polymerization reactions. Still others are used as antioxidants, surface reactive agents, colorants, inert fillers, and reinforcing fillers. Inert fillers like clay and talc often make a long-chain polymer easier to shape, and some like wood flour and chopped cotton fibers are used to lower the cost of the finished product. Reinforcing fillers (e.g., carbon black in rubbers and inorganic glass fibers in various polymers) improve the strength characteristics. In addition, compounds called plasticizers are added to a long-chain polymer to improve the flow characteristics and thus the moldability. Plasticizers typically are substances with smaller molecules than the polymer and are dissolved in the polymer. They lower the glass transition temperatures of some polymers to below room temperature, thereby making them more flexible and impact resistant. Further discussion of the effect of additives on polymeric structure and properties is presented in Chapter 18.

REFERENCES

1. Alfrey, T. and Gurnee, E. F. *Organic Polymers*, Prentice-Hall, Englewood Cliffs, N.J., 1967.

2. Billmeyer, F. W. *Textbook of Polymer Science*, 2nd edition, Wiley-Interscience, New York, 1971.

3. Mark, H. F. The Nature of Polymeric Materials, *Scientific American*, Vol. 217, No. 3, September 1967, pp. 148–156.

QUESTIONS AND PROBLEMS

6.1 The length of a linear polymer chain is commonly expressed by the degree of polymerization ($\bar{x}$), which is defined as the number of repeating units in a chain.

(a) Calculate the molecular weight of polystyrene having $\bar{x} = 100{,}000$.

(b) Calculate the approximate stretched-out chain length of one of the molecules, taking the C—C distance to be 0.154 nm.

6.2 Distinguish between long-chain polymers, network polymers, and elastomers in terms of structure and characteristics.

6.3 (a) Give the structural formula for vinyl chloride.
(b) Show how polyvinyl chloride can be prepared by chain polymerization.
(c) Schematically show how the monomer concentration varies with time, relative to completion of the chain polymerization process.

6.4 What characteristics must a monomer have in order to form a network polymer by chain polymerization?

6.5 Many condensation polymers can be partially polymerized, stored after the reaction is halted, and then completely polymerized at a later date.
(a) Explain how this can be done.
(b) Schematically show how the monomer concentration varies with time relative to the stepwise polymerization process.

6.6 (a) Can fully cured Bakelite be ground up and reused? Explain.
(b) Can polyethylene be ground up and reused? Explain.

6.7 (a) Most vinyl polymers produced by chain polymerization tend to be either completely or largely noncrystalline. Explain.
(b) When ethylene or vinylidene chloride (CH_2CCl_2) undergo chain polymerization, the resulting polymer tends to be crystalline. Explain.
(c) Chain polymerization of vinylidene chloride with a small amount of vinyl chloride results in a noncrystalline polymer (saran), which is much more pliable than the homopolymer polyvinylidene chloride. Explain.

6.8 Shown are the characteristics of polyethylene as a function of % crystallinity (determined from room temperature density) and molecular weight (determined from viscosity of the melt). The % crystallinity, in turn, is primarily a function of the amount of chain branching. Justify the characteristic behavior in each region in terms of polymer structure.

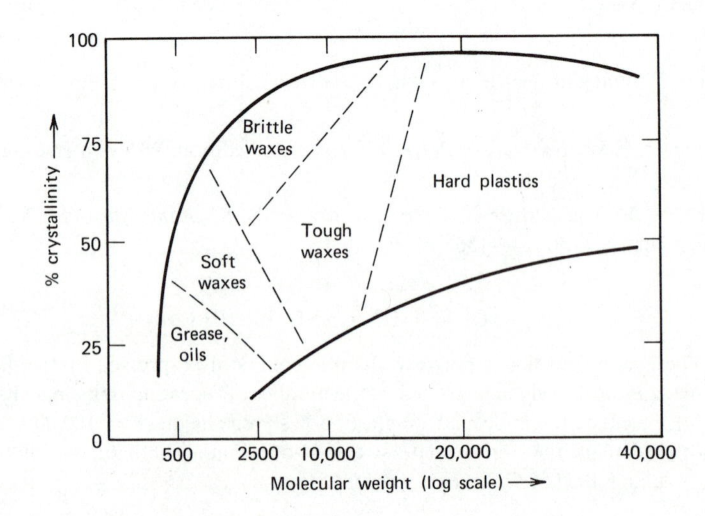

6.9 Certain long-chain polymers, such as isotactic polypropylene and many biological polymers, assume a helical chain conformation because of geometric hindrance of the side groups. Is crystallinity favored or disfavored because of this? Why?

6.10 Copolymerization is useful for decreasing crystallinity and for altering T_m and T_g somewhat independently (see Fig. 6.5). The glass transition temperature, T_g, marks the onset of chain segment motion for a polymer as temperature increases.

(a) For a random copolymer, T_g can be approximated by linear interpolation between the T_g values for the two homopolymers. Justify this statement.

(b) For a block copolymer, the brittle temperature will be essentially the glass transition temperature of the lower melting homopolymer (if the individual blocks are sufficiently long and a regular tacticity exists within the individual blocks). Furthermore, the softening temperature will essentially be the melting temperature of the higher melting homopolymer. Justify these statements and point out their limitations.

6.11 A polymer is said to be plasticized if it is made more pliable. Copolymerization can serve this purpose [see part (c) of problem 6.7]. Alternatively, additives, called plasticizers, can be used. A plasticizer is a lower molecular weight substance added to the polymer, usually a nonvolatile solvent that partially remains in the system in its ultimate use.

(a) What effect should a plasticizer have on T_g? Why?

(b) How will a plasticizer affect crystallinity? Why?

(c) Can a network polymer be plasticized?

6.12 The molecules within a polymer may be spatially oriented or not depending upon the fabrication process. Suppose a polymer is either stretched, or drawn through a die, above its glass transition temperature.

(a) Compare the molecular configuration with that of the same polymer solidified from the melt in a container.

(b) Why might stretching or drawing promote crystallization? Would this occur for all long-chain polymers? Explain.

6.13 What makes it possible to cross-link natural or synthetic rubber with sulfur?

6.14 (a) Unvulcanized *cis*-polyisoprene has a crystalline melting point above 30°C. Subsequent to vulcanization, the tendency to crystallize is very low. Why?

(b) On stretching, a vulcanized rubber may partially crystallize. Explain.

(c) Suggest why it might be desirable to have a vulcanized rubber whose crystalline melting point is below room temperature.

6.15 A *cis*-polyisoprene rubber gains 5 weight percent oxygen by oxidation. Assume that the oxygen produces cross-links.

(a) What fraction of the possible anchor points contain oxygen atoms?

(b) How does oxidation degrade the rubber?

6.16 Silicones are generally noncrystalline. In terms of molecular architecture, explain why.

7

Inorganic Solids

This chapter deals with the bonding and structure of inorganic solids.

7.1 INORGANIC GLASSES

The structure of inorganic glasses bears resemblance to the structures of some of the hydrocarbon solids discussed in the last two chapters. Most inorganic glasses, commonly encountered in the form of windows, containers, and lenses, are based on silica (SiO_2) or various silicates. Consequently, it is not surprising that silica glasses and silicone polymers have rather similar structural units. The basic structural unit in solid and liquid silica is a tetrahedron containing a small central silicon atom bonded to four larger oxygen atoms situated at the tetrahedron corners (Fig. 7.1). This geometry is a natural result of the four sp^3 hybrid orbitals of Si that overlap with a hybridized orbital of each oxygen. Although this suggests purely covalent bonding, in actuality the Si—0 bond is a mixture of covalent and ionic. In a tetrahedral unit, viewed by itself, the oxygen atoms are one electron short of having a filled shell (i.e., they have one unsatisfied orbital), and each is capable of bonding to one additional silicon atom, which in turn is bonded to three other oxygen atoms. Various groups of these tetrahedra can be found in crystalline forms of silica, in crystalline silicates, and in glasses.

If each tetrahedron shares corners only with other tetrahedra and the tetrahedra are arranged in an otherwise random manner, then the resulting network structure is silica glass or vitreous silica, as shown schematically in Fig. 7.2*a*. This is an inorganic network polymer with a three-dimensional network of polar covalent bonds. Inorganic glasses form readily from a number of compounds that have polar covalent bonding and that fulfill three empirical rules: (1) each "anion" forms a bonding bridge between no more than two "cations"; (2) each "cation" is surrounded by no more than four "anions," the grouping being a structural unit; and (3) each structural unit shares corners only, not edges or faces, with other structural units in the crystalline solid and in the glass. The "anions" and "cations" referred to above actually are electronegative atoms, such as O, S, and F, and relatively electropositive atoms, such as Si, B, and Be.

In addition to SiO_2, both BeF_2 and GeO_2 also have tetrahedral structural units and meet the conditions for being glass formers. On the other hand, B_2O_3 has an almost triangular structural unit consisting of three oxygen atoms that from a very flat pyramid with one boron atom. Again, the conditions for glass formation are met. When different but compatible glass formers are combined, their structural units fit into the same network. Thus, oxide glass formers may be combined, and the properties of the glass vary accordingly. Rather high temperatures must be achieved in order to reshape pure silica glass, and even higher temperatures are necessary to melt the raw material (e.g., sand) to form silica glass in the first place. The glassy state (see Sect. 1.4 and Fig. 1.5) results when crystallization from the melt occurs only with great difficulty. The specific volume change on cooling from above the equilibrium melting temperature is comparable to that for a completely noncrystalline long-chain polymer (i.e., *ABCD* in Fig. 6.4). Glasslike behavior occurs below the glass transition temperature range (or fictive temperature range), where the internal motion of the basic structural units is severely limited, that is, the structure is essentially independent of time and temperature below T_g.

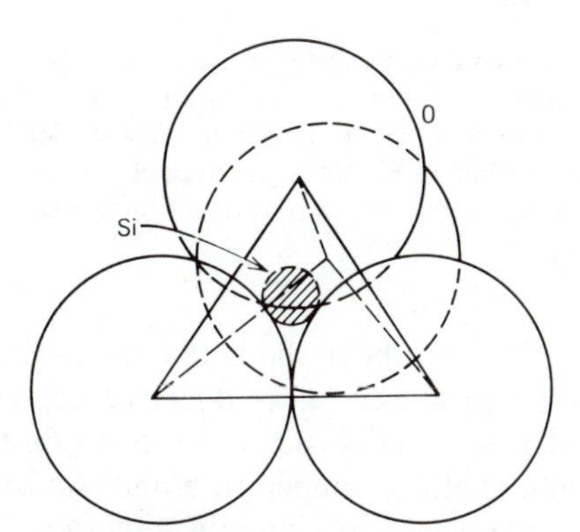

FIGURE 7.1 The basic structural unit in silica-based materials is a tetrahedron in which silicon is located at the center and oxygen at the corners of the tetrahedron.

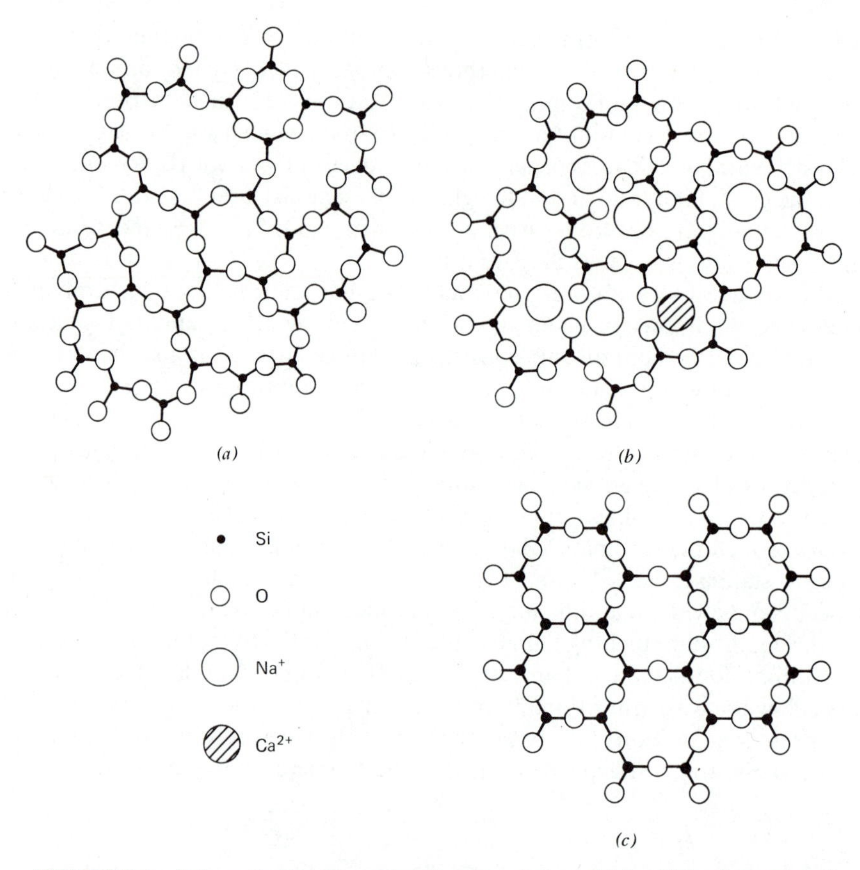

FIGURE 7.2 Schematic two-dimensional representation of (*a*) a SiO_2 glass network structure; (*b*) an interrupted glass network structure for a soda-lime-silica glass; and (*c*) a SiO_2 crystalline network structure. In (*b*) the oxygen ions from the modifiers (Na_2O and CaO) have attached to Si atoms, interrupting the Si—O network, and the cations remain in the interstices in order to provide electrical neutrality.

Shaping of glass is carried out above the fictive temperature range, which depends upon composition and structure, as does the crystalline melting point. Addition of certain oxides, such as Na_2O, K_2O, or CaO, to silica significantly lowers both the crystalline melting temperature and the softening temperature. Common window and container glasses contain soda (Na_2O) and lime (CaO) as well as silica and can be fabricated at relatively low temperatures. These oxide additives, which were present in ancient Egyptian and Roman glasses, are called modifiers because they interrupt the Si—O network (Fig. 7.2*b*). If too much of a modifier is present, then the network will be sufficiently broken up in the melt so that crystallization can occur readily. Alternatively, if a structure containing a large amount of a modifier does permit a glass to form on cooling, crystallization (termed devitrification) may occur if the glass is subsequently reheated above its softening point. Each oxygen atom from a modifier bonds covalently to a silicon atom such that two SiO_4 tetrahedra are broken apart, each having one unshared oxygen (Fig. 7.2*b*). The modifier cations fit into holes or interstices in the network in order to provide electrical neutrality. The resulting increased tendency toward crystallization, lower softening temperature, and decreased melt viscosity follow directly.

Some oxide additives to SiO_2, called intermediates, may act either as modifiers or network formers, although they are not glass formers by themselves. Examples are Al_2O_3 and PbO, which build extensively into a Si—O network. The presence of intermediates or of glass formers usually decreases any tendency toward devitrification. Modern glasses (Table 7.1) consist of a few to several glass formers, modifiers, and intermediates mixed together to achieve desired fabrication characteristics and properties.

TABLE 7.1 Some Technically Important Glasses

Type	*Typical Composition (wt. %)*	*Characteristics*
Soda-lime-silica glasses		
(a) Window glass	72% SiO_2–14% Na_2O–9% CaO–4% MgO–l% Al_2O_3	Relatively low melting; relatively high coefficient of thermal expansion.
(b) Container glass	73% SiO_2–15% Na_2O–5% CaO–4% MgO–2% Al_2O_3–1% K_2O	
High-silica glasses		
(a) Vitreous silica	>99.5% SiO_2	Relatively high melting; very low coefficient of thermal expansion; chemically resistant.
(b) Vycor	96% SiO_2–4% B_2O_3	
Borosilicate glasses		
Pyrex	80.5% SiO_2–13% B_2O_3–3.8% Na_2O–2.2% Al_2O_3–0.5% K_2O	Chemical and heat resistant; intermediate coefficient of thermal expansion.
Lead glasses		
Light flint optical glass	54% SiO_2–37% PbO–8% K_2O–1% Na_2O	Low softening temperature; high density; high index of refraction.

7.2 GELS

Another noncrystalline substance in which silica plays a role is silica gel. This can be produced, for example, as follows:

$$SiCl_4 + \text{water} \longrightarrow Si(OH)_4 \longrightarrow \text{silica gel (hydrous silica)}$$

The result is a rigid, three-dimensional silica network that has trapped water molecules, resulting from the reaction, in its interstices. A gel is a two-phase colloidal mixture of liquid in a solid, or sometimes gas in a solid, in which the component phases are mixed so intimately that the material behaves as if it were a noncrystalline solid even though the solid phase may be crystalline. Solid-liquid gels can be formed from sols [i.e., colloidal suspensions of very fine solid particles that have average dimensions less than about 100 nm (1000 Å) in a liquid] if the colloidal particles link together to form a solid network and the liquid is contained either in very fine capillaries between particles or in very small holes in the framework. Solid-liquid or solid-gas gels may also be formed if a fluid penetrates the solid along very fine capillaries, which is usually accompanied by swelling.

The classification of gels is similar to that used for noncrystalline solids. When the solid framework that immobilizes the liquid consists of long-chain molecules or particles bonded together only at a few points, the gel is called an elastic gel. In a rigid gel the framework is more comparable to the structure of a network polymer. Elastic gels become more liquid like as temperature increases because the bonds between particles or molecules are relatively weak. Rigid gels are similar to thermosetting polymers in that they do not soften appreciably as temperature is raised.

In gelatine, the most familiar elastic gel, long-chain protein molecules form the solid framework that traps water. Asphalt is another elastic gel, albeit a very complex one. It starts as a very viscous sol of high-molecular-weight hydrocarbons in an oily constituent. Gelling occurs below room temperature or as a result of oxidation, the framework being formed by the heavy hydrocarbons. Depending upon the preparation and conditions, therefore, asphalt may be a sol, a gel, or in between; its properties vary accordingly.

Portland cement gel is a common rigid gel produced during the setting of hydrated cement. Dry cement powder mainly consists of mixtures of calcium silicates, calcium aluminate, and calcium aluminoferrite. After the dry powder is mixed with water, hydration reactions occur that yield a complex hydrated calcium silicate gel and fine crystals of $Ca(OH)_2$ as a by-product. Initially, water molecules are adsorbed on the sheetlike silicate particles and serve to separate them in the cement paste. Once hydration starts, any other substances in the concrete aggregate mix (e.g., sand and gravel) are bound into a stiff mass by the cement, which *sets* into a rigid structure as the gelling reaction proceeds. Excess water must be kept to a minimum for the best properties, since it remains in large residual pores. Further chemical reactions, particularly continued hydration, proceed over a period of time before the final cement structure is achieved. Thus, the nature of a Portland cement gel depends upon the proportion of water added initially and upon the aging time.

7.3 SILICATES

Silica and a wide variety of silicates make up a large fraction of the earth's crust, such that oxygen and silicon are its two most abundant elements. Silicate minerals serve as some of the primary raw materials used to prepare numerous ceramics. Silicates exhibit very diverse crystallographic forms because various Si—O arrays are possible. As noted previously, Si does not bond extensively with itself in the same way that C does; thus, Si does not form a set of compounds comparable to organic compounds. Nevertheless, the combination of Si and O together does form the basis for silicone polymers, silicate glasses, and crystalline silicates. The same tetrahedral structural subunit (Fig. 7.1) that exists in silicate glasses also is found in crystalline silicates. Because of the polar nature of the Si—O bond, with the resulting positive charge on Si, and because of the two directed orbitals that can be used for bonding by each oxygen atom, the tetrahedral subunits in crystalline silicates usually share corners (shown schematically in Fig. 7.2*c*), seldom share edges, and never share faces.

Pure crystalline silica assumes a number of different crystal structures by the formation of ordered three-dimensional networks consisting of tetrahedral subunits which share corners only. Different crystalline forms of a compound are called polymorphs and, in some respects, are analogous to isomers in hydrocarbons. The simplest polymorph of SiO_2, stable at high temperatures, is cristobalite (Fig. 7.3). Tridymite and quartz are two other polymorphs of silica, stable at sequentially lower temperatures. Quartz, the densest of these three polymorphs, is the most common mineral in the crust of the earth.

Many silicate minerals also have framework (or network) structures; however, their networks are more open than those of silica. Almost invariably in framework silicates, Al atoms substitute in SiO_4 tetrahedra for a definite fraction of Si atoms. This, in itself, would result in unsatisfied bonding orbitals for some oxygen atoms (or, alternatively, a lack of electrical neutrality if we consider Si^{4+} being replaced by Al^{3+}). As a consequence, alkali or other electropositive atoms are present as ions in the framework interstices, their electrons having gone into the available oxygen orbitals. The feldspars, very common rock-forming minerals, comprise a very important group of framework silicates. An example is orthoclase, $KAlSi_3O_8$, in which one-quarter of the Si atoms have been replaced by Al atoms, and there is one K atom per Al atom.

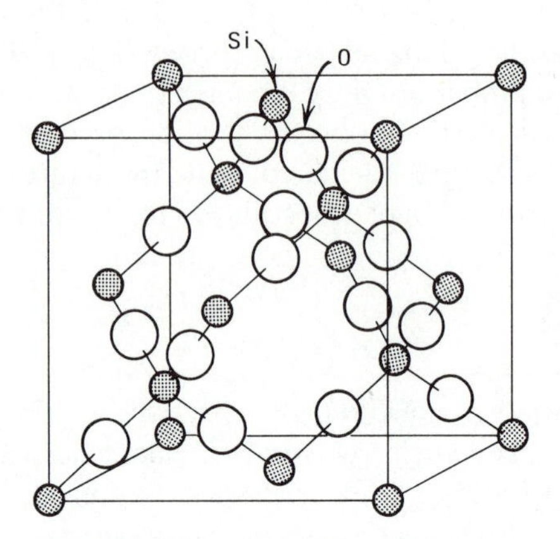

FIGURE 7.3 Idealized arrangement of silicon and oxygen atoms in the SiO_2 polymorph cristobalite. The cubic unit cell shown is a representative repeating unit. Oxygen atoms are essentially centered between pairs of silicon atoms, and SiO_4 tetrahedral subunits share corners.

Other silicates are classified as island, chain, or sheet structures, depending upon how the SiO_4 or $(Al, Si)O_4$ subunits are linked together. In the simplest island silicates, the tetrahedra are isolated from one another and exist as anionic polyhedra, $(SiO_4)^{4-}$, as shown in Fig. 7.4*a*. The cations present in these silicates provide charge neutrality as well as the ionic bonding that holds the charged polyhedra together. An example from the olivine minerals is forsterite, Mg_2SiO_4, in which each Mg^{2+} cation has six equidistant oxygen neighbors. Forsterite is present in some ceramics used as dielectric (i.e., electrically insulating) materials. Various other island silicates have isolated $(SiO_4)^{4-}$ anions or noncyclic isolated groupings of tetrahedra that share some corners.

When two corners of each tetrahedron are shared with other tetrahedra, the result is either an anionic single chain or an anionic ring, with the repeating unit formula $(SiO_3)^{2-}$, as shown in Fig. 7.4*b*. In the pyroxene mineral enstatite, $MgSiO_3$, essentially infinite $(SiO_3)_n^{2n-}$ chains are held together by Mg^{2+} cations, each of which has six equidistant oxygen neighbors. Enstatite is found in steatite ceramics, which are used as high-frequency electrical insulators. A double-chain silicate results if, on the average, two and one-half corners of each tetrahedron are shared with other tetrahedra (Fig. 7.4*c*). This double chain has the repeating unit formula $(Si_4O_{11})^{6-}$, which can be obtained by noting that the repeating unit in Fig. 7.4*c* contains four silicon atoms, six unshared oxygen atoms, and 10 shared oxygen

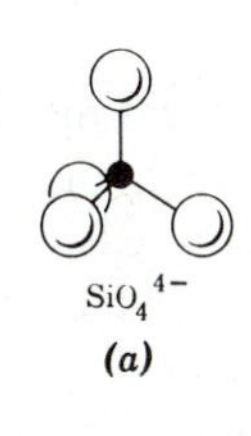

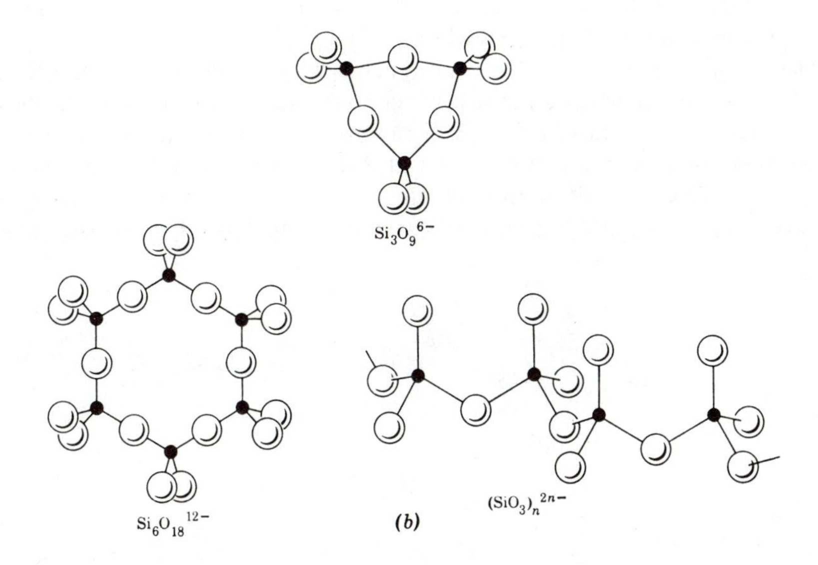

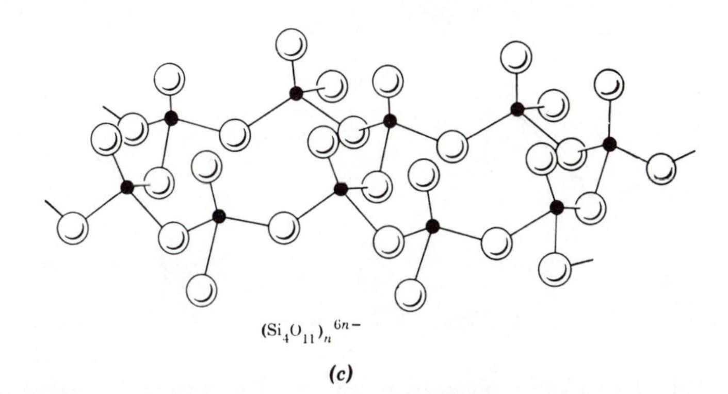

(c)

FIGURE 7.4 Some arrangements of SiO_4 tetrahedra that are found in silicate minerals. (*a*) Isolated tetrahedra, $(SiO_4)^{4-}$; (*b*) single-chain or ring structures with two corners of each tetrahedron shared by other tetrahedra, $(SiO_3)_n^{2n-}$; (*c*) double-chain structure with two and one-half corners of a tetrahedron shared, on the average, by other tetrahedra, $(Si_4O_{11})_n^{6n-}$.

atoms (half of which belong to the repeating unit). Once more, cations fit between the anionic double chains and hold them together. In all chain silicates, the weaker interchain bonds can be broken more easily than the intrachain polar covalent bonds. Consequently, cleavage between chains is relatively easy, and definite angles occur between cleavage surfaces (Figs. 7.5 and 7.6).

If three out of four oxygens in each tetrahedron are shared with other tetrahedra, anionic sheets result with a repeating unit formula $(Si_2O_5)^{2-}$ (Fig. 7.7). Sheet or layer silicates include talc $[Mg_3(OH)_2(Si_2O_5)_2]$, the micas [e.g., muscovite, $KAl_2(OH)_2(Si_3Al)O_{10}$], and the clay minerals [e.g., kaolinite, $Al_2(OH)_4Si_2O_5$], all of which are important ceramic raw materials. As with the framework silicates, Al commonly substitutes for Si in a Si_2O_5 sheet. The layers in sheet silicates are comprised not only of a Si_2O_5 sheet, but also incorporate $(OH)^-$ ions and any divalent or trivalent cations. In kaolinite and talc, van der Waals bonding exists between the layers, such that the layers can cleave readily and, in some cases, slide

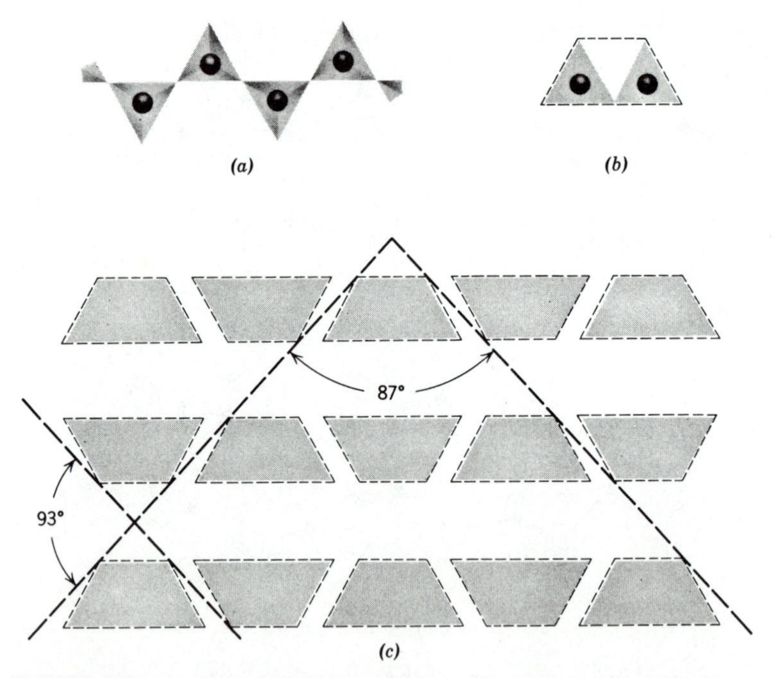

FIGURE 7.5 Another representation of SiO_4 tetrahedra in a single-chain silicate showing one chain viewed (*a*) from the top and (*b*) end on. (*c*) In a crystal, single chains pack together to form layers between which cations are located. Cleavage plane traces and their angles of intersection are shown in (*c*). Note that cleavage occurs along planes where interchain bonds are relatively weak. (From *The Structure and Properties of Materials,* Vol. I, by W. G. Moffatt, G. W. Pearsall, and J. Wulff, Wiley, New York, 1964.)

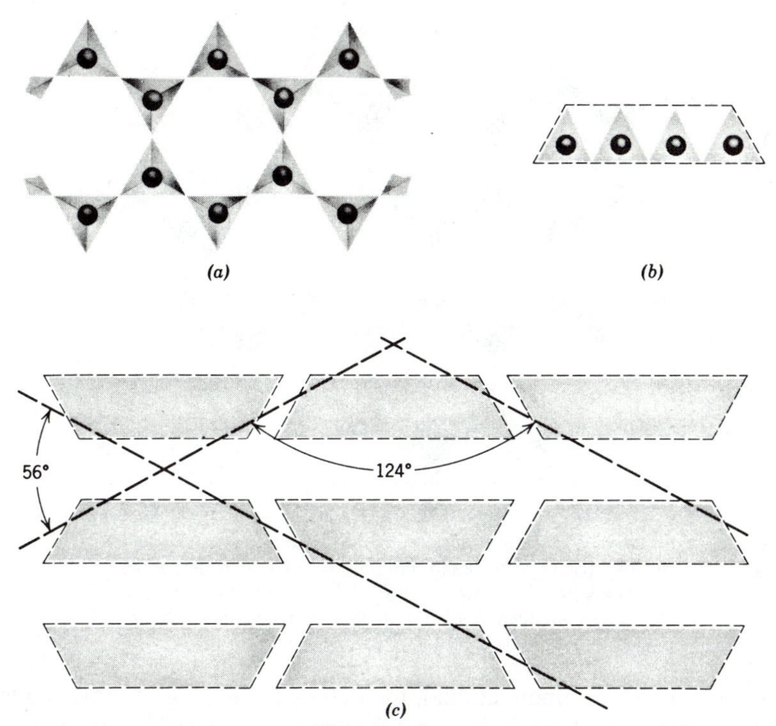

FIGURE 7.6 Another representation of SiO_4 tetrahedra in a double-chain silicate showing one chain viewed (*a*) from the top and (*b*) end on. (*c*) In a crystal, double chains pack together to form layers between which cations are located; cleavage plane traces and their angles of intersection are shown. (From *The Structure and Properties of Materials,* Vol. I, by W. G. Moffatt, G. W. Pearsall, and J. Wulff, Wiley, New York, 1964.)

with respect to one another. The layers of muscovite mica are bonded together by K^+ cations and, consequently, are more difficult to cleave.

A mixture of covalent and ionic bonding occurs in all silicates, and van der Waals bonding is important in some silicates. The resulting structures are among the most complex of all inorganic solids. For those inorganic solids where the bonding is either largely ionic, largely covalent, or largely metallic, the crystal structures often are much simpler.

7.4 IONIC SOLIDS

A large number of oxides, nitrides, carbides, sulfides, and halides exist as ionically bonded crystals. Although very few crystals are completely ionic, many have a significant degree of ionic bonding and can be classified, broadly, as ionic solids.

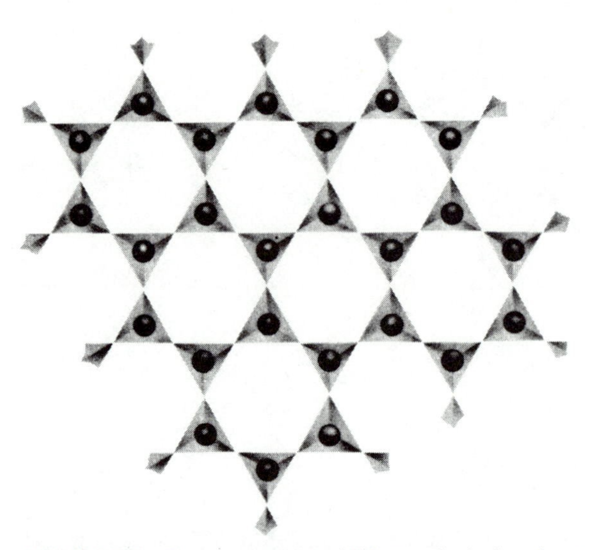

FIGURE 7.7 The arrangement of SiO_4 tetrahedra in a sheet silicate, $(Si_2O_5)_n^{2n-}$. Each tetrahedron shares three of its corners with other tetrahedra. (From *The Structure and Properties of Materials,* Vol. I, by W. G. Moffatt, G. W. Pearsall, and J. Wulff, Wiley, New York, 1964.)

Since the ionic bond is nondirectional, the details of packing in ionic solids are determined primarily by the relative ionic charges and the relative sizes of the ions. The relative number of anions to cations is dictated by the requirements of electrical neutrality. Within this general constraint, the packing of ions depends on the radius ratio, that is, the ratio of the smaller cation radius, r_C, to the larger anion radius, r_A (cf. Table 2.6). Each cation tends to be surrounded by the largest possible number of anions, subject to the restriction that the anions do not overlap with each other yet are in *contact* with the central cation. In a way we can visualize an ionic crystal as an orderly three-dimensional array of anions in which cations are distributed in a regular fashion in the interstices or holes between the anions. Very often the anion array is identical or close to one of the atomic arrays found in elemental metals (see Section 7.6 and Chapter 8) or it is even simpler.

Examples are provided by the unit cells of CsCl (Fig. 7.8*a*) and NaCl (Fig. 7.8*b*, cf. Fig. 3.1). A unit cell is defined as a small repeating unit representative of the whole crystal structure. Although both Cs^+ and Na^+ are monovalent, Cs^+ (radius = 0.169 nm) is coordinated or surrounded by eight Cl^- anions (radius = 0.181 nm) in the shape of a cube, and Na^+ (radius = 0.095 nm) is coordinated by six Cl^- anions in the shape of a regular octahedron. If NaCl had the same structure as CsCl, either the Cl^- anions would overlap or the Na^+ cations would not be as close as possible to the surrounding anions, and the resulting energy state would be much higher than that for the NaCl structure. The ranges of expected radius ratios,

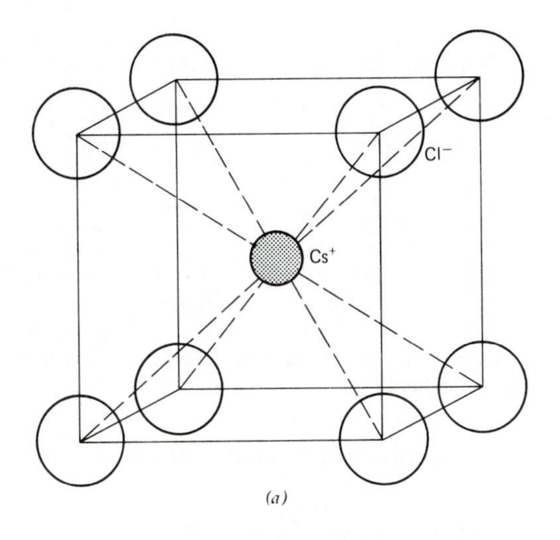

(a)

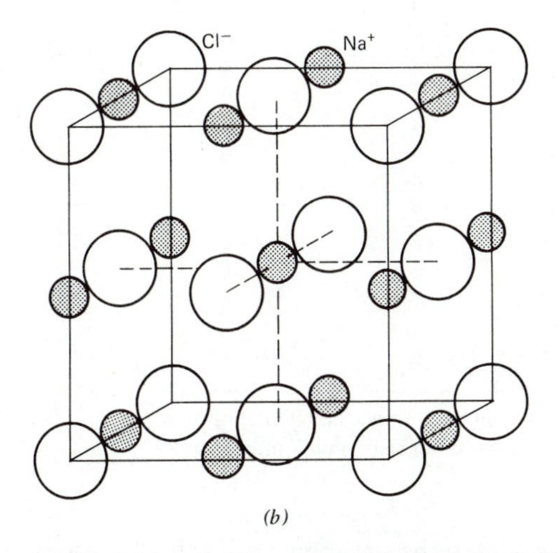

(b)

FIGURE 7.8 (*a*) The cubic unit cell of the CsCl structure. Each Cs^+ cation is coordinated by eight Cl^- anions in the form of a cube. (*b*) The cubic unit cell of the NaCl structure (cf. Fig. 3.1). Each Na^+ cation is coordinated by six Cl^- anions in the form of an octahedron.

TABLE 7.2 Expected Radius Ratio as a Function of Anion Coordination Number

Anion Coordination Number	*Expected Radius Ratio Range*	*Anion Coordination Polyhedron*[a]
2	0 to 0.155	Line
3	0.155 to 0.225	Equilateral triangle
4	0.225 to 0.414	Regular tetrahedron
6	0.414 to 0.732	Regular octahedron
8	0.732 to 1.000	Cube

[a]The polyhedron obtained by connecting the centers of anions that are coordinated to a central cation.

TABLE 7.3 Predicted and Observed Anion Coordination Numbers (CN) for Selected Ionic Solids

Compound	r_C/r_A	*Predicted CN*	*Observed CN*
BeS	0.17	3	4
BeO	0.22	3 or 4	4
SiO_2[a]	0.29	4	4
LiBr	0.31	4	6
Al_2O_3	0.36	4	6
MgO	0.46	6	6
MgF_2	0.48	6	6
TiO_2	0.49	6	6
NaCl	0.53	6	6
CaO	0.71	6	6
KCl	0.73	6 or 8	6
CaF_2	0.73	6 or 8	8
CsCl	0.93	8	8

[a]Although SiO_2 has predominately covalent bonding, the ionic radius ratio scheme is consistent with the observed coordination number.

for given anion coordinations, are shown in Table 7.2, and comparisons of observed and predicted anion coordination numbers are presented in Table 7.3. The discrepancies arise because ionic radii depend not only on the coordination but also on the partial covalent character of the bonding.

7.5 COVALENT SOLIDS

Inorganic solids that bond covalently include the nonmetallic elements and a large number of compounds. Whereas ionic solids have fairly dense ionic packing, covalent solids are generally more open, basically because the number of near neighbors to a particular atom is limited by the availability and directionality of bonding orbitals. This is apparent in diamond and in silica (Sect. 7.3). A commonly observed rule for covalent elemental solids (and often applicable to covalent compounds) is that the number of near or bonding neighbors equals $(8-N)$, where N is the group number in which the element is found in the periodic table. This rule reflects the maximum number of orbitals available for bonding in the elemental state. For example, carbon in group IVB should have four bonding neighbors, as is the case in diamond.

The group IVB elements C (as diamond), Si, Ge, and Sn (below room temperature) all have the diamond cubic structure (Fig. 7.9, cf. Fig. 3.2). Each of these is a relatively rigid substance because of the three-dimensional network of discrete, directional covalent bonds. Several compounds, such as zincblende (a ZnS polymorph), GaAs, and InSb, assume a structure similar to the diamond cubic structure, although the atoms are ordered such that, for example, all Zn atoms are bonded only to S atoms and vice versa. This type of ordered structure is called the zincblende structure. Many of the most important modern semiconductors have either the zincblende or the diamond cubic structure.

Graphite, the carbon allotrope stable under normal conditions, has an altogether different structure. It is made up of layers of hexagonally arrayed C atoms, in which each C atom bonds covalently with three coplanar neighbors. In addition to the localized bonding orbitals within the layer, a nonlocalized molecular orbital, which contributes somewhat to intralayer bonding, is located above and below each layer. This nonlocalized orbital is responsible for the relatively large electrical conductivity exhibited by graphite, but it does not lead to significant interlayer

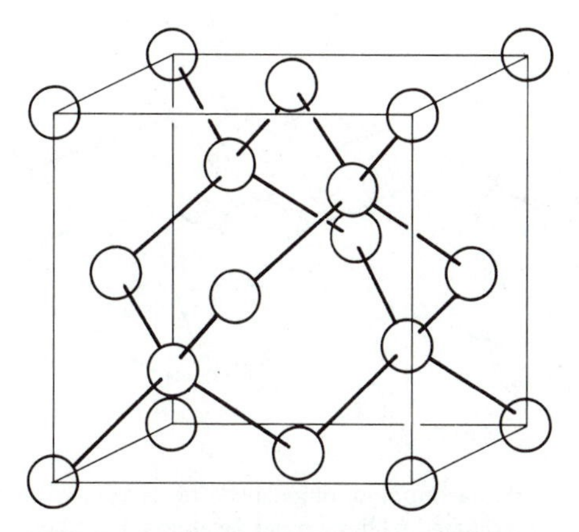

FIGURE 7.9 The unit cell of the diamond cubic structure (cf. Fig. 3.2). Each atom is coordinated by and bonded to four other atoms as indicated by the heavy lines.

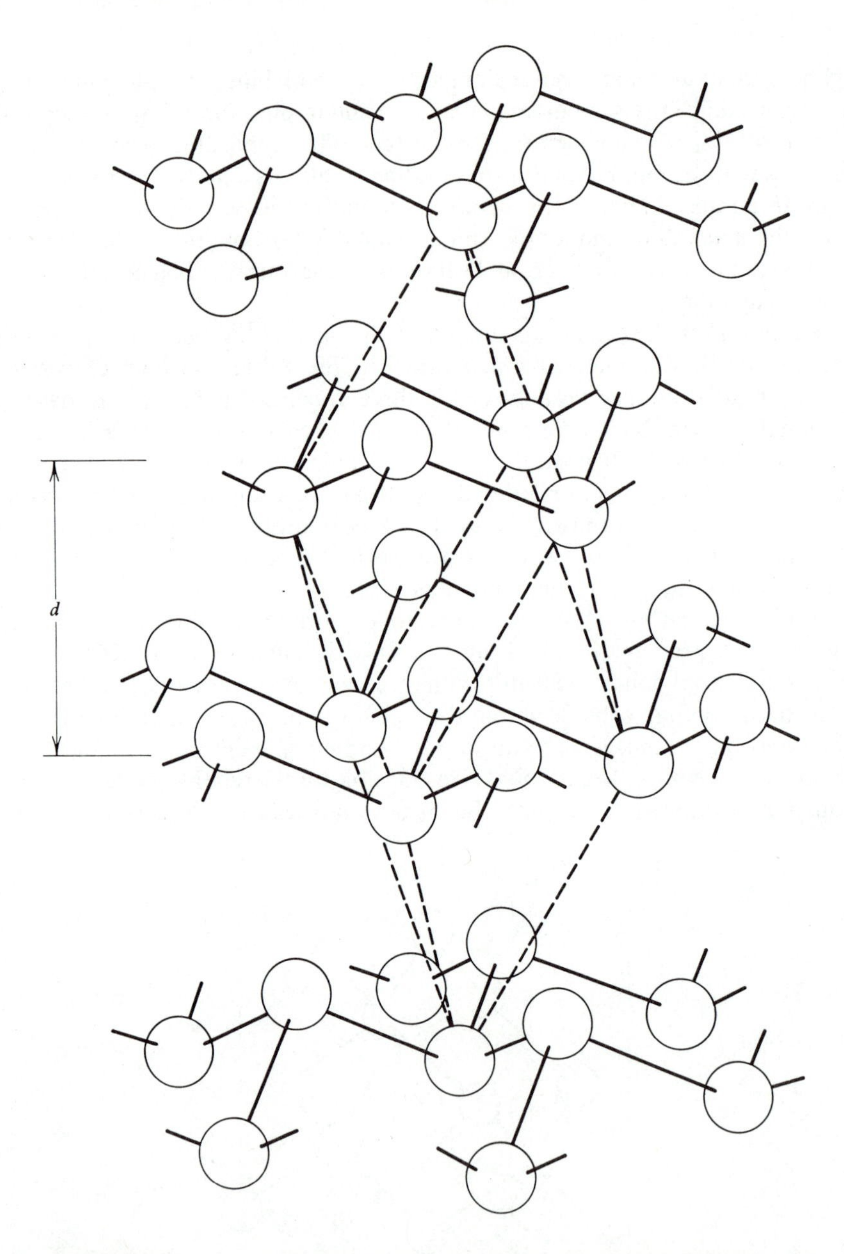

FIGURE 7.10 The crystal structure of arsenic. This is a puckered layer structure with each As atom coordinated by three other As atoms. The bonding within layers is strongly covalent (heavy lines), but weaker between layers; similarly the distance between layers, *d*, is greater than the interatomic distance within the layers. The unit cell, described further in Chapter 9, is indicated by the broken lines.

bonding. Rather, the layers are held together largely by van der Waals forces. Since the interlayer bonding is weak, graphite may be cleaved easily or layers may slide by one another. Talc, which has a structure similar to graphite, behaves in a like manner.

The group VB elements As, Sb, and Bi also crystallize in layer structures; however, their layers are somewhat puckered because the bond angles are less than 120°. The separation between layers is greater than the interatomic bond length (Fig. 7.10). The group VIB elements S, Se, and Te form chain or ring molecules in the liquid state. Indeed, rapid cooling of a liquid consisting of chain molecules results in a solid comparable to long-chain polymers (e.g., plastic sulfur). In the crystalline state each atom bonds covalently to two neighbors, and the molecules pack in an orderly array, being held together by van der Waals bonds.

Crystalline solids such as P and S are really molecular crystals, consisting of P_4 and S_8 molecules, which resemble solid methane (CH_4). That is, interatomic bonding is covalent, and intermolecular bonding is van der Waals in nature. Likewise, N_2, O_2, and the halogens form crystalline molecular solids at low temperatures.

7.6 OTHER ATOMIC SOLIDS

Solid inert gas elements and solid metallic elements can be considered together, even though their modes of bonding differ greatly. Each has nondirectional bonding which favors relatively dense atomic packing. Metallic bonding involves valence electrons that are shared by the crystal as a whole, such that we can visualize a solid metal as consisting of an orderly array of ion cores in a sea of free electrons (see Chapters 3 and 4). Although the electrons in a metal crystal repel one another, they are attracted by the positive ion cores of the crystalline array, and this interaction holds the solid together. In many instances, the ion cores are essentially spherical, and they may be considered as *hard balls* that cannot penetrate each other. Inert gas atoms are also spherical, since they have closed electronic shells. For the purpose of describing the spatial arrangements of atoms in metals and in solid inert gases, it is convenient to utilize the hard-ball model for atoms.

The lack of directional bonding leads to ion core or atomic packing analogous to the orderly packing of spheres. Most elemental metals crystallize in one of three common arrays. For copper or the solid inert gases, an atom has 12 nearest neighbors arranged as shown in Fig. 7.11*a*. In magnesium each atom also has 12 nearest neighbors, but they are arranged differently (Fig. 7.11*b*). In tungsten there are only 8 nearest neighbors (Fig. 7.11*c*). The crystal structures that result in each case are commonly called face-centered cubic (fcc), hexagonal close-packed (hcp), and body-centered cubic (bcc), respectively, and will be discussed more fully in Chapter 8. If hard spheres of equal size are packed together as efficiently as possible,

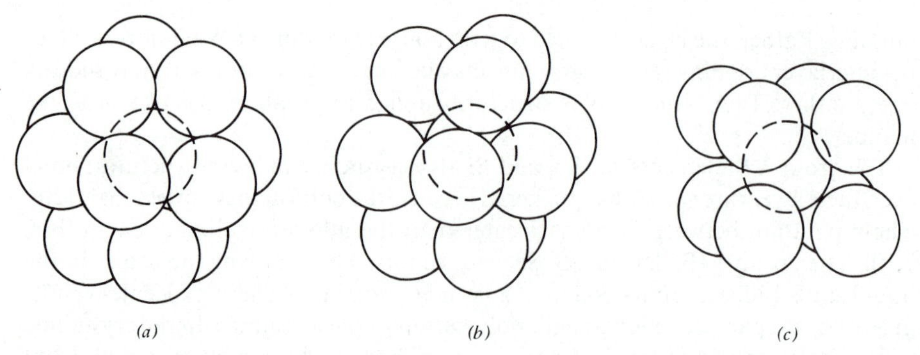

FIGURE 7.11 The arrangements of nearest neighbor atoms about a central atom for (*a*) the Cu structure, (*b*) the Mg structure, and (*c*) the W structure.

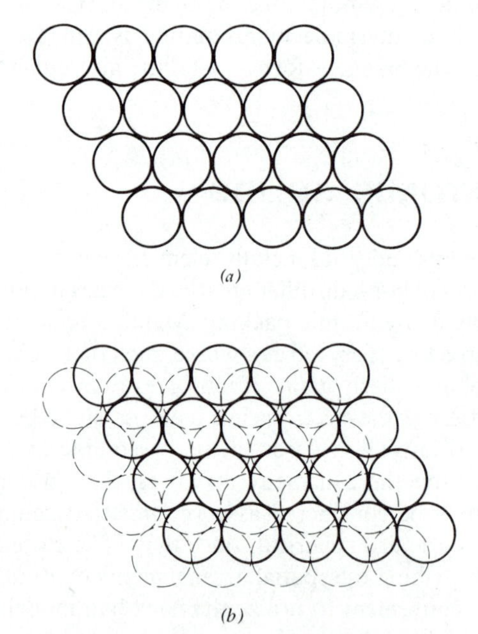

FIGURE 7.12 (*a*) Plan view of a close-packed plane of spheres. (*b*) Plan view of two close-packed planes of spheres, with spheres in the top plane (solid circles) situated in one set of triangular valleys in the bottom plane (broken circles). Note that a third close-packed plane may be positioned either so that its spheres lie directly above spheres in the first plane or so that its spheres lie above unoccupied triangular valleys in the first plane.

it is found that each sphere touches 12 neighboring spheres, either in the same manner as Cu atoms or in the same manner as Mg atoms. Hence, these most efficient atomic packing arrays are termed close-packed. In both of these cases the packing fraction (the fraction of space occupied by contacting spheres) is 0.74. The packing in W, in terms of contacting spheres, is less efficient, since the packing fraction is 0.68. Each of these values, however, represents quite efficient packing as compared to a covalent crystal such as diamond, which has a packing fraction of only 0.34.

The complete structures of Cu and Mg are better described in terms of close-packed plane stacking sequences. In a close-packed plane of similar atoms, the atoms are arrayed like billiard balls, each having six nearest neighbors within the plane (Fig. 7.12*a*). Close-packed planes stack together well, such that each atom in the second plane, for example, is situated in a triangular valley made by three atoms in the first plane. Figure 7.12*b* shows a close-packed plane sitting upon another close-packed plane. When a third close-packed plane is introduced, either the Cu structure or the Mg structure is represented partially. If atoms in the third plane lie directly above atoms in the first plane, those in the fourth plane above those in the second plane, and so forth, then the Mg structure (hcp) is achieved. In this case

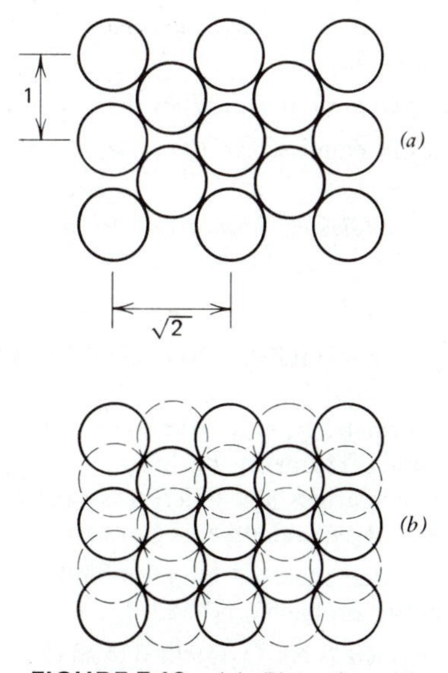

FIGURE 7.13 (*a*) Plan view of spheres representing the closest-packed plane in the W structure. (*b*) Plan view of spheres showing how the closest-packed planes stack in the W structure.

the stacking sequence is referred to as $ABABAB\cdots$. On the other hand, if atoms in the third plane lie above unoccupied triangular valleys in the first plane, those in the fourth plane directly above atoms in the first plane, and so forth, then the Cu structure (fcc) is achieved. Here the stacking sequence is referred to as $ABCABC\cdots$. Metallic structures having stacking sequences with a period greater than three exist, but are encountered less commonly.

The structure of W (bcc) is not close-packed, in the sense of hard-ball packing. Nevertheless, it is an efficient and symmetric packing scheme and is encountered in a large number of metallic elements. Although the W structure has no close-packed planes, there are nearly close-packed planes (Fig. 7.13*a*), in which each atom, viewed as a sphere, touches four nearest neighbors. If a second identical plane is positioned as shown in Fig. 7.13*b*, a third plane such that its atoms are directly above those in the first plane, and so forth, then the W structure is achieved. A more complete discussion of the three common metallic crystal structures will be presented in Chapter 8.

REFERENCES

1. Van Vlack, L. H. *Elements of Materials Science and Engineering*, 3rd edition, Addison-Wesley, Reading, Mass., 1975.
2. Kingery, W. D. *Introduction to Ceramics*, Wiley, New York, 1960.
3. Gilman, J. J. The Nature of Ceramics, *Scientific American*, Vol. 217, No. 3, September 1967, pp. 112–124.
4. Charles, R. J. The Nature of Glasses, *Scientific American*, Vol. 217, No. 3, September 1967, pp. 126–136.

QUESTIONS AND PROBLEMS

7.1 (a) With respect to formability, how do inorganic glasses differ (i) from organic network polymers, and (ii) from organic long-chain polymers?
(b) What structural similarities and differences exist between inorganic glasses and network polymers? Between inorganic glasses and long-chain polymers?

7.2 Compare and contrast the effects of modifiers added to an inorganic glass with the effects of plasticizers (see problem 6.11) added to a thermoplastic polymer.

7.3 As an oxide modifier (such as Na_2O) is added to silica glass the oxygen-to-silicon ratio increases, and it is empirically observed that the limit of glass formation is reached when $O/Si \approx 2.5$ to 3. Explain, in terms of structure, why a soda-silica mixture such that $2 < O/Si < 2.5$ will form a glass, whereas a soda-silica mixture such that $O/Si = 3$ will crystallize rather than forming a glass.

7.4 Sketch a plan view of the $(SiO_3)_6^{12-}$ anion ring for the mineral beryl, $Be_3Al_2(SiO_3)_6$.

7.5 For the sheet silicate, shown in Fig. 7.7, show that a repeating unit has the formula $(Si_2O_5)^{2-}$.

7.6 A view parallel to the layers of (1) talc, (2) muscovite mica, and (3) kaolinite can be represented schematically:

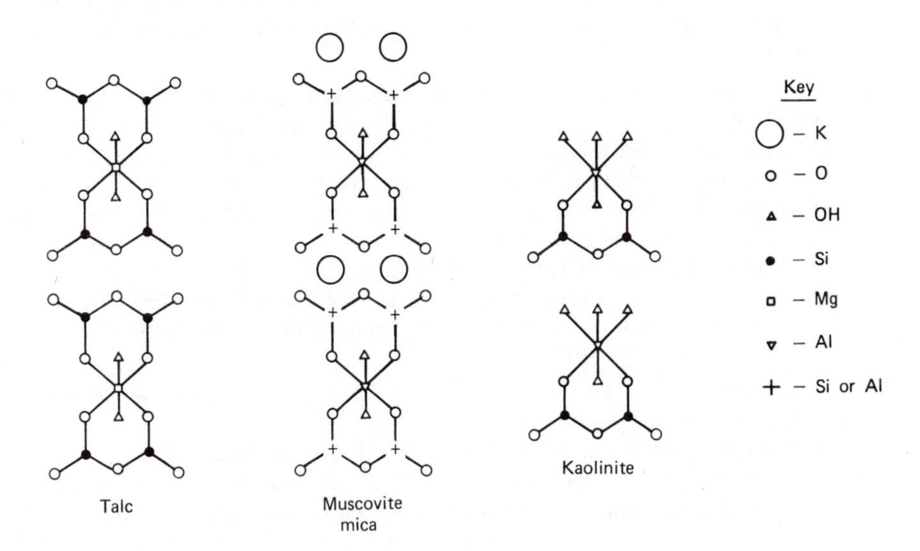

(a) Justify why these minerals cleave as they do.
(b) Explain why talc is much softer than mica.
(c) Explain why clay (e.g., kaolinite) absorbs water readily, whereas talc is relatively impervious to water.

7.7 The unit cell of the cesium chloride structure is shown in Fig. 7.8*a*. If the radius of the Cs^+ cation is 0.169 nm and that of the Cl^- anion is 0.181 nm, calculate the fractional volume of space occupied by spherical ions (packing fraction). Assume that the Cs^+ and Cl^- ions touch along cube body diagonals.

7.8 (a) MgO has the NaCl structure. Using radii of 0.140 nm for O^{2-} and 0.070 nm for Mg^{2+}, calculate the fractional volume of space occupied by spherical ions (packing fraction).
(b) Calculate the density of MgO.

7.9 The critical radius ratio is the ratio of cation radius to anion radius for the condition where the surrounding anions are just touching each other as well as the central cation. This represents the smallest radius ratio for which a particular anion coordination number should occur. Calculate the critical radius ratio for (a) cubic coordination, (b) octahedral coordination, (c) tetrahedral coordination, and (d) triangular coordination.

7.10 Inscribe a regular tetrahedron in a cube such that the tetrahedron corners coincide with some of the cube corners and the tetrahedron edges are cube face diagonals.
(a) Calculate the tetrahedron center-to-vertex distance (cube center-to-vertex distance) in terms of the tetrahedron edge length (cube face diagonal length).
(b) Assuming that equal size spheres of radius R are centered on the tetrahedron vertices and just contact along the tetrahedron edges, show that the largest sphere that can just fit between them (centered on the tetrahedron center) has a radius equal to $0.225R$.

7.11 The fluorite (CaF_2) structure unit cell can be generated by stacking eight cubic subunits (edge length $= a/2$) together to form a larger cube (edge length $= a$). Two different cubic subunits are necessary: one has anions and cations in the same positions as the CsCl unit cell; the other has anions in the same positions as the CsCl unit cell, but lacks the central cation. Four of each of these subunits are stacked such that like subunits share edges and corners only but not faces.
(a) Draw the CaF_2 unit cell and identify the ions.
(b) ThO_2 assumes the fluorite structure. The radius of Th^{4+} is 0.095 nm, and the radius of O^{2-} is 0.140 nm. Is the observed anion coordination about a Th^{4+} cation what you would predict?

7.12 Germanium assumes the diamond cubic structure (Fig. 7.9) with an interatomic distance (bond length) of 0.245 nm.
(a) For spheres packed in this manner, what is the packing fraction?
(b) Gallium arsenide (GaAs) has the cubic ZnS structure, which can be achieved by substituting Ga and As atoms for atoms in the diamond cubic unit cell such that each Ga bonds to four As atoms and each As bonds to four Ga atoms. Sketch the unit cell for GaAs.
(c) Is it possible that the packing fraction for GaAs is larger than value obtained in part (a)? Explain.

7.13 Explain why SiO_2 melts at a much higher temperature (1710°C) than SiF_4 (−77°C).

7.14 (a) How does the type of bonding affect local atomic packing? (Consider ionic, covalent, and metallic bonding.)
(b) What factors control the density of a crystal?

7.15 (a) Draw a cube with an atom at each corner and one in the center of each face. This unit cell is another description of the atomic packing in the Cu or face-centered cubic structure.
(b) These unit cells are packed face-to-face to form a crystal (only those portions of atoms contained with the cube belong to one unit cell). Calculate the number of atoms per unit cell.
(c) With spheres packed in this manner, verify that the packing fraction is 0.74.

7.16 (a) Redraw Fig. 7.12*b* and denote one tetrahedral interstice and one octahedral interstice between the two close-packed planes.
(b) For each atom in the first close-packed plane, show that there are two tetrahedral insterstices and one octahedral interstice between the two close-packed planes.

Metals

The crystal structures and characteristics of metals described in this chapter are largely dependent on the delocalized nature of the metallic bond.

8.1 BONDING

Almost three-quarters of the elements are metals. The malleability, opacity, and high thermal and electrical conductivities of metals all result from the character of the metallic bond. All bonding electrons do not reside in the same energy state because this would violate the Pauli exclusion principle. Rather, the energy states are very closely spaced, forming an almost continuous band that has empty states just above the highest occupied energy level (see Sect. 4.5). This partially accounts for the free nature of the bonding electrons in a metal and is a consequence of the delocalized bonding. Nondirectional bonding leads to crystal structures with dense atomic packing and, furthermore, permits most metals to deform plastically prior to failure. Conversely, because of their bonding and the resulting crystal structures, covalent and ionic crystals usually deform only to a very limited extent before they fracture.

The cohesive energy of a metallic solid is related to the number and type of valence electrons that are available for bonding. Stronger bonding results in a higher melting point or, more precisely, a higher boiling point. Figure 8.1 shows the melting points and electronic structures for metals in the sixth period of the periodic table (cf. Table 4.3). The melting point strongly depends on the electronic structure, which, in turn, determines the bonding. For example, the alkali metal cesium

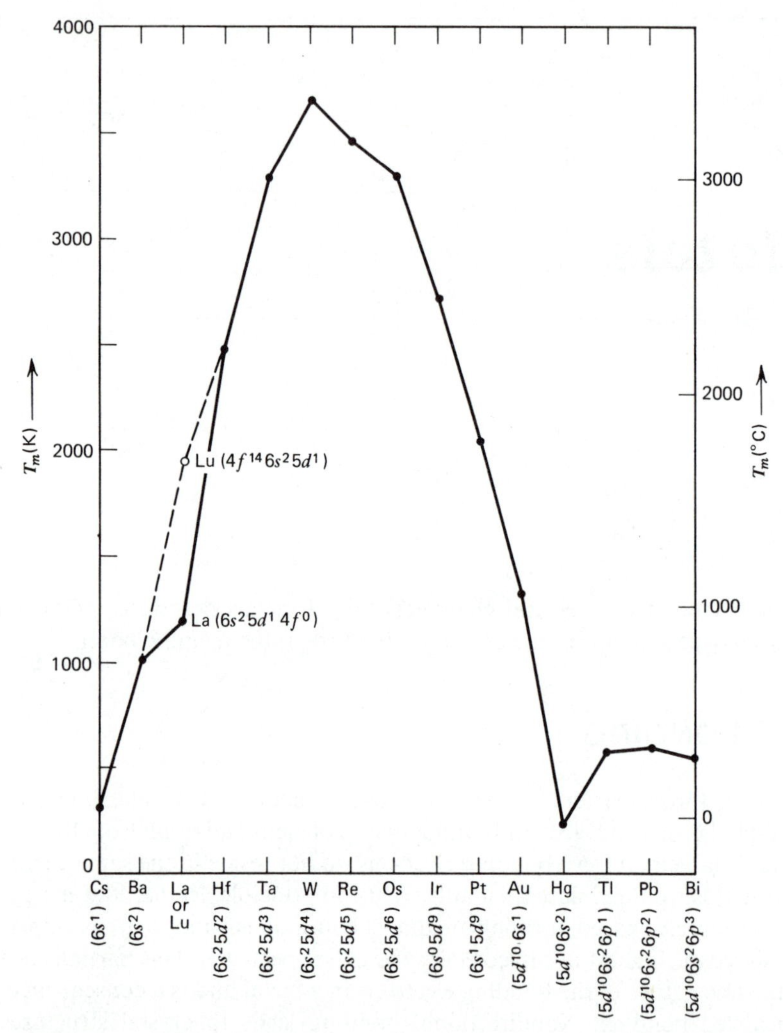

FIGURE 8.1 Melting point variation and electronic configurations for metals in the sixth period of the periodic table.

with only one valence electron per atom available for bonding, has a low melting point, whereas barium, with two valence electrons per atom, melts at a significantly higher temperature.

Elements in groups IIIA to VIII, characterized by incomplete *d* subshells, are called transition metals and have relatively high melting points due to the involvement of *d* electrons in bonding. Unlike *s* electrons, *d* electrons are not completely free in a metal, a fact that is partially reflected by the narrow and high density of states curve for the *d* electrons (Sect. 4.7). To a certain extent, we can ascribe a directionality to *d* electron bonding and thus consider the bonding in transition metals to be a mixture of metallic and covalent. Actually, metallic bonding predominates, but *d* electrons lead to a distinctive bond character such that many properties of transition metals differ from those of other metals. On the basis of melting point and of chemical behavior, we shall include the elements of group IB (Cu, Ag, Au) with the transition metals, even though Cu, Ag, and Au have filled *d* bands in the metallic state.

The number of *d* electrons involved in bonding equals the number of atomic *d* orbitals that contain an unpaired electron when the atom is in the bonding state. This bonding state might involve more available *d* orbitals than would be expected from the gaseous ground state electronic configuration since greater bonding effectiveness may result. In solid *W* we can consider that five *d* electrons and one *s* electron per atom participate in bonding, although the gaseous ground state configuration ($5d^4 6s^2$) indicates that only four *d* electrons are capable of bonding. Thus, an excited gaseous state configuration ($5d^5 6s^1$) essentially corresponds to the bonding state. The d^5 configuration represents the maximum number of *d* electrons that can participate in bonding, since this corresponds to a half-filled *d* subshell (see Hund's rule, Sect. 2.6, and Sect. 2.5). As a result, *W* has the highest melting and boiling points of any element in the sixth period. For elements with more than five electrons in the *d* subshell, fewer *d* electrons contribute to bonding because some occupy antibonding levels (see Fig. 4.17). The number of occupied antibonding levels per atom essentially equals the number of full atomic orbitals when a free atom is in the bonding state. In the case of Re with a $5d^6 6s^1$ configuration, for example, we would expect that only four (i.e., $10 - 6 = 4$) *d* electrons would be available for bonding; that is, five *d* electrons per atom occupy bonding states and one occupies an antibonding state so that a net of four *d* electrons and one *s* electron per Re atom contribute to bonding. Thus it is reasonable that the melting point of Re is lower than that of W. This simple description works reasonably well for transition elements in the fifth and sixth periods, but discrepancies occur in the fourth period for some elements that are antiferromagnetic or ferromagnetic (Chapter 25).

The rare earth metals (atomic numbers 57 to 71) and the actinium series metals (atomic numbers 89 to 103), which have incomplete *f* subshells, display a somewhat different behavior. The *f* electrons do not participate very effectively in metallic bonding, as judged by the relatively small change in melting point with each additional *f* electron. The accepted view is that *f* electrons are rather localized, remaining associated with individual atoms. It is important to reiterate that transition metals and rare earth metals are definitely metallic in character even though some covalent nature can be attributed to their bonding. Indeed, many of the most important engineering materials in use today are based on transition metals such as Fe, Co, Ni, and W.

8.2 UNIT CELLS AND CRYSTAL STRUCTURES

The three common metallic crystal structures that were discussed in Sect. 7.6 deserve further consideration since their geometries are very important relative to many properties. Each of these structures can be depicted conveniently by a small representative repeating unit termed a unit cell. A unit cell displays the relative atomic positions for a given crystal structure and the overall symmetry. The geometry of a unit cell is characterized by lattice parameters, that is, the edge lengths and the angles between edges. When a large number of identical unit cells are stacked together so as to fill space, the result is a simulation of the crystal structure. Strictly, the stacking of identical unit cells must be done in such a way that all are oriented in the same manner. A crystal structure can then be generated by repeated translations of unit cells along the three axes defined by the non-parallel edges of the cell.

The Cu structure, partially represented by nearest neighbors in Fig. 7.11*a* and discussed in terms of close-packed planes (Sect. 7.6) can also be portrayed as shown in Fig. 8.2. In Fig. 8.2*a* the dots represent the atom centers, whereas in Fig. 8.2*b* contacting spheres of equal size essentially represent the ion cores (cf. Fig. 3.3). The term face-centered cubic is descriptive, since the unit cell is cubic and atom centers are positioned at the cube corners and also the centers of the cube faces. Al, Ag, Cu, and Ni are among the metals that have this face-centered cubic (fcc) or Cu structure.

Each fcc unit cell contains four atoms, one-eighth of an atom at each corner and one-half of an atom on each face (Fig. 8.2*b*). Sometimes the unit cell is represented conventionally with whole atoms at each position (Fig. 3.3), but it is important to realize that the corner and face atoms actually are shared by neighboring unit cells in the crystal. The hard-ball model (Fig. 8.2*b*) shows that atoms are in contact along face diagonals, but not along cube edges. The radius of an atom, R, in an fcc

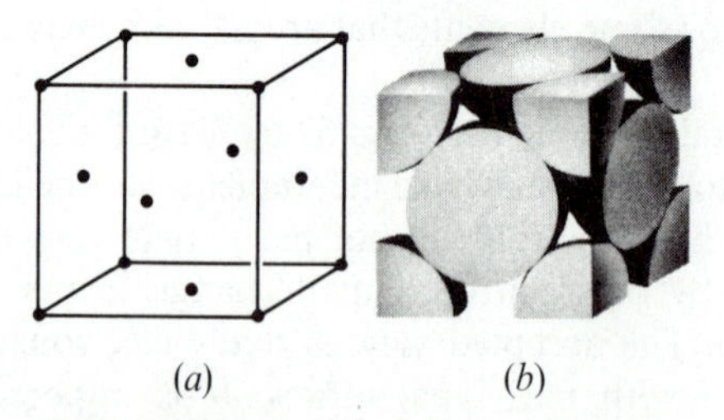

(*a*) (*b*)

FIGURE 8.2 The unit cell for the fcc or Cu structure (cf. Fig. 3.3). (*a*) The positions of atom centers are indicated by dots; (*b*) the fcc unit cell cut out of an fcc array of spheres.

metal is taken as half the distance of the closest interatomic approach; hence, $4R = a\sqrt{2}$, where a is the lattice parameter or cube edge length of the fcc unit cell. The face diagonals of an fcc unit cell represent very small portions of linear rows of contacting atoms referred to as close-packed directions. In Fig. 8.3 a corner of an fcc array has been sliced off, and a plane that includes three unit cell face diagonals is shown. This is simply a close-packed plane. By viewing the fcc unit cell along any axis perpendicular to a close-packed plane (and, incidentally, parallel to a cube body diagonal), we would see that the *ABC*-type stacking sequence discussed in Sect. 7.6 exists.

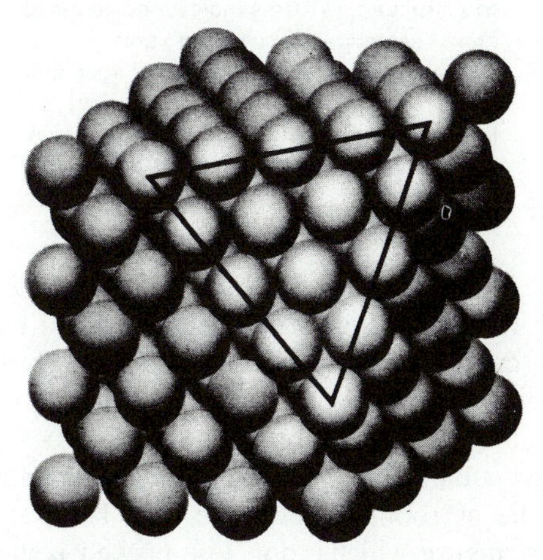

FIGURE 8.3 An fcc array of spheres sliced so as to expose a close-packed plane (outlined by the triangle). Note the orientation of this close-packed plane with respect to the unit cell edges. (From *The Structure and Properties of Materials,* Vol. I, by W. G. Moffatt, G. W. Pearsall, and J. Wulff, Wiley, New York, 1964.)

The hexagonal close-packed (hcp) or Mg structure is shown in Fig. 8.4*a*, where the dots represent atom centers. Here, two alternative unit cells are represented. The smallest possible unit cell contains two atoms, and the larger unit cell, which clearly displays hexagonal symmetry, contains six atoms. The larger unit cell, with contacting spheres, is shown in Fig. 8.4*b*. The only close-packed planes are parallel to the unit cell base and have an *AB*-type stacking sequence, and the only close-packed directions are contained within the close-packed planes. Be, Mg, Zn and, at room temperature, Ti are among the metals that have the hcp structure. If we visualize equal-size spheres packed in an hcp manner, then the ratio of the unit cell height, c, to the base edge, a, is uniquely defined by simple geometry as

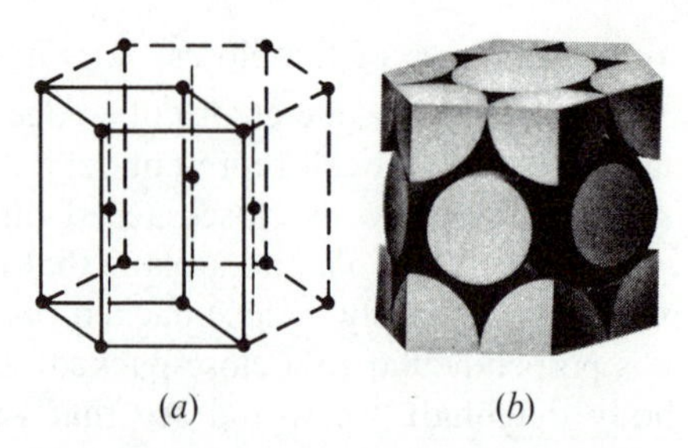

(*a*) (*b*)

FIGURE 8.4 The unit cells for the hcp or Mg structure. (*a*) Dots indicate positions of atom centers; the smallest unit cell is defined by solid lines, and the larger unit cell is defined by the whole figure; (*b*) the larger hcp unit cell cut out of an hcp array of spheres.

$(c/a) = \sqrt{8/3} = 1.633$ and the atomic radius is $R = a/2 = (c/2)\sqrt{3/8}$. Many metals that assume this structure do not have the ideal c/a ratio, as shown in Table 8.1. In those cases where c/a deviates significantly from the ideal value (e.g., Zn), we must modify the hard-ball model of the ion core, the true shape being more nearly spheroidal.

The body-centered cubic (bcc) or W structure is encountered in many metals, including Mo, W, Fe at room temperature and Ti at elevated temperatures. The unit cell of this structure, which is not close-packed, is shown in Fig. 8.5, where atom centers and hard balls are shown. The term body-centered cubic is descriptive of the unit cell, which contains two atoms. The closest-packed atomic planes (discussed in Sect. 7.6) cut through opposite, parallel edges of the unit cell and through the cell center. Thus, a closest-packed plane contains two opposite cell

TABLE 8.1 Axial Ratios for Some hcp Metals

Metal	*c/a*
Be, Y	1.57
Hf, Os, Ru, Ti	1.58
Sc, Zr	1.59
Tc	1.60
Co, Re	1.62
Mg	1.63
Zn	1.85
Cd	1.89
Ideal (sphere packing)	1.633

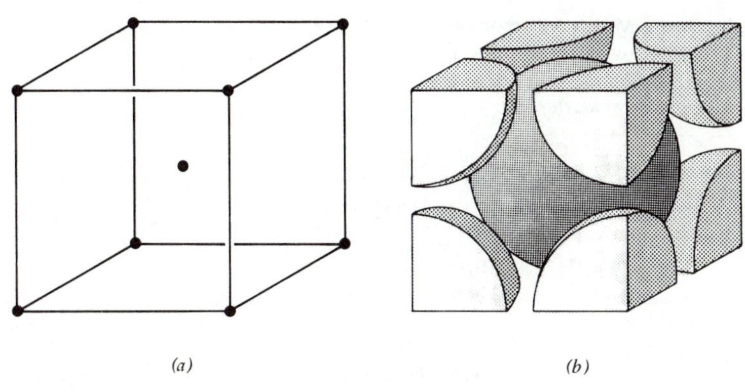

FIGURE 8.5 The unit cell for the bcc or W structure. (*a*) The positions of atom centers are indicated by dots; (*b*) the bcc unit cell cut out of a bcc array of spheres.

edges and two opposite face diagonals. The directions of atom contact, which coincide with cube body diagonals, are close-packed directions. Two of these are found in each of the closest-packed planes (see Fig. 7.12*a*). Since four atomic radii in a bcc metal coincide with the cube body diagonal, the atomic radius, R, is related to the lattice parameter of the bcc unit cell, a, as $4R = a\sqrt{3}$. Although a hard-ball model would lead to the conclusion that bcc packing is less efficient than either fcc or hcp packing, density and atomic volume comparisons for a number of metals clearly indicate that bcc, fcc, and hcp structures are about equally efficient in terms of atomic packing. Such comparisons are best made with a metal that can assume more than one crystal structure.

8.3 ALLOTROPY

Elements that occur in more than one crystallographic form are said to exhibit allotropy. Titanium, for example, is bcc above 883°C and hcp below. Many other metals, such as Na, Mn, Fe, and Co, undergo allotropic transformations (Table 8.2). This type of transformation is a phase change in the same sense as that of the liquid to solid transformation. Allotropic transformations with decreasing temperature are accompanied by heat evolution and with increasing temperature, by heat absorption. Although these enthalpies of transformation are always positive on heating and negative on cooling, the concurrent specific volume changes during transformation may be either positive or negative on heating or cooling. In the large majority of cases, metals that transform allotropically undergo volume changes on the order of 1% or less, and, as often as not, the change from a close-packed structure to bcc results in a volume decrease.

TABLE 8.2 Some Metal Allotropes

Metal	*Crystal Structure at Room Temperature*	*At Other Temperatures*
Ca	fcc	bcc (>447 °C)
Co	hcp	fcc (>427 °C)
Hf	hcp	bcc (>1742 °C)
Fe	bcc	fcc (912–1394 °C); bcc (>1394 °C)
Li	bcc	hcp (< −193 °C)
Na	bcc	hcp (< −233 °C)
Tl	hcp	bcc (>234 °C)
Ti	hcp	bcc (>883 °C)
Y	hcp	bcc (>1481 °C)
Zr	hcp	bcc (>872 °C)

The type of solid–solid transformation encountered is usually predictable. With the exception of Fe (which transforms from bcc to fcc to bcc again on cooling), most metals with allotropes tend to be bcc at elevated temperatures and to transform to hcp or fcc at lower temperatures. Examples are Li, Ti, and Zr (bcc to hcp) and Ca (bcc to fcc). Unlike the liquid–solid transformation where a more disordered, usually less dense phase goes to a more ordered, usually denser phase on cooling, the close-packed product of an allotropic transformation may be more dense or less dense than the bcc phase. Nevertheless, the high-temperature bcc phase can be considered as having less order in that the bcc structure permits larger atomic vibrations, with a correspondingly greater entropy. Regardless of the structures involved, in all cases the higher temperature phase has a higher entropy.

High pressures also may induce allotropic transformations, and the high-pressure allotrope always is more dense than the low-pressure allotrope. However, the corresponding enthalpy of transformation may be positive or negative. At present, high-pressure allotropic and polymorphic effects are receiving increasing attention, although much remains unknown because of experimental difficulties. Pressure-induced phase transformations, however, have been long recognized, and many of them are important. For example, industrial diamonds are presently prepared from graphite at high pressures and temperature with the aid of a catalyst.

8.4 ALLOYS

When a metal melts and its long-range order is destroyed, volume expansion on the order of 5% occurs. Since the ordinary geometrical limitations no longer apply, liquid metals usually are good solvents for other metals and some other elements. In fact, few pairs of the more common metals are insoluble in the liquid state. Most form homogeneous atomic mixtures or solutions that are single phase. Some pairs of metals, especially those whose atoms are of similar size ($\lesssim 15\%$ difference in radius or, equivalently, $\lesssim 50\%$ difference in atomic volume) and have similar outer electronic structures, are able to solidify as single-phase solids called solid solution alloys. For example, Cu and Ni form both liquid solutions and solid solutions at all compositions.

If the two metals do not satisfy the above conditions, solid solubility is likely to be quite limited, and solidification of the liquid leads to a two-phase mixture in the solid state called a two-phase alloy. The existence of one or two phases in a solid alloy can be readily ascertained by microscopic examination of a cut, ground, polished, and chemically etched section of the alloy. In a single-phase alloy only grain boundaries appear on etching (cf. Fig. 1.1), whereas in a two-phase alloy separate phases are revealed (cf. Fig. 1.2) because they are attacked differently by chemicals. The simplest form of plain carbon steel is a prime example of a two-phase alloy. The two phases present are called alpha or ferrite, which is the bcc form of Fe containing a small amount of carbon in solid solution (0.02 wt. % C maximum) and iron carbide or cementite, Fe_3C (6.69 wt. % C). The fractional amounts of these two phases in the microstructure depend upon the carbon content of the steel. For example, in a steel containing 0.4 wt. % C, there will be 5.7% Fe_3C and 94.3% ferrite by weight. For a steel of this composition, these amounts of Fe_3C and ferrite cannot be altered under equilibrium conditions, but the size and distribution of the two phases can be varied, either by heat treatment or by controlled solidification, to produce a broad range of useful mechanical properties.

8.5 NUCLEATION OF METAL CRYSTALS

When a pure metal solidifies in a crucible, it does so at a constant temperature equal to its equilibrium freezing point. An idealized typical plot of temperature versus cooling time is represented by the solid line in Fig. 8.6. In reality, a liquid metal will supercool below its freezing point (dotted line in Fig. 8.6) before any solid starts to form. The supercooling is caused by the difficulty in nucleating a particle of crystalline solid within the liquid, because an activation energy (see Sect. 1.6) is associated with the nucleation process. After some solid has nucleated, the temperature quickly approaches the equilibrium freezing point, and more solid grows at the expense of the liquid as a function of time. Under certain controlled laboratory conditions, isolated droplets of a pure liquid metal can be supercooled several hundred degrees Celsius before homogeneous nucleation of the solid commences. In almost all practical cases, however, heterogeneous nucleation occurs below, but very near to, the equilibrium freezing point, since inhomogeneities exist in the melt or on the walls of the container. The activation

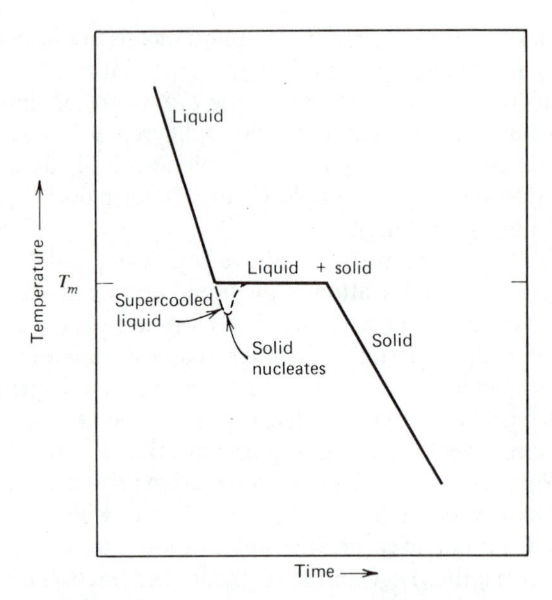

FIGURE 8.6 Cooling curve for a pure metal going from the liquid to the solid state. Under ideal equilibrium conditions (solid line), a thermal arrest occurs at the equilibrium freezing temperature when the liquid crystallizes, and the temperature remains constant until all liquid is converted to solid. In actual cases, the liquid supercools with respect to the equilibrium freezing point (dotted line) until some solid nucleates, and then the evolved heat of fusion raises the temperature to the equilibrium freezing point where further solidification proceeds.

energy for heterogeneous nucleation is much lower than that for homogeneous nucleation. Hence, the former occurs much more rapidly than the latter, and very limited supercooling is observed for normal cooling rates of a melt in a crucible or a mold. The processes of nucleation and of growth during crystallization from a melt affect the structure and properties of the resulting solid.

8.6 INGOT SOLIDIFICATION

Crystalline solids usually consist of a continuous aggregate of small crystals or grains, as can be confirmed by microscopic examination. In metals having very large grains the granular structure is often apparent to the unaided eye. Most engineering metals and alloys are prepared from the liquid either in final or near final shapes (castings) or in simple shapes (ingots). Ingots can be remelted or mechanically deformed in order to produce final products. Since the properties of a casting or final product depend to some extent on the nature of the cast structure, it is worthwhile to examine the solidification process for an ingot. This is most easily approached through consideration of the structure of a large single-phase ingot that has been sectioned (Fig. 8.7). As shown in this figure, there are three distinct grain-structure regions, the chill zone, the columnar zone, and the equiaxed zone, which developed sequentially during solidification.

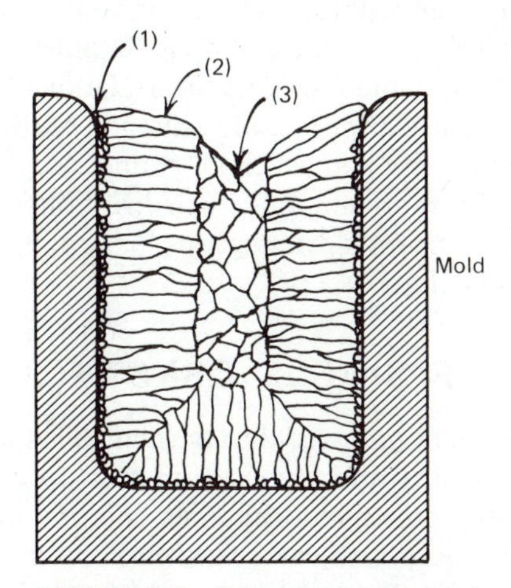

FIGURE 8.7 Schematic of the grain structure of an ingot. Three distinct regions are shown: (1) the chill zone at the edge of the mold wall, (2) the columnar zone, and (3) the central equiaxed zone. The formation of a "pipe" or shrinkage cavity near the center of the ingot is caused by the volume decrease that usually accompanies solidification.

The grains in the chill zone are small and equiaxed because the initially cold mold walls provide numerous heterogeneous nucleation sites. Initial crystallization, in the chill zone, is accompanied by evolution of heat associated with the freezing process ($-\Delta H_m$). This leads to a temperature distribution from the solid–liquid interface to the center of the mold that facilitates growth of the most favorably oriented grains. These grow into the melt in a direction opposite to that of the heat flow, thus resulting in columnar shapes. In a pure metal, especially one heated well above its melting point prior to being poured into a mold, these columnar grains may extend from the edge of the chill zone to the center of the ingot, and no central equiaxed zone occurs. In an alloy or an impure metal, the growth of columnar grains is arrested by the growth of equiaxed grains in the center of the ingot. The extent of the equiaxed zone depends upon the purity of the metal and upon the amount of *superheat* in the melt when it is poured. In general, the more impure the metal and the lower the pouring temperature, the greater is the extent of the equiaxed zone.

Unless the mold is fed continuously with molten metal, shrinkage during solidification will lead to undesirable cavities or voids. Sometimes these can be reduced to a minimum if the top of the mold is maintained at a higher temperature than the remainder of the mold by suitable insulation or external heating. By cooling the mold bottom, insulating the side walls and heating the top, it is possible in some cases to obtain a casting in which all grains except those in the chill zone have similar crystallographic orientations. Such directionally solidified ingots have anisotropic properties which are advantageous for certain purposes, for example, as turbine blades in jet engines.

8.7 GROWTH OF SINGLE CRYSTALS

Metal single crystals can be prepared through controlled solidification. Although many single crystals grown without constraint tend to exhibit regular external faces, metal single crystals grown in a container conform to the shape of the container, if it is simple. Single crystals can be grown in several ways. In one method, a crucible containing the liquid is withdrawn at a fixed, slow rate from a furnace (Fig. 8.8). Of the several grains formed at the pointed end of the crucible, only the most favorably oriented one survives to produce the single crystal. In a variation of this method, only a fraction of the material is molten at any one time, and this molten zone is moved slowly from one end of the material to the other, with a single crystal resulting as above. Such a molten zone can be achieved by a concentrated heat source in the form of an external electrical induction coil or an electrical resistance heating coil. In either of the above techniques, a small single crystal *seed*, which is kept solid at one end of the crucible, may be utilized to achieve a specific crystallographic orientation of the single crystal prepared.

Another widely used technique is that of crystal pulling. The material is melted in a crucible and held at a controlled temperature, and a small single crystal seed is brought into contact with the melt (Fig. 8.9). As the seed is slowly withdrawn and rotated simultaneously, a single crystal grows. The diameter of the resulting single crystal is controlled by varying the melt temperature and/or the pulling rate. Although a holder is used for the seed, the crystal–liquid interface is unconstrained.

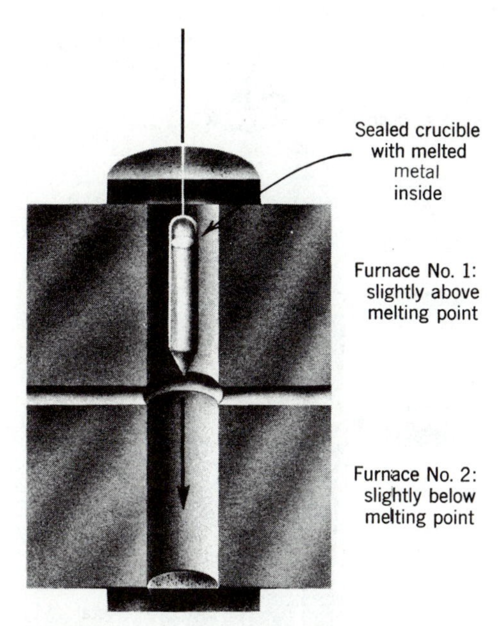

FIGURE 8.8 Diagram of apparatus used to grow single crystals by the Bridgman–Stockbarger technique. Molten metal contained in a pointed crucible is slowly lowered through a two-zone tube furnace. Several grains are usually formed at the pointed end, but only one grows to form the single crystal. (From *The Structure and Properties of Materials,* Vol. IV, by R. M. Rose, L. A. Shepard, and J. Wulff, Wiley, New York, 1966.)

Property anisotropy of single crystals, whether they are pure metals, solid solutions, or compounds, makes them useful for many purposes. Since anisotropy is a direct consequence of their crystalline nature, it is not surprising that a knowledge of their crystal structures, packing, and symmetry aids in understanding why properties vary systematically with different directions in a single crystal. Accordingly, considerations of crystal geometry facilitate not only a discussion of anisotropy but also the development of concepts that relate to the engineering behavior of crystalline solids.

8.8 DEFORMATION PROCESSING OF METALS

Metals, like all solids when subjected to a small applied load or stress, deform elastically and recover their original shape when the load is removed. Unlike ionically and covalently bonded solids, they can frequently be permanently

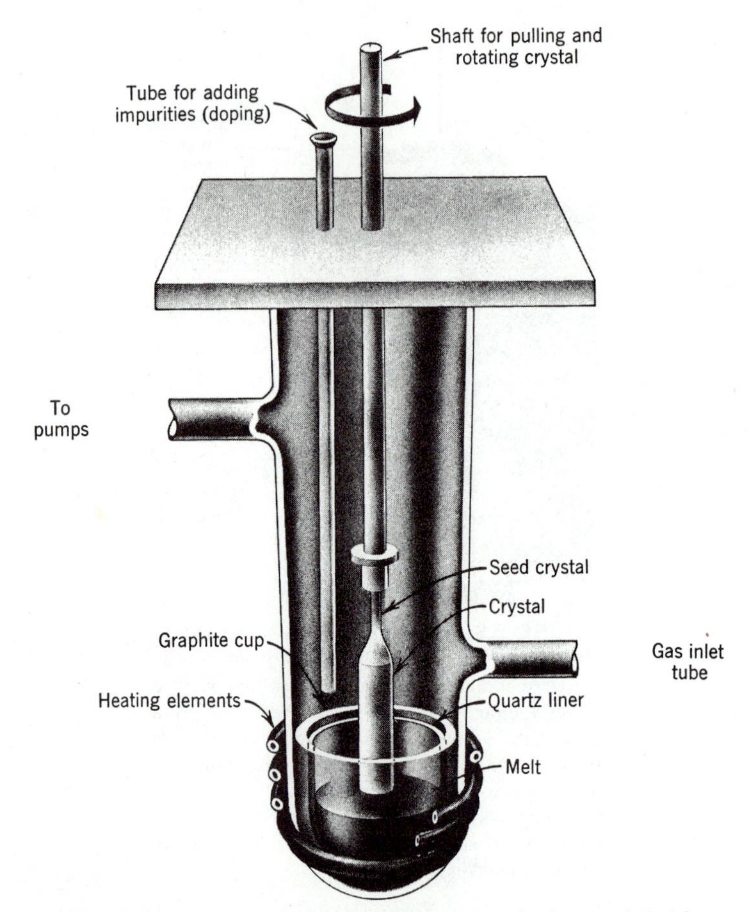

FIGURE 8.9 Diagram of apparatus used to grow single crystals by the Czochralski or Kyropoulos pulling technique. A seed crystal in contact with the melt is slowly withdrawn and rotated. This technique is widely utilized for the preparation of semiconductor single crystals that are used in the electronics industry. (From *The Structure and Properties of Materials* Vol. IV, by R. M. Rose, L. A. Shepard, and J. Wulff, Wiley, New York, 1966.)

deformed by large stresses without fracturing. Such changes in shape are possible because of the inherent plasticity of metals that occurs by slippage along densely packed crystallographic planes. Single crystals of metals are softer than polycrystalline metals of the same composition and can therefore be plastically deformed more easily. Single crystals of copper of appropriate size, for example, can be bent plastically with one hand. If deformed sufficiently, they cannot be bent back into their original shape even with both hands because cold plastic deformation causes work or strain hardening. Such hardening is attributed to the structural disorder induced in the atomic arrangements on slip planes and also at grain boundaries in polycrystalline metals. Face-centered cubic metals, however, are generally more malleable than hcp or bcc metals.

When a sheet of any metal is reduced in thickness by successively working it by hammering or rolling at room temperature, it becomes increasingly more difficult to deform plastically as mentioned above. Its inherent plasticity may be restored, however, at any stage by a heat-treatment called annealing. The heat treatment must be carried out above a certain critical temperature (called the recrystallization temperature) in order to restore ductility and malleability. This heat treatment produces a microstructure consisting of equiaxed grains that are strain-free. The recrystallization temperature is that temperature at which such new grains are visible under a microscope and is found to increase with the metal or alloy's melting point. The recrystallization temperature is not always the same for a given metal but is lower:

1. The greater the amount of prior cold-working.
2. The lower the temperature of prior working.
3. The purer the metal.
4. The smaller the grain size before cold-working.
5. The longer the time at the heat-treatment temperature.

In the shaping of a metal by mechanical deformation, it is customary to call deformation carried out above the recrystallization temperature hot-working to distinguish it from working below the recrystallization temperature or cold-working. Hot-working requires less energy than cold-working because the metal *anneals* itself during or immediately following deformation and therefore remains malleable.

The mechanisms invoked to explain the observed mechanical, thermal, electrical, and magnetic behavior of solids are described at greater length in Chapters 19 to 27. Before attempting such discussions it is worthwhile to examine in greater detail further aspects of the structure of solids that arise from short- or long-range atomic displacements or movement of atoms and groups of atoms that account for structural change in solids induced by different variables (mechanical, thermal, magnetic, etc.). First, it is necessary to introduce the crystallographic nomenclature commonly used to geometrically describe crystalline solids, and this topic will be treated in Chapter 9.

REFERENCES

1. Schlenker, B. R. *Introduction to Materials Science*, Wiley, Sydney, Australia, 1969.
2. Chalmers, B. *Physical Metallurgy*, Wiley, New York, 1959.
3. Cottrell, A. H. The Nature of Metals, *Scientific American*, Vol. 217, No. 3, September 1967, pp. 90–100.

QUESTIONS AND PROBLEMS

8.1 Consult Volume 1 of *Metals Handbook* and plot Young's modulus (modulus of elasticity) and molar volume for elements in the sixth period from La to Bi. Compare your curves with Fig. 8.1, and discuss the results in terms of the bonding characteristics of the metals.

8.2 Consult Volume 8 of *Metals Handbook* or the *Handbook of Chemistry and Physics* and tabulate the melting points of group IA elements (Li through Cs), group IIB elements (Zn through Hg), group VA elements (V through Ta), and the group VIII elements Co, Rh, and Ir. Discuss the trend in each group, and contrast transition metal groups with nontransition metal groups in terms of bonding.

8.3 The density of Cu (fcc) is 8.94 g/cm^3. Calculate its lattice parameter and its interatomic distance.

8.4 Calculate the volumes of an fcc unit cell, an ideal hcp unit cell, and a bcc unit cell in terms of the atomic radius, R, and in terms of the lattice parameters.

8.5 Sketch a conventional fcc unit cell (i.e., one with complete atoms at each position).
(a) Show that this unit cell contains four atoms.
(b) Show that each atom has 12 nearest neighbors.
(c) How many distinct (i.e., nonparallel) close-packed directions are there?
(d) Show that the packing fraction equals 0.74.

8.6 The atomic volume of an elemental metal is the volume of a piece of metal divided by the number of atoms it contains. Alternatively, the atomic volume is the unit cell volume divided by the number of atoms per unit cell.
(a) The lattice parameter of W (bcc) is $a = 0.3165$ nm. Calculate the atomic volume and the density of W.
(b) Calculate the volume of a sphere having the same radius as a W atom, and compare this to its atomic volume.
(c) Show that the packing fraction for W (0.68) is the answer to part (b) divided by the atomic volume.

8.7 (a) Show that the ideal c/a ratio for hexagonal close-packing is $\sqrt{8/3} = 1.633$.
(b) Show why an atom in a nonideal hcp metal ($c/a \neq 1.633$) does not have a unique radius. (*Hint*. Compare the interatomic distance within a close-packed plane to that between atoms in neighboring close-packed planes.)

8.8 (a) Sketch a portion of a cube face plane in an fcc metal and compute the number of atoms per unit area, in terms of the lattice parameter, a.
(b) Compute the number of atoms per unit area, in terms of the lattice parameter, for a close-packed plane in an fcc metal. Compare this to the answer of part (a).
(c) How many distinct (i.e., nonparallel) close-packed planes are there in an fcc metal?

8.9 Lanthanum at room temperature has a hexagonal crystal structure ($c/a = 3.22$) that can be visualized in terms of a *ABAC* close-packed plane stacking sequence. Sketch a small hexagonal unit cell for the La structure and show how this differs from the small hcp unit cell (Fig. 8.4*a*).

8.10 An interstice or interstitial site in a solid metal is a position between metal atoms that can be occupied by a very small impurity atom or a very small alloying element atom such as H, C, N, or O. The fcc, hcp, and bcc structures have octahedral interstices (surrounded by six metal atoms) and tetrahedral interstices (surrounded by four metal atoms). The octahedral interstices in the fcc structure are located at the unit cell center and at the centers of the unit cell edges.

(a) For an fcc array of contacting spheres of radius R, what is the radius, r, of the largest interstitial sphere that will just fit into an octahedral interstice?

(b) If Ti atoms assume an fcc array and all octahedral interstices are occupied by C atoms, we have the structure of TiC. Sketch the unit cell of TiC and compare this with the NaCl structure unit cell (Fig. 7.8*b*). TiC is largely metallic with some covalent bonding character, whereas NaCl is ionic. In terms of bonding and atomic or ionic sizes, explain why identical structures result for TiC and NaCl. (*Note*. Ti metal does not have an fcc allotrope, but some of its compounds are closely related to the fcc structure.)

8.11 (a) Sketch two fcc unit cells. On one indicate all octahedral interstices, and on the other indicate all tetrahedral interstices.

(b) How many of each type of interstice are there per unit cell?

(c) How many of each type of interstice are there per metal atom? (The answers to this question also apply to a hcp metal.)

8.12 It is sometimes useful to define an effective coordination number as follows. Connect the center of a given atom to the centers of all near neighbor atoms (*not* just nearest neighbors), and then draw planes bisecting these lines. This generates an atomic domain polyhedron about the reference atom having a volume equal to the atomic volume. The effective CN is the number of faces on the polyhedron. The resulting effective CN is 12 for fcc or hcp, even if nonideal.

(a) Construct the atomic domain for a bcc metal atom.

(b) What is the effective CN for a bcc metal atom?

(c) In terms of the bcc lattice parameter, a, what are the interatomic distances between a reference atom and its coordinated neighbors?

8.13 (a) For a metal with fcc and bcc allotropes, compute the atomic radius for the bcc form in terms of the atomic radius for the fcc form, assuming atomic volume remains constant.

(b) Pure iron transforms from bcc to fcc at 912°C with a 1.06% decrease in volume. In terms of the fcc atomic radius, what is the bcc atomic radius?

(c) Pure Ti transforms from hcp to bcc at 833°C with a 0.55% decrease in volume. Does the atomic radius increase or decrease?

8.14 Why should grain boundaries etch more readily than grain interiors?

8.15 Indicate in a simple sketch the grain structure of a directionally solidified ingot. Show how this structure might be obtained.

8.16 What effect might gas evolution during solidification have on the structure of an ingot?

Crystal Geometry

The concepts and notation commonly used to describe the internal arrangement of atoms and molecules in crystals are described in this chapter.

9.1 CRYSTAL SYSTEMS AND BRAVAIS LATTICES

Any crystalline material possesses a definite long-range orderly arrangement of atoms or ions that can be completely described by its unit cell. As noted previously, a number of substances may have the same crystal structure and, therefore, the same geometrical forms for their unit cells. Thus, different substances may be conveniently classed together. Various other geometrical features, amenable to grouping, aid in the orderly classification of crystalline materials. Crystalline substances have different degrees of symmetry, for example. A cubic material is most symmetrical, yet different crystal structures with overall cubic symmetry may have distinctly different symmetries with respect to local atomic arrangements.

On one level, it is possible to focus upon the overall symmetry of the unit cell itself. This allows the arrangement of all crystal structures into seven crystal systems, each of which is defined by a system of axes parallel to the unit cell edges and having natural units of measure along each axis equal to the three unit cell edge lengths or lattice parameters, a, b, and c. The crystal axes form the edges of a parallelepiped called the lattice unit cell (Fig. 9.1). The symmetry of each crystal system depends on the relations between the lattice parameters a, b, and c and between the interaxial angles α, β, and γ. In the cubic system $a = b = c$ and $\alpha = \beta = \gamma = 90°$, which is simply a Cartesian coordinate system. The least

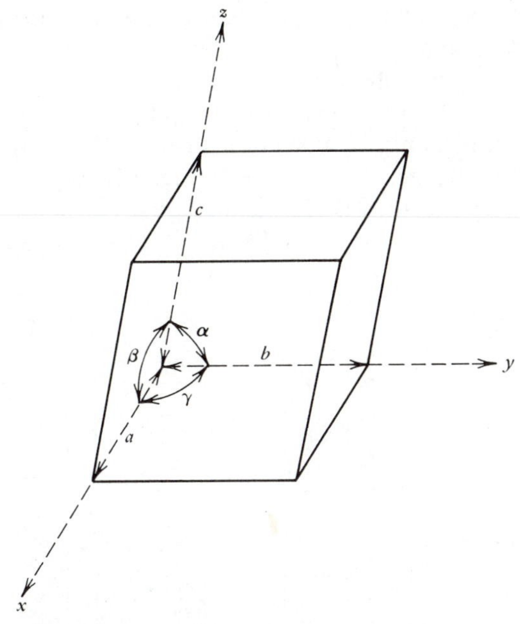

FIGURE 9.1 Lattice unit cell showing the axes and the lattice parameters, a, b, and c. In the most general case, the x, y, and z axes are not orthogonal, that is, $\alpha \neq \beta \neq \gamma \neq 90°$.

symmetrical crystal system is called triclinic, having $a \neq b \neq c$ and $\alpha \neq \beta \neq \gamma$, with none of the angles being 90°. Table 9.1 presents the relationships for the seven crystal systems.

For each of the crystal systems, there may be further ways in which symmetry can be described geometrically. For example, body-centered cubic is distinct from face-centered cubic, although both belong to the cubic system. Since every crystal

TABLE 9.1 The Seven Crystal Systems

System	*Axes and Interaxial Angles*	
Triclinic	$a \neq b \neq c$	$\alpha \neq \beta \neq \gamma$
Monoclinic	$a \neq b \neq c$	$\alpha = \gamma = 90° \neq \beta$
Orthorhombic	$a \neq b \neq c$	$\alpha = \beta = \gamma = 90°$
Tetragonal	$a = b \neq c$	$\alpha = \beta = \gamma = 90°$
Rhombohedral	$a = b = c$	$\alpha = \beta = \gamma \neq 90°$
Hexagonal	$a = b \neq c$	$\alpha = \beta = 90°, \gamma = 120°$
Cubic	$a = b = c$	$\alpha = \beta = \gamma = 90°$

has a periodic arrangement of atoms, it is convenient to relate the atomic arrangement to a network of points in space called the space lattice or Bravais lattice. Such a network, which can be subdivided into identical lattice unit cells, consists of lattice points, each of which has identical surroundings with respect to other lattice points. Mathematically, only 14 different networks of lattice points are possible, and these are the 14 distinct Bravais lattices (Fig. 9.2). Any crystalline solid must possess the symmetry and, thus, the lattice unit cell of one of the 14 Bravais lattices. It is very important to note that only the overall crystallographic symmetry, but

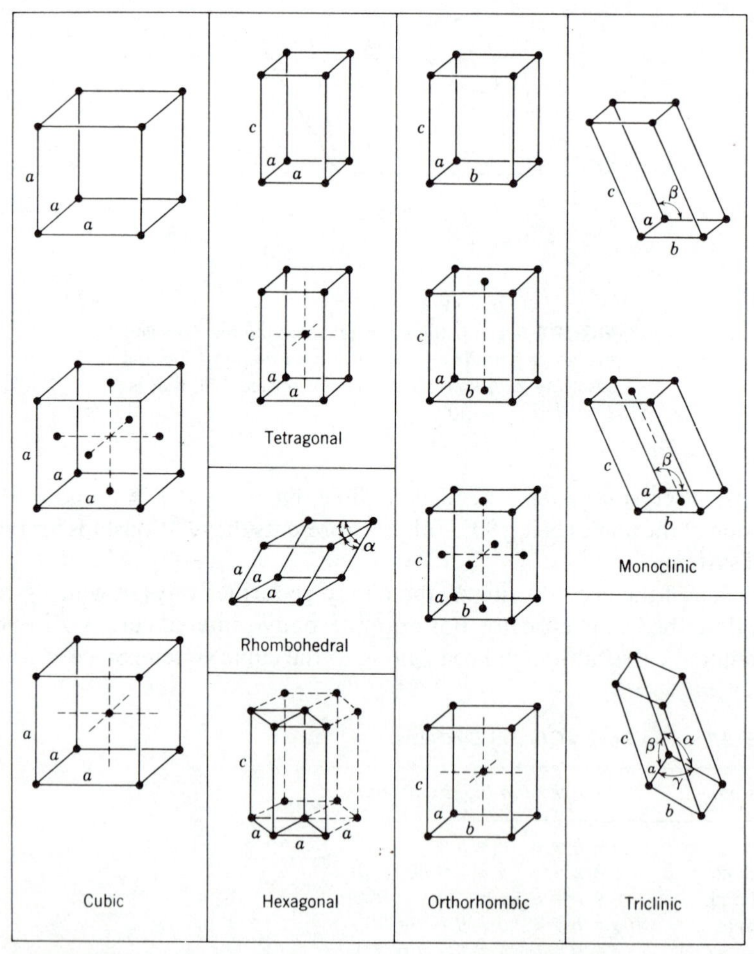

FIGURE 9.2 Conventional lattice unit cells of the 14 Bravais lattices grouped according to crystal system. The dots indicate lattice points that, when located on faces or at corners, are shared by other identical lattice unit cells.

not the chemical bonding characteristics, of a structure is reflected by the Bravais lattice.

Any lattice unit cell must have lattice points at its corners and must be a repeating unit representative of a Bravais lattice. Thus, when identical unit cells having the same orientation are stacked together, they will fill space and generate the Bravais lattice. One Bravais lattice can have a number of different unit cells. Conventionally, a lattice unit cell is chosen that has a simple geometry, contains only a few lattice points, and displays the greatest symmetry. Thus, a number of unit cells shown in Fig. 9.2 contains other lattice points in addition to those at the cell corners. The face-centered cubic Bravais lattice, although best visualized in terms of the conventional unit cell (Fig. 9.2), can be represented by a smaller unit cell having lattice points at its corners only. A lattice unit cell with lattice points only at its corners is called a primitive cell.

In most metals the relationship between the Bravais lattice and atomic packing is particularly simple since the basic building block is the atom itself and bonding

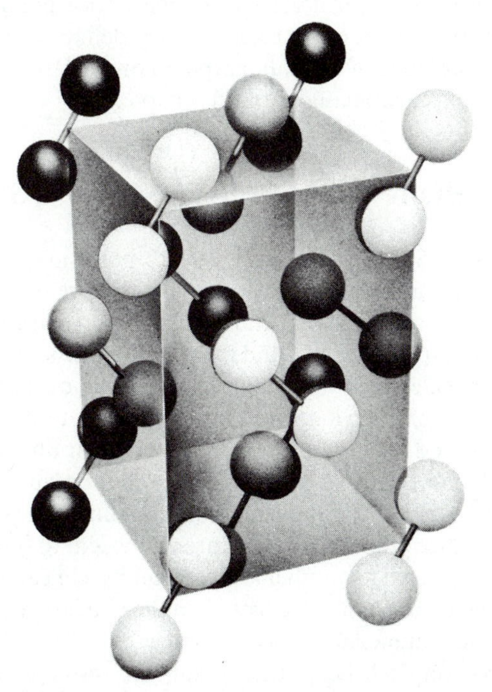

FIGURE 9.3 The crystal structure of solid iodine referred to the face-centered orthorhombic lattice unit cell. (From *The Structure and Properties of Materials,* Vol. I, by W. G. Moffatt, G. W. Pearsall, and J. Wulff, Wiley, New York, 1964.)

is nondirectional. This is not true for covalent or ionic solids. An example is provided by the crystal structure of arsenic, shown previously in Fig. 7.10, where the rhombohedral lattice unit cell is also indicated. As another example, the crystalline arrangement of I_2 molecules is face-centered orthorhombic (Fig. 9.3). In each of these cases, the most important structural feature, reflecting the covalent bonding, is not represented by the lattice unit cell. On the other hand, the crystallographic description and the structural description of a metal often are highly complementary, and it is necessary to be able to interchange between one description and the other at will.

Quite diverse crystal structures may have the same Bravais lattice. For example, Ni with the fcc or Cu structure (Fig. 8.2), MgO with the NaCl structure (Fig. 7.8*b*), Si with the diamond cubic structure (Fig. 7.9), and crystalline methane (Fig. 5.4) all have fcc Bravais lattices, yet their bonding characteristics and properties differ widely. In each of these cases the arrangement of atoms, ions, or molecules is uniquely related to the arrangement of fcc Bravais lattice points. However, the atomic positions must be specified relative to the lattice points and the lattice parameters must be specified in order to completely define each crystal structure. To do this, a certain amount of crystallographic nomenclature is necessary, and this will be developed in the remainder of this chapter.

9.2 LATTICE POINTS AND ATOMIC POSITIONS

It was pointed out in Sect. 9.1 that every lattice point in a Bravais lattice has identical surroundings. Accordingly, we can position one atom or an identical group of atoms at each lattice point and, thus, build up a crystal structure. Similarly, this can be done with a lattice unit cell, and the result will be a crystal structure unit cell such as those considered in Chapters 7 and 8. The coordinates of a lattice point or of an atom center within a unit cell are specified in terms of fractions of the lattice parameters a, b, and c. Thus, the body-centered cubic unit cell has two lattice points, one at 0, 0, 0 and the other at $\frac{1}{2}, \frac{1}{2}, \frac{1}{2}$. Note that because the conventional lattice unit cell contains redundant information, it is unnecessary to specify 1, 0, 0; 1, 1, 0; etc., as being lattice points since these are equivalent to 0, 0, 0. That is, if any lattice point can be connected to another lattice point by a linear, integral combination of vectors defined by the unit cell edges, then the two lattice points are equivalent. Table 9.2 lists the unique lattice points for each of the 14 Bravais lattices. If one atom is centered on each bcc lattice point, the result is the bcc structure, providing all atoms are identical. Lattice points need not coincide with all atom centers nor with any atom centers, however, as can be seen in the structure of solid I_2 (Fig. 9.3). The coordinates of lattice points in noncubic structures are described similarly to those in cubic structures, but this is accomplished in terms of the crystal axes, which are not necessarily orthogonal and whose units of

TABLE 9.2 Coordinates of the Unique Lattice Points for the 14 Bravais Lattice Unit Cells

Bravais Lattice	*Lattice Points in a Unit Cell*
Triclinic[a]	0, 0, 0
Simple monoclinic[a]	0, 0, 0
Base-centered monoclinic	0, 0, 0; $\frac{1}{2}$, $\frac{1}{2}$, 0
Simple orthorhombic[a]	0, 0, 0
Base-centered orthorhombic	0, 0, 0; $\frac{1}{2}$, $\frac{1}{2}$, 0
Face-centered orthorhombic	0, 0, 0; $\frac{1}{2}$, $\frac{1}{2}$, 0; $\frac{1}{2}$, 0, $\frac{1}{2}$; 0, $\frac{1}{2}$, $\frac{1}{2}$
Body-centered orthorhombic	0, 0, 0; $\frac{1}{2}$, $\frac{1}{2}$, $\frac{1}{2}$
Simple tetragonal[a]	0, 0, 0
Body-centered tetragonal	0, 0, 0; $\frac{1}{2}$, $\frac{1}{2}$, $\frac{1}{2}$
Rhombohedral[a]	0, 0, 0
Hexagonal[a]	0, 0, 0
Simple cubic[a]	0, 0, 0
Face-centered cubic	0, 0, 0; $\frac{1}{2}$, $\frac{1}{2}$, 0; $\frac{1}{2}$, 0, $\frac{1}{2}$; 0, $\frac{1}{2}$, $\frac{1}{2}$
Body-centered cubic	0, 0, 0; $\frac{1}{2}$, $\frac{1}{2}$, $\frac{1}{2}$

[a] These Bravais lattices have primitive unit cells, that is, there is one lattice point per unit cell.

measure need not be equal. Thus, the lattice points in the base-centered monoclinic lattice unit cell are at 0, 0, 0 and $\frac{1}{2}$, $\frac{1}{2}$, 0.

Atomic positions and any other points within a unit cell are specified in the same manner. For a hexagonal close-packed metal, the atoms are located at 0, 0, 0 and at $\frac{2}{3}$, $\frac{1}{3}$, $\frac{1}{2}$ in the structure unit cell (Fig. 8.4*a*), whereas the lattice unit cell has a single lattice point at 0, 0, 0. Thus, the hcp crystal structure has two atoms per lattice point. In the same hcp structure, one of the tetrahedral interstices is located at $\frac{2}{3}$, $\frac{1}{3}$, $\frac{1}{8}$.

9.3 ATOMIC PLANES

It is convenient to have a shorthand means of denoting various atomic planes in a crystal. Such a useful notation is based on the equation for a plane in three-dimensional space:

$$\frac{1}{A}\frac{x}{a} + \frac{1}{B}\frac{y}{b} + \frac{1}{C}\frac{z}{c} = 1 \tag{9.1}$$

The numerical intercepts of this plane along the three axes x, y, and z are A, B, and C, respectively, when expressed in terms of the lattice parameter units. These intercepts, when converted to a set of small integers $h = n/A$, $k = n/B$, and $l = n/C$, are called the Miller indices of the plane, and the plane is denoted as (hkl).

In describing crystallographic planes, crystal axes are used in conjunction with a reference origin on one lattice point, and the intercepts are determined as pure numbers in terms of the unit of measure (the lattice parameter) appropriate to each crystal axis regardless of the actual dimension. For example, consider Fig. 9.4, which shows the planes (220) and (110) for an orthorhombic unit cell. The respective intercepts are $\frac{1}{2}, \frac{1}{2}, \infty$; $1, 1, \infty$; and $2, 2, \infty$ along the *a*, *b*, and *c* axes. It is a matter of convention that the intercept reciprocals are multiplied by the smallest integer that will convert them into a set of integers having the same ratios. Thus, the planes with intercepts $1,1,\infty$ and $2,2,\infty$ are both (110) planes, whereas the plane having intercepts $\frac{1}{2}, \frac{1}{2}, \infty$ is a (220) plane. This notation provides a means by which equally spaced, parallel atomic planes have the same Miller indices. Furthermore, if the set of Miller indices is the smallest possible, then the planes it represents have identical atomic packing.

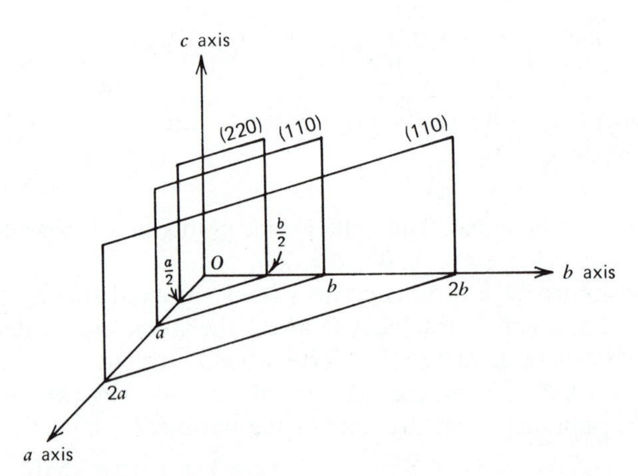

FIGURE 9.4 The (220) and (110) planes in an orthorhombic lattice. Note that the designation (110) applies to two of the planes shown.

Figure 9.5 shows the $(\bar{1}\bar{1}1)$ plane in the rhombohedral system. The intercepts of this plane along the *a*, *b*, and *c* axes are -1, -1, and 1, respectively. Negative Miller indices are denoted by a bar above the particular index. If the signs of all indices are reversed, for example $(\bar{1}\bar{1}1)$ to $(11\bar{1})$, an identical plane parallel to the first, but on an opposite side of the origin, is denoted. Thus, $(\bar{1}\bar{1}1)$ and $(11\bar{1})$ actually represent the same plane, and this redundancy is a general one. One further comment requires attention. Any plane passing through the origin has indeterminate indices; however, its Miller indices can be obtained either by translating the origin to an equivalent lattice point or by determining the Miller indices of a parallel, equivalent plane.

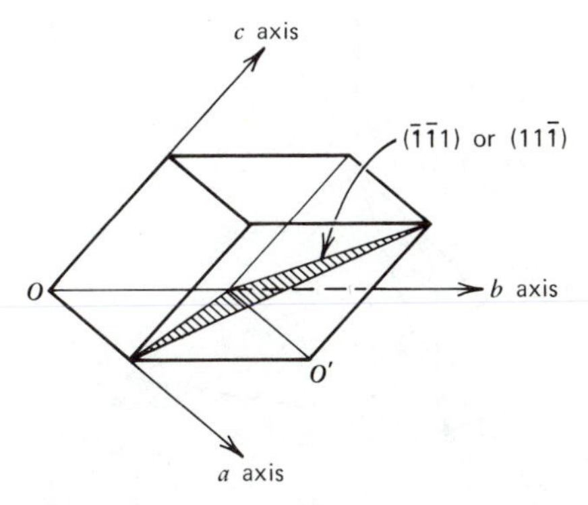

FIGURE 9.5 The $(\bar{1}\bar{1}1)$ plane in a rhombohedral lattice unit cell. That this is a $(\bar{1}\bar{1}1)$ plane is best seen by shifting the origin from point O to point O' so that axial intercepts can be obtained. Shifting the origin from one lattice point to an equivalent lattice point is permissible, providing the crystal axes remain parallel.

The method of obtaining the Miller indices (hkl) of a plane is summarized as follows:

1. Remove any indeterminacy.
2. Find the intercepts along the three axes of the crystal system.
3. Take the reciprocals of these numbers.
4. Multiply these reciprocals by the smallest integer necessary to convert them into a set of integers.
5. Enclose the resulting integers in parentheses, (hkl), without commas.

In many cases the atomic packing is the same on a number of nonparallel planes in a given crystal structure. This depends, of course, on the symmetry of the structure. Figure 9.6 shows the (111), $(11\bar{1})$, $(1\bar{1}1)$, and $(\bar{1}11)$ close-packed planes of the fcc structure. These four distinct close-packed planes, although unique in three-dimensional space, are crystallographically equivalent in terms of atomic packing. Crystallographically equivalent, nonparallel planes are termed planes of a form, or simply a family of planes, and are designated by braces rather than parentheses. Thus, the close-packed planes in an fcc metal can be referred to as the {111} family of planes. In a strict sense, {111} planes in an fcc metal include (111), $(11\bar{1})$, $(1\bar{1}1)$, $(\bar{1}11)$, $(\bar{1}\bar{1}\bar{1})$, $(\bar{1}\bar{1}1)$, $(\bar{1}1\bar{1})$, and $(1\bar{1}\bar{1})$; however, the last four listed are equivalent to the first four since one of each of the planes in the second group is parallel to one plane in the first.

Only in the cubic system do all planes having the same set of Miller indices have the same type of atomic packing. Thus, for a cubic crystal, the (135) plane and the

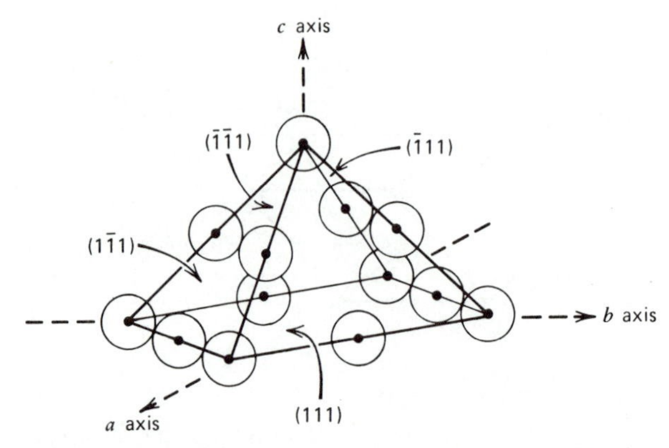

FIGURE 9.6 The family of close-packed planes, {111}, in a face-centered cubic metal. Each plane is uniquely oriented in space relative to the other planes, but all have identical atomic packing.

($\bar{1}$53) plane belong to the same family. On the other hand, for a tetragonal crystal, the (001) plane differs from the (100) and (010) planes, but the latter two are crystallographically equivalent (cf. Table 9.1).

9.4 CRYSTALLOGRAPHIC DIRECTIONS

Directions in a crystal are specified in a shorthand vector notation. The direction of a line from the origin of the unit cell to a point whose coordinates are the integers u, v, and w is denoted as [uvw]. Any direction parallel to this one is denoted the same since it is identical. Once more, it is conventional to avoid fractional indices and, furthermore, to use the smallest possible set of integers. Hence, lines from 0, 0, 0 to $\frac{1}{2}$, $\frac{1}{2}$, $\frac{1}{2}$ and from 0, 0, 0 to 2, 2, 2 also pass through 1, 1, 1 and the direction is designated as [111]. As with planes, any negative index is indicated by a bar above it. If the signs of all indices are reversed, for example, [110] to [$\bar{1}\bar{1}$0], the result is a direction antiparallel to the former. Consequently, if the sense of a direction is taken into account, the redundancy encountered with Miller indices for a plane is not found with direction indices.

Another way to visualize direction indices is in terms of vectors referred to as the crystal axes, with the respective unit vectors being **a**, **b**, and **c**, which define the unit cell. A vector, $\mathbf{r} = D\mathbf{a} + E\mathbf{b} + F\mathbf{c}$, in the crystal points in a direction [uvw], where u, v, and w are the smallest set of integers having the same ratios of the unit vector coefficients D, E, and F.

Crystallographically equivalent directions occur frequently, particularly in symmetrical crystal structures. Consider the three close-packed directions in the

(111) plane shown in Fig. 9.6. One of these goes from the atomic position at 1, 0, 0 to that at 0, 0, 1 and its direction indices are $[\bar{1}01]$. The other two distinct close-packed directions in the (111) plane are $[01\bar{1}]$ and $[1\bar{1}0]$. Indeed, for an fcc metal, each plane in the {111} family contains three close-packed directions. Since these are crystallographically equivalent, they are called directions of a form, or simply a family of directions, and are designated ⟨110⟩. Carets are used to denote a family of crystallographically equivalent directions, whereas brackets denote a specific direction. In the cubic system, directions having the same set of indices are crystallographically equivalent. For example, the [135] direction and the $[\bar{1}53]$ direction belong to the same family.

In general, there is no simple geometrical relationship between a particular direction and a plane with the same indices. In the cubic system, however, a direction is normal to a plane having identical indices. Thus, the $[1\bar{1}0]$ direction is normal to the $(1\bar{1}0)$ plane in a cubic crystal. Conversely, in an orthorhombic crystal the normal to the $(1\bar{1}0)$ plane is not parallel to the $[1\bar{1}0]$ direction.

9.5 MILLER–BRAVAIS INDICES FOR HEXAGONAL SYSTEMS

Consider an hcp unit cell as illustrated in Fig. 9.7. According to the indexing system we have introduced, the three equivalent atomic planes indicated in Fig. 9.7 have Miller indices (100), (010), and $(1\bar{1}0)$. It is both desirable and convenient for

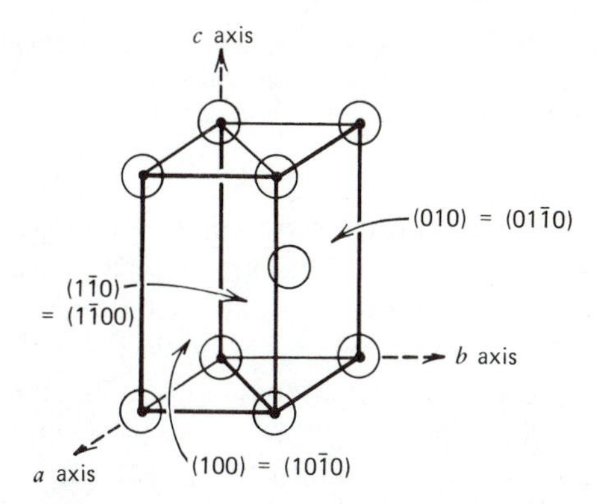

FIGURE 9.7 The hexagonal close-packed unit cell. Note that the three crystallographically equivalent planes indicated have different sets of Miller indices. Using the Miller–Bravais scheme, we see that these planes have the same set of indices.

crystallographically equivalent planes in the hexagonal system to have the same set of indices. An alternative indexing system that achieves this purpose utilizes four indices to denote a plane, $(hkil)$. These are called the Miller–Bravais indices. The indices h, k, and l are obtained as before, and the fourth index is given by $i = -(h + k)$. What this amounts to, essentially, is the introduction of a third axis in the basal plane (Fig. 9.8). Here, rather than labeling the initial two axes in the basal plane a and b, we shall label them a_1 and a_2, and the third axis, a_3. The latter is redundant as far as the coordinate system is concerned, since it provides no new information; however, it does reflect the inherent hexagonal symmetry. Under the Miller–Bravais indexing system, the equivalent planes (100), (010), and $(1\bar{1}0)$ are transformed to $(10\bar{1}0)$, $(01\bar{1}0)$, and $(1\bar{1}00)$, which now have the same set of indices and are members of the $\{1\bar{1}00\}$ family of planes. Note that the c axis is unaffected by this scheme.

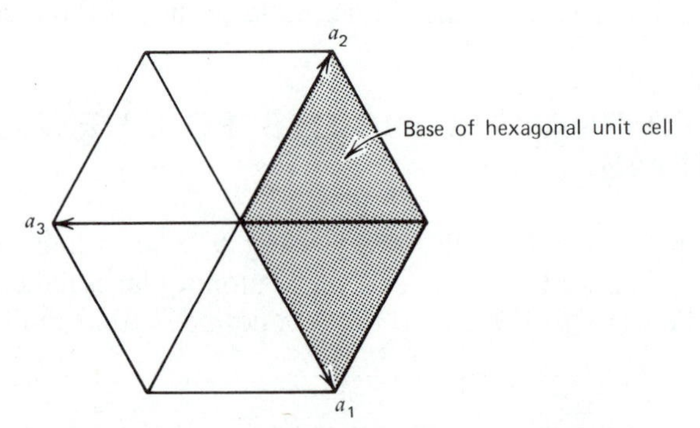

FIGURE 9.8 A hexagonal system basal plane showing the three coplanar axes used for obtaining the first three Miller–Bravais indices. The axis a_3 is actually redundant.

Direction indices also can be constructed so that crystallographically equivalent directions have the same set of indices. Thus, a specific direction is denoted as $[uvtw]$, where $t = -(u + v)$. The procedure for determining direction indices is not as straightforward as it is for planes. For example, the [010] direction does not simply become the $[01\bar{1}0]$ since the latter does not lie along the a_2 axis. Rather, the indices u, v, and t must be chosen by a proper vector summation such that the correct vector direction results, subject to the constraint $t = -(u + v)$. This is illustrated in Fig. 9.9, where the direction along the a_2 axis is shown to be $[\bar{1}2\bar{1}0]$. The procedure can be reduced to mathematical formulas:

$$u = \frac{n}{3}(2u' - v'), \qquad v = \frac{n}{3}(2v' - u'), \qquad t = -(u + v), \qquad w = nw' \quad (9.2)$$

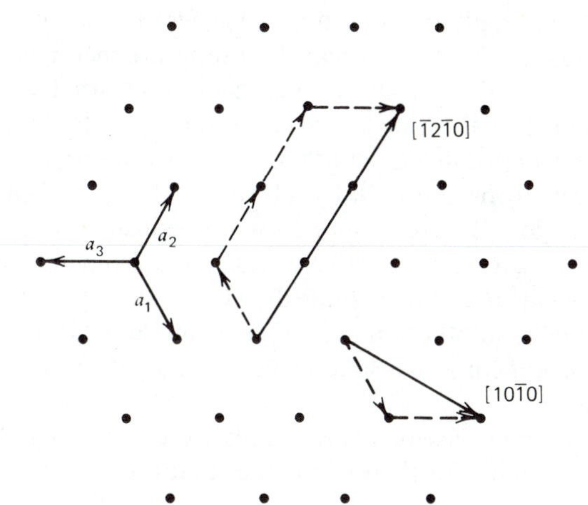

FIGURE 9.9 The four-index scheme applied to directions in hexagonal systems.

where $[u'v'w']$ would be obtained as explained in Sect. 9.4, $[uvtw]$ is the four index designation, and n is a factor necessary to obtain the set of smallest integers for u, v, t, and w. Hence, to find the four indices for a direction in a hexagonal crystal, first find the convential three index notation (Sect. 9.4) and then convert these according to Eq. 9.2.

9.6 INTERPLANAR SPACINGS AND THE DIFFRACTION OF X RAYS BY CUBIC CRYSTALS

Much of the current knowledge about crystal structures is a result of studies involving the diffraction of X rays by crystals. Diffraction is the preferential scattering of an incident X ray beam at definite angles by a crystal. This is similar, in many respects, to the interference effects that result when waves in the ocean impinge upon a number of equally spaced obstacles such as pilings. Such a diffraction phenomenon occurs only when the obstacle spacing is comparable to the wavelength. Since the interatomic spacing is comparable to the wavelength of X rays but much smaller than the wavelength of light (see Chapter 27), X rays can be diffracted by crystals whereas light cannot.

An X ray beam containing a spectrum of wavelengths will be partially absorbed and partially transmitted by a single crystal and a small fraction of the beam intensity will be diffracted at various definite angles. The latter is a direct consequence of the periodic array of atoms or ions in a crystal. On the other hand, a beam of fixed wavelength (monochromatic) generally will not be diffracted by a single crystal unless specific geometrical conditions are met. Basically, an incident X ray beam interacts with the electrons of each atom such that each atom scatters a portion of the beam in all directions. A situation analogous to this is provided by the effects of a pebble dropped in a quiet pond. An individual pebble causes a series of circular waves that propagate radially. However, if a large number of pebbles, arrayed two-dimensionally, are dropped simultaneously, each pebble causes the emission of circular waves, yet collectively these waves from individual pebbles add together and propagate in certain definite directions and subtract from one another so as to not propagate in other directions. In a simple sense, the interaction of X rays with atoms or ion cores in a solid is much the same, only the atomic array is three-dimensional.

We can develop some useful relationships by considering the diffraction of X rays from one set of identical atomic planes that are equally spaced (Fig. 9.10).

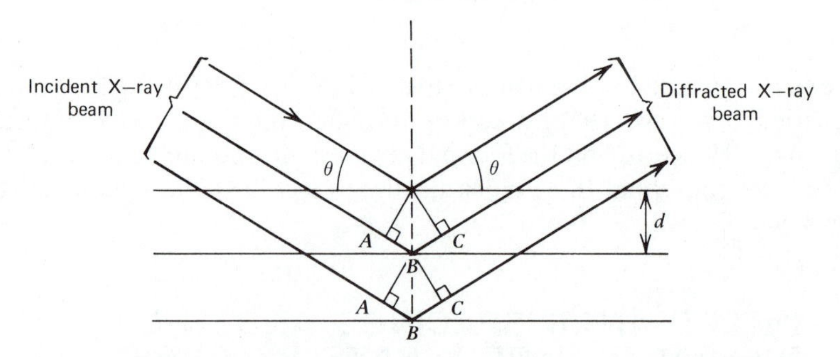

FIGURE 9.10 The geometrical conditions necessary for diffraction of monochromatic X rays from atomic planes. Atoms have been omitted for the sake of clarity. The path difference $\overline{AB} + \overline{BC}$ must contain an integral number of wavelengths if diffraction is to occur.

For an monochromatic X ray beam incident to a single crystal, diffraction occurs when the X ray path difference $\overline{AB} + \overline{BC}$, relative to neighboring planes, just equals one wavelength, λ. This corresponds to diffracted beams from any two neighboring planes, and, hence, from all equally spaced parallel planes, being in phase and, therefore, reinforcing each other. The incident and diffracted beams, making the diffraction angle 2θ, are always symmetrically oriented such that each makes an angle of θ with the diffracting planes. Within this constraint, the same atomic planes (Fig. 9.10) can diffract the X ray beam at other values of 2θ if the comparable X ray path differences equal an integral number of wavelengths, $n\lambda$.

This geometrical approach yields the Bragg law, which relates the X ray wavelength, λ, the interplanar spacing, d, and one-half of the diffraction angle, θ, as

$$n\lambda = 2d \sin \theta \tag{9.3}$$

If the incident beam makes an angle other than one that satisfies Eq. 9.3, no diffraction occurs from these planes.

In a cubic crystal, the interplanar spacing, d_{hkl}, is related to the Miller indices of the plane and to the lattice parameter by

$$d_{hkl} = \frac{a}{\sqrt{h^2 + k^2 + l^2}} \tag{9.4}$$

(For noncubic crystals, more complex relationships exist.) It is possible to illustrate a generality through consideration of a simple cubic crystal having one atom per lattice point. If we take the planes shown in Fig. 9.10 to be (100) atomic planes, then the smallest diffraction angle for these planes will correspond to $n = 1$, being given by $\sin \theta_1 = \lambda/(2d_{100})$. These same planes will also diffract when $n = 2$, or $\sin \theta_2 = 2\lambda/(2d_{100})$, which is actually the same as $\sin \theta_2 = \lambda/(2d_{200})$. This process, which can be continued such that $\sin \theta_n = n\lambda/(2d_{100}) = \lambda/(2d_{n00})$ where n is an integer, can be carried out for any diffracting plane in any crystal system. Accordingly, the Bragg law can be rewritten in a more general form, as

$$\lambda = 2d_{hkl} \sin \theta \tag{9.5}$$

where $d_{hkl} = d/n$. For cubic crystals, this becomes

$$\lambda = \frac{2a \sin \theta}{\sqrt{h^2 + k^2 + l^2}}$$

or

$$\sin^2\theta = \left(\frac{\lambda^2}{4a^2}\right)(h^2 + k^2 + l^2) \tag{9.6}$$

In a given experiment, the quantity $(\lambda^2/4a^2)$ is a constant, and discrete values of θ will correspond to discrete values of $(h^2 + k^2 + l^2)$. Thus, from measured values of θ, it is possible to determine the lattice parameter, a, providing λ and the Miller indices of the diffracting plane are known. Which (hkl) planes will actually diffract depends upon the particular Bravais lattice of the crystal in question and upon the atomic positions (Table 9.3).

TABLE 9.3 Diffracting Planes for Some Cubic Structure

Crystal Structure	*Miller Indices of Diffracting Planes*
Simple cubic	All values of $(h^2 + k^2 + l^2)$[a]
Body-centered cubic	$(h^2 + k^2 + l^2)$ = an even number
Face-centered cubic	h, k, l all even or all odd
Diamond cubic	$(h + k + l)$ = an odd number or an even multiple of two

[a] Note that certain numbers, such as 7, 15 and 23, are not possible.

Experimentally, it is possible to determine the diffraction angles of monochromatic X radiation either from a single crystal or from a polycrystalline or powder specimen. These data, when used in conjunction with the Bragg law and with information such as that given in Table 9.3, permit not only the calculation of lattice parameters, but also the determination of Bravais lattices, and additionally they provide information about atomic positions in a unit cell. Truly, X ray diffraction techniques provide powerful tools for the study of matter.

Up to this point, crystalline materials have been treated as if they were perfect. In reality almost all crystalline materials have microscopic defects that have a great influence on many of their properties. The types of deviation from crystalline perfection are discussed in the next chapter.

REFERENCES

1. Azároff, L. V., *Introduction to Solids*, McGraw-Hill, New York, 1960.
2. Buerger, M. J. *Elementary Crystallography*, Wiley, New York, 1956.

QUESTIONS AND PROBLEMS

9.1 (a) For a crystal having $a \neq b \neq c$ and $\alpha = \beta = \gamma = 90°$, what is the crystal system?
(b) For a crystal having $a = b \neq c$ and $\alpha \neq \beta \neq \gamma$, what is the crystal system? (*Hint.* Many of the inequalities shown in Table 9.1 mean that lattice parameters are not necessarily equal to, and generally different from each other.)
(c) Can you specify the Bravais lattices for parts (a) and (b)? Explain.

9.2 Draw a hcp structure unit cell and a bcc structure unit cell. Identify the atom sites and lattice points in each.

9.3 (a) Draw an fcc lattice unit cell. Connect the face-centered lattice points and two corner lattice points (at opposite ends of a cube body diagonal) and show that the result is a primitive cell.
(b) What is the Bravais lattice for the primitive cell obtained in part (a)?

9.4 (a) Draw an fcc structure unit cell and denote the unique coordinates of the four atoms that belong to the unit cell. Separately, list the coordinates of all atoms that are shown, and indicate which of these are redundant.

(b) Using the same unit cell, list the coordinates of all octahedral interstices and of all tetrahedral interstices.

9.5 (a) Calculate the size of the largest sphere, r, that can fit interstitially in a bcc array of contacting spheres of radius R. The largest interstitial sites have coordinates such as $0, \frac{1}{2}, \frac{1}{4}$ in a bcc unit cell.

(b) Repeat part (a) for an interstitial site located at $0, \frac{1}{2}, \frac{1}{2}$.

9.6 Sketch the following planes and directions in separate cubic unit cells: (123), [123]; (112), [112]; $(\bar{1}10)$, $[\bar{1}10]$. Show that all of these planes contain the $[11\bar{1}]$ direction. (Planes of a zone are all planes parallel to one line, called the zone axis.)

9.7 (a) Draw a (111) plane in a tetragonal unit cell.

(b) Draw a $(1\bar{1}0)$ plane in an orthorhombic unit cell.

(c) Draw a (100) plane in a monoclinic unit cell.

9.8 (a) What are the direction indices of the line of intersection of a (111) plane and a $(11\bar{1})$ plane in a cubic crystal? (Determine the answer by sketching the planes such that they intersect.)

(b) What are the direction indices of the lines normal to the planes given in part (a)?

9.9 (a) Determine the direction indices for a line extending from $\frac{1}{4}, \frac{1}{2}, 0$ to $\frac{3}{4}, \frac{1}{2}, \frac{1}{2}$ in a cubic material.

(b) Repeat part (a) for an orthorhombic material.

9.10 (a) Calculate the acute angle between the [100] and [111] directions in a cubic crystal.

(b) Repeat part (a) for the [100] and [110] directions.

9.11 The interplanar spacing, d_{hkl}, of (hkl) planes for a crystal having orthogonal axes is given by:

$$\frac{1}{d_{hkl}^2} = \frac{h^2}{a^2} + \frac{k^2}{b^2} + \frac{l^2}{c^2}$$

where a, b, and c are the lattice parameters.

(a) What does this equation reduce to for a cubic crystal?

(b) Calculate the spacing between close-packed planes in terms of the lattice parameter for an fcc metal. How does this compare with the nearest neighbor interatomic spacing?

9.12 Planar density of packing is defined as the fraction of total area occupied by contacting spheres whose centers lie in the plane.

(a) Sketch contacting circles so as to represent the (111) and (100) planes for an fcc array of spheres and the (110) plane for a bcc array of spheres.

(b) Calculate the planar density in each case.

(c) Show the close-packed directions contained in each plane and specify their direction indices.

9.13 (a) Give the Miller indices for the specific planes in the {120} family for an orthorhombic crystal.

(b) Repeat part (a) for a cubic crystal.

9.14 (a) Give the direction indices for the specific directions in the $\langle 011 \rangle$ family for a tetragonal crystal.

(b) Repeat part (a) for a cubic crystal.

9.15 Given that the lattice parameter of Ni (fcc) is $a = 0.3524$ nm, calculate the values of θ corresponding to the first six diffraction peaks. For the X ray wavelength, use $\lambda = 0.154$ nm.

9.16 Diffraction peaks were determined at the following values of θ for a powdered sample of a cubic metal bombarded by X rays with $\lambda = 0.154$ nm: 20°, 29°, 36.5°, 43.4°, 50.2°, 57.35°, and 65.55°.

(a) Determine the crystal structure and calculate the lattice parameter.

(b) Identify the metal by referring to tabulated data for the elements, as may be found in the *Handbook of Chemistry and Physics*.

10

Structural Imperfections

The local irregularities in the structures of crystals that are responsible for certain aspects of chemical and physical behavior are discussed in this chapter.

10.1 TYPES OF IMPERFECTIONS

In the preceding chapters on crystal geometry and structure it was implied that the regular crystalline order described by a unit cell exists throughout an entire crystal. In fact, numerous slight deviations from the ideal atomic array occur in every crystalline material. Two of these deviations, grain boundaries and impurity atoms, have been mentioned previously. In this chapter, we shall consider the structural defects that have one or more of their dimensions on the order of size of an atomic diameter. Such imperfections are classified on the basis of geometry as either point, line, or surface defects, depending upon whether their extent can be described best in terms of zero, one, or two dimensions. Each of these types of imperfections has an important bearing on the properties and behavior of crystalline materials.

10.2 IMPURITY ATOMS

What we normally consider as pure materials actually contain some dissolved impurities. In many cases, solid solutions are prepared intentionally so as to alter certain properties of a material. This effectively represents the controlled use of

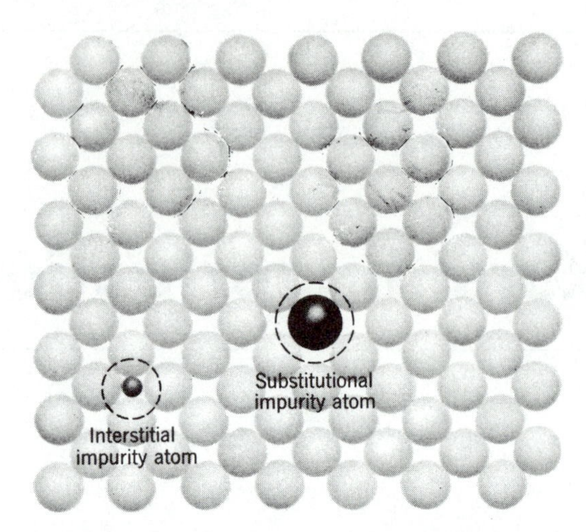

FIGURE 10.1 Two-dimensional representation of a simple crystalline solid containing interstitial and substitutional impurity atoms. (Adapted from *The Structure and Properties of Materials*, Vol. I, by W. G. Moffatt, G. W. Pearsall, and J. Wulff, Wiley, New York, 1964.)

potential impurities and, consequently, the concepts discussed below need only be extended and further developed for intentional solid solutions. Solute atoms, whether they are impurities or desired additions, lead to local structural disturbances in the host crystalline solid. Solute atoms are classified as point defects, but are of sufficient importance to be discussed separately from other point defects.

Because of bonding characteristics, solid metals usually can act as solvents and dissolve at least a limited amount of most other elements. As Fig. 10.1 shows, impurity atoms may occupy either solvent atom positions or small interstitial sites between the solvent atoms. In the former situation, an impurity atom substitutes for a host atom, and the result is called a substitutional solid solution. In the latter situation, an impurity atom dissolves intersititally, giving rise to an interstitial solid solution. Since atoms in metals are densely packed, the spaces available for impurity atoms in interstitial sites are limited, and only small atoms, such as hydrogen, carbon, nitrogen, and oxygen, dissolve interstitially in metals. Even for these atoms there is considerable distortion of the crystal structure near the interstitial position. This and chemical differences often lead to restricted interstitial solubilities. A striking exception is hcp Ti, which can dissolve up to 33 atomic percent oxygen. In transition metals, even very low interstitial contents can have a profound effect on mechanical and electrical properties.

Geometrical and chemical limitations also determine the extent to which substitutional impurities may dissolve in a metal and in ionic and covalent crystalline solids as well. If solute and solvent atoms have similar atomic sizes, then larger solubilities are favored. For extensive solid solubility, it is also necessary that the chemical natures of the solvent and solute be similar. With some metals (e.g., Cu–Ni) complete substitutional solubility occurs. Another example of this is provided by Nb–Ta alloys. Niobium and tantalum have nearly identical atomic atomic sizes and nearly identical chemical behavior and, therefore, form very nearly *ideal* solid solutions. Since niobium and tantalum ores are found together in nature, it is extremely difficult to prepare one of these pure metals without having some atoms of the other present as an impurity. In most other metallic substitutional solid solutions, differences in solute and solvent atom sizes cause local distortions of the structure that can be related to a strain energy. Furthermore, interactions between solute and solvent atoms usually lead to a preference for like or unlike nearest neighbors, such that in the first instance clustering of like atoms occurs and in the second case local ordering of unlike atoms takes place.

Substitutional impurities in covalent crystals often give rise to concurrent electronic defects that markedly alter the electrical conductivity of the material. In germanium, for example, the presence of arsenic with five valence electrons leaves one extra electron per As atom, since only four are required for bonding in the diamond cubic structure. If gallium is present rather than As, a deficiency of bonding electrons results in a *hole* in one bonding orbital. These situations are represented schematically in Fig. 10.2. The electrical conductivity in both *doped* Ge crystals

```
   ••    ••    ••    ••                ••    ••    ••    ••
: Ge : Ge : Ge : Ge :              : Ge : Ge : Ge : Ge :
   ••    ••    ••    ••                ••    ••    ••    ••
          •
: Ge : As : Ge : Ge :              : Ge : Ge : Ge : Ge :
   ••    ••    ••    ••                ••    ••    •o    ••
: Ge : Ge : Ge : Ge :              : Ge : Ge : Ga : Ge :
   ••    ••    ••    ••                ••    ••    ••    ••
: Ge : Ge : Ge·:: Ge :             : Ge : Ge : Ge : Ge :
   ••    ••    ••    ••                ••    ••    ••    ••
            (a)                                 (b)
```

FIGURE 10.2 Schematic representation of the effect of substitutional impurities in germanium, a covalent solid whose atoms have a valence of four. (*a*) The presence of an arsenic impurity atom, with a valence of five, leaves an extra electron that is free to move in the crystal. (*b*) The presence of a gallium impurity atom, with a valence of three, leaves one bond unsaturated, or a hole in the bond structure that can move about in the crystal.

is increased by the presence of the impurity atoms. With As as the impurity, each extra electron is not restricted to a discrete bonding orbital and, therefore, can move almost freely under the influence of an applied electric field, that is, it behaves very nearly like an electron in the conduction band (see Sect. 4.6). Likewise, when Ga is the impurity, the resulting *hole* can jump from one discrete bonding orbital to another, if an electric field is applied, and a net flow of electrons occurs in the opposite direction.

Substitutional impurities in ionic crystals also are permissible, providing electrical neutrality is maintained. Thus, similar anions or cations may substitute for host anions or cations. Usually, it is found that impurity cations are more common. When the impurity cation has a charge that differs from the host's, substitution necessitates the creation of another defect. For example, when Ca^{2+} is substituted for Na^+ in a NaCl crystal, charge neutrality can be preserved if a neighboring Na^+ cation site is vacant. In other words, the Ca^{2+} cation effectively substitutes for two Na^+ cations but can occupy only one of the sites. Thus, this impurity creates another defect, termed a cation vacancy, which affects the electrical properties of the ionic material. Since the cation vacancy can exchange positions repeatedly with neighboring cations when an electric field is applied, an electric current can be conducted. In this case, of course, conduction corresponds to the motion of positively charged cations.

The implications of these impurity defects with respect to electrical properties, particularly for covalent and ionic solids, will be explored in subsequent chapters. At this point, the vacant cation site mentioned above deserves further attention, for this type of defect may occur when impurities are absent.

10.3 POINT DEFECTS

Independent of impurity atoms, the two most common structural point defects in metallic or covalent solids are illustrated in Fig. 10.3. A vacancy is an atomic position that is not occupied in the crystal, and a self-interstitial is an atom from the crystal that occupies an interstitial site. Self-interstitials are much less common than vacancies because too great a structural distortion is necessary in metals and an unbonded atom in a covalent crystal is highly unfavorable. Calculations and experiments show that the fraction of vacant sites, X_V, is given by

$$X_V \simeq e^{-(\Delta H_v/RT)} \tag{10.1}$$

where R is the universal gas constant, T is absolute temperature, and ΔH_v is an enthalpy which, when divided by Avogadro's number, represents the increase in the

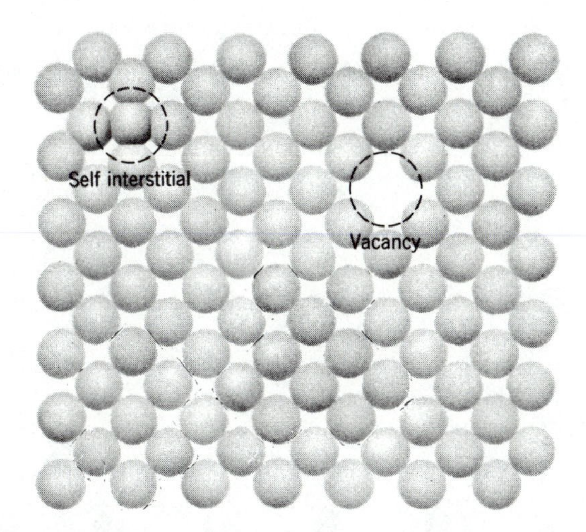

FIGURE 10.3 Two-dimensional representation of a simple crystalline solid, illustrating a vacancy and a self-interstitial. (Adapted from *The Structure and Properties of Materials,* Vol. I, by W. G. Moffatt, G. W. Pearsall, and J. Wulff, Wiley, New York, 1964.)

energy of the crystal because of one vacancy. As temperature is increased, X_V rises very rapidly, but even near the melting point of a metal the vacancy concentration only reaches a value on the order of 10^{-4}. A very important aspect of vacancies is that they can move around by exchanging positions with neighboring atoms. This provides a means by which atoms can migrate or diffuse in the solid state at elevated temperatures.

Equation 10.1 actually gives the approximate equilibrium vacancy concentration that may be exceeded if vacancies are generated by other than thermal means. For ionic solids, equations of the same form give the temperature dependence of the equilibrium point defect concentration. These point defects are slightly more complex, however, since electrical charge balance must exist locally. For example, a cation vacancy may require a nearby anion vacancy as shown in Fig. 10.4. Such a cation-anion vacancy pair, called a Schottky defect, is common in many ionic crystals. Alternatively, another type of point defect involves an interstitial cation in the vicinity of a cation vacancy (Fig. 10.4). This is called a Frenkel defect. Both of these defects contribute to the conductivity of ionic solids. In Ag I, for example, the Ag^+ ion is small enough to move into interstitial holes in the I^- ion lattice. On the other hand, Schottky defects are found in NaCl because the Na^+ and Cl^- ions are too large to be contained interstitially.

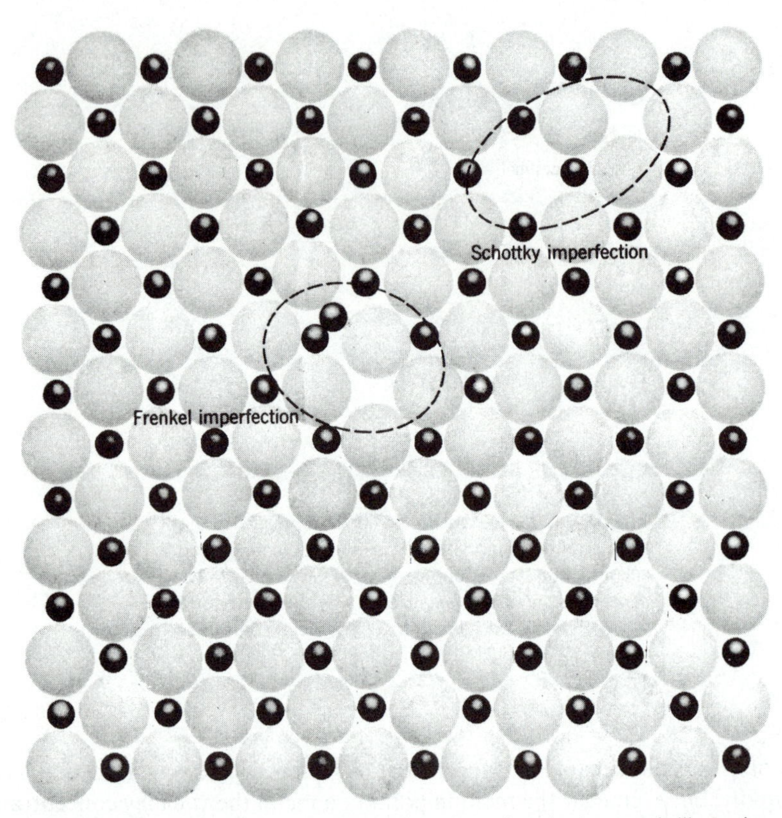

FIGURE 10.4 Two-dimensional representation of an ionic crystal, illustrating a Schottky defect and a Frenkel defect. (Adapted from *The Structure and Properties of Materials*, Vol. I, by W. G. Moffatt, G. W. Pearsall, and J. Wulff, Wiley, New York, 1964.)

10.4 LINE DEFECTS: THE EDGE DISLOCATION

The concept of a line defect in a solid was first postulated in 1934 in an attempt to explain the ease with which metals deform plastically (i.e., permanently). Prior theoretical calculations had indicated that the process of plastic deformation in a perfect crystal (see Fig. 10.5) should require very high shear forces. The experimentally observed shear forces leading to permanent deformation of pure single crystals were orders of magnitude lower than those predicted. Hence, a line defect called an edge dislocation was postulated to explain this discrepancy. An edge dislocation, viewed end on as shown in Fig. 10.6, corresponds to the edge of an atomic plane that terminates within the crystal rather than passing all the way

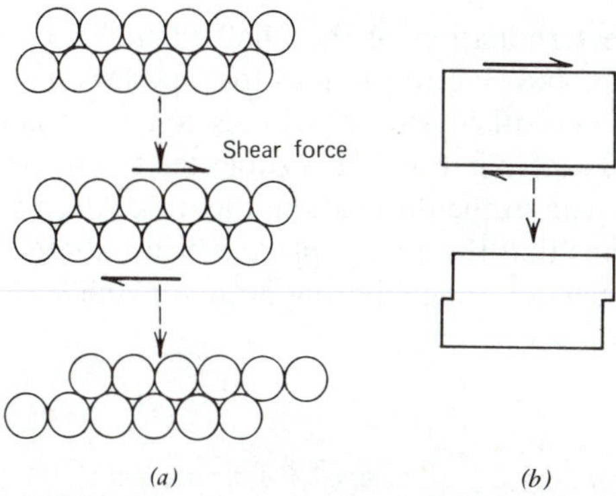

FIGURE 10.5 The process of permanent deformation in a perfect crystal. (*a*) An applied shear force causes one atomic plane in a crystal to be displaced by one interatomic distance with respect to a neighboring atomic plane. (*b*) On a macroscopic scale, the resulting permanent deformation involves the displacement of one portion of the crystal relative to the other. The shear force necessary to produce permanent deformation in this manner is much greater than that required for the permanent deformation of most real materials.

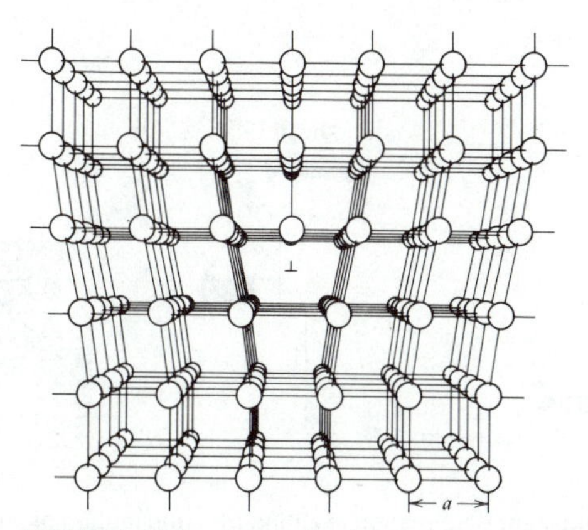

FIGURE 10.6 An end view of an edge dislocation in a simple cubic structure. The dislocation line extends normal to the figure and along the edge of a plane that terminates within the crystal. An edge dislocation is denoted by ⊥. (Adapted from A. G. Guy, *Elements of Physical Metallurgy*, Second Edition, 1959, Addison-Wesley, Reading, Mass.)

through. Since the termination of the "half plane" shown is a line extending normal to the figure and since deviations from perfect crystallinity occur only in the vicinity of this line, an edge dislocation is conventionally called a line defect.

When the crystal shown in Fig. 10.6 is subjected to sufficiently large shear forces directed perpendicular to the dislocation, the dislocation will move as shown in Fig. 10.7. This means that the atomic plane which initially terminated within the crystal joins with part of a neighboring atomic plane such that another atomic

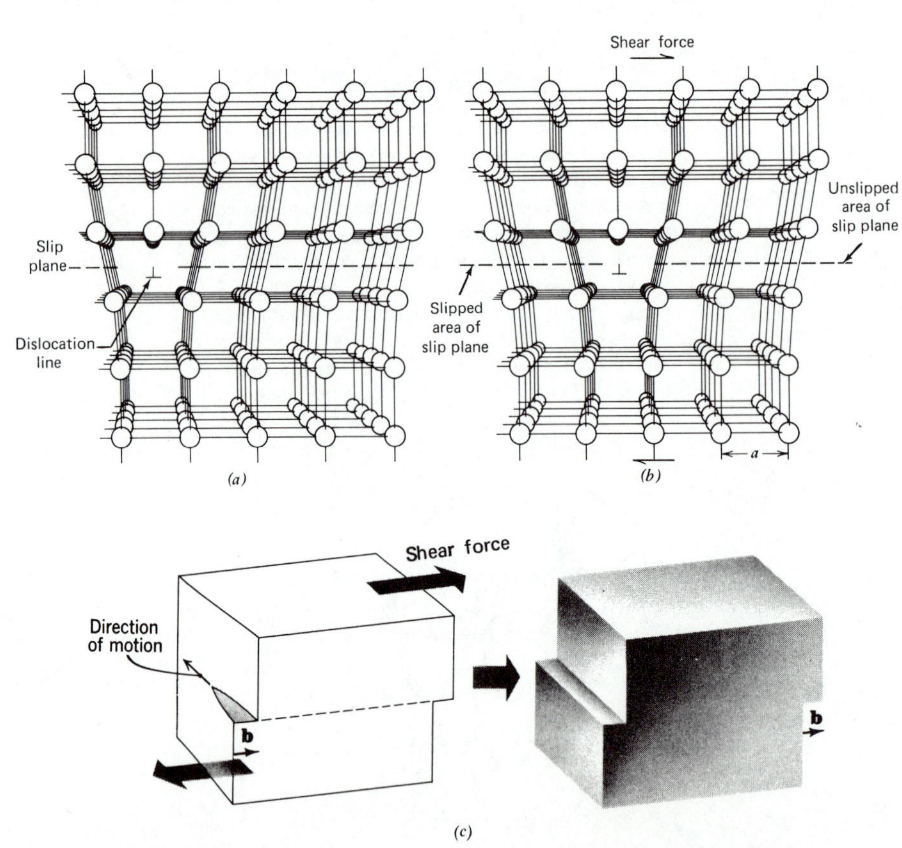

FIGURE 10.7 Permanent deformation resulting from the motion of an edge dislocation. Shown are the atomic arrangements in the vicinity of an edge dislocation (*a*) before and (*b*) after an applied shear force has caused it to move by one interatomic distance. (Adapted from A. G. Guy, *Elements of Physical Metallurgy,* Second Edition, 1959, Addison-Wesley, Reading, Mass.) If the dislocation continues to move to the right, (*c*) permanent offset of the top half of the crystal with respect to the bottom half is achieved. The shear forces necessary to produce dislocation motion are much lower than those required for the process illustrated in Fig. 10.5. (Adapted from *The Structure and Properties of Materials,* Vol. III, by H. W. Hayden, W. G. Moffatt, and J. Wulff, Wiley, New York, 1965.)

plane now terminates within the crystal. Because of the applied shear force, the edge dislocation has moved, and through repetition of this process it can continue to move or slip. If the dislocation moves through an entire crystal (Fig. 10.7*c*), then the upper half of the crystal will be shifted by one slip vector or interatomic spacing relative to the lower half. It is through the motion of dislocations that plastic deformation occurs, and this requires much smaller forces than those necessary to permanently deform a perfect crystal. Although much indirect experimental evidence indicated the existence of dislocations, it was not until 1956 that they were directly observed with the aid of the electron microscope.

Geometrically, an edge dislocation can be characterized by the direction of its line relative to its Burgers vector, which is the closure gap that ensues when the dislocation line is encircled, as shown in Fig. 10.8. The circuit followed would give closure in a perfect region of the crystal. The resulting Burgers vector, **b**, is perpendicular to the line of the edge dislocation, and this may be taken as the general definition of an edge dislocation. For an edge dislocation the Burgers vector, which is identical to the slip vector mentioned above, and the dislocation line define the plane in which the dislocation can slip. The slip plane is usually a densely packed atomic plane.

In metals and alloys, the Burgers vector always points in a close-packed direction and has a magnitude equal to the nearest neighbor interatomic spacing. Accordingly, for the simple cubic structure used in Figs. 10.6 and 10.8, the Burgers vector in crystallographic notation is given by $\mathbf{b} = a\langle 100 \rangle$; in bcc metals, $\mathbf{b} = (a/2)\langle 111 \rangle$, and in fcc metals, $\mathbf{b} = (a/2)\langle 110 \rangle$. (In the cubic system this notation,

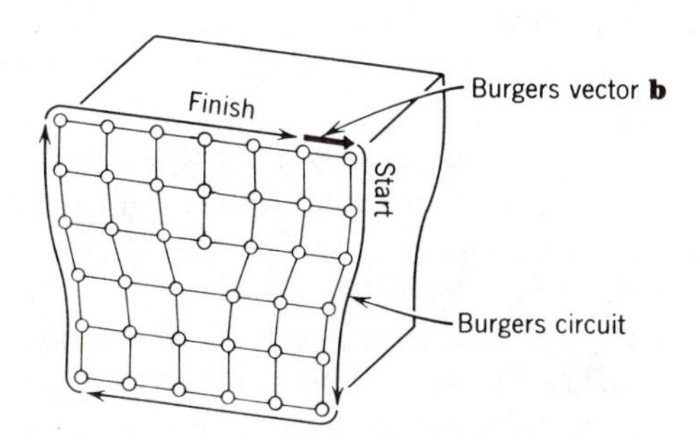

FIGURE 10.8 Burgers circuit around an edge dislocation in a simple cubic structure. The circuit followed would result in closure if made in a perfect region of the crystal or in a region containing a point defect; however, when this circuit is made around a dislocation, it must be completed by a vector, **b**, which is the Burgers vector of the dislocation.

e.g., $(a/2)[110]$, represents a vector pointing in the $[110]$ direction and having a magnitude equal to $(a/2)\sqrt{1^2 + 1^2 + 0^2} = a\sqrt{2}/2$.) This general rule for Burgers vectors is a consequence of the self-energy of a dislocation. Since the crystal structure is severely distorted in the vicinity of a dislocation line, the energy state of the crystal is higher than that of a perfect crystal. The increase in energy resulting from a dislocation (i.e., the dislocation self-energy) is directly proportional to the length of dislocation line and to the square of the Burgers vector magnitude. Thus, the energy is minimized when the Burgers vector extends from one atom to the next in a close-packed direction.

Dislocations occur in all crystalline materials; however, their energies in nonmetallic solids usually are significantly higher than in metals. This relates to the greater disturbance of bonding caused by dislocations in covalent and ionic solids. Furthermore, nonmetallic solids and metal compounds often have larger Burgers vectors than do metals. These factors and other geometrical limitations restrict the amount of dislocation motion in nonmetallic solids and in many metallic compounds such that they are capable of very little plastic deformation prior to fracture at room temperature.

10.5 INTERACTIONS OF EDGE DISLOCATIONS

Edge dislocations interact with various point defects as well as with each other. Near a solitary edge dislocation, the atoms are pushed together on one side of the slip plane (above the dislocation shown in Fig. 10.6) and are pulled apart on the other side (below the dislocation in Fig. 10.6). These regions of the crystal are in *compression* (C) and in *tension* (T), respectively, relative to a perfect region. Thus, substitutional solute atoms having larger diameters than the solvent atoms will tend to segregate to the tension side of an edge dislocation, where the space available for an atom is larger. This results in a decrease of total strain energy of the solid solution. Conversely, smaller substitutional solute atoms tend to segregate to the compression side of an edge dislocation. Interstitial atoms have an even greater tendency to segregate to the tension side of an edge dislocation, because the interstitial sites are considerably larger there than in a region of perfect crystal. These interactions suggest that impurity atoms will be weakly *bound* to a dislocation and that it will be more difficult to move a dislocation in an impure material than in a pure material. Thus, greater shear forces are necessary to plastically deform an impure material.

As an example of interactions between edge dislocations consider Fig. 10.9*a*, which shows two parallel dislocations with the same Burgers vectors. These dislocations will tend to move apart spontaneously because they exert a force on each other. The force arises from the interaction of atomic distortions and displacements

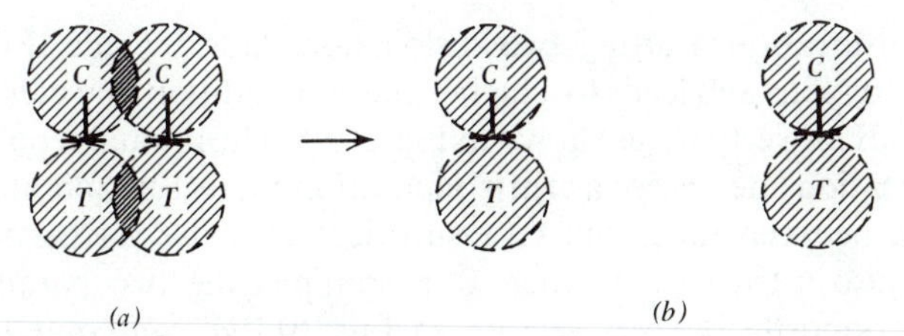

FIGURE 10.9 Two parallel edge dislocations that have identical Burgers vectors and that are on the same slip plane. These dislocations exert a repulsive force on each other and will move from (*a*) their initial positions to (*b*) more widely separated positions because of the interaction of crystal structure distortions about each dislocation. (*C* denotes a region of compression and *T* denotes a region of tension.)

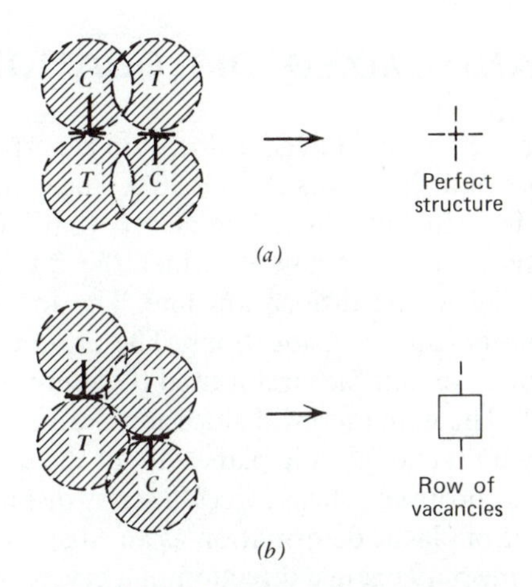

FIGURE 10.10 Interactions between parallel edge dislocations that have opposite Burgers vectors. (*a*) If the dislocations are on the same slip plane, they can come together and annihilate each other. (*b*) If they are on neighboring slip planes, a row of vacancies is formed when they come together. (Regions of compression and tension are indicated by *C* and *T*, respectively.)

around each dislocation. Putting two dislocations side by side (Fig. 10.9*a*) intensifies the distortions and leads to a higher energy condition that can be alleviated if the dislocations move apart, as shown in Fig. 10.9*b*. The mutual repulsion of these two edge dislocations decreases as their separation increases, and at some separation the force between them will be insufficient to cause further motion. Dislocations can also attract each other. For example, the two parallel edge dislocations with opposite Burgers vectors, in Fig. 10.10*a*, can come together and annihilate each other, thus producing a region of perfect crystal. Alternatively, if these dislocations are on neighboring slip planes, they will still annihilate each other, but a row of vacancies will result (Fig. 10.10*b*).

Many other interactions between dislocations are possible, including those involving nonparallel edge dislocations and those involving other types of dislocations. In general, the motion of any dislocation is impeded by the presence of other dislocations, and the forces necessary to deform a metal by slip increase as the number of dislocations increases.

10.6 SCREW AND MIXED DISLOCATIONS

Although the Burgers vector of an edge dislocation is perpendicular to the dislocation line, other types of dislocations are possible. The concept of a dislocation can be generalized so as to include those line defects that have a Burgers vector parallel to the dislocation line and those for which the Burgers vector is neither perpendicular nor parallel to the dislocation line. The former are called screw dislocations, and the latter, mixed dislocations. The atomic arrangement in the vicinity of a screw dislocation and the circuit used to determine its Burgers vector are shown in Fig. 10.11. The name screw dislocation derives from the continuous helical ramp formed by one set of atomic planes about the dislocation line. When a crystal is subjected to appropriate shear forces, a screw dislocation can move and give rise to the same type of plastic deformation as an edge dislocation (Fig. 10.12).

A single dislocation line may change direction in a crystal such that one part is pure edge, another part is mixed, and yet another part is pure screw (Fig. 10.13) since the Burgers vector must be the same at any point along a dislocation line. Sufficiently large shear forces, applied on planes parallel to the slip plane and in the direction of the Burgers vector, will cause the dislocation shown in Fig. 10.13 to move perpendicular to its line. The resulting slip between the top and bottom portions of the crystal will be the same (i.e., equal to the slip vector, which is identical to the Burgers vector) regardless of which part of the dislocation line sweeps by.

It is important to realize that, in a metal, the Burgers vector for any type of dislocation points in a close-packed direction and has a magnitude equal to the interatomic spacing. This is a consequence of energy considerations, since the

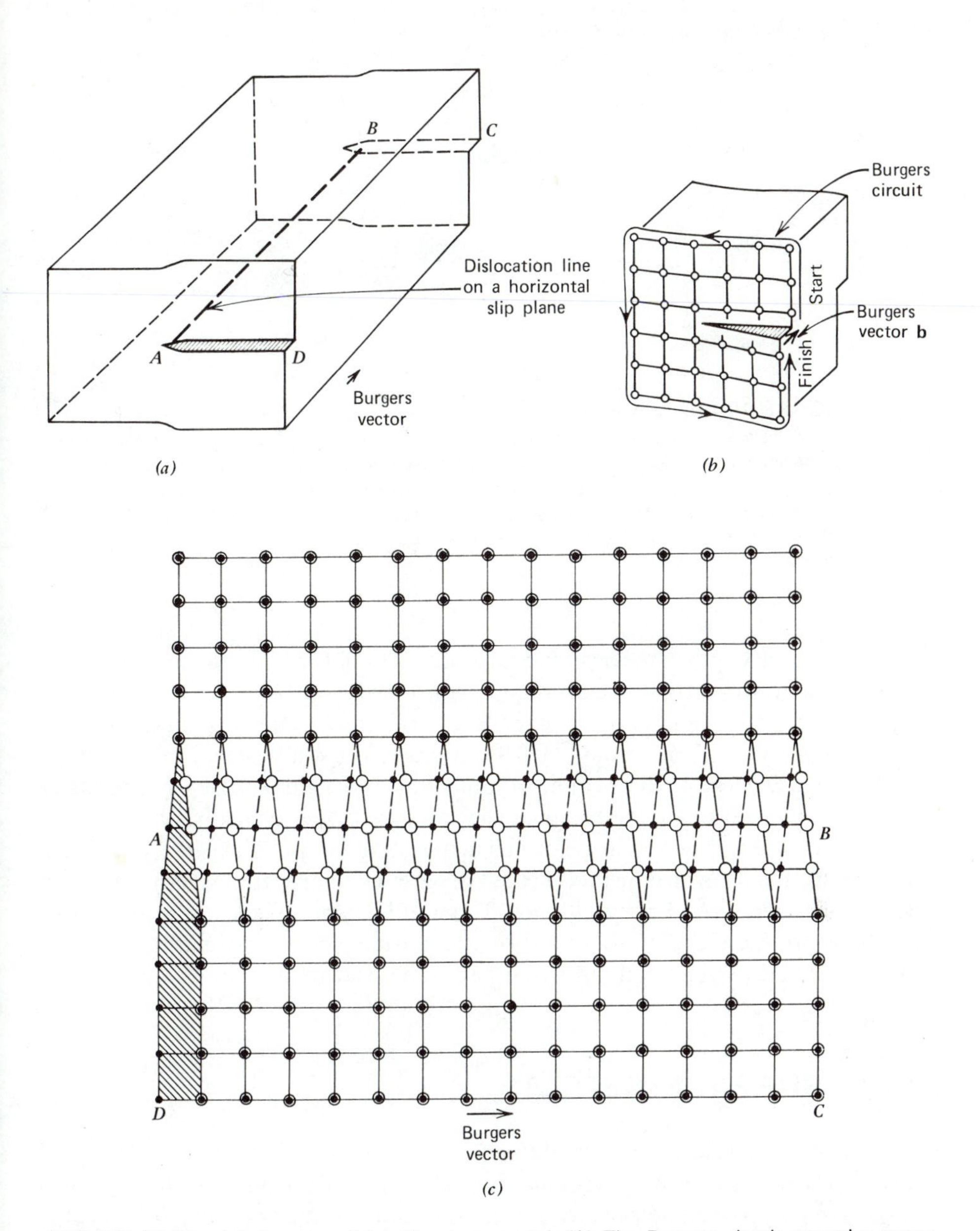

FIGURE 10.11 (*a*) A screw dislocation in a crystal. (*b*) The Burgers circuit around a screw dislocation in a simple cubic structure. The Burgers vector, **b**, is parallel to the dislocation line. (*c*) Plan view showing the atomic positions above (○) and below (●) the slip plane of a screw dislocation. The dislocation line extends from *A* to *B*.

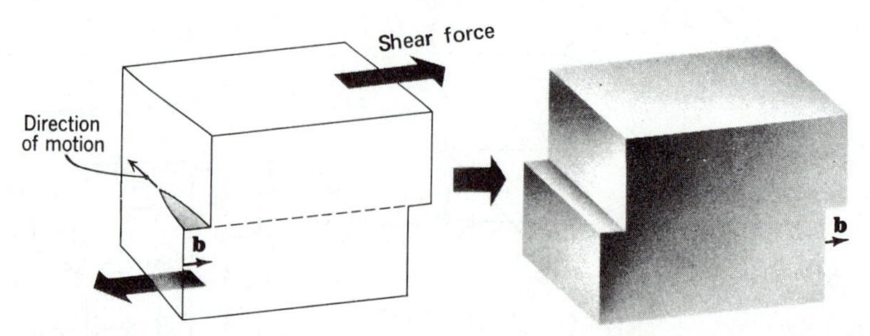

FIGURE 10.12 Permanent deformation achieved through the motion of a screw dislocation. Application of a shear force parallel to the Burgers vector and to the slip plane causes the screw dislocation to move in a direction perpendicular to its line, but the final offset of the crystal is the same as that shown in Fig. 10.7*c*. (Adapted from *The Structure and Properties of Materials,* Vol. III, by H. W. Hayden, W. G. Moffatt, and J. Wulff, Wiley, New York, 1965.)

energy per unit length of dislocation line is proportional to b^2. In the case of a screw dislocation, however, the proportionality constant is somewhat smaller than that for an edge dislocation, and a greater length of screw dislocation line usually is observed in crystals. The amount of dislocations in a crystal is measured in terms of the length of dislocation line per unit volume, which is called the dislocation density. In as-grown metal single crystals the dislocation density is typically 10^5 to 10^7 cm of line/cm^3, whereas in heavily deformed metals it can be as high as 10^{11} to 10^{12} cm of line/cm^3. If a deformed metal is heated to elevated temperatures, the dislocation density will decrease to about 10^7 to 10^8 cm of line/cm^3. Thus, not only are dislocations produced during solidification and during plastic deformation, but they may be partially eliminated during an annealing heat treatment.

10.7 SURFACE DEFECTS

During crystallization of a solid from vapor, solution, or melt, a single crystal forms only under certain conditions such as when a single seed is present. Most solids are polycrystalline because nucleation occurs at many points. In the solidification of an ingot, for example, many grains are formed. For a slowly solidified casting, the grains can sometimes be seen with the unaided eye (Fig. 10.14). If the solidification rate is rapid, the grain size will be much smaller, and the grains can be seen only with the aid of an optical microscope.

A polycrystalline solid has essentially the same density as a single crystal. Each individual grain is a single crystal, but neighboring grains have different, unrelated

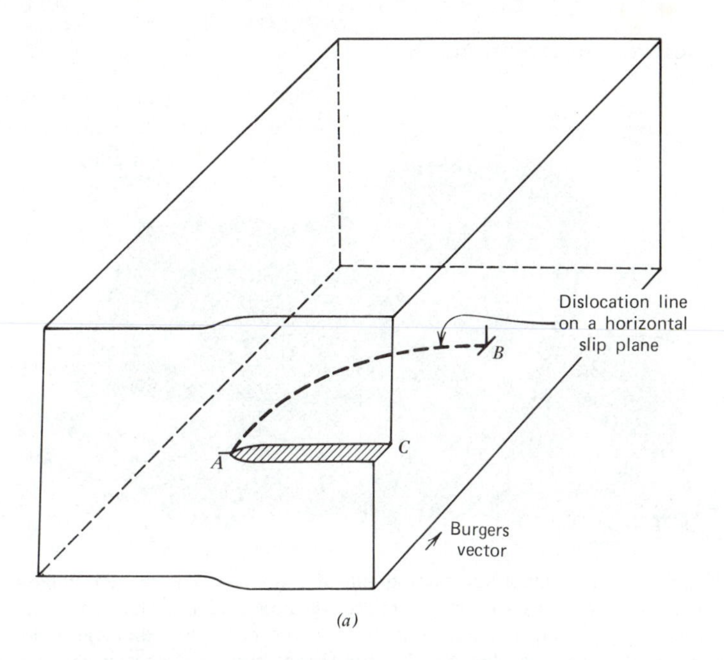

(a)

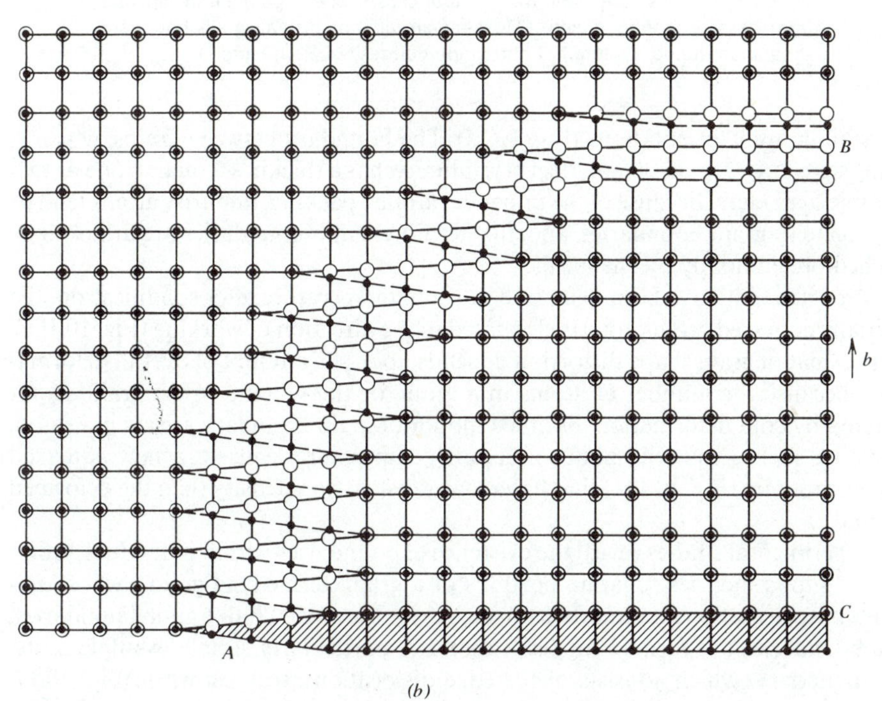

(b)

FIGURE 10.13 (*a*) Schematic representation of a curved dislocation line in a crystal. The Burgers vector is the same at any point along this dislocation line. (*b*) Plan view showing the atomic positions above (○) and below (●) the slip plane for a curved dislocation in a simple cubic structure.

FIGURE 10.14 Actual-size photograph of a polished and etched molybdenum ingot that has very large grains. The upper edge is the trace of a fracture surface, and it is apparent that fracture occurred along the grain boundaries in this case. Note that fracture caused one large grain to separate from its neighboring grains. (Most metals do not display grain boundary separation during fracture.) (Photograph courtesy of R. J. Farrell.)

crystallographic orientations (Fig. 10.15). The boundary between grains, which is a higher energy region of imperfect crystallinity, has a thickness equal to one or two atomic diameters. Because of the imperfect atomic packing, impurity atoms tend to segregate to grain boundaries, and, furthermore, grain boundaries are attacked or etched preferentially by chemicals.

In metals that have been deformed at room temperature after solidification, the initially equiaxed grains appear elongated in the direction of working (Fig. 10.16*a*). The extent of grain shape distortion depends upon the amount of overall deformation because the number of atoms in a grain, or the volume of the grain, is not altered by cold deformation, but the amount of grain boundary area is increased. Heating such a material above a certain temperature leads to a new equiaxed grain structure (Fig. 10.16*b*) that has a lower dislocation density than the deformed metal.

Experimental studies reveal the existence of boundaries within individual grains. These subboundaries separate portions of a grain, termed subgrains, whose relative misorientation amounts to a few degrees at the most. Unlike grain boundaries, subboundaries are formed by dislocations. A particularly simple example is the tilt boundary, which consists of the edge dislocation array shown in Fig. 10.17. Other types of low-angle boundaries may be produced by other dislocation arrays.

In many materials, incorrect stacking of atomic planes may occur during crystal growth. Consider an fcc metal, in which the normal stacking sequence of close-

FIGURE 10.15 Schematic representation of the orientations of individual grains in a single-phase polycrystalline aggregate. Within one grain, a given set of atomic planes has the same orientation in space. At a grain boundary, the orientation changes abruptly, and these atomic planes are not continuous from one grain to the next.

packed planes is $ABCABC\cdots$. If, when the crystal is growing, a B layer inadvertently forms on top of a C layer and then the reverse stacking sequence proceeds, the resulting configuration will be $\cdots ABCAB\boxed{C}BACBA\cdots$. Thus, the crystal to one side of the plane denoted as $\boxed{C}$ will be a mirror image or twin of the other side. The $\boxed{C}$ plane that delineates the twin boundary is called a twin plane. Twins and twin boundaries also can be generated by deformation, especially in some hcp metals and in bcc metals at low temperatures. Twin boundaries are recognizable in the microstructures of some metals (Fig. 10.18).

Closely related to a twinning fault is a stacking fault, which is another irregularity in the stacking of atomic planes. As distinct from a twinning fault where the stacking sequence is reversed on either side of the twin plane, the stacking sequence is the same on either side of a stacking fault. This defect commonly occurs in nonmetals that have layer structures, such as graphite, mica, and talc, as well as in many metals. In hcp cobalt, for example, the ideal $ABAB$ sequence of close-packed planes has faults every 10 to 14 layers. Two distinct stacking faults occur in some fcc metals:

$$\cdots ABCAB|ABCBC\cdots \qquad \text{and} \qquad \cdots ABCA|C|BCABC\cdots$$

The former, called an intrinsic stacking fault, is equivalent to one close-packed plane being missing from the stacking sequence, and the fault may be viewed as a

(b)
100 μm

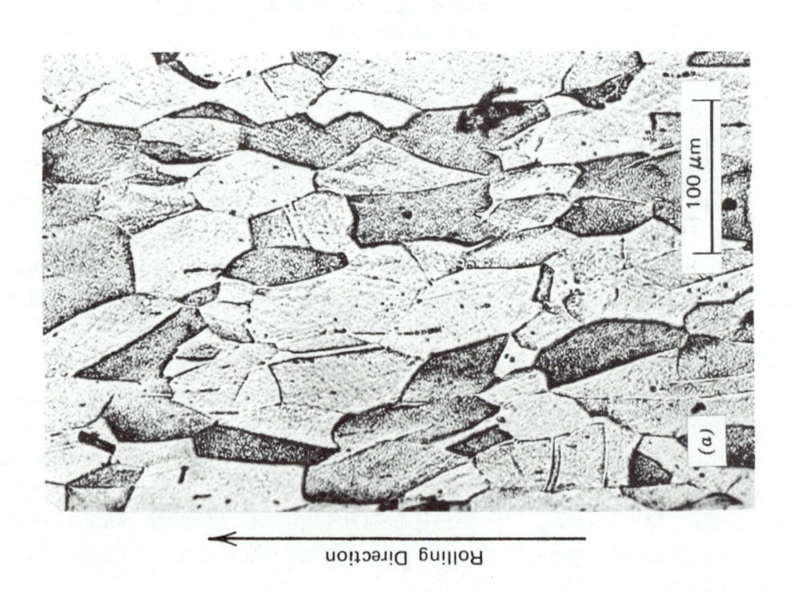
(a)
100 μm
Rolling Direction

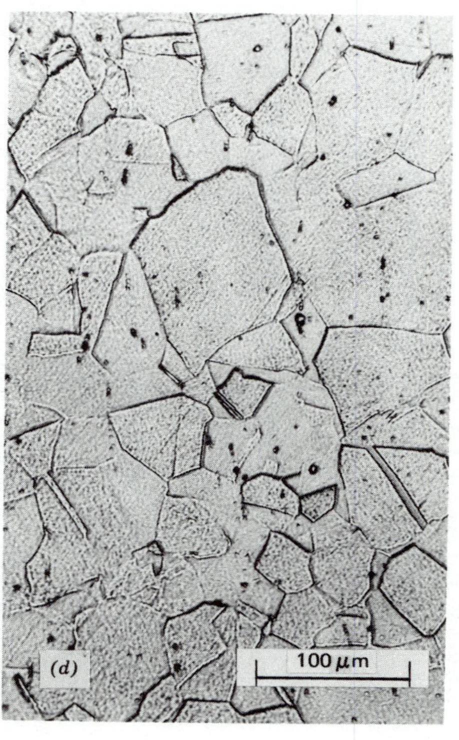

FIGURE 10.16 Recrystallization of a cold-worked metal. (*a*) Following 50% reduction by cold rolling, the grains are elongated in the rolling direction. (*b*) Little change in the microstructure is discernible after a heat treatment for 15 min at 600°C, but recrystallization has just begun. (*c*) A completely recrystallized structure consisting of equiaxed grains has formed after one hour at 600°C. (*d*) Continued annealing for two hours causes grain growth. (Annealing twins, evident in *c* and *d*, should not be confused with grains; see the text and Fig. 10.18.)

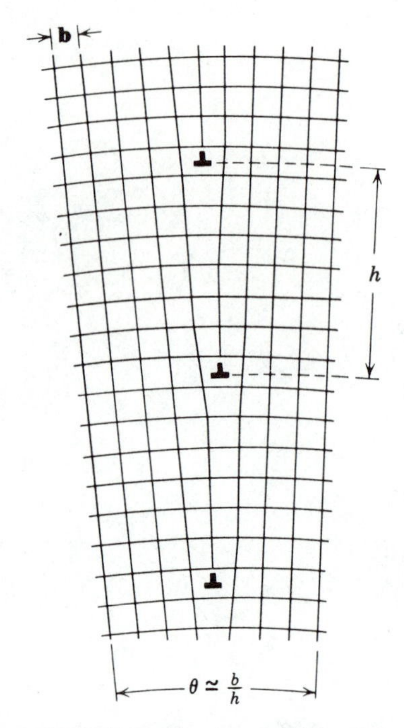

FIGURE 10.17 An example of a low-angle boundary in a simple cubic structure. This simple tilt boundary consists of an array of equally spaced edge dislocations. The angle of misorientation, θ, across the subboundary is given by $\theta \simeq b/h$, where b is the Burgers vector magnitude and h is the vertical separation between edge dislocations.

very thin layer of hcp stacking (i.e., *ABAB*) in the fcc crystal. The latter, called an extrinsic stacking fault, is equivalent to one additional close-packed plane being inserted into the stacking sequence, and the fault may be viewed as a very thin twin (e.g., $|C|BC$) in the fcc crystal. Intrinsic stacking faults are observed most frequently as thin ribbons of faulted material (Fig. 10.19). Since they do not reach across an entire grain, in general, they are necessarily bounded by *partial dislocations*. The width of these stacking faults in an fcc metal is determined by their energy, which is a measure of how much greater the energy is for hcp packing as compared to fcc packing.

FIGURE 10.18 The microstructure of annealed cartridge brass (70% Cu–30% Zn). Within many grains, twins (the regions with parallel sides) and twin boundaries are apparent. The contrast between twinned regions of an individual grain is a result of differing attack by the chemical etchant acting on different orientations.

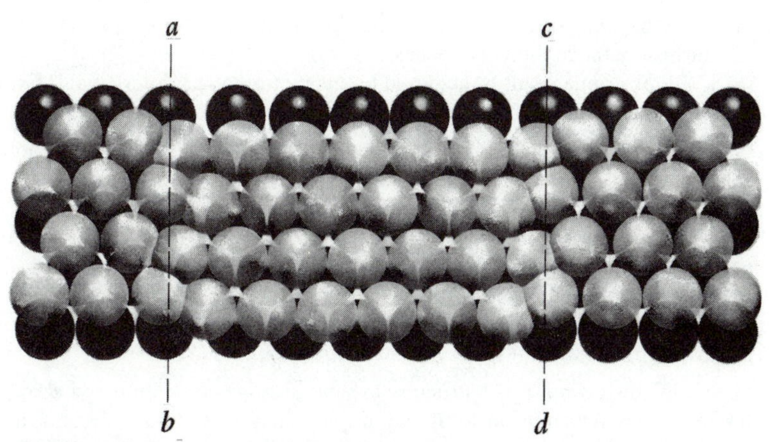

FIGURE 10.19 Plan view of a stacking fault in an fcc metal. To the left of *ab* and to the right of *cd*, the close-packed plane stacking sequence in and out of the plane of the figure is *ABC* · · ·. Between *ab* and *cd*, there is a thin region of hcp stacking, and the lines that bound this region (i.e., *ab* and *cd*) are partial dislocations. (From *The Structure and Properties of Materials,* Vol. I, by W. G. Moffatt, G. W. Pearsall, and J. Wulff, Wiley, New York, 1964.)

REFERENCES

1. Azároff, L. V. *Introduction to Solids*, McGraw-Hill, New York, 1960.

2. Cottrell, A. H. The Nature of Metals, *Scientific American*, Vol. 217, No. 3, September 1967, pp. 90–100.

QUESTIONS AND PROBLEMS

10.1 Calculate, in terms of the metal atom radius, R, the radius, r, of the largest atom that could fit interstitially in an fcc metal without distortion. Such an interstitial site is located at 0, 0, $\frac{1}{2}$, for example. How many metal atoms coordinate or surround the interstitial atom?

10.2 A substitutional impurity in an fcc metal may cause an increase or a decrease in the lattice parameter (and, in some cases, there may be no change), but an interstitial impurity in the same metal invariably causes an increase in lattice parameter. Justify this statement.

10.3 (a) Calculate, in terms of the metal atom radius, R, the radii of the largest atoms which can fit into (i) a tetrahedral interstitial site (at 0, $\frac{1}{4}$, $\frac{1}{2}$) and (ii) an octahedral interstitial site (at 0, $\frac{1}{2}$, $\frac{1}{2}$) in a bcc metal without distortion.
(b) Suggest a reason why interstitial impurities usually occupy octahedral interstices in a bcc metal.

10.4 If $\Delta H_v = 84$ kJ/mol, what is the fraction of vacant sites at 1000 K? At 1500 K?

10.5 (a) Describe, with sketches, how the motion of atoms in a crystal might be accomplished by the motion of vacancies.
(b) Could the same result be achieved by the motion of self-interstitials? Explain.

10.6 When substitutional atoms do not occupy random sites in a crystal structure, an ordered structure results. For example, even though Cu and Au show complete solid solubility at elevated temperatures, at lower temperatures ordering takes place at both 25 atomic percent Au and 50 atomic percent Au.
(a) Sketch a fcc unit cell, and designate the most likely sites for the Cu and Au atoms in Cu_3Au, such that as much symmetry as possible results.
(b) Repeat part (a) for CuAu.
(c) In terms of entropy, why does ordered Cu_3Au become a disordered solid solution at elevated temperatures?

10.7 Sketch the distortion of the structure around an edge dislocation and show the preferred regions for large substitutional atoms, small substitutional atoms, and interstitial atoms.

10.8 Sketch a schematic of two parallel screw dislocations with opposite Burgers vectors on the same slip plane in a simple cubic structure. Show how these dislocations can come together and annihilate each other.

10.9 Distinguish among the direction of the dislocation line, the Burgers vector, and the direction of dislocation motion for (*i*) an edge and (*ii*) a screw dislocation.

10.10 How would you expect dislocation motion to be affected by grain boundaries? Explain.

10.11 Grain boundaries are sometimes called "high-angle boundaries." Can they be described in terms of dislocation arrays? Explain.

10.12 Calculate the dislocation spacing in a symmetric 2° tilt boundary in an fcc metal crystal that has a lattice parameter equal to 0.361 nm.

Surfaces and Interfaces

Many chemical and mechanical interactions between materials depend upon the nature of their surfaces. In this chapter some important concepts relating to the characteristics of liquid and solid surfaces are introduced.

11.1 SURFACE ENERGY AND SURFACE TENSION

It is well known that liquids have a propensity to form spherical droplets unless constrained by external forces. This occurs because spheres have the smallest ratio of surface area to volume of all geometrical shapes. Furthermore, it is frequently observed that two spherical droplets of water or of mercury, when placed in contact, coalesce to form one larger spherical droplet. These tendencies of surfaces to assume the smallest possible total area are a consequence of total surface energy tending toward a minimum. Total surface energy is the product of surface energy and area. By definition, surface energy is the work required to create a unit area of additional surface and has units such as J/m^2 or erg/cm^2. The atomistic basis for surface energy relates to the fact that an atom or molecule at the surface of a liquid in contact with its vapor is bonded to fewer neighbors than one in the interior of the liquid. Thus, the energy of a surface atom is greater than that of an atom in the interior. Since decreasing the surface area of a liquid decreases the relative proportion of surface atoms without altering the nature of the surface, total surface energy is directly proportional to surface area.

The tendency of surfaces to contract also can be considered to be a result of a surface tension. For liquids, surface energy and surface tension have the same numerical value and dimensions, but the units of the latter are given by force per unit length (e.g., N/m). The equivalence of surface energy and surface tension results because the energy expended in creating a unit area of new surface equals the force per unit length required to extend the surface. This relationship may be illustrated by considering an expandable wire frame containing a soap film as shown in Fig. 11.1. If γ is the surface energy, then the increase in total surface energy on expanding the soap film by a distance Δx is $2l\,\Delta x \cdot \gamma$ (since there are two soap-air interfaces). This term equals the work ($f\ \Delta x$) done by a force, f, acting through the distance Δx; consequently, the surface tension ($f/2l$) is identical in magnitude to γ.

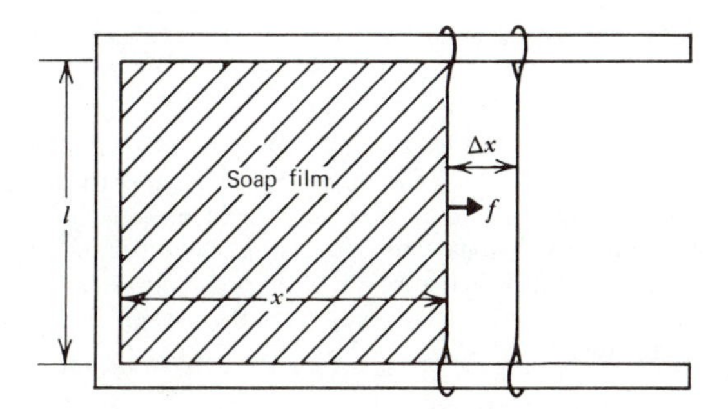

FIGURE 11.1 A soap film in an expandable wire frame. Expanding the film by a distance Δx increases the total surface energy by an amount $(2l\Delta x) \cdot \gamma$, where γ is the surface energy of the soap film. The force per unit length necessary to expand the film is given by $(f/2l) = \gamma$; this corresponds to the surface tension.

11.2 MEASUREMENT OF LIQUID SURFACE TENSION

One method used to determine the surface tension of a liquid is to measure its rise in a capillary tube and the contact angle it makes with the wall of the capillary tube (Fig. 11.2). The liquid is said to *wet* the tube if the meniscus is concave upward as shown in Fig. 11.2*a*. If the liquid does not wet the tube, the meniscus will be concave downward. At equilibrium, in the former case, the surface tension pulling liquid up to the tube is balanced by the weight of the liquid in the tube, and a force balance may be applied as shown in Fig. 11.2*b*. For a liquid of density ρ, the weight per unit volume is given by ρg, where g is the acceleration due to gravity. The weight, w, of liquid in the capillary tube is given by the product of ρg and $\pi r^2 h$,

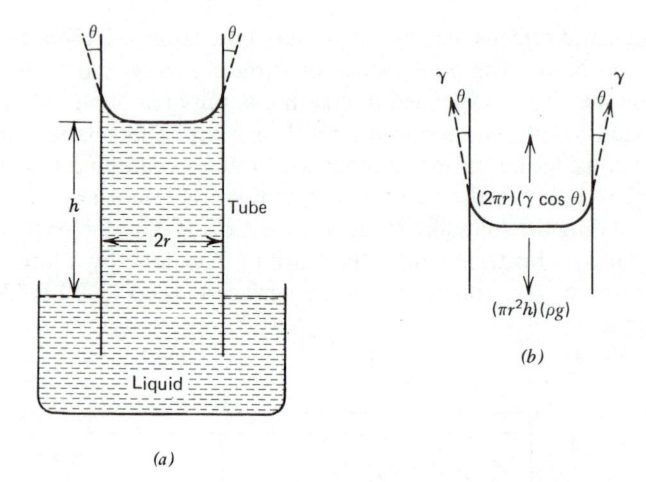

FIGURE 11.2 (*a*) The rise in a capillary tube of a liquid that wets the surface of the capillary tube. (*b*) A force balance between the surface tension and the weight of liquid in the capillary tube. The vertical component of total surface tension, $(2\pi r)(\gamma \cos\theta)$, balances the weight of the liquid drawn into the capillary tube, $(\pi r^2 h)(\rho g)$, where ρg is the weight per unit volume of liquid.

the liquid volume. The surface tension γ acts around the tube circumference with a resolved component $\gamma \cos\theta$ in the vertical direction, where θ is the contact angle. Balancing forces gives

$$(2\pi r)(\gamma \cos\theta) = (\pi r^2 h)(\rho g) \tag{11.1a}$$

or

$$\gamma = \frac{\rho g r h}{2} \cos\theta \tag{11.1b}$$

For liquids that readily wet the capillary tube material, θ is very small, such that $\cos\theta \simeq 1$, and

$$\gamma \simeq \frac{\rho g r h}{2} \tag{11.2}$$

Thus, for such liquids, surface tension can be determined directly by measuring the height to which the liquid rises in the capillary tube.

Different values of surface tension are obtained when the vapor in contact with the liquid differs chemically from the liquid. This is due to contamination of the surface by adsorbed gases which invariably reduce the measured surface tension. Consequently, measurements of the surface tensions of liquid metals, for example, are carried out either in a vacuum or in an inert gas atmosphere. The values so obtained represent surface tensions between the liquid and a vacuum rather than between a liquid and its own vapor; however, for most purposes these two values are sufficiently close that they may be considered interchangeable.

Liquid surface tension decreases with increasing temperature because thermal agitation reduces the interatomic or intermolecular attractive forces. For example, the surface tension of water decreases from 0.072 N/m (72 dyne/cm) at 25°C to 0.059 N/m (59 dyne/cm) at 100°C. Since water molecules in the liquid are attracted to each other primarily by physical forces, water has a relatively low surface energy. The chemical bonding in liquid metals and molten salts gives rise to much larger surface energies (Table 11.1).

Addition of a solute to a solvent liquid may lower the surface tension appreciably. Soaps and detergents effectively lower the surface tension of water. They form surface films on dirt and grease spots in wash water. The addition of soap or other compounds that are soluble in both phases of a mixture of immiscible liquids and that reduce the interfacial tension between them makes it possible to finely disperse one of the liquids in the other and thereby form an emulsion. The small

TABLE 11.1 Surface Energies of Liquids

Liquid	*Temperature (°C)*	*Surface Energy (J/m²)*[a]
Hg	20	0.48
Zn	650	0.75
Au	1130	1.10
Cu	1120–1150	1.10–1.27
Steel (Fe–0.40%C)	1600	1.56
Pb	350	0.442
Ag	1000	0.92
Pt	1770	1.865
H_2O	0	0.076
H_2O	100	0.059
NaCl	800	0.114
NaCl	910	0.106
$NaPO_3$	620	0.209
B_2O_3	900	0.08
FeO	1420	0.585
Al_2O_3	2080	0.70

[a] $1\ J/m^2 = 10^3\ erg/cm^2$.

drops of the one phase become coated with the emulsifying agents and therefore do not coalesce. Oil in water emulsions are stabilized by sodium soaps whereas water in oil emulsions are stabilized by calcium and magnesium soaps.

11.3 SURFACE ENERGIES OF SOLIDS

A solid in contact with its vapor usually has a somewhat higher surface tension than a liquid because the number of near neighbors of an atom *in the interior* of a solid is greater than in a liquid. Direct measurement of the surface tensions of solid metals can be made by suspending small weights from identical fine wires held at high temperatures in a vacuum or in an inert gas atmosphere. The weights tend to extend the wires and, since the wire volume remains constant, the surface area tends to increase. However, the surface tension opposes the extending force and, by measuring the weight at which one of the wires neither shortens nor lengthens, the surface tension can be determined. This technique can be used only at elevated temperatures where atomic mobility is sufficiently high to permit dimensional changes. Values of surface tension determined in this manner are shown in Table 11.2 along with other experimental data for surface energies of ionic and covalent solids.

If the wires used in the above experiment are polycrystalline, then grooves will develop at the grain boundaries as a result of thermal etching, which is a result of a surface energy equilibrium between the grain boundary surface energy and the vapor–solid surface energy as shown in Fig. 11.3. Such grooving, which reduces

TABLE 11.2 Surface Energies of Solid Materials

Material	*Surface Energy (J/m^2)*[a]
Au	1.40
Cu	1.43–1.70
Ag	1.14–1.20
Ni	1.90
NaCl	0.30
Al_2O_3	0.905
MgO	1.00–1.20
TiC	1.19
LiF	0.34
CaF_2	0.45
BaF_2	0.28
$CaCO_3$	0.23
Si	1.24

[a] $1\ J/m^2 = 10^3\ erg/cm^2$.

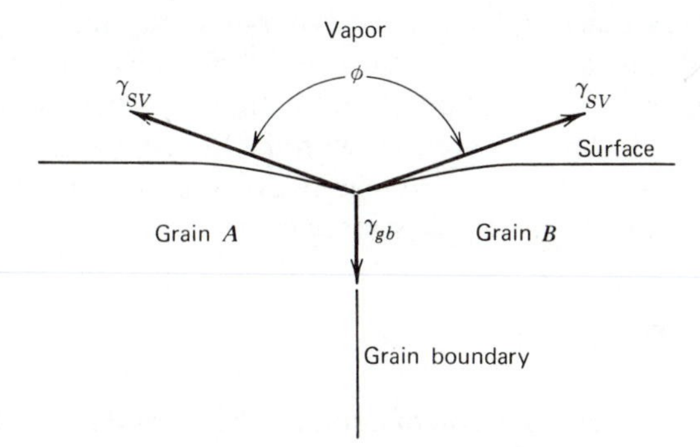

FIGURE 11.3 A grain boundary groove formed on the surface of a polycrystalline metal as a result of thermal etching. The ratio of grain boundary energy to solid–vapor surface energy can be determined with a surface tension balance. In this case, a vertical force balance gives $\gamma_{gb} = 2\gamma_{SV} \cos(\phi/2)$.

the total free energy, facilitates the determination of the ratio of grain boundary energy to surface energy by means of a surface tension balance (Fig. 11.3). In this manner, it is found that grain boundary energies are on the order of one-third of the solid–vapor surface energies, reflecting the fact that atoms in the vicinity of a grain boundary have more bonding neighbors than those at a free surface. It is worth noting at this point that microstructural observations of second-phase particles located in the grain boundaries of an alloy often can be used to ascertain the interfacial energy between the second-phase particle and the matrix phase in terms of the matrix-phase grain boundary energy (problem 11.5).

Mention was made in Sect. 10.7 that impurity atoms tend to segregate to grain boundaries. This occurs because the total energy is lowered. Likewise, impurity atoms that lower the surface energy of a free surface will tend to migrate to a free surface, and those that raise the surface energy will tend to migrate away from the free surface. In addition, since most solid surfaces are neither clean nor atomically smooth, surface energy depends upon both geometry and environment. As with liquids, surfaces of solids are readily coated by a thin layer of contaminants from the environment, resulting in a lower surface energy. The adsorbed atoms and molecules, and any surface compounds that form, greatly affect the manner in which contacting solid surfaces interact.

In contrast to liquid surfaces, crystal surfaces are not isotropic, and the surface energy depends upon which crystallographic plane is exposed. For example, densely packed atomic planes have relatively low surface energies, and a metal single crystal held at elevated temperatures for long times will always assume a shape bounded by crystallographic planes of minimum surface energy rather than becoming spherical. The anisotropy of surface energy is even more significant in ionic or covalent materials. Upon sharp impact, materials such as sodium chloride or diamond cleave and form new surfaces with the lowest possible surface energy. Thus, definite cleavage facet planes are readily observed with diamond and other materials that have highly anisotropic surface energies.

11.4 ADSORPTION

As mentioned above, the attraction of a solid or liquid surface for foreign atoms, ions, or molecules leads to adsorption. It is convenient to characterize weakly bonded films of foreign atoms or molecules as physically adsorbed. The bonding is then of the van der Waals type and the energy liberated during adsorption is less than about 40 kJ/mol (10 kcal/mol). Physical adsorption of a gas by a solid is readily decreased by lowering the pressure or raising the temperature of the substrate. In contrast to physically adsorbed films, which are several monolayers thick, chemisorbed films are usually monatomic or monomolecular. Furthermore the heat liberated during chemisorption can be of the order of 400 kJ/mol. Chemisorption is more apt to occur at high temperatures than at low temperatures; the rate of chemisorption follows an Arrhenius-type relationship. Most interphase chemical reactions of clean metals start by chemisorption, and the nucleation and growth of stable reaction products at preferred sites, such as dislocations and grain boundaries, follow.

Adsorption plays a significant role in suspensions of finely divided solids and liquids in gaseous or liquid media. If the particles of the suspended phase have a diameter of 1 to 100 nm they are called colloids. Such colloids can be agglomerated or floculated from colloidal suspension by raising the temperature or by addition of ions that neutralize the surface charges adsorbed on the colloidal particles. Colloidal dispersions of liquids in gases are commonly called aerosols while smokes are usually dispersions of solids in gases and foams are dispersions of gas bubbles in liquids or solids. Gels are generally colloidal mixtures of solids and liquids in which the solid particles have become linked together to form a continuous framework (see Section 7.2). If the interparticle bonds in the gel are weak an *elastic* gel results; if strong, a *rigid* gel. Jello and asphalt are elastic gels and Portland cement, when set, is a rigid gel.

11.5 WETTING

The shape assumed by a liquid in contact with a solid depends upon the relative values of surface energy as illustrated in Fig. 11.4*a*. In the plane of the the solid surface, force equilibrium must exist between the various surface tensions because

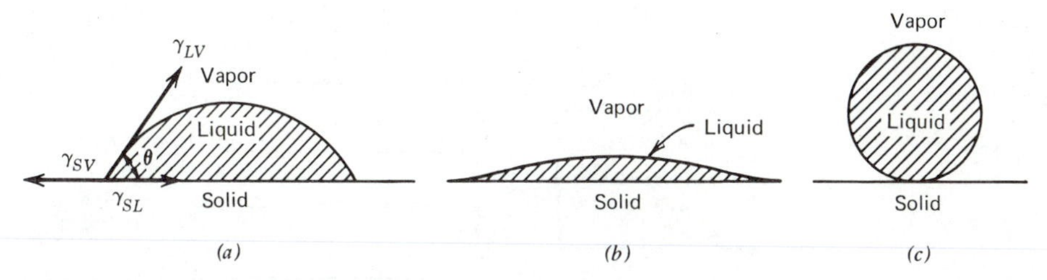

FIGURE 11.4 (*a*) Partial wetting of a solid substrate by a liquid. Balancing the surface tensions in the horizontal plane gives $\gamma_{SV} = \gamma_{LV} \cos\theta + \gamma_{SL}$. (*b*) Complete wetting (i.e., $\theta = 0°$) of the solid substrate occurs when the two new surfaces formed have less energy than the original solid–vapor interface, that is, $\gamma_{SV} > \gamma_{LV} + \gamma_{SL}$. (*c*) No wetting occurs when the solid–liquid interfacial energy is greater than the surface energies of the original interfaces, that is, $\gamma_{SL} < \gamma_{SV} + \gamma_{LV}$.

the liquid droplet is free to move until equilibrium is established. For partial wetting with a contact angle θ, a horizontal surface tension balance gives

$$\gamma_{SV} = \gamma_{LV} \cos\theta + \gamma_{SL} \tag{11.3}$$

where γ_{SV}, γ_{LV}, and γ_{SL} are the solid–vapor, liquid–vapor, and solid–liquid surface tensions, respectively. As shown in Fig. 11.4*b*, the liquid will completely wet ($\theta = 0°$) the solid surface if the two new surfaces formed have less energy than the original solid–vapor interface (i.e., $\gamma_{LV} + \gamma_{SL} < \gamma_{SV}$). A complete lack of wetting ($\theta = 180°$, Fig. 11.4*c*) occurs under the converse condition (i.e., $\gamma_{SL} > \gamma_{SV} + \gamma_{LV}$).

It is often found that solid–liquid surface energies are smaller than the corresponding solid–vapor or liquid–vapor surface energies since fewer bonds are unfulfilled at a solid–liquid interface. Indeed, wetting is generally favored if the solid and liquid are chemically compatible, providing adsorbed surface contaminants are absent. Such compatibility exists when compounds are formed at the interface or when the liquid is capable of alloying with the solid substrate. Thus, liquid copper will wet solid nickel because it can form a solid solution with nickel, whereas liquid lead, which is immiscible in liquid or solid iron, wets an iron surface only with great difficulty. Alteration of liquid lead by the addition of tin or antimony promotes wetting since the latter elements form compounds with iron while lead does not. This is one reason why Pb–Sn or Pb–Sb alloys are used as solders.

11.6 ADHESION

Surface energy plays an important role in the adhesion between solids. Adhesion refers to the attraction between two surfaces brought into contact. This can be represented by the work of adhesion, which is the work or energy required to

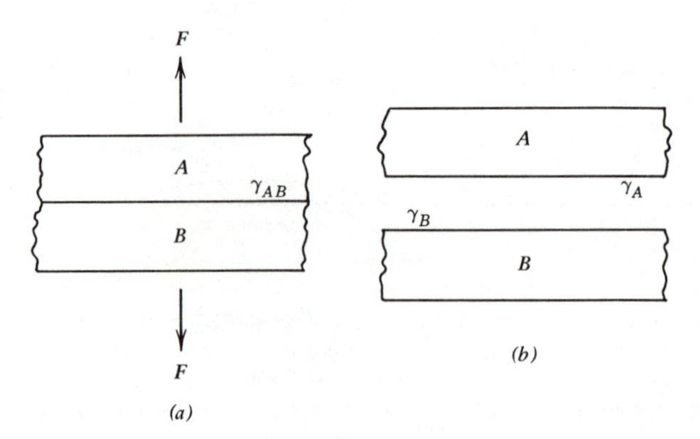

FIGURE 11.5 The work of adhesion in terms of surface energies. The energy increase between (*a*) and (*b*) is the work of adhesion, which is given by $W_{AB} = \gamma_A + \gamma_B - \gamma_{AB}$.

separate a unit area of the adhering surfaces. As shown in Fig. 11.5, the work of adhesion, W_{AB}, is given by

$$W_{AB} = \gamma_A + \gamma_B - \gamma_{AB} \tag{11.4}$$

where γ_A and γ_B are the surface energies of *A* and *B* (with respect to the final vapor or vacuum with which they are in contact) and γ_{AB} is the interfacial energy between *A* and *B*. Since γ_{AB} is small when two like or compatible surfaces are in contact, W_{AB} will be large in this case. Two totally unlike or incompatible surfaces—usually those between substances that do not form compounds or solid solutions with one another—have high values of γ_{AB}, and W_{AB} is small. Thus, if no adsorbed contaminants are present, compatible materials will adhere more strongly than incompatible materials. Perhaps the classic example is mica which, when cleaved in a vacuum and brought together again, adheres nearly as firmly as before cleavage. The high strength of grain boundaries in a polycrystalline solid provides another example.

That solids with nominally flat surfaces usually do not adhere very well when pressed together in air can be attributed to several factors: (1) the presence of surface contaminants limits intimate contact, (2) even nominally flat surfaces are rough on an atomic scale, and the true area of contact is therefore small, and (3) when the contact pressure is removed, residual elastic stresses in the solids rupture junctions where adhesion has taken place. Soft metals such as indium can be made to adhere to other metals with clean surfaces because the soft metals flow so as to give fairly good contact and the elastic stresses in the regions of contact are relieved by deformation of the soft metal rather than by rupture of the junctions. Strong adhesion joints, called *cold welds*, are readily made between ductile metals such as gold on gold or aluminum on aluminum, provided there is enough plastic deformation during the joining to sweep away adsorbed gases and to break up surface oxide films that may be present. Even with limited adhesion when two metals are pressed into contact, some metal-to-metal junctions are always formed. This has been demonstrated by bringing a radioactive metal into contact with a nonradioactive metal. After separation, radioactivity is invariably detected on the metal that was nonradioactive initially.

Organic adhesives such as animal glues have long been used to bond wood and other cellulose products. Inorganic polymeric cements such as silicic acid [$Si(OH)_4$] and combinations of a metal oxide and phosphoric acid have also proven useful in bonding inorganic materials. Currently, a wide variety of synthetic organic polymers are employed as adhesives. Some of them provide sufficient adhesion to bond metallic structural parts. An adhesive must wet both surfaces that are to be joined and subsequently *set* or harden without cracking. Although the adhesive material usually is much weaker than the solids joined, the large contact area and small thickness of the adhesive layer often afford considerable strength to the joint.

11.7 FRICTION AND WEAR

Junctions or cold welds formed between contacting surfaces are largely responsible for the phenomenon of friction. The coefficient of friction, μ, can be determined by an experiment of the type illustrated in Fig. 11.6. Here the horizontal

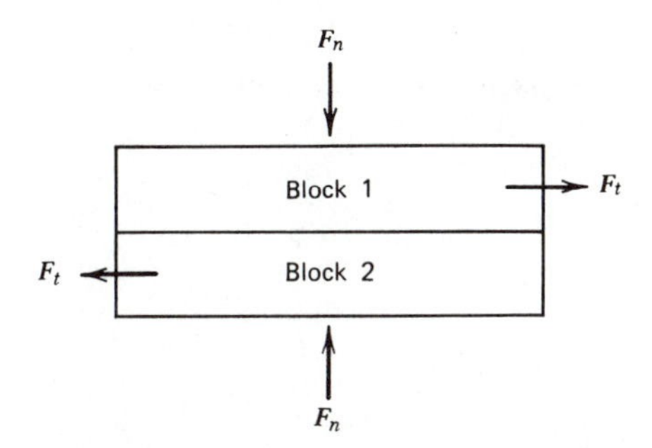

FIGURE 11.6 A procedure to measure the coefficient of friction between two surfaces. The coefficient of friction is the ratio of the horizontal force required to slide the blocks over one another, to the normal force pushing them together (i.e., $\mu = F_t/F_n$).

or tangential force, F_t, required to slide one block over another is measured and compared to the normal force, F_n. The ratio of these forces defines the coefficient of friction:

$$\mu = \frac{F_t}{F_n} \tag{11.5}$$

The greater the adhesion between the sliding surfaces, the greater will be the coefficient of friction. Experimental studies of friction indicate that the coefficient of friction between two solids is approximately independent of normal load, sliding velocity, surface geometry and, within limits, surface roughness. In part, this is a consequence of the real contact area of junctions between the two surfaces being directly proportional to the normal load and, in turn, the tangential force is directly proportional to the real contact area. Under the extreme conditions encountered in certain metalworking operations with very high normal loads, the real contact area approaches the actual area of the two surfaces, and the coefficient of friction no longer is constant. Rather, F_t divided by the area of contact remains constant, and μ varies with the normal force. Coefficients of friction between various materials in air are given in Table 11.3.

Boundary lubrication by a thin layer of soft material (liquid or solid) between two harder surfaces reduces μ significantly (Table 11.3) and has an even more marked effect in reducing adhesive wear. During the sliding of two surfaces, adhesive wear results because particles are pulled out of one surface (usually the softer surface) and adhere to the other surface. If some of these particles become oxidized and loose between the two surfaces, then they can cause one form of abrasive wear, since oxide particles typically are harder than metals. Another

TABLE 11.3 Typical Values of Coefficient of Static Friction

Material	μ
Steel on aluminum	1.2
Titanium on titanium	0.55
Wood on steel	0.45
Lead on lead	1.0
Soft steel on soft steel	0.3
Hard steel on hard steel	0.15
Soft steel on soft steel (thin film of lead between)	0.14
Soft steel on soft steel (lubricating oil)	0.08
Copper on copper	1.6
Stainless steel on nickel	0.4
Graphite on graphite	0.1
Polytetrafluoroethylene on self	0.04

form of abrasive wear occurs when a hard, rough surface scratches a softer surface. This leads to loose particles of the softer material between the two surfaces, and these particles can cause more abrasive wear if they oxidize. Unlike adhesive wear, abrasive wear is affected little by lubrication, except insofar as the lubricant inhibits oxidation or corrosion.

11.8 VITRIFICATION AND SINTERING

The starting materials used to prepare many ceramic objects are either fine powders or particulate substances like clay. By suitable means these materials can be molded into a desired shape, but to convert them into a cohesive body requires a

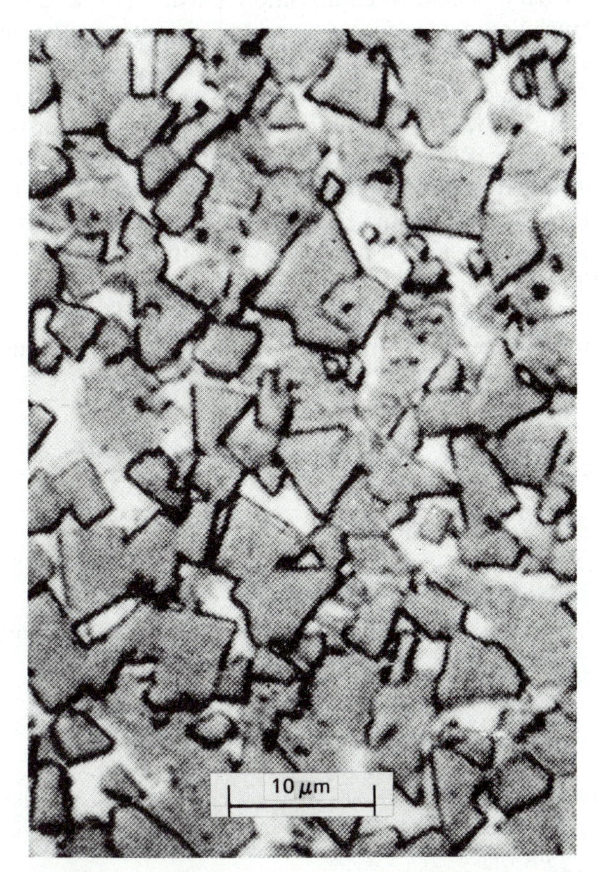

FIGURE 11.7 Microstructure of a cemented tungsten carbide containing 13% cobalt by weight. The dark gray particles are WC, and the lighter colored phase is Co.

firing treatment at an elevated temperature. Both interparticle bonding and densification of the body result from firing. Both processes take place by a mechanism of mass transport that involves either viscous flow or diffusion (see Chapter 12); the driving force is the reduction of surface free energy that accompanies the reduction of total surface area as the particles bond together.

The great majority of silicate-base ceramics are converted into viable bodies by a firing process called *vitrification.* During firing, a predetermined amount of the viscous liquid silicate-phase forms, and this phase wets the remaining crystalline particles and, by virtue of viscous flow, pulls them together. Upon subsequent cooling, the liquid phase becomes a glass that binds the crystalline particles to each other, and the final microstructure usually contains some residual porosity. (In order to avoid excessive sagging or shape distortion of the body during vitrification, it is necessary to control composition, particle size, and firing temperature so that too much liquid does not form and so that its viscosity is not too low.)

Certain ceramic oxides and ceramic-metal combinations (*cermets*) are fired by a process termed liquid-phase sintering. Again, the liquid formed on firing wets the crystalline particles but on cooling it crystallizes rather than becoming a glass. Cemented carbide cutting tools, such as those prepared from fine powders of tungsten carbide and cobalt, provide an example; a typical microstructure is shown in Fig. 11.7. Often, liquid-phase sintering leads to a body with no residual porosity, a feature that gives it greater strength than is present in a ceramic with porosity. When single-phase oxide or metal powders are sintered in the absence of a liquid phase, it is very difficult to produce a product without some residual porosity. This topic, dealing with solid state sintering, will be discussed in Sect. 12.7.

REFERENCES

1. Gomer, R. and Smith, C. S. (editors). *Structure and Properties of Solid Surfaces*, University of Chicago Press, Chicago, 1953.
2. Adamson, A. W. *Physical Chemistry of Surfaces*, Interscience, New York, 1960.

QUESTIONS AND PROBLEMS

11.1 Compare the surface to volume ratios of a cube, sphere, and infinitely long cylinder. Express your answer in terms of the radius of the sphere or cylinder and the edge length of the cube.

11.2 Metals that form solid solutions or compounds show a higher coefficient of friction and greater wear when in sliding contact in a vacuum than in air. Briefly explain.

11.3 (a) Consider the general case of three phases in surface equilibrium shown below.

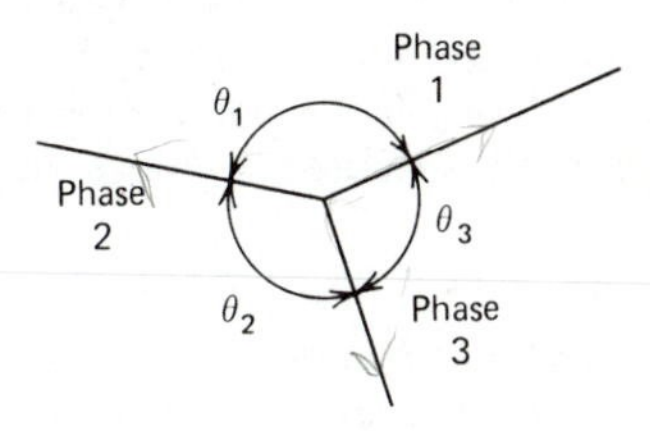

Perform a surface tension balance and show that

$$\frac{\gamma_{12}}{\sin \theta_3} = \frac{\gamma_{13}}{\sin \theta_2} = \frac{\gamma_{23}}{\sin \theta_1}$$

(b) Show that the equation you derived reduces to the relations given in Fig. 11.3 and by Eq. 11.3 under the special conditions considered in these cases.

11.4 If $\gamma_{gb} = \gamma_{SV}/3$, what would be the groove angle observed on thermal etching?

11.5 Pb is often added to copper and its alloys for better machining properties. The lead usually is found at the copper grain boundaries as shown below.

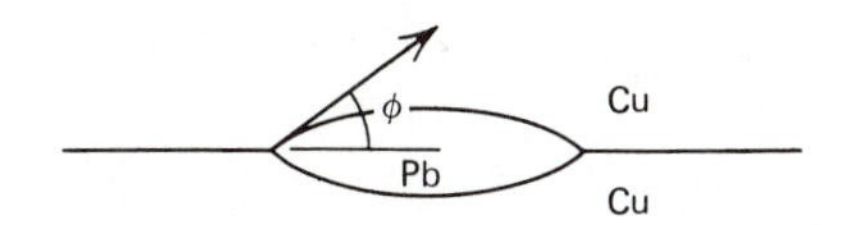

Find an expression for the Cu—Pb interfacial energy in terms of the angle ϕ and γ_{gb} for copper. Do this by a direct surface tension balance and by reduction of the equations in problem 11.3.

11.6 (a) Would you expect the twin boundaries and stacking faults described in Chapter 10 to have related values of surface energy? Explain.

(b) Would you expect these surface energies to be large or small in comparison with the solid–vapor surface energy?

11.7 (a) The surface energy of NaCl is 0.3 J/m^2. Calculate the size of salt particles at which the surface energy is 10% of the magnitude of the binding energy of NaCl.

(b) Explain why salt is hard to pour when it becomes moist.

11.8 If a polycrystalline material is heated to an elevated temperature, the average grain size increases. Explain why this process takes place.

11.9 What chemical and structural characteristics make graphite suitable as a lubricant?

11.10 Do you expect that crystallographic planes of densest packing to have the lowest values of solid–vapor surface energy? Explain your answer.

11.11 Thermodynamically, surface energy is actually a free energy; that is, $\gamma = H - TS$, where H and S are the surface enthalpy and surface entropy, respectively.

(a) If you have data for surface energy as a function of temperature, how can you determine surface enthalpy and surface entropy?

(b) Consult the *Handbook of Chemistry and Physics* for the surface energy of water as a function of temperature and compute the surface enthalpy and surface entropy at 25°C.

11.12 Surface enthalpy (see problem 11.11) can be estimated by counting the *fraction* of missing nearest neighbors per surface atom, multiplying this number by the binding energy per atom, and dividing the product by the surface area per atom.

(a) Estimate the surface enthalpy for copper surfaces consisting of (111), (100), and (110) atomic planes.

(b) Should the surface energy be greater or less than the surface enthalpy? Why? (Compare the surface enthalpies obtained in part (a) with the surface energy of copper given in Table 11.2.)

12

Diffusion

This chapter introduces some of the useful concepts and ideas concerning diffusion. Many reactions in solids and liquids are diffusion dependent and these will be discussed in subsequent chapters.

12.1 DIFFUSION PHENOMENA

In order for chemical reactions to occur efficiently and effectively, it is usually important that the reactants be well mixed. The mixing of gases or of liquids to form a solution can result from flow due to mechanical agitation such as stirring or due to thermal convection currents. Both of these means are very effective in producing compositional uniformity or homogeneity in gaseous solutions or liquid solutions. Even under quiescent conditions, mixing will proceed by diffusion, in which the thermal motion of individual atoms or molecules leads to homogeneity (Fig. 12.1). Diffusive mixing in liquids is much slower than in gases, and diffusion in solids is even slower because less freedom for atomic motion exists.

Diffusion in solids may be demonstrated experimentally by making a very clean joint, such as that shown in Fig. 12.2*a*, between pure gold and pure nickel. After this couple has been held at an elevated temperature (e.g., 900°C) for a long period of time, the extent of mixing of gold and nickel can be determined by suitable chemical analyses, and a composition profile can be plotted (Fig. 12.2*b*), showing that gold atoms have diffused into the nickel and nickel atoms, into the gold. While interdiffusion of gold and nickel atoms is occurring, nickel atoms are moving about within the nickel and gold atoms move within the gold also. This phenomenon,

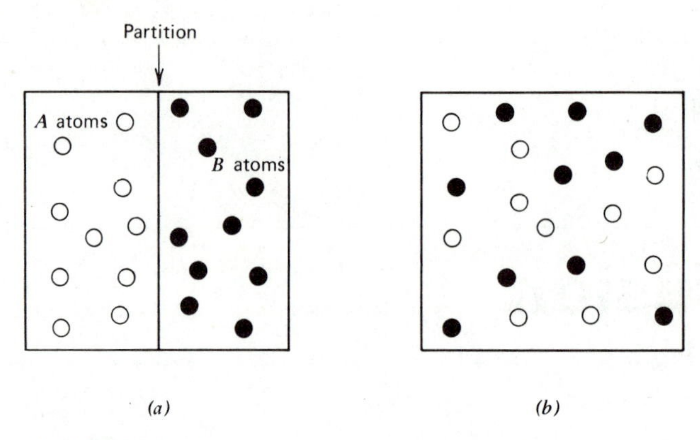

FIGURE 12.1 Mixing of gases by gaseous diffusion. (*a*) Gas atoms of species *A* and *B* are separated by a partition. (*b*) Upon removal of the partition, the gases mix, resulting in a homogeneous distribution.

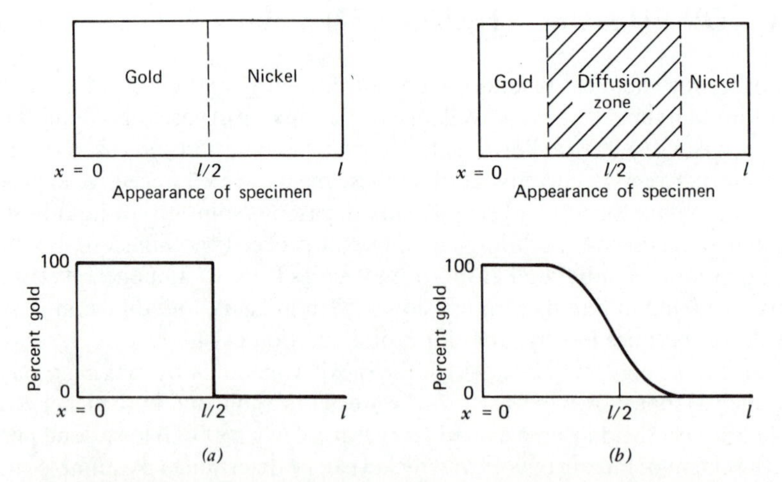

FIGURE 12.2 An Au–Ni diffusion couple. (*a*) Before the diffusion heat treatment, pure gold is on the left-hand side of the couple and pure nickel is on the right-hand side. (*b*) The same couple after a diffusion heat treatment shows that Au atoms have penetrated into the nickel and vice versa.

called self-diffusion, can be demonstrated by depositing a thin layer of radioactive nickel on nonradioactive nickel. After this chemically homogeneous specimen has been given an appropriate diffusion heat treatment, a suitable radiation detector can be used to measure radioactivity as a function of distance into the originally nonradioactive nickel. The penetration plot obtained is shown schematically in Fig. 12.3. Such experiments reveal, for a solid metal held just below its melting point, that over three years are required for an atom to move a distance of one centimeter by self-diffusion.

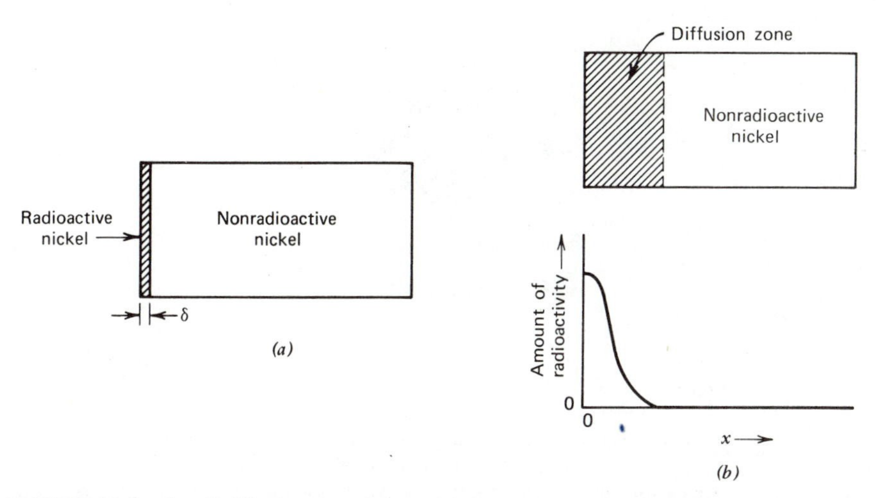

FIGURE 12.3 A self-diffusion experiment. (*a*) A thin layer of radioactive nickel has been deposited on one surface of a nonradioactive nickel specimen. (*b*) A diffusion heat treatment has resulted in the diffusion of radioactive Ni atoms into the initially nonradioactive material.

Atoms are able to move throughout solids because they are not stationary but execute rapid, small-amplitude vibrations about their equilibrium positions. Such vibrations increase with temperature, and at any temperature a very small fraction of atoms has sufficient amplitude to move from one atomic position to an adjacent one. The fraction of atoms possessing this amplitude increases markedly with rising temperature. In jumping from one equilibrium position to another, an atom passes through a higher energy state since atomic bonds are distorted and broken, and the increase in energy (i.e., the activation energy) is supplied by thermal vibrations. As might be expected, defects, especially vacancies, are quite instrumental in affecting the diffusion process in crystalline solids. In a particular material, diffusion depends on the type and number of defects that are present, as well as the thermal vibrations of atoms.

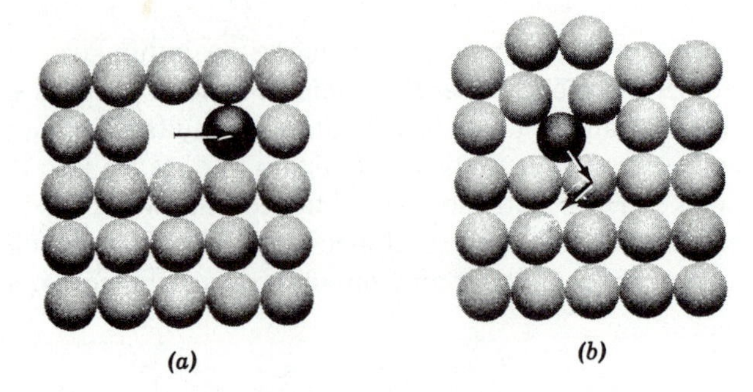

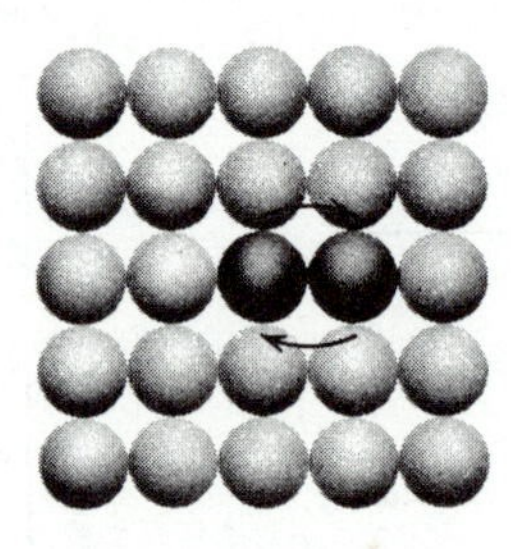

FIGURE 12.4 Possible mechanisms of diffusion. (*a*) The vacancy mechanism, in which an atom and an adjacent vacancy exchange positions, is the mechanism for diffusion in most metals and substitutional solid-solution alloys. The activation energy for self-diffusion by this mechanism is the sum of the energy required to form a vacancy and that required to move it. (*b*) The interstitial mechanism illustrated results in atomic motion since a self-interstitial atom can displace an adjacent atom to an interstitial position. Although the activation energy for such motion is low, self-interstitials occur so infrequently that they do not contribute to diffusion in most metals and substitutional solid-solution alloys. Interstitial impurity atoms and solute atoms in interstitial solid-solution alloys can move fairly readily by this mechanism. (*c*) The direct exchange mechanism, in which two adjacent atoms interchange positions, is highly unlikely because the associated activation energy is extremely high. (Adapted from *The Structure and Properties of Materials,* Vol. II, by J. H. Brophy, R. M. Rose, and J. Wulff, Wiley, New York, 1964.)

12.2 MECHANISMS OF VOLUME DIFFUSION IN METALS

The transfer of a metal atom from one equilibrium position to another is the basic step in the volume diffusion process, that is, mass transport that occurs by the movement of atoms within the bulk of the material. In the case of self-diffusion, there is no net mass transport, but atoms migrate in a random manner throughout the crystal. In interdiffusion, the mass transport almost always occurs so as to minimize compositional differences. Various atomic mechanisms for self-diffusion and interdiffusion have been proposed, and three of them are illustrated in Fig. 12.4. The most energetically favorable process, and one that has been experimentally shown to predominate in most metals, involves an interchange of places by an atom and a neighboring vacancy (Fig. 12.4*a*). For this process, the energy of the atom depends on its position, as shown in Fig. 12.5. The intermediate or activated state, in which the atom is halfway between the initial and final positions, has a much higher energy associated with it, and the increase in energy corresponds to the activation energy necessary to move the atom (or the vacancy). The activation energy for diffusion by a vacancy mechanism is the sum of the energy required to form a vacancy, ΔH_v, (see Sect. 10.3) and that required to move a vacancy from one equilibrium position to another, ΔH_m. That is, diffusion by this mechanism depends not only on the motion of vacancies (and the corresponding motion of atoms) but also on the fraction of vacant sites present.

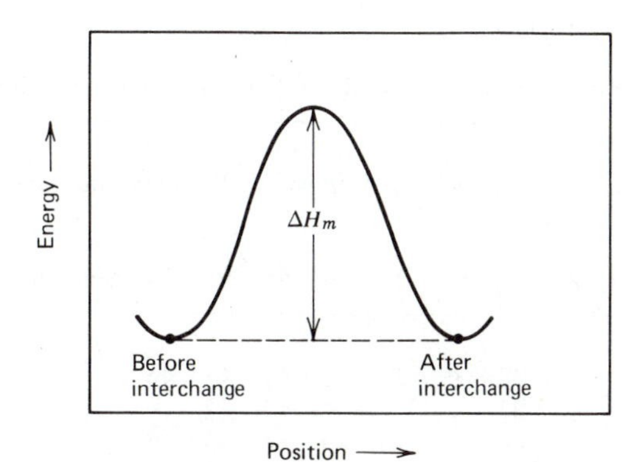

FIGURE 12.5 The energy associated with motion of atoms by a vacancy mechanism (see Fig. 12.4*a*). Halfway between equilibrium positions, the energy of an atom is greatest. ΔH_m represents the activation energy for such atomic motion.

A self-interstitial is more mobile than a vacancy in that a smaller activation energy for motion is encountered as the self-interstitial atom moves to an equilibrium atomic position and simultaneously displaces a neighboring atom into an interstitial site. Nevertheless, the equilibrium number of self-interstitial atoms present at any temperature is negligible in comparison to the number of vacancies, and the activation energy for diffusion by a self-interstitial mechanism (Fig. 12.4*b*) is much higher than that for a vacancy mechanism because the energy to form a self-interstitial is extremely large. Similarly, a direct interchange mechanism (Fig. 12.4*c*), which does not depend upon defects, is untenable because an exceptionally high activation energy would be required.

In alloys where solute atoms are sufficiently small in comparison with host atoms such that the former occupy interstitial sites, the interstitial mechanism predominates for interdiffusion. Thus, hydrogen, carbon, nitrogen, and oxygen diffuse interstitially in most metals, and the activation energy for diffusion is only that associated with motion since the number of unoccupied, adjacent interstitial sites usually is large. As a consequence, interstitial solute atoms diffuse much more rapidly than do substitutional solute atoms in metals.

For volume self-diffusion in a metal, the diffusivity or diffusion coefficient, which provides a measure of the rate at which diffusion occurs, is given by

$$D = D_0 e^{-(\Delta H_v + \Delta H_m)/RT} \tag{12.1a}$$

where D_0 is an experimental constant depending on the metal and $\Delta H_v + \Delta H_m$ is the activation energy for diffusion. In general, the diffusivity can be expressed as

$$D = D_0 e^{-Q/RT} \tag{12.1b}$$

TABLE 12.1 Activation Energies (in kcal/mol)[a] for Diffusion in Metals and Alloys

Diffusing Substance / *Solvent* (↓ *Increasing Melting Point of Solvent*)	*Increasing Melting Point of Diffusing Substance* →												
	Hg	Sn	Tl	Zn	Ag	Au	Cu	Mn	Ni	Co	Fe	Cr	Nb
Al				31				29					
Ag	38	39	38	42	<u>44</u>	48	46		55	60	49		
Au	37				40	<u>42</u>			42	42			
Cu	44		43	46	47	50	<u>48</u>	44	57	55	51		
Ni						65	61		<u>70</u>		55	65	
α-Fe (bcc)						62			59	65	<u>57</u>		
β-Ti (bcc)		69						65	60	61	61	66	70

Underlined activation energy values are for self-diffusion.

[a] 1 kcal/mol = 4.184 kJ/mol.

Table 12.1 lists some experimentally determined values of the activation energy for diffusion, Q, for several metals and alloys. In self-diffusion, it should be noted (problem 12.9) that Q increases with increasing melting point, indicating that the bonding energy between atoms strongly influences the rate at which diffusion occurs.

12.3 VOLUME DIFFUSION IN IONIC AND COVALENT SOLIDS

Diffusion in most ionic solids occurs by a vacancy mechanism. As a rule, the large anion can move only if anion vacancies are present, and the smaller cation, while in an interstitial position *with respect to the anions*, can diffuse only if cation vacancies are present. However, in certain open crystal structures, such as that of fluorite (CaF_2) or UO_2, anion diffusion does occur by an interstitial mechanism. Imperfections in ionic materials that influence diffusion (see Sects. 10.2 and 10.3) arise in two ways: (1) intrinsic point defects, such as Schottky and Frenkel defects, whose number depends upon temperature, and (2) extrinsic point defects, whose presence is due to impurity ions of different valence than the host ions. The former give rise to a temperature dependence of diffusion similar to that for self-diffusion in metals, and the latter result in a temperature dependence of diffusion which is similar to that for interstitial solute diffusion in metals.

The diffusivity of Na^+ cations in pure NaCl is comparable to self-diffusivities found in metals because Schottky defects are formed relatively easily in NaCl. In extremely pure stoichiometric metal oxides, on the other hand, the energy associated with intrinsic point defects is so high that their concentration becomes sufficient for significant diffusion only at very high temperatures (Table 12.2). A small

TABLE 12.2 Activation Energies for Diffusion in Some Ionic Materials

Diffuser	*Q (kcal/mol)*[a]
Fe in FeO	23
Na in NaCl	41
O in UO_2	36
U in UO_2	76
Co in CoO	25
Fe in Fe_3O_4	48
Cr in $NiCr_2O_4$	76
Ni in $NiCr_2O_4$	65
O in $NiCr_2O_4$	54
Mg in MgO	83
Ca in CaO	77

[a] 1 kcal/mol = 4.184 kJ/mol.

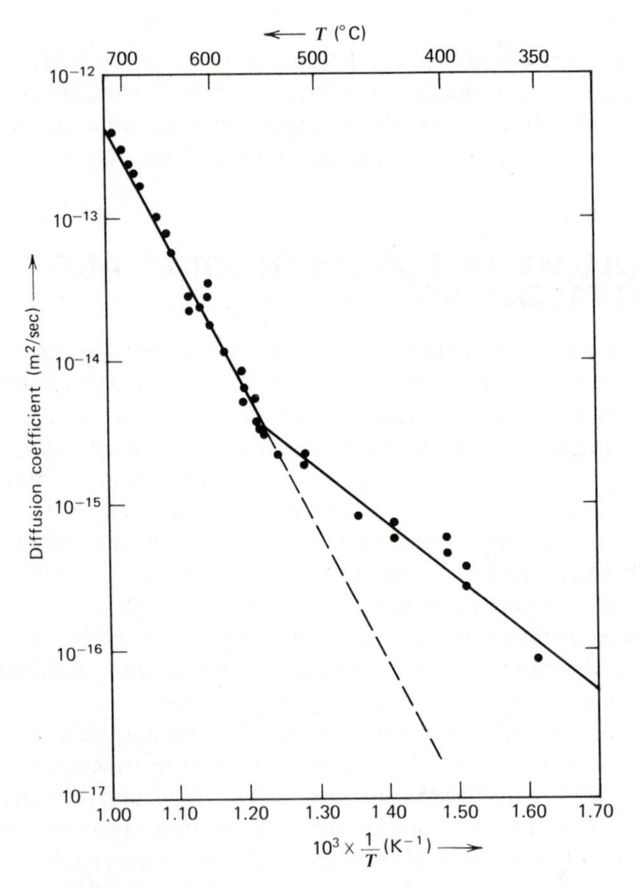

FIGURE 12.6 Self-diffusion coefficient of Na^+ in NaCl containing a small amount of dissolved $CdCl_2$. At high temperatures, the number of Na^+ vacancies associated with Schottky defects is much larger than the number associated with Cd^{2+} ions, and intrinsic diffusion dominates. At low temperatures, the reverse is true, and the vacancies resulting from the presence of Cd^{2+} ions enhance the diffusion of Na^+ ions. The extrapolated dotted line represents the diffusivity of Na^+ ions in the absence of Cd^{2+} ions. [After D. E. Mapother, H. N. Crooks, and R. J. Maurer, *J. Chem. Phys., 18,* 1231 (1950).]

amount of impurities can greatly enhance diffusion at moderate temperatures. An experimental example of this effect is provided by NaCl containing a trace of Cd^{2+} as an impurity (Fig. 12.6). At low temperatures where the number of Na^+ cation vacancies equals the number of Cd^{2+} cation impurities, the activation energy for Na^+ diffusion is associated with motion alone and the diffusivity is proportional to the impurity content. At high temperatures, the activation energy for Na^+ diffusion is larger since it involves the formation of Schottky defects in addition to the motion of Na^+ cations. It is of interest to note that the electrical conductivity of an ionic solid is directly related to the diffusion coefficient because ions are the charge carriers and electrical conduction corresponds to preferential ionic diffusion because of an applied voltage. For such materials, diffusivities can be obtained from electrical conductivity measurements.

Although most covalent solids have relatively open crystal structures (due to directional bonding) and thus larger sites for self-interstitials than metals and most ionic solids, the vacancy mechanism still predominates for self-diffusion and interdiffusion. In the diamond cubic structure (Fig. 3.2), for example, the self-interstitial sites have about the same volume as the atomic sites. Nevertheless, self-interstitials are unfavorable energetically because the covalent bond geometry of directed bonding orbitals cannot be fulfilled. Because of the directional bonding, the activation energies for self-diffusion in covalent solids generally are higher than in metals with comparable melting temperatures; Q for self-diffusion in germanium, for example, is 290 kJ/mol (69 kcal/mol) as compared to 186 kJ/mol (44 kcal/mol) for self-diffusion in silver, even though the melting points of Ag and Ge are within several degrees Celsius of each other.

12.4 OTHER TYPES OF DIFFUSION

In addition to diffusion through the bulk of a solid, atoms may migrate along external or internal paths that afford lower energy barriers to motion. Thus, diffusion can occur along dislocations, grain boundaries, or external surfaces. The rates of diffusion along such *short-circuit* paths are significantly higher than for volume diffusion since the associated activation energies are much lower than for volume diffusion. It is physically plausible that atoms should be most mobile on a free surface, and the atomic disarray at internal line or surface defects also results in relatively high atomic mobility. Nevertheless, most cases of mass transport are due to volume diffusion because the effective cross-sectional areas encountered in short-circuit processes are much smaller than those for volume diffusion. Exceptions to this general rule do occur, however. In fine-grained metals at relatively low temperatures, grain boundary diffusion is more important than volume diffusion (Fig. 12.7). This is so because grain boundary diffusivity, with a lower activation energy, varies less with temperature than does volume diffusivity,

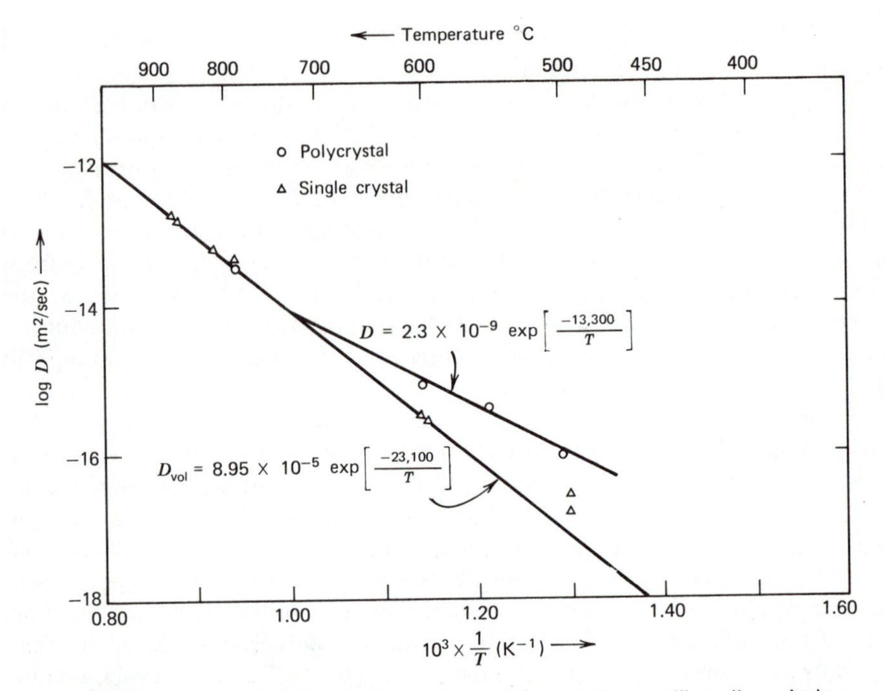

FIGURE 12.7 Self-diffusion coefficient in single and polycrystalline silver. At low temperatures, diffusion is faster in the polycrystalline material because the grain boundaries act as "short-circuit" paths for atomic motion. The activation energy is lower for grain boundary diffusion than it is for volume diffusion, which predominates at high temperatures. (Reproduced by permission from *Atom Movements,* American Society for Metals, 1951.)

with a higher activation energy. Furthermore, in certain powder processes involving sintering of metals or ceramics, the high surface-to-volume ratio of fine powder particles favors surface diffusion.

12.5 DIFFUSION IN NONCRYSTALLINE SOLIDS

Long-chain polymers, which are characterized by strong covalent bonds within molecules and weaker secondary bonds between molecules, exhibit sluggish diffusion because of their large molecular size. Self-diffusion in such materials involves the motion of molecular chain segments and is closely related to their viscous flow behavior. A greater degree of polymerization, with correspondingly larger molecules, results in lower diffusion rates. This can be observed experimentally when a polymer is dissolved in a solvent and interdiffusion is studied.

The more technologically important case of diffusion in polymers is that of foreign molecules, for this relates to the permeability and absorption characteristics displayed by polymers. During absorption, smaller molecules enter into the polymer causing swelling and, perhaps, chemical reactions, both of which can alter mechanical and physical properties. During permeation, smaller molecules diffuse through the polymer. Small molecules diffuse much more rapidly than do larger molecules, and only those atoms and molecules that are soluble yet remain essentially inert with respect to the polymer can diffuse readily. The diffusion path is almost invariably through noncrystalline regions within the polymer since crystalline regions present a much larger barrier to the migration of small foreign molecules. The rate of diffusion depends not only on the diffusing species but also on the state of the polymer. Below the glass transition temperature, diffusion is much less rapid than above it.

Somewhat analogous considerations hold for diffusion in inorganic glasses. In silicate glasses, the silicon atoms are so strongly bonded to their oxygen neighbors that their diffusivity remains low even at high temperatures, where the SiO_4 unit actually migrates. The rather large holes in the silicate network allow small atoms such as hydrogen or helium to permeate through glass quite readily. In addition, their chemical inertness with respect to the constituents of glass enhances their diffusivity. Such considerations explain the apparent *transparency* of glass to hydrogen and helium atoms and indicate the limited usefulness of glass for certain high-vacuum applications. Sodium and potassium ions also diffuse through glass with relative ease because of their small sizes. Their rates of diffusion, however, are considerably lower than those for hydrogen and helium since the cations experience a peripheral electrostatic attraction to the oxygen atoms in the Si—O network. This interaction is, nevertheless, much less restrictive than that sustained by silicon atoms.

12.6 MATHEMATICAL DESCRIPTION OF DIFFUSION

Diffusional processes can be classified as steady state or nonsteady state. In steady state diffusion the net number of atoms crossing a unit area perpendicular to a given direction per unit time is constant with time. In nonsteady state diffusion this quantity, called the flux, varies with time. Steady state diffusion is described by Fick's first law, which states that the flux, J, is proportional to the concentration gradient and the diffusion coefficient. For the one-dimensional case illustrated in Fig. 12.8, Fick's first law is given by

$$J_x = -D \frac{\Delta c}{\Delta x} = -D \frac{dc}{dx} \tag{12.2}$$

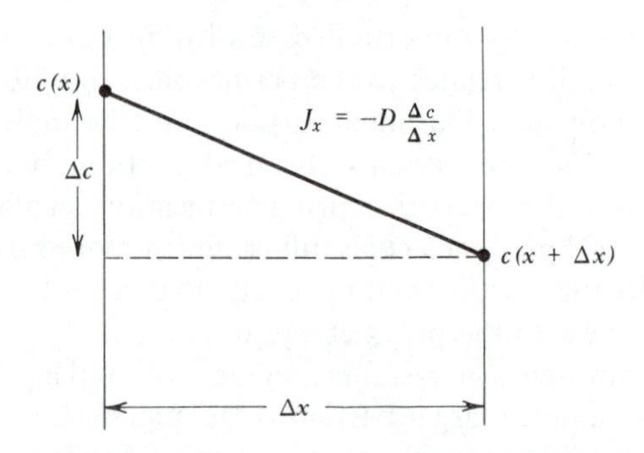

FIGURE 12.8 Steady state diffusion. If the concentration profile of the diffusing species remains constant with time, then steady state diffusion will occur. For the linear concentration gradient illustrated, the net number of atoms of the diffusing species that crosses a unit area per unit time (i.e., the flux) is given by Fick's first law, $J_x = -D\Delta c/\Delta x$. According to Fick's first law, atoms diffuse down a concentration gradient.

where $\Delta c/\Delta x$ (or dc/dx) is the concentration gradient. For diffusion considerations, convenient concentration units are atoms/cm^3 or g/cm^3 (either referred to one atomic species, for example, B atoms/cm^3), and the corresponding flux units are atoms/cm$^2 \cdot$ sec or g/cm$^2 \cdot$ sec, respectively. In both cases, the diffusion coefficient D has units of cm^2/sec. Equation 12.2 indicates that atoms tend to flow *down* a chemical concentration gradient. Equations analogous to Eq. 12.2 are used to describe steady state heat fluxes due to temperature gradients and steady state electrical currents due to electric fields (or potential gradients). An example of steady state diffusion is provided by the permeation of hydrogen atoms through a sheet of palladium with different imposed hydrogen gas pressures on either side of the slab. This process can be used to purify hydrogen gas since palladium is impermeable to other gases such as nitrogen, oxygen, and water vapor.

Most interesting cases of diffusion are not steady state, since the concentration at a given position changes with time and, therefore, the flux changes with time also. To treat such situations, a mass balance is applied to the diffusing species, and this gives rise to Fick's second law, which is used to analyze nonsteady state diffusion. In one dimension, Fick's second law is expressed as:

$$\frac{dc}{dt} = \frac{d}{dx}\left(D\,\frac{dc}{dx}\right) \tag{12.3}$$

where dc/dt is the time rate of change of concentration at a particular position x. If D is assumed to be constant (i.e., independent of concentration), then Fick's second law becomes

$$\frac{dc}{dt} = D\frac{d^2c}{dx^2} \tag{12.4}$$

Solutions to this equation have been obtained for a large number of boundary conditions and initial conditions. For example, in the case of a radioactive tracer deposited on the surface of a metal, the solution to Eq. 12.4 gives the concentration of the radioactive species in the originally nonradioactive metal as a function of time and position, x, measured from the original interface. Thus, for the particular case illustrated in Fig. 12.3, $c(x, t)$ is given by

$$c(x, t) = \frac{M}{\sqrt{\pi Dt}} e^{-(x^2/4Dt)} \tag{12.5}$$

where M is the mass of radioactive material originally deposited per unit area. When ln $c(x, t)$ is plotted as a function of x^2 after a given diffusion time at a constant heat treatment temperature, the slope equals $-1/4Dt$, and from this D can be determined. Concentration as a function of distance (Eq. 12.5) is plotted for two diffusion times in Fig. 12.9. It is noteworthy that to double the penetration distance effectively requires that the diffusion time be quadrupled. This arises from the exponent $x^2/4Dt$ in Eq. 12.5. It is generally found, regardless of the initial conditions and boundary conditions or of the exact form of the solution to Eq. 12.4, that the ratio x^2/Dt is the important parameter in determining the extent of diffusion. Accordingly, the order of magnitude of the effective diffusion distance in time t is given by $\sqrt{Dt}$, and the order of magnitude of time necessary for diffusion over a distance x is given by x^2/D. Thus, temperature has a large influence on diffusion distance and diffusion time by way of the diffusivity.

Brief comment on interdiffusion in a binary or two-component mixture is warranted because each component may migrate by the same mechanism, yet one component may diffuse more rapidly than the other. In the case of interdiffusion between gold and nickel, for example, gold atoms typically move faster than nickel atoms. One consequence of this is that the interdiffusion coefficient must depend upon the intrinsic diffusivities in binary alloys as well as the relative fractional amounts of each type of atom. For interdiffusion, the diffusion coefficient appearing in the Fick's law expressions (Eqs. 12.2 and 12.3) is actually the interdiffusion coefficient,

$$\tilde{D} = X_B D_A + X_A D_B \tag{12.6}$$

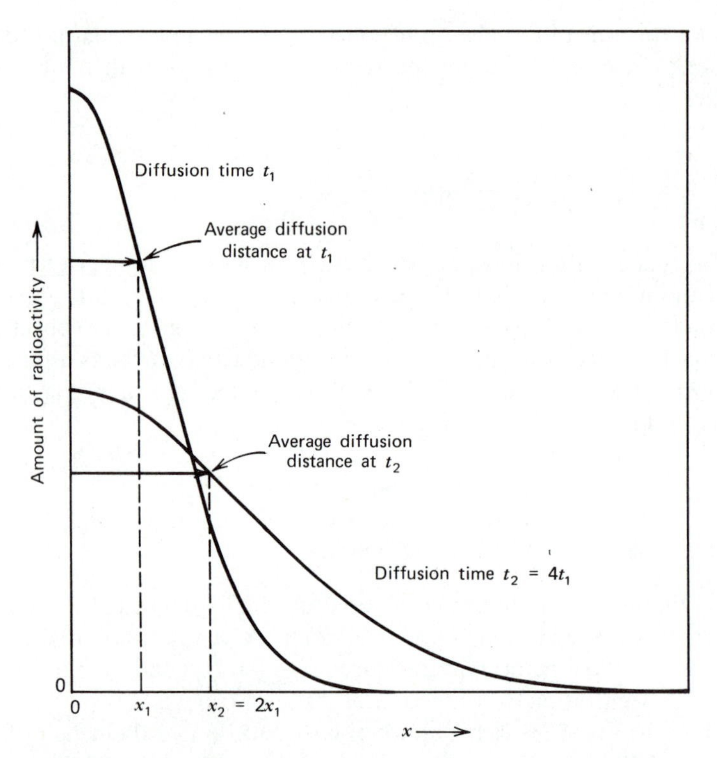

FIGURE 12.9 Solutions to Fick's second law for the self-diffusion experiment shown in Fig. 12.3. In this example of nonsteady state diffusion, it is necessary to quadruple the diffusion time, at a given temperature, in order to double the average diffusion distance of radioactive atoms into the nonradioactive material.

where D_A and D_B are the intrinsic diffusivities associated with the diffusion of A atoms and B atoms, respectively, in an A-B substitutional solid solution, and X_A and X_B are the respective mole fractions. Clearly, if D_A and D_B differ, then $\tilde{D}$ depends on composition. In most actual cases, D_A and D_B also depend on composition. It is important to note that, for interdiffusion in a dilute solution (i.e., $X_A \approx 1$ and $X_B \approx 0$), the intrinsic diffusivity of the dilute component gives the interdiffusion coefficient (i.e., $\tilde{D} \approx D_B$). Similarly, if one component is much more mobile than the other (i.e., $D_B \gg D_A$), this component determines the interdiffusion coefficient in alloys (i.e., $\tilde{D} \approx X_A D_B$, providing $X_A D_B \gg X_B D_A$). In the interdiffusion of an interstitial solute and a metal, each component diffuses by a different mechanism, and the diffusion coefficient in Fick's laws is that of the interstitial, since its diffusivity is so much greater than that of the substitutional atom.

Although Fick's first law (Eq. 12.2) indicates that the driving force for interdiffusion is a concentration gradient, it is more correct to consider the reduction in total free energy as the driving force. This coincides with equalization of the free energy throughout a phase, which corresponds to equalization of concentration. In most cases, elimination of concentration gradients does reduce the free energy; however, in some cases, diffusion occurs up, rather than down, a concentration gradient. Such uphill diffusion can be illustrated by the simple experiment shown in Fig. 12.10, which depicts carbon diffusion in a (Fe-Si)-steel diffusion couple. Here, more carbon accumulates on the steel side than would be expected if the concentration gradient were the driving force for diffusion. The presence of dissolved silicon on one side of the couple alters the equilibrium relationships, and, from a free energy point of view, carbon is more welcome in the steel than in the silicon iron.

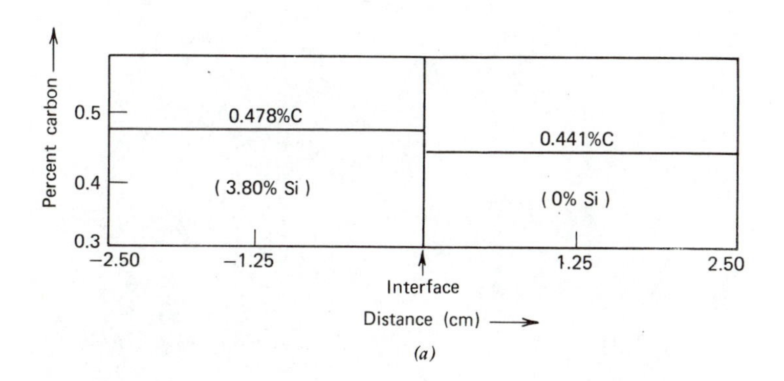

(*a*)

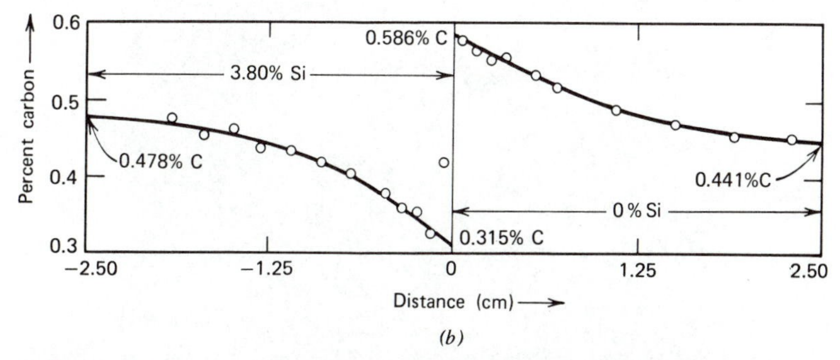

(*b*)

FIGURE 12.10 An example of uphill diffusion. (*a*) The initial carbon distribution is shown for a diffusion couple consisting of a steel with silicon and one without silicon. (*b*) After a diffusion heat treatment at 1050°C for 13 days, carbon has segregated preferentially into the steel that does not contain silicon. Thus, uphill diffusion has occurred. [After L. S. Darken, *Trans. AIME, 180,* 430 (1949).]

12.7 SOLID STATE SINTERING

Interparticle bonding and densification of powder compacts fired in the solid state depend on solid state diffusion. When fine oxide powders are compacted in steel dies at room temperature, between 25 and 60% porosity by volume remains; however, metal powders like copper or silver can deform during compaction and their compacts contain less porosity. A sintering treatment, conducted at elevated temperatures where solid state diffusion proceeds readily, causes pores to become smaller in volume and more spherical in shape. This leads to an increased density of the body (*densification*). The rate of densification increases at higher sintering temperatures because the diffusion coefficient increases.

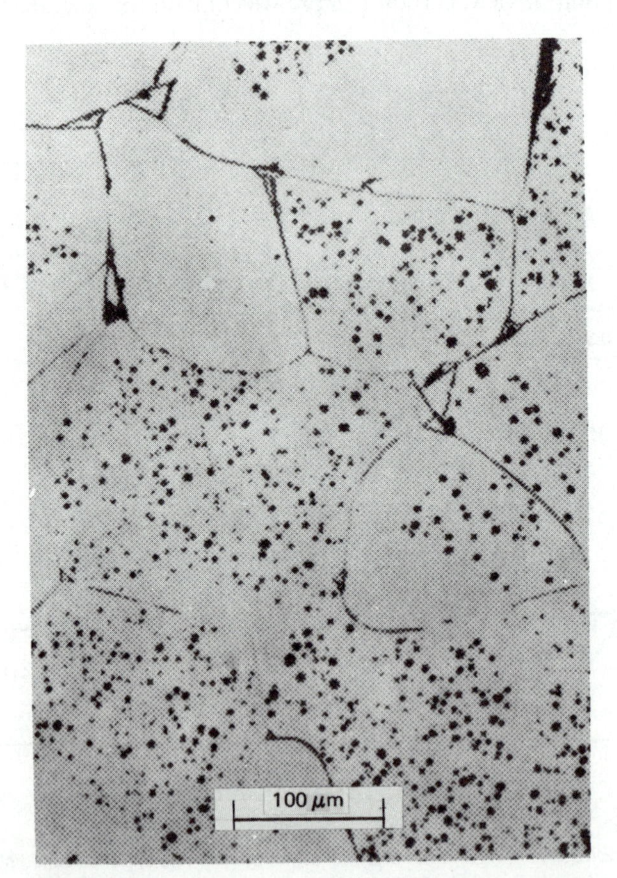

FIGURE 12.11 Microstructure of sintered Al_2O_3 showing the absence of porosity adjacent to grain boundaries and residual porosity within the grains. (Courtesy of J. E. Burke, General Electric Co.)

Although the mechanism of densification during solid state sintering is solid state diffusion, the driving force is the free energy decrease that accompanies the reduction in total surface area (see Sect. 11.8). Thus, for powder compacts of very fine initial particle size, the driving force is large and a high rate of densification can be achieved. Initially, the interparticle interfaces become grain boundaries that serve as sinks for vacancies originating at the pores. As long as grain boundaries intersect each pore, the pore volume decreases fairly rapidly with increasing sintering time. However, when grain growth takes place, grain boundaries move away from pores leaving them trapped within grains, and further densification becomes insignificant because volume diffusion, rather than grain boundary diffusion, must occur. As shown in Fig. 12.11, pure sintered Al_2O_3 has little porosity adjacent to grain boundaries but exhibits considerable porosity in the grain interiors.

Conventionally sintered products usually contain about 5% residual porosity by volume. Since residual porosity adversely affects mechanical strength, it is desirable to achieve complete densification. This requires more expensive processing procedures, however. One technique involves the application of mechanical pressure at the sintering temperature (termed *hot pressing*). Hot pressing is restricted to bodies of relatively simple shape and is limited by die materials. Another technique involves the addition of certain soluble substances that segregate to grain boundaries or the addition of very fine inert particles to the starting powder. These serve to inhibit grain boundary motion and, thus, restrict grain growth so that a fully dense, porefree, and fine-grained solid is produced by sintering.

REFERENCES

1. Girifalco, L. A. *Atomic Migration in Crystals*, Blaisdell, New York, 1964.
2. Shewmon, P. G. *Diffusion in Solids*, McGraw-Hill, New York, 1963.

QUESTIONS AND PROBLEMS

12.1 Calculate the diffusion coefficient for self-diffusion in copper at 1000°C if $Q = 48{,}000$ cal/mol and $D_0 = 0.2\ \text{cm}^2/\text{sec}$.

12.2 Consider a *random walk* problem by assuming a forward step is taken every time you throw a head in a head-tails process and a backward step every time a tail is thrown. Assume one second lapses for every time a step is taken. Take a coin and throw 20 flips and determine where you have walked to after 20 seconds, that is, if 11 heads and 9 tails are thrown your distance is $+2$, if 8 heads and 12 tails are thrown your distance is -4. Repeat this for a total of 200 flips and after each 20 flips determine your new position, $\bar{x}$. After you are all through plot the position, $(\bar{x})^2$ versus time. Do your results approximate a straight line? Such an experiment reveals the random nature of diffusion and illustrates how an atom wanders about in a crystal.

12.3 The diffusivities of carbon in α iron (bcc) and γ iron (fcc) are given by

$$D = 0.0079 \exp\left[-\frac{20{,}000 \text{ cal/mol}}{RT}\right] \text{cm}^2/\text{sec}$$

$$D = 0.21 \exp\left[-\frac{33{,}800 \text{ cal/mol}}{RT}\right] \text{cm}^2/\text{sec}$$

Calculate the respective diffusion coefficients at 800°C and explain the difference.

12.4 Make a schematic plot of $\ln D$ vs. $1/T$ for volume diffusion and grain boundary diffusion assuming $Q_{gb} \cong \frac{1}{2}Q_{\text{vol}}$. At which temperature regimes does grain boundary diffusion predominate over volume diffusion. (Your answer illustrates the general principle that high activation energy processes dominate at high temperatures and low activation energy ones at low temperatures.)

12.5 By considering a mass balance per unit time in a slab of thickness Δx and recognizing that the input of atoms per unit time per unit area is $J(x)$ and the corresponding output is $J(x + \Delta x) = J(x) + dJ/dx \cdot \Delta x$, derive Fick's second law.

12.6 Explain why, in Fig. 12.10*b*, there has been little or no net transfer of silicon from the left-hand side of the diffusion couple to the right-hand side. Sketch what you expect the concentration of silicon and carbon to be after diffusing for a very long time at 1200°C.

12.7 The diffusivity of hydrogen in fcc iron is three orders of magnitude greater than that of carbon in fcc iron at 1100°C. Explain.

12.8 What is the thermodynamic driving force for the mixing of gases illustrated in Fig. 12.1? For the mixing of radioactive and nonradioactive metals in Fig. 12.3?

12.9 Using the activation energies for self-diffusion in Table 12.1 and those supplied below, plot activation energy for diffusion vs. melting temperature in K. Discuss your results. *Data*: Self-diffusion activation energies (kcal/mol): Na(10), Nb(105), Cd(19), Pb(24), Cr(73), Ge(69), Co(68).

12.10 The activation energies for carbon, nitrogen, and hydrogen diffusion in bcc iron are 20, 18, and 3 kcal/mol, respectively. Explain the variation.

12.11 The activation energies for diffusion of Co in CoO and Fe in FeO are anomalously low (Table 12.2). Explain why. (*Hint*. Both Fe and Co are multivalent.)

13

Chemical Equilibrium and Reaction Rates

The conditions of chemical equilibrium and the rates of chemical reactions are of great technological importance. In this chapter the general principles of these topics are presented.

13.1 EQUILIBRIUM AND FREE ENERGY

Surfaces and interfaces are especially important in the consideration of chemical reactions involving solids and liquids since it is usually at a surface or interface where a chemical reaction is initiated. This is particularly true for corrosion and oxidation, which can cause deterioration of solids. These topics will be discussed in Chapters 14 and 15. It is first necessary, however, to introduce some pertinent concepts relating to chemical equilibrium and chemical reaction rates.

The brief discussion of phase changes, equilibrium, and rate of change presented in Chapter 1 can be extended directly to the topic of chemical reactions, since the principles involved are similar. A chemical system is said to be in stable equilibrium if it is not susceptible to a spontaneous change. Any change in temperature, concentration, or pressure can alter the equilibrium state and may produce a spontaneous chemical change. At constant temperature and pressure the thermodynamic criterion for a spontaneous change is a concurrent decrease in the free energy of the system. That is, a chemical reaction can take place only if ΔG is negative; in such a case, $-\Delta G$, is termed the *driving force* of the reaction. If ΔG is positive for a reaction as written, then the reaction can proceed only in the opposite direction. When a chemical system is at equilibrium under conditions of constant temperature and pressure, $\Delta G = 0$ for the reaction and the total free energy is a minimum. A word of caution is in order. Free energy is simply a useful quantity that serves as a measure of the possibility for a change in a system. Furthermore, free energy is suitably chosen so that it can be used to describe equilibrium at constant temperature and pressure. Water at room temperature and in contact with the atmosphere (saturated with water vapor) is stable, and this is called a stable equilibrium state since the free energy is at the lowest possible value for the given conditions of temperature and pressure. On the other hand, a gaseous mixture of hydrogen and oxygen may persist indefinitely at room temperature as a metastable system, although a reaction that forms water would lower the free energy. For a metastable equilibrium state any change to a more stable state necessitates an initial increase in free energy followed by a subsequently greater decrease. The required increase in free energy (activation free energy) largely determines the rate of a chemical reaction. Before discussing reaction rates, some thermodynamic concepts relating to equilibrium need to be developed.

13.2 FREE ENERGY AND THE EQUILIBRIUM CONSTANT

The free energy of a pure solid or a pure liquid depends primarily on temperature and only secondarily on pressure. The free energy of a solid or liquid solution involving two or more components also depends on the concentrations of the components. In contrast, the free energy of a gas or a solution of gases is strongly dependent on pressure as well as temperature and concentration of components. The latter, however, mainly influences the entropy of gaseous mixtures and has only a very slight effect on their internal energy or enthalpy.

Free energies have been determined experimentally for the elements and for a large number of chemical compounds under standardized conditions. These standard state free energies are designated G° and usually correspond to the state of a pure component (either element or compound) at one atmosphere pressure and the temperature of interest. Standard state free energies provide a reference level to which free energies of solutions and free energy changes associated with chemical reactions can be compared. For example, the free energy of one mole of an ideal gas as a function of pressure is given by

$$G = G^\circ + RT \ln P \tag{13.1}$$

where R is the universal gas constant, T is the absolute temperature, and P is the pressure in atmospheres. Similarly, for a mole of *one component* in a mixture of ideal gases, the free energy is

$$G_A = G_A^\circ + RT \ln p_A \tag{13.2}$$

Here, p_A is the partial pressure of component A, that is, the pressure that A would exert if all other components were removed from a given volume. Clearly, when P in Eq. 13.1 or p_A in Eq. 13.2 equals 1 atm, the free energy of the pure gas or of gaseous component A equals the standard state free energy. In effect, P and p_A represent *concentrations*, and a relationship analogous to the above equations gives the free energy of a mole of one component in an ideal liquid solution or an ideal solid solution:

$$G_A = G_A^\circ + RT \ln [A] \tag{13.3}$$

where $[A]$ designates the concentration of component A. Here unit concentration corresponds to the standard state. Although concentration can be specified in many different ways, the number of moles of a given component in a condensed-phase solution divided by the total number of moles of all components in the solution (i.e., the mole fraction) often is the most useful.

Equations 13.1 and 13.2 are applicable only for ideal gases. They can, however, be used for real gases at normal and low pressures because deviations from ideal behavior usually are slight under such conditions. On the other hand, Eq. 13.3 may be quite inaccurate if applied to real condensed-phase solutions (i.e., an *ideal* solution is by definition one for which Eq. 13.3 is true). Nevertheless, because of its simple form, we shall prefer to utilize Eq. 13.3 as necessary in this and subsequent chapters.

The formalism employed above can be used to advantage in ascertaining whether a chemical reaction can occur and what the relative proportions of reactants and products will be at equilibrium conditions. The oxidation of fine silver powder exposed to an atmosphere containing oxygen affords an example:

$$2\text{Ag}(s) + \tfrac{1}{2}\text{O}_2(g) \longrightarrow \text{Ag}_2\text{O}(s) \tag{13.4}$$

At any temperature, the free energy change for this reaction is

$$\Delta G = G_{\text{Ag}_2\text{O}} - 2G_{\text{Ag}} - \tfrac{1}{2}G_{\text{O}_2} \tag{13.5a}$$

and this may be rewritten as

$$\begin{aligned}\Delta G &= G^\circ_{\text{Ag}_2\text{O}} + RT \ln [\text{Ag}_2\text{O}] - 2G^\circ_{\text{Ag}} - 2RT \ln [\text{Ag}] - \tfrac{1}{2}G^\circ_{\text{O}_2} \\ &\quad - \tfrac{1}{2}RT \ln p_{\text{O}_2} \\ &= \Delta G^\circ + RT \ln \frac{[\text{Ag}_2\text{O}]}{[\text{Ag}]^2 p_{\text{O}_2}^{1/2}}\end{aligned} \tag{13.5b}$$

where $\Delta G^\circ = G^\circ_{Ag_2O} - 2G^\circ_{Ag} - \frac{1}{2}G^\circ_{O_2}$ = the standard state free energy change. Since we are considering pure Ag and pure Ag_2O, their respective concentrations in Eq. 13.5b are unity (i.e., these substances are in their standard states), and the free energy of reaction becomes

$$\Delta G = \Delta G^\circ + RT \ln \frac{1}{p_{O_2}^{1/2}} \tag{13.5c}$$

Whether or not the reaction (Eq. 13.4) is thermodynamically possible depends on the partial pressure of oxygen and the temperature, which appears explicitly in Eq. 13.5c and also implicitly since ΔG° varies with temperature. At 300 K, $\Delta G^\circ = -10500$ J per mole of Ag_2O formed (Fig. 13.1); thus, in terms of joules per mole of Ag_2O formed, Eq. 13.5c becomes

$$\Delta G(300\text{ K}) = -10500 + (8.314)(300) \ln \frac{1}{p_{O_2}^{1/2}} \tag{13.6a}$$

At constant temperature, ΔG will be greater than, equal to, or less than zero, depending on the partial pressure of oxygen. The equilibrium value of p_{O_2} at 300K (Fig. 13.1) occurs when $\Delta G = 0$, which corresponds to an oxygen partial pressure of $p_{O_2} \simeq 2.2 \times 10^{-4}$ atm. If p_{O_2} is less than this value (i.e., $\Delta G > 0$), Ag will not oxidize; if p_{O_2} is greater than this value, Ag *may* oxidize, providing the reaction rate is finite.

Chemical equilibrium is characterized not only by $\Delta G = 0$ but also by time-independent concentrations for each component in a system. A wealth of experimental evidence has demonstrated that there is a constant ratio of the concentrations of products to the concentrations of reactants when a chemical system is in equilibrium at a fixed temperature and pressure. For the general case of a chemical reaction having the form

$$aA + bB + cC + \cdots \longrightarrow lL + mM + nN + \cdots \tag{13.7}$$

the equilibrium relationship is written as

$$\left\{\frac{[L]^l[M]^m[N]^n \cdots}{[A]^a[B]^b[C]^c \cdots}\right\}_{\text{eqm}} = K_{\text{eqm}} \tag{13.8}$$

That is, the equilibrium concentrations of reactants and products are raised to the power corresponding to the number of moles of each species involved in the reaction, as indicated by the stoichiometric coefficients in Eq. 13.7, and Eq. 13.8

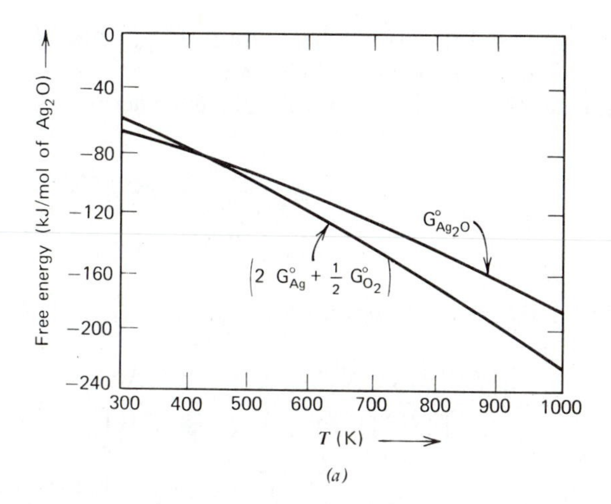

(*a*)

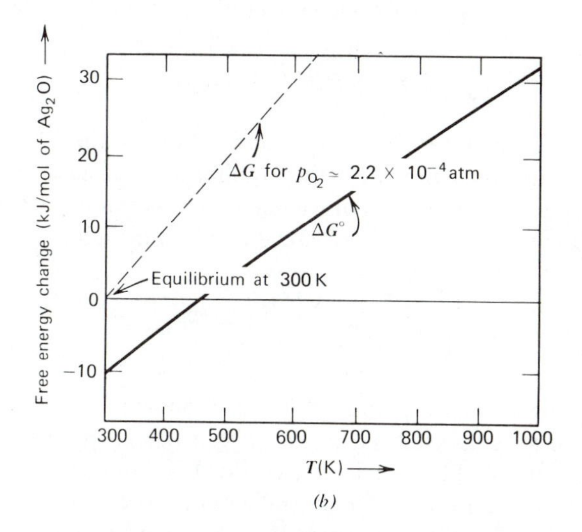

(*b*)

FIGURE 13.1 (*a*) The variation with temperature of the standard state free energies of the reactants and product for the oxidation of silver. (*b*) The free energy change associated with the formation of Ag_2O as a function of temperature. The solid line gives the standard state free energy change (i.e., for $p_{O_2} = 1$ atm), which is the difference between the curves shown in (*a*). The broken line gives the free energy change when $p_{O_2} \simeq 2.2 \times 10^{-4}$ atm, the equilibrium partial pressure of oxygen at room temperature (~300 K).

must be satisfied at equilibrium. Here, the quantity K_{eqm}, called the equilibrium constant for the reaction, is a function of temperature only.

For the general chemical reaction (Eq. 13.7), the free energy change at constant temperature is

$$\Delta G = \Delta G^\circ + RT \ln\left\{\frac{[L]^l[M]^m[N]^n \cdots}{[A]^a[B]^b[C]^c \cdots}\right\} \tag{13.9a}$$

which at equilibrium (i.e., $\Delta G = 0$) becomes

$$\ln\left\{\frac{[L]^l[M]^m[N]^n \cdots}{[A]^a[B]^b[C]^c \cdots}\right\}_{eqm} = -\frac{\Delta G^\circ}{RT} \tag{13.9b}$$

By identification of Eq. 13.9b with Eq. 13.8, it is apparent that the standard state free energy change and the equilibrium constant are uniquely related by

$$\Delta G^\circ = -RT \ln K_{eqm} \tag{13.10}$$

and a comparison of the ratio of actual concentrations of the products to actual concentrations of the reactants with the equilibrium constant can be used to ascertain whether equilibrium exists or if a reaction is thermodynamically possible. For example, if the ratio of actual concentrations at some temperature is

$$\frac{[L]^l[M]^m[N]^n \cdots}{[A]^a[B]^b[C]^c \cdots} < K_{eqm} \tag{13.11}$$

then $\Delta G < 0$ for Eq. 13.7 as written, and the reaction will tend to proceed toward the right. Thus, if silver is exposed to an oxygen partial pressure of 10^{-3} atm at room temperature, it should oxidize (see Eq. 13.6b and Fig. 13.2). In reality, silver can be exposed to air ($p_{O_2} = 0.21$ atm) without oxidation at room temperature. This is because the above thermodynamic considerations do not provide information about the rate of a reaction.

The distinction between the thermodynamics of a reaction and the kinetics or rate of a reaction is very important for the former relates solely to whether or not a reaction is possible and to what the equilibrium state is, whereas the latter relates to how fast the reaction occurs, that is, to how fast the equilibrium state is approached. It is observed experimentally that the rates of chemical reactions depend on four factors: (1) the concentrations of the reacting species and of the product species, (2) the temperature (and pressure, if gases are involved), (3) the nature of the reactants, and (4) the presence or absence of catalysts. These concepts are discussed in the remainder of this chapter.

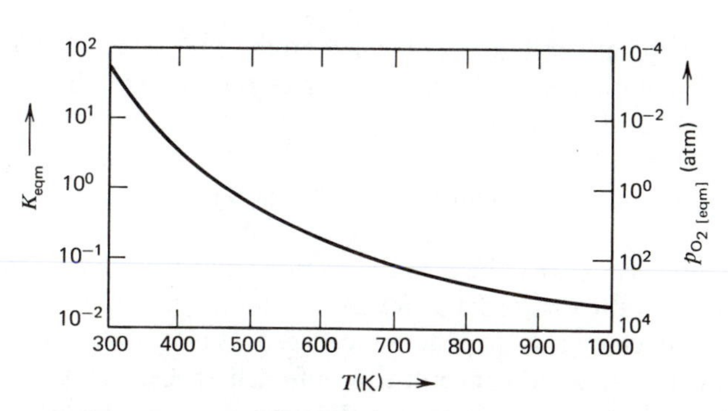

FIGURE 13.2 The equilibrium constant as a function of temperature for the reaction $2Ag + \frac{1}{2}O_2 \rightleftharpoons Ag_2O$. Compare the temperature variation of K_{eqm} with that of $\Delta G°$, shown in Fig. 13.1*b*. The right-hand ordinate gives the equilibrium partial pressure of oxygen, which is related to the equilibrium constant as $p_{O_2[eqm]} = K_{eqm}^{-2}$, in this case.

13.3 RATES OF CHEMICAL REACTIONS—EFFECT OF CONCENTRATION

The influence of concentration can be described by considering the following hypothetical reaction between two gaseous species at constant temperature:

$$A + BC \longrightarrow AC + B \tag{13.12}$$

The reaction, as read from left to right, can be physically visualized by noting that an A atom and a BC molecule must collide for the reaction to take place. The probability of such collisions per unit volume depends upon the number of A atoms and the number of BC molecules per unit volume, that is, their concentrations (or partial pressures). Doubling the concentration of A atoms will double this probability; alternatively, doubling the concentration of BC also doubles this probability. If the concentrations of both are doubled, the collision rate and, hence, the reaction rate quadruples. Thus the reaction rate for Eq. 13.12, going from left to right, may be written as

$$(\text{rate})_f = k_f[A][BC] \tag{13.13}$$

where $[A]$ and $[BC]$ represent the concentrations of the reactants and k_f is the reaction rate constant that depends on temperature, the nature of the reactants, and catalysts.

While A and BC are colliding to form B atoms and AC molecules, the reverse reaction occurs simultaneously. Using the same arguments as above, the rate of the reverse reaction can be written as

$$(\text{rate})_r = k_r[AC][B] \tag{13.14}$$

If equilibrium does not exist, the overall reaction will proceed at a rate given by the difference between the forward and reverse reaction rates, and the concentrations of the species will change as the reaction proceeds (Fig. 13.3). At equilibrium, there is no net reaction and the concentration of each species does not change with time (Fig. 13.3). Equilibrium is thus a dynamic state in which the rates of the forward and reverse reactions are equal:

$$k_f[A][BC] = k_r[AC][B] \tag{13.15a}$$

and the equilibrium ratio of the concentrations of the products to those of the reactants equals the ratio of the reaction rate constants:

$$\left\{\frac{[AC][B]}{[A][BC]}\right\}_{\text{eqm}} = \frac{k_f}{k_r} \tag{13.15b}$$

Comparison of Eq. 13.15b with Eq. 13.8 leads to the conclusion that the ratio of the forward and reverse reaction rate constants, k_f/k_r, is identical to the equilibrium constant K_{eqm}.

If a reaction involving gaseous species has the form

$$2A + BC \longrightarrow A_2C + B \tag{13.16}$$

then the probability concepts employed above show that the rate of the forward reaction is proportional to the square of the concentration of A; that is,

$$(\text{rate})_f = k_f[A]^2[BC] \tag{13.17}$$

The rate of the reverse reaction is

$$(\text{rate})_r = k_r[A_2C][B] \tag{13.18}$$

and the corresponding equilibrium relationship is given by

$$\left\{\frac{[A_2C][B]}{[A]^2[BC]}\right\}_{\text{eqm}} = \frac{k_f}{k_r} = K_{\text{eqm}} \tag{13.19}$$

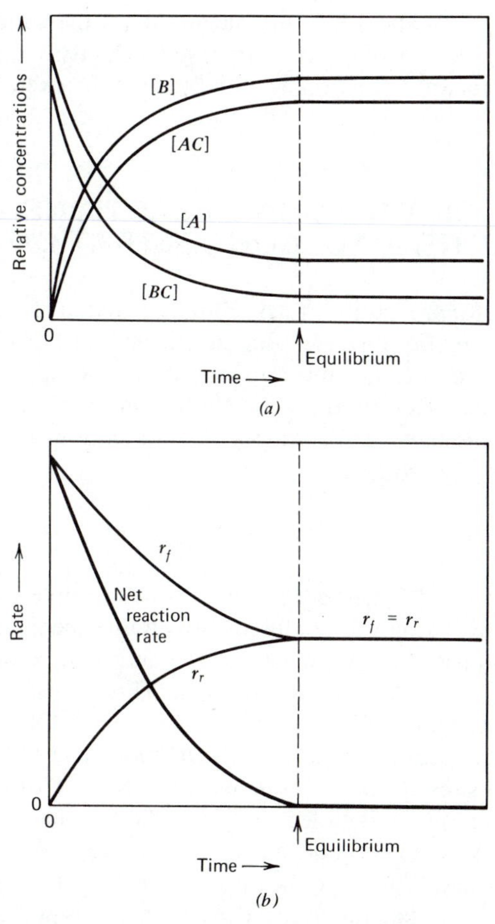

FIGURE 13.3 (*a*) Relative concentrations of reactants (*A* and *BC*) and products (*B* and *AC*) as a function of time during an isothermal reaction. Note that the concentrations remain constant after equilibrium is reached. (*b*) Rates of the forward reaction (r_f), the reverse reaction (r_r), and the overall reaction (net reaction rate = $r_f - r_r$) as a function of time at constant temperature. Note that the net reaction rate goes to zero (i.e., the forward and reverse reaction rates become equal) when equilibrium is attained.

According to Eqs. 13.15b and 13.19, the magnitude of the equilibrium constant depends on the relative sizes of the reaction rate constants, k_f and k_r. If $k_f \gg k_r$, then the rate of the forward reaction essentially gives the overall or net reaction rate.

13.4 RATES OF CHEMICAL REACTIONS—FACTORS IN ADDITION TO CONCENTRATION

The importance of temperature in determining reaction rates may be demonstrated by reconsidering Eq. 13.12. In order for the forward reaction to occur, not only must A and BC collide, they must do so with sufficient kinetic energy so that the BC bond can be broken to allow for the formation of an AC bond. In this regard, Eq. 13.12 may be written so as to show the intermediate, unstable state in which the BC bond is broken:

$$A + BC \longrightarrow (ABC)^* \longrightarrow AC + B \tag{13.20}$$

The intermediate state, $(ABC)^*$, is termed the activated complex, and exists for only about 10^{-13} sec; it should not be confused with an intermediate chemical species produced and consumed during a reaction. The activated complex has a higher energy than the original A atom and BC molecule. This increase in energy, denoted as the activation enthalpy ΔH^* and commonly called the activation energy, represents the energy necessary to disrupt the BC bond and to form the activated complex, which can subsequently decompose into $AC + B$ or revert to $A + BC$. Most of the gaseous atoms and molecules lack sufficient kinetic energy to achieve the activated complex state on collision. Thus, most collisions between A atoms and BC molecules will not produce the forward reaction (Eq. 13.20). However, at any finite temperature, a small fraction of constituents does possess kinetic energies greater than or equal to ΔH^*, and this fraction increases significantly as the gas temperature increases (Fig. 13.4). It is known that this fraction varies with temperature according to an Arrhenius relationship, that is, as $e^{-\Delta H^*/RT}$ (see Sect. 1.6).

If we assume that Eq. 13.20 represents the reaction mechanism, then an absolute reaction rate expression that includes the temperature dependence explicitly may be developed. For this chemical reaction, the net rate is given by the difference between Eq. 13.13 and Eq. 13.14:

$$\begin{aligned}(\text{rate})_{\text{net}} &= k_f[A][BC] - k_r[AC][B] \\ &= k_f[A][BC]\left(1 - \frac{k_r}{k_f}\frac{[AC][B]}{[A][BC]}\right)\end{aligned} \tag{13.21}$$

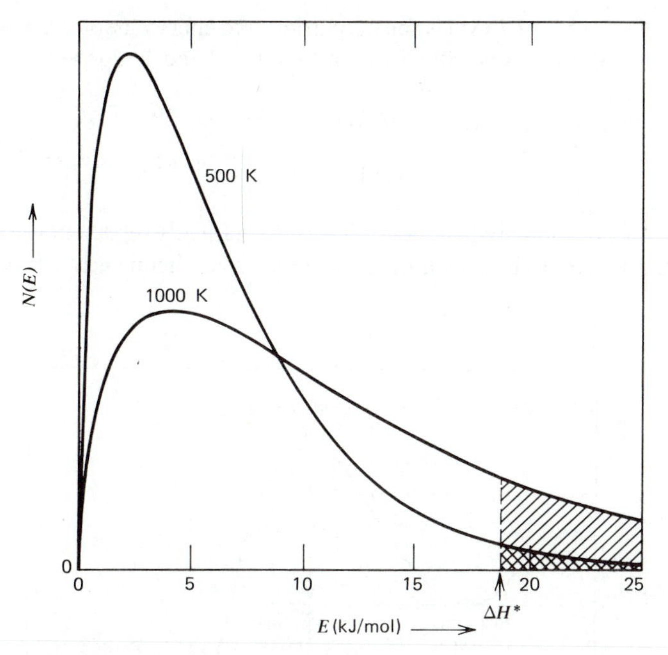

FIGURE 13.4 The distribution of translational kinetic energies for gaseous molecules at 500 K and 1000 K (cf. Fig. 1.3). This curve can be interpreted in the same manner as that for the density of electronic states in a metal (Fig. 4.15); that is, $N(E)\,dE$ is the fraction of molecules with energy between E and $E + dE$. The total fraction of molecules possessing energies greater than a hypothetical activation energy (ΔH^*) is given by the area under the distribution curve from $E = \Delta H^*$ to $E \to \infty$; note that this area (partially represented by the shaded regions) increases markedly as the temperature rises.

Through Eqs. 13.15b, 13.10, and 13.9a, this expression can be rewritten as

$$(\text{rate})_{\text{net}} = k_f[A][BC]\left(1 - \frac{[AC][B]}{[A][BC]}\, e^{\Delta G^\circ/RT}\right)$$
$$= k_f[A][BC](1 - e^{\Delta G/RT}) \tag{13.22}$$

As mentioned above, only a certain fraction of A atoms and BC molecules have sufficient kinetic energies to form activated complexes on collision, and this can be taken into account in the forward reaction rate constant, k_f:

$$k_f = ke^{-\Delta G^*/RT} = (ke^{\Delta S^*/R})e^{-\Delta H^*/RT} \tag{13.23}$$

where $\Delta G^*(=\Delta H^* - T\,\Delta S^*)$ is the activation free energy associated with forming the activated complex. A combination of Eqs. 13.22 and 13.23 yields

$$(\text{rate})_{\text{net}} = k[A][BC](1 - e^{\Delta G/RT})e^{-\Delta G^*/RT} \tag{13.24a}$$

$$= ke^{\Delta S^*/R}[A][BC](1 - e^{\Delta G/RT})e^{-\Delta H^*/RT} \tag{13.24b}$$

Equations 13.29a and b make it possible to schematically represent the relationship between the thermodynamics and the kinetics of a chemical reaction (Fig. 13.5).

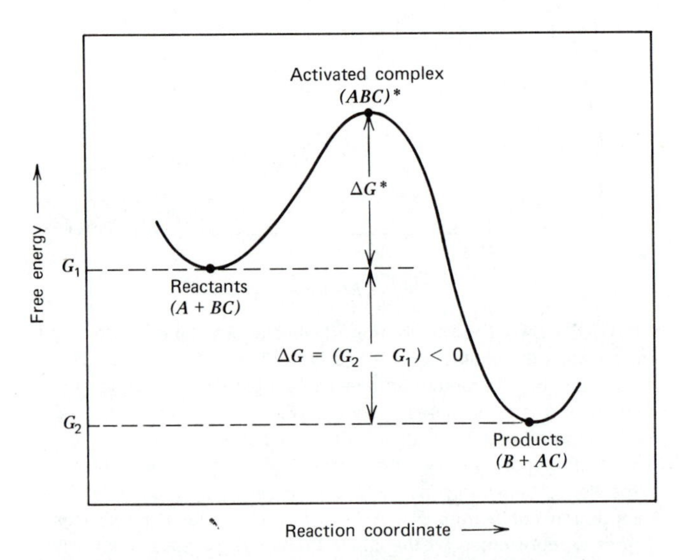

FIGURE 13.5 The variation of free energy for an isothermal chemical reaction as a function of reaction coordinate. For the reaction to occur, there must be sufficient thermal energy available so that the reactants (A and BC) can form the activated complex $(ABC)^*$, which can then produce the products (B and AC). During the course of the reaction, ΔG decreases in magnitude and becomes zero at equilibrium, whereas, ΔG^* remains essentially unchanged.

This type of representation is also valid for phase changes. The form of Eq. 13.24b is instructive because it shows that the main temperature dependence of the reaction rate is contributed by the exponential involving the activation energy ΔH^*.

Since the activation energy is sensitive to the nature of the reactants and to any catalysts, the rate of a reaction will depend on these factors. The effect of the nature of the reactants on the activation energy can be demonstrated as follows. For a simple reaction such as

$$A + A \longrightarrow A_2 \tag{13.25}$$

it is likely that the activation energy will be lower than in the case of the reaction in Eq. 13.20. In the present case, no chemical bonds need to be broken in order to initiate the reaction.

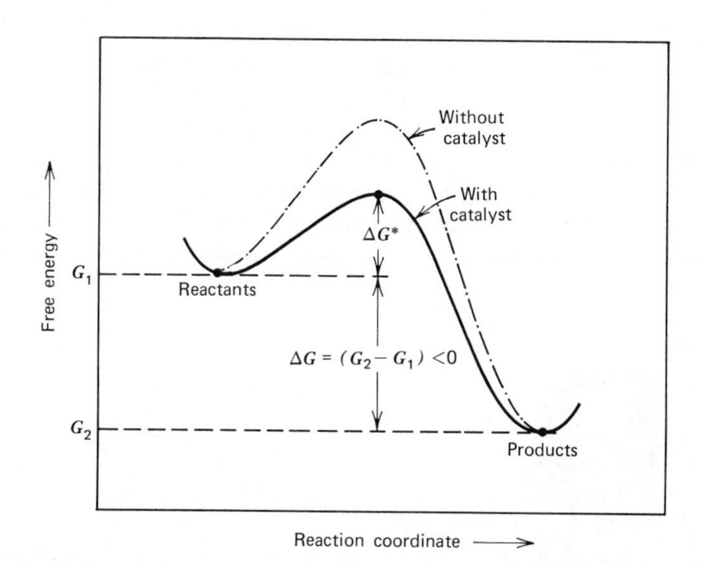

FIGURE 13.6 The effect of a catalyst on the variation of free energy for an isothermal chemical reaction as a function of reaction coordinate. The presence of a catalyst lowers the activation free energy and, hence, increases the reaction rate, but the catalyst does not alter ΔG, the free energy change associated with the reaction.

Catalysis increases reaction rates by lowering the activation energy barrier (or activation free energy barrier) without altering the equilibrium state. The precise way in which catalysts work is usually not completely known, but they generally alter the mechanism of a reaction by facilitating bond breaking or formation. The path of a reaction with and without a catalyst is shown schematically in Fig. 13.6.

13.5 KINETICS OF SOLIDIFICATION AND MELTING

The general principles of reaction-rate theory can be applied to an easily visualized physical process—that of freezing or melting of a pure material. The rate of melting is taken to be equal to the rate at which atoms migrate from the solid to the liquid across the solid–liquid interface, and the rate of freezing is the converse (Fig. 13.7*a*). The net rate of melting is simply the difference between the two rates of these elementary processes. If the rate of melting is greater than the rate of freezing, the material melts; if the opposite is true, the material freezes. At equilibrium, the rates are equal, and liquid and solid coexist.

Figure 13.7*b* shows that the activation free energy associated with the freezing process is ΔG^* and that associated with the melting process is $(\Delta G^* - \Delta G)$, where ΔG is the free energy change for the liquid to solid transformation. The activated state is one in which an atom is neither part of the liquid nor part of the solid, but in between. In addition to thermal activation, the rate of freezing depends on the concentration of atoms in the liquid at the solid–liquid interface, which can be taken as unity in this case, and on geometrical factors. The latter include the probability that an atom at the interface with sufficient thermal energy to overcome the activation barrier is moving in the direction of the solid, as well as

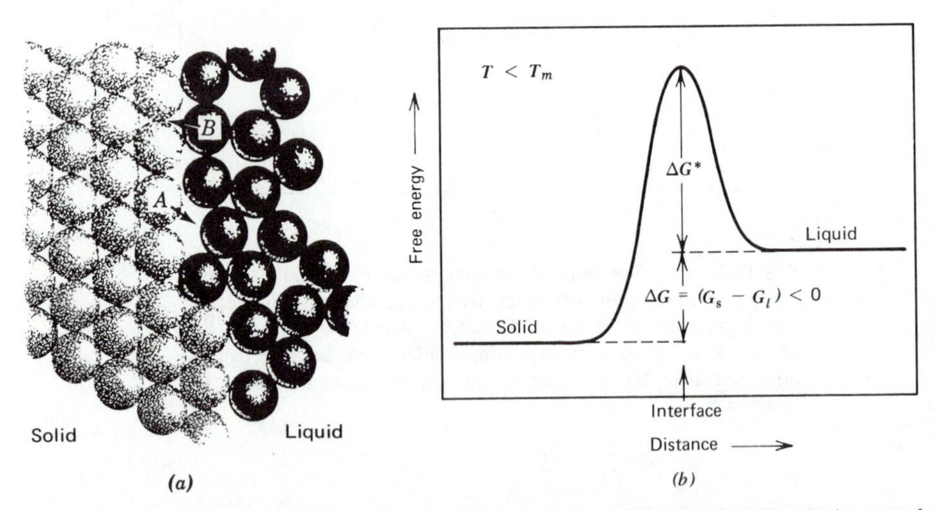

FIGURE 13.7 (*a*) Solidification viewed as a dynamic process. The rate of melting is the rate of transfer of atoms from the solid to the liquid at the solid–liquid interface (see atom *A*). The rate of freezing is the rate of transfer of atoms from the liquid to the solid at the interface (see atom *B*). (*b*) The free energy variation for an atom in the vicinity of the solid–liquid interface at a temperature below T_m. The free energy barrier for an atom going from the liquid to the solid is ΔG^*; the barrier for the opposite direction is $\Delta G^* - \Delta G$. For the case illustrated here, $\Delta G (= G_s - G_l)$ is negative.

the probability that such an atom can be accommodated into the structure of the crystalline solid. These two terms can be combined into a constant, α_f, and the freezing rate can be expressed as

$$r_f = \alpha_f \, e^{\Delta S^*/R} e^{-\Delta H^*/RT} \tag{13.26}$$

Similar considerations hold for the melting rate, which is given by

$$r_m = \alpha_m e^{(\Delta S^* - \Delta S)/R} e^{-(\Delta H^* - \Delta H)/RT} \tag{13.27}$$

where ΔS and ΔH are the entropy and enthalpy changes of the liquid to solid transformation. If ΔS^* and ΔS are insensitive to temperature (which is a valid approximation, in general), Eqs. 13.26 and 13.27 can be written as

$$r_f = k_f e^{-\Delta H^*/RT} \tag{13.28}$$

and

$$r_m = k_m e^{-(\Delta H^* - \Delta H)/RT} \tag{13.29}$$

where

$$k_f = \alpha_f e^{\Delta S^*/R}$$

and

$$k_m = \alpha_m \, e^{(\Delta S^* - \Delta S)/R}$$

The ratio of k_f to k_m can be calculated by realizing that r_f and r_m are equal at the equilibrium melting temperature, T_m. This means that solid and liquid can coexist indefinitely at this temperature and

$$k_f = k_m e^{\Delta H/RT_m} \tag{13.30}$$

Since ΔH and T_m can be measured experimentally, k_f/k_m is readily calculated. Knowledge of this parameter, in turn, allows calculation of the ratio r_f/r_m. Results of such calculations for copper are shown in Fig. 13.8; r_m rises more rapidly with increasing temperature than r_f since the former has a higher activation energy [i.e., $(\Delta H^* - \Delta H) > \Delta H^*$, because $\Delta H < 0$ for the liquid to solid transformation].

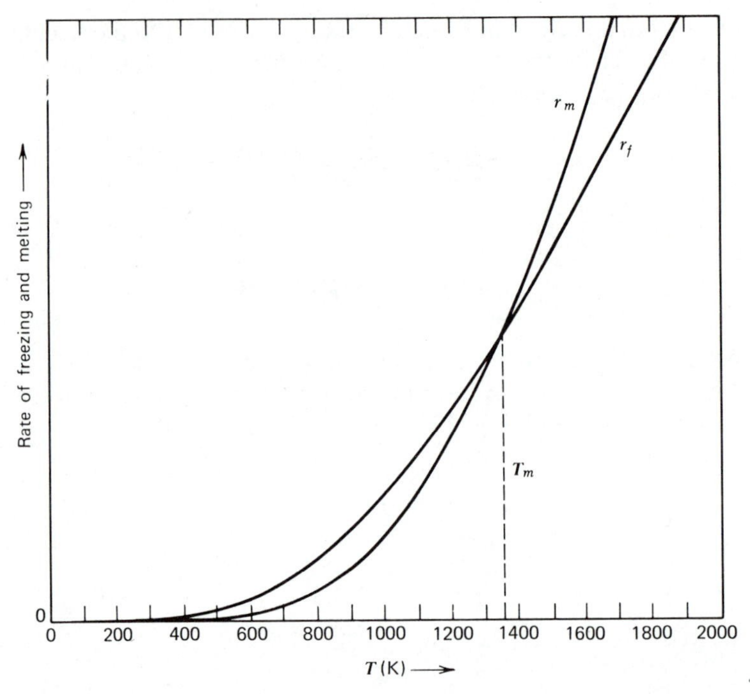

FIGURE 13.8 Schematic variation of the freezing and melting rates for copper. At T_m, $r_m = r_f$, corresponding to dynamic equilibrium. Above T_m, melting dominates; below T_m, freezing dominates. [After K. A. Jackson and B. Chalmers, *Can. J. Phys.*, *34*, 473 (1956).]

The net rate of melting (or the negative of the net rate of freezing), $\bar{r} = r_m - r_f$, is shown in Fig. 13.9. Above the equilibrium melting temperature, $\bar{r}$ is positive and melting dominates; below T_m, $\bar{r}$ is negative and freezing dominates. It is especially interesting to note that the net freezing rate reaches a maximum at a finite value of undercooling (Fig. 13.9). As will be shown in Chapter 17, such behavior, which is commonly observed for transformations occurring on cooling in condensed-phase systems, arises from the transformation being thermally activated. At low temperatures, even though the liquid to solid transformation is highly favored thermodynamically, thermal agitation is insufficient to activate the transformation. This characteristic is one of the important features that allows certain liquids to become glasses without crystallization. In contrast, the net rate of melting increases monotonically with superheating since thermal agitation and the thermodynamic driving force both increase with increasing temperature.

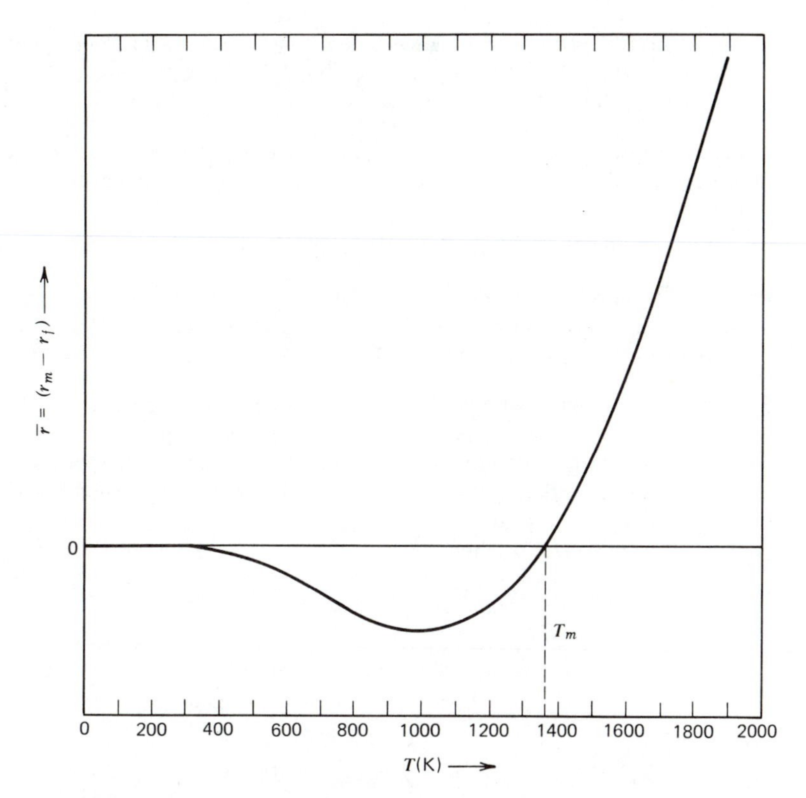

FIGURE 13.9 The net rate of melting, $\bar{r} = r_m - r_f$, for copper as a function of temperature. Positive values of $\bar{r}$ correspond to melting, and negative values correspond to freezing. The net rate of freezing reaches a maximum at a temperature near 1000 K. [After K. A. Jackson and B. Chalmers, *Can. J. Phys., 34,* 473 (1956).]

REFERENCES

1. Campbell, J. A. *Why Do Chemical Reactions Occur*? Prentice-Hall, Englewood Cliffs, N.J., 1965.
2. Mahan, B. H. *College Chemistry*, Addison-Wesley, Reading, Mass., 1969.

QUESTIONS AND PROBLEMS

13.1 The standard state entropies for iron, FeO and O_2 at 298 K (and 1 atm) are 6.49 cal/K·mol, 13.74 cal/K·mol and 49.01 cal/K·mol, respectively.

(a) Why should a gas such as oxygen have higher entropy values than solids (e.g., Fe and FeO)?

(b) Since $\Delta G^\circ = \Delta H^\circ - T\Delta S^\circ$, how does the stability of FeO change with temperature? (For FeO, $\Delta H^\circ = -63.8$ kcal/mol at 298 K; assume ΔH° and ΔS° are not functions of temperature.)

13.2 In the case of FeO (problem 13.1, part b), the enthalpy of formation is negative. This indicates that, when Fe and O_2 react to form FeO, heat is evolved (i.e., the reaction is exothermic).

(a) Using the relationship between ΔG° and K_{eqm}, show that $d \ln K_{\text{eqm}}/dT = \Delta H^\circ/RT^2$.

(b) Briefly discuss (use words and sketches) the temperature dependence of the equilibrium constant for (i) an exothermic reaction (i.e., $\Delta H^\circ < 0$) and (ii) an endothermic reaction (i.e., $\Delta H^\circ > 0$).

13.3 For reactions involving mixtures of gases in a container of fixed volume, it is sometimes convenient to express concentrations as the number of moles per unit volume. In the case of ideal gases, the concentration for a component is related to the partial pressure of that component by the gas law, that is, $p_i = c_i RT$, where c_i is the number of moles of component i per unit volume.

(a) Consider the general reaction

$$aA + bB \longrightarrow dD + eE$$

and show that the concentration equilibrium constant,

$$K_c = \left\{\frac{c_D^d c_E^e}{c_A^a c_B^b}\right\}_{\text{eqm}}$$

is related to the equilibrium constant defined by Eq. 13.8

$$K_{\text{eqm}} = \left\{\frac{[D]^d[E]^e}{[A]^a[B]^b}\right\}_{\text{eqm}} = \left\{\frac{p_D^d p_E^e}{p_A^a p_B^b}\right\}_{\text{eqm}}$$

as

$$K_{\text{eqm}} = K_c(RT)^{\Delta n}$$

where $R = 82$ atm cm^3/K·mol and Δn is the change in the number of moles of gas during the reaction. [In this case, $\Delta n = (d + e) - (a + b)$.]

(b) Because of the definition of standard state, $\Delta G^\circ = -RT \ln K_{\text{eqm}}$. Using this relationship and $\Delta G^\circ = \Delta H^\circ - T\Delta S^\circ$, show that $d \ln K_{\text{eqm}}/dT = \Delta H^\circ/RT^2$.

(c) Develop an expression for $d \ln K_c/dT$.

(d) Under what conditions will K_c be identical to K_{eqm}?

13.4 Nitrogen and hydrogen can react to form ammonia as follows:

$$N_2 + 3H_2 \longrightarrow 2NH_3$$

Thermodynamic data for this reaction are:

T (K)	$\Delta G°$ *(cal/mol of NH_3)*
298	−4000
400	−1600
500	+1200
600	+3700
700	+6350

(a) Is this reaction exothermic or endothermic? Explain.

(b) Ammonia is an important substance used, for example, in the manufacture of fertilizers and in the nitriding of steel in order to develop a hard surface layer (one type of case hardening). The direct synthesis of ammonia from the elements is not very practical. Using the above thermodynamic data, explain why.

13.5 (a) Starting with 1 mole of N_2 and 3 moles of H_2 in a container of volume $V = 1.31 \times 10^5$ cm^3 at 400 K, how many moles of NH_3 will exist at equilibrium? Use thermodynamic data given in problem 13.4. [*Hint.* If x moles of NH_3 form, then there will be $(1\text{-}x/2)$ moles of N_2 and $(3\text{-}3x/2)$ moles of H_2 remaining.]

(b) If the container volume in part (a) is reduced to $V = 1.31 \times 10^3$ cm^3 (corresponding to an initial total pressure change from 1 to 100 atm), how many moles of NH_3 will exist at equilibrium?

(c) For the same conditions as in part (a), except at a temperature of 600 K, will any NH_3 form? Explain.

13.6 An unconstrained exothermic reaction ($\Delta H < 0$) will cause the chemical system and the surroundings to heat up, whereas the opposite is true for an unconstrained endothermic reaction ($\Delta H > 0$).

(a) Does the temperature rise accompanying an unconstrained exothermic reaction favor or retard the reaction in terms of the extent to which the reaction can proceed? (Consider the equilibrium constant as a function of temperature.)

(b) Repeat part (a) for an endothermic reaction. [The above parts of this problem are examples of LeChatelier's principle: when a system at equilibrium is subjected to a change in temperature or pressure, the equilibrium state shifts in a direction that tends to counteract the change in temperature or pressure.]

(c) In terms of reaction rate, what can you say about the effect of the temperature change accompanying an exothermic reaction? An endothermic reaction?

13.7 It is observed that fine aluminum powder exposed to air oxidizes rapidly at 600°C, whereas a solid block of aluminum oxidizes slowly at the same temperature. Explain.

13.8 As temperature rises the dissociation of Br_2 gas molecules into Br gas atoms becomes more favorable thermodynamically. At a specific elevated temperature the rate of dissociation is slower for pure Br_2 gas than for a gaseous mixture of Br_2 and Ar having an equivalent Br_2 partial pressure. Discuss this observation.

13.9 (a) Draw an enthalpy diagram of a spontaneous exothermic reaction at constant temperature, indicating the following.

(1) Enthalpy of reactants
(2) Enthalpy of products
(3) Activation energy, ΔH^*
(4) The activated state
(5) ΔH for the reaction

(b) Do the same for a spontaneous endothermic reaction.

13.10 The rate of a spontaneous chemical reaction is inversely proportional to the time required for the reaction to occur. Listed below are the reaction times at different temperatures. Determine the activation energy for this reaction.

Reaction Time	*Temperature*
78 min	327°C
13.8 sec	427°C
0.316 sec	527°C
0.001 sec	727°C

13.11 In terms of absolute reaction rate theory (Sect. 13.4), discuss how the rate of an isothermal chemical reaction varies with the extent of the reaction.

13.12 Letting G_s, G_L, H_S, H_L, and S_S, S_L represent the free energies, enthalpies, and entropies of the solid and liquid phases of a pure material, show that the change in free energy accompanying the solid to liquid transformation is given by

$$\Delta G_{S\to L} = \Delta H_{S\to L}\left(1 - \frac{T}{T_m}\right)$$

where T_m is the equilibrium absolute melting temperature. (Assume $\Delta H_{S\to L}$ and $\Delta S_{S\to L}$ are independent of temperature.)

14

Electrochemical Reactions

The relationships between chemical change and electrical energy are discussed in this chapter.

14.1 ELECTRICAL CONDUCTION

The flow of electrical charge through a conducting substance such as an ionized gas, a metal, or an electrolyte constitutes an electric current I, which is defined as the rate of positive charge passage ($I = dq/dt$). On a microscopic scale, an electric current corresponds to a net flow of charge carriers in one direction. For charge carriers to move preferentially in one direction, there must be a voltage or potential difference. Thus, when the ends of a metal wire are attached to the terminals of a battery, a voltage is impressed along the wire and a current flows through the wire. Metals are good electrical conductors because their bonding electrons are not localized but can move throughout the entire solid. Since electrons are negatively charged, their net flow in one direction is exactly opposite to the conventionally defined current.

The relationship between voltage and current depends on the nature of the conductor. In the case of a metal, there is a direct proportionality known as Ohm's law:

$$I = \frac{V}{R} \tag{14.1}$$

The electrical resistance R is a constant for a given metal of fixed dimensions at a specific temperature. The units of current, voltage, and resistance are the ampere, volt, and ohm, respectively, and one ohm equals one volt per ampere. One ampere is defined as one coulomb per second, where one coulomb, the unit of electrical charge, is equivalent to the *magnitude* of charge on 6.24×10^{18} electrons.

The electrical resistance of a metal depends on its dimensions and, therefore, cannot be considered a property. If, however, the variables are expressed as current per unit area or current density, J, and voltage per unit length V/l (or electric field), then Ohm's law is expressed as

$$J = \frac{(V/l)}{\rho} \tag{14.2}$$

and the quantity ρ (electrical resistivity) is a property of the metal, independent of dimensions. Since current density is current per unit cross-sectional area (i.e., $J = I/A$), resistance and resistivity are related as

$$R = \frac{\rho l}{A} \tag{14.3}$$

where l and A are the conductor's length and cross-sectional area, respectively. Resistivity is determined by the nature of the conductor and by factors such as temperature. For a metal, structural defects and alloying elements raise the resistivity, and an increase in temperature also raises the resistivity because each of these interferes with conduction electron motion (Chapter 22).

Ohm's law holds for most simple conductors such as metals, semiconductors, and electrolytes. The simplest kind of electrolyte consists of an ionic substance dissolved in an aqueous solution, and current is conducted by ions, with cations and anions moving in opposite directions. In addition, solid-state electrolytes (i.e., ionic solids) and molten salts also can conduct electricity by the same means. Returning to the consideration of a metal wire connected to the terminals of a battery, we see that electrons enter at one end of the wire and exit at the other end when a current flows, and the battery must be part of the electrical circuit. In other words, charge must pass through the battery at the same rate as it passes through the wire. Since the electrolyte within the battery obeys Ohm's law, we might expect the voltage output to be a linear function of the current supplied. This not the case, however, because charge transfer also must occur at electrode-electrolyte interfaces and their current-voltage behavior is highly nonlinear. Indeed, with the exception of metal-metal junctions, most junctions between conductors do not obey Ohm's law, and this has great importance relative to semiconductor devices (Chapter 23). The topic of charge transfer at an electrode will be pursued after relationships between chemical and electrical energy are developed.

14.2 ELECTROLYSIS

An electrolysis cell provides an example of the equivalence of electrical and chemical energy. Electrical energy can be used to produce a chemical change in such a cell. An example is a fused-salt electrolysis cell, represented schematically in Fig. 14.1, which consists of two chemically inert electrodes immersed in molten NaCl. If the electrodes are connected externally to a direct current power supply that is capable of impressing a suitable voltage, the sodium cations will migrate toward the negative electrode where they can combine with electrons supplied by the external circuit to form neutral sodium atoms. Since the cell temperature is above the melting point of sodium, the product of this reaction is liquid sodium. As this process occurs, chloride anions simultaneously migrate to the positive electrode where each gives up one electron and is converted into a neutral chlorine

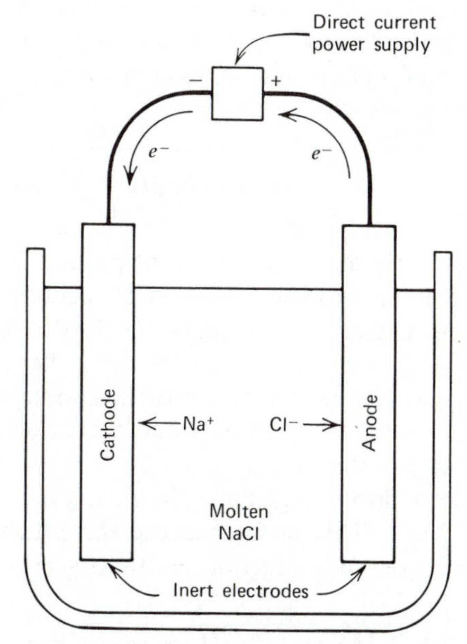

FIGURE 14.1 A fused salt electrolysis cell. The application of external electrical energy (e.g., by a direct current power source) effects a chemical change within the cell. In this case, metallic sodium is generated by reduction at the cathode (i.e., $Na^+ + e^- \rightarrow Na$), and chlorine gas is generated by oxidation at the anode (i.e., $Cl^- \rightarrow \frac{1}{2}Cl_2 + e^-$).

atom. In turn, the chlorine atoms combine to form Cl_2 molecules, which are evolved as a gas. The flow of charge through the external circuit necessitates that for every electron captured by a Na^+ cation at the cathode, one electron must be liberated from a Cl^- anion at the anode. Thus, the chemical reaction taking place in the cell may be written as two separate reactions:

$$Na^+ + e^- \longrightarrow Na(l) \quad \text{(cathode reaction)} \tag{14.3a}$$

$$Cl^- \longrightarrow \tfrac{1}{2}Cl_2(g) + e^- \quad \text{(anode reaction)} \tag{14.3b}$$

In chemical terminology, the cathode is defined as the electrode at which a reduction reaction occurs (i.e., where the oxidation state of the reacting species is reduced through the acquisition of one or more electrons), and the anode is defined as the electrode at which an oxidation reaction occurs (i.e., where the reacting species is oxidized through the loss of one or more electrons). Oxidation in this sense does not imply chemical combination of the species with oxygen.

The overall cell reaction for electrolysis of molten NaCl is the sum of the anode and cathode half-cell reactions:

$$Na^+ + Cl^- \longrightarrow Na(l) + \tfrac{1}{2}Cl_2(g) \tag{14.4}$$

Equations 14.3 and 14.4 show that while electrons participate in reactions at the electrodes, they do not enter into the overall cell reaction. This is true for all electrochemical reactions. Under normal conditions the decomposition of molten NaCl into the component elements is unfavorable thermodynamically (i.e., $\Delta G > 0$ for Eq. 14.4). However, the externally applied voltage shifts the chemical equilibrium so that $\Delta G < 0$ for the reaction, and electrical energy is expended in raising the free energy state of this system.

Electrolysis of sodium chloride aqueous solutions is more complex than the electrolysis of molten NaCl. This arises because the presence of water offers possible alternative oxidation and reduction reactions such as

$$2H_2O + 2e^- \longrightarrow H_2(g) + 2OH^- \quad \text{(reduction reaction)} \tag{14.5a}$$

and

$$H_2O \longrightarrow \tfrac{1}{2}O_2(g) + 2H^+ + 2e^- \quad \text{(oxidation reaction)} \tag{14.5b}$$

The electrode reactions requiring the smallest applied voltage (i.e., the least expenditure of energy) are the ones that dominate. In relatively concentrated NaCl solutions, Cl^- is oxidized more easily than H_2O, and H_2O is reduced more easily than Na^+. Consequently, the predominant half-cell reactions are

$$2H_2O + 2e^- \longrightarrow H_2(g) + 2OH^- \quad \text{(cathode)} \tag{14.6a}$$

$$2Cl^- \longrightarrow Cl_2(g) + 2e^- \quad \text{(anode)} \tag{14.6b}$$

The overall cell reaction is given by

$$2Cl^- + 2H_2O \longrightarrow Cl_2(g) + H_2(g) + 2OH^- \tag{14.7}$$

Because hydroxyl ions are a reaction product and Na^+ is present in the solution, sodium hydroxide can be precipitated and the above hydrolysis reaction serves as the basis for the industrial preparation of NaOH, with chlorine gas and hydrogen gas as by-products.

Another important electrolysis reaction involves copper sulfate aqueous solutions. The various reactions are

$$2Cu^{2+} + 4e^- \longrightarrow 2Cu(s) \quad \text{(cathode)} \tag{14.8a}$$

$$2H_2O \longrightarrow 4H^+ + O_2(g) + 4e^- \quad \text{(anode)} \tag{14.8b}$$

and

$$2Cu^{2+} + 2H_2O \longrightarrow 2Cu(s) + 4H^+ + O_2(g) \quad \text{(cell reaction)} \tag{14.8c}$$

Solid copper is deposited or plated on the cathode. This principle is utilized in the copper plating industry and also in the electrolytic refining of copper. For the latter, a block of impure copper serves as the anode and a sheet of pure copper serves as the cathode. Both electrodes are immersed in a sulfuric acid aqueous solution and connected to a power supply. Electrolysis causes oxidation at the anode [$Cu(s) \rightarrow Cu^{2+} + 2e^-$] and reduction at the cathode [$Cu^{2+} + 2e^- \rightarrow Cu(s)$], with the resulting electrode deposit being purer than the original copper. The initial impurities either remain dissolved in the electrolyte solution or form insoluble precipitates.

The rate of copper deposition in plating or refining equals the rate of the overall cell reaction, which can be calculated in terms of the cell current:

$$\text{Deposition rate [g/sec]} = \frac{I[\text{A}] \times \text{atomic weight [g/mol]}}{nF} \tag{14.9}$$

where n is the number of faradays to deposit one mole and F is the Faraday constant. In electrochemistry, the convenient unit of charge is the faraday, which equals the magnitude of charge on one mole (i.e., 6.022×10^{23}) of electrons; the conversion factor $F = 96{,}490$ C/mol of electrons is called the Faraday constant. For divalent copper, two faradays will plate out one mole (i.e., $n = 2$) and the atomic weight is 63.54 g/mol; hence,

$$\text{copper deposition rate[g/sec]} \simeq (3.3 \times 10^{-4}) \times I[\text{A}] \tag{14.10}$$

From a practical standpoint, deposition rate per unit area of cathode surface is important, and this can be obtained by substituting current density $J(= I/A_{\text{cathode surface}})$ for I in Eqns. 14.9 and 14.10.

14.3 CONCENTRATION UNITS

In order to partially characterize a solution, it is necessary to specify the concentrations of the components. Several standard ways exist for doing this. For materials systems, which will be discussed in subsequent chapters, compositions (or concentrations) in terms of mole fractions or weight fractions prove most useful. The mole fractions of the components in a two-component liquid or solid solution are given by

$$X_A = \frac{n_A}{n_A + n_B} \tag{14.11a}$$

and

$$X_B = \frac{n_B}{n_A + n_B} \tag{14.11b}$$

where n_A and n_B are the number of moles of A and B, respectively, in the solution. Similarly, the weight fractions are given by

$$w_A = \frac{m_A}{m_A + m_B} \tag{14.12a}$$

and

$$w_B = \frac{m_B}{m_A + m_B} \tag{14.12b}$$

where m_A and m_B are the respective masses of A and B in the solution.

For solutions involving liquid solvents, which are used frequently in chemistry, molarity and molality are convenient units of concentration. Molarity is defined as the number of moles of solute (i.e., the substance that dissolves) per 1000 cm^3 of solution. Molality is the number of moles of solute per kilogram of solvent. If 0.6 mole of NaCl (35.1 g) is dissolved in 0.3 kg of H_2O, for example, the result is a 2 molal NaCl aqueous solution. Although molality is a more fundamental concentration unit in electrolyte solutions because it is independent of temperature whereas molarity varies with temperature because of thermal expansion, molarity is used commonly. Molality and molarity become numerically equal for very dilute aqueous solutions.

14.4 GALVANIC CELLS

Whereas external electrical energy produces chemical change in an electrolysis cell, the exact opposite takes place in a galvanic cell, where the energy of a chemical reaction is converted into electrical energy. In the example of a galvanic cell shown in Fig. 14.2, the two half cells are connected internally by a salt bridge, which is an ionic conductor that permits current flow between the half cells but prevents mixing of the electrolytes. As indicated, the Zn and Cu electrodes are immersed in one molar zinc sulfate and one molar copper sulfate, respectively. The possible oxidation reactions that may occur at each electrode are $Zn \rightarrow Zn^{2+} + 2e^-$ and $Cu \rightarrow Cu^{2+} + 2e^-$. Since zinc is oxidized more readily in one molar $ZnSO_4$ than is copper in one molar $CuSO_4$, the electron density tends to build up in the zinc electrode. An external metallic connection between the electrodes allows electrons to flow from the Zn electrode to the Cu electrode where they reduce Cu^{2+} ions in the electrolyte. These events correspond to the half-cell reactions

$$Zn \longrightarrow Zn^{2+} + 2e^-(\text{anode}) \tag{14.13a}$$

and

$$Cu^{2+} + 2e^- \longrightarrow Cu \quad (\text{cathode}) \tag{14.13b}$$

and the overall cell reaction is

$$Zn + Cu^{2+} \longrightarrow Zn^{2+} + Cu \tag{14.13c}$$

Such a chemical reaction, which occurs spontaneously when the electrodes are connected through an external circuit, is the result of a reduction in the free energy of the chemcial system. This is the type of process that occurs in batteries.

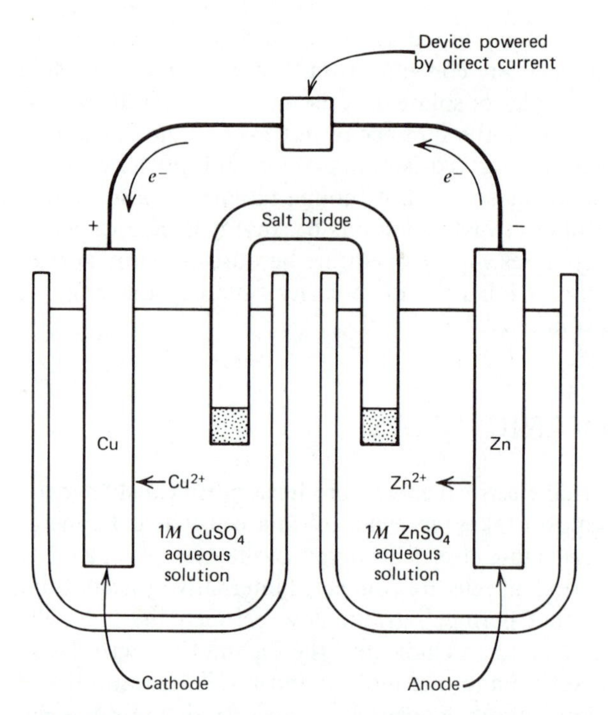

FIGURE 14.2 A galvanic cell. Since metallic zinc gives up electrons more readily than copper, the two half-cell reactions are $Zn \rightarrow Zn^{2+} + 2e^-$ (anode) and $Cu^{2+} + 2e^- \rightarrow Cu$ (cathode). The electrons liberated by the anode reaction travel through the external circuit to the cathode, where they are used in the cathode reaction. The electrical energy generated can be used to power a device operated by direct current.

A galvanic cell possesses a limited capacity for supplying electrical energy because of the finite amount of reactants. Thus, batteries often are rated in terms of ampere-hours, a designation that is related simply to the number of coulombs available. Other energy-producing cells such as fuel cells (see Sect. 14.7) do not share the limitations that batteries do. Before proceeding to fuel cells, it is useful to examine galvanic cells further.

14.5 STANDARD ELECTRODE POTENTIALS

The voltage between the electrodes of a galvanic cell depends on the electrolyte concentrations and composition, the electrode materials, and the temperature. For the cell described in the previous section, the measured voltage at 25°C is

1.10 V under conditions of very small external current. This voltage, also termed electromotive force (emf), is a direct measure of the free energy decrease accompanying the chemical reaction (Eq. 14.13c). The relationship between the free energy change ΔG and the cell voltage ΔV is given by

$$\Delta G = -q\,\Delta V \tag{14.14}$$

where q is the charge transferred per mole of substance undergoing the reaction. That is, $q = nF$, where n is the number of electrons participating in either half-cell reaction and F is the Faraday constant. ΔG will have units of J/mol since F is 96,490 C/mol of electrons and n equals the number of moles of electrons per mole of substance involved in the overall chemical reaction.

When the voltage is measured under standard conditions (i.e., $T = 25°C$, 1 atm pressure and 1 molar electrolyte concentrations) and vanishingly small current, it is called the standard state voltage and designated $\Delta V°$. (Note that the standard state concentrations used here differ from those defined in Sect. 13.2; for any thermodynamic treatment of a chemical system, it is advantageous to adopt a convenient standard state of reference). Under standard conditions, Eq. 14.14 becomes

$$\Delta G° = -\,nF\,\Delta V° \tag{14.15}$$

Necessarily, $\Delta V°$ must be positive if $\Delta G°$ is to be negative.

Since an overall cell reaction can be written as the sum of two half-cell reactions, it is useful to consider each half cell as having a half-cell potential such that the difference between the cathode and anode potentials equals the cell voltage:

$$V_c^{\circ} - V_a^{\circ} = \Delta V° \tag{14.16}$$

Implicit in this equation is the convention that the cathode potential be algebraically larger than the anode potential. In addition, although an electrode potential is independent of the direction in which a half-cell reaction proceeds, it is conventional to write all half-cell reactions as reduction reactions.

Individual electrode potentials are measured relative to a standard hydrogen electrode (Figure 14.3). This hydrogen electrode is arbitrarily *assigned* a half-cell potential of zero when the inert platinum electrode is immersed in an acid solution of unit hydrogen ion concentration (i.e., 1 molar H^+) at 25°C and hydrogen gas at 1 atm pressure is bubbled over its surface. In other words, a half-cell potential of zero is taken for the reaction $2H^+ + 2e^- \rightleftarrows H_2$ under standard conditions, and this serves as a reference for the determination of other half-cell or electrode potentials. For example, when a copper electrode is immersed in one molar $CuSO_4$,

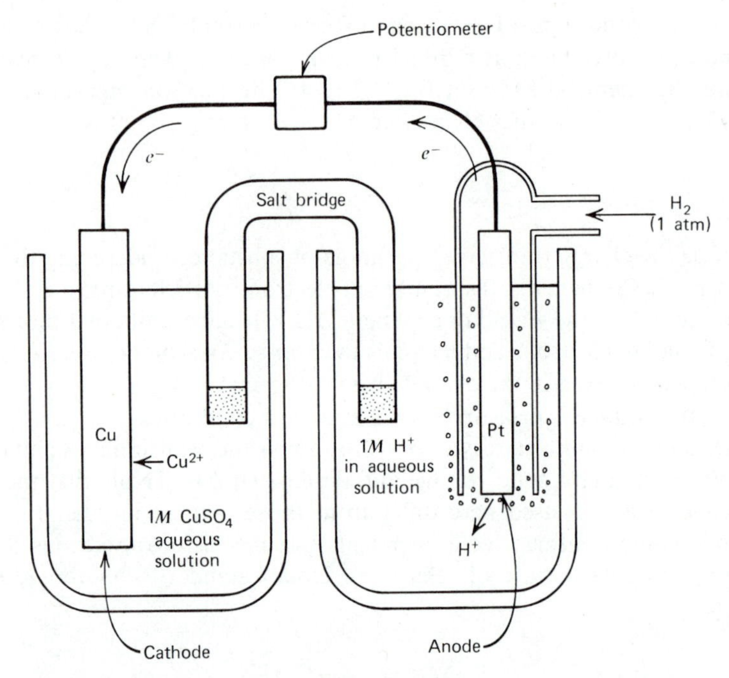

FIGURE 14.3 A copper–hydrogen galvanic cell. Gaseous hydrogen gives up electrons more readily than copper, and the two half-cell reactions are: $Cu^{2+} + 2e^- \rightarrow Cu$ (cathode) and $H_2 \rightarrow 2H^+ + 2e^-$ (anode). Since, under standard conditions, the half-cell potential of the hydrogen reaction is arbitrarily assigned a value of zero, measurement of the total cell voltage with a potentiometer yields the copper reaction half-cell potential.

the cell voltage with respect to a standard hydrogen electrode is 0.34 V, and the observed cell reaction is

$$Cu^{2+} + H_2 \longrightarrow Cu + 2H^+ \qquad \Delta V^\circ = 0.34 \text{ V} \tag{14.17}$$

Since hydrogen is oxidized (anode reaction) and copper is reduced (cathode reaction), we may use Eq. 14.16 to deduce the standard electrode potential for copper:

$$\Delta V^\circ = 0.34 \text{ V} = V_c^\circ - V_a^\circ = V_c^\circ - 0 \tag{14.18}$$

$$Cu^{2+} + 2e^- \rightleftharpoons Cu \qquad V^\circ = 0.34 \text{ V} \tag{14.19}$$

The positive sign indicates that Cu^{2+} ions accept electrons more readily than do H^+ ions. If a zinc electrode is immersed in one molar $ZnSO_4$, the cell voltage relative to a standard hydrogen electrode is 0.76 V, and the observed cell reaction is

$$Zn + 2H^+ \longrightarrow Zn^{2+} + H_2 \qquad \Delta V^\circ = 0.76\ \text{V} \tag{14.20}$$

Proceeding as before, we obtain

$$Zn^{2+} + 2e^- \rightleftharpoons Zn \qquad V^\circ = -0.76\ \text{V} \tag{14.21}$$

The negative sign indicates that Zn^{2+} ions accept electrons less readily than do H^+ ions. Appropriate summation of Eqs. 14.19 and 14.21 (or the addition of Eqs. 14.17 and 14.20) yields

$$Cu^{2+} + Zn \longrightarrow Cu + Zn^{2+} \qquad \Delta V^\circ = 1.10\ \text{V} \tag{14.22}$$

which agrees with the observed voltage for the galvanic cell considered in Sect. 14.4.

Every half-cell reaction can be assigned a standard electrode potential with respect to the standard hydrogen electrode, and some of these values are listed in Table 14.1. If two elements in the table are used as electrodes in a galvanic cell under standard conditions, the one having the algebraically smaller electrode potential will become the anode and the other, the cathode. (It is appropriate at this time to recall that the terms anode and cathode identify the electrodes at which oxidation and reduction occur, respectively. Consideration of Fig. 14.2, for example, should demonstrate that the chemically defined anode actually is the negative terminal of a galvanic cell and the chemically defined cathode is the positive terminal.)

14.6 FREE ENERGY CHANGES AND CONCENTRATION CELLS

Not all galvanic cells operate under standard state conditions, and the electrode potentials may differ considerably from those given in Table 14.1. Often this is primarily a consequence of nonstandard electrolyte concentrations whose effect can be handled in the same manner as was used in Sect. 13.2 for other chemical reactions. The free energy change for the reaction given in Eq. 14.22 is

$$\Delta G = \Delta G^\circ + RT \ln \frac{[Zn^{2+}]}{[Cu^{2+}]} \tag{14.23}$$

TABLE 14.1 Standard Electrochemical Potentials of Some Elements at 25°C

Reaction	*V° (volt)*	
$F_2 + 2e^- \rightarrow 2F^-$	2.76	
$Au^+ + e^- \rightarrow Au$	1.68	
$Au^{3+} + 3e^- \rightarrow Au$	1.46	
$Cl_2 + 2e^- \rightarrow 2Cl^-$	1.36	
$Pt^{2+} + 2e^- \rightarrow Pt$	1.20	↑
$Br_2 + 2e^- \rightarrow 2Br^-$	1.07	
$Hg^{2+} + 2e^- \rightarrow Hg$	0.86	
$Ag^+ + e^- \rightarrow Ag$	0.800	More noble
$Fe^{3+} + e^- \rightarrow Fe^{2+}$	0.771	(cathodic)
$I_2 + 2e^- \rightarrow 2I^-$	0.536	
$Cu^+ + e^- \rightarrow Cu$	0.522	
$Cu^{2+} + 2e^- \rightarrow Cu$	0.341	
$Sb^{3+} + 3e^- \rightarrow Sb$	0.11	
$2H^+ + 2e^- \rightarrow H_2$	0.00	
$Pb^{2+} + 2e^- \rightarrow Pb$	−0.126	
$Sn^{2+} + 2e^- \rightarrow Sn$	−0.136	
$Ni^{2+} + 2e^- \rightarrow Ni$	−0.24	
$Co^{2+} + 2e^- \rightarrow Co$	−0.29	
$Cd^{2+} + 2e^- \rightarrow Cd$	−0.4	
$Fe^{2+} + 2e^- \rightarrow Fe$	−0.42	
$Cr^{2+} + 2e^- \rightarrow Cr$	−0.53	
$Cr^{3+} + 3e^- \rightarrow Cr$	−0.68	
$Zn^{2+} + 2e^- \rightarrow Zn$	−0.762	
$Ti^{2+} + 2e^- \rightarrow Ti$	−1.63	
$Al^{3+} + 3e^- \rightarrow Al$	−1.67	More active
$Mg^{2+} + 2e^- \rightarrow Mg$	−2.36	(anodic)
$Na^+ + e^- \rightarrow Na$	−2.71	
$Ca^{2+} + 2e^- \rightarrow Ca$	−2.71	
$K^+ + e^- \rightarrow K$	−2.92	↓
$Rb^+ + e^- \rightarrow Rb$	−2.96	
$Cs^+ + e^- \rightarrow Cs$	−2.97	
$Li^+ + e^- \rightarrow Li$	−3.04	

where $[Zn^{2+}]$ and $[Cu^{2+}]$ are the concentrations of Zn^{2+} and Cu^{2+} ions, respectively, in terms of molarity, and $[Zn] = 1$ and $[Cu] = 1$. Substitution of $\Delta G = -nF\,\Delta V$ and $\Delta G^\circ = -nF\,\Delta V^\circ$ with $n = 2$ gives

$$\Delta V = \Delta V^\circ - \frac{RT}{2F} \ln \frac{[Zn^{2+}]}{[Cu^{2+}]} \tag{14.24}$$

Thus, the galvanic cell voltage ΔV will differ from ΔV° when the electrolyte concentrations are not unity. One implication of this is as follows. If the terminals of the galvanic cell shown in Fig. 14.2 are connected externally so that a current is allowed to flow, the initial voltage will be $\Delta V^\circ = 1.10$ V, but the voltage will decrease with time as $[Zn^{2+}]$ increases and $[Cu^{2+}]$ decreases. Eventually, current flow will stop and the cell voltage will become zero when the ratio of Zn^{2+} to Cu^{2+} concentrations satisfies

$$\frac{RT}{2F} \ln \frac{[Zn^{2+}]}{[Cu^{2+}]} = \Delta V^\circ \tag{14.25}$$

This, of course, represents a state of chemical equilibrium where free energy is a minimum and the chemical system no longer is capable of supplying energy or doing work. As will be shown in Sect. 14.8, other factors relating to the rate of reaction also decrease the cell voltage.

In analogy with Eq. 14.24, half-cell potentials can be written in terms of standard electrode potentials and ion concentrations. For copper with the half-cell reaction given by

$$Cu^{2+} + 2e^- \rightleftharpoons Cu \tag{14.26}$$

the electrode potential is

$$V = V^\circ - \frac{RT}{2F} \ln \frac{1}{[Cu^{2+}]} \tag{14.27}$$

In the general case

$$M^{n+} + ne^- \rightleftharpoons M \tag{14.28}$$

the electrode potential is

$$V = V^\circ - \frac{RT}{nF} \ln \frac{1}{[M^{n+}]} \tag{14.29}$$

Since an electrode potential depends on the ion concentration in the electrolyte solution, a voltage can be developed by a cell having identical metal electrodes that are immersed in electrolytes of different concentrations (i.e., a concentration cell). As shown in Fig. 14.4, a copper electrode in 0.01 molar $CuSO_4$ becomes the anode with respect to a copper electrode in 0.1 molar $CuSO_4$. If these electrodes are connected externally and a current flows, Cu atoms will be oxidized at the anode and Cu^{2+} ions will be reduced at the cathode. That is, the electrolyte concentrations will tend to equalize. In order that an energy-producing electrochemical cell have a long life, either large amounts of electrolyte must be present or there must be some means by which fresh electrolyte of fixed concentration can be supplied to each electrode.

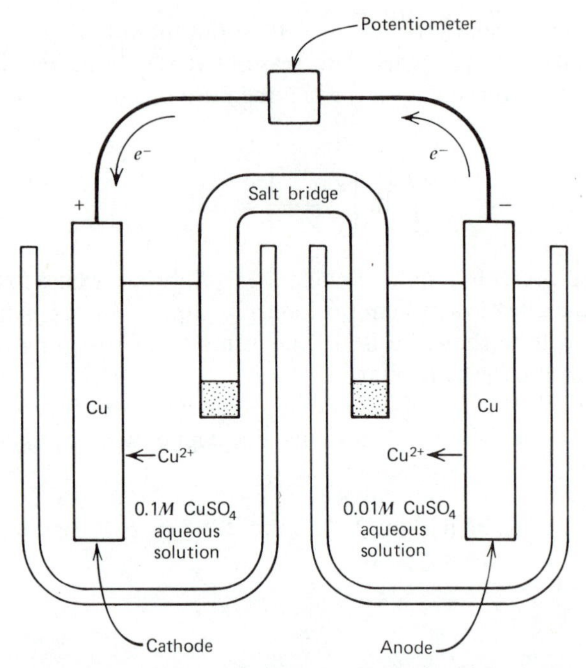

FIGURE 14.4 A concentration cell. Because the copper ion concentration is less in the 0.01 *M* solution than in the 0.1 *M* solution, the copper electrode in the more dilute electrolyte solution becomes the anode and that in the more concentrated solution becomes the cathode. The voltage developed by the cell is given by $V = (RT/2F) \ln 10$ (cf. Eq. 14.27).

14.7 FUEL CELLS

Electrical cells that convert the energy of combustion of fuels such as H_2, CO, or CH_4 directly into electrical energy are called fuel cells. Fuel cells possess about twice the efficiency of conversion as compared to conventional electrical generators that utilize an intermediate step involving mechanical energy. At present most electrical power is generated by steam turbines in which the combustion of fuel heats water to form steam (thermal energy), the steam drives a turbine (mechanical energy), and the turbine runs the electrical generator. The overall efficiency for this process is less than 35%; that is, less than 35% of the thermal energy is converted into electrical energy.

With fuel supplies becoming more scarce in the world, it is advantageous to employ energy conversion schemes that are more efficient, and fuel cells represent one type of energy conversion system that is being investigated. In a simple version of a fuel cell, hydrogen gas is bubbled through one of two porous carbon electrodes and oxygen gas through the other when both electrodes are immersed in a concentrated aqueous solution of KOH (Fig. 14.5). The reactions are

$$O_2(g) + 2H_2O(l) + 4e^- \longrightarrow 4OH^- \quad \text{(cathode)} \tag{14.30a}$$

$$2H_2(g) + 4OH^- \longrightarrow 4H_2O(l) + 4e^- \quad \text{(anode)} \tag{14.30b}$$

$$2H_2(g) + O_2(g) \longrightarrow 2H_2O(l) \quad \text{(overall)} \tag{14.30c}$$

Since the reaction rate is limited in part by how fast OH^- ions can migrate through the electrolyte, the cell is maintained at as high a temperature as possible. A more inherent and universal limitation, which exists for all electrochemical cells, results in restricted reaction rates at the electrodes. Present fuel cells, accordingly, are limited by the catalytic activity and short-life span of the electrodes and by the inability to utilize commercial fuels that do not have high purities.

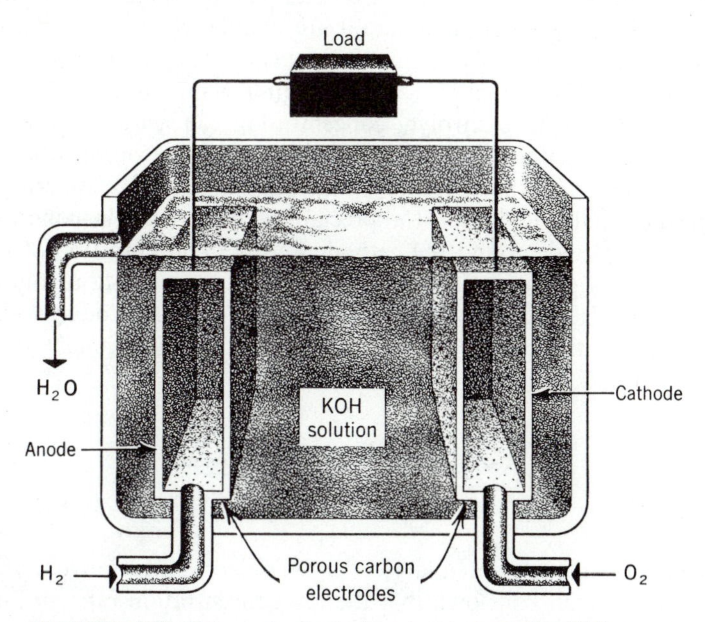

FIGURE 14.5 Schematic of a hydrogen-oxygen fuel cell. The overall reaction, $2H_2(g) + O_2(g) \rightarrow 2H_2O(l)$, is the sum of the cathode $[O_2(g) + 2H_2O(l) + 4e^- \rightarrow 4OH^-]$ and anode $[2H_2(g) + 4OH^- \rightarrow 4H_2O(l) + 4e^-]$ reactions. The chemical energy liberated by this reaction is converted directly to electrical energy, which can power an external device (the "load"). (From *General Chemistry*, Second Edition, by W. H. Slabaugh and T. D. Parsons, Wiley, New York, 1971.)

14.8 NONEQUILIBRIUM EFFECTS—POLARIZATION

Up to now electrode reactions have been considered under reversible or near-equilibrium conditions. Reversibility hinges on just balancing or stopping a chemical reaction, and this can be accomplished with an electrochemical cell through the application of a voltage that exactly counters the cell voltage. Equilibrium cell voltages and equilibrium electrode potentials are measured under the condition where an applied voltage just prevents current flow. It is found experimentally that when an energy-producing cell (e.g., a galvanic cell) delivers an electric current, the cell voltage is less than the difference between the equilibrium electrode potentials. In addition, when an electrolysis cell is forced to operate, the applied voltage must exceed the difference between the equilibrium electrode potentials. These observations are related to irreversibility or nonequilibrium effects associated with current flow. Irreversible effects at the electrodes result from two factors: (1) changes in electrolyte concentration around an electrode (concentration polarization) and (2) restricted reaction rates at an electrode (activation polarization). Both types of polarization are manifested by a change in the individual electrode potentials, and this change, called *overpotential* or *overvoltage*, is the actual electrode potential minus the equilibrium value.

A locally nonuniform electrolyte concentration can arise because ions must diffuse in order to maintain a uniform electrolyte concentration when electrode reactions take place. The importance of this can be seen by considering the zinc-copper galvanic cell (Fig. 14.2). At appreciable cell currents the concentration of Cu^{2+} ions is depleted around the cathode and the concentration of Zn^{2+} ions builds up around the anode. In effect, the resulting concentration gradient in the vicinity of either electrode acts like a concentration cell and produces a back voltage that opposes the cell voltage. Concentration polarization at the cathode changes the cathode potential by an overpotential

$$\eta_c^{\text{conc}} = -\frac{RT}{2F}\ln\frac{[Cu^{2+}]_{\text{solution}}}{[Cu^{2+}]_{\text{cathode}}} \tag{14.31}$$

that is, the actual cathode potential ($V + \eta_c^{\text{conc}}$) is decreased relative to the equilibrium value (V). It can be shown that the ion concentration ratio in Eq. 14.31 is related to the current density at the cathode surface, and the cathode overpotential due to concentration polarization can be generally written as

$$\eta_c^{\text{conc}} = \frac{RT}{nF}\ln\left(1 - \frac{J}{J_L}\right) \tag{14.32}$$

where J is the actual current density and J_L is a limiting value of current density, determined by the Cu^{2+} ion diffusion rate. The comparable expression for concentration polarization at the anode is

$$\eta_a^{\text{conc}} = -\frac{RT}{nF} \ln\left(1 - \frac{J}{J_L}\right) \tag{14.33}$$

Usually J_L is smaller for the cathode reaction than for the anode reaction, and the effect of concentration polarization is more severe at the cathode. Concentration polarization decreases as temperature rises because diffusion proceeds more rapidly at higher temperatures, and this leads to a significant increase in J_L. Furthermore, vigorous stirring of the electrolyte can minimize or virtually eliminate concentration polarization.

Activation polarization arises because electrode reaction rates inherently are restricted by activation energy barriers. The activation polarization overpotential associated with the deposition of metals usually is small, but that required for the discharge of a gas may be appreciable. At a cathode where hydrogen is evolved, for example, H^+ ions must acquire electrons to form neutral hydrogen atoms which, in turn, must combine to form H_2 molecules that can be evolved as a gas. If any step in this sequence is slow, then the electrode reaction rate will be limited by that step. This, of course, depends on the activation energy for each step. On most metals other than the platinum group (i.e., Pt, Pd, Ir, and Rh), the overpotential associated with H_2 evolution is sizable, and competing cathode reactions proceed preferentially. This is responsible for the observation that many metals can be electroplated in acidic electrolyte solutions even though hydrogen evolution would be expected. On the other hand, if the cathode material catalyzes the reaction to form H_2 (e.g., Pt), the amount of activation polarization will be small. This is the primary reason why platinum is used in a standard hydrogen electrode.

For many practical cases the activation polarization overpotential at a cathode can be written as

$$\eta_c^{\text{act}} = -\beta_c \ln\left(\frac{J}{J_0}\right) \tag{14.34}$$

and that at the anode, as

$$\eta_a^{\text{act}} = \beta_a \ln\left(\frac{J}{J_0}\right) \tag{14.35}$$

providing J/J_0 is greater than about 3. In either equation, J is the actual current density at the electrode surface, β is an empirical constant (typically about 0.05 V at room temperature), and J_0, called the exchange current density, is a parameter

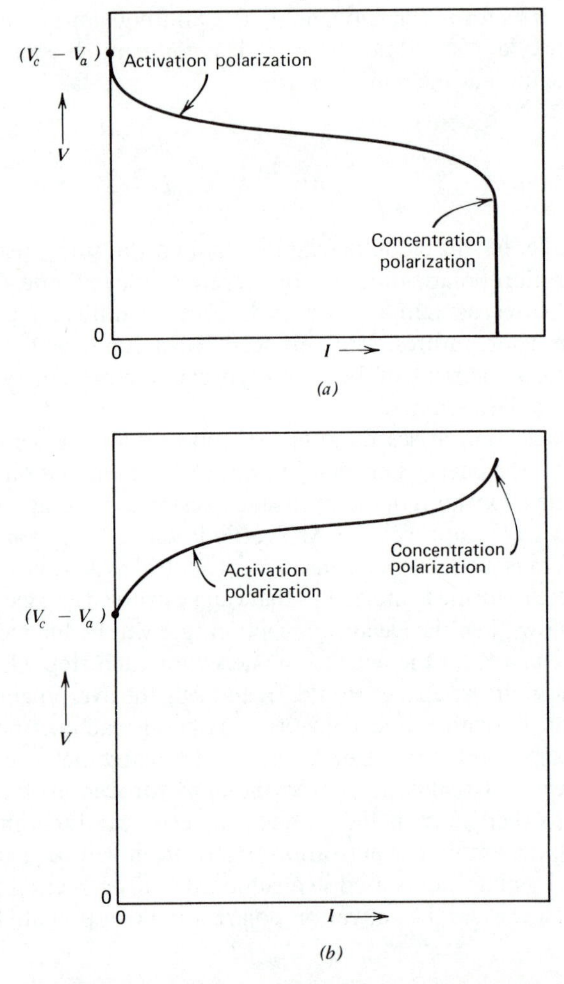

FIGURE 14.6 (*a*) Variation of cell voltage with current for a galvanic cell. The nonlinear *V–I* variation is due to activation polarization at low currents and concentration polarization at high currents (see Eqs. 14.32 to 14.36). (*b*) Variation of cell voltage with current for an electrolysis cell. Since the external voltage produces a chemical change opposite to that occurring in a galvanic cell, the resultant *V–I* relationship is almost a mirror image of that for a galvanic cell.

related to the equilibrium forward and reverse reactions rates (i.e., $J_0 = nFr_{\text{forward}} = nFr_{\text{reverse}}$). Equations 14.34 and 14.35 are limiting forms of a more general electrode kinetics expression, which will be discussed in the next section.

Even if the electrode reactions of a cell proceed readily without significant overpotentials, a voltage drop ($V = IR$) occurs across the electrolyte when current flows. This is so because the electrical resistance of the electrolyte leads to energy dissipation in the form of heat. Physically, the charge carriers (i.e., moving ions) continually bump into other atoms or molecules and lose kinetic energy in the process, and a voltage drop must exist if the current is to be maintained. The transferred kinetic energy corresponds to heat because it involves random atomic or molecular motion, and the rate of heating is proportional to IV or I^2R.

The various nonequilibrium effects can be incorporated into an expression for the voltage provided by a galvanic cell:

$$V_{\text{cell}} = \Delta V + \Delta\eta^{\text{conc}} + \Delta\eta^{\text{act}} - IR = (V_c - V_a) + (\eta_c^{\text{conc}} - \eta_a^{\text{conc}}) + (\eta_c^{\text{act}} - \eta_a^{\text{act}}) - IR \tag{14.36}$$

Only when $I = 0$ does $V_{\text{cell}} = \Delta V$. The variation of cell voltage with current for a galvanic cell is shown schematically in Fig. 14.6*a*, and Fig. 14.6*b* presents the corresponding variation for an electrolysis cell.

14.9 ELECTRODE KINETICS AND ACTIVATION POLARIZATION

The overpotential due to activation polarization is an inherent feature of electrode reactions, and the rate of interfacial charge transfer is the central aspect of electrochemistry. With the exception of superconducting materials (Chapter 22), whenever a current flows there is a corresponding voltage or potential difference. In the case of an electrode-electrolyte interface this is the activation polarization overpotential. There is a general relationship between current density J and overpotential η for an oxidation reaction ($\eta > 0$):

$$J = J_0(e^{\eta/\beta_a} - e^{-\eta/\beta_c}) \tag{14.37a}$$

and for a reduction reaction ($\eta < 0$):

$$J = J_0(e^{-\eta/\beta_c} - e^{\eta/\beta_a}) \tag{14.37b}$$

In these equations J_0, the exchange current density, and β_a and β_c are empirical constants for a particular electrode and specific electrode reaction. Equations 14.37a and b differ only in sign, and this is due to the convention that J be positive at any electrode in an electrochemical cell. For large positive η, Eq. 14.37a reduces to Eq. 14.35, and for large negative η, Eq. 14.37b reduces to Eq. 14.34.

The origin of the general electrode relationships can be seen as follows. When a metal electrode is inserted into an electrolyte, *electrochemical* equilibrium is established quickly by the development of a thin layer in the electrolyte, adjacent to the electrode surface, across which there is essentially a linear variation in electrical potential (Fig. 14.7). This equilibrium potential difference causes no current flow. It cannot be measured directly for a single electrode; however, the equilibrium voltage developed between two electrodes in a galvanic cell, for example, equals the

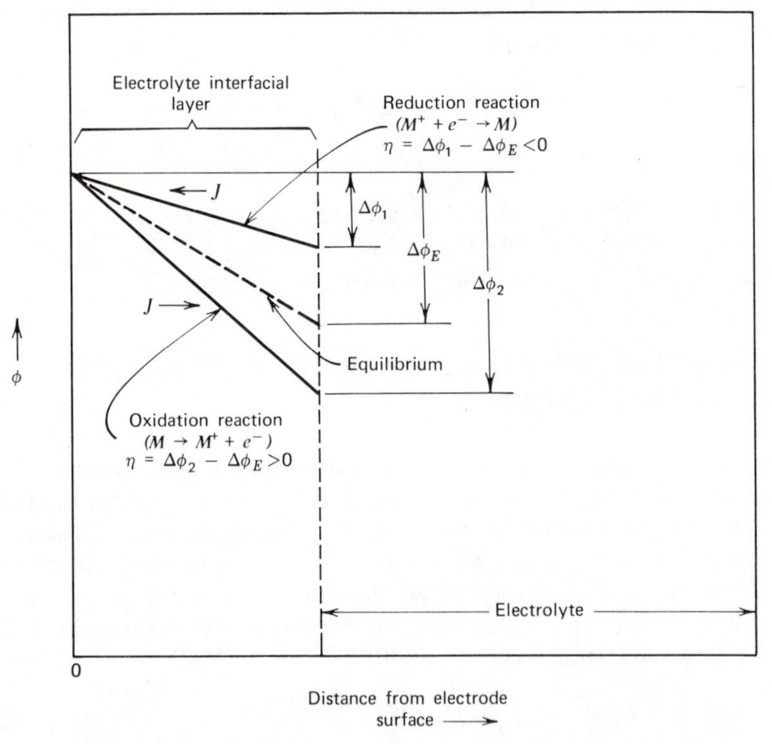

FIGURE 14.7 The variation of electric potential in the vicinity of an electrode–electrolyte interface. In the limiting case of zero current flow, the variation of the potential is denoted by the dotted line and the potential difference across the electrolyte interfacial layer is $\Delta\phi_E$. The potential variation depends upon the direction of current flow. If the current is flowing toward the electrode (i.e., it is the cathode), the potential difference ($\Delta\phi_1$) is less than $\Delta\phi_E$ and the overpotential ($\eta = \Delta\phi_1 - \Delta\phi_E$) is negative. When the electrode acts as the anode, the situation is reversed.

algebraic sum of the equilibrium interface potential differences of the two electrodes.

A free energy barrier exists in the interfacial electrolyte layer, and the rate of interfacial charge transfer depends on its size. For the oxidation reaction

$$M \longrightarrow M^+ + e^- \tag{14.38}$$

there is a free energy minimum when an M atom is on the electrode surface and another free energy minimum when an M^+ ion is hydrated (i.e., coordinated by polar H_2O molecules) just across the interfacial layer (Fig. 14.8). In between these two positions the M^+ ion cannot be completely hydrated, and the free energy passes through a maximum (Fig. 14.8). At electrode equilibrium the rates of the forward

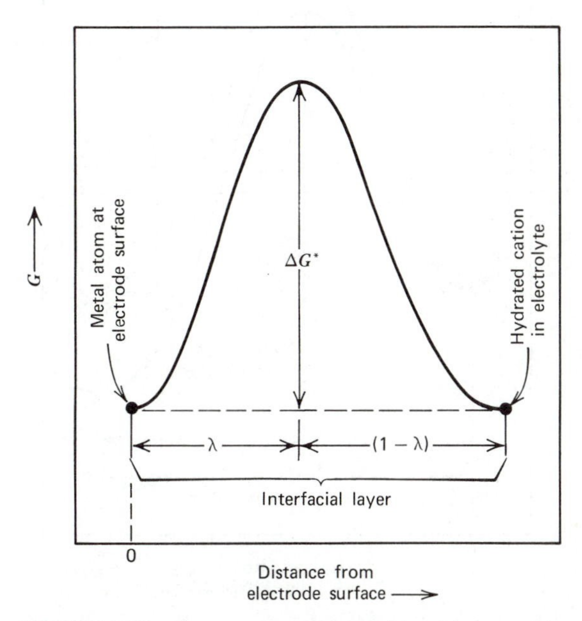

FIGURE 14.8 Free energy versus position for a metal ion in the vicinity of an electrode–electrolyte interface. The diagram shown is for electrode equilibrium, since a metal atom at the electrode surface has the same free energy as its corresponding hydrated ion (i.e., ion coordinated by polar H_2O molecules) in the electrolyte. Under this condition the forward and reverse reaction rates are identical because the activation free energy is the same for both directions. (λ represents the fractional position within the electrolyte interfacial layer at which a metal ion has a maximum free energy.)

and reverse reactions (Eq. 14.38) are identical because the activation free energy is the same in both directions and the free energy minima are at the same value. If this electrode is part of an electrochemical cell and the reaction (Eq. 14.38) is proceeding to the right as current flows, the interface potential difference is increased above the equilibrium value by an amount equal to the overpotential (i.e., $\eta > 0$). Current passage through the interfacial layer and the concurrent development of finite η are accompanied by an *electrochemical* driving force ($-\Delta G = F\eta > 0$) across the layer and a reduced activation free energy in the

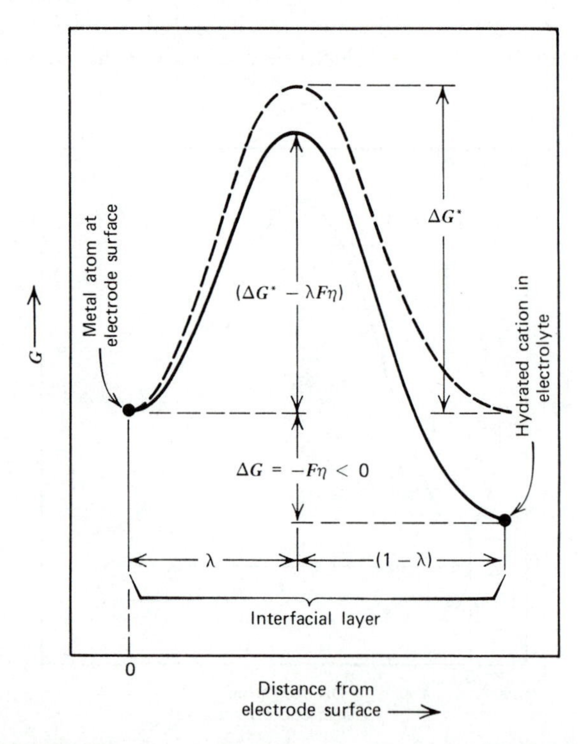

FIGURE 14.9 Free energy versus position in the vicinity of an electrode–electrolyte interface when an electric current is flowing. The example illustrated corresponds to an oxidation reaction (e.g., $M \rightarrow M^{+} + e^{-}$) where $\eta > 0$ (see Fig. 14.7). This reaction proceeds to the right because a hydrated cation has a lower free energy than a metal atom on the surface of the electrode (i.e., $\Delta G = -F\eta < 0$). The activation free energy in the forward direction ($\Delta G^{*} - \lambda F\eta$) is less than that in the reverse direction $[\Delta G^{*} + (1 - \lambda)F\eta]$. λ represents the fractional position within the electrolyte interfacial layer at which a metal ion has the maximum free energy.

forward direction (Fig. 14.9). The net rate of the reaction can be obtained from Eq. 13.27 (see Sect. 13.4):

$$\begin{aligned}(\text{rate})_{\text{net}} &= k_f[M](1 - e^{\Delta G/RT}) \\ &= k[M](1 - e^{-F\eta/RT})e^{-(\Delta G^* - \lambda F\eta)/RT}\end{aligned} \tag{14.39}$$

where the activation free energy is now $\Delta G^* - \lambda F\eta$ in the equation $k_f = k\,\exp[-(\Delta G^* - \lambda F\eta)/RT]$; that is, it is reduced for the case $\eta > 0$ (compare

TABLE 14.2 Selected Values of J_o and β at 25°C

Electrode Reaction	*Electrode*	log J_o *(A/cm²)*	λ	$\beta_a(V)$	$\beta_c(V)$
$Ce^{4+} + e^- \rightarrow Ce^{3+}$	Pt	−4.4	0.25	0.103	0.034
$Cr^{3+} + e^- \rightarrow Cr^{2+}$	Hg	−6	—	—	—
$Fe^{3+} + e^- \rightarrow Fe^{2+}$	Ir	−2.8	—	—	—
	Pd	−2.2	—	—	—
	Pt	−2.6	0.42	0.061	0.044
	Rh	−2.76	—	—	—
$2H^+ + 2e^- \rightarrow H_2$	Al	−10	—	—	—
	Au	−4.5	—	—	—
	Cd	−10.8	—	—	—
	Cu	−6.7	—	—	—
	Fe	−6.0	—	—	—
	Hg	−12.2	0.50	0.051	0.051
	Ir	−3.7	—	—	—
	Mn	−10.9	—	—	—
	Nb	−6.8	—	—	—
	Ni	−5.2	0.42	0.061	0.044
	Pb	−11.6	—	—	—
	Pd	−3.0	—	—	—
	Pt	−3.1	—	—	—
	Sn	−8.0	—	—	—
	Rh	−3.6	—	—	—
	Ti	−8.2	—	—	—
	Tl	−11.0	—	—	—
	W	−5.9	—	—	—
$O_2 + 4H^+ + 4e^- \rightarrow 2H_2O$	Au	−12.3	—	—	—
	Ir	−11	—	—	—
	Pt	−9	—	—	—
	Rh	−8.2	—	—	—

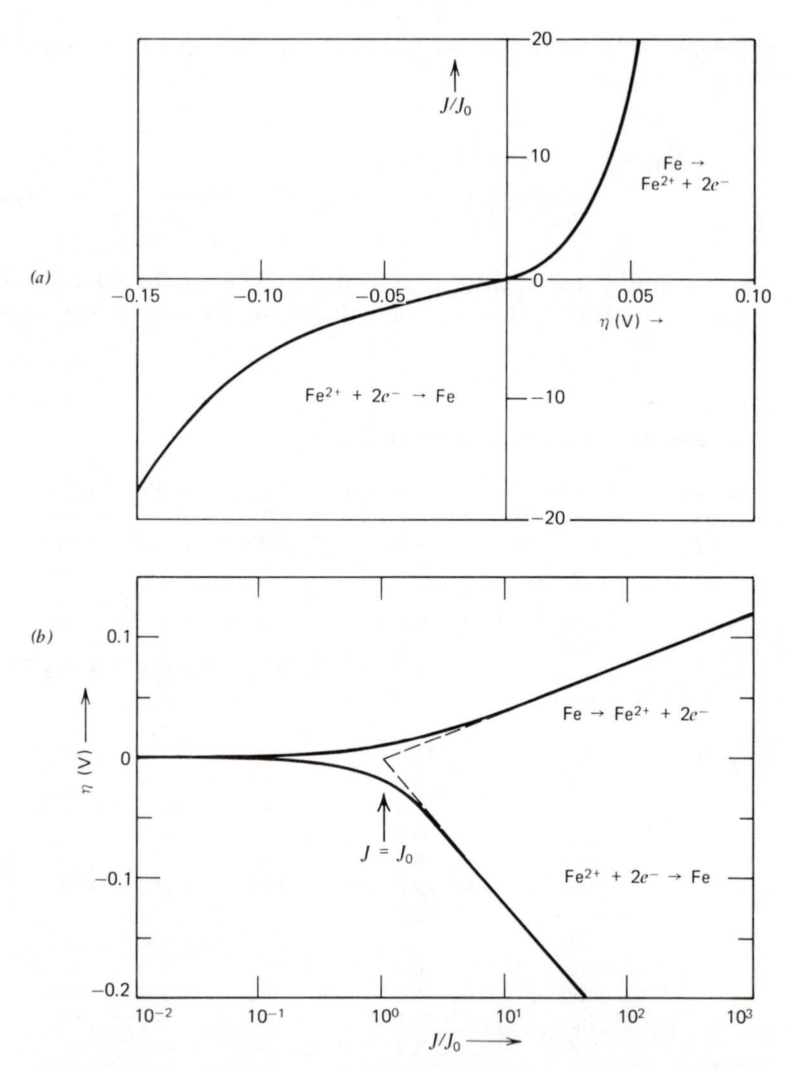

FIGURE 14.10 The effect of activation polarization on the variation of current density vs. overpotential for an iron electrode. (*a*) Unlike metals, an electrode–electrolyte junction displays a nonlinear current density–voltage relationship. The J–η variation also depends upon whether the electrode acts as an anode or as a cathode. (*Note.* In this part, the sign of J has been taken to correspond to that of η.) (*b*) A plot of overpotential versus the logarithm of current density demonstrates that the electrode potential remains constant (i.e., $\eta = 0$) when $J \ll J_o$, whereas either Eq. 14.34 or 14.35 describes the behavior when $J \gg J_o$. Note that the transition region occurs in the vicinity of $J = J_o$. This type of representation—that is, η vs. $\log J$ (or $\log I$)—is particularly useful for portraying electrochemical corrosion reactions (see Chapter 15).

Figs. 14.8 and 14.9). This expression can be rewritten as

$$
\begin{aligned}
(\text{rate})_{\text{net}} &= k[M]e^{-\Delta G^*/RT}[e^{\lambda F\eta/RT} - e^{-(1-\lambda)F\eta/RT}] \\
&= (\text{rate})_{f,\,\text{at eqm}}[e^{\lambda F\eta/RT} - e^{-(1-\lambda)F\eta/RT}]
\end{aligned}
\qquad (14.40)
$$

Since the net reaction rate is directly proportional to current density J and the forward reaction rate at equilibrium is directly proportional to the exchange current density J_0, we see that Eq. 14.40 has the same form as Eq. 14.37a. We can identify β_a, for example, with $RT/\lambda F$; however, because Eq. 14.40 was derived for a simplified case, this identification is not general. Nevertheless, both β_a and β_c are proportional to RT/F and the proportionality factor can be obtained by experiment.

Unlike metals, where there is a linear relationship between voltage and current density, electrodes display very nonlinear J vs. η behavior (Fig. 14.10) because of the accompanying chemical reaction. The operation of an electrochemical cell is related intimately to the values of the parameters J_0 and β_a (or β_c) for each electrode (Table 14.2). In particular, the polarization of an electrode interface is likely to be small if J_0 is large, whereas the polarization is likely to be large if J_0 is small. That is, the electrode overpotential is small when J is much less than J_0, and it becomes increasingly larger as J increases above J_0. The restrictions imposed by electrode reaction rates play a significant role in the operation of electrochemical cells and also in the electrochemical corrosion of metals, as discussed in the next chapter.

REFERENCES

1. Scully, J. G. *Fundamentals of Corrosion*, Pergamon Press, Oxford, 1966.
2. Bockris, J. O'M. and Reddy, A. K. N. *Modern Electrochemistry*, Plenum Press, New York, 1970, Vols. 1 and 2.

QUESTIONS AND PROBLEMS

14.1 Consider the galvanic cell illustrated in Fig. 14.2. If the cell is electrically connected and current flows until equilibrium is established, what will be the final ratio of Zn^{2+} ion concentration to Cu^{2+} ion concentration?

14.2 What electrolyte concentrations would be necessary to render copper the anode in the galvanic cell of Fig. 14.2?

14.3 Consider the copper concentration cell of Fig. 14.4. What will be the Cu^{2+} ion concentrations after the cell is electrically connected and current flows until equilibrium is established?

14.4 How would you experimentally measure the voltage between the Zn and Cu electrodes (Fig. 14.2) under conditions of near zero current flow?

14.5 A piece of iron and a piece of copper are placed in a beaker of dilute sodium chloride water solution and then externally connected by a wire. Sketch the cell and indicate the following:
(a) The anode
(b) The cathode
(c) The direction of external electron flow
(d) The anode half-reaction
(e) The cathode half-reaction
(f) The metal that dissolves (corrodes)

14.6 What effect will each of the following have on the cell of problem 14.5?
(a) A new dilute HCl electrolyte
(b) A new $FeSO_4$ electrolyte
(c) Inserting a battery in the external circuit (both ways)
(d) Complete deionization of the water
(e) Disconnecting the wire

14.7 With a cell consisting of an iron electrode electrically connected externally to a copper electrode immersed in distilled water, a very low current is measured. If ferrous sulfate is added to the water and dissolved slowly, the current initially increases and then decreases. Explain why.

14.8 What would be the current required to plate out 100 g of Cu in 10 minutes from an aqueous solution of $CuSO_4$?

14.9 Predict the anode and cathode reactions for the electrolysis of molten NaOH; repeat for aqueous NaOH.

14.10 Refer to Table 14.1. Calculate the standard potential for the reaction

$$Cu^+ + e^- \longrightarrow Cu^{2+} + 2e^-$$

Compare with the second ionization potential of copper, Chapter 2.

14.11 Refer to Table 14.1. Give the standard potential for the reaction

$$Fe^{3+} + 3e^- \longrightarrow Fe$$

Justify your answer.

15

Corrosion and Oxidation

Chemical degradation of materials and how it can be alleviated are discussed in this chapter.

15.1 CORROSION

The deterioration of a material resulting from chemical attack by its environment is called corrosion. As with all chemical reactions, the rate of corrosion depends on temperature and the concentrations of reactants and products. Additional factors such as mechanical stress, abrasion, or irradiation may accelerate the rate of degradation. Corrosion of a metal can result from direct chemical reaction with a nonmetal such as an acid, from dissolution by a liquid metal, or from an electrochemical reaction when dissimilar metals are in electrical contact in the presence of an electrolyte. Oxide formation, or oxidation in the limited sense, fits into the category of direct attack; however, as a well-defined type of materials deterioration that occurs typically at elevated temperatures, oxidation will be considered separately.

Nonmetallic materials such as ceramics and polymers usually are immune to electrochemical attack but not to direct chemical attack. For example, water on the surface of common glass will combine with alkali metal ions to initiate corrosion and crack formation. Similarly, ground water containing sulfate ions will react adversely with tricalcium aluminate in concrete. At high temperatures, ceramic materials may be attacked by molten salts, oxide slags and, in some cases, liquid metals. Deterioration of organic polymers can result from contact with organic solvents. Besides dissolution, a polymeric solid may change dimensions and properties by absorbing some of the solvent. The combined action of oxygen and ultraviolet radiation causes some polymers to deteriorate by depolymerization at normal temperatures and, in the absence of ultraviolet radiation, depolymerization occurs at elevated temperatures.

Corrosion reduces the usefulness of a solid, usually through the removal of material, and may cause complete destruction. In a somewhat limited but appropriate view, corrosion may be thought of as one or more uncontrolled chemical reactions that proceed in a direction opposite to those used to prepare or refine the material initially. That is, the tendency for corrosion to occur coincides with a reduction in free energy. Because aqueous corrosion of metals and alloys is extremely important economically, this topic will be considered further.

15.2 TYPES OF ELECTROCHEMICAL CORROSION

Corrosion of most metals and alloys is electrochemical in nature and phenomenologically similar to the action of a galvanic cell (Sect. 14.4). In this regard, the corrosion of a two-phase alloy such as a mild steel is easiest to visualize. When a piece of such steel is immersed in a corrosive medium, one phase (the compound Fe_3C, cementite) acts as a cathode and the other phase (primarily iron) acts as an anode. Being electrically coupled, these phases function as the electrodes in many microscopic galvanic cells, and corrosion proceeds at the anodes (ferrite regions) by the conversion of iron atoms to Fe^{2+} ions. The most common corrosive media are aqueous salt solutions.

Microscopic galvanic cells also are present on single-phase metals. Grain boundaries can serve as anodes and grain interiors as cathodes because atoms at a grain boundary can be removed and oxidized more easily than those in the interior of a grain. Other things being equal, therefore, a metal with fine grains is more susceptible to corrosive attack than a metal with coarser grains. In addition, regions of high dislocation density in a cold worked metal are anodic compared to regions of lower dislocation density, and a cold worked metal displays a greater tendency for corrosion than an annealed metal. Thus, even a pure metal or a single-phase alloy has a heterogeneous structure from an electrochemical standpoint, although it is not entirely appropriate to attribute corrosion only to spatially fixed microscopic cells. That is, oxidation and reduction reactions can occur randomly over the surface of a single-phase metallic material, and this type of corrosion is called uniform attack.

On a macroscopic scale, dissimilar metals in electrical contact are subject to *galvanic corrosion*. Here, one metal becomes the anode and the other becomes the cathode, in direct analogy with a galvanic cell. However, which metal corrodes preferentially (i.e., which is anodic) may differ from simple considerations of equilibrium electrode potentials (see Table 14.1) because of nonequilibrium effects that relate to a specific corrosive medium. Accordingly, it is useful to have empirical information pertaining to a common environment such as sea water. Such data, in tabular form, are called a *galvanic series* (Table 15.1). If a pair of metals from this list is connected electrically in sea water, the metal higher on the list will be the anode. For environments other than sea water, the galvanic series may differ significantly from Table 15.1.

Galvanic action equivalent to what takes place in a concentration cell (Sect. 14.6) also can lead to corrosion. Concentration cells may arise from differences in ion concentrations or from differences in dissolved gas concentrations. An example is represented schematically in Fig. 15.1 which illustrates *crevice corrosion* of iron.

TABLE 15.1 A Portion of the Galvanic Series in Sea Water

Corroded End (Anodic, or Least Noble)
Magnesium
Magnesium alloys
Zinc
Galvanized steel
Aluminum and its alloys
Mild steel
Cast iron
50% Pb–50% Sn alloy
Nonpassivated 304 (18% Cr–8% Ni) stainless steel
Nonpassivated 316 (18% Cr–8% Ni–3% Mo) stainless steel
Lead
Tin
Muntz metal
Naval brass
Yellow brass
Admirality brass
Aluminum bronze
Red brass
Copper
Silicon bronze
70–30 Copper–nickel
Passivated nickel
Passivated inconel
Monel
Passivated 304 (18% Cr–8% Ni) stainless steel
Passivated 316 (18% Cr–8% Ni–3% Mo) stainless steel
Protected End (cathodic, or Most Noble)

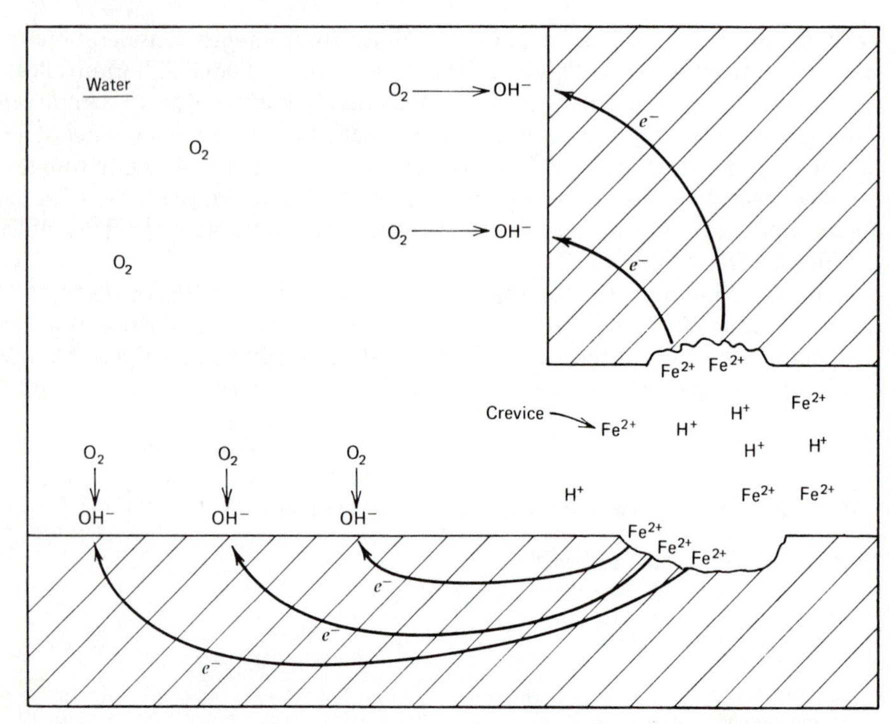

FIGURE 15.1 Crevice corrosion of iron. Because the oxygen is depleted rapidly in the crevice, an oxygen concentration cell is set up between the crevice and the more exposed areas adjacent to it. The crevice region is anodic, with the corrosion half-cell reaction Fe → Fe^{2+} + $2e^-$.

The cathode reaction that accompanies the corrosion of iron in an alkaline or neutral solution is

$$\tfrac{1}{2}O_2 + H_2O + 2e^- \longrightarrow 2OH^- \tag{15.1}$$

Within a crevice the oxygen dissolved in water is depleted rather quickly by uniform corrosion reactions, and its replenishment takes place slowly because diffusion is necessary. Outside the crevice, however, sufficient oxygen remains available so that the cathode reaction can proceed. As a consequence, the crevice region becomes anodic, and corrosion occurs according to the reaction

$$Fe \longrightarrow Fe^{2+} + 2e^- \tag{15.2}$$

If the aqueous solution contains chloride ions (e.g., sea water), then the Fe^{2+} ions within the crevice electrostatically attract Cl^- ions from regions outside of the crevice. With the exception of alkali metal salts, most metal chlorides (and sulfates) undergo hydrolysis in water such that metal cations are converted into insoluble hydroxides:

$$M^+ + Cl^- + H_2O \longrightarrow MOH + H^+ + Cl^- \qquad (15.3)$$

In the case of iron the insoluble hydroxides are known commonly as rust. Although rust can and does form when chloride ions are absent, these anions have a catalytic effect on the rate of rust formation and on the corrosion rate because as hydrolysis proceeds (Eq. 15.3) the anode reaction (Eq. 15.2) in the crevice is favored and accelerates.

The effect of oxygen concentration, resulting from differential aeration, can be observed readily with steel pilings in the ocean. The oxygen concentration in sea water is highest near the surface. Consequently, corrosion of steel pilings takes place most readily slightly below the ocean surface at locations that are anodic relative to the locations with a plentiful supply of dissolved oxygen. These examples of crevice corrosion and corrosion of pilings illustrate the insidious nature of corrosion in that the greatest damage often occurs at positions that are not readily visible.

A form of extremely localized attack that results in holes in a metal is called *pitting*. Stainless steels in particular are susceptible to pitting when exposed to aerated electrolytes containing chloride anions. The initiation of pits occurs slowly and is related to the penetration of the protective hydrated oxide surface film by Cl^- ions. Once pits start, however, their growth is self-stimulated or autocatalytic, and the mechanism of pit penetration is essentially identical to that of crevice corrosion in the presence of Cl^- ions (Fig. 15.2).

TABLE 15.2 Various Forms of Corrosion

Corrosion Type	*Example and/or Comment*
Uniform attack	Steel immersed in dilute sulfuric acid.
Galvanic corrosion	Dissimilar metals (e.g., Zn and Fe) immersed in a corrosive media and connected electrically.
Crevice corrosion	Intense localized corrosion at a crevice under a bolt head.
Pitting	Localized attack, in the form of pits, such as observed on stainless steel immersed in chloride ion solutions.
Intergranular corrosion	Corrosion occurring in the vicinity of grain boundaries such as is observed in the vicinity of some weldments of stainless steel.
Erosion corrosion	Acceleration of corrosion because of relative movement between corrosive fluid and material as may occur in pumping equipment.
Fretting corrosion	Corrosion occurring at contact areas between materials under load subjected to vibration or sliding.
Stress corrosion	Cracking caused by simultaneous presence of tensile stress and particular corrosive medium.

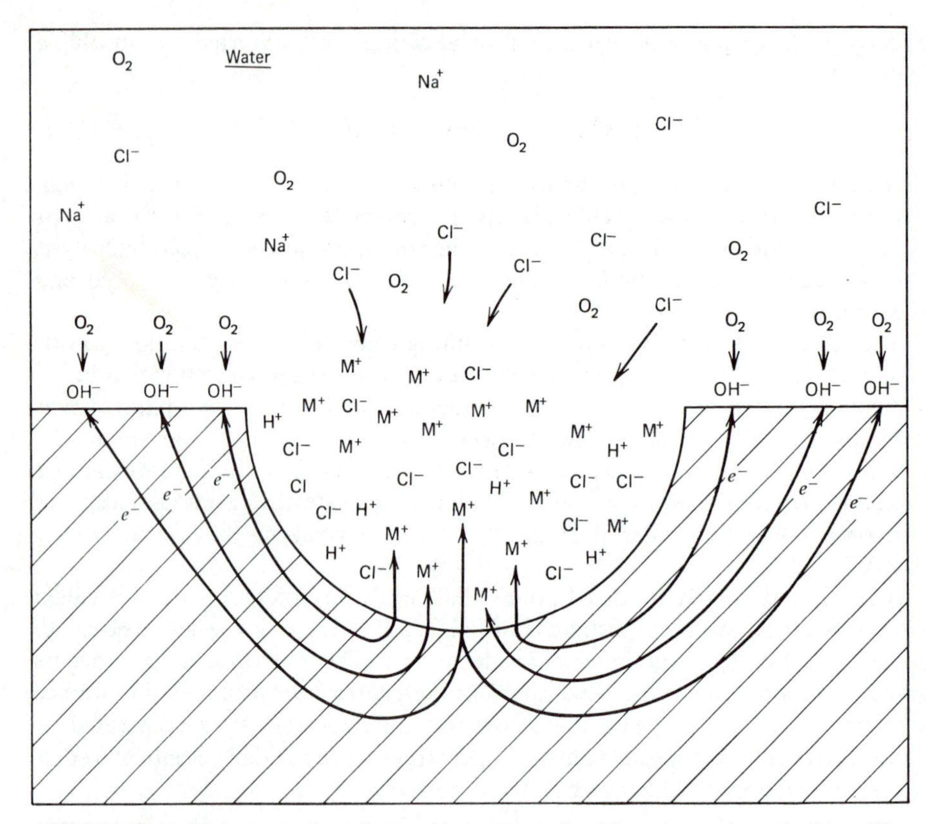

FIGURE 15.2 Pitting corrosion of a metal in a salt solution. The oxygen concentration cell in pitting is similar to that associated with a crevice (Fig. 15.1). The pitting process tends to be autocatalytic in nature because, as the pit grows, the region formed is depleted of oxygen, leading to an increasing corrosion rate. (Adapted from *Corrosion Engineering* by M. G. Fontana and N. D. Greene. Copyright 1967 by McGraw-Hill Book Co. Used with permission of McGraw-Hill Book Co.)

Table 15.2 lists various forms of corrosion including those discussed above. It is important to realize that most metals and alloys in addition to iron and steel exhibit many of the forms of corrosion.

15.3 RUST

In general, corrosion of a metal involves removal of material by virtue of a chemical reaction, and this invariably corresponds to an anode reaction. There may or may not be a visible, solid corrosion product. Iron exposed to an alkaline or neutral solution does have a definite corrosion product, rust, which is a result of several chemical steps. The appropriate cathode and anode reactions for the corrosion of iron in alkaline or neutral solutions are given by Eq. 15.1 and Eq. 15.2, respectively. The overall reaction is a summation of these half-cell reactions:

$$Fe + \tfrac{1}{2}O_2 + H_2O \longrightarrow Fe^{2+} + 2OH^- \longrightarrow Fe(OH)_2 \quad (15.4)$$

Because ferrous hydroxide is relatively insoluble in water it precipitates out of solution. The presence of dissolved O_2 causes ferrous hydroxide to be oxidized to ferric hydroxide, the principal component of rust:

$$2Fe(OH)_2 + H_2O + \tfrac{1}{2}O_2 \longrightarrow 2Fe(OH)_3 \quad (15.5)$$

Strictly speaking, rust consists of hydrated iron oxides (principally hydrated Fe_2O_3). Since iron oxides comprise the main ores that serve as a source for iron metal, it is clear that the corrosion process proceeds in a direction opposite to the refining process.

It is a common misconception that the location of rust coincides with the position where corrosion occurs (i.e., the anodic regions). For uniform corrosion this is true in essence; however, in other cases the rust deposit may be somewhat removed from the anodic regions. In part, this arises because Fe^{2+} cations produced at the anode and OH^- anions produced at the cathode must migrate or diffuse toward each other through the electrolyte before they can combine and lead to a rust deposit. The relative mobilities of the ions and also the presence of Cl^-, which migrates faster than OH^-, influence where rust forms and how fast iron or steel corrodes.

15.4 CORROSION REACTIONS

A corroding metal is analogous to a short-circuited galvanic cell possessing a a very small electrolyte resistance. If the metal is isolated electrically, the total rate of oxidation must equal the total rate of reduction so that no charge accumulation results. Consequently, the anode current or corrosion current must equal the cathode current. This can be portrayed with potential-current diagrams based on the concepts developed in Sects. 14.8 and 14.9. Such diagrams, shown schematically in Fig. 15.3, indicate that the cathode potential decreases and the anode potential increases as current increases because of activation polarization for each reaction. The intersection of the two curves defines the corrosion current and the so-called corrosion potential. For example, when zinc is immersed in a hydrochloric acid

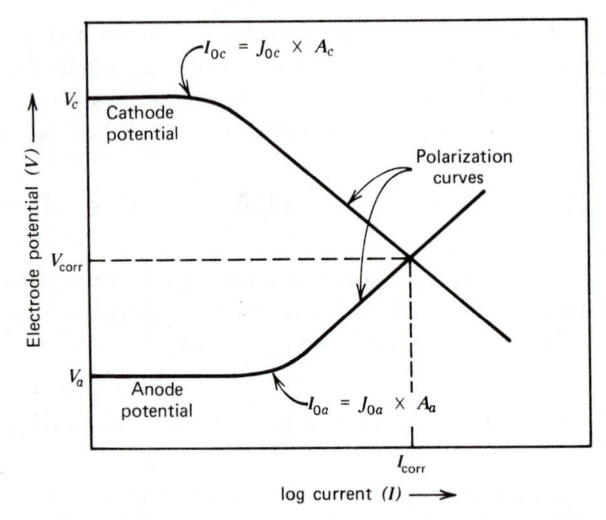

FIGURE 15.3 The potential–current relationship for anode and cathode reactions during corrosion (cf. Fig. 14.10*b*). Activation polarization decreases the cathode potential and increases the anode potential at finite currents. The current increases until the cathode and anode potential curves intersect. The intersection point defines the steady state corrosion current (I_{corr}) and the corrosion potential (V_{corr}). [*Note.* Activation polarization becomes significant when I_0 is exceeded for either electrode reaction (cf. Eqs. 14.34 and 14.35); I_0 is the product of the electrode area, A, and the exchange current density, J_0, which was discussed in Sects. 14.8 and 14.9.]

aqueous solution, zinc oxidation (i.e., $Zn \rightarrow Zn^{2+} + 2e^-$) and hydrogen ion reduction (i.e., $2H^+ + 2e^- \rightarrow H_2$) begin, and the zinc itself serves as the *external* metallic connection between the electrodes on its surface. Since the microscopic anode and cathode regions virtually superimpose, the resistance to electron flow in the zinc is negligible; hence, the current increases to a maximum value (i.e., the corrosion current) that is attained when the difference between cathode and anode potentials goes to zero. In principle, the corrosion current density (i.e., anode current divided by anode area) provides a quantitative measure of corrosion rate per unit area. In actual situations, however, the prediction of corrosion rates on this basis is hampered because pertinent electrode parameters usually are unknown. Nevertheless, qualitative application of potential-current diagrams facilitates the understanding of electrochemical corrosion.

Galvanic coupling of zinc with a cathodic metal like platinum greatly enhances the zinc dissolution rate in acid solutions because the exchange current density

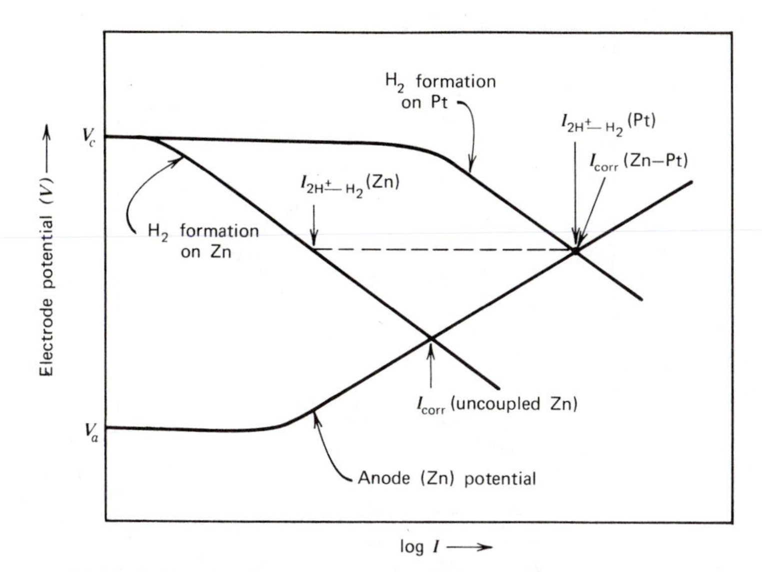

FIGURE 15.4 The effect of galvanic coupling of platinum with zinc on the corrosion rate of zinc. The intersection of the anode potential curve with the cathode potential curve for the hydrogen reaction on zinc defines the corrosion rate of uncoupled zinc in an acid solution. When the zinc is coupled to platinum in the same acid solution, the corrosion current increases to I_{corr}(Zn–Pt) = $I_{2H^+ \rightarrow H_2}$(Zn) + $I_{2H^+ \rightarrow H_2}$(Pt). Because the exchange current density is much greater for the hydrogen reaction on Pt, the total corrosion rate is approximately $I_{2H^+ \rightarrow H_2}$(Pt).

for hydrogen evolution is much greater on Pt than on Zn (Fig. 15.4). In addition, the relative anode and cathode areas have an important effect on corrosion rate. The corrosion current density is small when the anode area is large compared to that of the cathode, and it is large when the anode area is relatively small (Fig. 15.5). Thus, an iron rivet in a large sheet of copper corrodes rapidly, but a copper rivet in a large sheet of iron contributes insignificantly to the corrosion rate of the iron sheets. It is interesting to note that Pt, when coupled with Ti or Cr, does not accelerate the corrosion rate of either of the latter metals but protects them because it produces a protective hydrated oxide film on the exposed Cr or Ti.

Potential-current diagrams are especially useful in providing an explanation as to why one cathode reaction prevails over competing cathode reactions. Both hydrogen ion reduction,

$$2H^+ + 2e^- \longrightarrow H_2 \tag{15.6}$$

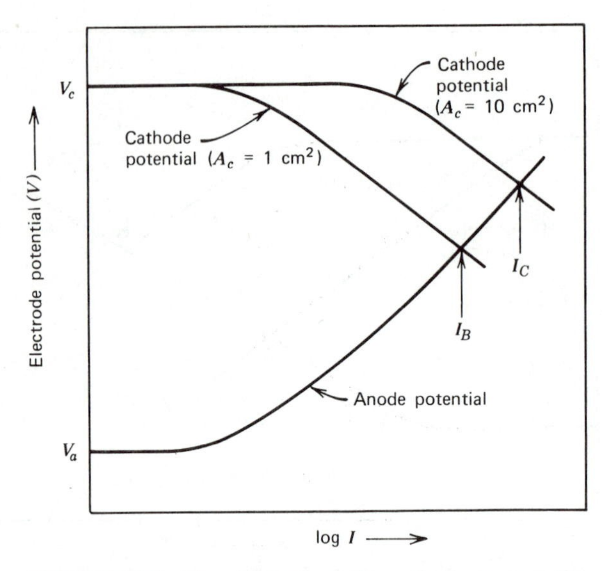

FIGURE 15.5 The effect of relative cathode area on the corrosion current. Increasing the cathode area from 1 to 10 cm^2 causes the corrosion current to increase from I_B to I_C because the cathode polarization is decreased. The latter is a direct consequence of the decrease in current density at the cathode.

and oxygen reduction,

$$O_2 + 4H^+ + 4e^- \longrightarrow 2H_2O \tag{15.7}$$

or

$$O_2 + 2H_2O + 4e^- \longrightarrow 4OH^- \tag{15.8}$$

should be possible in the case of aqueous corrosion. Thermodynamically, either oxygen reduction reaction is favored because the corresponding half-cell potential remains 1.23 V above that for hydrogen ion reduction regardless of the H^+ concentration or pH ($= -\log[H^+]$). Nevertheless, Eq. 15.6 predominates over Eq. 15.7 in acid solutions because the former is favored kinetically. That is, the exchange current density for H^+ reduction usually is much greater than that for O_2 reduction, and the oxygen reduction reaction also is prone to diffusion limitations (i.e., concentration polarization). These effects are shown schematically in Fig. 15.6. Only in neutral or alkaline solutions, where the H^+ ion concentration is small, does oxygen reduction (Eq. 15.8) prevail.

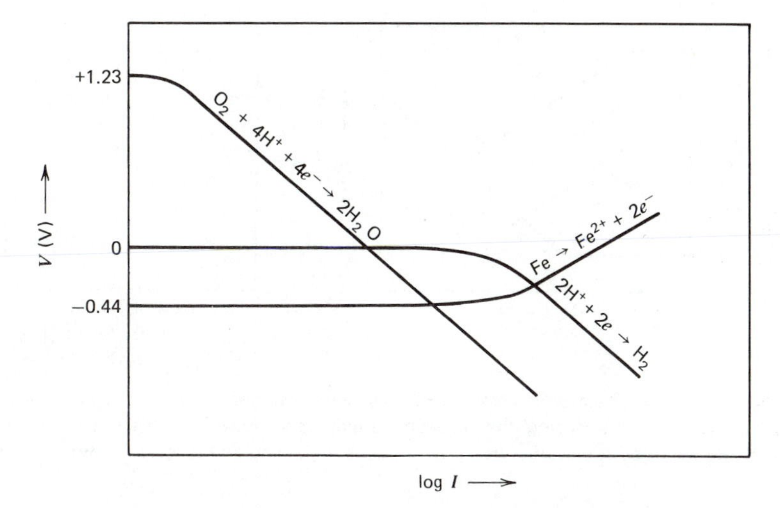

FIGURE 15.6 Potential–current diagram showing that the hydrogen reduction reaction is favored over the oxygen reduction reaction in the aqueous corrosion of iron. Although thermodynamics favors the oxygen reduction reaction, the higher exchange current density for the hydrogen reduction reaction makes this the predominant cathode reaction. (Adapted from *Modern Electrochemistry*, by J. O'M. Bockris and A. K. N. Reddy, Plenum Press, New York, 1970.)

Additional cathode reactions due to the presence of an oxidizing agent may have a profound effect on corrosion rate. An oxidizing agent is a substance that is capable of bringing about oxidation by undergoing reduction itself (e.g., $Fe^{3+} + e^- \rightarrow Fe^{2+}$). In the case of corrosion in an acid solution containing a ferric salt, the total corrosion current equals the sum of the two reduction currents:

$$I_{corr} = I_{Fe^{3+} \rightarrow Fe^{2+}} + I_{2H^+ \rightarrow H_2} \tag{15.9}$$

and corrosion is more severe because of the presence of the ferric salt.

15.5 PASSIVITY

For most active metals the rate of corrosion increases as the concentration of a dissolved oxidizing agent increases (Fig. 15.7). Contrary to this behavior, certain active metals such as chromium, iron, nickel, titanium, and many of their alloys become passive or more noble in behavior when exposed to suitable oxidizing agents at high enough concentrations. The active-passive transition exhibited by these metals is caused by the formation of a surface film of hydrated oxide that

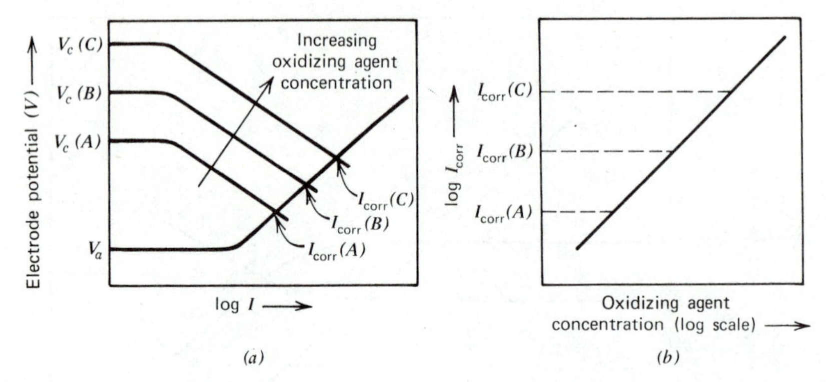

FIGURE 15.7 The effect of increasing oxidizing agent concentration on the corrosion rate of an active metal. (*a*) Increasing the oxidizing agent concentration raises both the corrosion potential and I_{corr}. (*b*) A plot of corrosion current vs. oxidizing agent concentration, as deduced from (*a*).

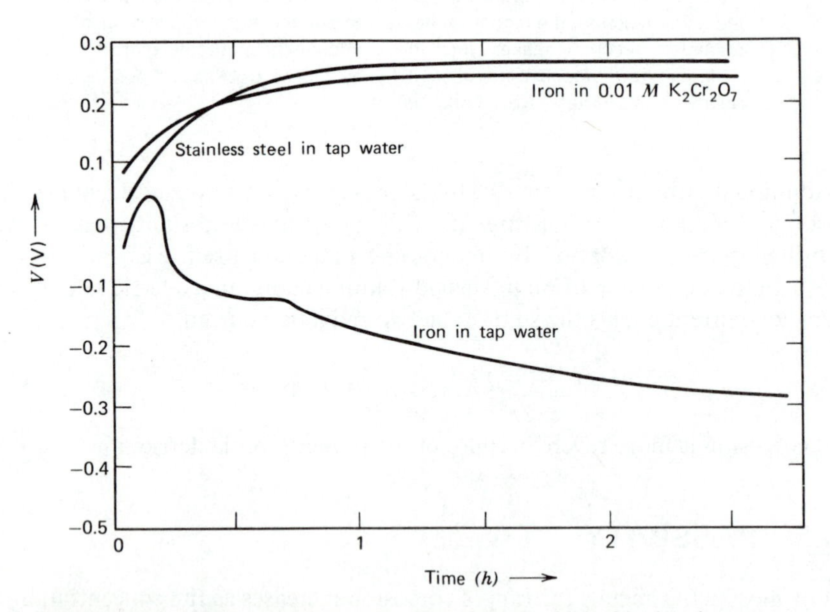

FIGURE 15.8 The potential, as a function of time, exhibited by iron in tap water and in a solution of $K_2Cr_2O_7$, and by stainless steel in tap water. Both stainless steel and iron in the potassium dichromate solution become passive as time proceeds because a protective film develops on their surfaces. [After R. M. Burns, *J. Appl. Phys.*, **8**, 398 (1937).]

interferes with the anode reaction and drastically lowers the corrosion rate. Thus an ordinary iron electrode develops a noble potential like that of stainless steel, when immersed in a tap water solution of potassium dichromate (Fig. 15.8). The $Cr_2O_7^{2-}$ ions are reduced at the iron surface and Fe is oxidized with the result being a protective film consisting of hydrated Cr_2O_3 and Fe_2O_3. When an iron electrode is immersed in tap water which does not contain $K_2Cr_2O_7$, it is initially anodic with respect to a standard hydrogen electrode, then becomes temporarily cathodic, and finally anodic as dissolved oxygen is used up. Stainless steel, on the other hand, becomes passive in tap water because dissolved oxygen is a suitable oxidizing agent to form a protective film on stainless steel, which is comparable to that developed on iron by $Cr_2O_7^{2-}$ ions.

Some metals and alloys, such as titanium and stainless steel, acquire a natural layer of oxide when exposed to air but a more protective one after prolonged exposure in an oxygenated aqueous solution. In halide solutions of low pH, however, the film on stainless steel can be penetrated by halide ions and the metal corrodes. Stainless steels containing more than about 12% Cr do not corrode uniformly in neutral salt solutions but do corrode generally in salt solutions

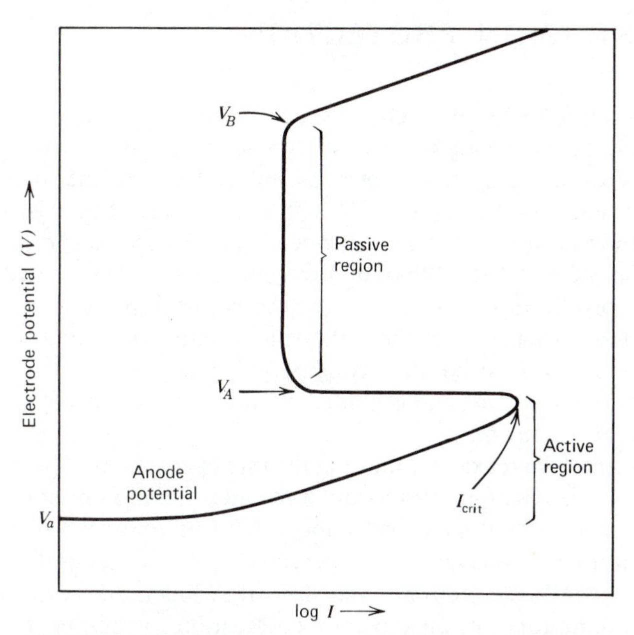

FIGURE 15.9 Passivation of an electrode by electrical means. When the electrode potential reaches a critical value (V_A), the current abruptly decreases as a result of the formation of a passive film, which is disrupted only at a much higher voltage (V_B).

having a pH less than about 2.7. When the pH is higher, stainless steels can corrode by pitting because Cl^- ions, for example, penetrate the film locally and repassivation at the base of a pit can occur only if sufficient oxygen is replenished from the exterior solution. The autocatalytic nature of pit growth (Sect. 15.2) and the low solubility and diffusivity of oxygen in water often preclude repassivation at the base of a narrow pit.

Passivation also can be effected electrically through anodic polarization in which a protective film forms when the overpotential reaches a critical value (Fig. 15.9). (The shape of the curve shown in Fig. 15.9 should be compared with that shown in Fig. 15.7a.) In the passive region the anode current density becomes much less than in the active region. Furthermore, it becomes relatively insensitive to the anode potential, and the constant metal dissolution rate with increasing potential may be attributed to a corresponding thickening of the film such that the electric field within the film remains constant. At sufficiently high anode potentials, either oxygen evolution or breakdown of the oxide film ensues.

15.6 CORROSION PROTECTION

Corrosion protection can be accomplished by physical, chemical, or electrical means. Painting or enameling a metal surface may prevent corrosive attack by virtue of physical shielding that prevents contact between the metal and any corrosive environment. Plating a metal with another metal that is more noble (i.e., more cathodic) utilizes the same principle of physical shielding; however, there are deficiencies to this kind of protection. Chromium-plated steel, for example, will corrode rapidly when the plating is punctured because the anode (steel) area is much smaller than the cathode (chromium) area and this situation favors a large corrosion current density. Any type of physical protection depends on the integrity of the coating and is limited because the coating may degrade with time in certain environments.

Two fundamental ways exist for limiting the rate of corrosion. The first involves decreasing electrode reaction rates through the addition of inhibitors to the corrosive environment. Inhibitors are substances that lead to greater polarization at the anode or cathode for a given current density (Fig. 15.10) by interfering with the reaction at either or both electrodes and thus by reducing the corrosion current density. Some inhibitors poison the cathode reaction by raising the activation energy, others block the cathode or anode surface by forming a film, and still others, particularly certain polar organic compounds, become strongly adsorbed on the metal surface. Necessarily, inhibitors should be effective when present in small concentrations and should not contaminate the environment. They are most

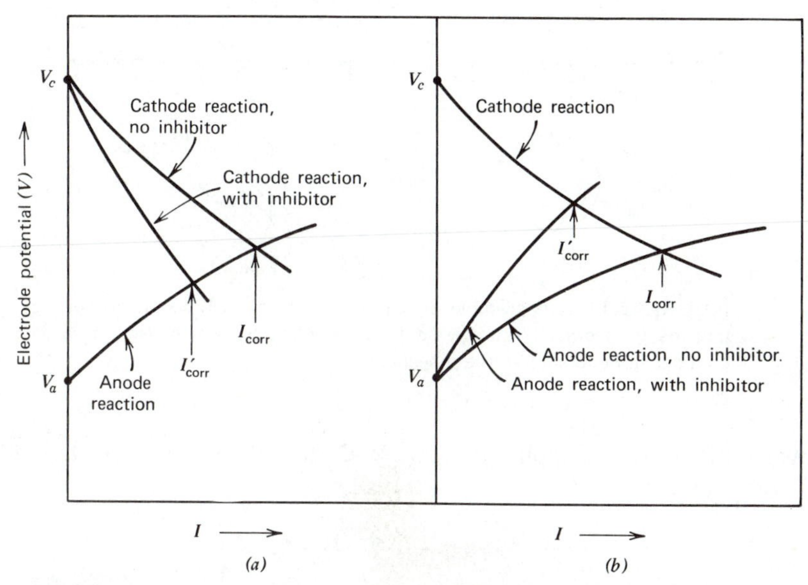

FIGURE 15.10 Schematic diagrams showing the effect of inhibitors on the rate of corrosion. (*a*) An inhibitor that interferes with the hydrogen reduction reaction increases the cathode polarization and reduces the corrosion current from I_{corr} to I'_{corr}. (*b*) The formation of an adsorbed layer on the anode surface increases the anode polarization and decreases the corrosion current from I_{corr} to I'_{corr}. (Note that linear, rather than logarithmic, current scales have been used.)

useful in lessening corrosion in closed systems such as automobile radiators, boilers and laundry equipment. Electrode reaction rates also can be reduced by the partial or complete removal of one necessary reactant. Thus, deaerated water lengthens the life of hot water heating systems.

A second fundamental way of limiting the rate of corrosion is to make a metal's potential cathodic relative to the corrosion potential. This is called cathodic protection and can be accomplished chemically through appropriate galvanic coupling or electrically through an applied current. For example, a steel pipe buried in the ground can be protected by magnesium or zinc anodes that are electrically connected to the pipe. Such active anodes corrode preferentially to the steel (Fig. 15.11) and, accordingly, are termed sacrificial anodes. Steel coated with zinc (galvanized) is protected in much the same manner. Even when the anodic zinc coating is punctured, galvanized steel corrodes only slowly in normal environments because the corrosion current is spread over a large anode area. The combination of an inert electrode and an external power supply that impresses a current produces the same effect as galvanic coupling to a sacrificial anode. As

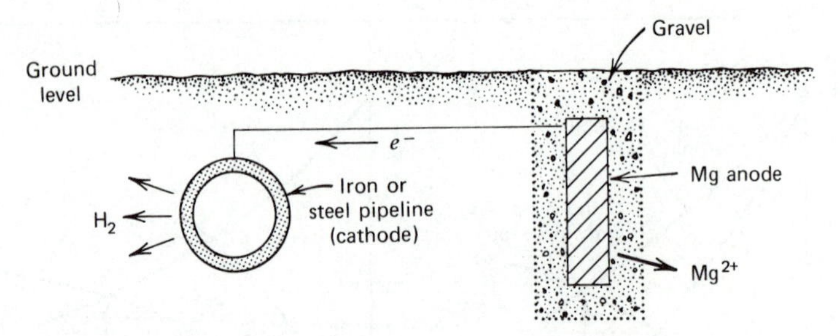

FIGURE 15.11 An example of cathodic protection. The iron or steel pipeline is electrically connected to an Mg anode, which sacrificially corrodes in preference to the pipeline.

shown in Fig. 15.12, the applied current equals the difference between the cathode and corrosion currents

$$I_{app} = I_c - I'_{corr} \tag{15.10}$$

and the corrosion current decreases as the corrosion potential is decreased.

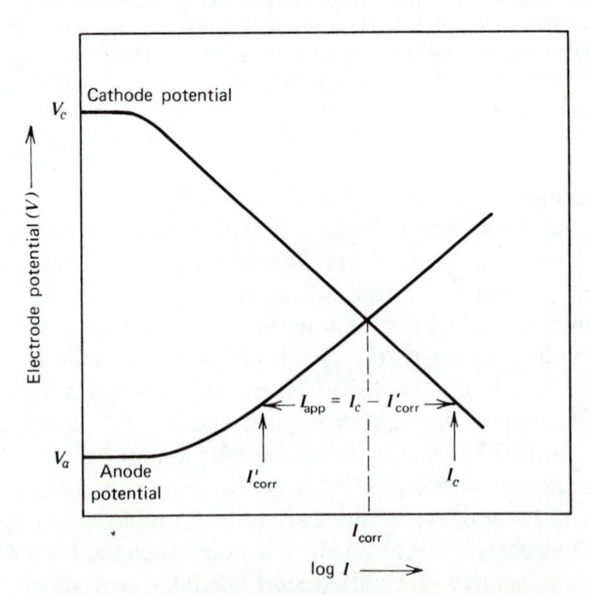

FIGURE 15.12 Corrosion protection can be afforded by external cathodic protection. To reduce I_{corr} to I'_{corr}, an external current (I_{app}) equal to the difference between the cathode current (I_c) and I'_{corr} must be impressed.

For metals that display an active-passive transition, the above methods of corrosion protection still apply, including the use of inhibitors that cause a passivating film to form (Fig. 15.13). Such inhibitors, however, must be present in sufficiently large concentrations to be effective because the presence of insufficient amounts of an oxidizing agent inhibitor may greatly accelerate the corrosion rate (see Fig. 15.7*b* and original portion of Fig. 15.13*b*). Alternatively, a metal that becomes passive can be protected electrically by raising its potential.

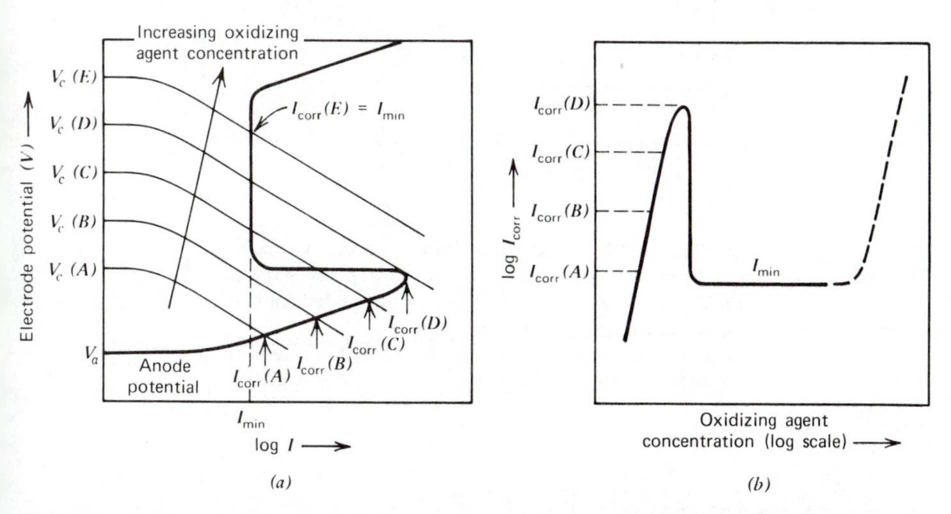

FIGURE 15.13 The effect of an oxidizing agent inhibitor on the corrosion behavior of a metal that exhibits a passive transition. (*a*) As the concentration of the oxidizing agent is increased, the corrosion potential increases. (*b*) In the active region, the corrosion current increases (see also Fig. 15.7*b*), but upon exceeding the level of inhibitor corresponding to curve *D*, the corrosion current is reduced markedly to I_{min} since the corrosion potential now lies in the passive region.

This is termed anodic protection and can be brought about with a device, called a potentiostat, that maintains the metal at a constant potential with respect to a reference electrode. When this potential lies in the passive range, corrosion protection results (Fig. 15.14). In this case the applied current required is given by

$$I_{app} = I_{min} - I_c \tag{15.11}$$

and is smaller typically than that needed for cathodic protection (compare Figs. 15.12 and 15.14).

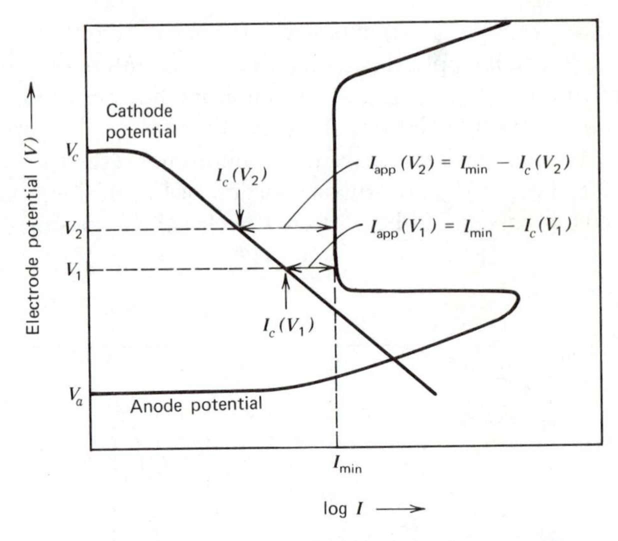

FIGURE 15.14 Anodic protection of a metal is achieved by raising its potential. For the active–passive metal shown here, the corrosion current can be reduced to I_{min} by raising the potential to values such as V_1 or V_2. The impressed current corresponding to either potential is simply $I_{app} = I_{min} - I_c$.

15.7 OXIDATION

Almost all metals and many nonmetallic materials react with the atmosphere to form oxides and, in some cases, sulfides and nitrides. Very few pure metals are found in nature. Rather, the great majority of metallic elements exist in ores as chemical compounds. Metals that have been extracted from ores are in a metastable state in that they tend to recombine with oxygen or other elements present in the atmosphere. Standard free energies of formation at room temperature for selected metal oxides are given in Table 15.3. Note that ΔG° is negative in all but one case, indicating that the respective metals are not stable in air at room temperature. In addition, the large, negative heats of formation (Table 15.3) indicate that oxidation is strongly exothermic. That a metal such as aluminum has wide uses and is commonplace even though its oxide is much more stable can be attributed to the protective nature of the thin oxide layer that forms on aluminum and completely isolates the metal from the air. Conversely, the extreme stability of solid Al_2O_3 itself accounts for its widespread use as a refractory material at high temperatures.

TABLE 15.3 Selected Values of Standard Free Energies ($\Delta G°$) and Enthalpies ($\Delta H°$) of Oxide Formation From the Elements

Oxide	$\Delta H° \left(\frac{kcal}{mol\ of\ oxide}\right)$	$\Delta G° \left(\frac{kcal}{mol\ of\ oxide}\right)$	$\Delta H° \left(\frac{kcal}{mol\ of\ O_2\ consumed}\right)$	$\Delta G° \left(\frac{kcal}{mol\ of\ O_2\ consumed}\right)$
CaO	−152	−144	−304	−288
BeO	−146	−139	−292	−278
MgO	−144	−136	−288	−272
BaO	−133	−126	−266	−252
Al_2O_3	−399	−377	−266	−251
ZrO_2	−258	−244	−258	−244
TiO_2	−218	−204	−218	−204
SiO_2 (quartz)	−205	−192	−205	−192
B_2O_3	−302	−283	−201	−188
Cr_2O_3	−270	−250	−180	−167
ZnO	−83	−76	−166	−152
V_2O_5	−373	−344	−149	−138
SnO_2	−139	−124	−139	−124
Fe_3O_4	−267	−242	−134	−121
WO_3	−201	−182	−134	−121
Fe_2O_3	−197	−177	−131	−118
FeO	−64	−58	−128	−116
MnO_2	−125	−111	−125	−111
CdO	−61	−54	−122	−108
MoO_3	−180	−162	−120	−108
NiO	−58	−52	−116	−104
Sb_2O_5	−234	−201	−94	−80
Bi_2O_3	−138	−119	−92	−79
TeO_2	−78	−65	−78	−65
PbO_2	−66	−52	−66	−52
HgO	−22	−14	−44	−28
Ag_2O	−7	−3	−14	−6
Au_2O_3	+19	+39	+13	+26

Whether a metal oxidizes readily at elevated temperatures depends on the partial pressure of oxygen in equilibrium with the metal and its oxide. If the ambient partial pressure of oxygen is greater than the equilibrium value, then the metal will tend to oxidize. If the converse holds, then oxidation will not occur or any existing oxide will tend to decompose into metal plus oxygen gas. The rate of either chemical reaction depends strongly on temperature and is determined by additional factors as described below.

15.8 INITIATION AND GROWTH OF OXIDE FILMS ON METALS

The first oxygen molecules to come into contact with a perfectly clean metal surface at low temperatures usually decompose to form a monolayer of adsorbed oxygen atoms. Since the oxygen atoms have a greater affinity for electrons than the metal, oxygen ions form and this gives rise to a strong electric field that repels valence electrons in the metal away from its surface, leaving metal cations next to the oxygen ions. As more oxygen molecules contact the surface and decompose, the process is repeated and metal ions move through the film to combine with O^{2-} ions, and nucleate a metal oxide layer. Although the initial oxidation rate may be high when the metal and oxygen molecules can achieve intimate contact, the subsequent rate often decreases because the oxide layer acts as a diffusion barrier. As shown in Fig. 15.15, oxide formation can be viewed as a *dry* electrochemical reaction. That is, the overall reaction

$$aM + \frac{b}{2}O_2 \longrightarrow M_aO_b \tag{15.12}$$

can be divided into an anode reaction

$$aM \longrightarrow aM^{n+} + nae^- \tag{15.13a}$$

which occurs at the metal-oxide interface, and a cathode reaction

$$\frac{b}{2}O_2 + 2be^- \longrightarrow bO^{2-} \tag{15.13b}$$

which occurs at the oxide-gas interface. In these equations, $n = 2b/a$ so that electrical neutrality is maintained. The oxide layer serves as both the electrolyte and the *external* conductor. Film growth according to this picture slows down rapidly because ion diffusion is reduced. Metal ions must migrate outward or

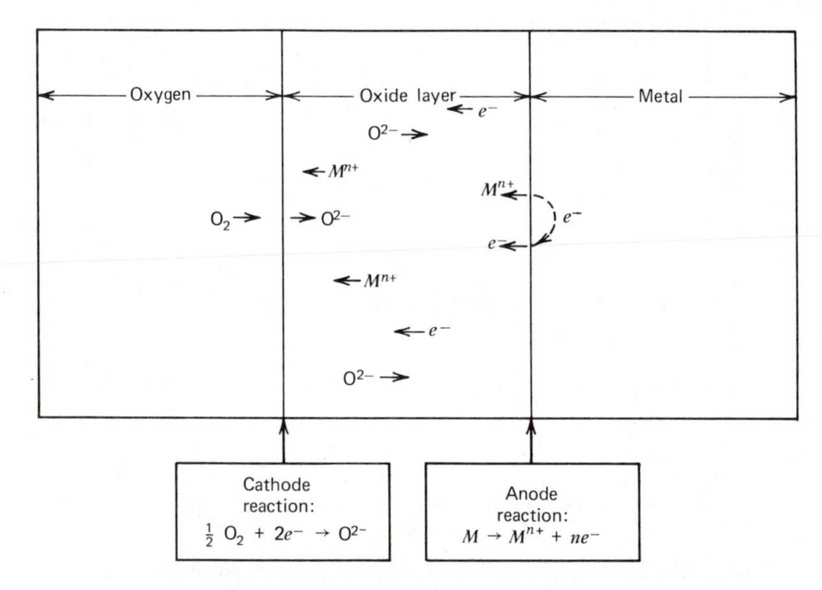

FIGURE 15.15 Oxide formation considered as a dry electrochemical reaction. The anode half-reaction can be considered to take place at the oxide–metal interface and the cathode half-reaction, at the oxide–oxygen interface. The reaction proceeds by diffusion of oxygen ions to the anode interface and/or by diffusion of metal ions to the cathode interface.

oxygen ions inward or both. Usually one or the other of these processes predominates. In addition, electrons must be conducted from the metal-oxide interface toward the oxide-gas interface if the half-cell reactions (Eqs. 15.13a and b) are to proceed. Thus, as an oxide layer thickens, the diffusion distance increases and the electric field across the layer decreases. Both of these contribute to a slower growth rate. The above picture does not apply to certain metals (e.g., the alkali metals) that form thick, porous oxide scales. In such a case the oxide is porous enough to permit rapid diffusion of molecular oxygen inward and direct combination at the metal surface.

The rate of oxidation after initial film formation usually is measured by weighing a test specimen exposed to air or oxygen in order to determine the amount of oxygen that has reacted. In general, this rate increases with temperature since diffusion proceeds faster at higher temperatures. The amount of oxide formed varies with time in one or more ways, depending on the metal, the temperature, and the structure of the oxide. As shown in Fig. 15.16, weight gain may be logarithmic, parabolic, linear, or discontinuous with time. The logarithmic relationship applies to certain protective oxide layers that usually have a high electrical resistivity.

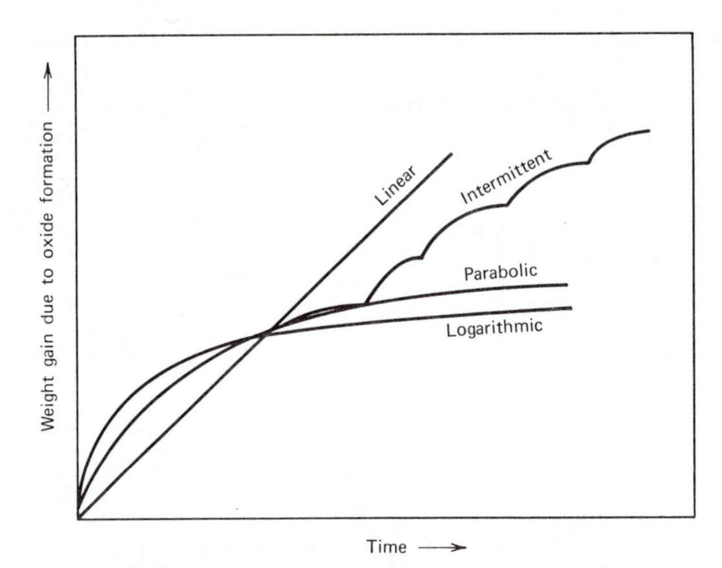

FIGURE 15.16 Various types of oxidation behavior. Logarithmic and parabolic oxidation are protective in that the oxidation rate decreases as the oxide layer thickness increases. Intermittent oxidation is often associated with the spalling or flaking off of an otherwise protective oxide, which leads to an intermittent increase in the oxidation rate. Linear oxidation occurs when a nonprotective oxide permits fresh metal surface to be in contact with oxygen.

Low-temperature oxidation of copper or iron is logarithmic, and so is the oxidation of aluminum that has a single cation valence state (i.e., Al^{3+}) and whose oxide maintains a high resistivity even at elevated temperatures. The parabolic relationship applies to other protective layers such as those formed on Cu and Fe at intermediate temperatures. Here, ion diffusion limits the growth. The linear relationships applies to all oxide films formed initially and to the formation of nonprotective scales, as for the alkali metals. Discontinuous or breakaway oxidation corresponds to cracking or spalling of an otherwise protective oxide layer. In such a case an initial parabolic rate is altered by oxide cracking or spalling, and at long times a roughly linear rate is observed.

The tendency for an oxide film to be protective can be approximately ascertained by a parameter called the Pilling–Bedworth ratio. This is the ratio of oxide volume to the volume of metal from which the oxide was formed. If the oxidation reaction is represented by Eq. 15.12, then the Pilling–Bedworth ratio is

$$\mathrm{PB} = \frac{\text{oxide volume per metal atom}}{\text{metal volume per atom}} = \frac{W\rho_M}{aw\rho_{\mathrm{ox}}} \tag{15.14}$$

where W is the formula weight of the oxide, w is the atomic weight of the metal, and ρ_M and ρ_{ox} are the metal and oxide densities, respectively. When $PB < 1$, the oxide tends to be porous and unprotective. When PB is equal to or slightly greater than unity, an intact and protective oxide layer should form. For values much greater than unity the brittle oxide layer spalls or flakes off the metal surface and breakaway oxidation occurs. Because oxidation proceeds more rapidly as temperature rises, even an adherent oxide layer with a parabolic growth rate may not afford sufficient protection at elevated temperatures.

Some techniques used for corrosion control also can be applied for oxidation control. For example, a protective coating that is oxidation resistant and that adheres well to a metal surface, yet does not react with the metal, provides one way to prevent oxidation. Silicide and aluminide coatings are of potential use with transition metals and their alloys because these refractory metals typically possess poor resistance to high-temperature oxidation. More commonly, oxidation resistance is improved through alloying. Effective alloying elements are strong oxide formers that favor the formation of a more protective oxide layer by changing the volume or crystal structure of the oxide. The addition of Cr or Al to Fe, Ni, or Co alloys improves their oxidation resistance, and there are a number of such alloys that have a combination of high strength and oxidation resistance at elevated temperatures.

In many cases there are complications that have not been discussed above. Grain boundary penetration of oxygen and accompanying oxide formation at grain boundaries can severely limit the usefulness of a metal or alloy at high temperatures. In addition, some alloys are susceptible to internal oxidation. An example is copper containing a small amount of beryllium. As an outer oxide scale forms, some oxygen dissolves interstitially in the copper and subsequently reacts with Be to form BeO particles. Depending upon the distribution of the oxide particles, the alloys may or may not suffer degradation. If internal oxidation is effected intentionally in a controlled manner, strengthening may result from the dispersion of hard oxide particles that interfere with dislocation motion. Often, however, internal oxidation is not controlled, and the oxide particle dispersion embrittles the alloy.

Finally, it should be mentioned that many metals have more than one distinct oxide, and the scale that forms often reflects this. Thus, it is possible for iron to have successive layers of FeO, Fe_3O_4, and Fe_2O_3 between the metal surface and the gaseous atmosphere if the temperature is above 560°C, whereas below this temperature the FeO layer is absent. The relative thickness of each phase depends on temperature and diffusion considerations, and the overall oxidation kinetics may be determined largely by the growth of a single oxide phase. That multiple oxide scales can form is related to metal–oxygen phase diagrams, and general considerations of phase diagrams will be presented in the next chapter.

REFERENCES

1. Scully, J. C. *Fundamentals of Corrosion*, Pergamon Press, Oxford, 1966.

2. Fontana, M. G. and Greene, N. D. *Corrosion Engineering*, McGraw-Hill, New York, 1967.

QUESTIONS AND PROBLEMS

15.1 Explain the difference between the emf series (Table 14.1) and the galvanic series (Table 15.1) of metals.

15.2 Why are very thin plates of copper followed by nickel recommended as base coats for chromium electroplate on steel?

15.3 Explain the following experimental observations.
(a) Alloying Ag with a small amount of Be and subsequently heat treating it in air increases its tarnish resistance.
(b) Dipping steel for a short time in molten aluminum increases the oxidation resistance of the steel.
(c) Alloying steel with more than about 20% Cr markedly increases its high temperature oxidation resistance and low temperature corrosion resistance in aqueous media.

15.4 Fe-Ni-Cr alloys are frequently used as high temperature oxidation resistant materials.
(a) The first oxide to form on the surface is the protective oxide, Cr_2O_3. Why is this so?
(b) As oxidation proceeds, the Cr_2O_3 layer grows in a protective manner. At a certain critical weight gain, depending on the overall alloy composition and temperature, M_3O_4 oxide begins to form and the oxide spalls off. Explain why M_3O_4 oxide eventually forms as opposed to continued formation of Cr_2O_3.

15.5 (a) Refer to Fig. 15.5. Draw a schematic diagram similar to this figure for Pt surface areas of 0.1, 1, and 10 cm^2 and a Zn anode surface area of 1 cm^2, and plot I_{corr} vs. cathode surface area.
(b) Repeat part (a) for a cathode surface area of 1 cm^2 and anode surface areas of 0.1, 1.0, and 10 cm^2.

15.6 (a) In Sect. 15.4, the statement is made that the oxygen reduction reaction is favored over the hydrogen reaction because the corresponding half-cell potential for the oxygen reaction remains 1.23 V above that for hydrogen reduction, irrespective of the pH of the solution. Using the relationship between half-cell potentials and ionic concentration (Chapter 14), prove this statement.
(b) Explain why the oxygen reduction reaction predominates when the H^+ ion concentration is small. [*Hint*. Schematically plot Fig. 15.6 as a function of H^+ ion concentration. Remember, from part (a), that the half-cell potential of the oxygen reaction always is 1.23 V above that for the hydrogen reaction, but also remember that the absolute potentials of the competing reduction reactions depend upon the pH of the solution].

15.7 In anodic protection of a metal, the voltage is usually raised to a level corresponding to V_2 (Fig. 15.14), although in theory a voltage as low as V_1 is all that is necessary for passivation. Explain why the higher voltage, with correspondingly higher external current and power dissipation, is used in real applications.

15.8 According to the Pilling–Bedworth ratio, which of the following oxides would *not* be protective?

Oxide	*Oxide Density (g/cm^3)*	*Metal Density (g/cm^3)*
Na_2O	2.27	0.97
FeO	5.70	7.87
Fe_3O_4	5.18	7.87
Fe_2O_3	5.24	7.87
Al_2O_3	3.70	2.70
U_3O_8	7.31	18.17

15.9 The parabolic relationship for oxide formation is $x^2 = kt$, where k is a temperature dependent constant. Show that when diffusion of ions controls oxide film growth, the parabolic relationship results. How is k related to the diffusion coefficient, D?

15.10 Which of the following will preferentially corrode when placed in electrical contact in sea water?

(i) Copper and zinc
(ii) Magnesium and lead
(iii) Lead and yellow brass
(iv) Zinc and cast iron
(v) Zinc and steel

15.11 According to Table 15.1, zinc should corrode preferentially to galvanized steel in sea water. Can you think of a reason why this should be so?

16

Phase Diagrams

An understanding of phase equilibria and of ways to represent phase equilibria is very important for comprehending many materials systems. In this chapter maps are used to indicate the phases present in a material at equilibrium.

16.1 SINGLE COMPONENT SUBSTANCES

A pure substance, under equilibrium conditions, may exist as vapor, liquid, or crystalline solid, depending upon the conditions of temperature and pressure. Phase diagrams provide a convenient way of representing which state of aggregation is stable for a particular set of conditions. Moreover, a phase diagram is a map that shows which *phase* is stable and not just the state of aggregation. It is useful, at this point, to recall that a phase is a structurally homogeneous portion of matter. There is only one vapor phase no matter how many constituents make it up. Similarly, for a pure substance, there is only one liquid phase; however, there may be more than one solid phase (i.e., allotropes or polymorphs) because of differences in crystal structure.

Pure substances such as H_2O, NaCl, CH_4, and Cu all may be regarded as one-component substances. A phase diagram for a pure substance, called a unary equilibrium diagram, shows which phases exist as a function of temperature and pressure. A portion of the unary phase diagram for H_2O (Fig. 16.1*a*) indicates that at 1 atm pressure ice is stable below 0°C ,water is stable between 0°C and 100°C, and vapor is stable above 100°C. The temperatures separating the single-phase regions are fixed points (i.e., invariants) at a given pressure, and they represent the

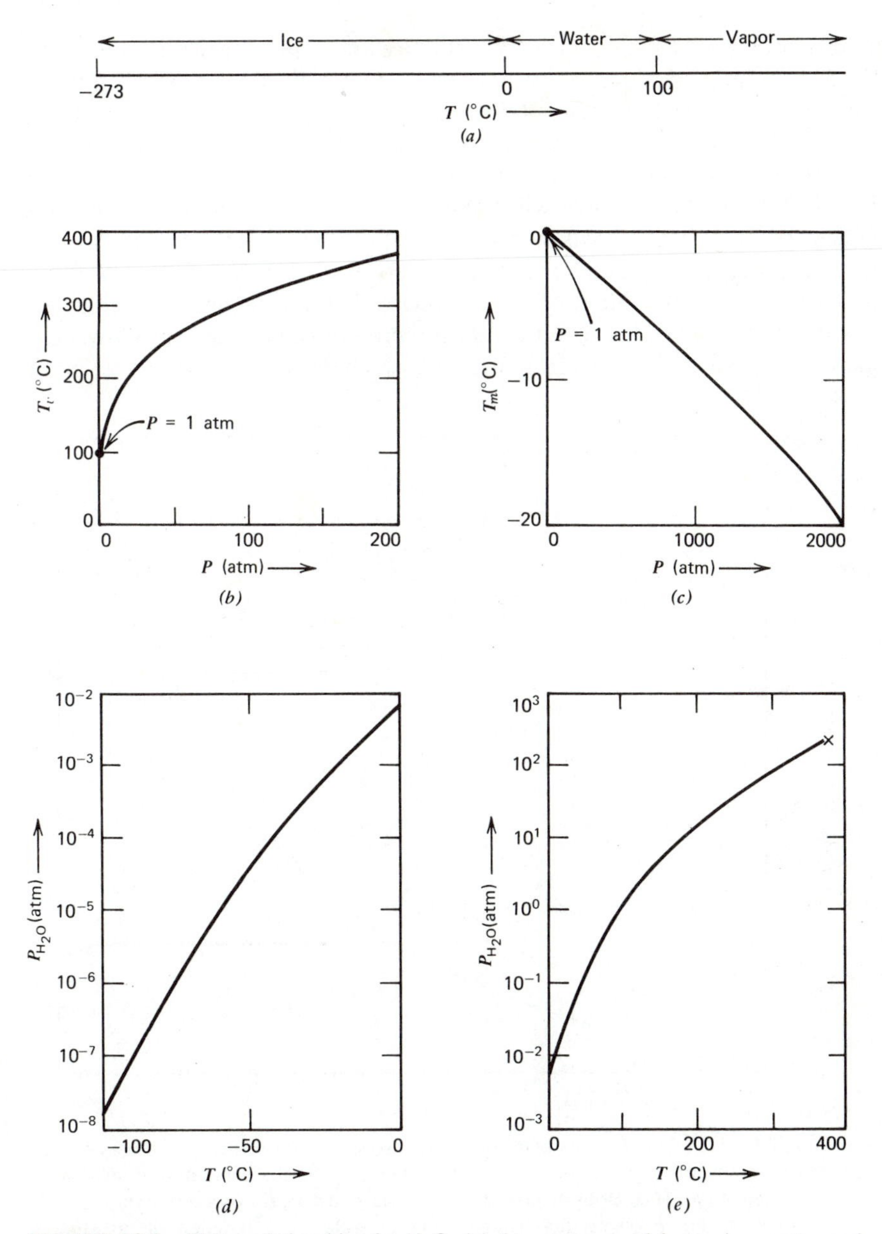

FIGURE 16.1 Phase relationships for H_2O. (*a*) At a pressure of 1 atm, ice, water, and vapor are stable over different temperature ranges. (*b*) The boiling point of water is a function of pressure. (*c*) The melting point of ice is a function of pressure. (*d*) The vapor pressure of ice is a function of temperature. (*e*) The vapor pressure of water is a function of temperature.

conditions of two-phase coexistence. Thus, ice and water coexist in equilibrium at 0°C and 1 atm, and water and vapor coexist in equilibrium at 100°C and 1 atm; at either point, the free energies of the two coexisting phases are equal (see Fig. 1.6).

It is well-known that the boiling point of water increases with increasing pressure (Fig. 16.1*b*); also, the melting point of ice decreases with increasing pressure (Fig. 16.1*c*). Similarly, water will vaporize or ice will sublimate if pressure is reduced sufficiently (i.e., so as to reach the vapor pressure of the phase) at a given temperature (Figs. 16.1*d* and *e*). All of the above are features of the *T–P* diagram for H_2O (Fig. 16.2), which is a unary phase diagram having temperature and pressure as variables. In such a diagram there are single-phase fields where one phase has the lowest free energy, lines along which two phases coexist (i.e., bounda-

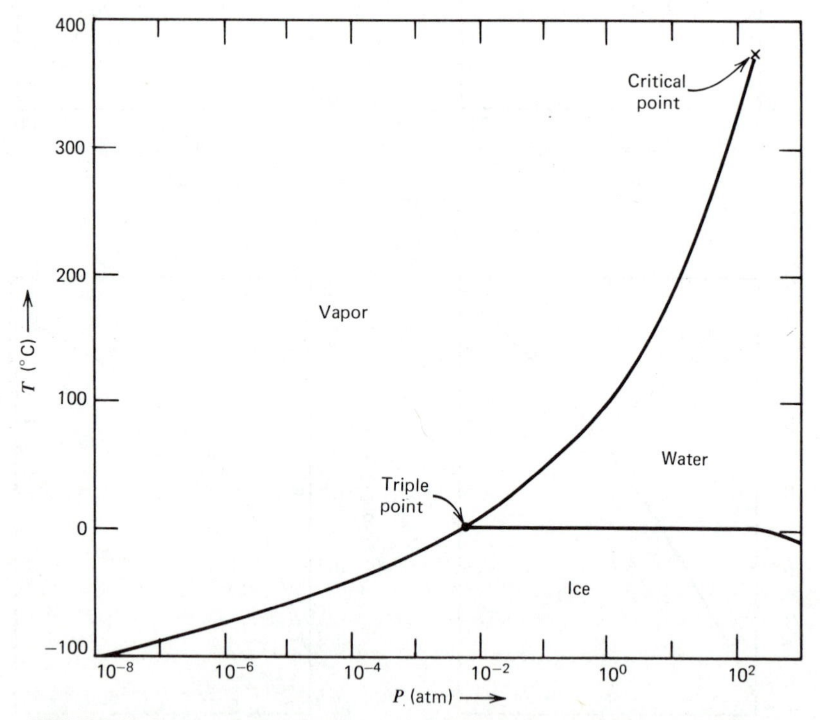

FIGURE 16.2 The *T–P* diagram for H_2O. The equilibrium unary phase diagram, as a function of temperature and pressure, incorporates the separate information presented in Fig. 16.1. Boundaries between single-phase fields represent two-phase coexistence. For example, the boiling point of water as a function of pressure (which is identical to the vapor pressure of water as a function of temperature) separates the liquid (i.e., water) and vapor fields. This boundary terminates at critical values of temperature and pressure (i.e., the critical point), above which liquid can be converted to vapor or vice versa without a distinct or sharp change in phase being observed.

ries between single-phase fields) where two phases have the same free energy, and one or more distinct points at which three phases coexist. These triple points are said to be invariants of the system, for each occurs at a definite, fixed combination of temperature and pressure for a given substance. Triple points provide absolute reference values for both temperature and pressure.

At moderate and low pressures, other one-component substances possess unary phase diagrams which are qualitatively similar to that for water; however, the melting point of most substances increases as pressure increases because most solids are more dense than their liquids. Melting temperature as a function of pressure and vapor pressure as a function of temperature (or, equivalently, sublimation temperature as a function of pressure) represent the extreme limits of

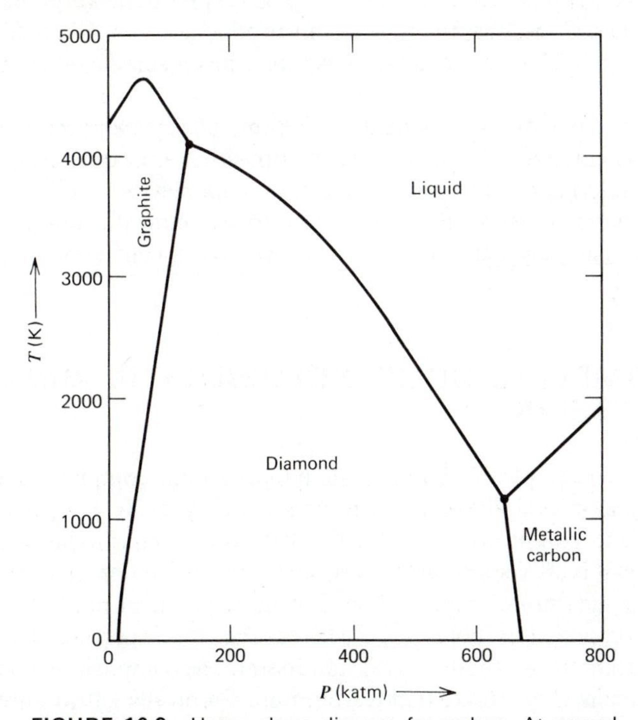

FIGURE 16.3 Unary phase diagram for carbon. At normal pressures graphite is the stable solid phase; at high pressures diamond is the stable solid phase. Thus, production of synthetic diamonds requires very high pressures. Note that metallic carbon can exist only at very high pressures. Phase equilibrium features for normal atmospheric pressure and reduced pressures are not portrayed on this diagram. [After G. C. Suits, *Amer. Scientist,* **52**, 395 (1964). Reproduced by permission, *The American Scientist,* journal of Sigma Xi, the Scientific Research Society of North America.]

usefulness of a crystalline solid for most engineering purposes. That is, if a crystalline solid begins to melt or vaporize, it loses mechanical integrity.

Changes in temperature and pressure can cause allotropic or polymorphic transformations, and such transformations occur at high pressures for many substances. High-pressure phase diagram studies have been rather limited, however, because of difficult experimental conditions. Nevertheless, such studies are important because they provide information about the state of matter in the earth's core and about phases that can be produced only at high pressures. As an example of the latter, Fig. 16.3 shows the unary phase diagram for carbon. It is apparent that graphite is the stable solid phase at atmospheric pressure, whereas diamond is stable at high pressures. Diamond is extremely valuable for cutting tool purposes and for gem purposes because it is very hard; however, its occurrence in nature is rare. Consequently, the current production of synthetic diamonds at high pressures and elevated temperatures represents a significant contribution to technology.

When more than one component is present, phase diagrams become more complex since temperature, pressure, and composition are variables. Because most engineering materials consist of more than one component but are used at pressures near atmospheric pressure, it is important to examine the phase equilibrium relationships that may exist in binary or two-component systems at constant pressure.

16.2 COMPLETE SOLID SOLUBILITY IN BINARY SYSTEMS

An important example of a binary system that exhibits complete solid solubility is that of copper-nickel alloys at 1 atm pressure (Fig. 16.4). The coordinates are temperature and composition, and in this case the latter is expressed as weight percent nickel. It is also common to use atom (or mole) fraction or atomic (or mole) percent as the measure of composition. For most practical purposes, the weight percent of a component is more useful, whereas the atomic percent of a component is more important from a fundamental standpoint. The composition coordinate for the Cu–Ni system (Fig. 16.4) extends from pure Cu on the left to pure Ni on the right.

In a simple view, any phase diagram can be considered as a map. If the alloy composition and temperature are specified, then the phase diagram allows determination of the phase or phases that will be present under equilibrium conditions. In Fig. 16.4, three regions are shown: a solid or α field, a liquid field, and a liquid plus solid two-phase region. If the temperature and composition correspond to point A, the alloy is a single-phase solid; at point B, it is a two-phase mixture of solid and liquid, and at point C, a single-phase liquid. Complete solid

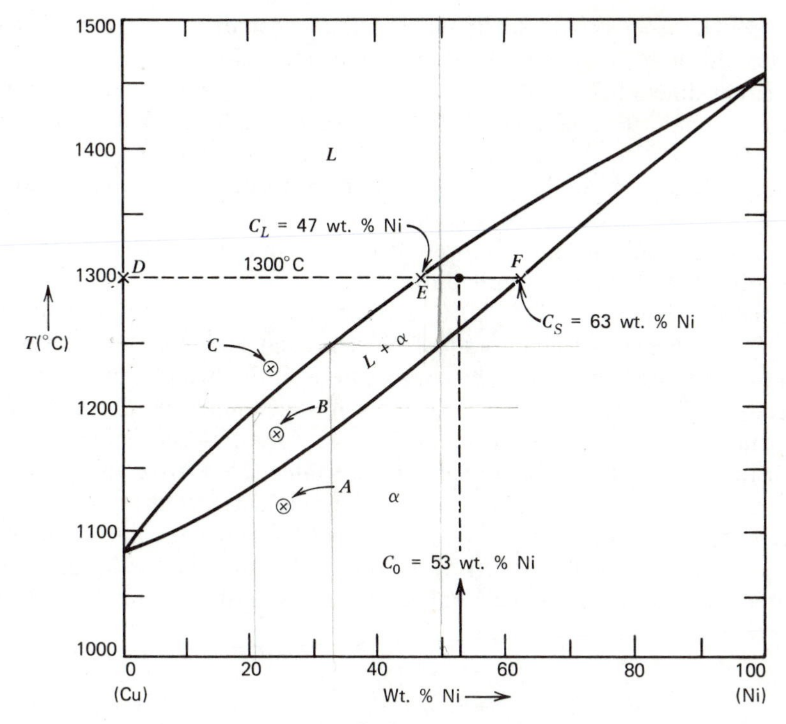

FIGURE 16.4 The copper–nickel binary phase diagram (at $P = 1$ atm). Complete liquid solubility or miscibility and complete solid solubility or miscibility occur because Cu and Ni are compatible elements. Cu–Ni solid solutions melt (or solidify) over a range of temperatures rather than at a fixed temperature. The temperature–composition points *A*, *B*, and *C* correspond to solid, solid and liquid, and liquid, respectively, in equilibrium.

In the experiment described in the text, copper is heated to 1300°C and nickel is slowly added, changing the composition along the line *DE*. At 1300°C the liquid becomes saturated with nickel when the Ni content reaches 47 wt. % Ni (point *E*); the addition of more nickel results in the appearance of a solid (α) phase containing 63 wt. % Ni (point *F*).

solubility is possible only when the two pure components have the same crystal structure and form a substitutional solid solution in all proportions.

The significance of the features of the Cu–Ni phase diagram may be illustrated in part by a simple experiment in which copper is heated above its melting point but below that of pure nickel, say 1300°C. If particles of pure Ni are added little by little to the molten copper, the composition of the resulting liquid alloy changes along the line *DE* in Fig. 16.4. At 1300°C it is found that Ni dissolves in molten Cu, producing a homogeneous liquid solution until the alloy composition reaches

47 wt. % Ni, whereupon the liquid solution becomes saturated with nickel. Further addition of Ni leads to the formation of a nickel-rich solid solution whose composition differs from that of the saturated liquid solution with which it coexists. The composition of this solid solution (63 wt. % Ni at 1300°C) is obtained by extending the line *DE* to point *F*, the boundary between the solid and solid plus liquid regions. The above considerations are rather analogous to the dissolution of sugar in hot tea. At a given tea temperature, sugar dissolves up to a certain amount, and addition of more sugar results in a sugar-rich solid that sinks to the bottom of the sugar-saturated tea.

For the Cu–Ni system at 1300°C, increasing the overall Ni content above 47 wt. % Ni increases the amount of Ni-rich solid and decreases the relative amount of saturated liquid, but the compositions of the two phases, given by the intersections of the isothermal line (tie line) with the boundaries of the two-phase field, remain the same. Since the compositions of the two coexisting phases are fixed at this temperature, it is possible to determine the relative amount of each phase if the alloy composition is known. For example, consider an alloy of composition $C_0 = 53$ wt. % Ni at 1300°C. Letting W_α and W_L be the fractional amounts by weight of solid and liquid, respectively, two conservation of mass equations must be satisfied. First, the sum of the fractional amounts must equal unity,

$$W_\alpha + W_L = 1 \tag{16.1}$$

since the alloy consists of a two-phase solid plus liquid mixture. Second, the mass of Ni distributed between the two phases must equal the mass of Ni in the overall alloy, and this can be expressed as

$$W_\alpha C_s + W_L C_L = C_0 \tag{16.2}$$

(Consideration of the mass of Cu would give an identical equation.) Simultaneous solution of Eqs. 16.1 and 16.2 gives the fractional amount of liquid by weight,

$$W_L = \frac{C_s - C_0}{C_s - C_L} \tag{16.3a}$$

and the fractional amount of solid by weight,

$$W_\alpha = \frac{C_0 - C_L}{C_s - C_L} \tag{16.3b}$$

For the example under consideration (C_0 = 53 wt. % Ni, C_s = 63 wt. % Ni, and C_L = 47 wt. % Ni), we have

$$W_L = \frac{63 - 53}{63 - 47} = 0.625 \quad (16.4)$$

That is, 62.5 % of the alloy mass is liquid, and 37.5 % is solid.

Equations 16.3a and b are referred to as the lever rule equations and can be used to determine the relative amounts of phases present in *any* two-phase region of a binary phase diagram, providing the alloy composition and temperature are known. For example, where two-phase equilibrium exists between phases α and β, the relative amount of α at a designated temperature is obtained by measuring the length of an isothermal line between the alloy composition (C_0) and the *opposite* phase-field boundary (C_β), and then dividing this by the total length of tie line within the two-phase region [i.e., $W_\alpha = (C_\beta - C_0)/(C_\beta - C_\alpha)$].

The type of experiment carried out with Cu and Ni at 1300°C can be repeated at other temperatures between 1084.5°C (T_m for Cu) and 1455°C (T_m for Ni), and a pair of composition points (C_L and C_s), representing the liquid and solid compositions for two-phase equilibrium, will be obtained at each temperature. The curved lines generated by connecting the C_L vs. T points and the C_s vs. T points are called the liquidus and solidus lines, respectively, and are important

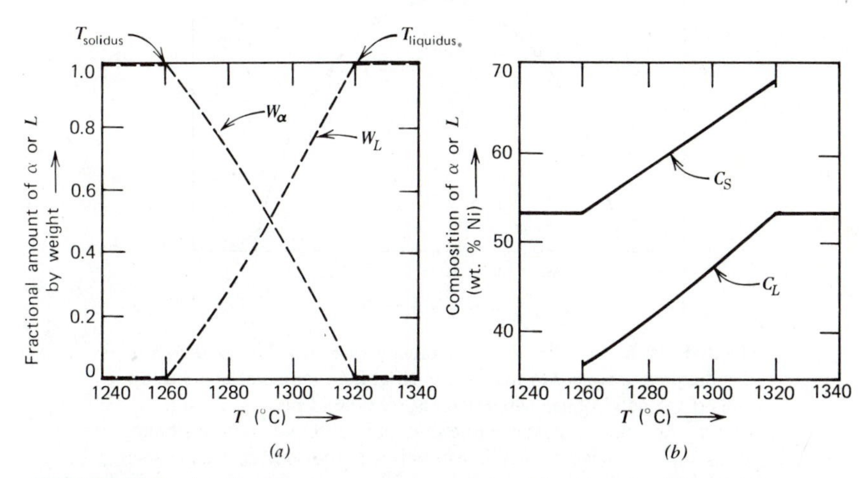

FIGURE 16.5 (*a*) Fractional amounts by weight of liquid and solid and (*b*) compositions of the phases as a function of temperature for a Cu–53 wt. % Ni alloy. At any temperature between the solidus and the liquidus, the equilibrium fraction of solid or liquid can be obtained with the lever rule.

features of the phase diagram since they define the limits of the two-phase region. In addition, the solidus and liquidus lines give the temperatures at which melting starts and is completed, respectively, for an alloy of fixed composition. Thus, if a solid alloy of composition C_0 is heated, melting begins when the solidus temperature for the alloy is reached. The melting process continues, as temperature is raised through the two-phase field, until all solid has been converted to liquid at the liquidus temperature for the alloy. For this binary alloy, the fractional amounts of liquid and solid vary continuously between zero and unity and between unity and zero, respectively, as temperature is raised from the solidus to the liquidus value (Fig. 16.5). This, of course, corresponds to the continuous change in the composition of each coexisting phase, given by the solidus or liquidus line as a function of temperature. The characteristic of melting over a range of temperatures or, conversely, of solidifying over a range of temperatures, which is common in

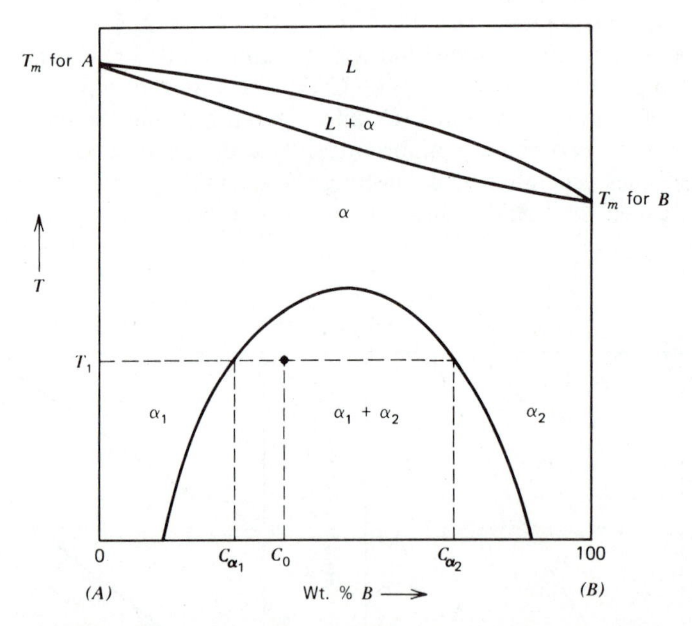

FIGURE 16.6 A solid state miscibility gap in a binary diagram. This diagram is similar to that of Cu–Ni at high temperatures, but on cooling the solid phase separates into two solid phases, α_1 and α_2, having the same crystal structure but different compositions. (Subscripts are used to distinguish between the solid phases in this case.) At temperature T_1, an alloy of composition C_0 consists of two phases coexisting in equilibrium and having compositions C_{α_1} and C_{α_2}. The lever rule can be used to determine the relative amounts of these phases.

binary systems, is quite distinct from the fixed-temperature melting or freezing of a single-component substance.

Cu and Ni exhibit complete solid solubility, but in many alloys only a certain amount of one component can dissolve into the solid form of the other component and vice versa. This is necessarily the case if the two pure components have different crystal structures. Even when they have the same crystal structure, differences in relative atomic size and in chemical nature often lead to limited solid solubility. An example is shown in Fig. 16.6, where complete solid solubility occurs at elevated temperatures, but immiscibility (or a miscibility gap) develops at lower temperatures. Thus, on cooling an alloy of composition C_0, the high-temperature α solid solution separates into two distinct solid phases, designated α_1 and α_2, which have the same crystal structure but different compositions. Since the lever rule may be used in any two-phase region, it is possible to calculate the relative amounts of α_1 and α_2 at a particular temperature. For the alloy C_0 at temperature T_1 (Fig. 16.6), the two coexisting solid phases have compositions C_{α_1} and C_{α_2}, and their fractional amounts by weight are $W_{\alpha_1} = (C_{\alpha_2} - C_0)/(C_{\alpha_2} - C_{\alpha_1})$ and $W_{\alpha_2} = (C_0 - C_{\alpha_1})/(C_{\alpha_2} - C_{\alpha_1})$. Liquid-state miscibility gaps also are found in certain binary systems.

16.3 THERMAL ANALYSIS AND PHASE DIAGRAMS

Before proceeding to other features of binary systems, it is instructive to outline another experimental procedure which, in principle, can be used to determine the boundaries in a binary phase diagram. The hypothetical set of experiments suggested in the previous section could be used to delineate the Cu–Ni phase diagram; however, in actual practice, it is more convenient to utilize an alternative technique called thermal analysis. With reference to Fig. 16.7*a*, consider an alloy of composition C_0 that has been heated well into the single-phase liquid region. After equilibration at the elevated temperature, heat is slowly extracted from the system at a constant rate, and the temperature is measured as a function of time. A typical cooling curve obtained in this manner is shown in Fig. 16.7*b*. The temperature initially decreases with time in an approximately linear manner (see Sect. 1.4) until solidification begins at the liquidus temperature. As liquid is converted to solid during cooling through the two-phase region, evolution of the heat of fusion decreases the rate at which temperature falls as compared to cooling through a single-phase region. This is reflected by a change in slope of the cooling curve at the liquidus temperature. When the solidus temperature is reached, solidification is complete, and the slope of the cooling curve becomes steeper again. Thus, for a given alloy, the liquidus and solidus temperatures can be determined by noting where the cooling curve displays discrete changes in slope.

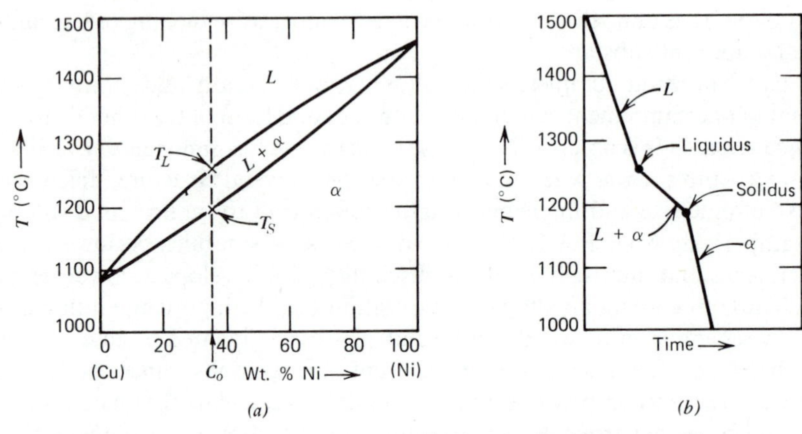

FIGURE 16.7 An alternate way of determining the liquidus and solidus curves. (*a*) An alloy of composition C_0 is heated into the liquid region (to 1500°C), and heat is slowly extracted. (*b*) The cooling curve (temperature vs. time) enables determination of the liquidus and solidus temperatures for this alloy because slope discontinuities occur at these points.

Repetition of this procedure with different alloys enables determination of the complete phase diagram as shown in Fig. 16.8. It should be noted that the liquidus and solidus temperatures coincide at the melting point of a pure component, and a *thermal arrest* results at T_m because liquid and solid can coexist in equilibrium only at this temperature (see Sect. 1.4). Cooling curve thermal arrests can also occur for alloys in many binary systems that are somewhat more complex than those discussed previously.

16.4 BINARY EUTECTIC SYSTEMS

A very large number of two-component substances consist of solid two-phase aggregates rather than a single solid solution phase. An especially important type of phase diagram in which two-phase solids are encountered is a simple eutectic system (Fig. 16.9). In this case, addition of B to pure A results in a substance that melts at temperatures lower than T_m for A, and addition of A to pure B results in decreased melting temperatures as well. Furthermore, there are certain compositions for which complete melting occurs at temperatures less than T_m for either pure component. Such behavior is important in the case of alloys used for soldering or brazing. It is also related to the action of permanent antifreeze when added to water, since the ethylene glycol-water system is a simple eutectic.

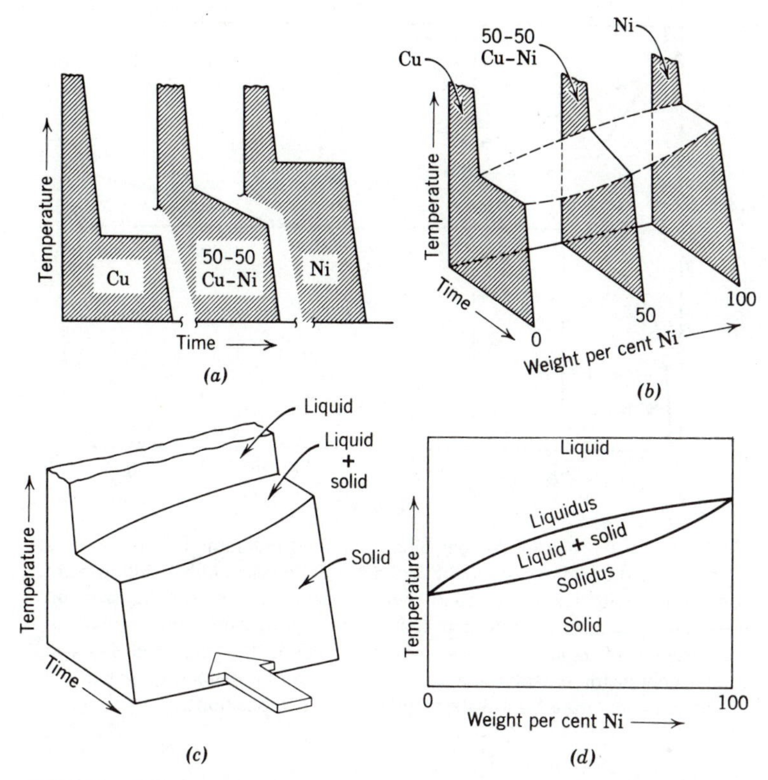

FIGURE 16.8 (*a*) A series of cooling curves for Cu–Ni alloys. (*b*) The cooling curves of (*a*) on time-temperature-composition coordinates. (*c*) The surfaces generated by cooling curves for all possible Cu–Ni alloys. (*d*) The Cu–Ni phase diagram, which is (*c*) viewed in the direction indicated by the arrow. (From *The Structure and Properties of Materials*, Vol. I, by W. G. Moffatt, G. W. Pearsall, and J. Wulff, Wiley, New York, 1964.)

Three different two-phase regions are found in a simple eutectic phase diagram. The unique features of such a diagram, however, are the eutectic temperature, T_E, and the eutectic composition, C_E. The latter corresponds to the composition at which the $L/L + \alpha$ and $L/L + \beta$ liquidus lines intersect with each other and with the horizontal line at T_E, called the eutectic invariant line. Any alloy in equilibrium at the eutectic temperature and having a composition lying on the eutectic invariant line consists of three coexisting phases, and these three phases (α, β, and L) have definite, fixed compositions corresponding to C_{α_E}, C_{β_E}, and C_E, respectively (Fig. 16.9). The compositions of the phases and the temperature for three-phase equilibrium in a binary system at constant pressure are invariants of the system, just as the melting point of a pure substance at constant pressure is invariant.

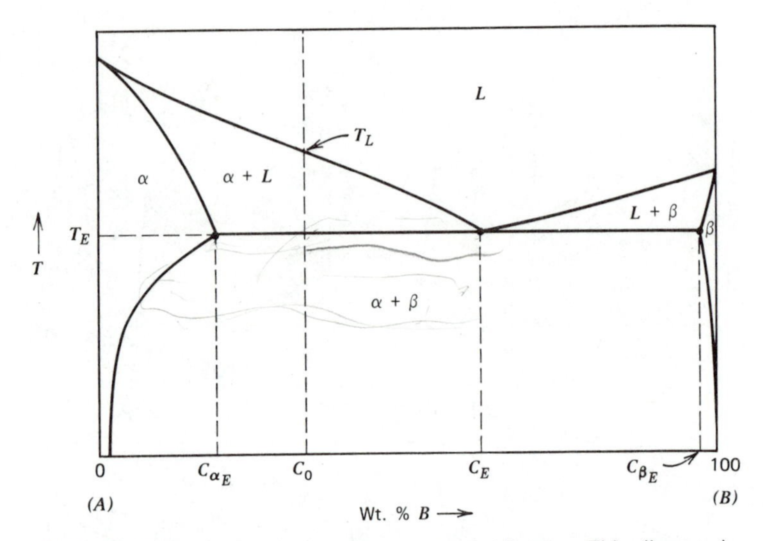

FIGURE 16.9 A hypothetical binary eutectic diagram. This diagram is characterized by limited solubility for each solid phase and by melting point depression with respect to each pure component. The eutectic invariant reaction is the most important feature of this system, and equilibrium coexistence of α, β, and L (having compositions $C_{\alpha E}$, $C_{\beta E}$, and C_E) can occur only at the eutectic temperature T_E. Note that melting begins at the eutectic temperature for a wide range of alloy compositions.

A cooling curve for an alloy of the eutectic composition is represented schematically in Fig. 16.10. The form of this curve is identical to that for a pure substance because a thermal arrest occurs at the solidification temperature, T_E; however, the process of solidification differs in these two cases. Note that the alloy is completely liquid at a temperature infinitesimally above T_E but is a solid two-phase mixture of α and β at a temperature infinitesimally below T_E. At the eutectic temperature, solidification proceeds according to the reaction

$$L(C_E) \xrightarrow{\text{cooling at } T_E} \alpha(C_{\alpha_E}) + \beta(C_{\beta_E}) \tag{16.5}$$

This equation is a shorthand way of representing the eutectic invariant reaction; it specifies that liquid of composition C_E decomposes, on cooling, at T_E into two solid phases (α and β) having compositions C_{α_E} and C_{β_E}. Evolution of heat of fusion accompanies the eutectic reaction, and since the three phases can coexist only at the eutectic temperature, a thermal arrest occurs. On heating, the exact

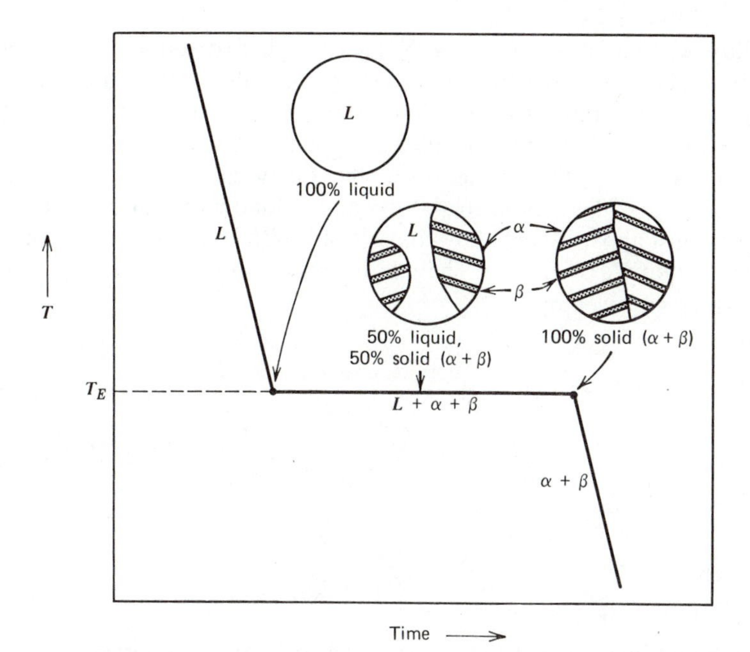

FIGURE 16.10 Cooling curve for the eutectic alloy (composition C_E in Fig. 16.9). A thermal arrest is observed at the eutectic temperature because equilibrium three-phase coexistence can occur only at this temperature. Two distinct solid phases form simultaneously from the liquid, and the heat of fusion that is evolved must be removed isothermally.

opposite occurs, and the solid alloy of composition C_E completely melts at T_E (i.e., Eq. 16.5 proceeds in the reverse direction).

Microstructures of the eutectic alloy before, during, and after solidification are represented schematically in Fig. 16.10. The distribution and morphology of phases in the eutectic solid are of considerable technological importance. Since α and β form simultaneously from the melt, they invariably are intermixed on a very intimate scale. In many eutectics, the arrangement is a lamellar array with alternating layers of α and β, as represented in Fig. 16.10. The intimate two-phase mixture in a eutectic solid is a consequence of the necessity of diffusion during the eutectic reaction, since the three phases involved have different compositions. The interlamellar spacing and the thicknesses of α and β lamellae depend upon cooling rate. A faster solidification rate (i.e., more rapid extraction of heat) imposes a limit on diffusion that results in a finer eutectic structure (i.e., smaller interlamellar spacing).

For alloys with compositions other than C_E, the phase diagram not only provides information about the solidification process but also information about the resulting microstructure. Consider an alloy of composition C_0 such that $C_{\alpha E} < C_0 < C_E$ (Fig. 16.9). This is called a hypoeutectic alloy. On cooling between the liquidus and eutectic temperatures, the fractional amount of α increases monotonically, and the compositions of α and L follow the solidus and liquidus lines, so that just above T_E the alloy consists of α with composition C_{α_E} and liquid with composition C_E (Fig. 16.9). The relative proportions of α and L in this two-phase mixture are given by

$$W_\alpha = \frac{C_E - C_0}{C_E - C_{\alpha E}} \tag{16.6a}$$

and

$$W_L = \frac{C_0 - C_{\alpha E}}{C_E - C_{\alpha E}} \tag{16.6b}$$

Since the liquid has reached the eutectic composition, it undergoes the eutectic reaction (Eq. 16.5) at T_E. The cooling curve for this hypoeutectic alloy exhibits a slope change at the liquidus temperature and a thermal arrest at T_E (Fig. 16.11). For a fixed mass of alloy, the duration of the thermal arrest varies from a maximum when $C_0 = C_E$ to zero when $C_0 = C_{\alpha E}$ because the amount of liquid that undergoes the eutectic reaction varies with alloy composition according to Eq. 16.6b.

During the eutectic reaction, any α which had formed above the eutectic temperature, termed proeutectic or primary α, remains unaffected. Thus, the microstructure of a completely solidified hypoeutectic alloy consists of a mixture of primary α and the two-phase $(\alpha + \beta)$ eutectic solid (Fig. 16.11). These morphologically distinct entities (i.e., primary α and eutectic solid), called microconstituents, are important because the properties of a hypoeutectic alloy often depend more upon the relative amounts of microconstituents than upon the relative amounts of phases. In this case, the fractional amounts of the microconstituents are identical to the fractional amounts of the two phases present just prior to the start of the eutectic invariant reaction:

$$W_{\text{primary }\alpha} = \frac{C_E - C_0}{C_E - C_{\alpha E}} \quad (\equiv W_\alpha \text{ at } T_E + \Delta T) \tag{16.7a}$$

$$W_{\text{eutectic solid}} = \frac{C_0 - C_{\alpha E}}{C_E - C_{\alpha E}} \quad (\equiv W_L \text{ at } T_E + \Delta T) \tag{16.7b}$$

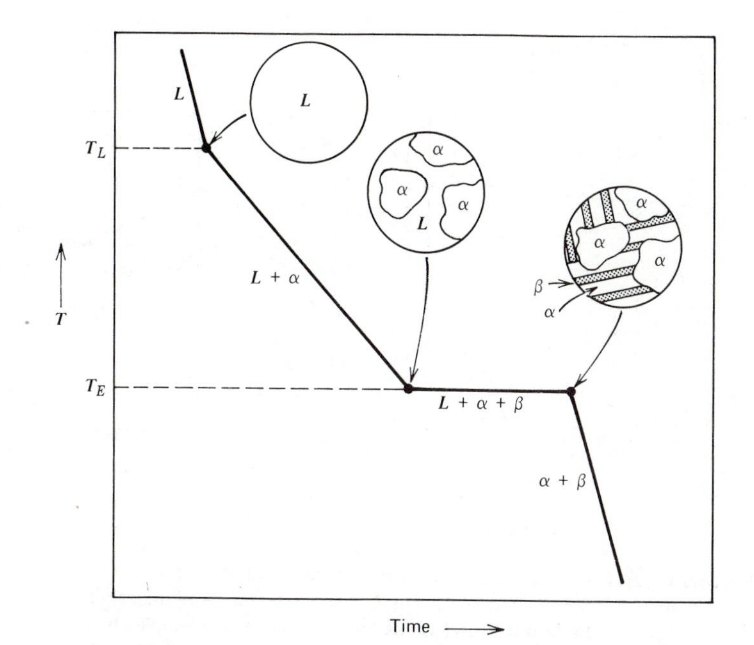

FIGURE 16.11 Cooling curve for a hypoeutectic alloy (composition C_0 in Fig. 16.9). Primary or proeutectic α begins to crystallize at the liquidus temperature where the cooling curve displays a slope change. Just above the eutectic temperature, liquid coexists with a maximum amount of primary α. At the eutectic temperature, this liquid (having the eutectic composition) undergoes the eutectic reaction with a corresponding thermal arrest. Once the eutectic reaction is completed, the solidified alloy consists of a mixture of primary α and eutectic solid, which is a mixture of α and β formed by the eutectic reaction.

On the other hand, the fractional amount of total α in the solid (i.e., primary α plus α in the eutectic solid) is given by

$$W_\alpha = \frac{C_{\beta_E} - C_0}{C_{\beta_E} - C_{\alpha_E}} \tag{16.8}$$

which is obviously greater than the fractional amount of primary α.

For an alloy whose composition lies between C_E and C_{β_E}, termed a hypereutectic alloy, the same considerations discussed above apply, but the final microstructure will contain primary β and eutectic solid as microconstituents. A complete set of thermal analysis experiments with alloys having different compositions facilitates determination of a simple eutectic phase diagram. Typical results and the corresponding phase diagram are shown in Fig. 16.12.

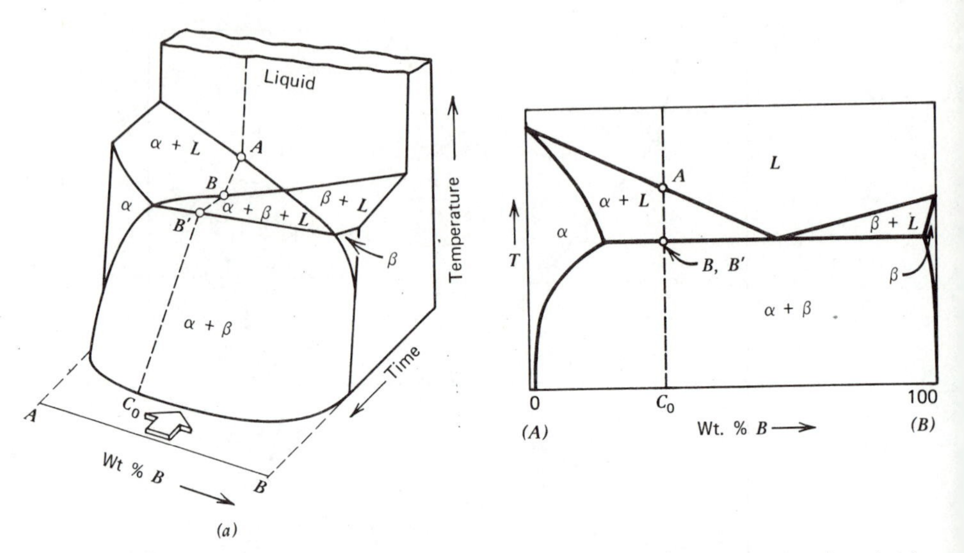

FIGURE 16.12 (*a*) The results of a series of thermal analysis experiments for alloys in the system shown in Fig. 16.9. The cooling curves delineate the liquidus, solidus, and eutectic temperatures as well as the boundaries (solvus lines) between the single-phase solid fields and the two-phase solid regions. The diagram in (*b*) is obtained by viewing (*a*) in the direction noted by the arrow.

16.5 OTHER THREE-PHASE REACTIONS IN BINARY SYSTEMS

The eutectic reaction is but one of a number of possible three-phase invariant reactions in binary systems. In the solid-state analog of a eutectic reaction, called a eutectoid reaction, one solid phase having the eutectoid composition decomposes on cooling to two different solid phases. A particularly important eutectoid reaction occurs in the Fe–Fe_3C system, which is shown in Fig. 16.13. At the eutectoid temperature (723°C), the fcc γ phase (austenite) decomposes on cooling to form the bcc α phase (ferrite) and Fe_3C (cementite), according to the reaction:

$$\gamma(0.80 \text{ wt. }\%\text{C}) \xrightarrow{\text{cooling at } 723^\circ\text{C}} \alpha(0.02 \text{ wt. }\%\text{C}) + \text{Fe}_3\text{C}(6.69 \text{ wt. }\%\text{C}) \tag{16.9}$$

The eutectoid decomposition product, whose microstructure is shown in Fig. 16.14*a*, is called pearlite and is a microconstituent consisting of alternating ferrite (α) and cementite (Fe_3C) lamellae. Pearlite is the only microconstituent present in a

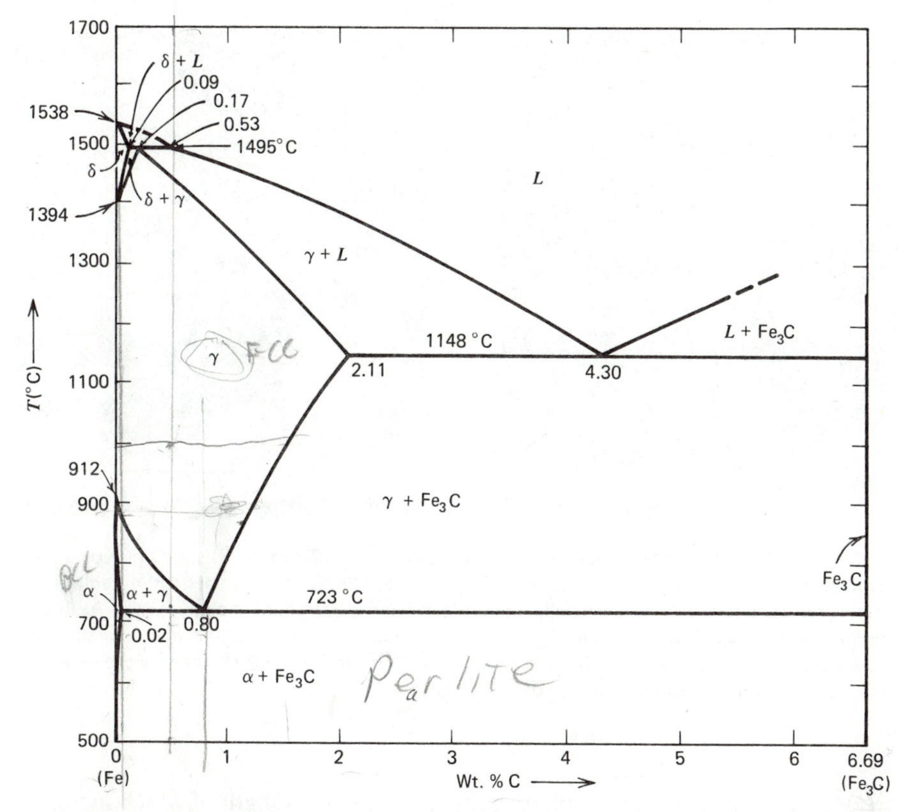

FIGURE 16.13 The Fe–Fe_3C system. In this important binary system, four solid phases participate in three invariant reactions (peritectic at 1495°C, eutectic at 1148°C, and eutectoid at 723°C).

slowly cooled alloy having the eutectoid composition. Alloys with compositions less than or greater than 0.80 wt. % C develop different microstructures. For example, when a hypoeutectoid alloy (e.g., Fe—0.5 wt. % C) cools from the austenite (γ) region into the $\alpha + \gamma$ region, ferrite begins to nucleate and grow at austenite grain boundaries. This process, which is similar to the formation of the primary microconstituent in a eutectic system, results in an increasing amount of ferrite and a decreasing amount of austenite that becomes enriched in carbon content. Just above the eutectoid temperature, the austenite approaches the eutectoid composition (0.80 wt. % C), and all of the remaining γ undergoes the eutectoid reaction at 723°C. The final microstructure contains two microconstituents, proeutectoid ferrite and pearlite (Fig. 16.14*b*), whose relative amounts may be determined with equations analogous to Eqs. 16.7a and b. Such iron-

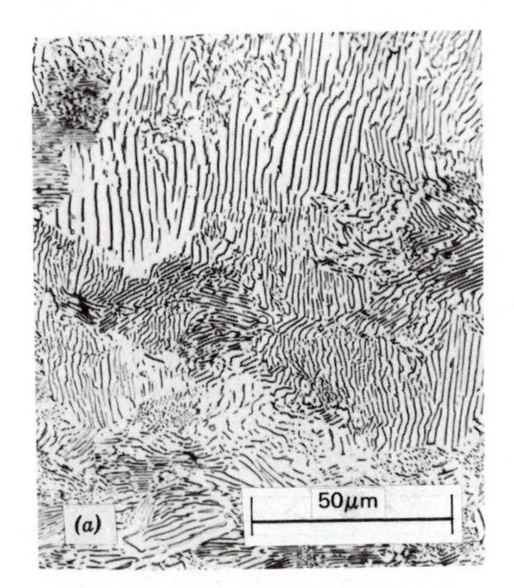

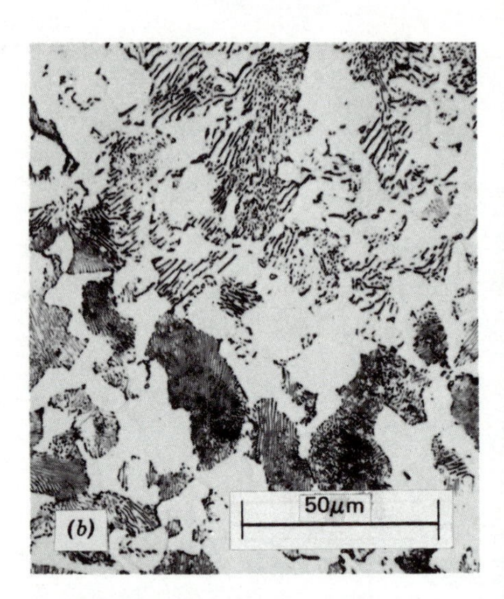

FIGURE 16.14 (*a*) Photomicrograph of a eutectoid steel (0.80 wt. % C). Shown is pearlite, a microconstituent consisting of ferrite (α) and cementite (Fe_3C) that formed by eutectoid decomposition of austenite (γ). This lamellar microconstituent is important in many steels. (*b*) Photomicrograph of a hypoeutectoid steel (0.5 wt. % C). This alloy contains two microconstituents: proeutectoid ferrite (formed prior to the eutectoid reaction) and pearlite (formed by the eutectoid reaction). (Photomicrographs courtesy of United States Steel Corp.)

carbon alloys are one form of plain carbon steels, the strength of which increases almost linearly with the fractional amount of pearlite present.

The Fe–Fe_3C system also contains a peritectic invariant reaction at 1495°C. In this case, a solid phase decomposes on heating into a liquid and another solid phase, and the reverse occurs on cooling. This peritectic reaction is

$$\delta(0.09 \text{ wt. } \%\text{C}) + L(0.53 \text{ wt. } \%\text{C}) \xrightarrow{\text{cooling at } 1495^\circ\text{C}} \gamma(0.17 \text{ wt. } \%\text{C}) \tag{16.10}$$

It is worth noting that many chemical compounds are said to decompose rather than melt at a specific temperature, and often this occurs by means of a peritectic invariant reaction. The solid-state analog of the peritectic reaction, called a peritectoid reaction, involves three different solid phases instead of a liquid and two solid phases. Peritectic and peritectoid reactions do not give rise to microconstituents as the eutectic and eutectoid reactions do.

The three-phase invariant reactions represented in Eqs. 16.5, 16.9, and 16.10 have been written for cooling processes. On heating, each of these reactions

TABLE 16.1 Types of Three-Phase Reactions Occurring Most Commonly in Condensed Systems

Name of Reaction	*Equation*	*Phase Diagram Characteristic*
Monotectic	$L_I \xrightarrow{\text{cooling}} \alpha + L_{II}$	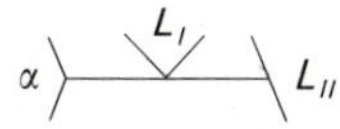 α L_I L_{II}
Monotectoid	$\alpha_1 \xrightarrow{\text{cooling}} \alpha_2 + \beta$	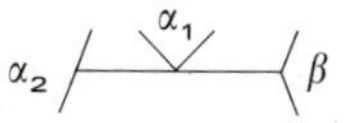 α_2 α_1 β
Eutectic	$L \xrightarrow{\text{cooling}} \alpha + \beta$	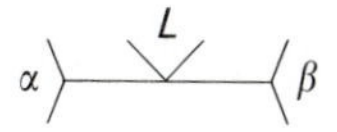 α L β
Eutectoid	$\alpha \xrightarrow{\text{cooling}} \beta + \gamma$	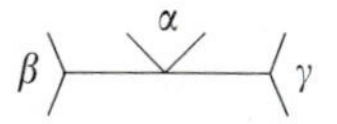 β α γ
Syntectic	$L_I + L_{II} \xrightarrow{\text{cooling}} \alpha$	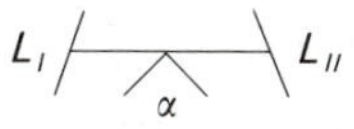 L_I L_{II} α
Peritectic	$\alpha + L \xrightarrow{\text{cooling}} \beta$	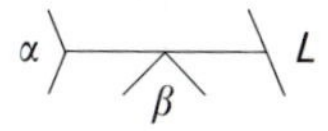 α L β
Peritectoid	$\alpha + \beta \xrightarrow{\text{cooling}} \gamma$	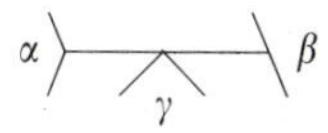 α β γ

proceeds in the reverse direction. This is also true for other three-phase reactions in binary systems that may involve vapor, liquid, and solid phases. Although a large number of three-phase reactions are possible, those involving condensed phases usually are of more practical significance in binary metallic and ceramic systems. The most important of these are summarized in Table 16.1.

16.6 INTERMEDIATE PHASES

It was mentioned in Sect. 16.2 that limited solid solubility is more common than complete solid solubility. Accordingly, in most cases, only a finite amount of one component will dissolve in another pure component. Such a solid phase, based on a pure component and extending to certain finite compositions into a binary phase diagram, is called a terminal solid solution, and the line representing the solubility limit of a terminal solid solution with respect to a two-phase solid region is called a solvus line. Many phase diagrams contain intermediate phases whose occurrence cannot be readily predicted from the nature of the pure components.

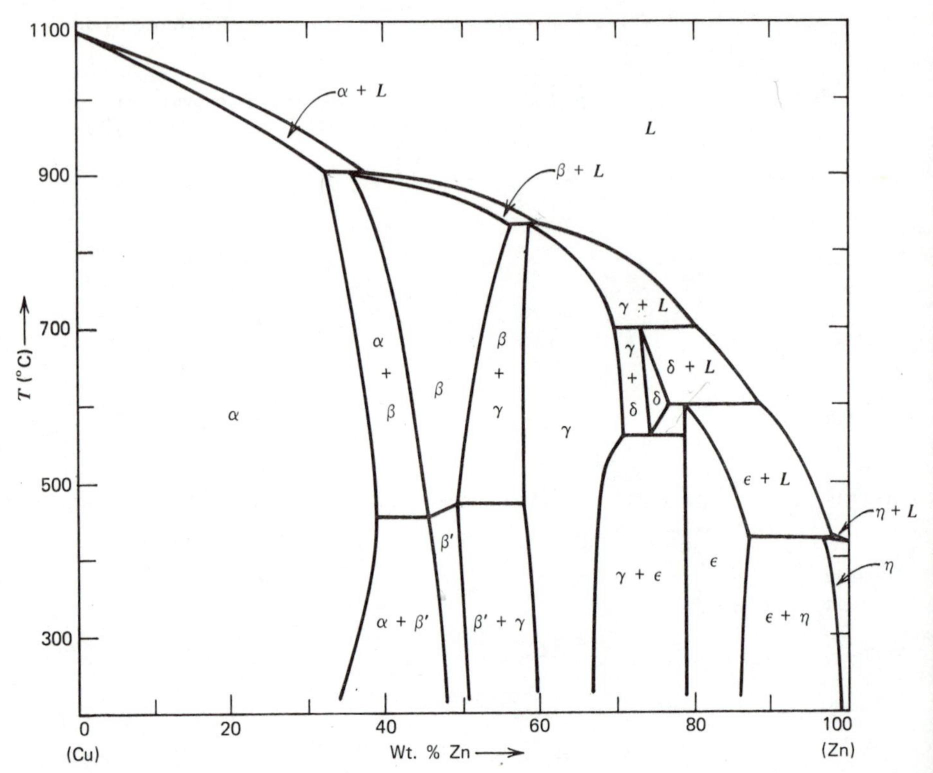

FIGURE 16.15 The Cu–Zn system. This system contains a number of intermediate phases (intermediate solid solutions) and a number of invariant reactions. Copper-rich alloys in this system constitute the important technological alloys known as brasses.

An intermediate phase may exist over a range of compositions (intermediate solid solution) or only at a relatively fixed composition (a compound, if there is a stoichiometric relationship between the components). The Cu–Zn system displays a number of intermediate solid solutions (Fig. 16.15). Intermediate solid solutions often have higher electrical resistivities and hardnesses than either of the two components. Examples of compounds are found in the Mg–Ni system (Fig. 16.16). Here, both Mg_2Ni and $MgNi_2$ have very narrow composition ranges and are designated intermetallic compounds. Intermetallic compounds differ from common chemical compounds in that the bonding is primarily metallic rather than ionic or covalent, as would be found with compounds in certain metal–nonmetal systems or ceramic systems. Some metal–nonmetal compounds, such as Fe_3C, are metallic in character, whereas others, such as MgO, are ceramic phases.

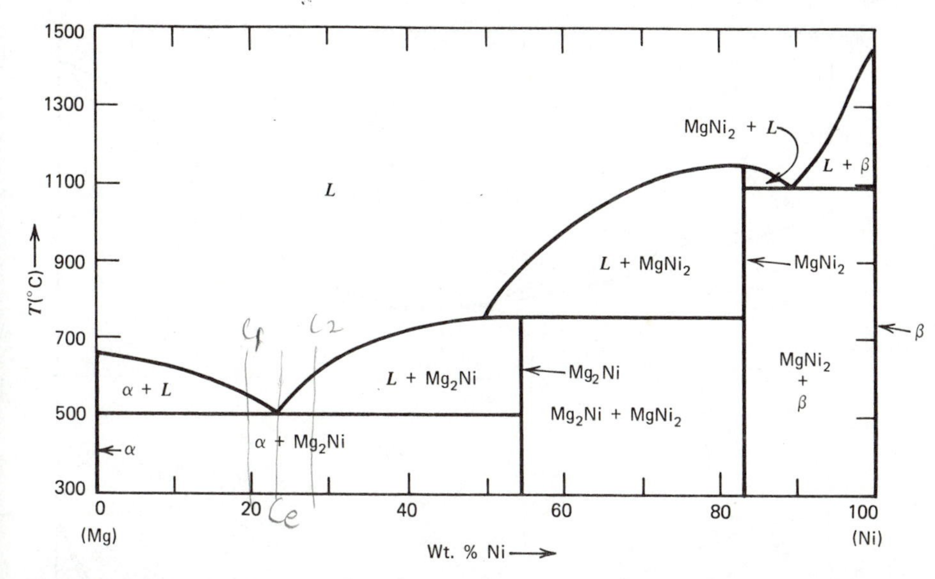

FIGURE 16.16 The Mg–Ni system. In this system, intermetallic compounds (having fixed compositions and definite stoichiometries) occur. Unlike covalent or ionic compounds, intermetallic compounds often have stoichiometries that are dictated by relative atomic sizes and atomic packing considerations.

Intermediate phases may undergo polymorphic transformations, and some may melt at a fixed temperature (e.g., $MgNi_2$). These are termed congruent transformations, in which one phase changes to another phase of the same composition at a definite temperature. Thus, $MgNi_2$ is said to be a congruently melting phase, whereas Mg_2Ni melts incongruently since it undergoes peritectic decomposition. A congruently melting compound can serve as a pure component in a binary system. For example, the Mg–Ni system (Fig. 16.16) can be subdivided into the Mg–$MgNi_2$ and $MgNi_2$–Ni systems. Congruent transformations also are observed at specific alloy compositions for some binary solid solutions. The Ti–Zr system (Fig. 16.17) contains two examples. The 50 wt. % Ti–50 wt. % Zr alloy melts congruently at 1537°C (minimum melting point), and the 34 wt. % Ti–66 wt. % Zr alloy undergoes a congruent solid state transformation at about 535°C.

16.7 MULTICOMPONENT SYSTEMS

Most commercially important alloys and ceramic materials contain more than two components. Although in some cases the proportion of the third or additional components is small, they can have a large effect on the properties of the material.

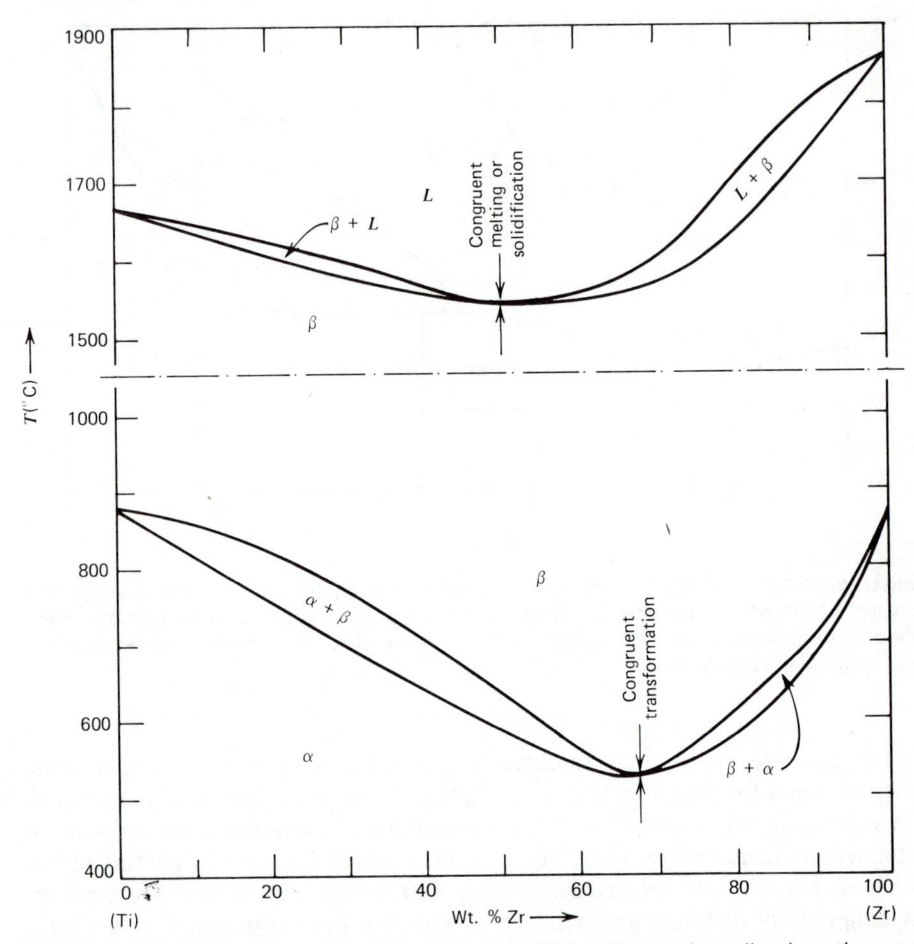

FIGURE 16.17 The Ti–Zr system. Titanium and zirconium are isomorphous (i.e., have the same crystal structure) at both high and low temperatures. This leads to a rather unique miscibility in the low-temperature (hcp) form as well as in the high-temperature (bcc) form. An alloy containing 50 wt. % Zr melts congruently at 1537°C and an alloy containing 66 wt. % Zr undergoes a solid state congruent transformation at 535°C.

Since it is difficult to delineate phase relationships for even a ternary system at constant pressure without recourse to a three-dimensional model, two-dimensional maps often are utilized. In such a case, one of two variables is usually fixed—either the temperature or the percentage of one of the components. For our purposes, it is sufficient to note that addition of a third component not only alters the composition of a particular phase in a binary system but also raises or lowers the temperatures at which phase changes take place. This is demonstrated in the plot

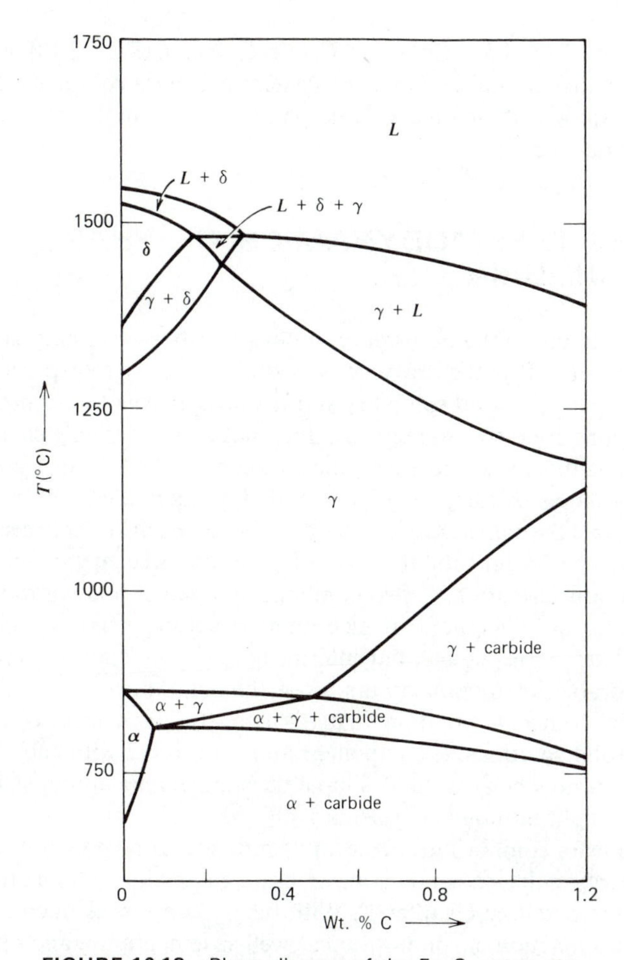

FIGURE 16.18 Phase diagram of the Fe–C system containing 5% Cr. In this "pseudobinary" diagram, the phase relationships between iron and carbon have been altered by the presence of chromium. Note that three phases can be present in a region of this diagram, unlike the case of a binary diagram. (Adapted from L. H. Van Vlack, *Materials Science for Engineers*, 1970, Addison-Wesley, Reading, Mass.)

that shows the effect of 5% Cr on the Fe–Fe_3C system (Fig. 16.18). Furthermore, invariant reactions in multicomponent systems are more complex than in binary systems, for example, three distinct solid phases form from the eutectic liquid in a ternary eutectic reaction.

16.8 THE THERMODYNAMICS OF PHASE EQUILIBRIA

Concepts relating to the occurrence of phases or phase stability are essentially chemical in nature. It is the province of crystal chemistry to explain and predict extensive or restricted solid solubility and the occurrence of intermediate phases. Such predictions are rather limited currently, but two underlying chemical factors are apparent: atomic size and electronic structure. If these factors are similar for two pure components, then changes in enthalpy upon solid solution formation will be small, and the increased entropy due to increased disorder in a solid solution will favor extensive solid solubility. That is, the total free energy will be lowered by solid solution formation. If two components interact unfavorably because of size differences or differences in electronic structure, solid solubility usually will be restricted or negligible, but intermediate phases may or may not occur. Truly unfavorable interaction occurs when the component atoms do not bond well with each other, a situation that disfavors the formation of intermediate phases and solid solutions. If component atoms do bond with each other, intermediate phases may be favored. A simple thermodynamic approach can provide some useful insight into binary systems.

The criterion for equilibrium at constant temperature and pressure is a minimum in the free energy of the system. This was considered previously for a pure substance at constant pressure (see Chapter 1). With binary substances, such as alloys, the free energy is a function of composition as well as temperature and often is lowest when a mixture of two phases of different compositions coexist. This is amply demonstrated by the large number of two-phase regions in binary phase diagrams. To understand the thermodynamic basis for binary equilibrium diagrams, it is necessary to develop expressions for the free energy of a phase. This can be done in terms of expressions for enthalpy and entropy.

When two substances that form a solid solution are mixed together, there invariably is a decrease in free energy that reflects both a change in enthalpy because of bonding differences and a change in entropy because of the increased disorder. The enthalpy change may be positive or negative, but it is dominated by the increase in entropy. The molar enthalpy of a solid solution at some temperature can be represented as

$$H = X_A H_A^\circ + X_B H_B^\circ + \Delta H_{\text{mix}} \tag{16.11}$$

where H_A° and H_B° are the standard state enthalpies of the pure components, $X_A (=1 - X_B)$ is the mole fraction of A in solution, X_B is the mole fraction of B in solution, and ΔH_{mix} is the enthalpy of mixing. Similarly, the molar entropy of a solid solution at some temperature can be represented as

$$S = X_A S_A^\circ + X_B S_B^\circ + \Delta S_{\text{mix}} \qquad (16.12)$$

A combination of Eqs. 16.11 and 16.12 provides the free energy of a solid solution:

$$\begin{aligned} G &= X_A(H_A^\circ - TS_A^\circ) + X_B(H_B^\circ - TS_B^\circ) + \Delta H_{\text{mix}} - T\,\Delta S_{\text{mix}} \\ &= X_A G_A^\circ + X_B G_B^\circ + \Delta G_{\text{mix}} \end{aligned} \qquad (16.13)$$

Since this equation affords no information about the chemical nature of the solid solution, it is useful to examine ΔH_{mix} and ΔS_{mix} more closely.

In certain cases, the enthalpy of mixing can be approximated by

$$\Delta H_{\text{mix}} \approx X_A X_B V \qquad (16.14)$$

where V is an interaction energy. If $V < 0$, an A atom bonds more strongly with B atoms than with other A atoms, and a B atom bonds more strongly with A atoms; that is, unlike nearest neighbors are favored. If $V > 0$, like nearest neighbors are favored, since like atoms bond more strongly than do unlike atoms. If $V = 0$, unlike and like neighbors are equally favored, and a truly random solid solution should occur.

An approximate form for the entropy of mixing, on a molar basis, is

$$\Delta S_{\text{mix}} \approx -R(X_A \ln X_A + X_B \ln X_B) \qquad (16.15)$$

where R is the universal gas constant. Since X_A and X_B must be smaller than or equal to unity, ΔS_{mix} as given by this equation necessarily will be positive. This result is largely independent of bonding, depending on randomness considerations alone. Substitution of the physically plausible Eqs. 16.14 and 16.15 into Eq. 16.13 gives

$$G \approx X_A G_A^\circ + X_B G_B^\circ + X_A X_B V + RT(X_A \ln X_A + X_B \ln X_B) \qquad (16.16)$$

(Comparison of this equation, with $V = 0$, should be made with Eq. 13.3.) Equation 16.16 provides two important results. First, a small amount of B added to A (or vice versa) causes a decrease in the free energy. Second, if $V \leq 0$, the free energy curve will always have a positive curvature and, accordingly, exhibit a single minimum. Features similar to those described above also pertain to liquid solutions. The case of intermediate phases is only slightly more complex, although the

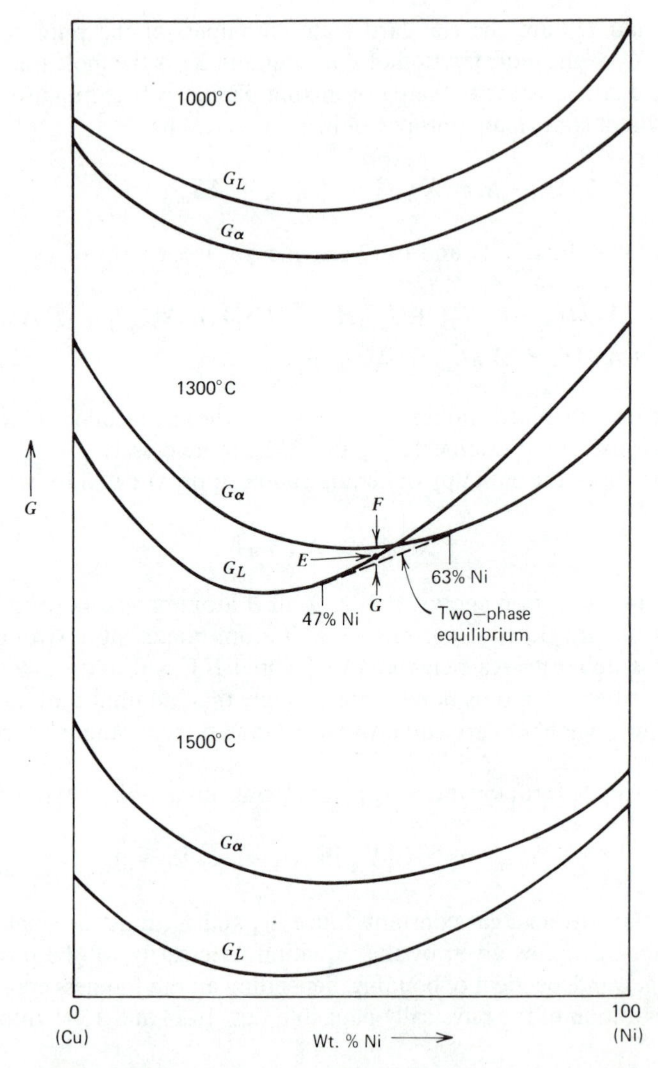

FIGURE 16.19 Schematic free energy vs. composition curves at various temperatures for the Cu–Ni system (see Fig. 16.4). For any composition, the lowest free energy represents stable equilibrium. When the free energy curves for two phases intersect, the lowest free energy (for a certain composition range) occurs for a two-phase mixture; this is represented by a common tangent line between the two free energy curves.

enthalpy of mixing almost invariably is negative, and the entropy of mixing may be very small for a stoichiometric or ordered phase.

Schematic free energy curves for the Cu–Ni system (Fig. 16.4) at various temperatures are presented in Fig. 16.19. When $T > T_m$ for Ni, the liquid is stable at all compositions, having a lower free energy than the phase α. When $T < T_m$ for Cu, α is stable at all compositions, since its free energy curve lies below that of the liquid. [Note that the standard state free energies used in representing the molar free energy of a phase (Eq. 16.13 or 16.16)—that is, the end points of the α free energy curves—are the free energies of the respective pure components existing as that phase.] At a temperature intermediate between the melting points of Cu and Ni,

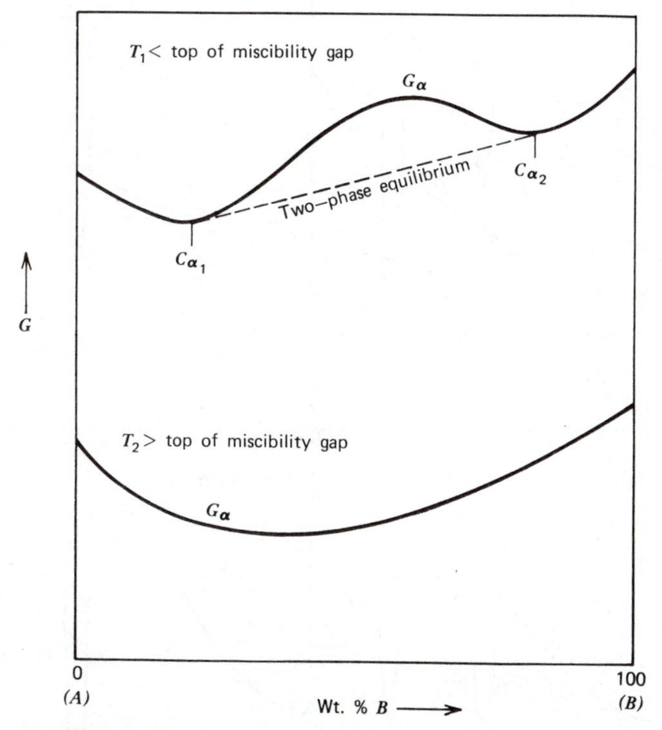

FIGURE 16.20 Schematic free energy vs. composition curves for the solid phase in a system that contains a solid state miscibility gap (see Fig. 16.6). At a temperature below the top of the miscibility gap (T_1), the free energy curve has two minima and a maximum because the enthalpy of mixing is positive and, as a consequence, two-phase equilibrium occurs over a certain range of compositions. At a temperature above the top of the miscibility gap (T_2), the free energy curve has only one minimum and a single phase exists at all compositions.

the liquid and α free energy curves intersect, with one being lower for certain compositions and the other being lower for other compositions. At 1300°C and an alloy composition of 53 wt. % Ni, the liquid (point E) is more stable than α (point F); however, the free energy of the alloy is lowest at point G on a common tangent line between the two free energy curves; that is, the points on the tangent line represent the free energy of a mixture of the α and liquid phases, and this, corresponds to two-phase equilibrium. The discrete points of common tangency define the compositions of the two coexisting phases (47 wt. % Ni and 63 wt. % Ni). Because

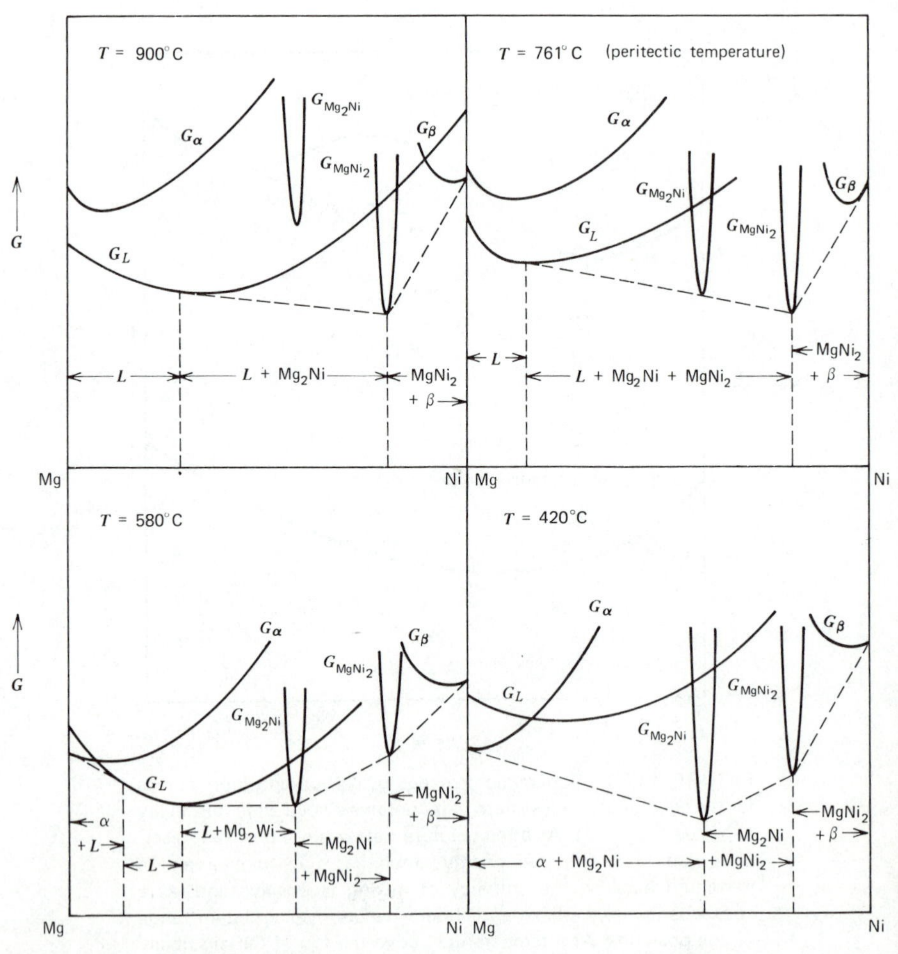

FIGURE 16.21 Schematic free energy vs. composition curves for the Mg–Ni system (Fig. 16.16). Intermediate phases or intermetallic compounds have discrete free energy curves.

only one common tangent line is possible, it should be apparent that α and L coexisting in equilibrium at a given temperature will necessarily have fixed compositions regardless of their relative amounts.

As another example, the binary phase diagram shown in Fig. 16.6 represents a case in which the enthalpy of mixing (or the interaction energy V) is positive for the solid phase (Fig. 16.20). Since $V > 0$, the free energy curve for the solid phase undergoes a change of shape from a positive curvature only and one minimum at higher temperatures to two minima and a maximum below a critical temperature. The latter is responsible for the development of a miscibility gap, since a common tangent can be drawn between two points on this single free energy curve.

When intermediate phases exist, as in the Mg–Ni system (Fig. 16.16), free energy diagrams are still straightforward (Fig. 16.21). Note that the free energy curve for a compound has a very sharp minimum and that three-phase coexistence (e.g., at the peritectic temperature) corresponds to simultaneous common tangency between the free energy curves for three phases.

16.9 THE GIBBS PHASE RULE

The thermodynamic conditions for phase equilibria lead to a simple equation that can be used to ascertain the number of coexisting phases. This is the Gibbs phase rule:

$$p + f = c + n \tag{16.17}$$

where p designates the number of phases in a materials system, c the number of components, n the number of noncompositional variables (e.g., T and P), and f the number of degrees of freedom. The latter is defined as the number of variables (temperature, pressure, and composition) that can be changed independently without changing the number of phases existing at equilibrium.

In the case of a unary system, the T–P diagram (e.g., Fig. 16.2) can be examined with the Gibbs phase rule. Here, $c = 1$ and $n = 2$, so that $p + f = 3$. In a single-phase field $p = 1$, and both temperature and pressure can be changed independently since $f = 2$. Along a two-phase coexistence line $p = 2$, and if temperature is changed, there must be a prescribed change in pressure in order to maintain two-phase equilibrium. At a triple point $p = 3$, and $f = 0$. No variable can be altered if three-phase equilibrium is to be maintained; $f = 0$ defines an invariance.

In the case of a simple binary eutectic at constant pressure (e.g., Fig. 16.9), $c = 2$ and $n = 1$, since temperature is the only noncompositional variable. Thus, $p + f = 3$, and three-phase coexistence between α, β, and L represents an invariant for which each phase has a definite composition and the temperature is fixed.

In the $\alpha + L$ two-phase region $p = 2$ and $f = 1$. Accordingly, a change in temperature must be accompanied by corresponding changes in the composition of each phase in accordance with the solidus and liquidus lines.

The Gibbs phase rule serves as a mnemonic device that allows easy computation of the maximum number of equilibrium phases in a materials system. Furthermore, it can be used to analyze a nonequilibrium situation, for example, the existence of three phases in the microstructure of a binary substance at constant pressure over a range of temperatures. Such situations often occur because phase transformations are limited kinetically by diffusion rates, a topic that will be discussed in the next chapter.

REFERENCES

1. Van Vlack, L. H. *Elements of Materials Science and Engineering*, 3rd edition, Addison-Wesley, Reading, Mass., 1975.
2. Guy, A. G., and Hren, J. J. *Elements of Physical Metallurgy*, 3rd edition, Addison-Wesley, Reading, Mass., 1975.

QUESTIONS AND PROBLEMS

16.1 Pure iron melts at 1538°C. Body-centered cubic iron is stable between 1538°C and 1394°C and also below 912°C. Face-centered cubic iron is stable between 1394°C and 912°C

(a) Sketch the cooling curve expected under equilibrium, for pure Fe cooled from 1600°C to room temperature.

(b) Sketch schematically the free energy curves as a function of temperature over the same temperature range for each phase.

16.2 It can be shown that the total differential of internal energy is given by

$$dE = T\,dS - P\,dV$$

for a phase of a pure substance.

(a) Using the definitions of enthalpy and free energy (Eqs. 1.5 and 1.7), show that $dG = -S\,dT + V\,dP$.

(b) At constant pressure, how does G vary with T? At constant temperature, how does G vary with P?

(c) For a pure substance, two-phase coexistence occurs when each phase has the same free energy on a molar basis. Alternatively, the differentials of their respective free energies must be equal if equilibrium exists. Based on this and using the expression in part (a), show that the slope of a phase field boundary in a unary diagram is

$$\frac{dT}{dP} = \frac{\Delta V_{\text{trans}}}{\Delta S_{\text{trans}}} = \frac{T_{\text{trans}}\,\Delta V_{\text{trans}}}{\Delta H_{\text{trans}}}$$

16.3 When a pure substance undergoes a phase change on heating at constant pressure, there is a discrete increase in enthalpy (i.e., $\Delta H_{trans} > 0$ for heating); however, the volume may increase or decrease, and this determines the sign of the slope of a boundary in a unary diagram [see relation in part (c) of problem 16.2].

(a) Demonstrate that this is so by considering the S/L and L/V boundaries for H_2O (Fig. 16.2).

(b) α-Fe, stable at 912°C and below, is bcc and has a lattice parameter of $a = 0.2904$ nm at 912°C; γ-Fe, stable at 912°C and above, is fcc and has a lattice parameter of $a = 0.3646$ nm at 912°C. At high pressures, will the α–γ transformation occur at higher or lower temperatures than 912°C? Explain.

(c) On heating, α-Ti transforms to β-Ti with a relative volume change of -0.55%. How does the α/β boundary vary as a function of pressure?

16.4 Describe in detail the process of equilibrium solidification in the case of a 50 wt. % Cu–50 wt. % Ni alloy (Fig. 16.4). Sketch the compositions of the liquid and solid as a function of temperature.

16.5 Consider the phase diagram shown in Fig. 16.6. For an alloy containing 40 wt. % B, calculate the fractional amounts (by weight) of α_1 and α_2 existing at T_1, T_2, and T_3 (where $T_1 > T_2 > T_3$).

T	C_{α_1} (wt. % B)	C_{α_2} (wt. % B)
T_1	48	52
T_2	36	64
T_3	12	88

16.6 Can the lever rule be applied at the temperature of a three-phase invariant reaction in a binary system? Why or why not?

16.7 The most convenient way to specify compositions in a materials system is either in terms of weight percent of each component or in terms of atomic (or mole) percent of each component (which is 100 times the mole fraction).

(a) For a binary alloy containing C_0 wt. % B, set up general expressions for at. % A and for at. % B in terms of C_0 and the atomic weights of each component.

(b) What at. % C is there in cementite (6.69 wt. % C)? What is the maximum at. % C that can dissolve in ferrite (at 723°C)?

(c) Suggest why compositions in terms of weight percent are more useful from a practical point of view than those in terms of atomic percent.

16.8 One eutectic reaction in the Mg–Ni system (Fig. 16.16) is

$$L\ (23.5\ \text{wt. \% Ni}) \xrightleftharpoons{507^\circ\text{C}} \alpha\ (0\ \text{wt. \% Ni}) + \text{Mg}_2\text{Ni}\ (54.6\ \text{wt. \% Ni})$$

Consider two alloys of composition C_1 and C_2 containing less Ni and more Ni, respectively, than the eutectic composition. The fractional amount of primary microconstituent by weight is the same in each alloy, but the fractional amount of total α in alloy C_1 is 2.5 times the fractional amount of total α in alloy C_2. Calculate the compositions C_1 and C_2 in terms of wt. % Ni.

16.9 Identify and completely write out all three-phase invariant reactions occurring in the Cu–Zn system (Fig. 16.15).

16.10 Part of the water-cane sugar ($C_{12}H_{22}O_{11}$) system is shown below.

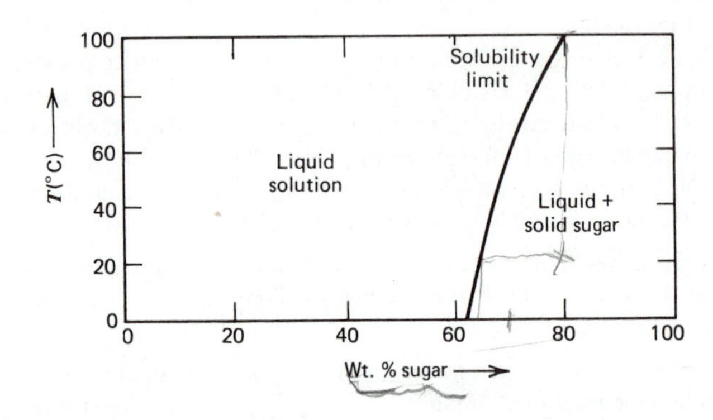

(a) What is the maximum amount of sugar that can be dissolved in boiling water at 100°C? Give both wt. % sugar and mole % sugar.

(b) If the above solution is cooled to 20°C, what will be the composition of the water solution? What will be the fractional amounts (by weight) of saturated water and solid sugar at 20°C?

16.11 The formation of substitutional solid solutions by pairs of metals usually is disfavored if (i) the solvent and solute atoms have atomic radii that differ by more than $\pm 15\%$ and (ii) the electronegativities of the respective elements differ by more than ± 0.4 (Table 3.5). The size factor often plays a dominant role in precluding substitutional solid solution formation.

(a) Solely on the basis of information provided below, indicate which of the following pairs of metals would *not* be likely to form a continuous series of solid solutions: Ta–W, Pt–Pb, Co–Ni, Co–Zn, and Ti–Ta.

Metal	*Elemental Atomic Radius (nm)*	*Crystal Structure*
Ti	0.1461	hcp ($T < 883°C$); bcc ($T > 883°C$)
Co	0.1251	hcp ($T < 427°C$); fcc ($T > 427°C$)
Ni	0.1246	fcc
Zn	0.1332	hcp
Ta	0.1430	bcc
W	0.1370	bcc
Pt	0.1387	fcc
Pb	0.1750	fcc

(b) Consult *Metals Handbook*, 8th edition, Volume 8, and check your predictions.

16.12 Explain why complete mutual solubility can occur between the two components of a substitutional solid solution but not for an interstitial solid solution.

16.13 The cation-to-anion radius ratios for MgO, Al_2O_3, and Cr_2O_3 are 0.47, 0.36, and 0.40, respectively. Al_2O_3 and Cr_2O_3 form a continuous series of solid solutions.
(a) Is this a surprising result? Why or why not?
(b) Would you expect to find limited or extensive solid solubility in the MgO–Cr_2O_3 system? Why or why not?

16.14 (a) Using free energy curves for the solid phases of a binary system based on two metallic components having different crystal structures, illustrate (i) the case in which there is very limited solid solubility in each terminal solid solution and (ii) the case in which there is extensive solid solubility in each terminal solid solution. (Assume that no intermediate phases exist.)
(b) What conclusions can you draw from part (a)?
(c) What effect does an intermediate phase have on the solid solubility limit of a terminal solid solution?

16.15 Cu (atomic radius, 0.1278 nm) melts at 1084.5°C. Bismuth (atomic radius, 0.1535 nm) melts at 271.4°C. Complete liquid miscibility exists in the Cu–Bi system, and a three-phase reaction takes place at 270°C. Sketch, in as much detail as you can, the expected phase diagram.

16.16 How much ferrite, austenite, and cementite by weight are in 0.1 kg of an Fe–C alloy containing 0.5 wt. % C at (i) 1000°C, (ii) 724°C, and (iii) room temperature?

16.17 How much pearlite by weight is there in 1.0 kg of an Fe–C alloy containing 0.4 wt. % C at 722°C?

16.18 For plain carbon steels (i.e., Fe–C alloys) that have been cooled slowly from 1000°C to room temperature, plot the fractional amounts by weight of (i) proeutectoid ferrite and pearlite and (ii) ferrite and cementite as a function of carbon content between 0.02 wt. % C and 0.8 wt. % C.

16.19 Draw schematic cooling curves for Fe—0.4 wt. % C and Fe—0.8 wt. % C cooled slowly from 1000°C to room temperature. Sketch the microstructures at selected temperatures, and identify phases and microconstituents.

16.20 Sketch schematic free energy vs. composition curves at various temperatures for the eutectic system shown in Fig. 16.9. Be sure to include the eutectic temperature.

16.21 (a) Sketch schematic free energy vs. wt. % C curves for Fe–Fe_3C alloys at various temperatures below 1200°C.
(b) Fe_3C is a metastable phase; that is, its free energy curve lies above the common tangent lines between the free energy curves of liquid and graphite, austenite and graphite, and ferrite and graphite. (The iron-graphite system is stable and the iron-cementite system is metastable.) If L, γ, and α, respectively, coexisted with graphite rather than Fe_3C, how would the eutectic and eutectoid temperatures be altered and how would the solubility limit of C in γ and of C in α be altered?

16.22 Sketch schematic free energy vs. composition curves at various temperatures, including 1537°C and 535°C, for the Ti–Zr system (Fig. 16.17).

17

Phase Transformations

The rate of approach to equilibrium in condensed systems is often sufficiently slow so that nonequilibrium or near-equilibrium structures are often observed in solids. Since such structures are frequently of technological importance, the basic concepts relating to the rates of transformations are presented in this chapter.

17.1 OBSTACLES TO PHASE TRANSFORMATIONS

An equilibrium diagram represents the phases and phase changes expected in a materials system when equilibrium conditions exist, but it provides no information about the rates of transformation. Although changes in pressure, composition, or temperature can cause phase transformations, it is temperature changes that are most important in materials systems. If a heating or cooling process is carried out very slowly, then a phase change will take place near the equilibrium temperature specified by the phase diagram. Even so, both the initiation and continuation of the transformation do require some small increment of superheating or supercooling relative to the equilibrium temperature. Under nonequilibrium conditions such as relatively rapid heating or cooling, it is found that a finite amount of superheating or supercooling precedes any measureable indication that the reaction is taking place. The kinetics of nucleation and growth, which are the fundamental processes of phase transformations, often are governed by thermal activation in much the same way as the rates of chemical reactions. Indeed, there is a parallelism between considerations of chemical equilibrium and phase equilibrium and between considerations of chemical kinetics and kinetics of phase changes.

Both nucleation (i.e., the initial formation of small particles of the product phase from the parent phase) and growth (i.e., the increase in size of the nucleated particles) require that the accompanying free energy change be negative. Consequently, the superheating or supercooling that is necessary for a phase change is to be expected. That is, a transformation cannot take place precisely at the equilibrium transformation temperature because, by definition, this is the temperature at which the free energies of the phases are equal. In addition to the necessity that $\Delta G < 0$, there are two basic impediments to most phase transformations that affect the transformation rate. First, every transformation involves a rearrangement of atoms because of compositional changes, crystal structure differences, or both. Since these rearrangements usually are accomplished by diffusive motion of atoms, the rate at which they proceed often is limited by how fast diffusion can occur. Diffusion limits both the nucleation and growth rates in many cases. The other impediment to a transformation is more subtle; this is the difficulty encountered in nucleating small particles of the product phase. The main reason for this difficulty is that there is a surface energy associated with the interface between the nucleated phase and the parent matrix that causes a local increase in free energy rather than the expected decrease.

An understanding of the obstacles to phase changes, how they relate to crystal structure, and how they enter into the fundamental processes of nucleation and growth is important for the understanding of near-equilibrium and nonequilibrium phase transformations.

17.2 NUCLEATION

Consider the formation of a spherical solid particle within a pure liquid cooled below the equilibrium freezing temperature. The formation of the solid particle leads to a decrease in free energy because of the lower free energy of the bulk solid as compared to that of the liquid (see Fig. 1.6, for example). Concurrent with this decrease in bulk free energy, however, is an increase in free energy because of the creation of an interface between the solid and the liquid. These two opposing tendencies can be represented by an equation that gives the free energy change associated with the formation of a spherical particle as a function of sphere radius r:

$$\Delta G = \left(\frac{4\pi}{3} r^3\right) \Delta G_v + (4\pi r^2)\gamma \qquad (17.1)$$

The first term in this equation gives the total reduction in volume free energy per particle, where ΔG_v (which is negative) is the free energy change per unit volume of solid forming from the liquid. The second term gives the total surface energy increase per particle, where γ is the solid–liquid surface energy per unit area. At very small particle sizes, the total surface energy is greater in magnitude than the total volume free energy change, and the free energy per particle increases with particle size. As shown in Figure 17.1, the free energy per particle increases until the value ΔG^* is reached at a critical radius r^*; this maximum free energy condition defines the critical nucleus. For $r > r^*$, the particle can spontaneously increase in size because this is accompanied by a decrease in total free energy, and the volume free energy term in Eq. 17.1 dominates. For $r < r^*$, the particle has a tendency to decrease in size since the free energy is decreased by this process.

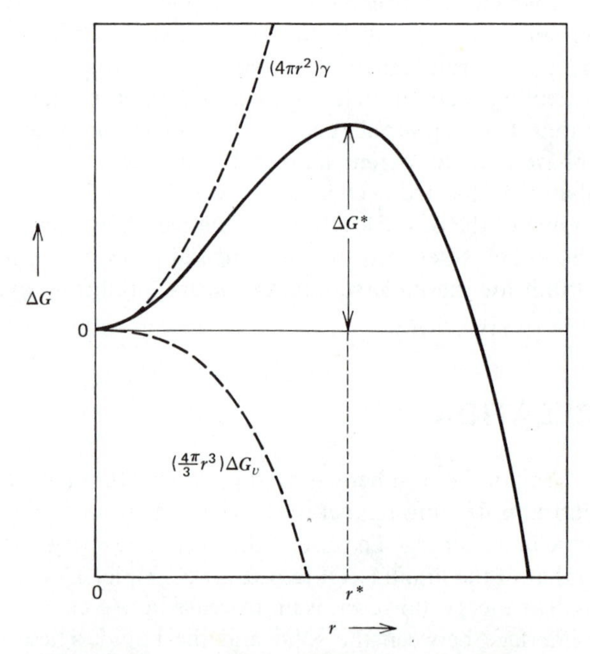

FIGURE 17.1 The change in free energy accompanying the formation of a solid droplet of radius r within a liquid at constant temperature. The initial increase in free energy is due to the surface energy ($4\pi r^2\gamma$) of the solid-liquid interface. As the surface-to-volume ratio decreases at larger sizes, the volume free energy term $[(4\pi r^3/3)\,\Delta G_v]$ becomes more important than the surface energy term. The maximum free energy, ΔG^*, which is the activation free energy barrier for nucleation, occurs at the critical radius, r^*. Particles of radius less than r^* tend to shrink in size, whereas those of radius greater than r^* can grow spontaneously.

In order to nucleate a particle, the free energy barrier must be surmounted. This activation free energy per critical nucleus is

$$\Delta G^* = \frac{16\pi}{3} \frac{\gamma^3}{(\Delta G_v)^2} \tag{17.2}$$

and the corresponding critical radius that must be achieved is

$$r^* = -\frac{2\gamma}{\Delta G_v}. \tag{17.3}$$

Note that r^* is a physically meaningful quantity only when $\Delta G_v < 0$, and that $\Delta G^* \rightarrow \infty$ and $r^* \rightarrow \infty$ at the equilibrium freezing temperature (T_m) where $\Delta G_v = 0$. Clearly, nucleation is impossible at the equilibrium transformation temperature.

Surface energy is not nearly as sensitive to temperature as is ΔG_v, which becomes increasingly negative as temperature decreases below T_m. Thus, both ΔG^* and r^* decrease with increasing supercooling (Fig. 17.2). As temperature is lowered, however, the probability of thermal activation becomes smaller, and there is a

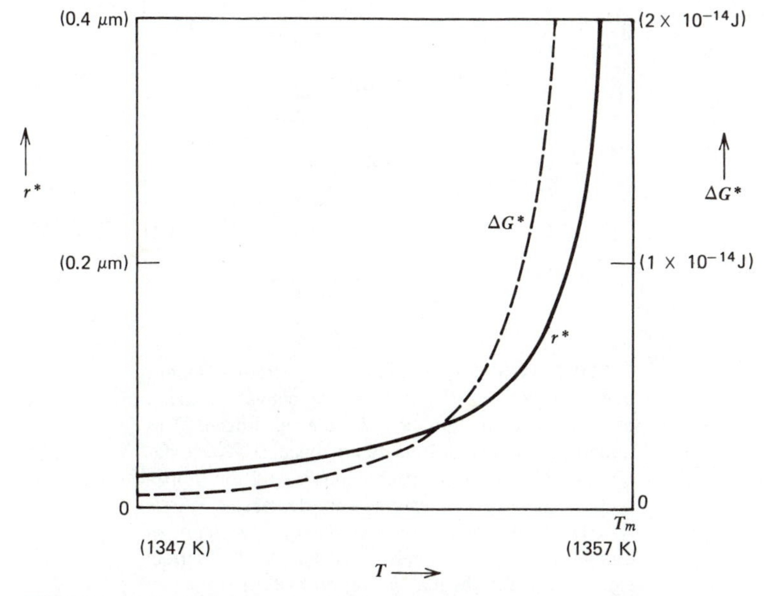

FIGURE 17.2 Variation of critical nucleus parameters (r^* and ΔG^*) as a function of temperature. At the equilibrium freezing temperature, T_m, both of these quantities approach infinity. The ordinate and abscissa scales provide approximate numerical values corresponding to the nucleation of solid copper from its melt.

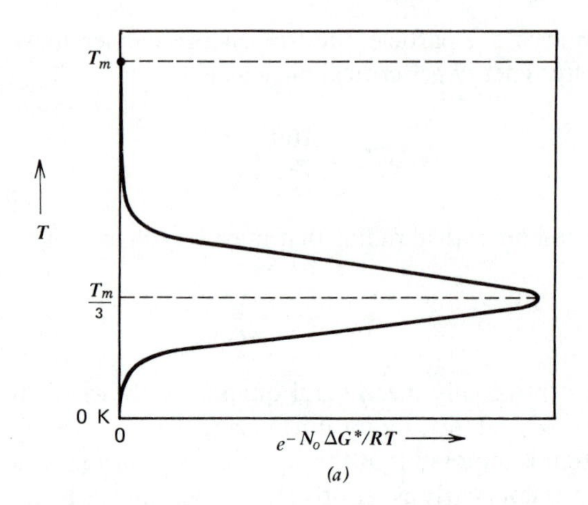

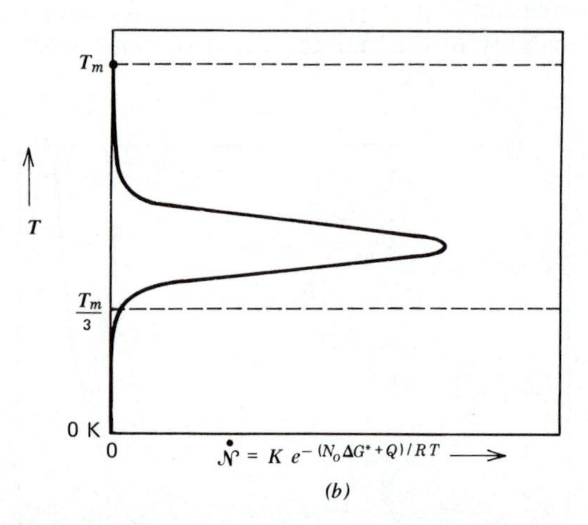

FIGURE 17.3 (*a*) Schematic temperature dependence of $e^{-N_0 \Delta G^*/RT}$, the thermal activation factor associated with forming N_0 critical nuclei. This quantity exhibits a maximum because ΔG^* decreases with decreasing temperature and the probability of thermal activation also decreases. (*Note.* $N_0 \, \Delta G^*$ is the activation free energy per mole of critical nuclei; this gives a dimensionally correct exponent in the thermal activation factor.)

(*b*) Schematic temperature dependence of the nucleation rate, $\dot{\mathcal{N}}$. For a critical nucleus to form, atoms must jump across the interface; hence, a thermally activated diffusion process is a necessary part of the nucleation.

temperature at which the value of $e^{-(N_0\Delta G^*/RT)}$ is a maximum (Fig. 17.3*a*). The nucleation rate depends not only on the activation free energy for forming a critical nucleus but also on the ability of atoms to jump from the liquid to the solid across the interface. This process, which is necessary for particles to increase in size, is thermally activated, and the activation energy Q is associated with diffusion in the vicinity of the liquid–solid interface. A combination of both activation energies gives the overall nucleation rate:

$$\dot{\mathscr{N}} = Ke^{-(N_0\Delta G^* + Q)/RT} \tag{17.4}$$

where $\dot{\mathscr{N}}$ is the number of critical nuclei that forms per unit volume of parent phase in unit time and K is a nucleation rate constant that is essentially independent of temperature. The temperature dependence of $\dot{\mathscr{N}}$ is shown in Fig. 17.3*b*.

The above discussion provides a description of *homogeneous* nucleation, that is, the formation of a critical nucleus without the aid of a catalyst. In practice, most transformations in condensed-phase systems are aided by the presence of catalysts, and the corresponding nucleation process is termed *heterogeneous* nucleation. In solidification the catalysts usually are foreign agents such as a mold wall or particles of oxide. These nucleating agents decrease the activation free energy for nucleation as well as the critical nucleus volume. This situation arises, for example, when the solid partially wets the nucleating agent, as illustrated in Fig. 17.4. The contact angle θ is determined by a surface tension balance (see Sect. 11.5), and in all cases where $0 \leq \theta < 180°$, ΔG^* is less for heterogeneous nucleation than for homogeneous nucleation. Indeed, ΔG^* becomes progressively smaller as θ decreases, and for complete wetting (i.e., $\theta = 0$), ΔG^* approaches zero since the new interfaces have less total surface energy than the original liquid-nucleating agent interface. In almost all real ingots or castings, all nucleation is heterogeneous. In fact, homogeneous nucleation during solidification can be observed only under very closely controlled laboratory conditions.

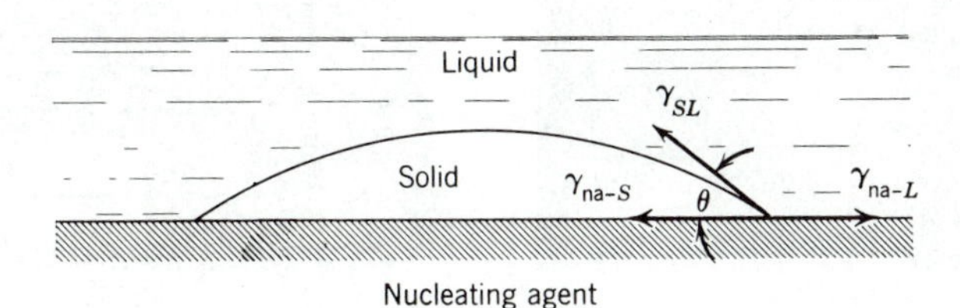

FIGURE 17.4 Heterogeneous nucleation of a solid from its melt. Partial wetting of the nucleating agent (e.g., mold wall) lowers the activation free energy for nucleation and also lowers the volume of the critical nucleus. The contact angle θ can be determined from a surface tension balance, $\gamma_{na-L} = \gamma_{na-S} + \gamma_{SL} \cos\theta$.

Similar considerations, albeit with different catalytic agents, account for heterogeneous nucleation in solids. Although it is believed that homogeneous nucleation does occur in certain solid–solid transformations, heterogeneous nucleation at grain boundaries, dislocations, and other defects such as inclusions is common. Grain boundaries are particularly effective as nucleating agents at high temperatures. The example shown in Fig. 17.5*a* illustrates the initial stages of precipitation of the compound Fe_3C from the face-centered cubic form of iron in a high carbon content steel. At lower temperatures, precipitation occurs within the grains as well as at grain boundaries as noted in the example of precipitation in a Ti alloy, Fig. 17.5*b*.

For most solid–solid transformations another positive energy term should be added to Eq. 17.1 in order to account for strain energy. This additional term originates because the precipitate typically has a different volume than the parent matrix from which it forms. When both strain energy and surface energy contribute to the activation free energy for nucleation, critical nuclei having other than

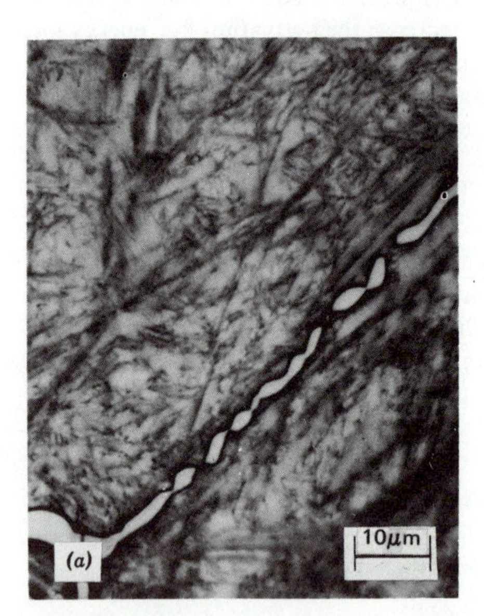

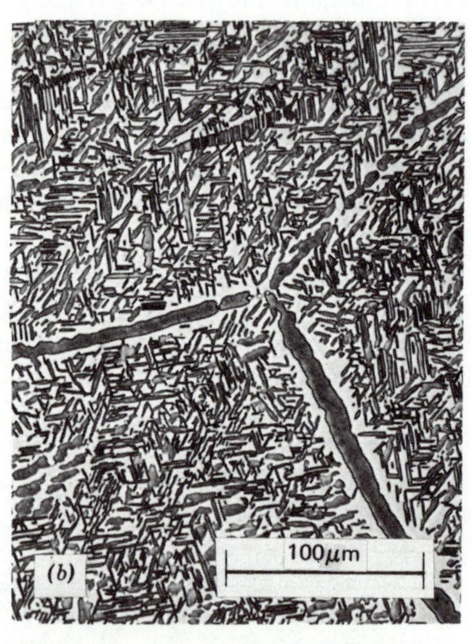

FIGURE 17.5 (*a*) The microstructure of an iron alloy with a high carbon content. Note the precipitation of Fe_3C at the grain boundaries of the parent face-centered cubic form of iron.

(*b*) Precipitation of hexagonal close-packed alpha phase in a matrix of body-centered cubic beta phase in a Ti–Cr alloy. The temperature of precipitation is such that the precipitate forms at both the beta phase grain boundaries and within the beta grains. The precipitate within the grains is plateletlike in form. (From W. Rostoker and J. R. Dvorak, *Interpretation of Metallographic Structures*, Academic Press, New York, 1965. Photomicrographs courtesy of W. Rostoker.)

spherical shapes may be favored even though this results in a greater interfacial area per unit volume of critical nucleus. Thus, critical nuclei may have the form of platelets or needles, which tend to minimize the effect of strain energy.

17.3 GROWTH AND THE OVERALL TRANSFORMATION RATE

Many transformations occur as a consequence of continuous formation of critical nuclei in the parent phase and the subsequent growth of the particles. Growth may proceed in two radically different manners. In one type of growth, individual atoms move independently from the parent to the product phase. That is, diffusion is important, and this type of growth is thermally activated. In the other type of growth, which can occur only in solid-state transformations, many atoms move cooperatively, and thermal activation is unimportant. The former will be treated in this section, while consideration of the latter will be deferred until Sect. 17.5.

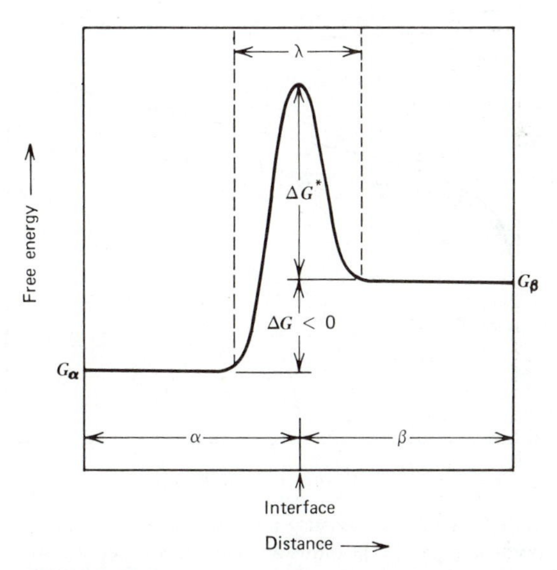

FIGURE 17.6 Variation of free energy in the vicinity of an α–β interface. Atoms in β have a higher free energy than those in α; thus, the interface advances to the right. The α phase grows at the expense of β because a lower activation free energy is encountered by atoms jumping from β to α than vice versa.

The nature of thermally activated growth can be elucidated for the simple case involving either the solidification of a pure substance or an allotropic or polymorphic transformation near equilibrium. In effect, this has been treated briefly in Sect. 13.5; however, it is instructive to look at the process once again. The growth of a product phase (α) from a parent phase (β) corresponds to atoms jumping across the interface. A net flux of atoms from β to α results in interface motion, and this is directly related to the growth rate $\mathcal{G} = dr/dt$ (or interface velocity). That is, the growth rate is taken as the rate of increase of a linear dimension of a growing particle. During growth, atoms jump form β to α and from α to β, but the free energy barrier differs for the two directions (Fig. 17.6). The flux of atoms from β to α is given by

$$J_{\beta \to \alpha} = s\nu e^{-\Delta G^*/RT} \tag{17.5}$$

where ν is an atomic vibration frequency factor and s is a geometric factor (see Sect. 13.5). Similarly, the flux of atoms from α to β is given by

$$J_{\alpha \to \beta} = s\nu e^{-(\Delta G^* - \Delta G)/RT} \tag{17.6}$$

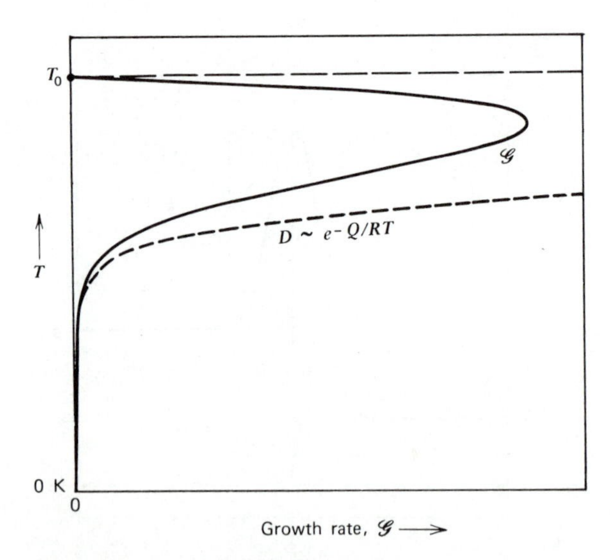

FIGURE 17.7 Schematic temperature dependence of growth rate, $\mathcal{G}$. Small diffusion rates at low temperatures limit the growth rate. At temperatures near the equilibrium transformation temperature T_0, the growth rate becomes small again because the activation free energy barrier for atoms jumping from the parent α phase to the product β phase approaches that for atoms jumping in the opposite direction.

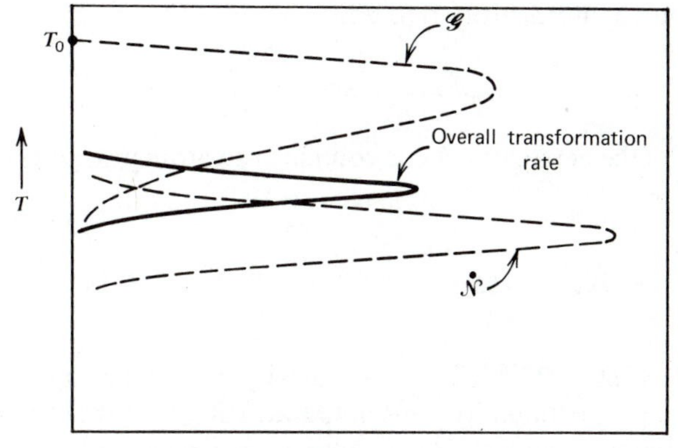

(a)

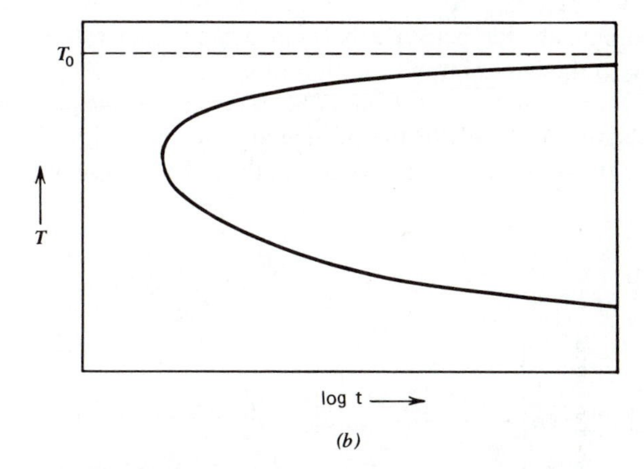

(b)

FIGURE 17.8 (*a*) Schematic variation of overall transformation rate with temperature. The overall transformation rate is a result of both nucleation rate $\dot{\mathcal{N}}$ and growth rate $\mathcal{G}$ and has a maximum value at a temperature between that for the maximum $\dot{\mathcal{N}}$ and that for the maximum $\mathcal{G}$.

(*b*) The time required for an isothermal transformation as a function of temperature. This *C*-curve, obtained from the reciprocal of the overall transformation rate, is characteristic of thermally activated nucleation and growth transformations.

Thus, the *net* flux of atoms from β to α is

$$J = J_{\beta \to \alpha} - J_{\alpha \to \beta} = sve^{-\Delta G^*/RT}(1 - e^{\Delta G/RT}) \tag{17.7}$$

Multiplication of the net flux by Ω, the volume of an atom in α, gives the growth rate of a particle of α:

$$\mathscr{G} = \frac{dr}{dt} = J\Omega = sv\Omega e^{-\Delta G^*/RT}(1 - e^{\Delta G/RT}) = KD(1 - e^{\Delta G/RT}) \tag{17.8}$$

In Eq. 17.8, $D = D_0 e^{-\Delta H^*/RT}$ is the diffusivity for atoms jumping across the interface, K is a proportionality constant, and $\Delta G(<0)$ is the difference in molar free energy between β and α. The temperature dependence of $\mathscr{G}$, illustrated in Fig. 17.7, is qualitatively correct for any thermally activated growth process and is not restricted to the example under consideration. At temperatures well below the equilibrium temperature T_0, growth rate is determined essentially by the diffusivity. At temperatures near T_0, $\mathscr{G}$ is dominated by the free energy difference between the two phases, and $\mathscr{G} = 0$ at $T = T_0$ because $\Delta G = 0$. Growth rate reaches a maximum at a temperature that is invariably higher than the temperature corresponding to the maximum nucleation rate.

The overall transformation rate (Fig. 17.8*a*) is given by some product of $\mathscr{G}$ and $\dot{\mathscr{N}}$, which depends on the details of the particular transformation. In turn, the time required for a transformation to proceed halfway to completion, which can be

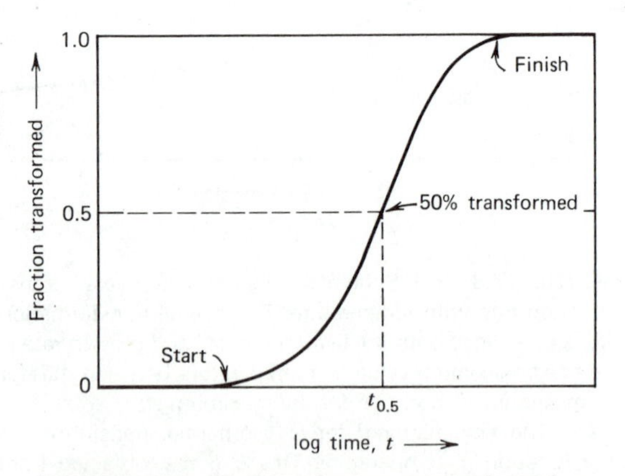

FIGURE 17.9 Fractional amount transformed as a function of time at constant temperature. Although the instantaneous transformation rate (i.e., the slope of the curve at any time) varies as the transformation proceeds, the reciprocal of the time required to take the transformation halfway provides a conventional measure of the overall transformation rate.

determined experimentally, has a reciprocal relationship to the overall transformation rate, and the temperature dependence of this time is shown in Fig. 17.8*b*. The *C-curve* kinetics illustrated in Fig. 17.8*b* are observed in all thermally activated nucleation and growth processes involving the transformation of a high-temperature phase to a low-temperature phase. It should be noted that the *instantaneous* rate varies with the course of a transformation (Fig. 17.9), largely because diffusion plays such an important role in both nucleation and growth. Furthermore, transformations may take place under conditions where growth is rapid and nucleation is slow, or vice versa, and these conditions greatly affect the nature of the resulting microstructure. At small degrees of supercooling, where slow nucleation and rapid growth prevail, relatively coarse particles result; at larger degrees of supercooling, relatively fine particles result.

17.4 DIFFUSION-CONTROLLED TRANSFORMATIONS IN SOLIDS

The simplest phase changes in solids do not involve a change in composition, that is, congruent transformations. These include allotropic transformations such as that found with titanium [β(bcc) $\rightarrow$ α(hcp) on cooling], where the change in atomic coordination requires little atomic motion. The polymorphic change of SiO_2 from cristobalite to tridymite at 1470°C, however, is much more sluggish because strong covalent bonds must be broken when atoms or groups of atoms rearrange themselves. In contrast, the rearrangement that occurs when either of these forms of SiO_2 is cooled quickly (e.g., β-cristobalite $\rightarrow$ α-cristobalite on cooling) only involves changes in bond angles, and this proceeds rapidly because diffusion is not necessary. Order–disorder reactions represent another type of diffusion-dependent congruent transformation. An example is provided by 75 at. % Cu–25 at. % Au, which is a random substitutional solid solution at temperatures above 410°C. Rapid cooling (i.e., quenching) of this alloy from an elevated temperature to room temperature results in the retention of the random fcc solid solution. If the alloy is then reheated to a temperature below 410°C, the Au atoms move to unit cell corners and the Cu atoms to face centers, giving the ordered Cu_3Au crystal structure. The fact that the high-temperature phase can be retained at room temperature by quenching is a consequence of the necessity for diffusion in this transformation and of the variation in overall transformation rate with temperature (Fig. 17.8).

Many important solid-state phase transformations are complicated by the fact that the products differ in composition from the parent phase. The additional diffusion problems that exist cause greater nucleation and growth difficulties than are encountered in congruent transformations. In spite of the complexities, precipitation and eutectoid reaction rates, as examples, can be explained with the concepts developed in Sects. 17.2 and 17.3. The discussion of overall transformation rate in the previous section provides the qualitative basis for understanding nucleation and growth in solids, yet the theory is insufficient to predict quantitative transformation rates. Thus, empirical studies must be utilized. Although many practical solid-state transformations may be effected under nonisothermal conditions, the sluggish nature of diffusion-controlled nucleation and growth in solids readily affords the opportunity to study isothermal transformations.

To illustrate this, consider the precipitation of α phase particles from the β phase in a Ti–Mo alloy (the Ti–Mo phase diagram is shown in Fig. 17.10). If the alloy is cooled rapidly from a temperature in the β single-phase region to a temperature in the $\alpha + \beta$ two-phase region and then held there, the progress of α precipitation may be followed by observation of the microstructure after various times at this temperature. Initially, supersaturated β is retained, but after a certain period of time some α begins to appear and the extent of the transformation as a function of time varies as shown in Fig. 17.9. By using different specimens of this alloy, it is possible to repeat this procedure at different temperatures in the two-phase region. Such an experimental schedule, called an isothermal transformation study, gives results that can be summarized in an isothermal transformation diagram, also known as a TTT (time-temperature-transformation) diagram, as shown in Fig. 17.11 (cf. Fig. 17.8*b*). Here, the time at which α is first observed and the time at which precipitation is complete are plotted as a function of isothermal transformation temperature. The resulting *C*-shaped curves are mirror images of the transformation rate (cf. Fig. 17.8*a*).

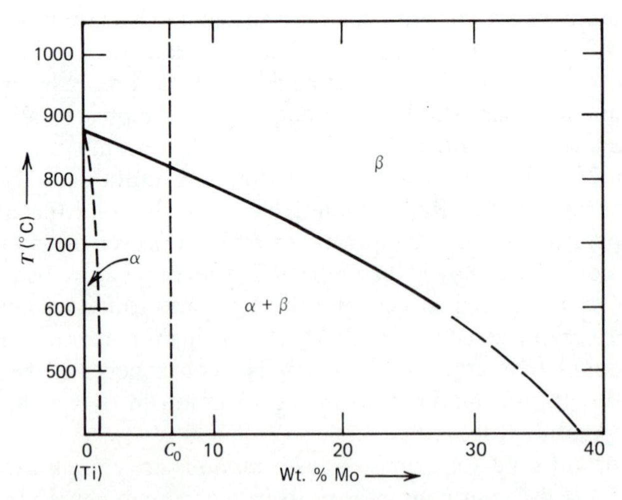

FIGURE 17.10 A portion of the Ti–Mo phase diagram. When the alloy having composition C_0 is cooled from the single-phase β region to the two-phase $\alpha + \beta$ region, α forms by nucleation and growth.

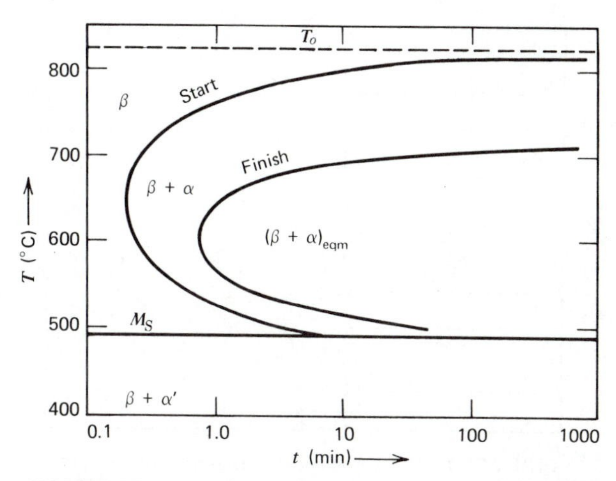

FIGURE 17.11 Experimental time-temperature-transformation (TTT) diagram for Ti–Mo. The start and finish times of the isothermal precipitation reaction vary with temperature as a result of the temperature dependence of the nucleation and growth processes. Precipitation is complete, at any temperature, when the equilibrium fraction of α is established in accordance with the lever rule. The solid horizontal line represents the athermal (or nonthermally activated) martensitic transformation that occurs when the β phase is quenched (see Sect. 17.5).

Solid transformations involving three-phase reactions fit into much the same scheme as that discussed above, even though they are slightly more complicated. The basic principles relating to nucleation and growth apply, and, once again, TTT diagrams prove useful. Eutectoid decomposition, which is the most important three-phase reaction in alloys, proceeds by nucleation and simultaneous growth of two product phases from the parent phase (Fig. 17.12). The eutectoid reaction in the Fe–C system (Fig. 16.13) will be used as an example. If an isothermal transformation study is conducted with an alloy having the eutectoid composition, the TTT diagram shown in Fig. 17.13 results. Here, the completion of the reaction corresponds to a complete transformation of γ into a two-phase mixture of α and Fe_3C. The intermixture of the two product phases, which is readily apparent in the microstructure (Fig. 17.14), is largely a consequence of the cooperative growth of these two phases during the eutectoid reaction. This intermixture becomes finer as the transformation temperature is reduced (compare Fig. 17.14*a* and Fig. 17.14*b*), a situation comparable to the observation that finer eutectic structures result as solidification rate is increased for eutectic alloys.

By virtue of the nature of diffusion-controlled nucleation and growth, it is often possible to retain a high-temperature solid solution at room temperature by

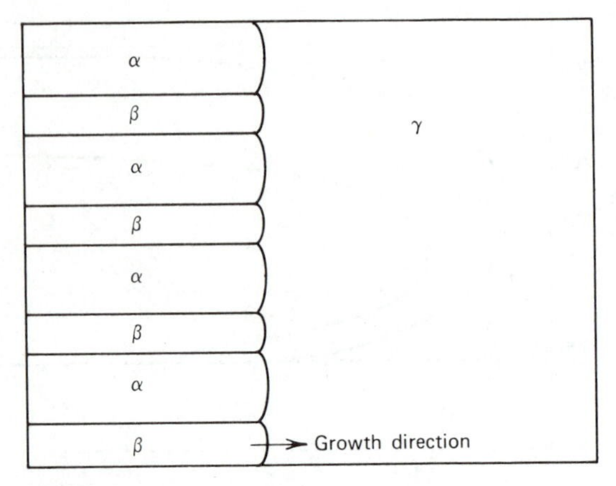

FIGURE 17.12 Schematic representation of the formation of the two-phase eutectoid solid from a third solid phase. The two product phases form and grow simultaneously into the parent phase. The idealized lamellar morphology shown here is often observed in eutectoid transformations.

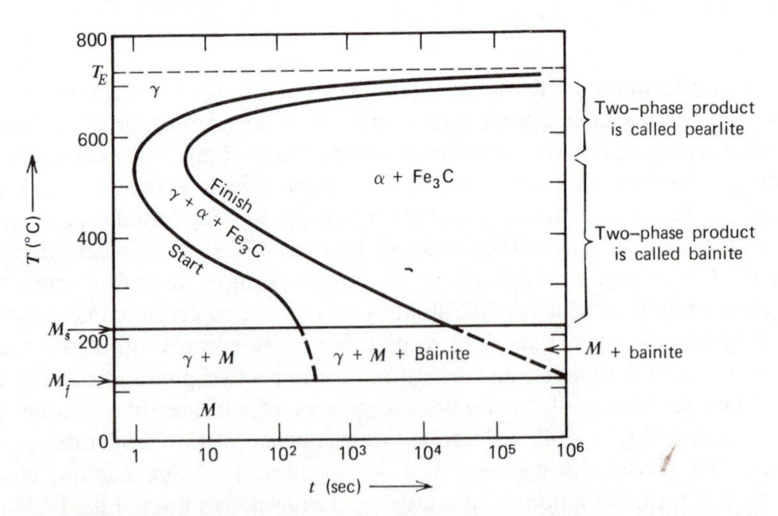

FIGURE 17.13 Experimental TTT diagram for a steel having the eutectoid composition (0.80 wt. % C). The start and finish curves for the isothermal eutectoid decomposition reaction are qualitatively similar to those for precipitation (cf. Fig. 17.11). (Adapted from *Atlas of Isothermal Transformation Diagrams*, United States Steel Corp., Pittsburgh, 1951.)

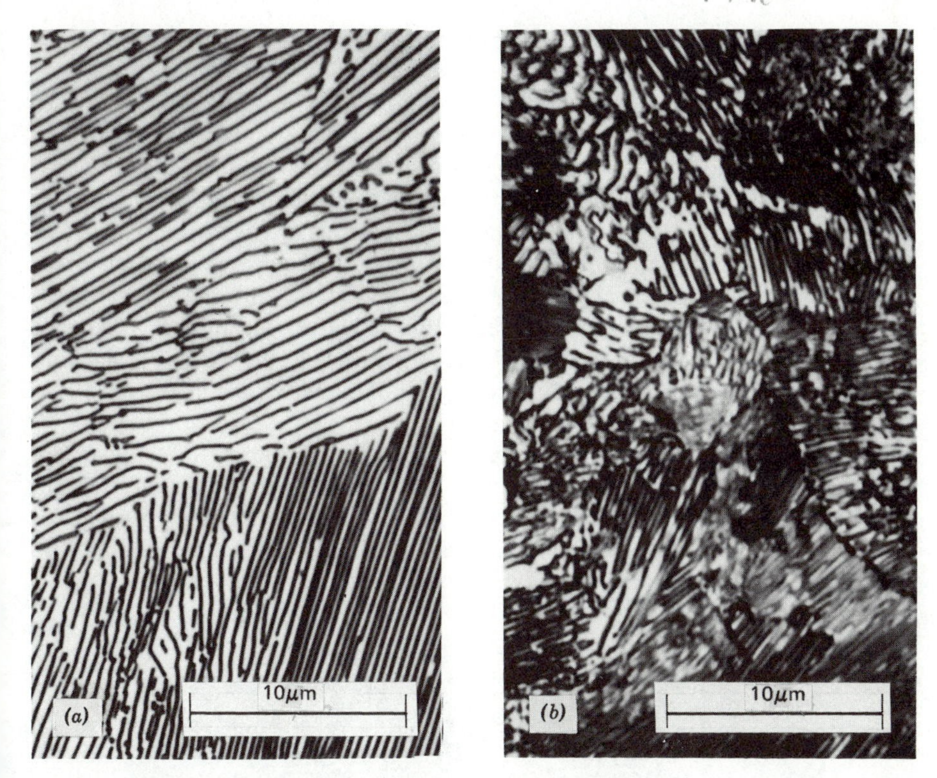

FIGURE 17.14 (*a*) Microstructure of a eutectoid steel transformed at 705°C to form coarse pearlite. (*b*) Microstructure of same steel transformed at 650°C to form fine pearlite. (Photomicrographs courtesy of United States Steel Corp.)

quenching. That is, the cooling rate during a quench is sufficiently rapid so that diffusional processes do not occur and any transformation is suppressed. This is a useful procedure for many alloys, for it permits subsequent, controlled isothermal transformation to be carried out. On the other hand, many high-temperature solid solutions based on a metal that has two allotropic forms cannot be retained upon quenching. This situation, arising because growth may proceed without diffusion in certain cases, will be discussed in the next section.

17.5 MARTENSITIC TRANSFORMATIONS IN SOLIDS

A congruent transformation such as the transformation of fcc cobalt to hcp cobalt on cooling below 427°C is considered diffusionless because it takes place at a rate approaching the speed of sound. It can be envisioned as a cooperative type of process in which, without the aid of thermal activation, atoms move into new locations because of the strain energy resulting from like movements of adjacent atoms. That is, the motion of a group of atoms to new positions causes their neighbors to move in a similar manner, and plate-shaped regions of hcp cobalt form rapidly in the fcc matrix (Fig. 17.15). The strains set up in the fcc matrix in the vicinity of an hcp plate tend to impede further transformation in this region. Thus, a lower temperature or mechanical deformation may be required to complete the transformation throughout the material. The cooperative displacement of atoms during such a *martensitic* transformation resembles a shear process in which (111) atomic planes in the fcc structure slide over one another to convert the $ABCABC \cdots$ close-packed plane stacking sequence to the $ABAB \cdots$ stacking sequence of the hcp structure. This merely requires that atoms in the fcc phase move a fraction of an interatomic distance.

Because of its crystallographic nature, a martensitic transformation only occurs in the solid state. Furthermore, the crystal structure of the product must be easily generated from that of the parent phase without diffusive motion of atoms, and the externally imposed conditions must be such that diffusion is limited. This is true for most allotropic transformations in metals that occur at low temperatures or

FIGURE 17.15 The microstructure of a rapidly cooled Co–15% Mo alloy. Shown are hcp platelets that have formed by a martensitic transformation in an fcc matrix. (From W. Rostoker and J. R. Dvorak, *Interpretation of Metallographic Structures*, Academic Press, New York, 1965. Photomicrograph courtesy of W. Rostoker.)

for high-temperature allotropic transformations of metals brought about by a quench. Thus, β-Ti, which is stable above 883°C, cannot be retained by a quench to room temperature because α-Ti forms martensitically. The fundamental reason for this and other martensitic transformations is twofold: (1) the free energy difference between the high-temperature and low-temperature phases becomes increasingly negative with decreasing temperature, and (2) the crystal structures of allotropes of a metal are relatively simple in most cases and share similar features with each other. Because the latter condition is not satisfied by the polymorphic forms of most compounds or by the allotropic forms of tin (bct Sn and diamond cubic Sn), for example, these substances can transform only by diffusion-controlled nucleation and growth. Even β-Ti will transform to α-Ti by nucleation and growth under conditions of slow cooling. Consequently, diffusion-controlled nucleation and growth and a martensitic change are competitive processes in many cases.

The martensitic transformation starts at a temperature designated M_s, which is generally below the equilibrium transformation temperature, T_0, and is completed at a lower temperature designated M_f. The amount of parent phase converted into product phase (called martensite) depends on temperature only (Fig. 17.16) and is independent of time. Furthermore, in most cases, the M_s temperature and the fractional amount of martensite as a function of temperature are independent of quenching rate, provided the cooling rate is sufficiently fast so that diffusion does

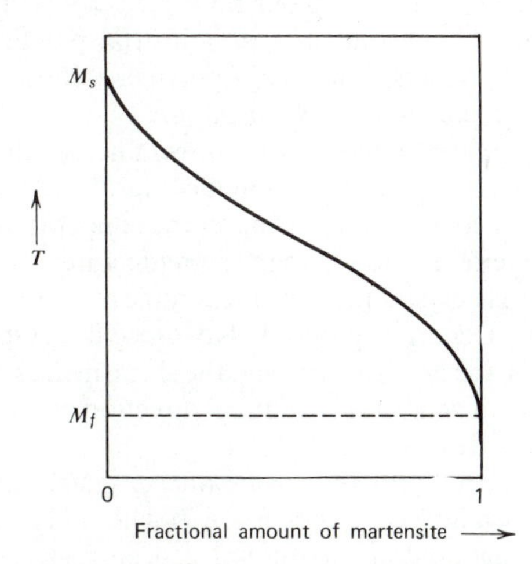

FIGURE 17.16 Fractional amount of martensite produced as a function of temperature. Martensite starts to form at the M_s temperature, and the transformation is complete at the M_f temperature.

not occur. Consequently, horizontal lines at the M_s and M_f temperatures serve to respresent a martensitic transformation on a TTT diagram (see Figs. 17.11 or 17.13). It should be noted that by making use of the "catalytic" effect of cold deformation, one can make the M_s temperature approach the equilibrium transformation temperature.

Martensitic transformations in titanium and especially in iron-carbon alloys are of great technological importance. Because of this a more complete description of the martensite transformation in steels is presented in Chapter 18.

17.6 SOLIDIFICATION

In the solidification of metals and alloys, nucleation presents no problem because heterogeneities invariably provide effective nucleation sites. Consequently, only a very small amount of supercooling is necessary for nucleation to occur, and the overall solidification rate is governed primarily by the growth rate. The growth rate, in turn, is determined by how fast heat is extracted. Physically, the ease of nucleation and growth of a metal from its melt results because atoms are packed almost as densely in the liquid as in the crystalline solid and because diffusion in the liquid is relatively rapid. Quite a different situation prevails in the case of many nonmetallic liquids, where nucleation and growth are restricted by the necessary structural rearrangements as well as complex structural units and low diffusivities. Thus, the liquid structure of a SiO_2 melt can be frozen in as a glass at the cooling rates normally encountered in industrial practice. Another extreme example is provided by atactic long-chain polymers whose irregular molecular structure completely precludes any crystallization.

When crystallization does proceed in most metallic or other inorganic liquids, dendritic growth often is the predominate mode (Figs. 17.17*a* and *b*). This situation arises with a pure metal, for example, whenever the temperature in the liquid ahead of the solid–liquid interface is lower than the temperature at the interface. Each dendrite is a small single crystal that advances rapidly into the melt and develops branches or sidearms. Dendrites generally have specific crystallographic orientations; for cubic metals, the dendrite axis and the sidearm axes are parallel to $\langle 100 \rangle$ directions. Thickening of dendrites in a lateral direction proceeds less rapidly than does the lengthening of dendrites.

Nucleation and dendritic growth provide further insight into the origin of the ingot structure discussed in Sect. 8.6 (see Fig. 8.7). The chill zone, consisting of fine, equiaxed grains having random orientations, occurs at the mold wall where heterogeneous nucleation takes place. The remainder of the ingot structure (i.e., the columnar zone and, with impure metals and alloys, the central equiaxed zone) arises from growth. In particular, the columnar zone forms because grains grow preferentially in a direction opposite to the direction of heat flow in the solid, and

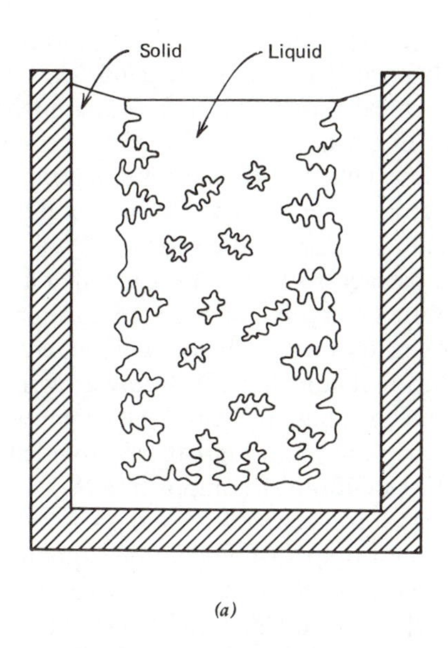

(a)

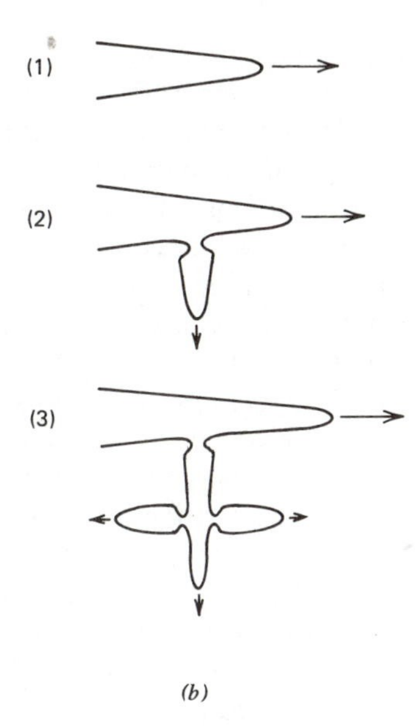

(b)

FIGURE 17.17 (*a*) A model of dendritic growth during the solidification of a metal ingot.
(*b*) The growth of an individual dendrite. (1) As the dendrite grows along its axis, (2) sidearms develop along the main axis and eventually (3) tertiary arms also form.
(*c*) Photomicrograph of a cast 70% Cu–30% Ni alloy showing dendrites. The inhomogeneous composition, which resulted from coring during solidification, reveals that solidification proceeded by dendritic growth. The etchant has preferentially attacked regions of varying composition. (Photomicrograph courtesy of T. F. Bower, Kennecott Copper Corp.)

the columnar grains posses a preferred crystallographic orientation; that is, certain of the chill zone grains are oriented favorably such that they grow dendritically into the melt, giving rise to columnar grains. Typically, columnar grains are much thicker than the spacing between individual dendrites, mainly because a number of dendrites originate from the same grain and, thus, have the same crystallographic orientation. Before considering the equiaxed zone, a very important nonequilibrium microstructural feature in ingots and castings deserves attention.

Dendrites are especially apparent in the microstructure of an impure metal or an alloy (Fig. 17.17*c*) because the impurity or solute has been concentrated in interdendritic spaces. Such compositional inhomogeneity in cast alloys, called microsegregation or coring, is a consequence of diffusional impediments during growth. Coring commonly arises whenever a solid that is forming must change its composition during the solidification process. A description of coring can be provided with reference to the solidification of a Si–Ge solution of composition X_0, as shown in Fig. 17.18, although the following discussion also pertains to alloys as well as nonmetallic solutions. As the solution is cooled to just under the liquidus temperature, small solid particles of composition X_1 nucleate. On further cooling to T_2, the solid particles grow, usually as dendrites, and the liquid composition shifts to X'_2. Under equilibrium conditions at T_2, the solid should have a uniform composition given by X_2. Indeed, equilibrium at any temperature necessitates that the compositions of solid and liquid be given by the solidus and liquidus lines, respectively. This situation can occur only if diffusion in the solid and in the liquid is sufficiently rapid so as to maintain compositional uniformity throughout each phase during solidification. In actuality, the diffusion rate in the solid is much slower than that in the liquid and much slower than the rate of solidification. Consequently, the solid that formed initially often very nearly retains the composition it had at that temperature (i.e., X_1), and subsequent growth of solid with decreasing temperature involves the addition of successive incremental layers, each having a composition given by the solidus line at the temperature where it formed. That is, as a dendrite and its sidearms thicken, the local composition varies.

While the composition at the surface of growing solid particles is given by the solidus line at any temperature (e.g., X_3 at T_3) and the solid surface is essentially in equilibrium with the liquid (e.g., composition X'_3 at T_3), the solid particles present at T_3 have a nonuniform composition that varies from X_3 at their surfaces to X_1 at their centers (i.e., along the axes of the dendrites and the sidearms). Hence, the average composition of the solid at T_3 lies between X_1 and X_3. This process continues as temperature decreases, leading to an inhomogeneous or cored solid whose average composition follows the broken line in Fig. 17.18. The position of this line, referred to as the nonequilibrium solidus line, depends upon the amount of diffusion in the solid during the solidification process as well as the overall alloy composition. Since the average composition of the solid is less than that predicted by the equilibrium diagram, more liquid is present at any temperature than would

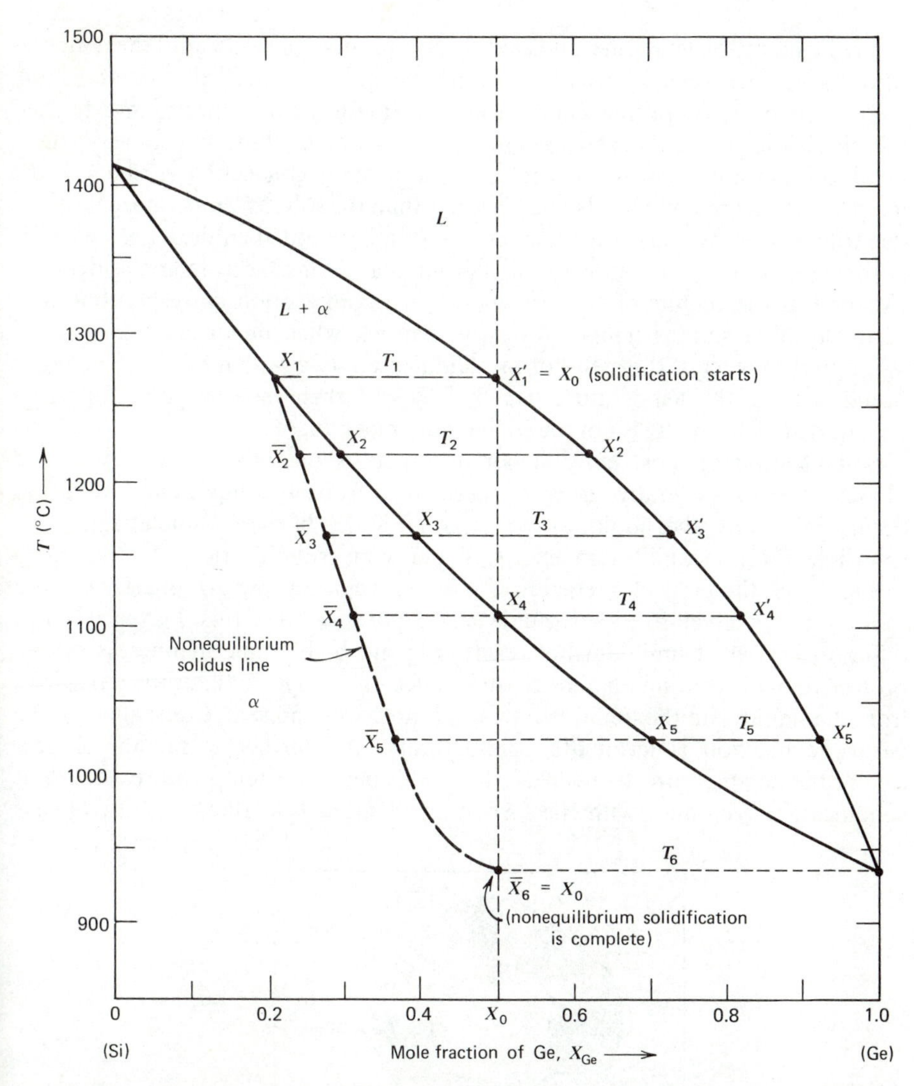

FIGURE 17.18 The Si–Ge phase diagram. As discussed in the text, nonequilibrium cooling of an alloy of composition X_0 produces the nonequilibrium solidus line (dotted line). The nonequilibrium solidus gives the average composition of the solid ($\overline{X}$) as a function of temperature, and solidification is not complete until the nonequilibrium solidus intersects the alloy composition X_0. [After C. D. Thurmond, *J. Phys. Chem.*, **57**, 827 (1953).]

be predicted by the lever rule, and solidification is not complete until the temperature at which the nonequilibrium solidus line reaches the overall alloy composition.

The above considerations can be used to explain the existence of an equiaxed zone in an alloy ingot. As dendrites grow inward from the chill zone, some sidearms break off and are swept to the center of the melt by convection currents. If the temperature in the center of the melt is lower than the solidus temperature for these dendrite fragments, they will grow into randomly oriented equiaxed grains. This is the case in alloys because the first solid that forms (i.e., along the dendrite axis and in the centers of the branches) has a composition corresponding to a relatively high solidus temperature. In contrast, when dendrite fragments are swept to the center of the melt during solidification of a pure metal, they revert to liquid because the temperature usually is above their melting point. Thus, an equiaxed zone normally is not found in a pure metal ingot.

Solidification of most eutectic liquids proceeds readily since the two solid phases forming generally grow cooperatively from the liquid and diffusion occurs rapidly in the liquid. However, solid state diffusion difficulties are responsible for nonequilibrium effects in peritectic solidification. As illustrated in Fig. 17.19, the peritectic reaction $L + \alpha \rightarrow \beta$ begins at the α-L interface, and a thin layer of β envelops or surrounds each α particle. Since this β layer acts as a diffusion barrier, it limits further reaction in much the same manner as does a protective oxide coating on a metal. In essence, the β layer isolates the α particles from the liquid, and the liquid behaves as if no α were present. Consequently, the peritectic reaction is terminated prematurely, and further extraction of heat causes the temperature to decrease from the peritectic temperature such that solidification continues with the formation of more β in the $\beta + L$ two-phase

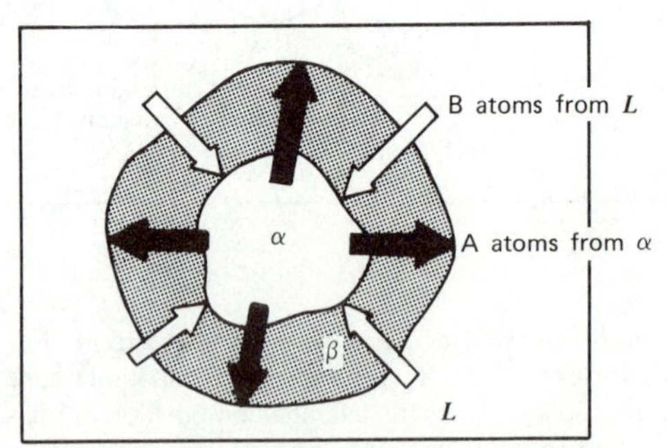

FIGURE 17.19 The phenomenon of *surrounding* in a peritectic reaction. The peritectic reaction $\alpha + L \rightarrow \beta$ begins at the α-L interface. The formation of β, however, reduces the rate of the peritectic reaction since further transformation can only occur by diffusion of atoms through the solid β phase.

field. The resulting microstructure consists of cored α phase particles surrounded by cored β phase.

Inhomogeneities in cast products usually are undesirable and can be eliminated by homogenization, that is, by heating the alloy to a temperature at which solid state diffusion occurs. For this situation, the solution to the diffusion equation (Fick's second law, Eq. 12.4) shows that the differences in composition within a cored alloy will decay with time according to the exponential relationship $e^{-(t/\tau)}$, where τ, the relaxation time, equals $l^2/\pi D$, D being the diffusivity and l, the characteristic distance over which composition variations occur. The latter factor is related to dendrite size or the spacing between dendrite arms, and homogenization is enhanced by producing castings with small dendrite arm spacings. Hot or cold working the alloy also speeds up the homogenization process. For the former, homogenization occurs simultaneously with the shape change due to deformation, whereas for the latter, it takes place after the change in shape. In both cases, the deformation provides some mechanical mixing and, thus, gives a smaller effective value of l. Additionally, working enhances diffusion due to the generation of defects (dislocations and vacancies) during deformation.

REFERENCES

1. Guy, A. G., and Hren, J. J. *Elements of Physical Metallurgy*, 3rd edition, Addison-Wesley, Reading, Mass., 1974.
2. Brophy, J. H., Rose, R. M., and Wulff, J. *Structure and Properties of Materials*, Vol. II, Wiley, New York, 1964.

QUESTIONS AND PROBLEMS

17.1 Differentiate Eq. 17.1 with respect to r to show that $r^* = -2\gamma/\Delta G_v$. Substitute this into the original equation to show that $\Delta G^* = (16\pi/3)\gamma^3/(\Delta G_v)^2$.

17.2 Repeat problem 17.1 for a cube of edge length a and find the "critical" cube edge length, a^*, and ΔG^* for such a cube. Why is ΔG^* higher for a cube than for a sphere?

17.3 Consider heterogeneous nucleation for a case where $\theta = 90°$ (i.e., a hemispherical drop). Show that ΔG^* for this case is one-half of ΔG^* for homogeneous nucleation.

17.4 (a) As mentioned in the text, nucleation of a particle within a solid is often accompanied by strain energy. Assume that, for a spherical precipitate, the strain energy is proportional to the sphere volume and show that for homogeneous nucleation, $\Delta G^* = (16\pi/3)\gamma^3/(\Delta G_v + E)^2$, where E is the coefficient of proportionality.
(b) Show that the above analysis predicts a vanishing nucleation rate not only at the bulk equilibrium temperature, but also for finite amounts of undercooling.

17.5 Calculate the critical radius of a body-centered cubic beta phase nucleus in a superheated alpha titanium matrix at 1175 K. Assume (i) a spherical nucleus, (ii) negligible strain effects, and (iii) ΔG_v (1175 K) $= -0.065$ J/m^3, $\gamma_{\alpha\beta} = 0.2$ J/m^2.

17.6 Look up the Mg–Zn phase diagram. Describe how an equiatomic alloy of Mg and Zn can show four phases in the "as-cast" state because of incomplete peritectic solidification.

17.7 If the liquidus and solidus boundaries in a phase diagram are approximately straight lines, the nonequilibrium weight fraction of liquid, at any given temperature, is given by

$$\ln W_L = \frac{1}{\left(\dfrac{C_s}{C_L} - 1\right)} \ln \frac{C_L}{C_0}$$

where C_0 is the overall alloy composition and C_L and C_s are the equilibrium liquidus and solidus compositions at the given temperature. Approximately linear solidus and liquidus lines are observed in the Al-rich end of the Al–Cu diagram. That is, the liquidus line is a straight line extending from the melting point of aluminum ($T = 660°\text{C}$) to the eutectic point ($T = 548°\text{C}$, $C_E = 33.2$ wt. % Cu), and the solidus line is a straight line from the melting point of aluminum to the point of maximum copper solubility in Al ($T = 548°\text{C}$, $C = 5.65$ wt. % Cu). In Al–Cu alloys containing less than 33.2 wt. % Cu, the liquid that is present just above the eutectic temperature transforms into the eutectic mixture of $\alpha + \theta$. Plot the weight fraction of eutectic in cast Al–Cu alloys as a function of copper content (0 to 33.2 wt. % Cu), for both equilibrium and nonequilibrium solidification.

17.8 (Library problem) What is "splat cooling," and what are typical cooling rates observed in splat cooling? What kinds of structure result from splat cooling of metals?

17.9 (Library problem) What is zone refining? Explain how the principles of nonequilibrium solidification are used to good avail in zone refining of silicon and germanium.

17.10 The Cu–Ni diagram, like that of Ge–Si, shows a complete range of solid solubility. Cu–Ni alloys are homogenized much more readily than Si–Ge alloys. Explain why.

17.11 (a) As a pure liquid is supercooled below its equilibrium freezing temperature (T_m), the free energy difference between solid and liquid becomes increasingly negative. The variation of ΔG_v with temperature is given approximately by $\Delta G_v = (\Delta H_m/T_m)(T_m - T)$, where ΔH_m (< 0) is the latent heat of solidification. Using this expression, show that $e^{-N_0\Delta G^*/RT}$ is a maximum when $T = T_m/3$ as indicated in Fig. 17.3*a*.

(b) Explain why the maximum nucleation rate occurs at a temperature greater than $T_m/3$ (see Fig. 17.3*b*).

17.12 (a) Using schematic free energy vs. composition diagrams (Chapter 16), show that the highest temperature at which α can form martensitically from β in titanium alloys containing molybdenum (Fig. 17.10) lies between the equilibrium $\beta/\alpha + \beta$ and $\alpha + \beta/\alpha$ boundaries. (*Hint.* The thermodynamic upper limit for martensite formation is the temperature at which the free energies of α and β, having the same molybdenum content, are equal.)

(b) The difference between the M_s temperature and the equilibrium temperature (M_0), as defined in part (a), usually is nearly a constant for a given alloy system. Using this along with schematic free energy vs. composition diagrams, show that M_s

should decrease with increasing molybdenum content in Ti–Mo alloys and with increasing carbon content in Fe–C alloys (Fig. 16.13).

(c) Sketch the form of a phase diagram (based on titanium, for example) in which the M_s temperature would *increase* with increasing alloy content.

17.13 (a) Consider a eutectic diagram for two components, one of which is metallic in nature and the other is covalent or semimetallic in nature (e.g., Al–Si). In such a system it is observed frequently that nucleation of the nonmetallic solid phase is difficult in comparison to nucleation of the metallic phase. Consequently, under conditions of rapid solidification, the effective liquidus temperature of the nonmetallic phase lies below the equilibrium liquidus. Since the eutectic point can be defined as the temperature at which the two liquidus lines (i.e., $L/L + \alpha$ and $L/L + \beta$) intersect, show that the effective eutectic temperature is decreased and that the effective eutectic composition is displaced toward compositions richer in the nonmetallic component if solidification is rapid. Show your result in terms of an effective phase diagram superimposed on the equilibrium diagram.

(b) Consider alloys containing more of the metallic component than is contained in the eutectic composition. With reference to your diagram in part (a), is the fraction of primary microconstituent resulting from rapid solidification greater or less than what would be present under equilibrium conditions? Compare your answer to that obtained in problem 17.7 and comment on the comparison.

17.14 A portion of the potassium disilicate [$K_2O \cdot 2SiO_2$(abbreviated KS_2)]–silica (SiO_2) phase diagram is shown below.

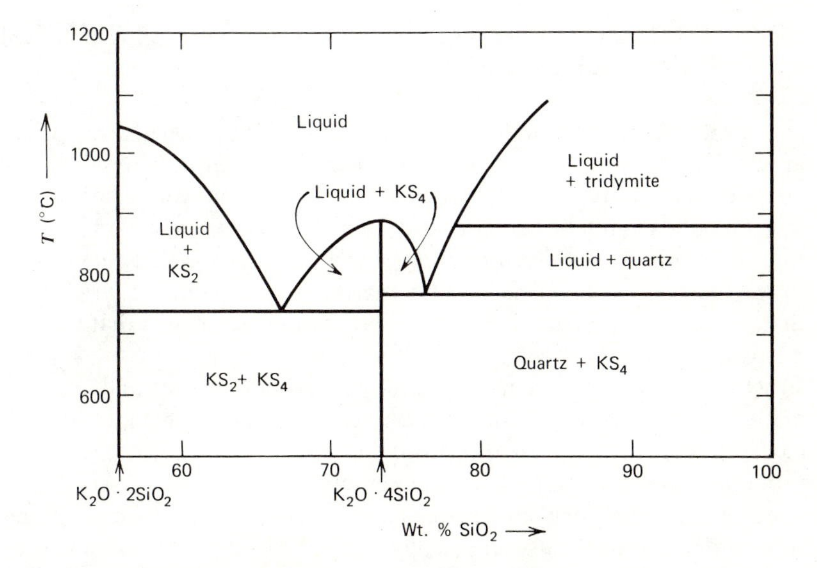

The phase $K_2O \cdot 4SiO_2$(abbreviated KS_4) is difficult to nucleate. Show that, under such circumstances, a nonequilibrium eutectic between KS_2 and quartz can occur at a temperature of about 550°C.

18

Structural and Property Changes

In this chapter the relationships between microstructure and properties of materials are discussed in terms of specific examples.

18.1 THE RELATIONSHIP BETWEEN STRUCTURE AND PROPERTIES

Almost without exception, physical, chemical, and mechanical properties are affected by changes in chemical composition, crystal structure, microstructure, and defects. Many properties depend primarily on composition and crystal structure and cannot be varied greatly by processing treatments. These are referred to as structure-insensitive properties, which only means that they are insensitive to microstructure and defects. Other properties are termed structure-sensitive; that is, they are influenced greatly by processing treatments that alter or control microstructure and defects. The difference between structure-insensitive and structure-sensitive properties is basically a matter of degree and, in some cases, a matter of circumstance. For example, single crystals typically are anisotropic (i.e., properties vary with direction) but polycrystals often are isotropic (i.e., properties are the same in any direction); however, polycrystals can become anisotropic when they are processed such that the individual grains assume similar orientations. This anisotropy applies to structure-insensitive as well as structure-sensitive properties.

Microstructural manipulation has a dramatic effect on many mechanical properties of metals. The widespread utilization of metallic materials in a large number of applications is a direct consequence of their ability to be shaped by deformation processing and their combination of strength and ductility. In addition, metals possess a variety of physical properties that make them preeminently useful for numerous engineering purposes. Nonmetallic materials also find widespread application. In part, this is because many ceramics and polymers are less expensive than metals and require simpler processing; however, they also possess certain unique properties. For example, a nonmetallic material may serve as an electrical or thermal insulator, whereas a metal usually cannot. Even in structural applications, particularly at elevated temperatures, ceramics may be superior to metallic alloys.

The very large number of materials employed for engineering purposes runs the gamut from stones, naturally occurring polymers, and simply processed ceramics based on clay, to common steels, complex synthetic polymers, transition metal carbides, and sophisticated semiconductors used in electronic devices. Some properties of each material are sensitive to microstructure, whereas others are determined largely by the interatomic or intermolecular bonding. Frequently, it is the combination of bonding structure and microstructure that gives rise to observed characteristics. Furthermore, the processing that a material receives often alters its microstructure and, as a consequence, its properties.

Although certain properties may dominate in the choice of a material for specific purposes, mechanical behavior is almost always significant. While the mechanical characteristics of a material are related intimately to bonding and crystal structure (or lack of crystallinity), the strength displayed often depends on its microstructure and the presence of defects and microstructural flaws. These aspects will be introduced in this chapter. More specific considerations of the mechanical and physical properties of both metals and nonmetals will be treated in subsequent chapters.

18.2 COLD WORK AND ANNEALING

During deformation, metals work harden and become stronger. In practice, many metallic materials are formed into shape by cold metalworking processes such as hammering (forging), rolling, or wire drawing. These processes utilize tools or dies that effect a permanent shape change of the metal while at the same time strengthening it. The macroscopic change of shape of a metal is reflected in the microscopic change of grain shape. Cold work causes grains to elongate in the flow direction and to become thinner in other directions since grain volume remains unaltered, and there is an accompanying increase in dislocation density. Thus, as a result of increased dislocation density and increased grain boundary area, a cold-worked metal is higher in energy. Concurrent with the structural changes, mechanical properties are changed. Cold working raises the strength and decreases the ductility of a metal.

A cold-worked structure can be returned to a strain-free structure, consisting of equiaxed grains and having a much lower dislocation density, through an elevated temperature heat treatment (see Sects. 10.6 and 10.7). This process, termed recrystallization (see Sect. 8.8), can be used to soften a cold-worked metal, to restore its ductility, and to increase its capacity for further cold deformation. Recrystallization involves the nucleation and growth of new strain-free grains (i.e., grains with a low dislocation density) at the expense of the deformed grains that possess a higher dislocation density. The thermodynamic driving force for recrystallization is the reduction of free energy that accompanies a reduction in dislocation density. The nucleation event is not well-characterized as in precipitation, but it involves local motion of grain boundaries and subboundaries. This is a thermally activated process as is the subsequent growth of the recrystallized grain wherein atoms jump from higher energy, strained grains across grain boundaries to lower-energy, strain-free grains. That is, the diffusive motion of atoms simultaneously eliminates elongated grains with a high dislocation density and forms equiaxed grains having a lower dislocation density. This continues until all deformed grains are consumed. Since thermal motion aids atomic jumping, elevated temperatures (often in the range of 0.4–0.5 T_m, where T_m is the absolute melting point) are required for recrystallization. In addition to the heat-treatment temperature, the time duration at temperature and the amount of prior cold work affect the extent and rate of recrystallization (Fig. 18.1).

Intermediate recrystallization treatments can be employed between metalworking operations to control fabrication parameters and final mechanical properties. Control of the latter is achieved by varying the point at which a final recrystallization heat treatment is conducted. Copper wire, rod and sheet, for example, are supplied in a number of conditions denoted as soft, 1/4 hard, 1/2 hard, etc. These correspond to recrystallization at final size ("soft") and at sizes increasingly removed from the final size.

Figure 18.2 provides a summary of the effect of cold work and annealing treatments on mechanical properties. As shown, cold work also alters certain physical properties such as electrical resistivity but to a lesser extent as compared to the change in mechanical properties. Heating a cold-worked metal to temperatures insufficient to cause recrystallization results in minor dislocation rearrangements and the elimination of excess vacancies that were generated during the working process. These processes, collectively referred to as recovery processes, hardly alter the mechanical properties of a cold-worked metal but tend to change physical properties toward values characteristic of the unworked condition (see Fig. 18.2*b*). Dislocation rearrangement during recovery, however, does serve to relieve any long-range elastic stresses or strains (stress relief) without softening the metal.

Some polycrystalline metals having low melting points (e.g., high-purity lead) do not work harden or display other facets of cold working when they are plastically deformed at room temperature. This is because they recrystallize either during the

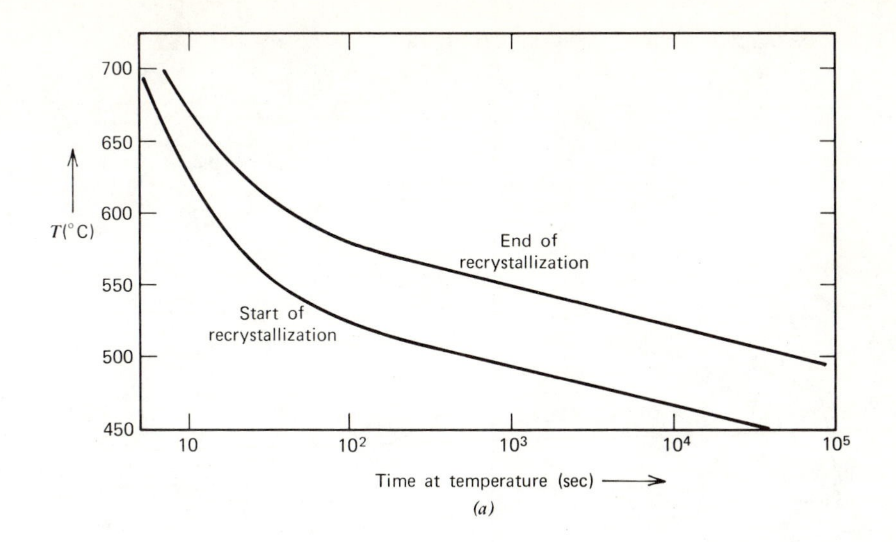

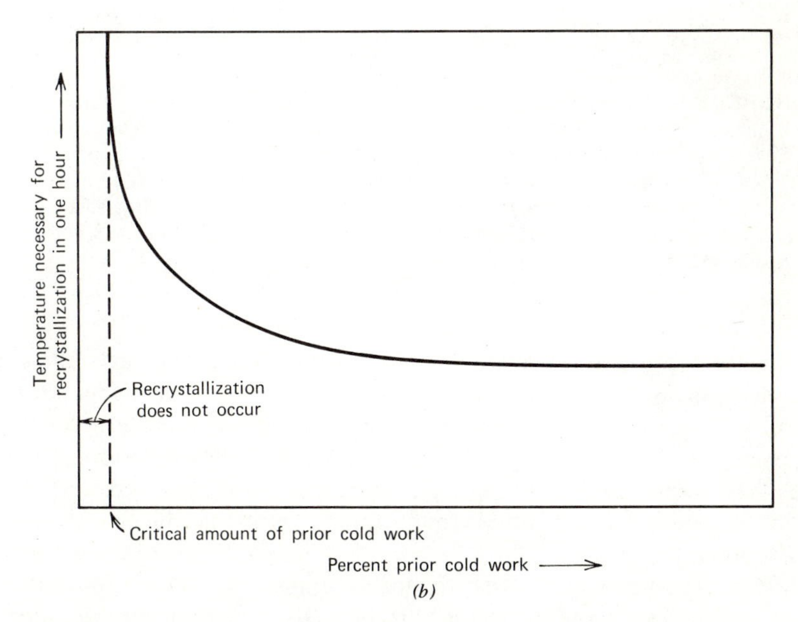

FIGURE 18.1 (*a*) A time-temperature-transformation diagram for recrystallization of rimmed steel. The curve illustrates the general principle that the higher the temperature, the less time required for recrystallization. [After R. H. Goodenow, *Trans. ASM*, **59**, 804 (1966).]

(*b*) Schematic illustration showing the variation of recrystallization temperature (one-hour heat treatment time) with percent prior cold work. Note that, regardless of temperature, a critical amount of cold work is required before recrystallization will take place, and the recrystallization temperature approaches a lower limit at high amounts of prior cold work.

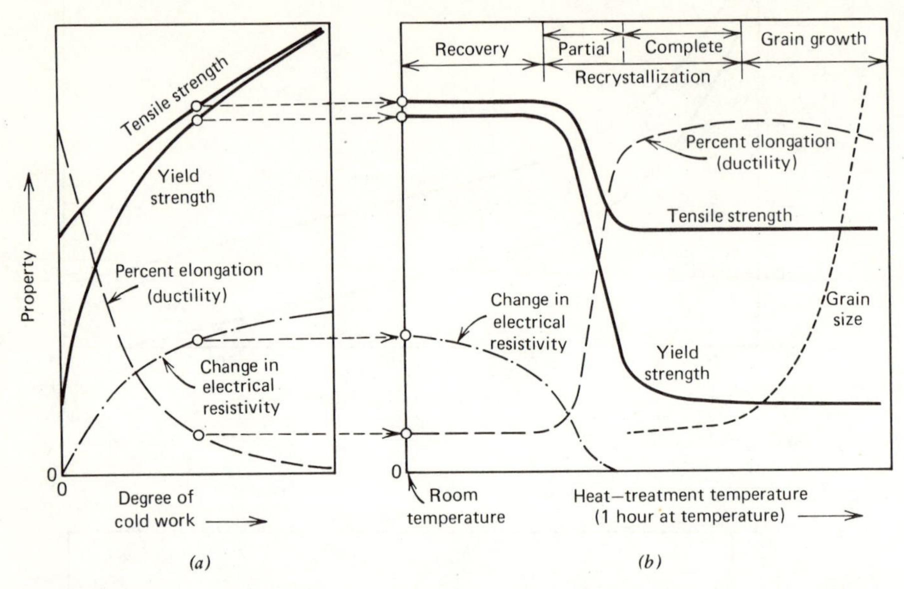

FIGURE 18.2 The effects of cold work and annealing after cold work on the mechanical and physical properties of metals: (*a*) Cold work markedly increases strength and decreases ductility; physical properties, such as electrical resistivity, are altered to a lesser degree. (*b*) During low-temperature heat treatment, physical properties change to a much greater extent than do mechanical properties, which only return to values characteristic of the unworked state after recrystallization occurs. Grain size increases rapidly at very high annealing temperatures.

deformation process or very shortly thereafter. Other polycrystalline metals exhibit behavior similar to this when deformed at elevated temperatures. Hot working is the designation applied to a forming process conducted at temperatures where recrystallization occurs on the same time scale as the forming operation itself. The result of hot working typically is a uniform, equiaxed, and strain-free grain structure quite comparable to the grain structure produced by recrystallization following cold work.

If a cold-worked metal is heat treated at temperatures much above the lowest temperature necessary for recrystallization, its microstructure is altered additionally by grain growth (Fig. 18.2*b*). Recovery, recrystallization, and grain growth processes occur over successively higher temperature ranges, but there is no sharp dividing line between the latter two. During grain growth, the average grain size increases and, necessarily, the number of grains per unit volume decreases. The thermodynamic driving force for grain growth is the reduction in free energy that accompanies a decrease in the grain boundary area per unit volume. The process proceeds by atoms jumping from one grain across a grain boundary to a neighboring larger grain, and the larger grains grow at the expense of their smaller neighbors. Although grain growth in metals is accompanied by slight decreases in strength (Fig. 18.2*b*), it generally is not used technologically to manipulate metal properties. Indeed, large grain sizes usually are undesirable. It should be noted that, unlike recovery and recrystallization, the process of grain growth need not be preceded by cold work. Consequently, grain growth occurs in ceramics, which are not amenable to cold work, and plays a role in their processing.

Many ceramic materials are produced by pressing and sintering processes in which a fine powder of the ceramic is cold pressed into a viable compact that contains about 30 to 35% porosity by volume. This resultant porosity is reduced markedly by a subsequent heat treatment process called sintering. During sintering, grain growth occurs coincidentally, and this is partly responsible for the difficulty of producing a final product without residual porosity. In addition, of course, it produces a larger grain size, and both of these effects are adverse insofar as ceramic mechanical properties are concerned.

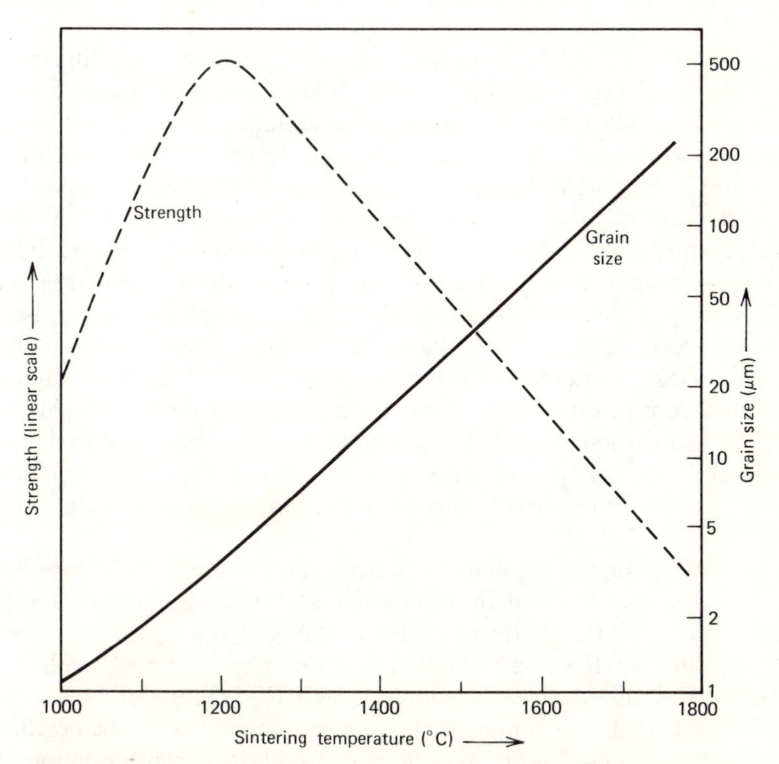

FIGURE 18.3 Schematic variation of room temperature strength and grain size of MgO as a function of sintering temperature. Optimum properties are obtained by the best combination of small grain size and low porosity.

Porosity has an especially adverse effect on strength, and ceramics with a fine grain size generally have superior mechanical properties as compared to those with a coarse grain size. Although grain size is minimized by a low firing temperature, this may lead to insufficient densification by the sintering process. Thus, there is an optimum firing temperature to produce the maximum room temperature strength (Fig. 18.3), and this corresponds to the best combination of low pore volume fraction and fine grain size.

Fully dense, porefree, and fine-grained ceramic bodies can be produced, in some cases, if grain boundary motion is inhibited by the presence of certain soluble substances that segregate to grain boundaries or of very fine inert particles added to the starting powder. An example is provided by Al_2O_3, containing a small amount of MgO, which becomes fully dense and is transparent rather than translucent or opaque, as in the case of sintered Al_2O_3 that has residual porosity.

18.3 PEARLITE AND BAINITE IN STEELS

The most technologically important and widely studied eutectoid reaction occurs in steels (see Fig. 16.13). The lamellar decomposition product, a two-phase microconstituent called pearlite, consists of alternating lamellae or plates of α and Fe_3C, and its structure becomes progressively finer as the transformation temperature is decreased. As was shown in Fig. 17.13, the decomposition of austenite displays *C*-curve behavior typical of transformations that go from a high-temperature phase to lower-temperature products and that require diffusion. At transformation temperatures above the knee of the *C*-curve (Fig. 17.13), pearlite is the product, whereas below the knee, a two-phase microconstituent called bainite forms from austenite. Basically, bainite is a transformation product that is not as close to equilibrium as pearlite. Furthermore, bainite is morphologically distinct from pearlite and consists of a dispersion of very small particles of Fe_3C (or of another less stable iron carbide) within or between fine ferrite plates. The scale of the carbide particle dispersion in bainite is sufficiently fine that the microstructure can be resolved completely only with the aid of an electron microscope (Fig. 18.4).

The transformations of austenite to pearlite and to bainite are competitive in that austenite will decompose to the product for which there is a higher transformation rate. Clearly, the bainite transformation is favored at a high degree of supercooling, and the pearlite transformation at a low degree of supercooling. It is important to note that the *C*-curve shown in the TTT diagram for a plain carbon eutectoid steel (Fig. 17.13) represents the overlap of *C*-curves for the pearlite and bainite transformations. That is, in principle, each of these transformations has a characteristic *C*-curve, and it is only coincidence that for plain carbon and some alloy steels the combination of the individual *C*-curves is itself a single *C*-curve.

In the case of many alloy steels the knees of the individual pearlite and bainite transformation curves are displaced sufficiently in temperature and in time so that the overall transformation curve consists of two *C*-curves joined in an intermediate region.

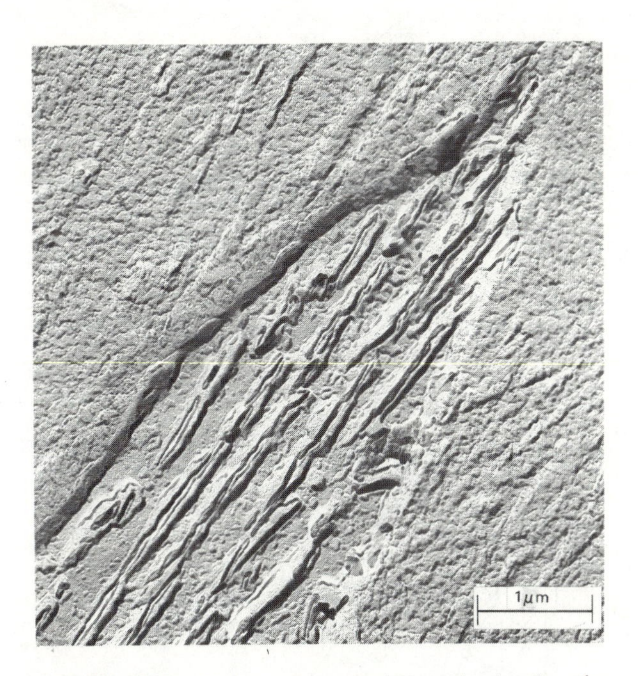

FIGURE 18.4 Replica electron micrograph showing the structure of bainite in a 4340 steel. A needle of bainite is shown in a martensite matrix. The needle consists of a ferrite matrix and elongated particles of Fe_3C. The specimen was austenitized, transformed at 450°C, and quenched. (Reproduced by permission from *Metals Handbook*, 8th Edition, Vol. 8, American Society for Metals, 1973.)

The mechanical properties of pearlitic and bainitic steels are related to their microstructures. Bainitic steels exhibit excellent combinations of strength and ductility. Pearlitic steels generally are not as strong as bainitic steels, although fine pearlite can result in high strengths. One notable example is piano wire, which is a pearlitic steel that has been subjected to extensive plastic deformation by wire drawing and is as strong as any steel yet produced.

The kinetics of austenite decomposition depend strongly on the composition of a steel. As a rule, increasing the carbon content or the content of alloying elements slows down the rates of pearlite and bainite formation, but not to the same

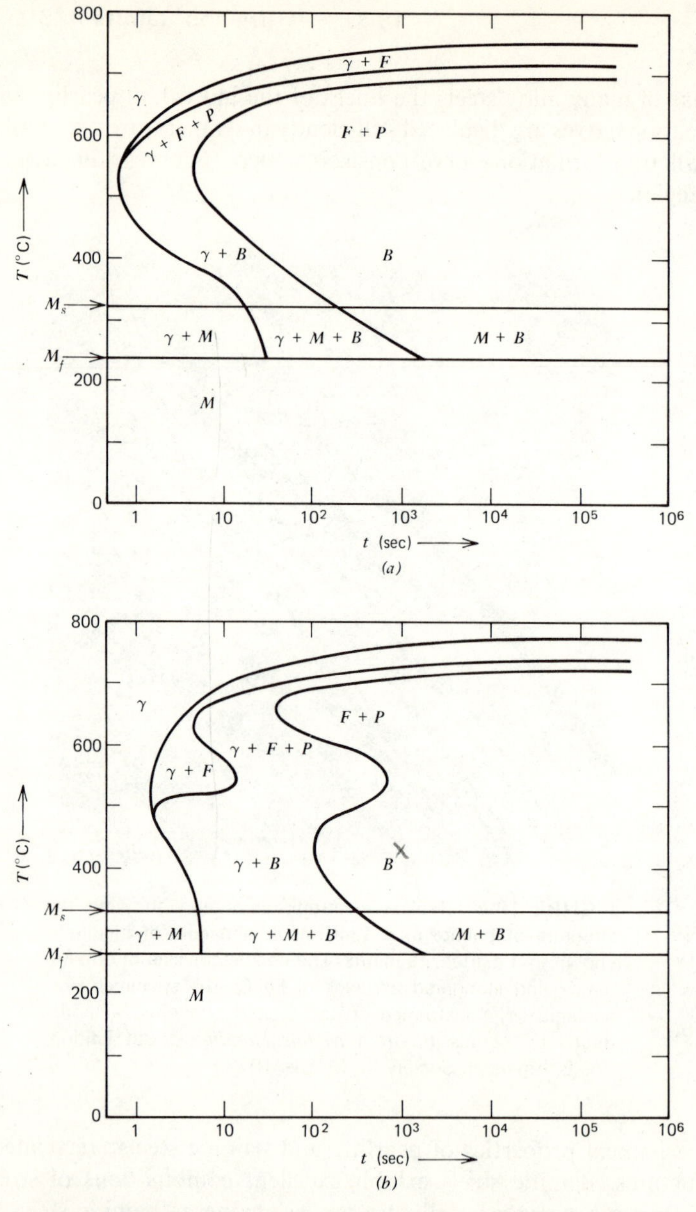

FIGURE 18.5 The isothermal transformation diagrams of (*a*) a plain carbon steel containing 0.50 wt. % carbon and (*b*) an alloy steel containing chromium. The pearlite and bainite reactions are accelerated by decreasing carbon content [compare (*a*) with Fig. 17.13] and retarded by increasing alloy content [0.93% Cr in the case shown in (*b*)]. In (*b*) two separate transformation *C*-curves for the pearlite and bainite reactions are apparent. [γ = austenite; F = proeutectoid ferrite; P = pearlite; B = bainite; M = martensite.] (Adapted from *Atlas of Isothermal Transformation Diagrams*, United States Steel Corp., Pittsburgh, 1951.)

degree. Substitutional alloying elements often are added to a steel in order to slow down transformations involving diffusion because this permits greater microstructural manipulation than would be possible otherwise. Isothermal transformation diagrams for a plain carbon steel and an alloy steel containing chromium are shown in Fig. 18.5. Decreasing the carbon content (compare Fig. 18.5*a* with Fig. 17.13) increases the rate of austenite decomposition, and adding an alloying element (compare Fig. 18.5*b* with Fig. 18.5*a*) makes the transformation more sluggish. Incidentally, this alloy steel TTT diagram (Fig. 18.5*b*) clearly shows separate *C*-curves for the pearlite and bainite transformations.

Since actual technological practice usually involves continuous cooling as opposed to isothermal transformation, "mixed" microstructures often are obtained. In plain carbon steels, the mixture usually is coarse and fine pearlite. In low-alloy steels, bainite frequently is present. In higher alloy steels, or in steels subjected to high cooling rates, both the pearlite and bainite reactions may be suppressed because insufficient time is spent in an appropriate temperature range for diffusion to occur. In these circumstances, austenite either may be retained at room temperature or, more commonly, may transform martensitically.

18.4 MARTENSITE IN STEELS

As noted in Chapter 17, a martensitic transformation can be considered competitive with diffusion-controlled nucleation and growth transformations. In steels, the formation of pearlite or bainite necessitates diffusion for nucleation and growth, and the martensitic transformation can dominate if a steel is cooled rapidly (i.e., quenched) so that it remains at temperatures favorable for austenite decomposition by diffusional processes for times that are insufficient for such processes to occur (recall Fig. 17.8). For a martensitic transformation to take place at a given temperature, the austenite must have a higher free energy than a body-centered cubic phase having the same carbon content. In steels the product of this transformation, which occurs by a shear process without any diffusion, is simply called martensite and is characterized by some of the features discussed in Sect. 17.5. One important exception is that steel martensite possesses a body-centered tetragonal crystal structure (Fig. 18.6) that is a slight distortion of the expected bcc crystal structure. This distorted structure results because interstitial carbon atoms necessarily occupy one set of preferred octahedral interstitial sites in the structure. That is, the martensitic transformation carries the carbon atoms residing in the octahedral interstitial sites in austenite to only one of three possible sets of octahedral interstitial sites in martensite, and this leads to a tetragonal distortion of the crystal structure. The extent of tetragonal distortion, as measured by the ratio of the longer *c* axis to the shorter *a* axis, increases with increasing carbon

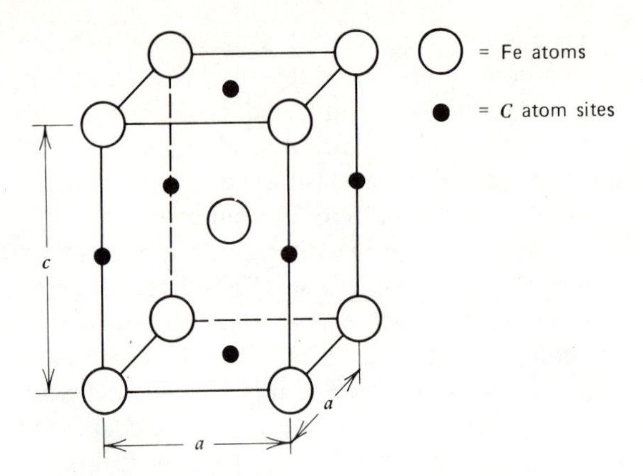

FIGURE 18.6 The body-centered tetragonal unit cell of steel martensite. The tetragonality arises from the preferential occupation of octahedral interstitial sites along the c axis by carbon atoms. The higher the carbon content, the higher the c/a ratio.

content (Fig. 18.7*a*), and the hardness of the martensite also increases with increasing carbon content (Fig. 18.7*b*).

Martensite formed directly by rapid cooling from the austenite region (i.e., as-quenched martensite) possesses mechanical properties that are similar to those of glass in some respects, that is, inherently high strength coupled with nil ductility (brittleness) and a tendency to fracture in a catastrophic manner when subjected to mechanical or thermal shock. The high strength and low ductility of martensite are related directly to the strengthening afforded by carbon in solid solution, with the accompanying tetragonality, and to a rather high dislocation density produced during the martensitic transformation. Steel martensite is relatively unique in these respects, and its characteristics render it unusable for structural purposes in the as-quenched form. Steel martensite may be altered, however, into a somewhat less strong but more ductile structure by tempering. Tempering is a heat treatment given to a martensitic steel in order to permit some diffusion to occur in a reasonable period of time. During tempering, small carbide particles precipitate from the supersaturated martensite, and this leads to both decreased tetragonality and decreased hardness with a concurrent gain in ductility and toughness. Extended tempering at temperatures slightly below the eutectoid temperature produces a structure called spheroidite. This two-phase structure consists of globular Fe_3C particles dispersed in a ferrite matrix having the equilibrium carbon content (Fig. 18.8). Spheroidite is the softest yet toughest structure that a steel may have. At lower tempering temperatures (or shorter heat treatment times), the structure obtained by tempering is intermediate between that of spheroidite

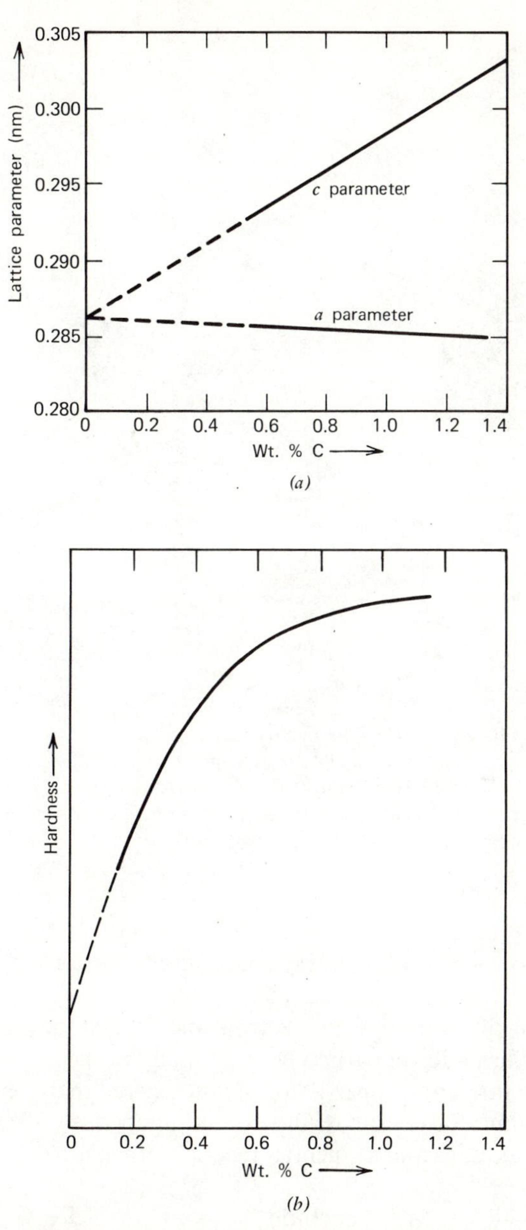

FIGURE 18.7 (*a*) How the degree of tetragonality of steel martensite varies with carbon content (see the unit cell in Fig. 18.6). (*b*) Schematic variation of the hardness of martensite with carbon content.

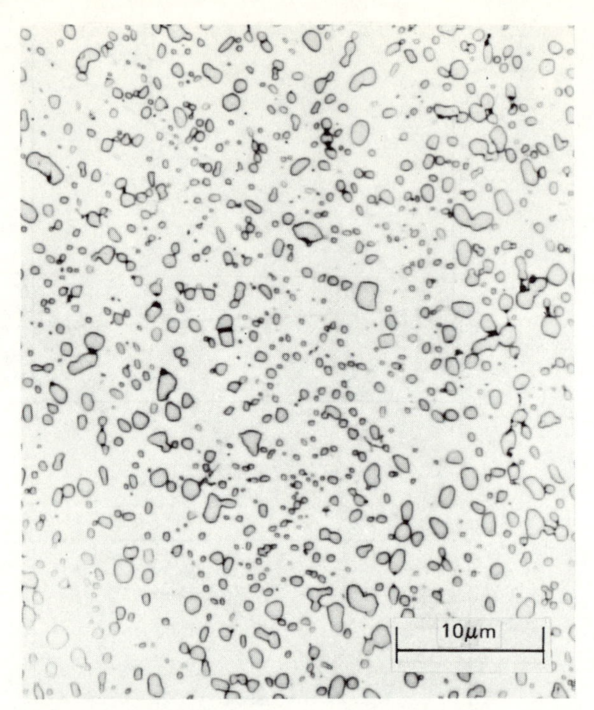

FIGURE 18.8 The microstructure of spheroidite obtained by tempering quenched martensite at a temperature near the eutectoid. The microstructure consists of globular Fe_3C particles dispersed in a matrix of ferrite. (Photomicrograph courtesy of United States Steel Corp.)

and that of martensite, and the mechanical properties are intermediate also. *Tempered martensite* often consists of very fine carbide particles, which can be resolved only with the aid of an electron microscope, dispersed in a matrix of martensite that has a lower carbon content than did the as-quenched martensite. (The microstructure and properties of such tempered martensite are very similar to those of bainite formed at relatively low temperatures.) Tempering may be viewed as a continuous and sequential reaction of the form

$$M \longrightarrow M' + \text{carbide} \longrightarrow \alpha + Fe_3C\ (\text{globular}) \quad (18.1)$$

where M and M' refer to the original and lower-carbon martensites, respectively. This is essentially a process of precipitation from supersaturated solid solution followed by a coarsening process. The optimum mechanical properties of a steel, combining excellent strength *with* good ductility and toughness, are obtained with martensite tempered to the intermediate stage (Fig. 18.9).

Inspection of Figs. 17.13 and 18.5*a* shows that very high cooling rates are needed to suppress pearlite formation in plain carbon steels. As a consequence, it is difficult to form martensite in plain carbon steels, particularly in the case of large section sizes where the cooling rates at the interior are fairly low. Indeed, even if the surface cools instantaneously when inserted into the quenchant medium, the interior

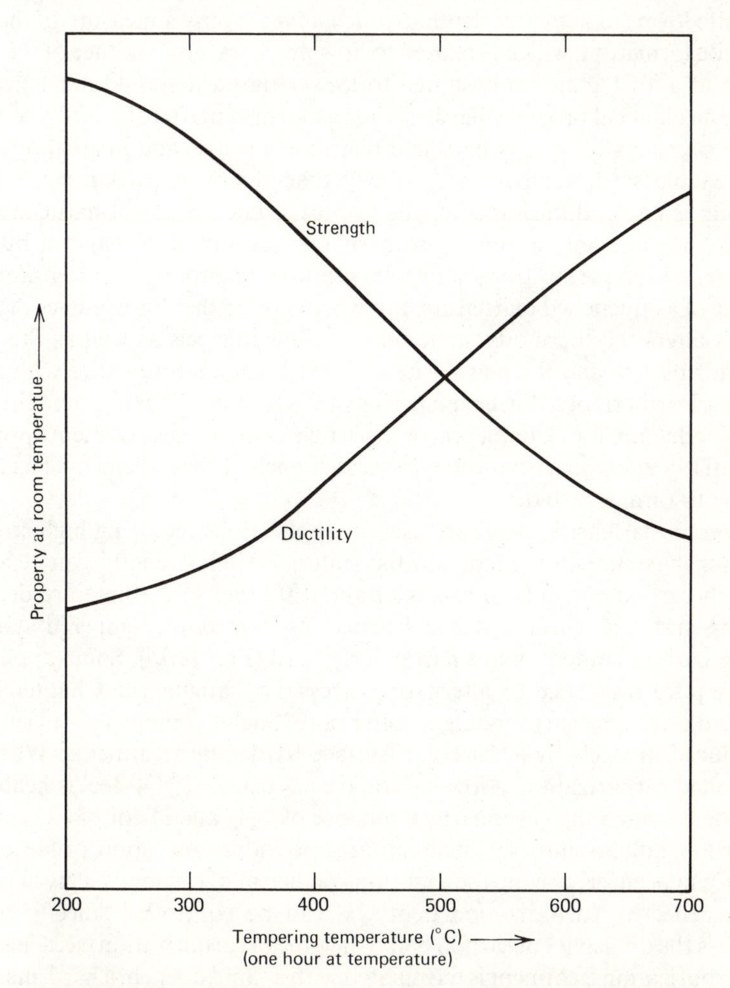

FIGURE 18.9 Schematic room temperature strength and ductility of martensite after tempering for one hour at the temperatures indicated. Optimum combinations of strength and ductility are obtained at intermediate tempering temperatures.

will cool at a rate governed by thermal conductivity, and among metals steels are not particularly good thermal conductors. The useful properties of tempered martensite can be realized only if martensite can form to a sufficient depth below the surface of a piece of steel. These considerations account for the use of some alloying elements in steels since the rates of bainite and pearlite formation are reduced markedly by their presence, and martensite can be formed at lower cooling rates. Alloy steels are said to be more *hardenable* than plain carbon steels because they are converted to martensite more easily or, for a given section thickness, martensite forms to a greater depth. *Hardenability* is only a measure of the ease of martensite formation, which is related to how much the nose or knee of the pearlite reaction in a TTT diagram is shifted to longer times; it should not be confused with the mechanical property hardness, for as-formed martensite of a given carbon content is essentially equally hard in a plain carbon steel and in an alloy steel.

Hardenable steels temper more slowly than do plain carbon steels because tempering is also a diffusion-controlled process. Each grade of hardenable steel, therefore, has not only a unique isothermal transformation diagram but also a unique set of tempering times and temperatures to produce a given amount of softening of as-quenched martensite. It is worth noting that the presence of alloying elements alters the thermodynamic relationships in steels as well as the kinetics of transformation, and the martensite start (M_s) temperature varies with alloying additions. Furthermore, during tempering of steels containing tungsten, chromium, and molybdenum, for example, carbide particles rich in these elements precipitate preferentially in comparison to Fe_3C because each of these elements has a strong tendency to form a carbide.

Tempered martensitic steels are used in applications requiring high strength—often over the entire cross section of the material. High strengths can be achieved only with carbon contents in excess of about 0.3 wt. % C (these are designated medium- and high-carbon steels) because as-formed or tempered martensite of lower carbon content is not particularly hard (Fig. 18.7*b*). Some applications, which require resistance to alternating stress (i.e., fatigue, see Chapter 21), call for a hard steel case surrounding a softer core. Such "composite" structures can be produced in steels by a variety of surface hardening treatments. With one of these, called carburization, a low-carbon (i.e., $\lesssim 0.3$ wt. % C) steel is heated in an atmosphere containing carbon (say a mixture of CH_4 and H_2 or of CO_2 and CO), and carbon diffuses into the steel surface, providing the appropriate chemical reaction can occur. Knowing the diffusion coefficient of carbon at a given temperature, the effective carburization depth, x, can be controlled since $x \simeq (Dt)^{1/2}$, where D is the diffusivity of carbon and t, the time of carburization (see Chapter 12). If the carburization treatment is terminated with a rapid quench, a hard, martensitic case is formed at the surface, and no martensite forms in the interior since it not only cools more slowly but also is composed of a less hardenable steel because of its lower carbon content. Martensitic cases can be produced on medium- and high-carbon steels, in addition, by flame or induction hardening, either of which locally heats the steel surface into the austenite region. Subsequent quenching gives a martensitic case around a ferrite-pearlite mixture, which is essentially what was present initially because the core was not heated above the eutectoid temperature.

Unlike steels, most engineering metals and alloys cannot be hardened by quenching, and alternative techniques, often involving cold work, are utilized to produce case-core properties similar to those discussed above. One of these techniques, called shot peening, involves sand blasting of the metal surface with hardened steel or similar hard shot. This produces local cold work that hardens the surface, which is desirable for applications involving cyclical stress.

18.5 CAST IRONS

Iron-carbon alloys containing less than about 1.2 wt. % C are classified as steels, and those containing more than about 2 wt. % C are classified as cast irons. Plain carbon steels actually contain minor amounts of silicon and manganese by virtue of necessary processing treatments. Similarly, most cast irons contain about 2 wt. % Si. Cast irons differ microstructurally from steels in that graphite usually is a constituent. Graphite is the equilibrium carbon-rich phase found in the Fe–C system. The equilibrium graphite phase is seldom, if ever, observed in steels because the solid state transformations that could give rise to it are so sluggish that transformations yielding cementite always predominate. Furthermore, nucleation of graphite within cementite does not occur, for all practical purposes, in steels. The situation differs in the lower melting cast irons, in that graphite can form directly from eutectic solidification and by solid state processes. Both of these are related to the silicon content because silicon favors graphite formation.

Whether carbon is present as graphite or tied up in Fe_3C in a cast iron depends on whether the solidification and cooling processes are carried out under conditions close to equilibrium (thus favoring graphite formation) or under conditions far from equilibrium (thus favoring Fe_3C formation). The similarity to austenite decomposition should be noted; the more nearly equilibrium pearlite forms in preference to the less stable bainite for a transformation close to equilibrium, and vice versa. Some of the types of structures that can be realized by varying processing treatments given to cast irons are shown schematically in Fig. 18.10. For slow cooling rates (or equivalently, high silicon contents), graphite forms initially in the eutectic reaction and continues to grow with decreasing temperature as the carbon content of the iron-rich phase (γ or α) decreases. Under these conditions, the phases present in zones I, II, and III (see Fig. 18.10) are austenite and liquid, austenite and graphite, and ferrite and graphite, respectively. Under normal

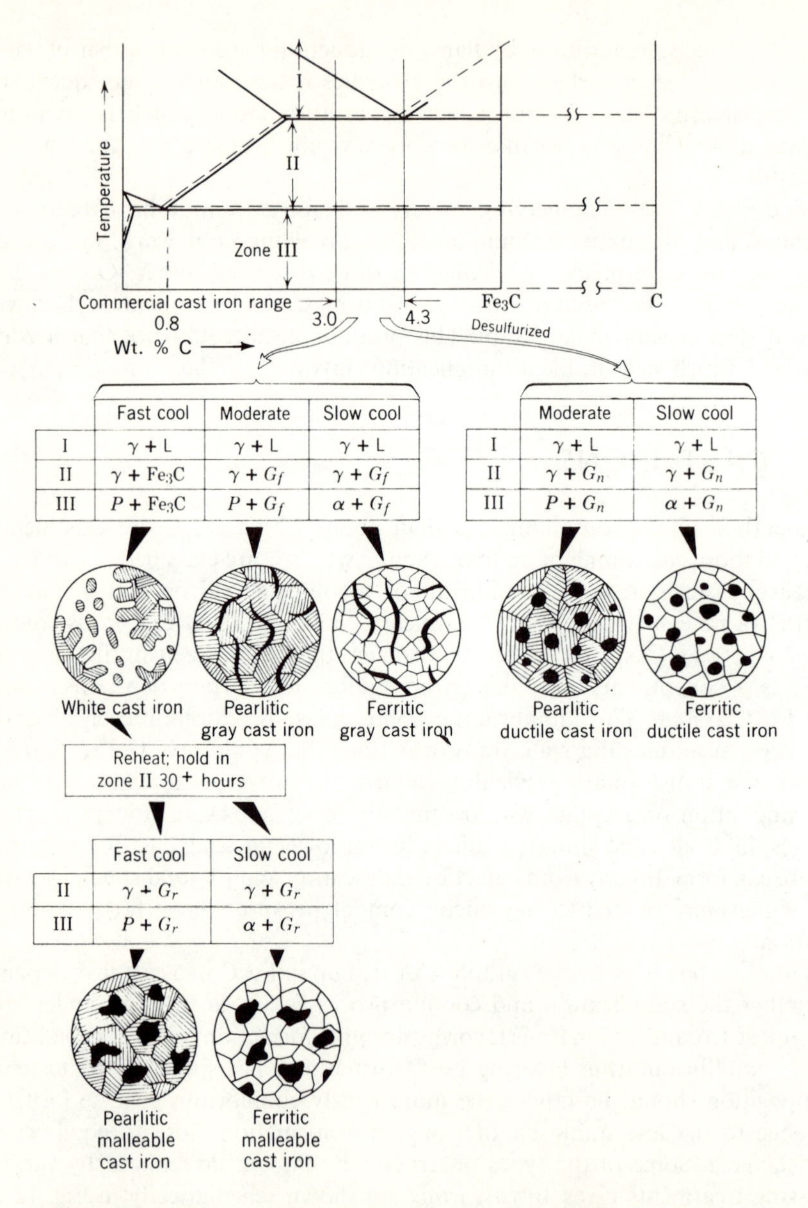

FIGURE 18.10 The iron-carbon phase diagram showing the composition range for commercial cast irons and schematic microstructures obtained by varying chemical and thermal treatment. [G_f = graphite flakes; G_r = graphite rosettes; G_n = graphite nodules; γ = austenite; P = pearlite; α = ferrite.] (Adapted from *The Structure and Properties of Materials*, Vol. I, by W. G. Moffatt, G. W. Pearsall, and J. Wulff, Wiley, New York, 1964.)

conditions, the graphite is present as flakes, but if the melt is desulfurized by "innoculation" with cerium or magnesium, the graphite will be present as nodules. The normal structure is called ferritic gray cast iron (see Fig. 1.2), and that for the innoculated material is called ferritic nodular (or ferritic ductile) cast iron.

With moderate cooling rates, graphite still forms in the eutectic reaction, but Fe_3C is one product of the eutectoid reaction. The final structure, consisting of graphite flakes (which are more or less interconnected locally) dispersed in a pearlitic matrix, is called pearlitic gray cast iron or, if innoculated, pearlitic nodular (or pearlite ductile) cast iron. The gray cast irons (ferritic or pearlitic) are noticeably less ductile than the nodular cast irons as a consequence of the difference in graphite particle shapes since the graphite flakes act as nuclei for cracks and promote premature mechanical failure. On the other hand, this same factor accounts for the good machinability of gray cast irons.

White cast irons, formed by rapid cooling, or by decreasing the silicon content to about 0.5%, have structures similar to steels in that no graphite is present. Thus, austenite and Fe_3C are present in zone II, and pearlite and Fe_3C are present in zone III. White cast irons are very brittle, reflecting the brittleness of their massive Fe_3C constituent and, accordingly, are used directly only in applications such as grinding ores in which this feature is not deleterious. White cast irons can be converted into more usable structures by heating into zone II for fairly long

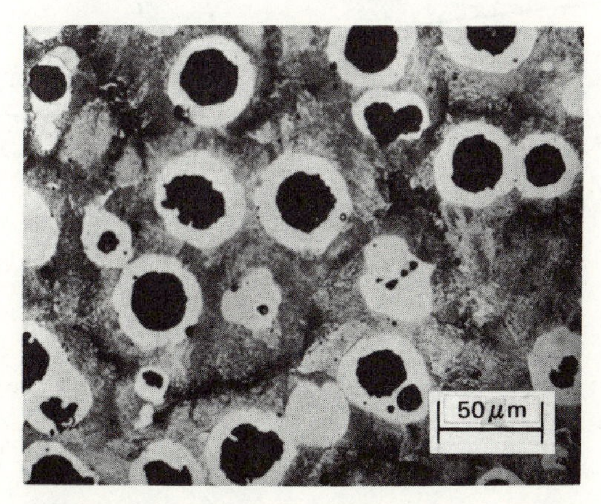

FIGURE 18.11 A mixed ferritic–pearlitic nodular cast iron. The graphite nodules are surrounded by free ferrite in a pearlitic matrix (the mottled area is pearlite; the magnification used is insufficient to resolve the individual ferrite and cementite lamellae in the pearlite). (Reproduced by permission from *Metals Handbook*, 8th Edition, Vol. 7, American Society for Metals, 1972.)

periods of time. This treatment causes graphite, in the shape of "rosettes," to precipitate and the cementite is eliminated. Pearlitic malleable and ferritic malleable cast irons produced in this manner are reasonably ductile, but the processing is more involved than that necessary to produce nodular cast irons.

It is important to note that cooling schedules normally encountered in practice render the above classifications (i.e., "slow," "moderate," and "rapid") somewhat arbitrary. Thus, cast irons often exhibit a mixture of the various microstructures noted above (Fig. 18.11).

18.6 PRECIPITATION PROCESSES

Steels are but one example of alloys that are particularly useful because of the existence of nonequilibrium phases and microconstituents. Because of the sluggish nature of solid state reactions that require diffusion, nonequilibrium microstructures can persist over extremely long periods of time, especially at room temperature. This "kinetically induced metastability" also is found with most high-strength aluminum alloys. The latter, on a microstructural level, consist

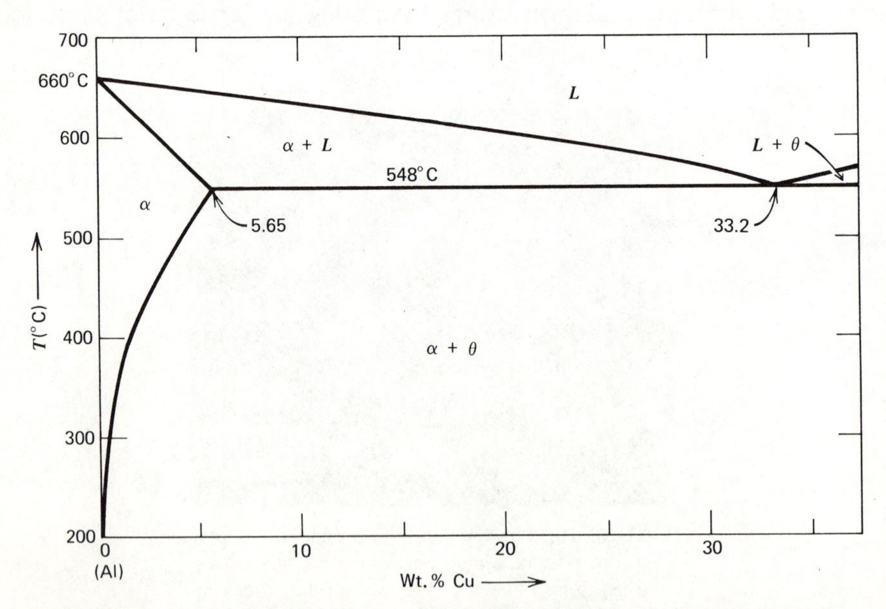

FIGURE 18.12 The Al-rich portion of the Al–Cu phase diagram. In this system, alloys containing less than 5.65 wt. % Cu can be heated into the single-phase α region (solutionized). Upon quenching to room temperature, θ precipitation is suppressed; subsequent heat treatment in the two-phase region results in the formation of a fine precipitate, leading to increased strength of the alloy.

of very fine, metastable precipitate particules dispersed within an aluminum-rich matrix phase. A variety of such useful precipitation-hardening aluminum alloys, containing a number of alloying elements in differing proportions, exist, but the thermal processing employed to effect strengthening can be illustrated with reference only to aluminum-rich Al–Cu alloys (Fig. 18.12).

A necessary (but not sufficient) condition for an alloy to be capable of being hardened by precipitation is that the solubility of the alloy element in the base material increase with increasing temperature, and the solvus line between the α field and the $\alpha + \theta$ two-phase region in the Al–Cu phase diagram meets this criterion. If an alloy containing less than 5.65 wt. % Cu is heated into the single-phase α region (solutionizing treatment) and then quenched rapidly, it is possible to prevent the precipitation of any θ phase particles. That is, the supersaturated solid solution can be retained at room temperature. Upon subsequent heating (aging treatment), a copper-rich phase will precipitate from the supersaturated α. With Al–Cu alloys, the approach to equilibrium is indirect in that the appearance of equilibrium θ precipitates is preceded by formation of copper-rich clusters in the α and by nonequilibrium or transition precipitates that can be seen only with the aid of an electron microscope. These intermediate stages in the precipitation sequence give associated mechanical strengths that are superior to those of the initial supersaturated solid solution and to those of the final equilibrium $(\alpha + \theta)$

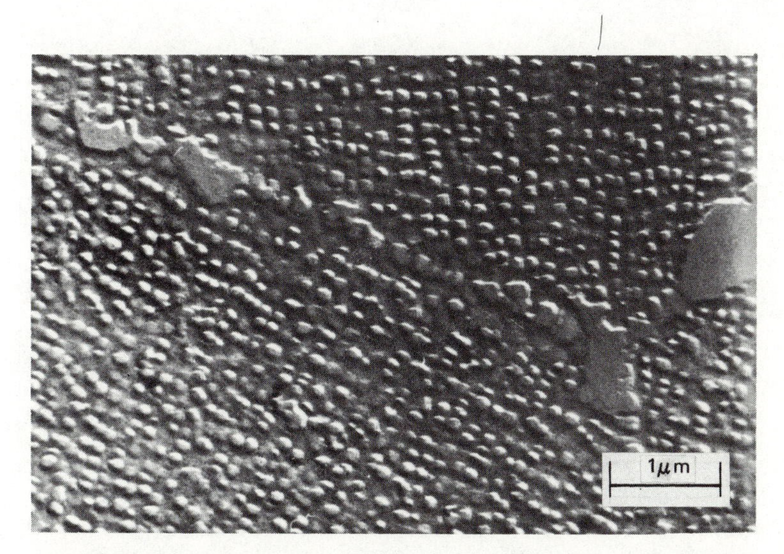

FIGURE 18.13 Replica electron micrograph of age-hardened Inconel X-750, a heat resistant nickel-base alloy. Hardening is provided by the small precipitate particles of $Ni_3(Al,Ti)$ dispersed in the matrix. The larger particles shown are a carbide that formed at the grain boundaries. (Reproduced by permission from *Metals Handbook*, 8th Edition, Vol. 7, American Society for Metals, 1972.)

two-phase mixture. An intermediate structure, with its attendant improved properties, can be retained simply by cooling the alloy from the aging temperature. It is important to note that the improved strength resulting from precipitation (or age) hardening is due to the very fine dispersion of the precipitate and, therefore, is not directly related to their nonequilibrium nature. That is, the additional sufficient condition alluded to earlier is that the precipitate that is formed be extremely fine. With some precipitation-hardening alloys, therefore, a fine dispersion of equilibrium precipitates achieves the same purpose. Examples include the "maraging steels" (Fe–Ni–Mo–Ti alloys), precipitation-hardening stainless steels, nickel- or cobalt-base alloys used for high temperature service (Fig. 18.13), and many two-phase $\alpha + \beta$ titanium alloys.

The appearance of transition precipitates is not predicted by the phase diagram, of course. Their occurrence, however, is proof that the free energy of the alloy at an intermediate stage of aging is lower than that of the original supersaturated solid solution. Whether or not the equilibrium state is preceded by a transition state mainly depends on certain kinetic factors that were discussed in Chapter 17. For the situation shown in Fig. 18.14*a*, a transition precipitate does form, whereas the equilibrium phase forms directly for the situation illustrated in Fig. 18.14*b*.

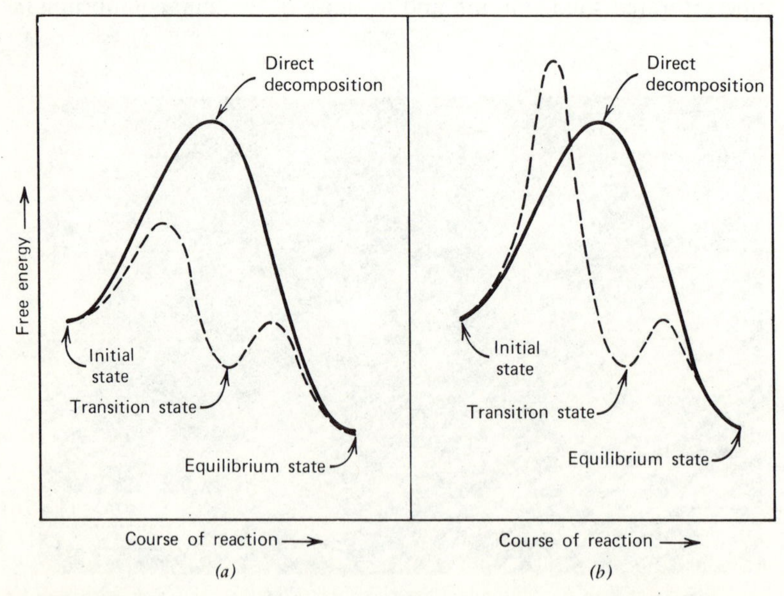

FIGURE 18.14 Schematic free energy versus course of reaction curves for the initial, final, and transition states in precipitation. (*a*) A transition precipitate can form initially because a lower activation energy barrier is associated with this process. (*b*) The initial state will decompose directly to the final equilibrium state because a lower activation energy barrier is associated with direct decomposition.

When transition precipitates are found, it means there is a lower activation energy for their formation than for the formation of equilibrium precipitates. Physically, this comes about because of definite structural similarities between the precipitate and the matrix, which result in a low value of the interfacial or strain energy terms that enter into the activation free energy. Quite often, a low value of interfacial energy is due to atomic plane matchup or coherency between the precipitate particle and the matrix.

Although considerations of strength do not exhaust the effect of microstructure on properties, in the case of metals, strength is particularly important in practice as well as particularly dependent upon microstructure. Virtually anything that impedes dislocation motion in a metal will strengthen that metal. Thus, microstructural alteration allows control of strength and other properties.

18.7 STRUCTURE AND PROPERTIES OF CERAMICS

The microstructure of ceramic materials profoundly affects their properties. The properties of crystalline ceramics produced by pressing and sintering, for example, depend strongly on the crystallite grain size and amount of residual porosity. Other ceramics, produced by vitrification, are two-phase and consist of crystalline particles embedded in a glassy matrix. While more economically produced than most crystalline ceramics, these materials have inferior mechanical properties in comparison. Such two-phase materials are especially prone to thermal shock. This arises because the individual phases have different thermal expansion coefficients [i.e., they expand (or contract) at different rates as the temperature increases (or decreases)]. On cooling from a high temperature, therefore, the different contraction in each phase produces internal stresses greater than those that would exist if the structure consisted only of one phase, and the possibility of thermal shock is increased.

Although inorganic glasses possess inherently high strengths, due primarily to the combination of strong covalent bonding and noncrystallinity, the presence of microscopic surface cracks leads to actual strengths that are rather low. Control of these surface cracks is virtually impossible since they develop as a result of chemical attack or abrasion. Nevertheless, certain chemical and thermal treatments, which cause the surface layers to be in residual compression, increase the strength of a glass. Composites consisting of glass and a polymer also are used either to render the glass resistant to shatter (e.g., safety glass) or to make use of its inherently high strength (e.g., fiberglass). In the latter case, glass fibers are embedded in a polymer matrix that protects them and prevents surface crack formation.

The extreme brittleness of inorganic glasses does not preclude applications. As pointed out previously, they are an important part of ceramics prepared by vitrification. Furthermore, the combination of high transparency and reasonable strength is advantageous for many uses. Glasses also are employed as glazes on the surface of ceramic bodies and as decorative enamels on steel. In either of these cases, high transparency is undesirable, and microstructural control provides a means of opacification and coloring. Controlled nucleation and growth of crystals below the equilibrium melting point produces opacification of enamels, for example. Most such enamels are borosilicate glasses containing small amounts of titania (TiO_2), which is soluble in the liquid. After the enamel is applied as a surface coating on the metal, it is fired at a relatively low temperature in order to precipitate small crystals of TiO_2. The firing temperature must be such that precipitation occurs in a reasonable length of time yet produces sufficiently fine titania particles (about 0.5 μm across) so that efficient opacification results. In effect, the fine particles scatter light to such an extent that the glass loses its transparency.

Controlled nucleation and growth also are employed in the case of crystallized-glass ceramics. These are fully dense crystalline solids characterized by a very fine

FIGURE 18.15 The microstructure of Pyroceram, a crystallized-glass ceramic, after crystallization. The very fine-grained microstructure is a result of controlled crystallization of a glass containing properly dispersed nucleating agents. The average grain diameter is 0.2 μm. (Photomicrograph courtesy G. B. Carrier, Corning Glass Works.)

grain size (Fig. 18.15) and by mechanical properties that are superior to most other polycrystalline ceramics and to glass. Such substances start off as viscous glasses and, thus, can be shaped by deformation processes. Compositions have been adjusted and small amounts of nucleating agents have been added so that, subsequent to shaping, firing at a lower temperature accomplishes widespread nucleation of crystallites and, then, increasing the temperature somewhat produces a growth rate sufficient to yield a porefree, fully crystalline product. The ensuing microstructure yields excellent strength and very good thermal shock resistance.

18.8 STRUCTURE-PROPERTY RELATIONSHIPS IN POLYMERS

Polymers, like ceramics, can exist in either the noncrystalline state, the crystalline state, or as a two-phase crystalline-noncrystalline mixture. The factors affecting whether or not a particular polymer crystallizes depend on the basic molecular architecture of the polymer and have been discussed in Chapter 6. For a given polymer, the degree of crystallinity affects properties, especially mechanical properties. For example, the crystalline regions in a slightly crystalline polymer serve much the same function as cross-links since they are well bonded, by way

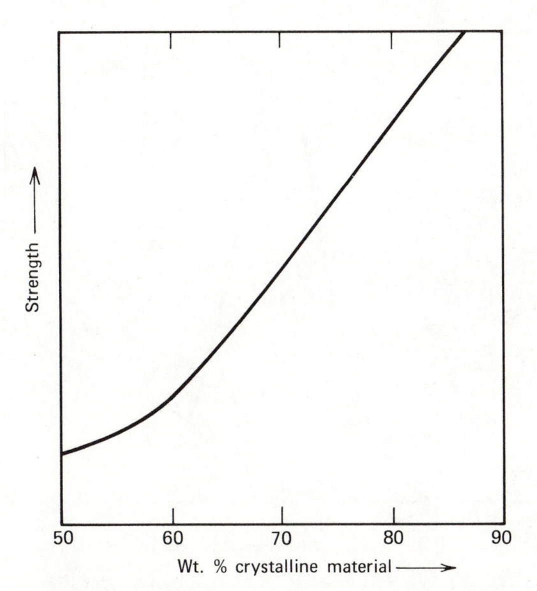

FIGURE 18.16 The strength of polyethylene shown schematically as a function of crystallinity. The strength increases with increasing crystallinity.

of polymer chains, to the amorphous regions. With higher degrees of crystallinity, the more rigid crystallites can directly support much of the applied load because they are interconnected to each other, and this leads to a stiff and rigid structure. Highly crystalline polymers deform very similarly to other crystalline materials. For pliable polymers, that is, polymers above their glass transition temperature, the degree of crystallinity is especially important in determining mechanical properties, as is illustrated schematically for polyethylene in Fig. 18.16.

Although the tendency for crystallization is determined largely by molecular architecture (see Sect. 6.4), the formation of crystalline regions in a polymer that does crystallize depends on the rates of crystal nucleation and growth, and the overall crystallization rate is a maximum at a temperature intermediate between the equilibrium melting temperature and the glass transition temperature (Fig. 18.17). Furthermore, the final crystallite size depends on the crystallization temperature; small spherulites correspond to lower crystallization temperatures (higher nucleation rates but lower growth rates), Fig. 18.18. As with metals and ceramics, a small grain size in polymeric materials usually is associated with improved mechanical properties, especially from the point of view of fracture and, as noted, this feature can be controlled by processing.

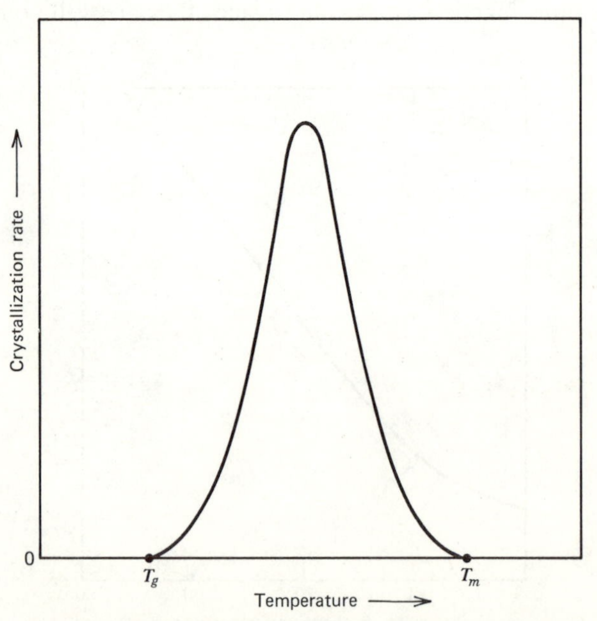

FIGURE 18.17 Schematic diagram showing that the rate of polymer crystallization reaches a maximum at a temperature between the equilibrium melting temperature and the glass transition temperature.

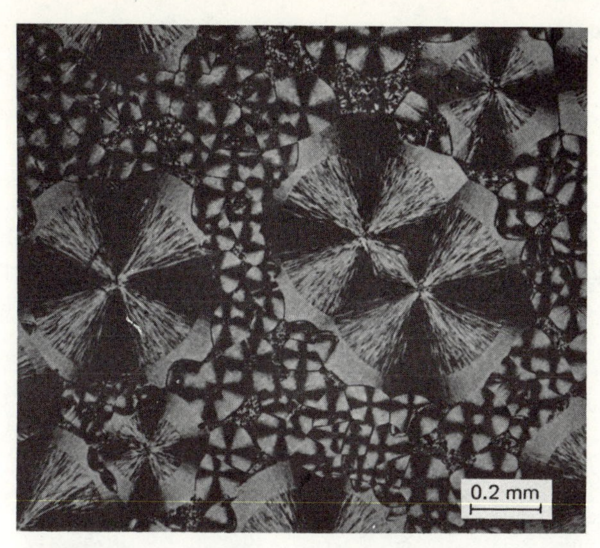

FIGURE 18.18 Spherulites in a siliconelike polymer. The large spherulites were formed by crystallization at a higher temperature than the small spherulites. (Photomicrograph courtesy of F. P. Price, General Electric Co.)

Addition of other substances to polymers markedly alters their structure and properties. The added substance may either dissolve into the base polymer or isolate itself into small regions. Alternatively, in the case of copolymers, different repeating units become integral parts of the basic molecular chains during the polymerization process. Usually, copolymer chains have decreased regularity as compared to homopolymer chains, and copolymers tend to be less crystalline than the respective homopolymers. The glass transition temperature of a copolymer usually is some suitable average of the glass transition temperatures of the homopolymers but, unless the individual repeating units are very similar structurally, the equilibrium melting temperature is decreased on copolymerization. The variation of T_m and T_g for varying degrees of constituents in a copolymer system may be summarized in a diagram of the type shown in Fig. 18.19. In some respects this figure is similar to a phase diagram but, unlike a true equilibrium diagram, it cannot be used to determine the relative amounts of the various phases present because each composition corresponds to the composition of individual molecules as polymerized, and this does not vary with changes in state. As noted in Fig. 18.19, the range of temperature between T_m and T_g is decreased by copolymerization and, in the system illustrated, there exists a range of copolymer compositions over which the polymer cannot be crystallized at all, so that the

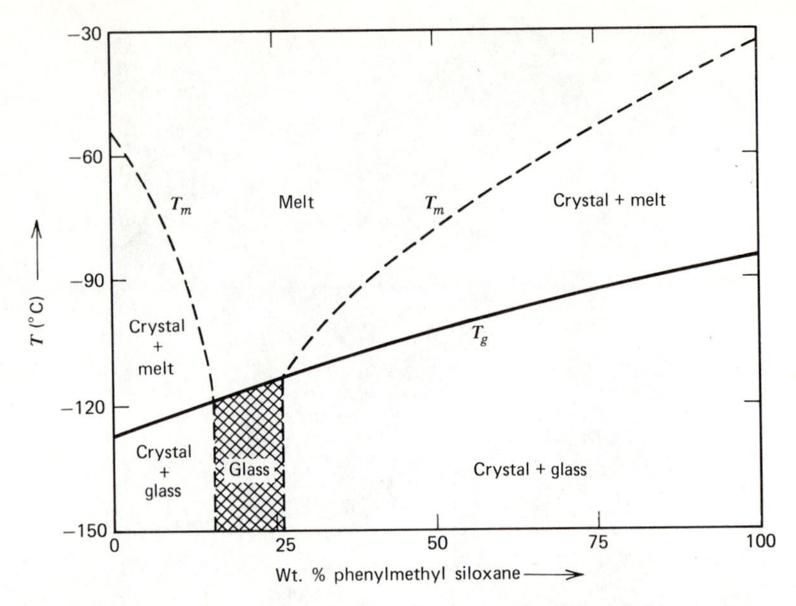

FIGURE 18.19 The effect of copolymerization on the glass transition temperature and equilibrium melting temperature in the phenylmethyl siloxane—dimethyl siloxane system. The glass transition temperature of these random copolymers is a suitable weighted average of the glass transition temperatures of the individual homopolymers; the melting temperature and the degree of crystallinity decrease with increasing amounts of the second constituent. This gives rise to a region (shaded) in which the polymer goes directly from a liquid to a completely amorphous, glassy polymer. [Regular copolymers, block copolymers, and graft copolymers (Fig. 6.3) often exhibit behavior that differs from the above.] [After K. E. Polmanteer and M. J. Hunter, *J. Appl. Polymer Sci.*, **1**, 3 (1959).]

copolymer changes directly from a liquid to an amorphous glassy substance on cooling.

Plasticizers, nonvolatile solvents added to polymers, are distinguished from a copolymer constituent in that they do not become an integral part of the polymeric chains, but rather remain as isolated dispersed molecules; the result is therefore a molecular solution. Plasticizers are restricted to use with amorphous polymers and certain semicrystalline polymers with a low degree of crystallinity because the smaller molecules of a plasticizer are not soluble in crystalline regions but do dissolve in noncrystalline regions. Plasticizers lower the polymer glass transition temperature and, thus, are used to make plastics more pliable and to afford greater control over properties. Polyvinyl chloride polymers are compatible with a wide range of plasticizers, and the properties of this class of material can thus be altered greatly by varying the type and amount of plasticizer. Commercial plasticizers

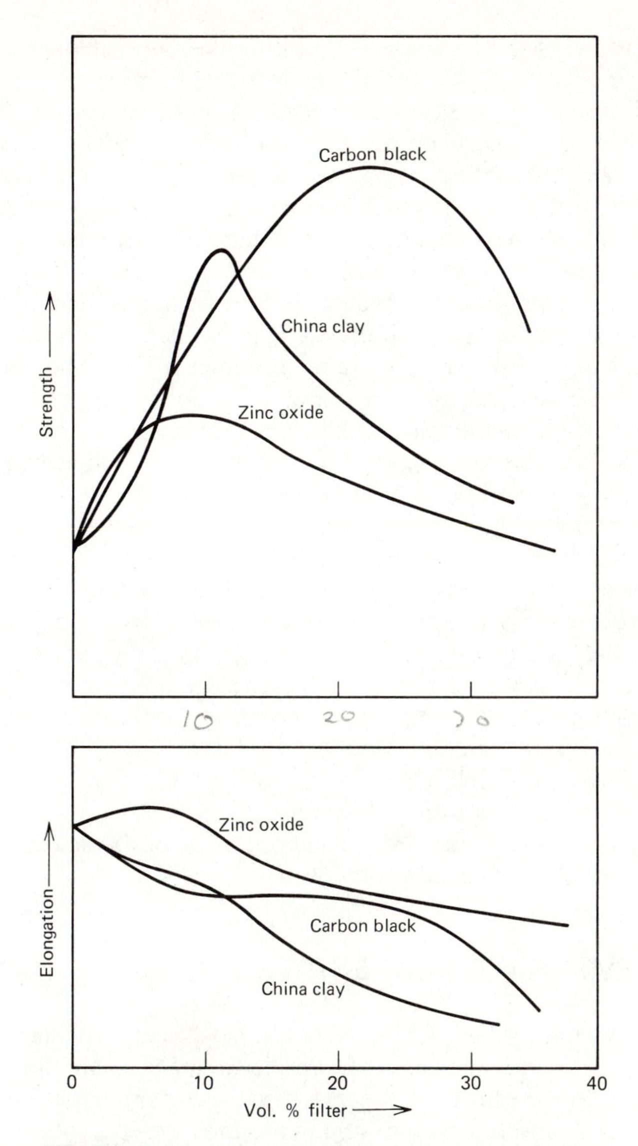

FIGURE 18.20 Schematic effect of the volume fraction of various fillers on the strength and the elongation at failure of natural rubber. Small amounts of filler increase strength without seriously reducing elongation. Above an optimum amount of filler, which depends on the filler characteristics, strength decreases.

include phthalate esters, adipates, azelates, oleates, and sulfonamides, as well as many other organic substances. The properties of a plasticized polymer may degrade with time if the plasticizer escapes to the surroundings.

Fillers, which are particulate additives with sizes typically in the range of thousands of nanometers on up, are also added to many polymers. In proper amounts, fillers significantly strengthen many plastics and are particularly beneficial to the properties of elastomers. In order for a small fraction of filler to function effectively, the filler must bond to the polymer and, by so doing, behave in a manner similar to a large cross-link. For a high volume fraction of filler, the filler itself carries part of an applied mechanical load. Typical fillers in plastics are asbestos, mica, quartz, and glass. Fillers also are used to improve other properties; quartz fillers, for example, generally enhance resistance to moisture penetration and mica fillers can improve the electrical insulation properties of polymers. Fillers may also serve to increase the toughness and flexibility of glassy and network polymers. Thus, elastomeric dispersions are commonly added to high-strength thermoplastics for the purpose of imparting ductility and toughness.

Perhaps nowhere is the importance of fillers as great as in the case of elastomers. Carbon black (graphite having particles of colloidal size, about 10 nm) is the most common filler used in elastomers. In combination with cross-linking by vulcanization, carbon black is responsible for the properties of the tough rubbers in present use. The addition of too much filler can result in a degradation of properties since, beyond the optimum filler addition, the filler merely acts as a diluent and, consequently, properties deteriorate (Fig. 18.20). Although particulate additives can increase the strength of metals as well as elastomers and other polymers, the most efficient means of increasing strength is when the second-phase additive is a high strength fiber or filament. Such "fiber composites," to be discussed in Sect. 18.10, can exhibit excellent mechanical properties.

18.9 COMPOSITE MATERIALS

Polymers containing fillers are but one example of composite materials. In the broadest sense, a composite may be considered to be a multiphase material exhibiting certain properties that are superior to those of the individual constituent phases. From this viewpoint, cast irons and precipitation-strengthened aluminum alloys can be classed as composites. By convention, however, the commonly accepted definition of a composite is a material whose structure is produced by an artificial, rather than naturally occurring, blending of two or more phases. Using this definition, we see that a carbon-black reinforced elastomer is a composite, whereas the metallic alloys mentioned above are not.

The properties of a composite often are related directly to the geometrical arrangement of the phases. Three of the most common types of phase arrangement

—particulate, fibrous, and lamellar—are illustrated schematically in Fig. 18.21. In the particularly important case of fibrous composites, the fibers can be arrayed randomly, as shown in Fig. 18.21, or can all be aligned in one direction. Generally, but not always, the minor phase by volume fraction consists of the particles in a particulate composite or of the fibers in a fibrous composite.

Most polymers containing fillers are particulate composites. Particulate composites are also manufactured from ceramic–metal combinations (cermets), and WC–Co alloys are prototypes of this class. This composite and others involving different refractory metal carbides are used widely as cutting tools for hardened steels. The suitability of these materials for such a purpose is due, in part, to the

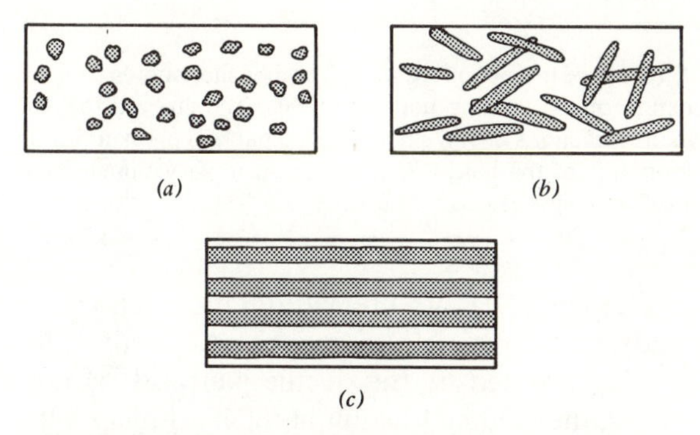

FIGURE 18.21 Phase arrangements in various types of composite materials: (*a*) particulate, (*b*) fibrous, (*c*) lamellar. In fibrous composites, the fibers may be arranged randomly in space as in (*b*) or the long axes of the fibers may all be aligned in one direction.

arrangement of the individual phases, which consists of carbide particles dispersed in a metallic matrix; the hard, brittle carbide particles serve as cutting edges, while the ductile metal matrix isolates them, thereby preventing easy crack propagation and brittle fracture. In monolithic form, such carbides cannot withstand the stresses that the cermet is capable of supporting. Besides cutting applications, certain cermets are useful as die materials for many metalworking operations because they possess good abrasion resistance and high stiffness.

The mechanical properties of a cermet such as WC–Co depend on the volume fractions of the particulate and matrix phases (Fig. 18.22). As with fillers in polymers, optimum properties are achieved at some intermediate volume fraction of the particulate phase. For cermets, however, the decrease in strength at high volume fractions of the brittle constituent is associated with continuity of the brittle phase (i.e., the ceramic particles contact each other rather than being isolated by

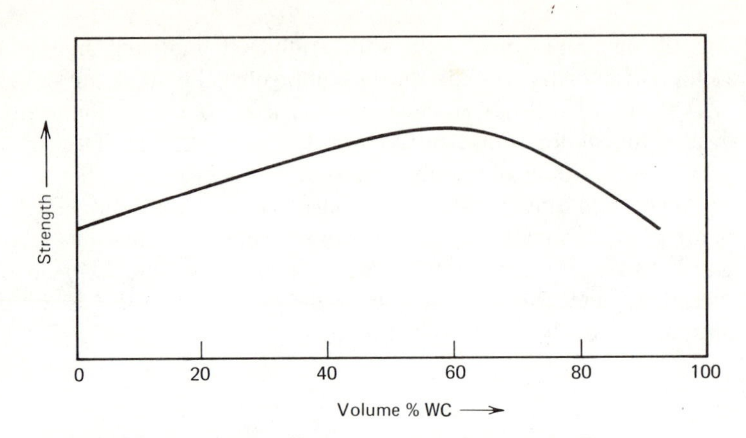

FIGURE 18.22 Strength of WC-cobalt composites shown schematically as a function of the volume fraction of WC. The strength decreases at higher volume fractions of WC since the material fails prematurely because of the continuity of the hard WC particles. At lower volume fractions of WC, the WC contributes less to the strength.

the matrix). Such a geometry favors premature failure since a crack formed in one tungsten carbide particle is able to propagate directly through other WC particles without being arrested by the ductile matrix. It is important to note that, in actual cermets, the volume fraction of the hard phase often exceeds 90% in order to maximize abrasion resistance. For most particulate composites, a significantly smaller volume fraction of reinforcing phase is used.

Dispersion-strengthened alloys represent another important type of particulate composite. In these, the microstructure of precipitation-strengthened alloys is emulated by dispersing a few volume percent of very fine ceramic particles in a metallic matrix. Examples include TD nickel and TD nichrome, where TD stands for thoria-dispersed. The fine, hard particles of ThO_2 impede dislocation motion in the nickel or Ni–Cr solid solution matrix and, thus, serve much the same purpose as precipitate particles in a precipitation-strengthened alloy (Sect. 18.6). The strengthening effect is not so great, however, because it is not possible to form an artificial dispersion of particles as fine or as closely spaced as particles resulting from solid-state precipitation. Nevertheless, dispersion-strengthened alloys are useful for structural applications at elevated temperatures because their microstructures remain very stable, whereas precipitate particles in a precipitation-strengthened alloy will either dissolve or coarsen at elevated temperatures. Further discussion of this topic will be found in Chapter 20.

In general, fibers provide more effective reinforcement than compact particles. For example, the stiffness of glass reinforced with aligned Al_2O_3 fibers is considerably greater than that of glass reinforced with spherical Al_2O_3 particles. It is necessary to point out, however, that the production and manufacture of fiber composites, particularly those in which fibers are aligned in one direction, costs much more than the production and manufacture of particulate composites. Consequently, aligned fiber composites are utilized only when the advantages of increased strength or stiffness outweigh the greater expense.

18.10 FIBER COMPOSITES

Undoubtedly the most familiar example of fiber composites is glass-reinforced plastic. In such a case, the reinforcing fiber adds strength to the plastic. Reinforcement by brittle fibers, especially inorganic glass, is of particular interest because brittle materials seldomly are employed in monolithic form to support loads applied in tension. One reason for the use of glass fibers is that glass can be prepared readily in such a form and will exhibit strengths much greater than in the bulk. The greater strength of glass fibers arises partially as a result of geometric considerations. That is, the chance of a serious surface flaw being present decreases as the surface area and cross-sectional area become smaller. As mentioned previously, incorporation of glass fibers into a composite also protects their surfaces from subsequent chemical attack or mechanical abrasion, either of which would lead to reduced strength. The matrix or binder phase also serves to reduce the tendency for catastrophic failure, normally associated with brittle materials, because it usually is capable of blunting or arresting cracks arising from the failure of individual fibers. In order for fiber strengthening to be effective, the fiber length to diameter (aspect) ratio must be large. Furthermore, while it is true that greater reinforcement is attained with larger fiber lengths, it is generally true that if the fiber aspect ratio is greater than about 20 to 100, depending upon the composite, reinforcement approaching that expected for infinitely long fibers results (the Wonka effect). This situation arises because fibers are bonded to the matrix, and with increasing interfacial bonding, significant fiber strengthening is achieved with smaller fiber aspect ratios. With composites utilizing a polymeric matrix, there usually is no chemical bonding at the interface but rather relatively weak physical bonding and interlocking. Nevertheless, most advanced composites used currently consist of polymeric matrixes containing glass, boron, or graphite filaments. Such composites have low densities and, therefore, good ratios of strength to density; furthermore, they are decidedly less expensive to produce than metal-matrix composites, which are characterized by strong chemical bonding at the fiber-matrix interface. Indeed, the relative expense of metal-matrix composities renders them noncompetitive with polymer matrix composites, except for elevated temperature applications where metal-matrix composites provide the only alternative.

At present, the most common metal-matrix composite consists of boron filaments embedded in aluminum. Preparation of this composite involves either hot pressing an assembly consisting of alternate layers of boron filaments and aluminum sheets or sequentially flame spraying aluminum directly onto layers of boron filaments. Another class of metal-matrix composites, which is being studied currently, is that of naturally occurring or in situ composites containing aligned fibers. An example of this, with TaC fibers in a nickel alloy matrix, is shown in Fig. 18.23. The fiber alignment results from directional solidification of a eutectic alloy in the TaC–(Ni–Cr) system. Although it may appear that naturally occurring (or in situ) composites, such as eutectic solids, violate the conventional definition of composites as set forth near the beginning of Sect. 18.9, the necessity of a specific processing treatment, such as directional solidification, to produce an artificial microstructure is consistent with the definition.

Strong fiber-matrix bonding characterizes in situ composites, which have the further advantage of being processed rather simply (e.g., by directional solidification). Conversely, one drawback of in situ composites is that the volume fraction

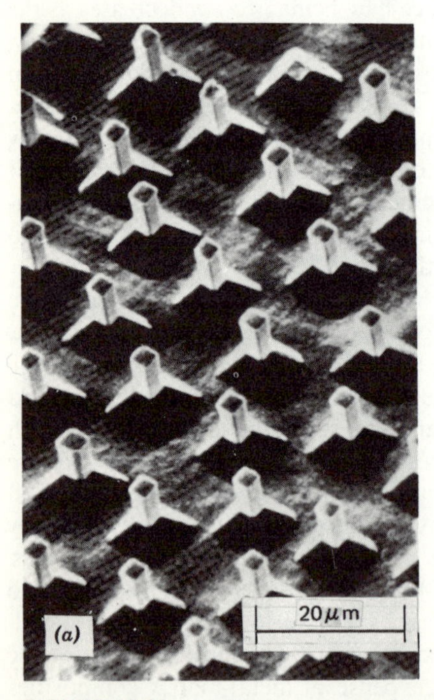

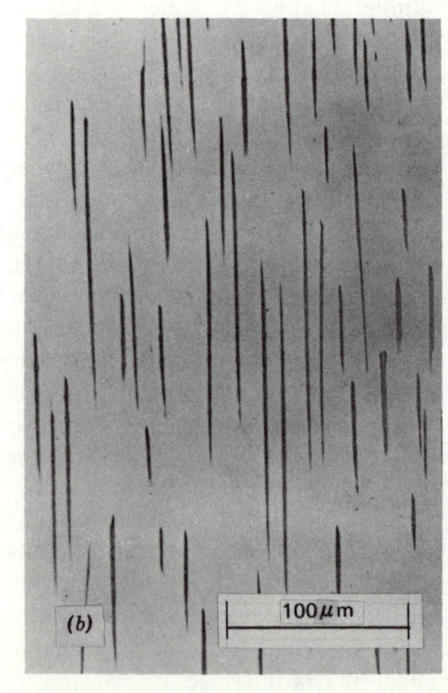

FIGURE 18.23 (*a*) Transverse and (*b*) longitudinal sections of the TaC–(Ni–Cr) directionally solidified fiber composite. Because the fibers weave in and out of the plane of polish, the lengths of the fibers are much greater than indicated in (*b*). (Photomicrographs courtesy of M. F. Henry, General Electric Co.)

of fibers appears to be limited to about 30%. For higher volume fractions of the second phase, a lamellar morphology grows preferentially. Because of this, recent work has been devoted to studies of aligned lamellar composites and to increasing the matrix contribution to the strength of fibrous composites.

Aligned fibrous composites are inherently anisotropic; their properties depend on the direction in which they are measured. Thus, maximum strength is achieved when an aligned composite is loaded along the fiber axis direction, whereas loading in other directions results in much lower strengths. In certain applications, such as turbine blades for a jet aircraft engine, high unidirectional strength is advantageous since loading occurs primarily in one direction. In a pressure vessel, on the other hand, biaxial loading occurs, and a special technique like cross-plying is necessary to produce composites suitable for such an application. If isotropic behavior is required or if the cost of producing an aligned fiber composite is excessive, then fibers can be oriented randomly within the matrix as they are in ordinary fiberglass and certain as-cast eutectics, even though greatly reduced strengthening is associated with the random alignment.

18.11 MACROSCOPIC COMPOSITES

A fine line of distinction often exists between an engineering structure whose members are connected together by other materials (e.g., lumber fastened by steel nails, in the framework of a house) and a macroscopic composite (e.g., plywood, which is composed of thin sheets of wood, having alternate grain directions, glued together). Nevertheless, macroscopic composites can be distinguished because their components are integral parts of the whole material. In addition to plywood, belted automobile tires and steel-reinforced concrete are commonplace, and such esoteric composites as graphite-reinforced epoxy golf club shafts are available commercially. The more mundane macroscopic composites are and will continue to be extremely important and useful for engineering purposes. The technology for producing them, the resulting properties, and the relatively low costs are very definite advantages.

Macroscopic composites and microscopic composites share a common basis in terms of structure-property relationships. Thus, concrete, consisting of an aggregate (usually a mixture of sand and gravel) held together by hydrated Portland cement gel (see Sect. 7.2), is the macroscopic counterpart of microscopic particulate composites. The fine and coarse portions of the aggregate comprise most of the volume (80 to 90%) and largely are responsible for the strength of concrete, providing a proper amount of water is mixed with the dry Portland cement. Too little water leads to an incomplete hydration reaction, and too much water results in excessive porosity, which lowers the strength of the resulting gel. In addition, high porosity permits subsequent water absorption, which, combined with freezing conditions in cold weather, can cause cracking and spalling. The amount of cement must be sufficient to coat aggregate particles, but the aggregate should occupy the majority of the volume because it is stronger than the hydrated cement and less expensive. Accordingly, fine aggregate usually constitutes about 33 to 40% of the *total aggregate volume* so that a good packing density of aggregate particles is achieved.

Concrete, like other brittle materials, is capable of sustaining only limited deformation prior to fracture. Steel bars or rods are incorporated into concrete in order to improve its load-bearing capability. Although the cement-steel bond is very weak, mechanical gripping and interlocking provide for effective load transfer to the steel reinforcing bars, and these macroscopic fibrous composites exhibit superior ability for supporting tensile loads as compared to unreinforced concrete. Indeed, the enhanced strength makes it possible to use lighter and less bulky members in structural applications.

Macroscopic composites illustrate the seemingly simple yet at the same time sophisticated approach to modern materials design engineering. Briefly, the intent of this approach is to produce a final structure with certain desired characteristics, and individual components are added, as necessary, to realize the objective. Glass-coated steels, solid state circuits for electronic purposes, and steel-belted radial tires are but the more common of many examples of this philosophy.

REFERENCES

1. Schlenker, R. B. *Introduction to Materials Science*, Wiley, Sydney, Australia, 1969.
2. Kingery, W. D. *Introduction to Ceramics*, Wiley, New York, 1960.
3. Smith, C. S. Materials, *Scientific American*, Vol. 217, No. 3, September, 1967, pp. 68–79.
4. Kelly, A. The Nature of Composite Materials, *Scientific American*, Vol. 217, No. 3, September, 1967, pp. 160–176.

QUESTIONS AND PROBLEMS

18.1 Microstructural examination of pure lead that has been deformed 50% by rolling at room temperature reveals an equiaxed grain structure. Explain this observation.

18.2 (a) The recrystallization temperature (for a 1 hour heat treatment) decreases with an increasing amount of prior cold work and reaches a fixed value at high amounts of prior cold work (Fig. 18.1). Justify this.

(b) With less than a certain amount of prior cold work, a metal cannot be recrystallized at any temperature (Fig. 18.1). Explain.

18.3 It is observed experimentally that the grain size, just after recrystallization is complete, decreases with an increasing amount of prior deformation (cold work). Explain this observation in terms of the effects of nucleation and growth on the overall transformation rate.

18.4 The grain size of copper can be refined (i.e., made smaller) only by heating after it has been cold worked. The grain size of iron, on the other hand, can be refined by heating without the necessity for prior cold work. Explain this difference in response.

18.5 A kinetic equation for grain growth at a constant temperature can be obtained by recognizing that the rate of grain growth is proportional ideally to the grain boundary energy per unit volume:

$$\frac{dD}{dt} \propto \frac{S}{V}\gamma$$

where D is the mean grain diameter, (S/V) is the grain boundary area per unit volume, and γ is the grain boundary free energy (per unit area). Since $(S/V) = (\text{constant}/D)$, the above differential equation can be written as

$$\frac{dD}{dt} = \frac{K}{D}$$

where K includes a thermal activation (or Arrhenius) factor. Solve the differential equation and explain briefly the time variation of D.

18.6 A fine dispersion of second phase particles can markedly reduce the rate of grain growth. Explain why this is so. (*Hint.* What happens to the total grain boundary energy of a metal when a grain boundary tries to move past a second phase particle?)

18.7 (a) Plot schematically a TTT diagram showing the start and finish curves for the recrystallization of a metal subjected to a fixed amount of prior cold work.
(b) Why does your diagram not exhibit C-curve kinetics? (Recall Sect. 17.4.)

18.8 In Fig. 18.5*a* there is a high-temperature range in which cementite does not form from the decomposing austenite. What is this temperature range and how is it related to the thermodynamics of Fe–Fe_3C equilibrium?

18.9 Justify why alloying elements shift the pearlite and bainite C-curves toward longer times in a TTT diagram.

18.10 In terms of microstructure, discuss why the mechanical properties of tempered martensite and bainite should be roughly comparable.

18.11 Which structure is more stable thermodynamically, pearlite or spheroidite? (*Hint.* Consider the interfacial free energy per unit volume of a lamellar versus a globular two-phase alloy.)

18.12 Sketch schematically the variation in hardness from the surface to the center for round bars of the following steels that have been quenched from the austenite region:
(a) A low-hardenability steel containing 0.6 wt. % C.
(b) A medium-hardenability steel containing 0.6 wt. % C.
(c) A high-hardenability steel containing 0.6 wt. % C.

18.13 Sketch schematically the variation in hardness from the surface to the center of a carburized and quenched steel rod that has a surface composition of 1.0 wt. % C and an interior composition of 0.2 wt. % C.

18.14 It is found empirically that the extent of softening due to tempering of martensite is almost a unique function of a tempering parameter, $P_H = T(\log t + C)$, where T is the absolute tempering temperature, t is the tempering time, and C is a constant for a given steel. Briefly restated, each hardness value corresponds to a specific value of P_H, regardless of what the particular temperature and time combination happens to be. Knowing that tempering depends on diffusion, justify the form of the tempering parameter.

18.15 Briefly describe a thermal process to produce a martensitic cast iron consisting of a martensite matrix that contains (a) graphite flakes, (b) graphite nodules, and (c) graphite rosettes. Which of these cast irons would be impractical? Why?

18.16 Compare white cast iron and as-formed martensite in terms of properties and in terms of what processing treatments are conducted to make these materials useful.

18.17 Discuss the similarities and differences between the tempering of steel and the age-hardening of Al–Cu alloys. Consider both structure and properties.

18.18 Cast age-hardening aluminum alloys seldom are solutionized at a temperature above the eutectic temperature. Explain why.

18.19 (a) Compare and contrast brittle behavior and ductile behavior.
(b) At normal temperatures, which types of materials (i.e., metals, ceramics, and polymers) are brittle and which are ductile? Qualify your answer if necessary.

18.20 Circular wheels containing abrasive grains (e.g., alumina, emory, or diamond) frequently are used to cut metals. The abrasive grains are bonded together by bronze or by a polymeric or vitreous material. Describe how such cutoff wheels could be made.

18.21 Give at least two reasons, in terms of processing and microstructure, why a crystallized-glass ceramic is preferable to a conventionally processed ceramic (e.g., one prepared by sintering).

18.22 Shown are schematic nucleation and growth rate curves for the crystallization of two different glasses as a function of temperature.

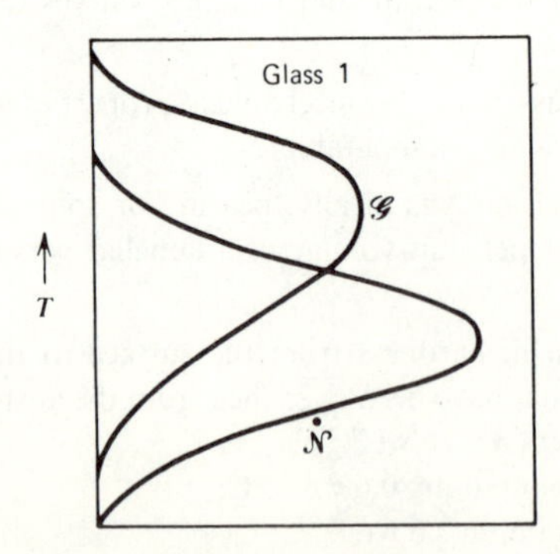

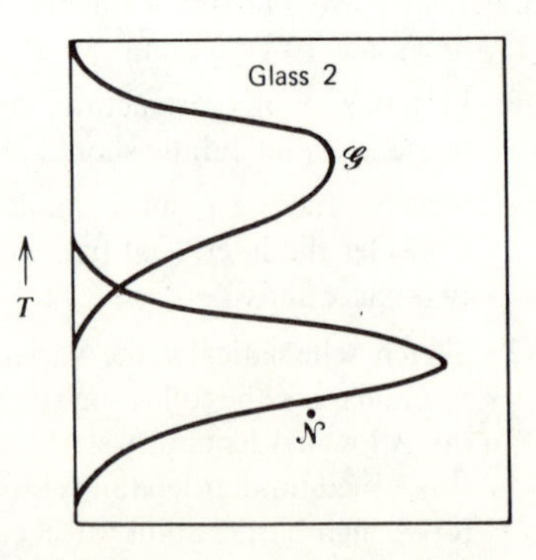

(a) Which glass could be converted into a *useful* crystallized-glass ceramic? Why?
(b) Briefly outline a heat-treatment schedule that would lead to good mechanical properties. Justify your schedule.

18.23 Explain how advantageous mechanical properties are attained by the cross-lamination process in the manufacture of plywood.

18.24 Suggest why a *mixture* of sand and gravel is used as the aggregate for concrete.

18.25 High speed steel (18W–4Cr–1V–0.36C, balance Fe) when quenched to form martensite retains about 5% austenite by volume. Describe briefly the type of tempering treatment you would use to eliminate the retained austenite, and describe the resulting microstructure.

18.26 Look up the Cu–Be phase diagram (*ASM Metals Handbook*, Vol. 8) and choose a Cu-rich alloy amenable to precipitation hardening. State how you would process a fine wire of this composition to attain a high strength.

19

Elastic Behavior

This chapter, dealing with reversible deformation, initiates our discussion of the properties of materials and their dependence on internal structure.

19.1 STRESS AND STRAIN

All solids undergo some change in shape when subjected to external forces or loads. If the shape change persists indefinitely after the load is removed, then the material has suffered permanent deformation. Alternatively, if the shape change is fully recovered after the load is removed, then the material has displayed elastic behavior. A metal coil spring provides examples of these two types of response. Elastic extension of the spring results from a relatively small applied force, since the spring reverts to its original length when the force is removed. A large applied force can cause the spring to be elastically extended and permanently stretched at the same time, so that the spring remains longer than its original length when the force is removed. In the purely elastic state, the extension of the coil spring ($l - l_0$) varies linearly with applied force P (i.e., $l - l_0 = kP$), and the spring constant, k, depends upon both the spring geometry and the properties of the spring material. Elastic properties provide the link in the relationship between what shape change a small element of the spring undergoes and how the force is transmitted through the element. Since a coil spring is unwieldy to consider, it is advantageous to choose a simpler example.

Consider the effect of hanging weights on the ends of the three aluminum rods illustrated in Fig. 19.1. If the loads are insufficient to cause permanent deformation, then the elongation, Δl, of each rod is directly proportional to the load, P, although the constant of proportionality differs for each rod. A single materials property is found to be the constant of proportionality for all three rods when the variables are expressed as relative elongation or strain, $\Delta l/l_0$, and intensity of force or stress, P/A_0:

$$\frac{\Delta l}{l_0} = \frac{l - l_0}{l_0} = \frac{1}{E}\frac{P}{A_0} \tag{19.1}$$

In this equation $E = 6.9 \times 10^{10}$ N/m^2($= 10 \times 10^6$ psi) is Young's modulus for aluminum. If the weights were placed on top of the three aluminum rods, then Eq. 19.1 would still apply, but both strain and stress would be negative.

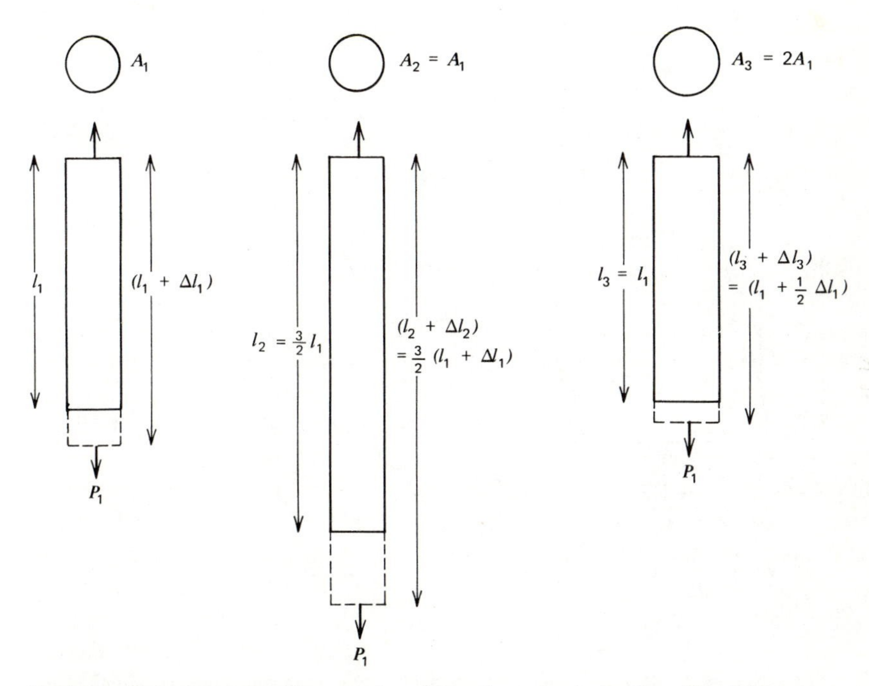

FIGURE 19.1 Three aluminum rods subjected to elastic loading. The initial cross-sectional areas are indicated at the top, and the initial lengths are shown to the left of each rod. Application of the same load to each rod causes length changes (dashed line portions). These length changes have been greatly exaggerated for the sake of clarity. The behavior of all three rods is described by Eq. 19.1; that is,

$$\frac{\Delta l_1/l_1}{P_1/A_1} = \frac{\Delta l_2/l_2}{P_1/A_2} = \frac{\Delta l_3/l_3}{P_1/A_3} = \frac{1}{E}$$

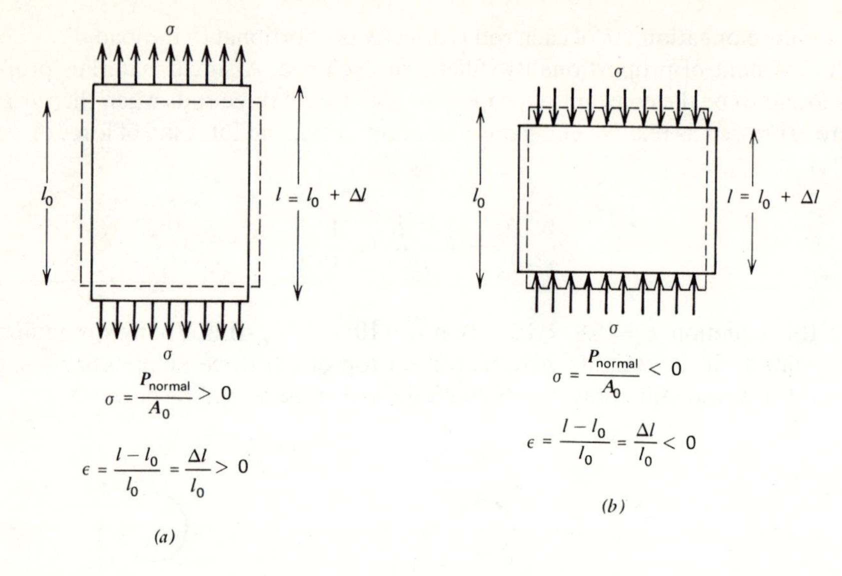

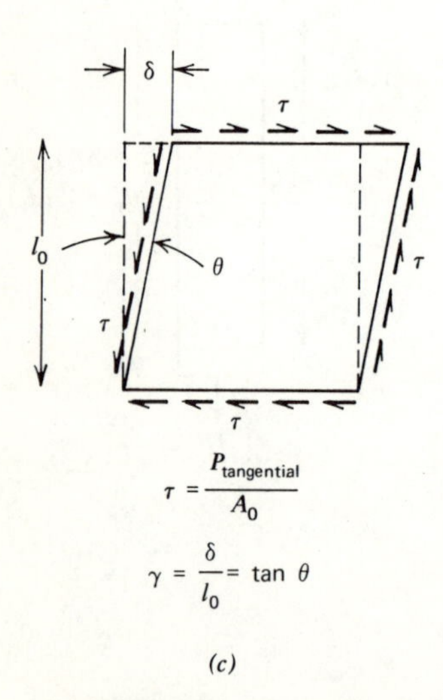

FIGURE 19.2 Types of stress and strain, illustrated for a solid cube as viewed from the side. The broken lines show the original shape before a stress is applied. The solid lines indicate the shape change: (*a*) a tensile stress causes a positive linear strain (as well as a lateral contraction); (*b*) a compressive stress causes a negative linear strain (as well as a lateral expansion); (*c*) a shear stress causes a shear strain. The lateral strains shown in (*a*) and (*b*) will be discussed in Sect. 19.3.

Since stress and strain are central to any discussion of the mechanical behavior of materials, these concepts shall be developed further. Stress is force per unit area [with units such as N/m^2 or lb/in^2 (psi)] acting on a surface, external or internal. For convenience, it is conventional to utilize stress components that are normal to and parallel to the area of interest, corresponding to normal and shear force components. Normal stresses (σ) are termed tensile stresses if they tend to pull a body apart or compressive stresses if they tend to push a body together (Figs. 19.2*a* and *b*). Tensile and compressive stresses are designated by positive and negative signs, respectively. Shear stresses (τ) tend to distort a body (Fig. 19.2*c*); their sign is a matter of arbitrary convention. Although forces are vector quantities with three orthogonal components, stresses are more complex (six components) and cannot be resolved vectorially.

Each type of stress has a conjugate type of strain that represents the relative shape change of a body; linear strain is conjugate to normal stress, and shear strain is conjugate to shear stress. Linear strains (ϵ), which represent fractional changes in length, can be tensile (positive) or compressive (negative) (Fig. 19.2*a*, *b*). Shear strains (γ) represent relative distortion, as shown in Fig. 19.2*c*. Although strains are dimensionless quantities, units such as in/in or m/m are sometimes used for linear strain as defined in Figs. 19.2*a* and *b*.

Strain may be the response of a material to an applied stress or, conversely, stress may be the response of a material to an applied strain. It is through the interrelationships between stress and strain that we shall initiate discussions of the mechanical properties of materials.

19.2 THE TENSION TEST

The tension test (Fig. 19.3) provides one convenient method for evaluating the mechanical behavior of materials. In one commonly used procedure, the specimen is pulled at a constant rate, and a load cell measures the tensile force P required to extend the specimen by an amount Δl, which is measured simultaneously by an extensometer. The resulting load-elongation curve can be readily converted to an engineering stress–strain curve with appropriate *scale factors*; that is, $\sigma = P/A_0$ and $\epsilon = \Delta l/l_0$, where A_0 is the initial cross-sectional area and l_0 is the initial gauge length between the arms of the extensometer.

Examples of tensile stress–strain curves for several different engineering materials are shown in Fig. 19.4. Rubber, Lucite, Bakelite, silica glass, and alumina are elastic all the way to the point of failure, in that if they were unloaded prior to failure, they would return to their original dimensions. Rubber, an elastomer with a characteristically nonlinear stress–strain curve, can have an extensibility of several hundred percent prior to failure. By comparison, the glassy polymer (Lucite), the network polymer (Bakelite), silica glass, and alumina are brittle

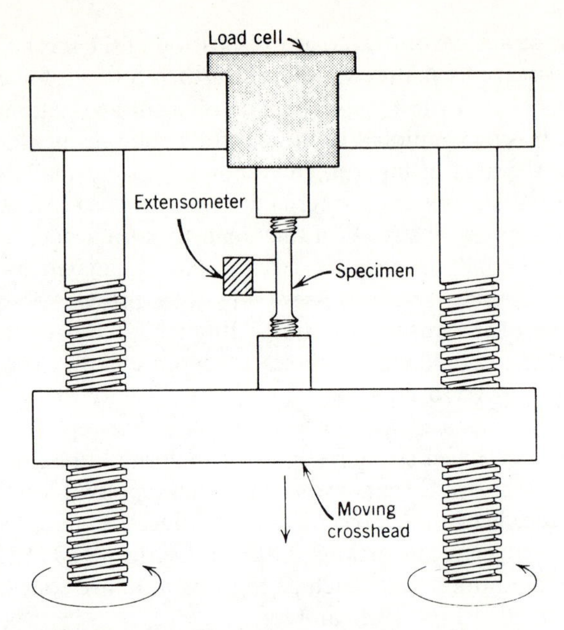

FIGURE 19.3 Schematic diagram of a tension test. The moving crosshead stretches the specimen, and the force required is measured by the load cell. The extensometer measures the extension, Δl, between two points along the gauge section. These raw data are converted to stress and strain to obtain the engineering stress–strain curve for the specimen. (Adapted from *The Structure and Properties of Materials*, Vol. III, by H. W. Hayden, W. G. Moffatt, and J. Wulff, Wiley, New York, 1965.)

materials that fail after only small strains and, except for Lucite, exhibit linear stress–strain curves. Polycrystalline aluminum and polyethylene display purely elastic behavior initially. After a very small amount of linear elastic elongation, polycrystalline aluminum begins to permanently deform by plastic flow. In the nonlinear portion of the stress–strain curve, which persists to failure, the strain is mainly plastic strain with a small amount of elastic strain. On the other hand, the more extensive initial nonlinear elastic response of polyethylene is followed by a combination of permanent deformation because of viscous flow and significant elastic strain up to the point of failure.

The remainder of this chapter will be devoted to elastic behavior, and the next chapter will treat those aspects of mechanical behavior that involve permanent deformation.

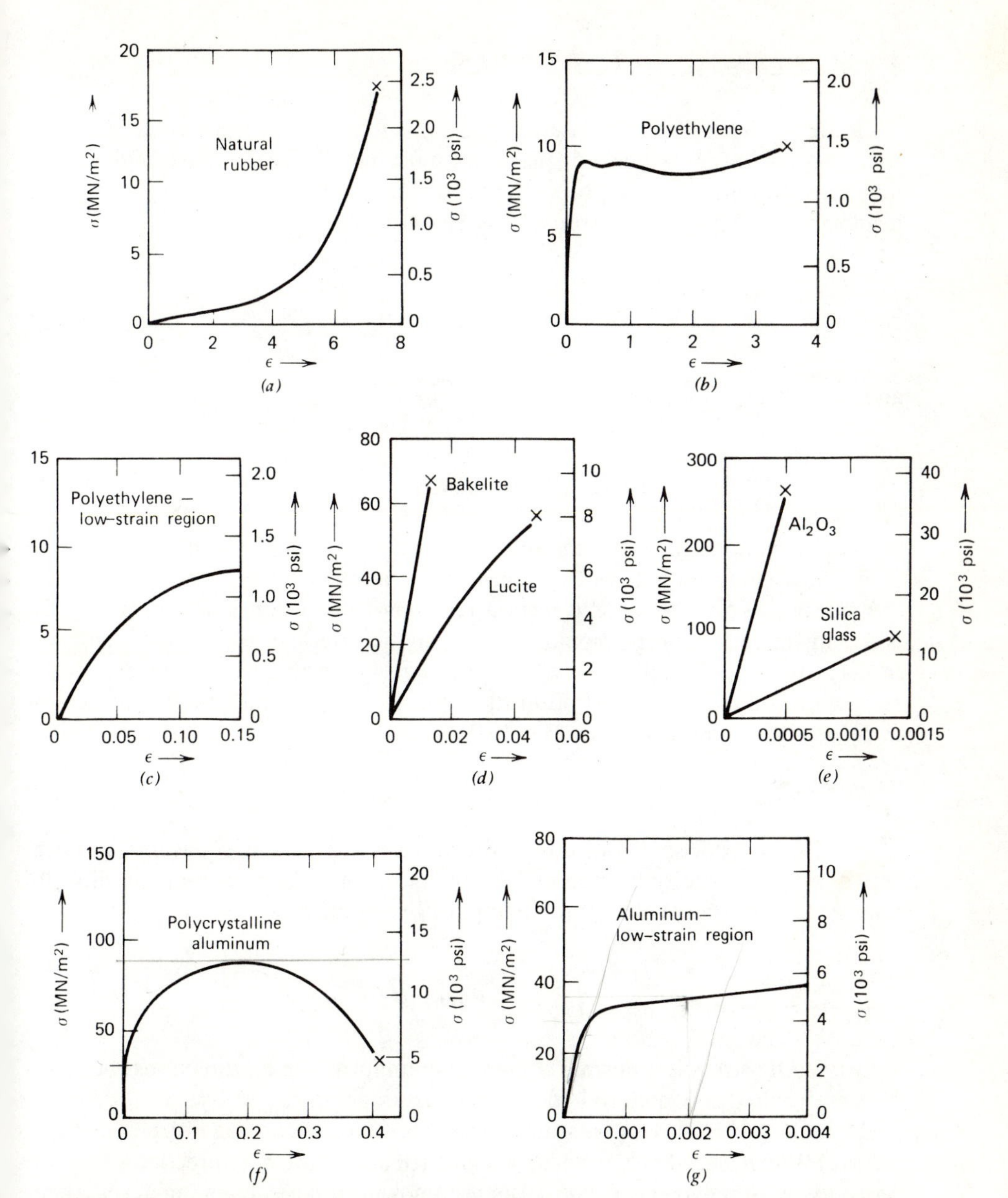

FIGURE 19.4 Engineering stress–strain curves for selected materials as determined by tension tests at room temperature. The point of fracture is denoted by ×. Note the different magnitudes of stress and strain in each case. (*a*) Natural rubber, an elastomer, (*b*) and (*c*) polyethylene, a ductile polymer, (*d*) Bakelite (phenol-formaldehyde), a network polymer; and Lucite (polymethyl methacrylate), a glassy polymer, (*e*) alumina, a crystalline ceramic; and silica glass, a ceramic glass, (*f*) and (*g*) polycrystalline aluminum.

19.3 ELASTIC PROPERTIES

Elastic properties are the materials constants that relate stress to elastic strain. In the case of linear elasticity, displayed by most metallic and ceramic engineering materials, strain is directly proportional to stress and the response is essentially instantaneous. This can be expressed by Hooke's law for a tensile or compressive stress as

$$\epsilon = \frac{1}{E}\sigma \tag{19.2}$$

and for a shear stress as

$$\gamma = \frac{1}{G}\tau \tag{19.3}$$

In these equations, E is the modulus of elasticity or Young's modulus and G is the shear modulus, both of which have stress units.

A linear strain in the axial direction resulting from a normal stress (Eq. 19.2) is accompanied by a proportional linear strain in a lateral direction that is smaller in magnitude and usually of the opposite sign. For example, an applied tensile stress causes a tensile elongation in the same direction and a lateral contraction in perpendicular directions (see Fig. 19.2*a*):

$$\epsilon_{\text{lateral}} = -\nu\epsilon_{\text{axial}} \tag{19.4}$$

where ν, called Poisson's ratio, is a property defined by this equation. Another elastic property, the bulk modulus K (or its reciprocal, the compressibility β), relates relative volume change to an applied hydrostatic stress, σ_{hyd}:

$$\frac{\Delta V}{V_0} = \frac{1}{K}\sigma_{\text{hyd}} = \beta\sigma_{\text{hyd}} \tag{19.5}$$

A hydrostatic stress is a normal stress acting uniformly in all directions; because of convention, it is related to hydrostatic pressure, p, as $\sigma_{\text{hyd}} = -p$.

Of the four elastic constants introduced above, only two are independent for an isotropic material (i.e., one in which properties are not directional). Consequently, if any two elastic constants are known for an isotropic material, then the others can be calculated through relationships such as

$$G = \frac{E}{2(1+\nu)} \tag{19.6}$$

and

$$K = \frac{1}{\beta} = \frac{E}{3(1 - 2\nu)} \tag{19.7}$$

Isotropy is common in inorganic glasses and in many polycrystalline materials; however, single crystals usually are anisotropic, as are polycrystalline solids whose grains have a preferred crystallographic orientation. The number of independent elastic constants necessary to relate stress and strain for an anisotropic solid ranges from 3 for cubic materials to 21 for triclinic materials, and the elasticity equations become more complicated.

19.4 BASIS FOR LINEAR ELASTICITY

The elastic strain suffered by a crystalline solid or an inorganic glass in response to an applied stress is simply a macroscopic manifestation of changes in bond lengths or in bond angles between atoms, ions, or molecules in the solid. Applied stress is transmitted through a solid by way of interatomic forces or internal stresses which, as a natural result of bonding, arise concurrently with interparticle distance changes or bond angle changes. Like the melting point, the magnitudes of E, K, and G depend on the strength of interparticle bonding. Values of Young's modulus for a number of different materials, including organic polymers, are listed in Table 19.1. Materials such as metals, with strong chemical bonds between atoms, have higher values of elastic moduli than do molecular solids, with weak van der Waals bonds between molecules.

The elastic response of a crystalline solid can be visualized with the aid of potential energy and internal stress curves as a function of interparticle separation (e.g., Fig. 3.4). Every crystalline solid possesses a characteristic potential energy curve with respect to volume (Fig. 19.5*a*), and the slope of this curve at any point is the hydrostatic stress necessary to change the volume of the solid relative to the equilibrium molar volume, V_0 (Fig. 19.5*b*). In turn, the slope of the hydrostatic stress curve at V_0 is related to the bulk modulus by $K = V_0 \sigma_{\text{hyd}}/\Delta V$ or, in terms of the energy-volume relationship, $K = V_0(d^2E/dV^2)_{V_0}$. For example, the relationship between potential energy and ionic spacing for an ionic solid (see Sect, 3.3) can be converted to an energy-volume relationship and K is found to vary as

$$K \sim \frac{U_0}{V_0} \tag{19.8}$$

TABLE 19.1 Young's Modulus for Isotropic Solids at Room Temperature, Melting Point, and Bond Type

Material	E (10⁶ psi)	E (10^{10} N/m²)	T_m (°C)	Bond Type
Aluminum	10	6.9	660	Metallic
α-Iron	30	20.7	1538	Metallic
Copper	16	11	1084.5	Metallic
Niobium	17	12	2471	Metallic
Tungsten	59	41	3387	Metallic
Osmium	80	55	3033	Metallic
Diamond	165	114	(>3800)	Covalent
Silicon	16	11	1414	Covalent
Titanium carbide (TiC)	45	31	3180	Metallic and covalent
Alumina (Al_2O_3)	58	40	2050	Covalent and ionic
Magnesia (MgO)	45	31	2900	Ionic and covalent
Rocksalt (NaCl)	5.3	3.7	800	Ionic
Silica glass	10	7	($Tg \sim 1150$)	Covalent and ionic
Phenol-formaldehyde (Bakelite)	0.7	0.5	—	Covalent
Hard rubber (ebonite)	0.6	0.4	—	Covalent
Polystyrene (atactic)	0.4[a]	0.3[a]	($Tg = 100$)	Van der Waals
Polyethylene (branched)	0.03[a]	0.02[a]	137	Van der Waals
Natural rubber	$\sim 10^{-4}$ to 10^{-3}	$\sim 10^{-4}$ to 10^{-3}	28	Van der Waals (plus covalent crosslinks)
Argon (solid)	0.5[b]	0.35[b]	−189	Van der Waals

[a] Relaxation moduli (see Sect. 19.5).
[b] At 0 K.

where U_0 is the binding energy per mole. The relationship, Eq. 19.8, although discussed in terms of an ionic solid is found to be generally valid for all crystalline solids.

In principle, Young's modulus (or the shear modulus) can be related to potential energy and internal stress in the same manner. This is shown in Fig. 19.6 for two materials, one with strong bonds and one with weak bonds. In both cases ϵ varies linearly with σ for small strains, but a nonlinear relationship exists at larger strains. Such nonlinear elasticity is observed rarely in strongly bonded solids because the applied stress reaches a sufficiently large value at small strains (typically $\epsilon < 0.01$) so that either fracture or permanent deformation intervenes. Only in a case such as fine metal whiskers of exceptionally high strength is nonlinear elastic behavior found. For weakly bonded molecular crystals, on the other hand, nonlinear

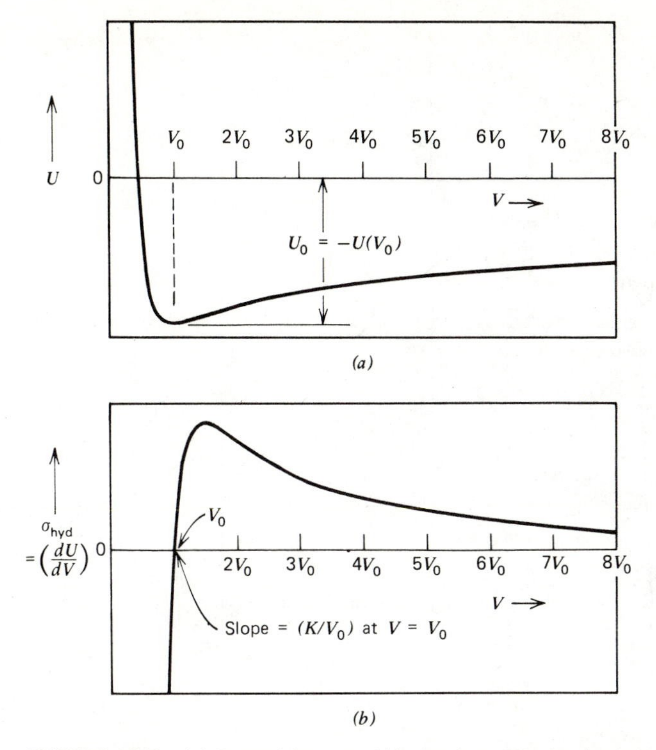

FIGURE 19.5 (*a*) Potential energy (U) as a function of volume for a crystalline solid. (*b*) The corresponding hydrostatic stress as a function of volume. This represents the applied hydrostatic stress that is just balanced by internal stresses resulting from interatomic forces at any volume. Note that the equilibrium volume, V_0, corresponds to the stressfree condition.

elasticity may be observed because of the greater curvature in the internal stress vs. strain relationship. Since bond strength decreases with increasing temperature, the elastic moduli also decrease with temperature. The variation of Young's modulus with temperature is shown for a number of common materials in Fig. 19.7.

Poisson's ratio differs between different materials, as shown in Table 19.2. A liquid has low resistance to shear (low G) and yet high resistance to volume change (high K). The ratio of Eq. 19.6 to Eq. 19.7 gives $(G/K) = 3(1 - 2\nu)/2(1 + \nu)$. Thus, if an isotropic material behaves like a liquid [i.e., $(G/K) \rightarrow 0$], then Poisson's ratio must approach 0.5. It can be shown that volume remains constant for any type of elastic loading if $\nu = 0.5$. Uniaxial elastic extension of a metal or ceramic results in a volume increase, however, since ν is usually in the range 0.2 to less than

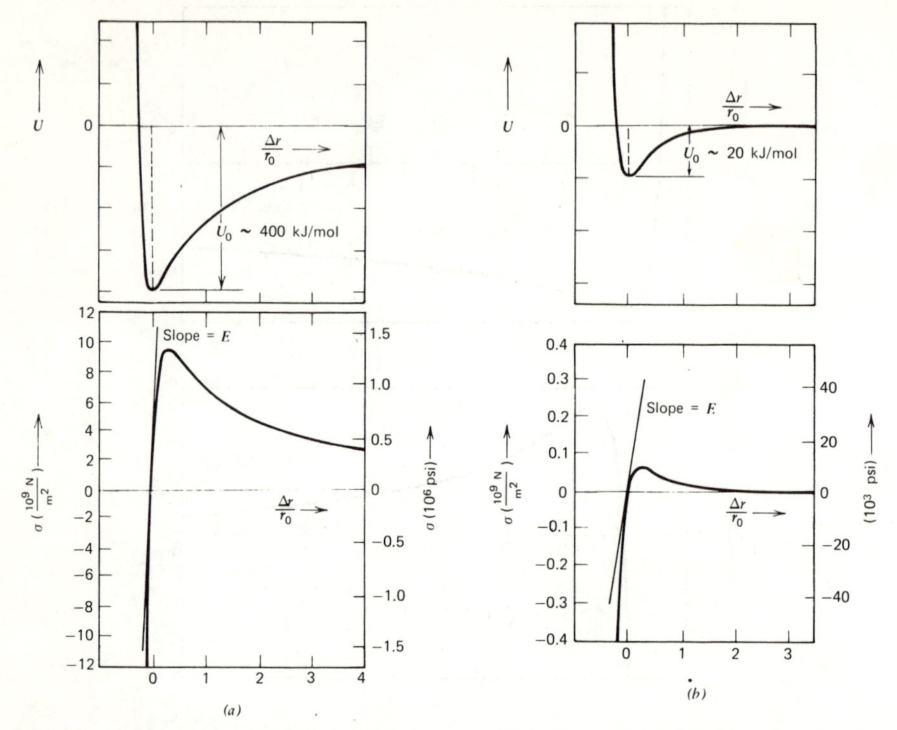

FIGURE 19.6 Schematic diagrams showing potential energy and applied stress versus linear strain for crystalline solids with (*a*) strong bonds and (*b*) weak bonds. Note the linear stress–strain relationship for small strains and the different slopes corresponding to Young's moduli.

0.5 for these materials. That is, the increased interatomic separation in the direction of an applied tensile stress favors a lateral contraction, but repulsive interatomic forces limit this lateral contraction to less than that necessary for no volume change.

The spacing of atoms or molecules in a crystalline solid depends on crystallographic direction, and the bond strength and elastic moduli also vary with direction. The orientation dependence of Young's modulus, for example, can be determined by performing tensile tests with single crystals having different crystallographic directions oriented along the tensile axis. Some results of such tests, along with Young's modulus for the respective isotropic polycrystalline materials, are shown in Table 19.3. The latter is a weighted average of the anisotropic moduli. With the exception of tungsten, which is inherently elastically isotropic, there is a marked anisotropy for different crystallographic directions.

Except for anisotropy considerations, inorganic glasses fit into much the same scheme as the crystalline solids discussed in this section. Most polymeric materials behave quite differently.

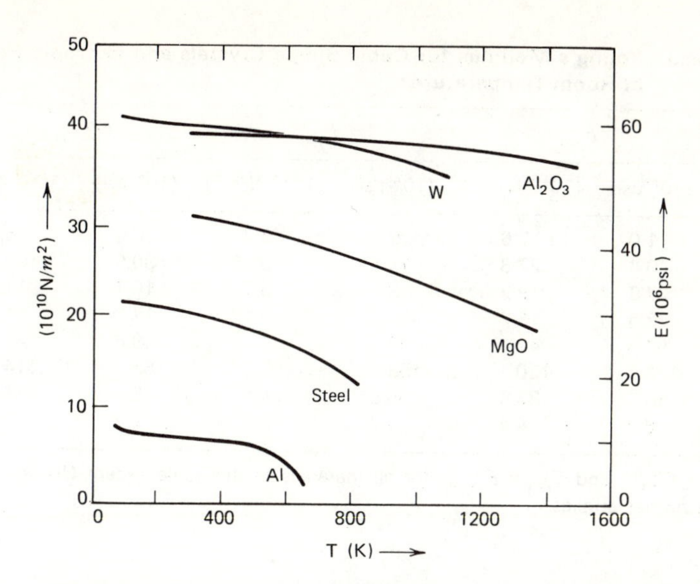

FIGURE 19.7 Variation of Young's modulus with temperature for some polycrystalline materials. Young's modulus decreases with temperature but is not markedly temperature dependent.

TABLE 19.2 Poisson's Ratio for Isotropic Solids at Room Temperature

Material	ν	*Structural Characteristics*
Aluminum	0.33	Crystalline metallic solid
α-Iron	0.29	Crystalline metallic solid
Copper	0.33	Crystalline metallic solid
Diamond	0.07	Crystalline covalent solid
Alumina (Al_2O_3)	0.23	Crystalline covalent/ionic solid
Magnesia (MgO)	0.19	Crystalline ionic/covalent solid
Silica glass	0.2	Covalent/ionic inorganic glass
Phenol-formaldehyde (Bakelite)	0.3	Tight network polymer
Hard rubber (ebonite)	0.39	Network polymer
Polystyrene (atactic)	0.33[a]	Glassy noncrystalline polymer
Polyethylene (branched)	0.4[a]	Supercooled liquid plus crystalline polymer
Natural rubber	0.49	Lightly cross-linked noncrystalline polymer
Argon (solid)	0.25[b]	Crystalline van der Waals solid

[a] Time dependent.
[b] At 0 K.

TABLE 19.3 Young's Modulus for Cubic Single Crystals and Isotropic Polycrystals at Room Temperature[a]

	E_{max}		E_{min}		$E_{polycrystalline}$	
Material	(10^6 psi)	(10^{10} N/m²)	(10^6 psi)	(10^{10} N/m²)	(10^6 psi)	(10^{10} N/m²)
Al	11.0	7.6	9.2	6.4	10.0	6.9
α-Fe	40.4	27.8	19.1	13.2	30.0	20.7
Cu	27.8	19.2	9.7	6.7	16.1	11.1
Nb	22.0	15.2	11.8	8.1	17.0	11.7
W	59.4	41	59.4	41	59.4	41
Diamond	175	120	153	105	165	114
MgO	48.7	33.6	35.6	24.5	45	31
NaCl	6.3	4.4	4.7	3.2	5.3	3.7

[a] $E_{max} = E_{\langle 111 \rangle}$ and $E_{min} = E_{\langle 100 \rangle}$ for all materials in this table except Nb and NaCl, where the reverse applies.

19.5 POLYMER ELASTICITY

The relationship between bonding and modulus described in the previous section has only limited application for describing elasticity in polymers. This situation arises from the characteristic molecular architecture of most polymeric substances. Nevertheless, the elastic response of network polymers and some glassy polymers can be described in terms of bonding since their internal structure is such that the elastic response to an applied stress is due to stretching of molecular and atomic bonds. Examples include Bakelite and polystyrene which are both relatively stiff (Table 19.1). Bakelite has a tight three-dimensional network of covalent bonds, and polystyrene is an organic glass at room temperature with tangled long-chain molecules held together by van der Waals bonds. The thermal motion of the molecular segments in polystyrene is limited at room temperature. On the other hand, in polymers where motion of the molecular segments occurs because of thermal agitation, time-dependent elasticity (viscoelasticity), which is distinct from linear elasticity, is observed. Such elasticity cannot be treated in terms of potential energy.

The mechanical response of elastomers (rubber elasticity) is an extreme example of elasticity arising from the capability of motion of molecular segments. (An elastomeric material would be liquidlike without cross-linking). As an elastomer is extended, the highly coiled, cross-linked molecules are straightened and aligned. Upon removal of the applied force the molecules coil up again so as to recover the elastic strain. This process happens readily because the molecular chain segments constantly undergo much thermal motion. Only at very large extensions on the order of 500% (Fig. 19.4*a*) does stretching of primary bonds occur and increase the "modulus." The response still remains elastic, however, since the cross-links prevent one molecule from sliding by another. The nonlinear nature of rubber elasticity necessitates an operational definition of stiffness distinct from the Young's modulus employed to characterize linearly elastic materials. One useful measure of stiffness is the tensile stress required to produce a 300% elongation; an almost equivalent specification is this stress divided by the extension ratio (i.e., by $l/l_0 = 1 + \epsilon = 4$). An interesting and unique feature of rubber elasticity is that the stiffness shows a small linear increase with increasing temperature over the temperature region where the material can be classified as an elastomer.

The elastic response at small strains of noncrystalline long-chain polymers at temperatures above their glass transition temperature is similar to that of elastomers in the sense that it corresponds to the uncoiling of molecules. These materials, however, possess a lesser mobility for molecular chain segments to move by one another, and the response is often sluggish. Larger strains lead to permanent deformation, which results in part because the molecular chain segments have too limited a mobility to completely recoil and in part because there are no chemical bond cross-links to prevent gross sliding of molecules. As the degree of crystallinity is increased in such materials they become stiffer. In polymers with a small degree of crystallinity the widely dispersed crystallites effectively serve as cross-links. When the degree of crystallinity exceeds 40 to 50%, crystalline regions are interconnected, forming a continuous load-bearing phase that is stiff compared to the noncrystalline regions.

The retarded elasticity of totally noncrystalline long-chain polymers, because of their sluggish configurational change, is observed in all polymers, although to a much lesser degree in crystalline, glassy, and network polymers. This *viscoelastic* behavior results, for example, in a decrease of stress with time after a constant elastic strain has been imposed, because of the continuing rearrangement of molecular segments. Under such circumstances, the "elastic modulus" will decrease with time. One useful way of specifying the elastic modulus is the relaxation modulus, $E_r(t)$, which is defined as the ratio of the stress to a small imposed strain after a fixed length of time t. For example, the ten second relaxation modulus is given by E_r (10 sec) $= \sigma(10 \text{ sec})/\epsilon$. The relaxation modulus is usually constant over a range of small strains but may be quite sensitive to the time employed. As shown in Fig. 19.8, time effects become increasingly important as the fraction of noncrystalline material increases; however, below the glass transition temperature, the relaxation modulus is also relatively insensitive to time for all polymers.

Some values of relaxation modulus at room temperature are listed in Table 19.1. For long-chain polymers, the temperature dependence of relaxation modulus is striking (Fig. 19.9). The primary effect of increasing temperature is to enhance the mobility of molecular chain segments, which facilitates elastic response particularly in noncrystalline regions. As shown in Fig. 19.9, depending upon temperature, a noncrystalline long-chain polymer may be glassy, leathery, rubbery, or

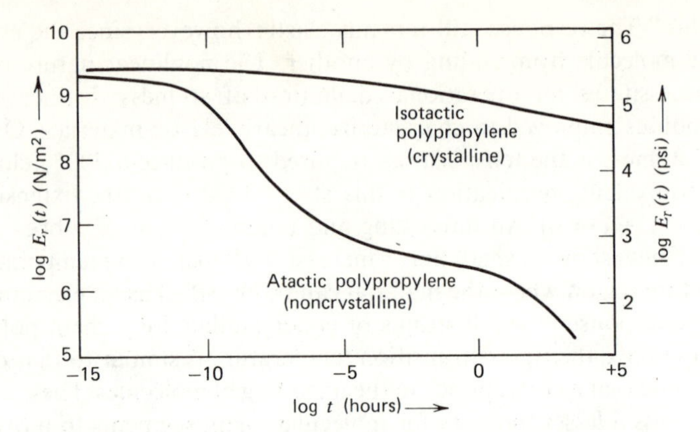

FIGURE 19.8 Relaxation modulus as a function of time at 25°C for crystalline and noncrystalline forms of polypropylene ($T_g \approx -18°C$, $T_m = 176°C$). The relaxation modulus is particularly sensitive to the time of measurement for noncrystalline polymers between T_g and T_m. The above data indicate that atactic polypropylene will have a glasslike response (high stiffness) to a rapidly applied load and rubbery response (low stiffness) to a slowly applied load. [After J. A. Faucher, *Trans. Soc. Rheol.*, **III**, 81 (1959).]

liquidlike in character. If such a polymer is lightly cross-linked, it exhibits elastomeric behavior over a relatively broad temperature range (e.g., lightly cross-linked atactic polystyrene, Fig. 19.9). The relaxation modulus of a crystalline polymer remains relatively high until the melting point is approached, although the effect of the glass transition is still evident (Fig. 19.9). By contrast, a totally cross-linked or network polymer displays glasslike behavior until a temperature is reached where chemical decomposition begins.

Crystalline polymers may be processed into thin filaments in which the molecular chain axes roughly coincide with the filament axis. Such filaments are markedly anisotropic and have greater values of relaxation modulus than bulk, isotropic materials.

It is interesting to note that the value of Poisson's ratio can serve as a rough guide to the mechanical characteristics of a polymer (Table 19.2). The Poisson's ratio of long-chain polymers and elastomers above their glass transition temperature is typically in the range 0.4 to 0.5. Thus, these substances behave as liquidlike solids, which is consistent with their structures. In contrast, glassy and network polymers have Poisson's ratios in the same range as metals and ceramics (Table 19.2). It is further observed that the value of ν decreases from that characteristics of a liquid toward values characteristic of inorganic solids as the cross-link density increases in a long-chain polymer.

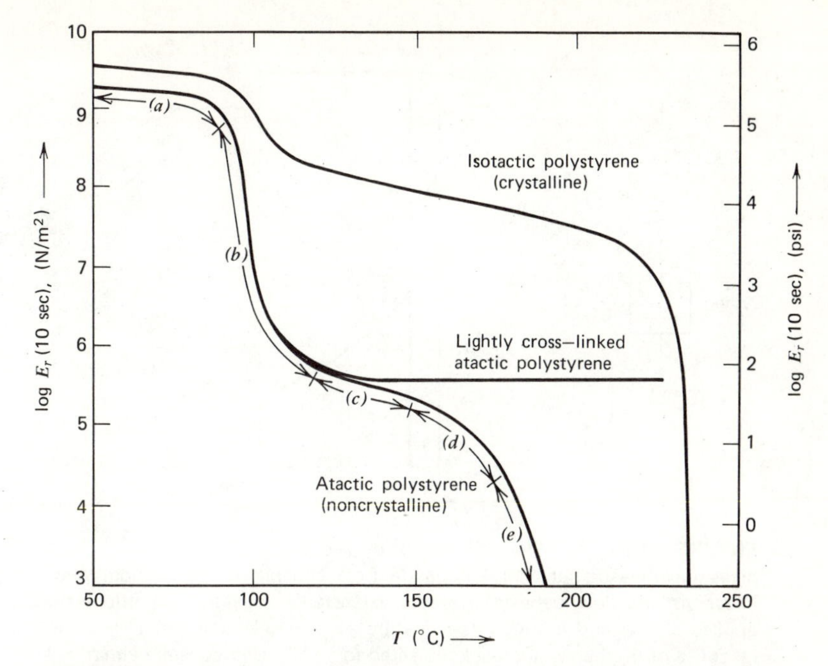

FIGURE 19.9 Relaxation modulus versus temperature for structurally different polystyrenes. Atactic polystyrene, which is totally noncrystalline, has five regions of behavior; (*a*) glassy, (*b*) leathery (in the glass transition region; T_g = 100°C), (*c*) rubbery, (*d*) rubbery flow (transition to liquid behavior), and (*e*) liquid flow. Lightly cross-linked atactic polystyrene, essentially an elastomer above T_g, also exhibits regions (*a*) and (*b*) as well as an extended region (*c*) in which the relaxation modulus increases linearly with absolute temperature, but this is not apparent in the figure because of the logarithmic scale used. Isotactic polystyrene, which is semicrystalline, shows a less marked decrease in relaxation modulus near T_g and maintains a much higher relaxation modulus until its melting point (T_m = 240°C) is approached. (After *Properties and Structure of Polymers,* by A. V. Tobolsky, Wiley, New York, 1960.)

19.6 VISCOELASTICITY: A MODEL

Many noncrystalline and sometimes even polycrystalline solids can exhibit both elastic and viscous behavior at relatively small stresses. Strictly, such viscoelastic behavior is displayed by most polymers discussed in the preceding section. A simple mechanical model of viscoelasticity, proposed by Maxwell in 1867, is shown in Fig. 19.10*a*. It consists of an elastic medium (the spring) and a viscous medium (the dashpot) in series. Numerous other mechanical models can be constructed with various arrangements of springs and dashpots in series and/or in parallel. None of these truly represent real behavior, but they are qualitatively

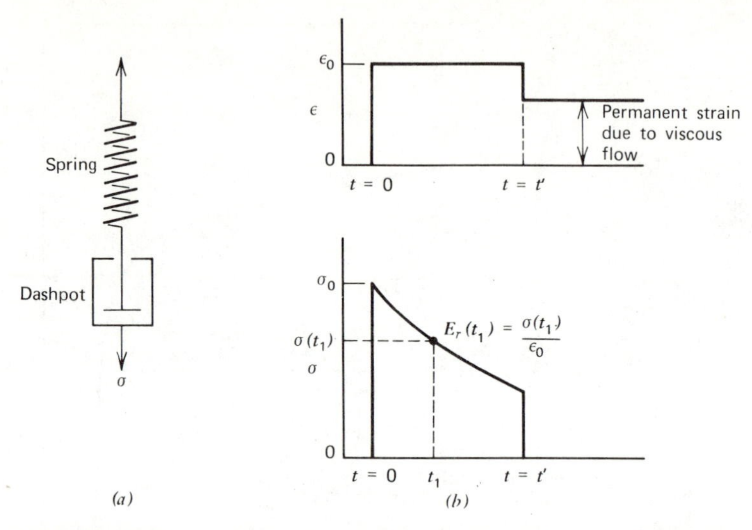

FIGURE 19.10 (*a*) Maxwell model for viscoelastic behavior. The spring gives linear elastic behavior, independent of temperature. The dashpot gives linear viscous flow behavior (rate of extension is directly proportional to applied stress) and is a very temperature sensitive element. (*b*) Relaxation response of the Maxwell model subjected to an imposed constant strain and then released. The relaxation modulus at time t_1 is indicated.

instructive. For the model in Fig. 19.10*a*, the stress response to a strain applied at $t = 0$ and held constant until $t = t'$ is shown in Fig. 19.10*b*. The instantaneous response is that of the elastic medium and subsequently the stress decreases with time up to t' because of the viscous member. Upon release of the imposed strain, the remaining elastic portion is instantaneously recovered, but the viscous strain persists indefinitely. Complete recovery of strain can result only in models where all dashpots are in parallel with springs.

By suitable combination of parallel and series arrangements of springs and dashpots, the mechanical response of such models can be made to simulate the viscoelastic response of polymers over the range from glassy to liquidlike behavior.

19.7 ANELASTICITY

In the ideal elastic solid, an applied stress produces an elastic strain instantaneously, and the strain disappears immediately when the stress is removed. On an atomistic basis, this means the atoms or molecules are instantaneously displaced from their equilibrium positions and then instantaneously return to the lowest energy positions or configurations on removal of the stress. In real solids the time of displacement or of return is finite. Time dependent elasticity of inorganic

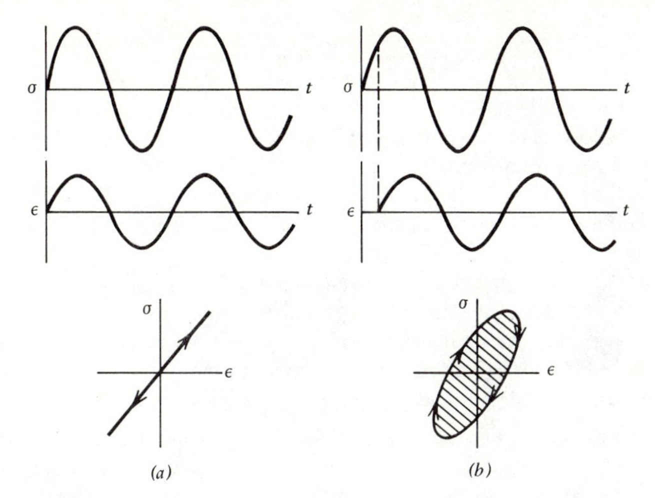

FIGURE 19.11 Stress–time, strain–time, and stress–strain relationships due to cyclic loading of (*a*) an ideal elastic material and (*b*) an anelastic material. Although most polymers are inherently anelastic (i.e., viscoelastic), metals and ceramics may be almost ideally elastic or quite anelastic, depending upon temperature and frequency of loading.

solids and metals is termed anelasticity. The viscoelasticity of polymers may be regarded simply as very exaggerated anelasticity.

The response of an ideal elastic solid to a cyclic stress such as alternating tension and compression is shown in Fig. 19.11*a*. For an anelastic solid, where strain is time dependent, strain is out of phase with respect to stress, and the stress–strain relation per cycle forms a hysteresis loop as shown in Fig. 19.11*b*. The area within the hysteresis loop represents the input of energy that is dissipated in heating up the material and its surroundings; this is the energy loss per unit volume for a cycle.

Anelastic effects are observed in a variety of materials under various conditions. For elastomers subject to cyclic stresses, the temperature increase can lead to degradation of the material with time. In the case of metals, cyclic stresses can lead to vibrations and, as described in Chapter 21, to fatigue failure. Conversely, anelasticity can be useful in damping out vibrations (elastic waves).

REFERENCES

1. Biggs, W. D. *The Mechanical Behaviour of Engineering Materials*, Pergamon Press, Oxford, 1965.
2. Hayden, H. W., Moffatt, W. G., and Wulff, J. *Structure and Properties of Materials*, Vol. III, Wiley, New York, 1965.

QUESTIONS AND PROBLEMS

19.1 Consider a fcc single crystal.

(a) Draw a portion of the (010) atomic plane with the [001] direction vertical and the [100] direction horizontal.

(b) Show the atomic positions that result when a tensile stress is applied in the [001] direction so as to cause elastic deformation. (Exaggerate the atomic displacements.)

(c) Show the atomic positions that result when a shear stress is applied on the (001) plane in the [100] direction. (Exaggerate the atomic displacements.)

(d) How large must the shear strain be in part (c) in order that each (001) atomic plane be shifted by one lattice parameter relative to the (001) atomic plane below it? Sketch the resulting atomic positions.

19.2 Even when a metal begins permanent deformation by plastic flow, part of the total strain under load is elastic strain given by Hooke's law (the modulus of elasticity being unaffected by plastic flow in a tension test).

(a) Refer to the engineering stress–strain curve for polycrystalline aluminum, (Figs. 19.4*f* and *g*) and calculate the elastic strain when the total tensile strain is 0.003.

(b) If a 20 cm long bar of aluminum is pulled in tension to a strain of 0.003 and then the load is released, how long will the bar be?

19.3 Poisson's ratio ν gives the negative of the ratio of transverse strain to axial strain in a uniaxial tensile test. For an isotropic rectangular bar of original length l_z and transverse dimensions l_x and l_y show that the fractional change in volume $(\Delta V/V_0)$ for elastic uniaxial tension in the z direction is equal to $\sigma_z(1 - 2\nu)/E$.

19.4 The nature of linear elasticity (i.e., small strains that are linearly proportional to stress) permits Hooke's law to be obtained for biaxial and triaxial stress states by the principle of superposition. This principle states that the total strain in any direction is the sum of strain components resulting from each stress acting on a body.

(a) For a cube of isotropic material subjected to three normal stresses (σ_x, σ_y, and σ_z) show that the resulting linear strains are given by the generalized Hooke's law relationships for an isotropic material; that is,

$$\varepsilon_x = \frac{1}{E}[\sigma_x - \nu(\sigma_y + \sigma_z)]$$

$$\varepsilon_y = \frac{1}{E}[\sigma_y - \nu(\sigma_z + \sigma_x)]$$

$$\varepsilon_z = \frac{1}{E}[\sigma_z - \nu(\sigma_x + \sigma_y)]$$

(b) If a shear stress were also applied to this cube, it would not appear in the above equations. Why?

19.5 (a) For an isotropic linear elastic material (small elastic strains), show that the relative elastic volume change is given by $\Delta V/V_0 = \varepsilon_x + \varepsilon_y + \varepsilon_z$.

(b) For any state of stress, the hydrostatic stress *component* is defined as $\sigma_{hyd} = (\sigma_x + \sigma_y + \sigma_z)/3$. Using this, your answer to part (a), and the generalized Hooke's law equations (problem 19.4), show that bulk modulus is given by $K = E/3(1 - 2\nu)$.

19.6 A state of biaxial stress involving a tensile stress and a compressive stress of equal *magnitude* is equivalent to a state of pure shear:

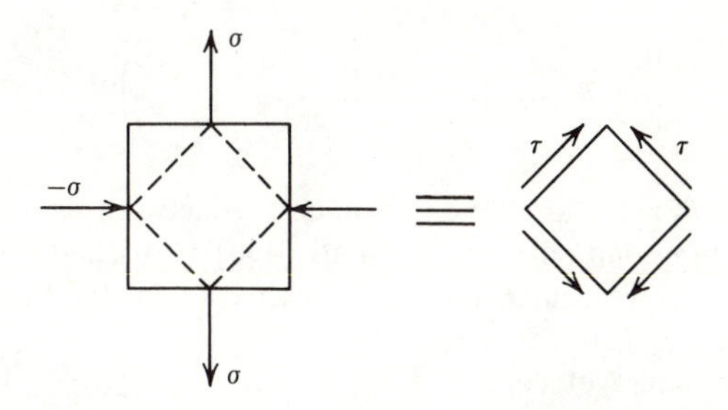

(a) Using this and the generalized Hooke's law equations (problem 19.4), show that for an isotropic linear elastic material $G = E/2(1 + \nu)$.

(b) Show that a state of pure shear stress leads to no elastic volume change.

19.7 (a) Young's modulus for steel is 20.7×10^{10} N/m^2. How much will a 2.5 mm diameter wire, 12 cm long, be strained when it supports a load of 450 N.

(b) What diameter aluminum wire, also 12 cm long, would be strained the same amount as the steel in part (a) by the same load? ($E = 6.9 \times 10^{10}$ N/m^2 for Al.)

19.8 Consider three aluminum and steel rods that are supporting a weight of 1000 lb:

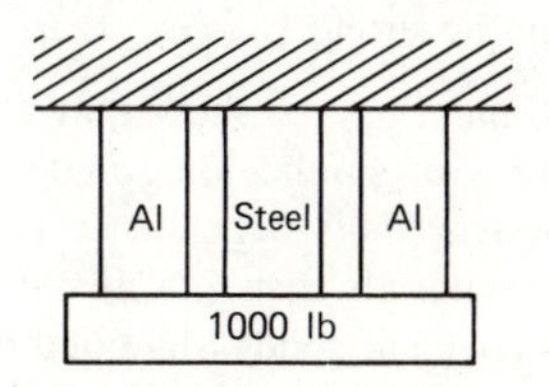

If the cross-sectional areas of the Al and steel rods are 0.2 in.^{2} each and the strains are equal, what is the stress on the Al and the stress on the steel? ($E_{Al} = 10 \times 10^6$ psi; $E_{steel} = 30 \times 10^6$ psi.)

19.9 Thermal stresses often are of considerable importance in engineering structures. Most solids (and liquids) expand as temperature increases. If this expansion cannot occur or is limited by constraints, then thermal stresses result, and these can be treated in terms of elasticity, providing they do not cause permanent deformation.

(a) By definition, the coefficient of linear thermal expansion is $\alpha_l = (1/l)(dl/dT) \simeq (1/l_0)(\Delta l/\Delta T)$. If a 10 ft long steel bar with 1 in^2 cross-sectional area just fits between two rigid abutments at 20°C, what stress will be induced in the bar when the temperature rises to 70°C? (For steel, $\alpha_l = 11.7 \times 10^{-6}\ {}^\circ C^{-1}$ and $E = 30 \times 10^6$ psi).

(b) The coefficient of volume thermal expansion is defined as $\alpha_v = (1/V)(dV/dT) \simeq (1/V_0)(\Delta V/\Delta T)$. Show that $\alpha_v = 3\alpha_l$.

19.10 Young's modulus for a cubic single crystal usually is anisotropic. The orientation dependence is given by

$$\frac{1}{E_{[uvw]}} = \frac{1}{E_{\langle 100\rangle}} - 3\left(\frac{1}{E_{\langle 100\rangle}} - \frac{1}{E_{\langle 111\rangle}}\right)(\alpha^2\beta^2 + \beta^2\gamma^2 + \gamma^2\alpha^2)$$

where α, β, and γ are the direction cosines between the direction of interest, $[uvw]$, and the three unit cell axial directions, [100], [010], and [001]. For most cubic metals $E_{\langle 111\rangle}$ is larger than $E_{\langle 100\rangle}$, and these represent the extremum values of Young's modulus (see Table 19.3).

(a) In the case of copper, $E_{\langle 111\rangle} = 27.8 \times 10^6$ psi and $E_{\langle 100\rangle} = 9.7 \times 10^6$ psi. Calculate Young's modulus for a copper single crystal having a [110] axial direction.

(b) Plot Young's modulus in copper as a function of direction in the $(0\bar{1}1)$ plane, starting with the [100] direction and ending with the [011] direction. [*Hint*. Sketch a unit cell and draw the $(0\bar{1}1)$ plane so that it passes through the origin. Then pick lines in this plane that start at the origin, so that you can conveniently compute the direction cosines with simple geometry. Plot E vs. θ, the angle between any chosen direction and the [100] direction.].

19.11 Many polycrystalline metals can be made elastically anisotropic by selected deformation processing schedules. List some advantages and disadvantages relative to the potential use of such a material in the form of a sheet whose Young's modulus varies with direction within the sheet.

19.12 (a) Using Eq. 19.8 and Eq. 3.5, show that $K \sim 1/r_0^4$, where r_0 is the equilibrium interionic spacing, for similar ionic materials (e.g., NaCl structure compounds with divalent ions). [This dependence ($K \sim r_0^{-4}$) is empirically observed for other classes of materials such as CaF_2 structure ionic solids, the alkali metals, and tetrahedral covalent solids (diamond cubic crystal structure and ZnS structure), but not for transition metals.]

(b) Shear modulus shows a similar dependence ($G \sim r_0^{-4}$) for divalent ionic solids and alkali metals, but a greater dependence on interatomic spacing is found for tetrahedral covalent solids ($G \sim r_0^{-5}$). Using physical reasoning, explain the latter relative to the former. In effect, justify why a covalent solid is more resistant to elastic shear (more rigid) than an ionic or metallic solid.

19.13 The simple theory of idealized rubber elasticity, in which an applied force raises the free energy by lowering the entropy of the rubber ($P = -T\,dS/dl$), leads to a constitutive equation for simple extension or compression:

$$\sigma = \frac{P}{A_0} = \frac{E'}{3}\left(\lambda - \frac{1}{\lambda^2}\right)$$

Here $\lambda = l/l_0$ is the extension ratio and E' is the initial slope of the stress–strain curve (i.e., the slope at $\lambda = 1$). This equation usually describes actual behavior well for λ between about 0.3 and 3.

(a) Using $E' = 133$ psi, plot σ vs. λ from $\lambda = 0.4$ to $\lambda = 6$, and indicate the linear strain scale ($\varepsilon = \lambda - 1$) as well. (Note that, unlike a linear elastic solid, an elastomer displays quite different responses to tensile and compressive stresses.) Compare your curve to that for natural rubber shown in Fig. 19.4.

(b) Compute the stiffness of this rubber in the two ways indicated in Sect. 19.5 and compare these to the slope of your curve in part (a) at $\lambda = 4$.

(c) The thermodynamic relation for idealized rubber elasticity gives the applied force P in terms of absolute temperature T and the decrease in entropy relative to extension dS/dl. This can be rewritten as $\sigma = -(T/V)(dS/d\lambda)$. Since $dS/d\lambda$ (< 0) depends on λ only and V is independent of λ, show that the stiffness of rubber increases linearly with absolute temperature.

19.14 For purposes of linear elasticity, normal stress as defined in this chapter (e.g., tensile load divided by *initial* cross-sectional area) is completely sufficient in terms of engineering usefulness and physical reality. For a material that undergoes sizable elastic elongation (e.g., an elastomer) or permanent deformation (e.g., polyethylene or aluminum), engineering stress (i.e., $\sigma = P/A_0$) remains useful (cf. Fig. 19.4), but the actual or true stress the material is subjected to is $\sigma_T = P/A$, where A is the actual cross-sectional area at any instant. In a tensile test, for example, $A < A_0$ due to lateral contraction, and the value of A depends on the extent and type of deformation.

(a) Show that $\sigma_T \approx \sigma$ for elastic tensile extension of a linear elastic material. (*Hint.* Make use of your knowledge of typical elastic properties.)

(b) The constitutive equation given in Problem 19.13 relates engineering stress to extension ratio for ideal rubber. For rubber, volume remains essentially constant ($\nu \approx 0.5$), and $\sigma_T = \sigma\lambda$, where $\lambda = l/l_0$. Derive this relationship and plot σ_T and σ for values of λ between 0.4 and 6.

(c) Comment on the physical basis of the difference in level found for the two curves in part (b) for $\lambda < 1$ and for $\lambda > 1$.

19.15 It can be contended that a noncrystalline long-chain polymer is a true solid below T_g but a supercooled liquid above T_g. In terms of elastic behavior, how can you support this? In particular, which polymers listed in Table 19.2 approach liquidlike behavior at room temperature?

19.16 (a) Sketch the relaxation modulus (logarithmic scale) versus temperature for unvulcanized natural rubber, for lightly vulcanized (elastomeric) natural rubber, and for hard rubber (completely cross-linked).

(b) Explain in terms of structure the different levels shown in your answer to part (a). In which condition will E_r be most time sensitive? Least time sensitive? Why?

19.17 The Voigt model of a viscoelastic material consists of a spring and dashpot in parallel:

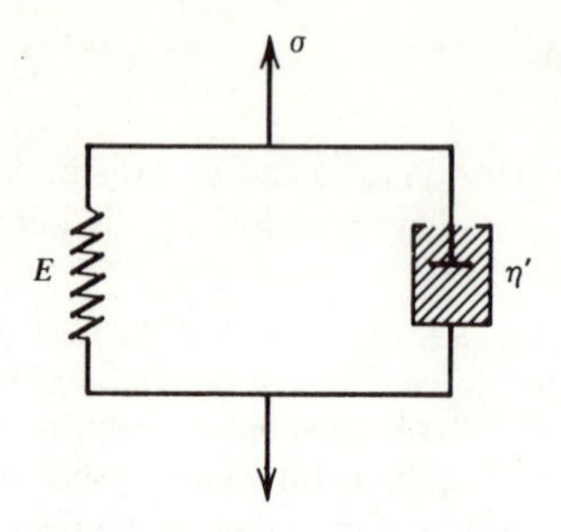

The equation that describes its viscoelastic response is

$$\sigma = E\varepsilon + \eta' \frac{d\varepsilon}{dt}$$

(a) If a stress is applied instantaneously at time $t = 0$ and held constant thereafter, sketch schematically the strain in the Voigt model as a function of time. (Use either the above equation or physical reasoning. If the equation is employed, explain the response in physical terms as well.)
(b) At $t = 0$, which member supports the applied stress?
(c) As $t \to \infty$, which member supports the applied stress?
(d) What is the limiting strain as $t \to \infty$?
(e) After the limiting strain is approached at constant stress, what response will occur if the stress is removed?

19.18 It is found experimentally that Poisson's ratio for a polymer subjected to a constant linear strain is a function of time, starting at some value and increasing toward a limiting value near 0.5 with increasing time. Explain this behavior.

20

Permanent Deformation

The flow of deformable materials, with an emphasis on the plasticity of crystalline solids, is discussed in this chapter.

20.1 TYPES OF DEFORMATION

While some materials are elastic up to the point of fracture (e.g., elastomers and most ceramics), many useful engineering materials such as metals and thermoplastic polymers can undergo substantial permanent deformation prior to fracture (see Fig. 19.4). This characteristic makes possible deformation processing operations that are widely used to shape and fabricate many metallic and polymeric materials and, in addition, it imposes some limitations on the engineering usefulness of such materials.

Permanent deformation is due to flow, a process of shear in which particles in a substance change neighbors. The shear nature of flow leads to the expectation that no volume change results, and this is essentially correct in most cases. Furthermore, it is reasonable to expect that interatomic or intermolecular forces and structure play an important role, although the former are much less significant in flow than they are in elastic behavior. The two most important types of permanent deformation are plastic flow, in the case of crystalline solids, and viscous flow, in the case of noncrystalline substances. Plastic flow involves the sliding of atomic planes relative to one another in a crystallographically organized fashion. Viscous flow, commonly associated with fluids, involves the switching of neighbors by small groups of atoms or molecules with a freedom that usually does not exist in a crystalline solid.

20.2 MECHANISMS OF PLASTIC FLOW

Slip is the predominant mechanism of plastic deformation. When a metal single crystal with a polished surface is pulled in tension, it begins to elongate plastically at a relatively low stress level, and blocks of the crystal slide over one another because of dislocation motion. Simultaneously, slip lines appear on the surface (Fig. 20.1*a*). These reflect the shear and the crystallographic natures of slip (Fig. 20.1*b*). A study of the orientation of slip lines, in conjunction with X ray analysis of a deformed metal single crystal, reveals that slip occurs most easily on planes with a high atomic density and a large interplanar spacing. These slip planes are often the closest-packed planes for a given crystal structure. In addition, the slip direction is parallel to the Burgers vector and invariably is a close-packed direction

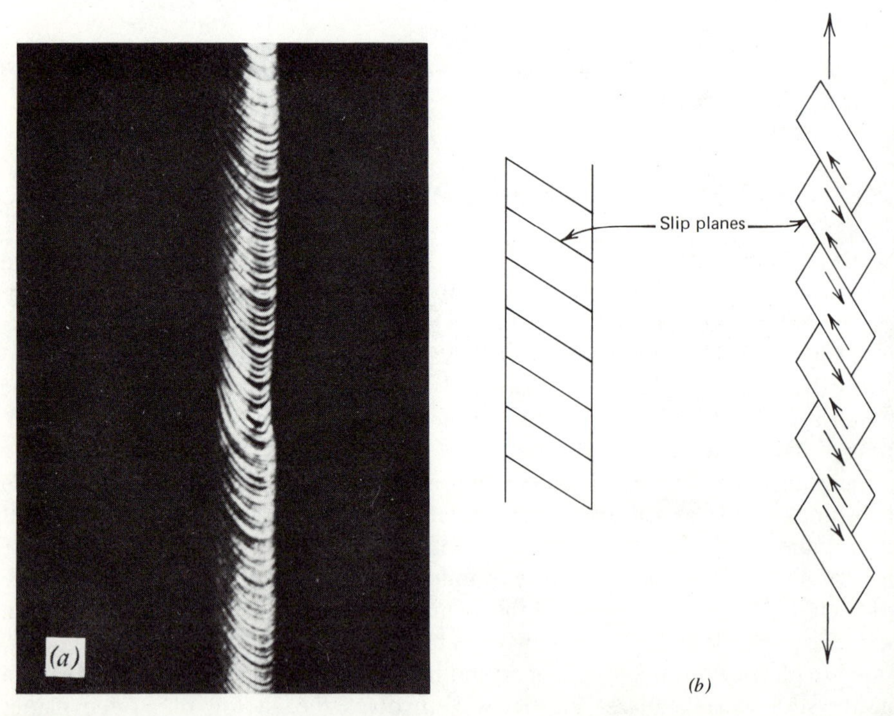

FIGURE 20.1 Plastic deformation by slip in a zinc single crystal. (*a*) The slip markings or slip lines indicate that flow occurs by shear on parallel planes. (*b*) Blocks of crystal sliding on slip planes give rise to the deformation observed in (*a*). Note that the continued tensile deformation causes a rotation of the slip planes, but volume remains constant. (Photograph from *The Distortion of Metal Crystals*, by C. F. Elam, Oxford University Press, London, 1935.)

(see Sects. 10.4 and 10.6). Table 20.1 lists the crystallographic planes and directions for slip in metals. It should be noted that a given slip plane can contain more than one slip direction and a given slip direction may be in more than one slip plane. The number of *slip systems* in a crystal is the number of possible combinations of slip planes and slip directions. Thus, the 12 slip systems in fcc metals exist because there are four geometrically distinct close-packed planes, each of which contains three distinct close-packed directions.

TABLE 20.1 Slip Systems in Metals

Crystal Structure	*Slip Planes*	*Slip Directions*	*Number of Slip Systems*	*Examples*
fcc	{111}	$\langle 1\bar{1}0 \rangle$	12	Al, Cu, Ni
bcc	{110}[a]	$\langle \bar{1}11 \rangle$	12	α-Fe, Ta, W
	{211}	$\langle \bar{1}11 \rangle$	12	α-Fe, Ta, W
	{321}	$\langle \bar{1}11 \rangle$	24	α-Fe
hcp	(0001)	$\langle 11\bar{2}0 \rangle$	3	Be, Mg, Zn[b]
	$\{10\bar{1}0\}$	$\langle 11\bar{2}0 \rangle$	3	α-Ti, α-Zr, Re[c]
	$\{10\bar{1}1\}$	$\langle 11\bar{2}0 \rangle$	6	

[a] Dominant slip plane at all temperatures.
[b] Other slip systems become operative at elevated temperatures.
[c] Other slip systems operate in polycrystals at room temperature.

The large number of slip systems (≥ 12) for most metals is one feature that makes possible the extensive plastic flow or ductility of metals. In bcc and hcp metals, slip always occurs in close-packed directions but may occur on crystallographically different planes. Which slip systems are active depends on temperature, with more being active at higher temperatures. For example, all slip systems in a bcc metal may participate in plastic deformation at high temperatures, but at low temperatures only those involving the most closely packed {110} planes do. Similarly, one type of slip plane is favored in hcp metals at low temperatures; however, it is not the same slip plane for all hcp metals.

In contrast to metals, ionic and covalent engineering solids normally are brittle at room temperature. This is not due to inherently greater bond strengths or due to a lack of dislocations; rather, these materials have fewer active slip systems that cannot accommodate significant plastic flow. Only at elevated temperatures will a material such as MgO deform plastically when subjected to an applied tensile stress. In compression, on the other hand, MgO single crystals can undergo more than 5% plastic strain at room temperature.

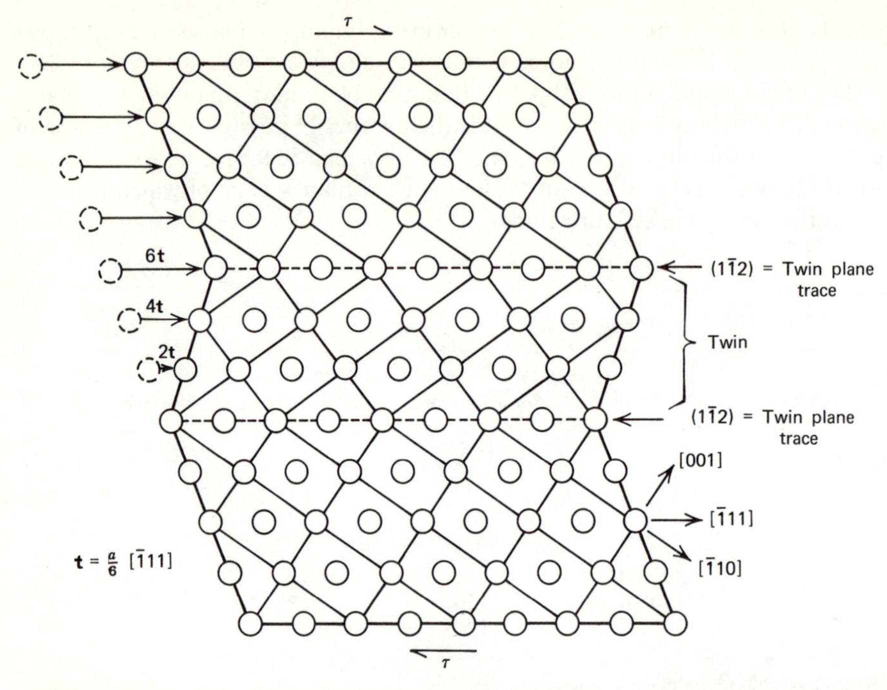

FIGURE 20.2 The crystallography of mechanical twinning in a bcc metal. A twin forms, because of the applied shear stress, when successive (1$\bar{1}$2) atomic planes shear over each other by an amount equal to **t**, the twinning vector. [In the (110) plane of view, only half of the intersecting (1$\bar{1}$2) atomic planes traces are apparent.] The cooperative motion of the atoms causes permanent deformation without any volume change and the crystal is a mirror image of itself across the twin plane trace.

A second mechanism of plastic deformation, which is important in some materials, is deformation twinning. Like slip, deformation twinning involves the shearing of atomic planes relative to one another, but the process is a cooperative one that leads to a twinned region of the crystal (see Sect. 10.7). Twin planes and twinning directions for metals are listed in Table 20.2; Fig. 20.2 shows the atomic movements associated with mechanical twinning in bcc metals. Twinning in metals

TABLE 20.2 Twinning Systems in Metals

Crystal Structure	*Twinning Planes*	*Twinning Directions*
fcc	$\{111\}$	$\langle 11\bar{2} \rangle$
bcc	$\{112\}$	$\langle 11\bar{1} \rangle$
hcp	$\{10\bar{1}2\}$	$\langle 10\bar{1}1 \rangle$

tends to occur only when slip is restricted, because the stress necessary is usually higher than that for slip. Thus, some hcp metals with a limited number of slip systems (e.g., Zn) may twin preferentially under certain conditions. Also, bcc metals twin at low temperatures because slip is difficult. Of course, twinning and slip may occur sequentially or even concurrently in some cases.

20.3 SLIP IN SINGLE CRYSTALS

The necessity of dislocation motion for plastic flow is well illustrated by the mechanical behavior of whiskers—fine fibers of metal or oxide having diameters on the order of 1 μm (10^{-4} cm). The conditions of whisker growth lead to essentially dislocation-free single crystals. When tested in tension, whiskers remain elastic to exceptionally high stress levels before fracture or plastic yielding occurs (Fig. 20.3). The onset of plastic flow (yielding) in this case corresponds to the generation of dislocations, after which plastic flow continues at a much lower stress than was necessary for yielding (Fig. 20.3*b*).

Most metal single crystals do not display the high strengths of whiskers, for they contain a relatively large number of dislocations that formed during their growth from liquid or vapor. For such ordinary single crystals of reasonable perfection, the tensile stress required to induce plastic deformation depends upon the crystallographic orientation of the crystal with respect to the tensile axis. Whether or not dislocations move is determined by the magnitude of the shear stress acting on a slip plane and in a slip direction. When an applied tensile stress causes this resolved shear stress to reach a critical value, then yielding occurs. The *critical resolved shear stress*, τ_0, is a property of the material that determines when slip starts. It varies with temperature (Fig. 20.4), and its exact value is strongly dependent on the purity of the material and its structural perfection, as well.

Since metals have a number of slip systems, each differing in orientation, it should be apparent that an applied tensile stress acting on a single crystal does not lead to the same resolved shear stress on each slip system. Rather, it is simply a problem of geometry to find which slip system is oriented most favorably, that is, which has the largest resolved shear stress when the crystal is elastically deformed. As the applied tensile stress is increased, the resolved shear stress will increase proportionately on all slip systems until the critical resolved shear stress is reached on the most favorably oriented slip system. Then plastic flow commences. To see the influence of crystal orientation on the magnitude of the tensile stress required for plastic yielding, it is necessary to examine how resolved shear stress is geometrically related to applied tensile stress. For the single crystal in Fig. 20.5, one slip system is shown; ϕ is the angle between the slip plane normal and the tensile axis, and λ is the angle between the slip direction and the tensile axis. (It is important to note that these three vectors are *not* coplanar in general, that is, $\phi + \lambda \neq 90°$.)

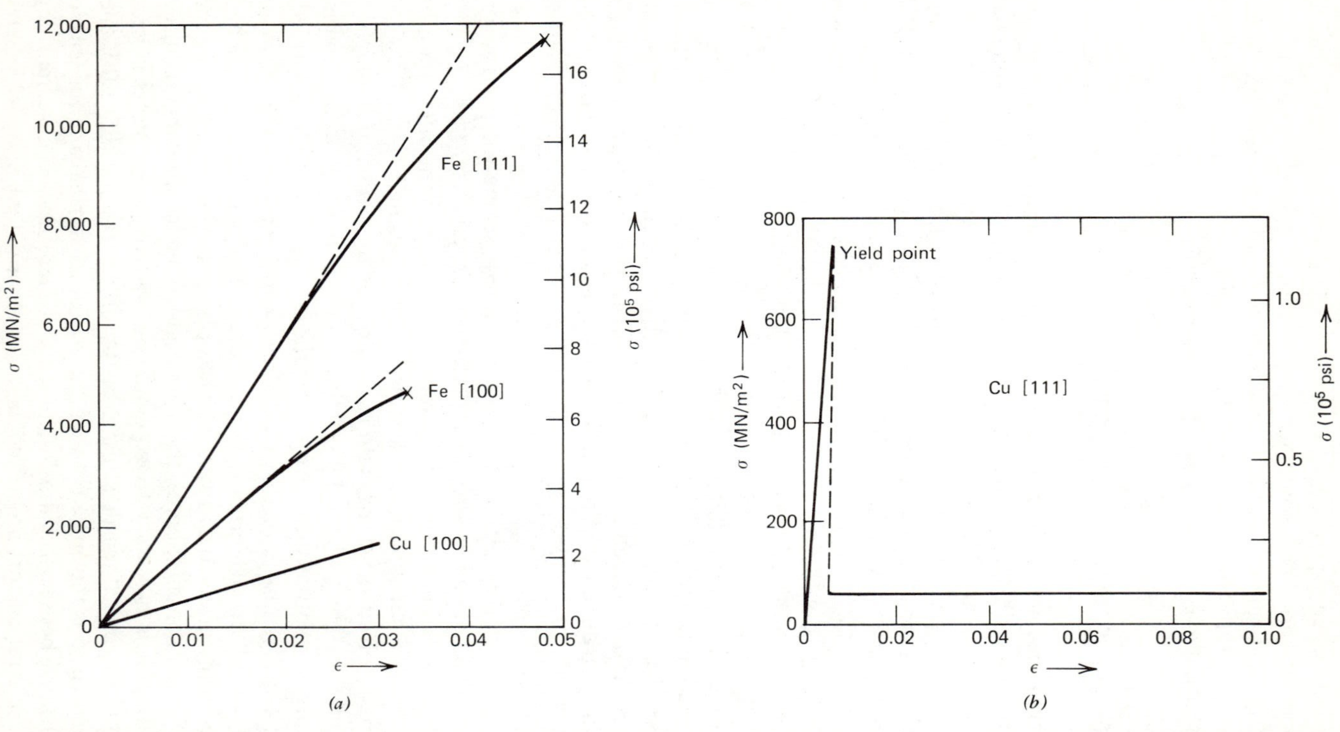

FIGURE 20.3 Tensile behavior of iron and copper single crystal whiskers. (*a*) The absence of dislocations permits elastic extension to the point of fracture (near the theoretical fracture strength) for some whiskers. Note the elastic anisotropy and the nonlinear elasticity displayed by the two Fe whiskers with different axial orientations. (*b*) Plastic flow occurs in some whiskers after dislocations are generated at high applied stress levels. A large, discontinuous drop in applied stress (broken line) accompanies yielding, and subsequent (plastic) deformation proceeds at a relatively low stress level. [After S. S. Brenner, *J. Appl. Phys.*, **27**, 1484 (1956); **28**, 1023 (1957).]

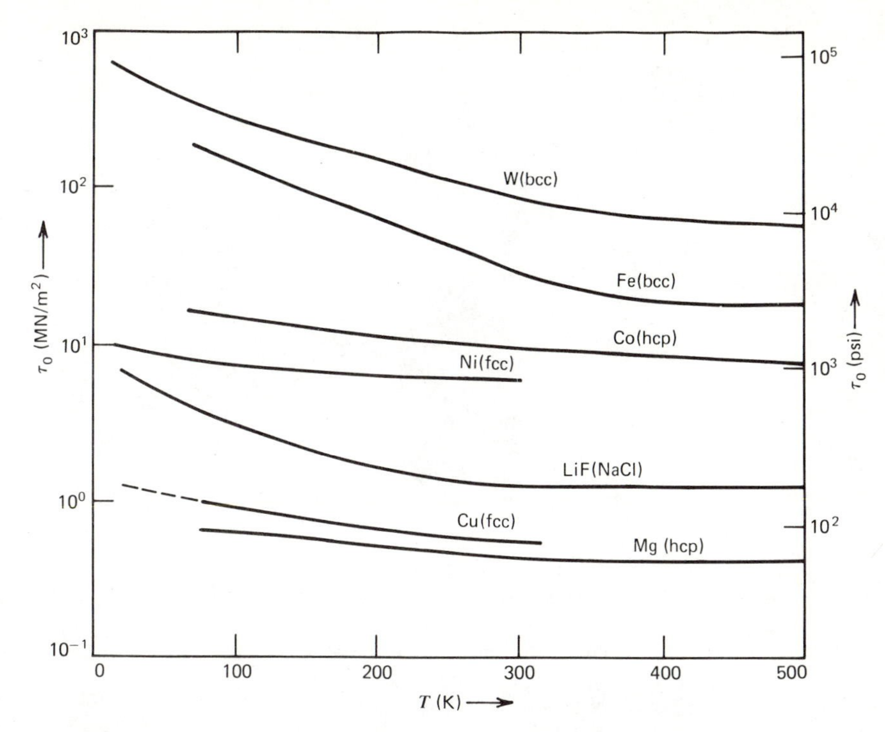

FIGURE 20.4 Temperature variation of critical resolved shear stress for single crystals of materials having different crystal structures. Note that τ_0 for the close-packed metals (i.e., Mg, Cu, Ni, and Co) only rises moderately as temperature decreases, whereas τ_0 increases significantly at low temperatures for the bcc metals and substances having the NaCl structure (e.g., LiF). [Data from: R. M. Rose, D. P. Ferriss, and J. Wulff, *Trans. Met. Soc. AIME,* **224**, 981 (1962); J. J. Cox, R. F. Mehl, and G. T. Horne, *Trans. ASM,* **49**, 118 (1957); K. G. Davis and E. Teghtsoonian, *Trans. Met. Soc. AIME,* **227**, 762 (1963); P. Haasen, *Phil. Mag.,* **3**, 384 (1958); W. G. Johnston, *J. Appl. Phys.,* **33**, 2050 (1962); J. Garstone and R. W. K. Honeycombe, in *Dislocations and Mechanical Properties of Crystals,* edited by J. C. Fisher, et al., Wiley, New York, 1957; W. F. Sheely and R. R. Nash, *Trans. Met. Soc. AIME,* **218**, 416 (1960).]

The component of applied force acting in the slip direction is $P \cos \lambda = \sigma A \cos \lambda$, and the area of the slip plane is $A/\cos \phi$, where A is the nominal cross-sectional area of the crystal. Thus, the resolved shear stress is given by

$$\tau = \frac{\sigma A \cos \lambda}{A/\cos \phi} = \sigma \cos \phi \cos \lambda \tag{20.1}$$

Since ϕ and λ are changed only imperceptibly by linear elastic deformation, it is clear that τ is directly proportional to σ. Furthermore, it is the geometrical factor

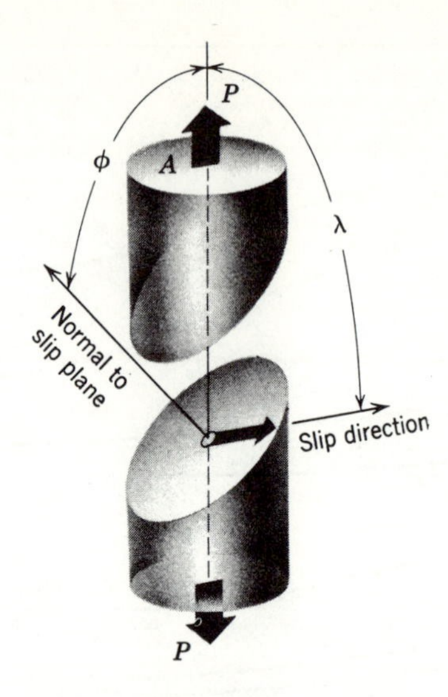

FIGURE 20.5 Geometry used for calculating the resolved shear stress (indicated by heavy arrow) acting on a slip plane and in a slip direction of a single crystal. (From *The Structure and Properties of Materials*, Vol. III, by H. W. Hayden, W. G. Moffatt, and J. Wulff, Wiley, New York, 1965.)

($\cos\phi\cos\lambda$), which depends on axial orientation, that determines the magnitude of τ relative to the magnitude of σ. The resolved shear stress on any slip system can be ascertained with Eq. 20.1; for the most favorably oriented slip system, $\cos\phi\cos\lambda$ is a maximum:

$$\tau_{\max} = \sigma(\cos\phi\cos\lambda)_{\max} \tag{20.2}$$

When $\tau_{\max} = \tau_0$, yielding occurs, and the applied tensile stress necessary for yielding is given by Schmid's law:

$$\sigma_{\text{yield}} = \frac{\tau_0}{(\cos\phi\cos\lambda)_{\max}} \tag{20.3}$$

This relationship is valid for an applied tensile or compressive stress and can also be used in the case of twinning. It is important to point out, however, that τ_0 for twinning usually is greater than τ_0 for slip.

Figure 20.6 shows some tensile stress–strain curves for copper single crystals at room temperature. Although each of these single crystals begins to plastically deform at the critical resolved shear stress, the tensile stress necessary for yielding depends on the crystal orientation, and the plastic flow behavior subsequent to yielding also has a marked dependence on the axial orientation. The general tendency for a crystalline solid to become stronger with increasing plastic deformation, called strain hardening or work hardening, is evident as well. Only in the case of the single crystal whose axial direction is such that a single slip system is favorably oriented initially does some plastic flow occur at an almost constant applied stress (*easy glide*), although after a certain amount of plastic strain, slip on other slip systems begins in this crystal, and strain hardening becomes significant. For the single crystals oriented such that multiple slip is favored even at yielding, strain hardening is significant over the entire range of plastic flow. The importance of these results is that marked strain hardening is closely related to interfering

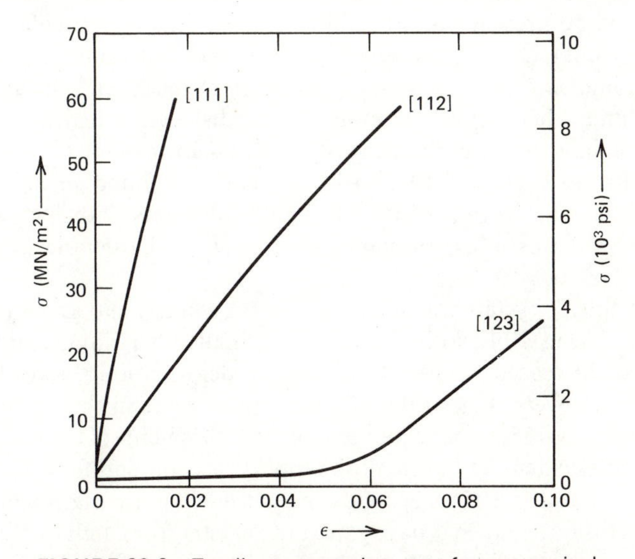

FIGURE 20.6 Tensile stress–strain curves for copper single crystals at small plastic strains. When two or more slip systems are equally favored at the onset of plastic flow (e.g., for [111] and [112] axial orientations), rapid strain hardening accompanies plastic deformation. When only one slip system is favorably oriented (e.g., for the [123] axial orientation), plastic deformation with very little strain hardening (i.e., easy glide) occurs initially. Crystallographic rotation during easy glide makes possible slip on other slip systems, which results in increased strain hardening beyond $\varepsilon \approx 0.05$ for this [123] crystal. [After J. Diehl, *Z. Metallk.*, **47**, 331 (1956).]

slip (i.e., slip on intersecting slip systems). It is interesting to note that the Zn single crystal shown in Fig. 20.1 can exhibit easy glide all the way to fracture, which occurs after an elongation of several hundred percent.

20.4 DISLOCATION MOTION AND GENERATION

The ability of dislocations to move readily in metals at small applied stress levels arises because bonds are not truly broken in the process, yet an inherent resistance to motion (*friction*) originates from the unfavorable atomic configurations and bond stretching concurrent with any motion. Accordingly, the critical resolved shear stress for a single crystal relates directly to crystal structure considerations and the type of bonding. These factors are outweighed, in most engineering metals, by crystal imperfections (including impurity atoms) and second-phase particles, all of which provide a more significant impediment to slip than does *friction*. Since dislocations interact with one another, their motion is hindered by the presence of other dislocations. This is particularly important in the case of interfering slip, where moving dislocations must cut through dislocations on intersecting slip systems. Observations made with the aid of electron microscopy and other techniques indicate that dislocation densities of the order of 10^6 cm of line/cm^3 are characteristic of undeformed metal single crystals, whereas values in excess of 10^9 cm of line/cm^3 are found in deformed single crystals. That is, dislocation density generally increases with increasing plastic deformation, and this is the primary cause of strain hardening in single-phase solids.

In certain single crystals where slip occurs on a single slip system (e.g., the Zn crystal shown in Fig. 20.1), strain hardening is small in magnitude, and dislocation density is not increased drastically by plastic deformation. Nevertheless, more dislocations than were present initially must participate in the slip process if the observed plastic strain is to be accounted for. Measurements of the size and number of slip steps after extensive plastic flow support this contention, because each slip line on the crystal surface is a step whose magnitude is a summation of the Burgers vectors of all dislocations that have exited the crystal from a given slip plane.

An applied stress usually cannot cause a dislocation to form in a perfect region of a crystal; however, it can activate dislocation sources that are present. An example of such a source, shown in Fig. 20.7*a*, is a segment of dislocation line on an active slip plane. Such segments are fairly common since dislocation lines are not restricted to one slip plane but can extend through the crystal on various slip planes in three dimensions. When the resolved shear stress reaches a sufficiently large value on the slip plane containing the dislocation segment (Fig, 20.7*a*), then the segment bows out (Figs. 20.7*b* and *c*), eventually forming a separate loop and regenerating itself (Figs. 20.7*d* and *e*). This process, which can occur repeatedly

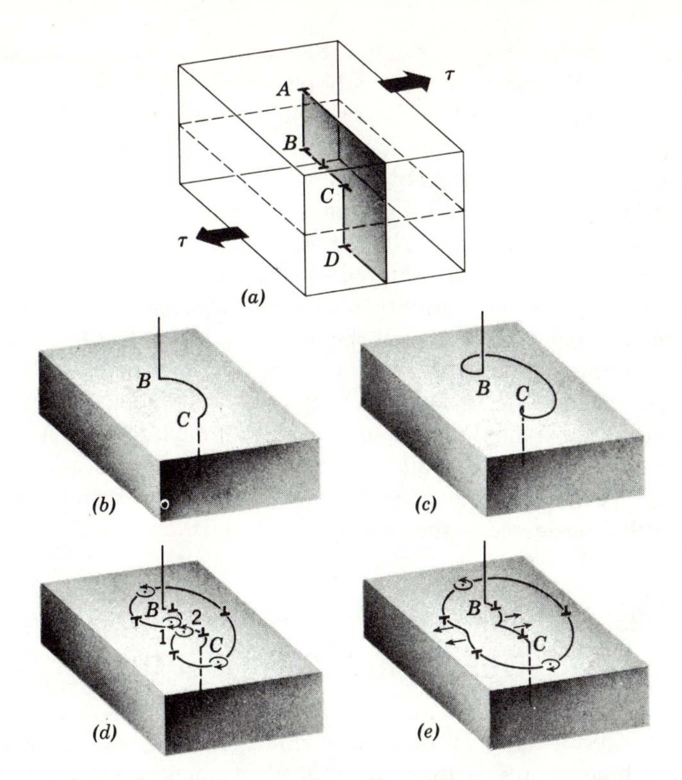

FIGURE 20.7 Example of a dislocation source in a crystal. (*a*) The dislocation line (*ABCD*) has a segment of length l (between *B* and *C*) on the slip plane subjected to the shear stress τ. The shear stress causes the dislocation segment to move as indicated in the sequence (*b*), (*c*), (*d*), and (*e*). The dislocation loop that is generated has the same Burgers vector as the initial segment. (Adapted from *The Structure and Properties of Materials*, Vol. III, by H. W. Hayden, W. G. Moffatt, and J. Wulff, Wiley, New York, 1965.)

and give rise to many dislocation loops in one slip plane, is largely responsible for the discrete slip lines on a deformed single crystal (Fig. 20.1). That is, although plastic deformation by slip is macroscopically homogeneous, it is microscopically inhomogeneous, with active dislocation sources existing on widely separated groups of parallel slip planes but not on each successive parallel slip plane. The resolved shear stress that will activate a dislocation source (Fig. 20.7) is given by

$$\tau_c \approx \frac{2Gb}{l} \tag{20.4}$$

where G is the shear modulus, b is the Burgers vector magnitude, and l is the distance between the ends of the dislocation segment.

Long dislocation segments begin to operate at shear stress levels that are lower than τ_0 (but necessarily greater than or equal to the friction stress); however, the number of dislocations generated by these sources is insufficient to provide plastic strains characteristic of yielding. Consequently, the applied stress level must be increased until sources are activated so as to produce a sufficient number of mobile dislocations, and this is achieved when the resolved shear stress reaches a critical value (i.e., τ_0). As sources operate and dislocation density increases, shorter segments may be activated as the higher applied stress levels necessitated by strain hardening. It may appear paradoxical that higher dislocation densities and, perhaps, more active dislocation sources correspond to a higher strength, but it must be borne in mind that these result in strain hardening such that further slip is more difficult to effect. While the above features of dislocation motion and generation are pertinent to polycrystals as well as to single crystals, other complications must be considered in the case of polycrystals.

20.5 PLASTIC DEFORMATION OF POLYCRYSTALS

To initiate plastic flow in pure, well-annealed metal polycrystals requires substantially higher stresses than for equivalent single crystals. Much of this increase is due to geometric reasons. Although some grains may be oriented favorably for slip, yielding cannot occur unless the unfavorably oriented neighboring grains can also slip. In a polycrystalline aggregate, individual grains provide a mutual geometrical constraint on one another, and this precludes plastic yielding at low applied stresses. Once yielding has occurred, continued plastic deformation is possible only if enough slip systems are simultaneously operative so as to accommodate grain shape changes while maintaining grain boundary integrity. It has been shown that at least five independent slip systems must be mutually operative for a polycrystalline solid to exhibit ductility. Thus, polycrystalline zinc, with only three slip systems at room temperature, fractures after a very small amount of plastic strain, whereas polycrystalline copper, with 12 slip systems, displays extensive plastic flow prior to fracture.

The yield strength of polycrystals increases with decreasing grain size since grain boundaries act as internal obstructions to dislocation motion. It has been theoretically proposed and experimentally verified that the yield strength, σ_Y, of many polycrystalline metals follows a Hall–Petch type relationship:

$$\sigma_y = \sigma_0 + kd^{-1/2} \tag{20.5}$$

where d is the average grain diameter and σ_0 and k are empirical constants (σ_0 may be identified with the yield strength of an imaginary polycrystal having infinite grain size).

The meaning of yield strength, as used above, warrants further comment. It might seem reasonable to choose the elastic limit—essentially where a stress–strain curve deviates from linearity—as the yield strength; however, such a value is difficult to determine experimentally. For engineering purposes, a conventional yield strength is generally used. Most commonly, the tensile or compressive stress level required to produce a permanent offset strain of 0.002 (or 0.2%) is taken to delineate the onset of appreciable plastic deformation. The 0.2% offset yield strength is determined, as shown in Fig. 20.8, by the interaction of the stress–strain

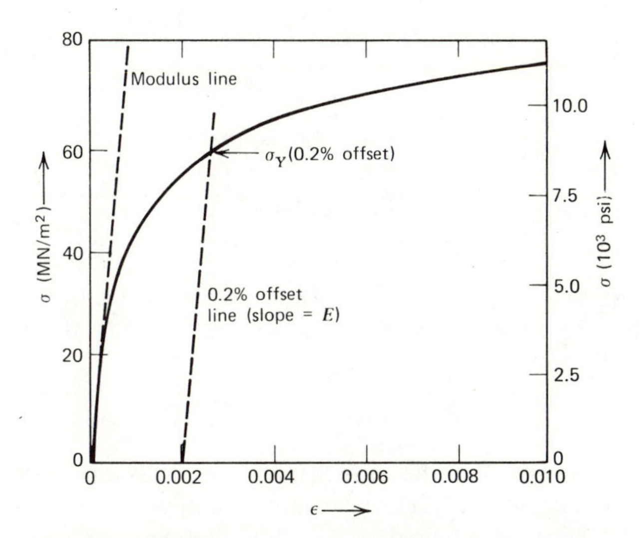

FIGURE 20.8 Initial portion of the stress–strain curve for polycrystalline copper. The 0.2% offset yield strength is indicated. If the specimen was loaded to the yield strength and then unloaded, it would suffer a permanent strain of 0.002.

curve with a line drawn parallel to the original modulus line and starting at an offset of 0.002 along the strain axis. Other conventions may be employed, and it is necessary to specify the convention when stating a yield strength value.

Strain hardening commences immediately when plastic flow begins in polycrystalline copper (Fig. 20.9*a*), and this is true for most other polycrystals as well. Some polycrystalline metals, such as mild steel, display a discrete yield point type of behavior (Fig. 20.9*b*) which is qualitatively similar to the behavior of the Cu

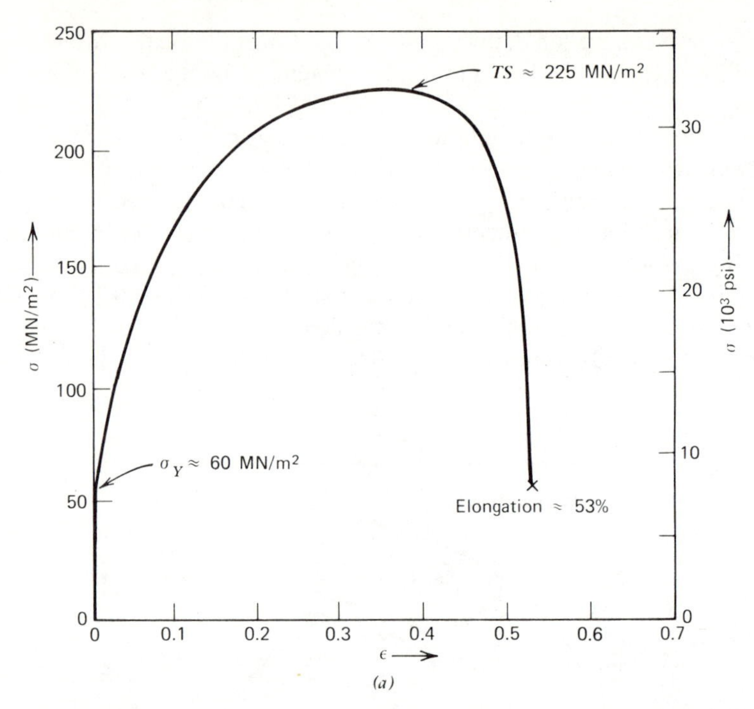

FIGURE 20.9 (*a*) Complete engineering stress–strain curves for annealed polycrystalline copper. σ_Y (taken from Fig. 20.8), tensile strength, and percent elongation are indicated.

whisker shown in Fig. 20.3*b* in the sense that a higher stress is necessary to initiate plastic flow than to continue it. Here, however, the upper yield point (Fig. 20.9*b*) corresponds to the stress required to begin the motion of existing dislocations rather than to generate dislocations, as in the case of the Cu whisker. A finite amount of inhomogeneous plastic strain (Lüder's strain), in the form of a band that propagates along the length of the specimen, occurs at the lower yield point, and uniform deformation with strain hardening commences subsequently. Since the lower yield point is well defined and characteristic of the material while the upper yield point varies with exact testing procedure, the lower yield point always should be identified with yield strength, and an offset convention is unnecessary for a polycrystalline material that exhibits a yield point phenomenon.

Because many metals are used to support tensile loads, their stress–strain behavior as ascertained with tension tests (see Sect. 19.2) is of interest. For certain engineering applications, the yield strength may represent the limiting stress that can be supported by a material; for other applications, the tensile strength repre-

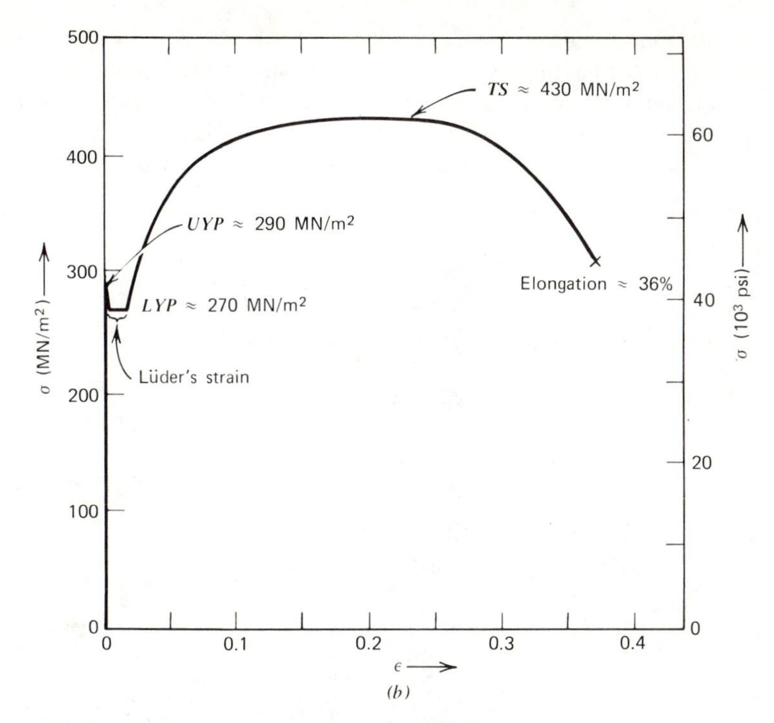

FIGURE 20.9 (*b*) Complete engineering stress–strain curves for mild steel (1018 hot-rolled steel). The upper and lower yield points (*UYP* and *LYP*), tensile strength, percent elongation, and Lüder's strain are indicated.

sents the limit. The maximum stress (P_{max}/A_0) that a material can withstand in a tensile test is called the tensile strength, *TS* (Fig. 20.9). The tensile load-bearing capability increases until the tensile strength is reached and then decreases continuously until fracture occurs. Such behavior does *not* mean that strain hardening occurs up to the tensile strength and "strain softening," thereafter. Actually, strain hardening continues to the point of fracture, but the geometry of deformation changes at the tensile strength. After yielding and up to the tensile strength, the axial elongation due to plastic flow is accompanied by a decrease in cross-sectional area that is uniform along the length of the specimen. Since volume remains constant during plastic deformation, the instantaneous cross-sectional area A_i can be related to the instantaneous specimen length l_i by

$$A_i l_i = A_0 l_0 \tag{20.6}$$

where A_0 and l_0 are the original values prior to tensile testing. Thus, the actual or *true stress* ($\sigma_T = P/A_i$) acting on the specimen becomes increasingly greater than the nominal or *engineering stress* ($\sigma = P/A_0$) with increasing tensile deformation, and both of these increase due to strain hardening. Ultimately, the strain-hardening ability of the material is unable to compensate for the decreasing cross-sectional

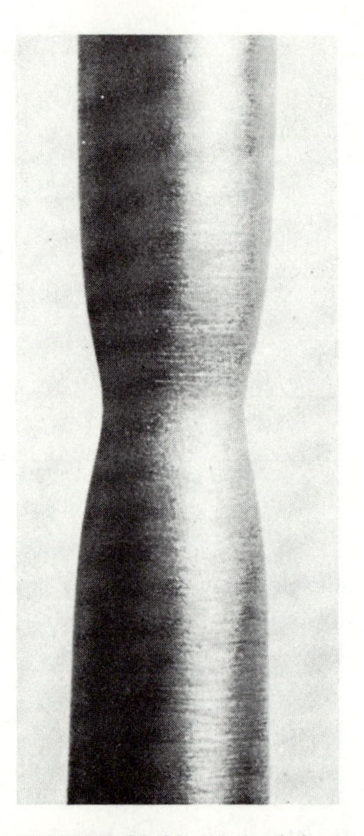

FIGURE 20.10 A neck in a tensile specimen. The neck starts to develop when the tensile strength of the material is reached, and becomes more and more pronounced as the tension test is continued.

area, and an instability in the form of a constriction or neck develops (Fig. 20.10). From this point on, further plastic deformation is largely restricted to the neck region, with an ever decreasing cross-sectional area, and the engineering stress ($\sigma = P/A_0$) decreases until fracture occurs but the true stress ($\sigma_T = P/A_{neck}$) continues to increase.

Just as true stress often is more meaningful than engineering stress, true strain represents large deformations better than engineering strain ($\epsilon = \Delta l/l_0$). True strain is given by

$$\epsilon_T = \ln\left(\frac{l_i}{l_0}\right) = \ln(1 + \epsilon) \tag{20.7a}$$

for uniform deformation and by

$$\epsilon_T = \ln\left(\frac{A_0}{A_{\text{neck}}}\right) \tag{20.7b}$$

after necking begins. The engineering stress–strain curve, which is identical in form to the load-elongation curve, and the true stress–true strain curve reflect different facets of the plastic flow behavior of a material. For example, it is the increase in true stress, also called tensile flow stress, with true plastic strain that is a real measure of strain hardening. True strain at fracture, computed with Eq. 20.7b, can be used as a measure of ductility or capacity for plastic deformation. More commonly, ductility is taken as the total engineering plastic strain to fracture, expressed as percent elongation (Fig. 20.9), or as the percent reduction in area. The latter is given by

$$\%\,\text{RA} = \frac{A_0 - A_f}{A_0} \times 100 \tag{20.8}$$

where A_0 and A_f are the initial and final (e.g., neck) cross-sectional areas. Both measures of ductility are useful, but % RA provides a better indication of capacity for plastic deformation.

20.6 STRENGTHENING MECHANISMS IN METALS

Any thermal, chemical, or mechanical process that increases the strength of a ductile crystalline material produces barriers to dislocation motion and thus decreases dislocation mobility. Ways of decreasing dislocation mobility can be classified into strengthening mechanisms, which depend on the nature of the barriers and how dislocations interact with them. Decreasing the grain size of a metal to increase its yield strength and cold working a metal to increase its flow stress are examples of strengthening processes.

Great differences in strain hardening are exhibited by single crystals and polycrystals of the same metal (e.g., Cu, Figs. 20.6 and 20.8). The basic mechanism involved in strain hardening is the increased difficulty of dislocation motion because of an increased dislocation density resulting from plastic flow. Strain hardening is related uniquely to dislocation density changes, and for many metals the resolved shear stress τ necessary to continue plastic flow varies with dislocation density ρ as

$$\tau = \tau_0 + A\sqrt{\rho} \tag{20.9}$$

where τ_0 is the resolved shear stress to move a dislocation in the absence of interfering dislocations, and A is an empirical constant that approximately equals the product of shear modulus G and Burgers vector magnitude b. As shown in Fig. 20.11, data for both single crystals and polycrystals of copper are in accord with Eq. 20.9, thus demonstrating the importance of dislocation density.

In addition to grain size control and strain hardening, a single-phase metal also can be strengthened by solid solution formation. Solute atoms tend to segregate to dislocations (see Chapter 10) because the total elastic strain energy associated with a dislocation and the solute atoms is reduced. This segregation

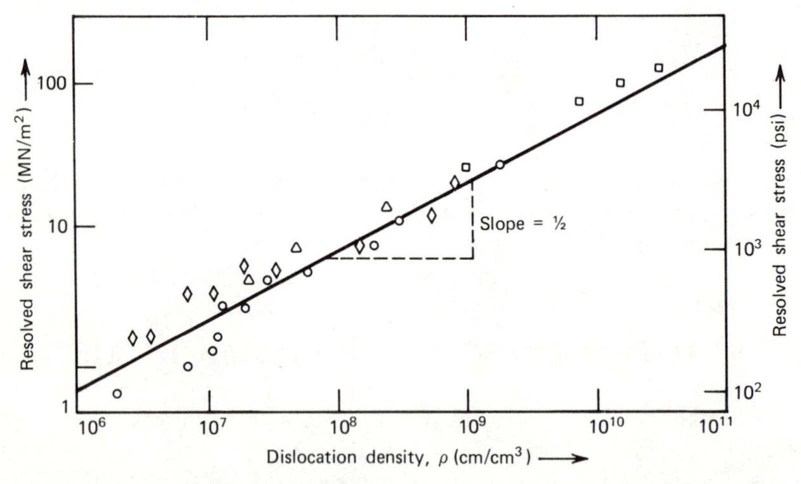

FIGURE 20.11 Resolved shear stress as a function of average dislocation density for copper. Data for single crystals having various orientations and for polycrystals follow the same trend and show that strain hardening is a direct result of increased dislocation density.

□ = polycrystalline copper. Single crystals of copper are oriented so that ○ means that one slip system is operative, ◇ that two slip systems are operative, and △ that six slip systems are operative. [After H. Wiedersich, *J. Metals,* **16,** 425 (1964).]

reduces dislocation mobility because of the binding between solute atoms and the dislocation; that is, a higher resolved shear stress is required to move the dislocation and produce plastic flow. Moving dislocations also interact with solute atoms, and a higher resolved shear stress must be acting in order to prevent a moving dislocation from being trapped by solute atoms.

The effectiveness of a solute atom for strengthening depends on the attractive interaction between it and a dislocation. Since the lattice strain energy associated with a solute increases with the disparity in size between the solvent and solute atoms, the binding energy between a solute and a dislocation increases in a similar manner. For example, interstitial atoms that create severe lattice distortions are particularly effective as solid-solution strengtheners (e.g., C in Fe and N in Nb, Fig. 20.12). Besides size factors, the effective elastic modulus of a solute atom also determines its effectiveness as a solid-solution strengthener. This is because the degree of stress relaxation associated with solute-dislocation segregation depends on the relative *hardness* or *softness* of the two atomic species. The size

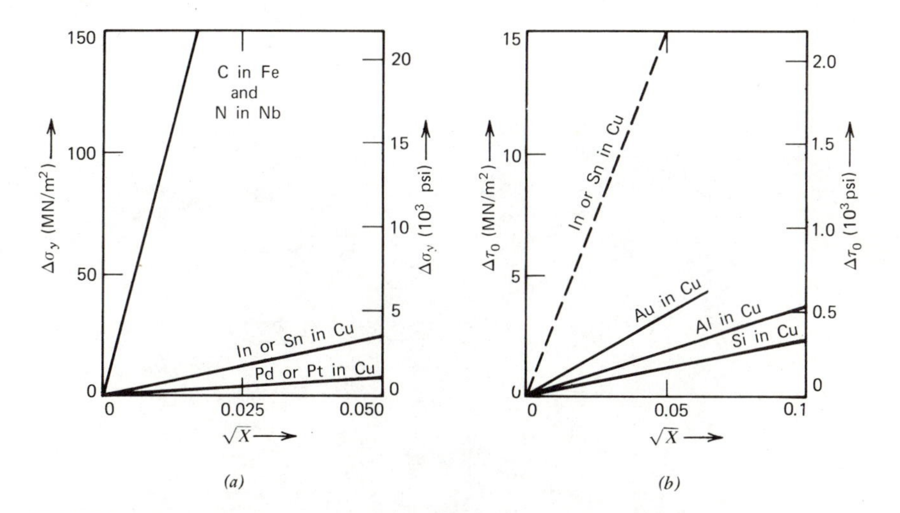

FIGURE 20.12 Variation of strength with the square root of solute mole fraction for dilute solid-solution alloys. (*a*) The yield strength of polycrystalline alloys increases much more when the solute causes an asymmetrical structural distortion (e.g., C in Fe and N in Nb) than when the solute results in a symmetrical structural distortion (e.g., the substitutional solutes in Cu). [Data from: C. Wert, *Trans. AIME,* **188**, 1242 (1950); P. R. V. Evans, *J. Less-Common Metals,* **4**, 78 (1962); R. L. Fleischer, *Acta Met.,* **11**, 203 (1963).]
(*b*) Values of critical resolved shear stress for alloy single crystals exhibit a trend similar to that for the yield strength of polycrystals. Note that the stress and composition scales differ from those in (*a*). [Data from: J. O. Linde and S. Edwards, *Arkiv Fysik,* **8**, 511 (1954); T. J. Koppenaal and M. E. Fine, *Trans. Met. Soc. AIME,* **224**, 347 (1962).]

and effective modulus contributions can be combined into a strengthening parameter K that is different for each solute-solvent combination, and the increase in critical resolved shear stress $\Delta\tau_0$ for dilute solutions can be expressed as

$$\Delta\tau_0 = KX^{1/2} \tag{20.10}$$

where X is the solute mole fraction (see Fig. 20.12). Although Eq. 20.10 is based on both firm theoretical and experimental grounds, it is not always obeyed. For example, at very low solute concentrations the critical resolved shear stress often is observed to increase linearly with solute content.

Deformation of two-phase or multiphase alloys generally is more complicated than that of single-phase materials. In many commercial alloys, the tensile stress–strain behavior can be approximated fairly accurately by a volumetric weighted average of the flow stresses of the individual phases. In the case of an alloy consisting of two ductile phases, for example, the stress supported by the alloy at strain ε is

$$\sigma_{\alpha+\beta}(\epsilon) = f_\alpha \sigma_\alpha(\epsilon) + f_\beta \sigma_\beta(\epsilon) \tag{20.11}$$

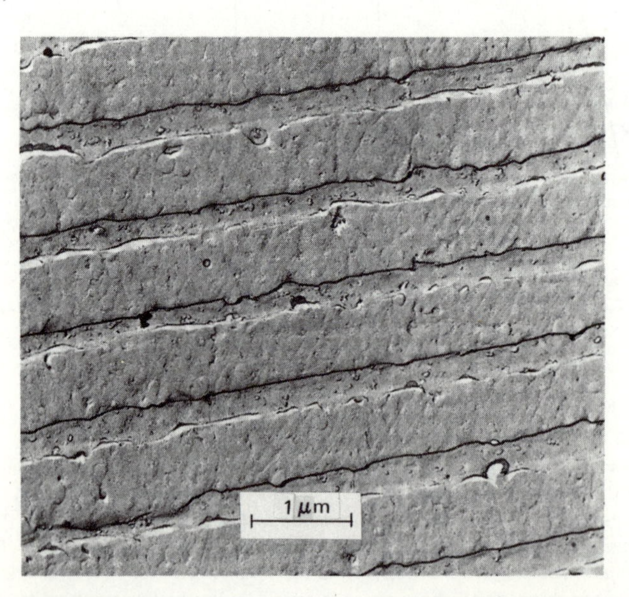

FIGURE 20.13 A micrograph of directionally solidified Ag–Cu eutectic. The individual silver-rich lamellae are 0.67 μm thick and the copper-rich lamellae are 0.27 μm thick. [From H. E. Cline and D. F. Stein, *Trans. TMS-AIME,* **245**, 841 (1969).]

where f_α and f_β are the volume fractions of the α and β phases and $\sigma_\alpha(\epsilon)$ and $\sigma_\beta(\epsilon)$ are the stresses at strain ϵ as determined from the individual stress–strain curves for each phase. This equation does not hold for small strains or beyond the tensile strength of the alloy and is further restricted to alloys where the strain in both phases is essentially equal. Many fiber composites (see Sect. 18.10) exhibit mechanical behavior that is consistent with the volume fraction rule (Eq. 20.11); however, if the phase dispersion is on a very fine scale, care must be taken in using Eq. 20.11. An illustrative example is provided by the directionally solidified, lamellar Cu–Ag eutectic. When solidified at suitable rates, quite thin individual lamellae result (Fig. 20.13), and the yield strength of the eutectic varies with interlamellar spacing according to a Hall–Petch relationship (Fig. 20.14). Thus, the volume fraction rule is applicable here only if spacing effects are taken into account for the individual phases. Similar considerations account for the differences in strength displayed by coarse and fine pearlite (Sect. 18.3).

Although not explicitly stated, the volume fraction rule is based on the assumption that the strains suffered by the individual phases in a multiphase alloy are

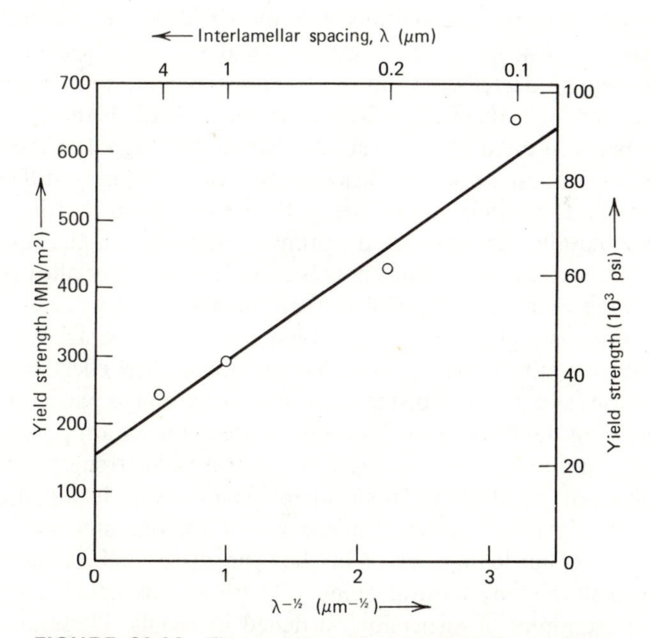

FIGURE 20.14 The variation of yield strength vs. (interlamellar spacing)$^{-1/2}$ for directionally solidified Ag–Cu eutectics. The interlamellar spacing λ is the sum of the individual Ag and Cu lamellar thicknesses. [From H. E. Cline and D. F. Stein, *Trans. TMS-AIME*, **245**, 841 (1969).]

equal. This is not the case in precipitation- or dispersion-strengthened alloys in which a fine dispersion of nondeforming (or only slightly deforming) particles is embedded in a much softer matrix. Here, the volume fraction of the stronger phase is relatively low, and the dispersoid impedes dislocation motion in the matrix, thereby strengthening the material.

The strength of an alloy may be increased as a result of precipitation from solid solution even though this process lowers the solute content of the solid solution. Strengthening occurs because of a fine dispersion of heterogeneities that may be either local regions rich in solute atoms (clusters) or actual second-phase precipitate particles. Both types of heterogeneity impede dislocation motion in the alloy, but second phase particles can be particularly effective. The processing techniques and the conditions necessary for precipitation hardening have been discussed previously (Sect. 18.6). The usual result of a precipitation hardening treatment is a fine dispersion of hard precipitate particles embedded in a relatively soft matrix phase. If the particles cannot deform, dislocation motion in the matrix can occur only by dislocation bowing or extrusion between the particles, a process that also leaves dislocation loops around the particles (Fig. 20.15). The stress necessary for dislocation bowing is given by Eq. 20.4, provided l is now taken as the interparticle spacing (or actually the matrix distance between particles). According to Eq. 20.4, increasingly greater yield strengths will be achieved as interparticle spacing decreases. In most common precipitation-hardening alloys, it becomes easier for a dislocation to cut or shear through the particles rather than to bow between them when the spacing is small enough. For these materials, the scale of dispersion at which dislocations find it equally difficult to bow around or to shear particles correlates with the maximum hardness and yield strength, and the alloy is considered optimally aged. When the microstructure is coarse enough so that dislocations bow between particles, the alloy is considered overaged and the yield strength is lower than the maximum value.

Although not distinguished by high yield strengths, overaged precipitation-hardening alloys strain harden rapidly. This is because the effective interparticle spacing decreases as successive dislocations bow between the particles (Fig. 20.15), thereby increasing the flow stress. By appropriate deformation processing, therefore, an overaged alloy can be produced that has a yield strength comparable to that of an optimally aged alloy. This principle is important in the utilization and preparation of dispersion-hardened alloys which, for our purposes, are distinct from precipitation-hardening alloys in that the dispersoid in them is formed artificially rather than by natural aging. TD (thoria-dispersed) nickel and TD nichrome are examples of dispersion-hardened materials. These alloys are prepared by suitable blending and subsequent compaction of colloidal-size ThO_2 (thoria) particles and a metal or alloy matrix material. Although the virgin strength of these alloys is not remarkable, appropriate thermomechanical treatments can markedly increase the flow stress. When properly processed, such alloys can

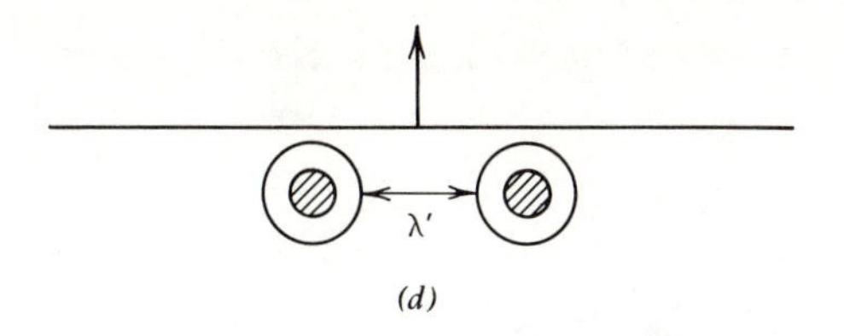

(d)

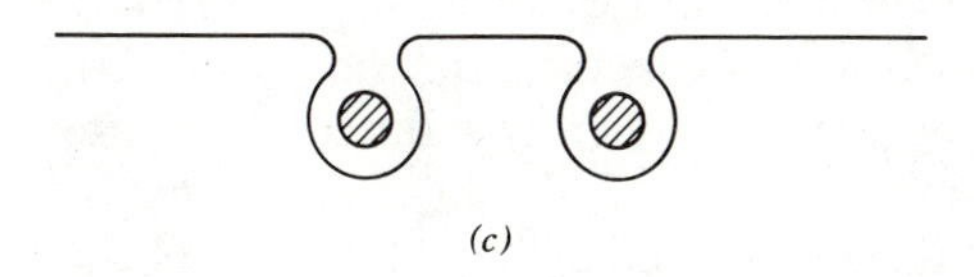
(c)

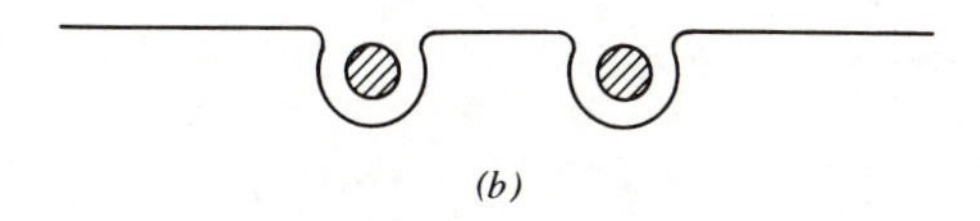
(b)

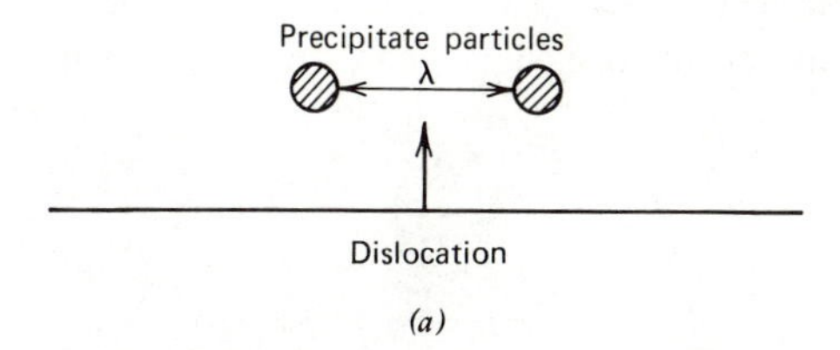

(a)

FIGURE 20.15 The sequence of events (*a*) to (*d*) occurs as a dislocation moves along a slip plane and encounters precipitate particles. In the situation shown, the dislocation bows around the particles leaving each encircled by a dislocation loop. A subsequent dislocation would require a higher stress for the same process to be repeated since the effective interparticle spacing has been reduced to λ' because of the passage of the first dislocation.

be used to very high temperatures without matrix recrystallization, and the stability of the dispersoid particles makes them effective as dislocation barriers even at elevated temperatures.

20.7 TEMPERATURE AND STRAIN RATE EFFECTS

The plastic behavior of crystalline materials depends on both temperature and strain rate. At elevated temperatures, thermal recovery processes that are related to recrystallization can occur concurrently with deformation, and both strength and strain hardening are consequently reduced. The latter effect results in decreasing the difference between yield and tensile strengths until, at sufficiently high temperatures, they are essentially equal. One important consequence of the thermal recovery processes is that viscoplastic deformation (creep) takes place at high temperatures. Creep refers to the continuing permanent elongation of a material under a constant tensile load and will be discussed further in Chapter 21.

Like recrystallization, thermal recovery processes depend on temperature and time. High strain rates can thus compensate somewhat for the reduced strength caused by elevated temperatures. This is true insofar as the thermal recovery time is longer than the time during which deformation takes place. The effect of strain rate on tensile strength is much more important at elevated temperatures than at normal temperatures (Fig. 20.16). The linear relationship on logarithmic coordinates, illustrated in Fig. 20.16, indicates that the tensile strength of copper depends on strain rate according to an equation of the form

$$TS = K'\left(\frac{d\epsilon}{dt}\right)^{m'} = K'(\dot{\epsilon})^{m'} \tag{20.12}$$

The exponent m', which depends on temperature, is a measure of the strain-rate sensitivity of the material. A relationship similar to Eq. 20.12 is observed to hold for the yield strength as a function of strain rate, that is,

$$\sigma_Y = K(\dot{\epsilon})^m \tag{20.13}$$

where m is the strain rate sensitivity exponent. This exponent usually is small at low temperatures but can be appreciable at high temperatures. In certain metals and alloys, m is on the order of 0.5 or greater at elevated temperatures and materials with such values of m behave very much like viscous solids (Sect. 20.8) and are capable of great extension.

Below the temperature at which thermal recovery processes are possible, temperature has a much greater effect on mechanical behavior than does strain rate.

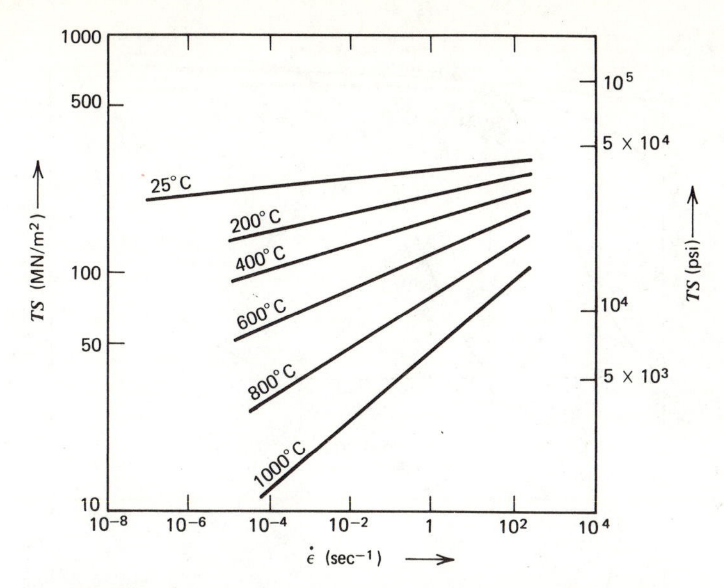

FIGURE 20.16 Tensile strength versus strain rate for polycrystalline copper (T_m = 1084.5°C) at various temperatures. Note that strain rate sensitivity is greater at higher temperatures. [After A. Nadai and M. J. Manjoine, *J. Appl. Mech.*, **8**, *A*77 (1941).]

Strain rate sensitivity exponents ordinarily are small at low temperatures where, for fcc metals, for example, m is of the order of 0.01. Body-centered cubic metals are more strain rate sensitive at low temperatures, but m is still typically only on the order of 0.1 at most. Since $m = 0$ corresponds to no strain rate sensitivity, it is apparent that at ordinary temperatures, the plastic behavior of metals is not greatly affected by strain rate.

Temperature has a marked influence on low temperature deformation of crystalline materials. Mention has been made of how temperature affects the number of active slip systems in some metals (Sect. 20.2), and this determines the ductility of polycrystalline metals such as Zn and Mg. Furthermore, dislocations move less readily as temperature decreases. Because of these factors, yield and tensile strengths of polycrystalline materials increase and their ductility, measured in terms of percent reduction in area, decreases with decreasing temperature (Fig. 20.17). As noted in Fig. 20.17, below a certain temperature—the ductile-to-brittle transition temperature—bcc transition metals are prone to fail prior to any measurable plastic deformation. On the other hand, face-centered cubic metals seldom show such a transition and their ductility decreases only slightly with decreasing temperature.

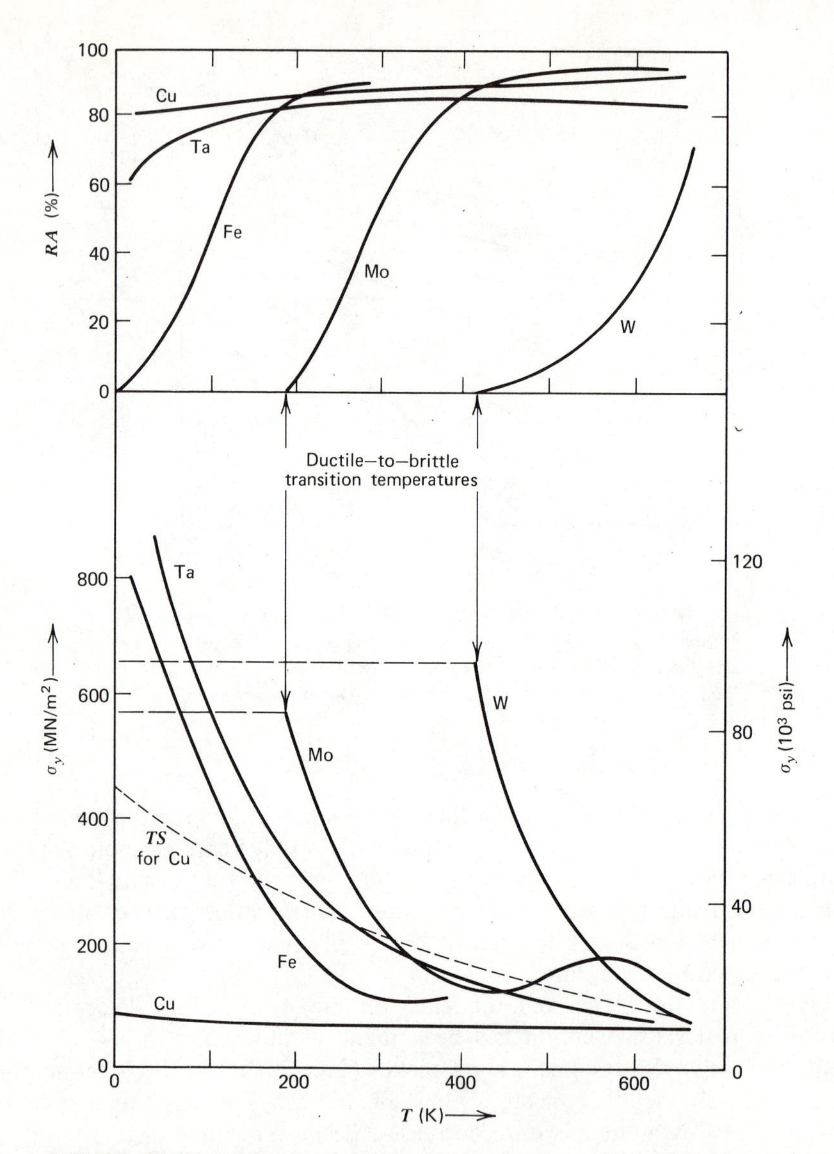

FIGURE 20.17 Variation of mechanical properties with temperature for constant strain rate conditions. Note that with the exception of tensile strength the properties of fcc metals are markedly less temperature sensitive than those of bcc metals. The ductile-to-brittle transition temperatures for some of the latter are noted. [Adapted in part from J. H. Bechtold, *Acta Met.*, **3**, 249 (1955), and from R. P. Carreker and W. R. Hibbard, *Acta Met.*, **1**, 654 (1953).]

20.8 VISCOUS FLOW

In contrast to an ideal crystalline solid, an ideal fluid adjusts its geometry continually and uniformly when subjected to an applied shear stress. The behavior of many common gases and liquids approaches that of the ideal fluid represented schematically in Fig. 20.18. Here, the amount of viscous flow, as measured in terms

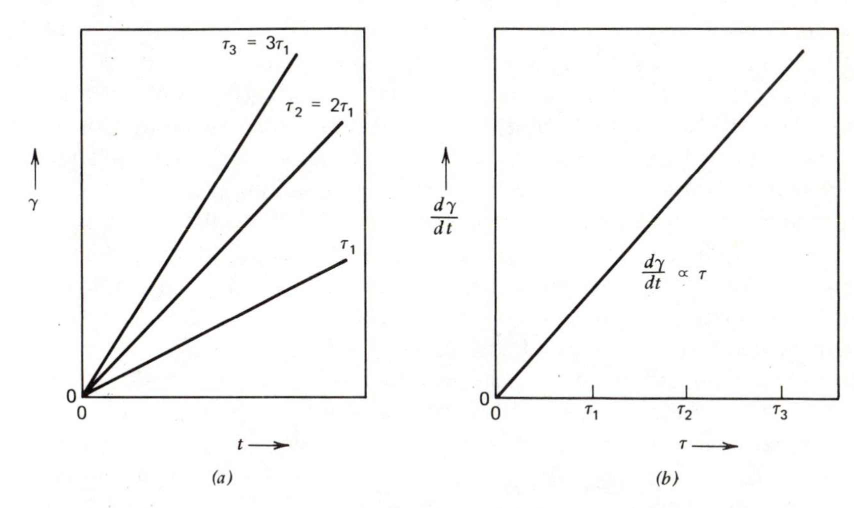

FIGURE 20.18 Phenomenological description of linear viscous flow. (*a*) Amount of flow versus time at different constant shear stress levels. (*b*) Shear strain rate, $(d\gamma/dt) = \dot{\gamma}$, varies linearly with applied shear stress.

of shear strain by which layers of atoms or molecules flow over one another, varies linearly with time for a constant applied shear stress. In addition, the rate of viscous flow or shear strain rate is directly proportional to the applied shear stress:

$$\frac{d\gamma}{dt} = \frac{1}{\eta}\tau \tag{20.14}$$

This equation defines viscosity η, which is a property of the fluid; the reciprocal of viscosity is called fluidity and hence the higher the viscosity the less *fluid* the material is and vice versa.

An inorganic glass or a thermoplastic polymer, above the glass transition temperature, can undergo significant viscous flow when subjected to sufficiently large stresses. Unlike an ideal fluid, however, these materials can withstand tensile stresses, and the tensile equivalent of Eq. 20.14 frequently is used to describe their behavior:

$$\sigma_T = 3\eta \frac{d\epsilon_T}{dt} = 3\eta\dot{\epsilon}_T \tag{20.15}$$

The ability to sustain normal stresses is due to the internal structure of these materials and is reflected in their high viscosities relative to simple fluids (Table 20.3). Such high viscosities are due to their relatively strong interparticle bonding that results in less atomic or molecular motion which is necessary for viscous flow. Above the glass transition temperature, continuous thermal motion of atomic groups occurs (e.g., SiO_4 units in a silica glass or molecular chain segments in a long-chain polymer move as a result of thermal agitation), and an applied stress biases local configurations so that the particles switch neighbors preferentially in order to produce a shape change consonant with the applied stress. Since many such local configurational rearrangements must occur for an observable macroscopic deformation, the viscosity is a measure of the ease and time scale for such rearrangements. Thus, simple fluids respond rapidly and have low viscosities, but noncrystalline solids are more sluggish. The molecular mechanism responsible for viscosity in these materials also implies a strain rate sensitivity with respect to deformation mode. For example, a ball of asphalt or of silicone putty will behave elastically if stressed quickly (brittle fracture can occur if a very high deformation rate is attempted) or viscously if stressed slowly.

TABLE 20.3 Viscosities of Selected Substances (at 20°C Unless Otherwise Noted)

Substance	η (Pa·sec[a])
Air	0.000019
Water	0.00100
Mercury	0.0156
Lead (liquid)	0.0026 (at 350°C)
Glycerin	1.49
Asphalt	$\sim 10^7$
Soda-lime-silica glass	10^{19} (at 20°C[b])
	10^7 (at 600°C)
	10 (at 1400°C)

[a] 1 Pascal · sec = 10 dyne · sec/cm² = 10 poise.

[b] This is well below the glass transition temperature ($T_g \approx 550$°C).

Since thermal motion is important in viscous flow, the process is highly dependent on temperature and for many substances the viscosity η varies according to an inverse Arrhenius relationship

$$\eta = \eta_0 e^{Q/RT} \tag{20.16}$$

As shown in Table 20.3, the viscosity of a glass changes by many orders of magnitude on cooling from a temperature where it is very fluid to its glass transition temperature where it can no longer properly be considered as viscouslike. The pre-exponential coefficient η_0 and Q, the activation energy in Eq. 20.16, depend on the bonding and structure of the material.

Viscous flow is of primary importance in the processing of many noncrystalline solids. The ready formability of inorganic glasses at elevated temperatures by glass blowing results because there is no tendency for necking to occur. This can be demonstrated easily by heating a soft glass rod uniformly in a Bunsen burner and then pulling it into a very long, fine filament. The elongation is accompanied by a uniform decrease in cross-sectional area. Long-chain polymers may be fabricated in much the same manner as inorganic glasses, but tearing because of molecular separation tends to occur after a certain amount of strain.

The tensile characteristics of semicrystalline polymers between T_g and T_m are also utilized in the production of strong polymeric filaments. When such a material is pulled in tension (e.g., polyethylene, Fig. 19.4) after yielding, permanent deformation proceeds at a comparable but lower engineering stress level. Subsequent to a certain amount of uniform permanent deformation (the amount depending on the material), a constriction or neck forms, and further flow continues at an approximately constant stress as the neck propagates along the specimen while the cross-sectional area of the neck itself remains constant. The formation of the neck does not lead to failure because the molecules within it have been reoriented parallel to the applied stress, making it stronger. During this process of *drawing*, which is used in the production of synthetic textile fibers, the original random spherulitic pattern is destroyed and replaced by a more ordered, anisotropic crystalline structure that affords greater strength parallel to the filament axis. Furthermore, such filaments also are considerably stiffer than their isotropic counterparts.

REFERENCES

1. Biggs, W. D. *The Mechanical Behaviour of Engineering Materials*, Pergamon Press, Oxford, 1965.
2. Hayden, H. W., Moffatt, W. G., Wulff, J. *The Structure and Properties of Materials*, Vol. III, Wiley, New York, 1965.

QUESTIONS AND PROBLEMS

20.1 Slip in NaCl-structure ionic crystals occurs most commonly on $\{110\}$ planes and in $\langle 1\bar{1}0\rangle$ directions.

(a) How many slip systems are there?

(b) In terms of the lattice parameter, what is the length of the Burgers vector?

(c) Sketch the (001) plane showing the ions in a square region containing at least 16 unit cell faces; (i) indicate the (110) slip plane trace and the $[\bar{1}10]$ slip direction; (ii) remove half of the anions on a $(1\bar{1}0)$ plane trace and half of the cations on a neighboring $(2\bar{2}0)$ plane trace, and show that the result is a edge dislocation having the correct Burgers vector.

(d) How are the anions and cations arrayed along the edge dislocation line (i.e., along the termination edge of the half plane)?

20.2 A zinc single crystal is oriented with the normal to the basal plane making an angle of 60° with the tensile axis, and the three slip directions making angles of 38°, 45°, and 85° with the tensile axis. If plastic deformation is first observed at a tensile stress of 330 psi, what is the critical resolved shear stress for Zn?

20.3 If the critical resolved shear stress for Al is 35 psi (0.24 MN/m^2) calculate the tensile stress required to cause yielding in a crystal having an [011] axial orientation. What compressive stress would cause this crystal to yield?

20.4 When an fcc single crystal is pulled in tension along the [001] axis, how many slip systems are equally favorable?

20.5 The theoretical shear strength of a crystalline solid can be roughly approximated by assuming $\tau = \tau_{max} \sin(2\pi x/b)$ describes the nonlinear elastic behavior in shear. In this equation, τ_{max} is the theoretical shear strength, x is the shear displacement of one atomic plane with respect to its neighboring atomic plane, and b is the interatomic distance along the direction of shear.

(a) Determine τ_{max}, in terms of the shear modulus G, when one close-packed plane is sheared in a close-packed direction over an adjacent close-packed plane in an fcc metal. (*Hint.* Remember that linear elastic behavior occurs for small strains, and shear strain is given by x/d, where d is the interplanar spacing.)

(b) Would your answer to part (a) differ significantly if you considered a bcc or an hcp metal? Explain.

(c) Using your answer to part (a), give some numerical values of theoretical shear strength for a few metals, taking $G = E/2(1 + \nu)$ for polycrystals.

20.6 Refer to Fig. 20.2 and calculate the shear strain associated with deformation twinning in a bcc metal.

20.7 It has been found experimentally that the resolved shear stress (or shear flow stress) during plastic flow for various fcc and bcc single crystals and polycrystals is related to dislocation density, ρ, as

$$\tau = \tau_0 + \alpha G b \sqrt{\rho}$$

In this equation, τ_0 is the stress to move a dislocation in the absence of interfering dislocations (i.e., essentially the critical resolved shear stress if the initial dislocation density is less than about 10^4 cm of line/cm^3), α is a constant (≈ 0.5), G is the shear modulus, and b is the Burgers vector magnitude.

(a) If $\tau_0 = 100$ psi for Cu single crystals with an initial dislocation density of $\rho = 10^5$ cm^{-2}, what will be the observed critical resolved shear stress? (Data for Cu: $G = 6 \times 10^6$ psi, $b = 0.256$ nm.)

(b) Repeat part (a) for initial dislocation densities of 10^6 cm^{-2} and 10^8 cm^{-2}.

(c) Using the information given in part (a) and with reference to Fig. 20.6, what is the dislocation density of the [111] Cu single crystal after 1% plastic strain? How much plastic strain must the [112] Cu single crystal suffer to have the same dislocation density? (*Hint.* Use the initial orientation relations to convert τ to σ so that an answer can be obtained.)

20.8 Consider moving dislocations being pinned at the mean interdislocation spacing $l \approx \rho^{-1/2}$, where ρ is the dislocation density in cm of line/cm^3. Assume that for motion to continue, dislocation segments must be extruded in the same manner as the segment of a dislocation source, so that the shear flow stress is given by Eq. 20.4. Calculate the dislocation density in a Cu single crystal that has been strain hardened to the stage where the applied tensile stress causes the resolved shear stress to be 14 MN/m^2. (For Cu, $G = 4.1 \times 10^{10}$ N/m^2 and $b = 0.256$ nm.)

20.9 For initially annealed metal polycrystals, the plastic portion of the true stress–strain curve can often be approximated by

$$\sigma_T = K\varepsilon_T^n$$

where K and n are empirically determined constants called the strength coefficient and the strain-hardening exponent, respectively. (*Note.* This equation should not be confused with Eq. 20.12, which gives the *strain rate* sensitivity.) For Cu, $n \approx 0.5$ and $n \approx 0.2$ for Al.

(a) Which metal has a greater strain-hardening capability? Why?

(b) For an equivalent amount of plastic strain, will Cu or Al have a greater dislocation density? Explain.

20.10 Necking begins when the tensile strength is reached in a tension test, that is, when the slope of the engineering stress–strain curve equals zero.

(a) Show that this is equivalent to $d\sigma_T/d\varepsilon_T = \sigma_T$.

(b) Using the equation in problem 20.9, what does the true strain equal at the onset of necking?

20.11 Explain why the level of the engineering stress–strain curve is lower than that of the true stress–strain curve in a tension test, while the opposite is true in a compression test. (*Note.* The true stress–strain curve is the same in tension and compression).

20.12 (a) Refer to Fig. 19.4, and provide the following information for polycrystalline aluminum: 0.2% offset yield strength, tensile strength, percent elongation.

(b) If this Al specimen had an initial diameter $d_0 = 0.505$ in., what is its diameter at the onset of necking? What is the true stress at the onset of necking?

(c) Is the information given in Fig. 20.4 sufficient to allow the calculation of percent reduction in area? Explain.

20.13 For the Cu specimen whose engineering stress–strain curve is shown in Fig. 20.9*a*, the true stress–strain curve can be approximately represented by $\sigma_T \approx 75000\, \varepsilon_T^{0.45}$ psi.

(a) Using this relationship, plot σ_T vs. ε_T up to $\varepsilon_T = 2.0$, the true strain at fracture.

(b) Compare this curve to that in Fig. 20.9*a*, and comment on the differences and similarities.

20.14 (a) Show that $\ln(l_i/l_0)$ and $\ln(A_0/A_i)$ are equivalent ways of representing true strain during plastic flow (i.e., when volume is constant).

(b) Comment on why only the latter form is appropriate after necking begins.

20.15 Hardness is a measure of the resistance to penetration by an indenter. In common hardness tests, a standard load is used to force an indenter into the surface of a specimen, and the hardness number is determined either in terms of the depth of penetration (e.g., Rockwell hardness) or in terms of the applied load divided by the area of contact after penetration (e.g., Brinell hardness). As a metal is cold worked, how does its hardness vary with the amount of cold work? Explain.

20.16 A severely cold-worked sample of polycrystalline copper has a dislocation density of 10^{12} cm of line/cm^3; when completely recrystallized, the same sample has a dislocation density of 10^7 cm of line/cm^3. The energy per unit length associated with a dislocation line is approximately Gb^2, where G and b are the shear modulus and the Burgers vector magnitude, respectively.

(a) Calculate (in MJ/m^3) the stored energy released when the Cu sample is recrystallized ($G = 4.1 \times 10^{10}$ N/m^2; $b = 0.256$ nm).

(b) Compare your answer in part (a) with the enthalpy of fusion for Cu ($\Delta H_m = 1840$ MJ/m^3), and calculate the upper limit of dislocation density by assuming ΔH_m is the maximum stored energy that is possible.

20.17 Schematically sketch the temperature variation of the yield strength for a bcc metal (i) with a fine grain size and (ii) with a coarse grain size. Assuming that the stress for brittle fracture is independent of temperature and grain size, use your sketch to show that the ductile-to-brittle transition temperature is higher for the fine-grained material than for the coarse-grained material.

20.18 Determine the strain rate sensitivity exponent (i.e., m' in $TS \approx K'\dot{\varepsilon}^{m'}$) for the tensile strength of copper, using the data in Fig. 20.16. Plot m' as a function of temperature.

20.19 Consider an inorganic glass rod having a constriction as shown:

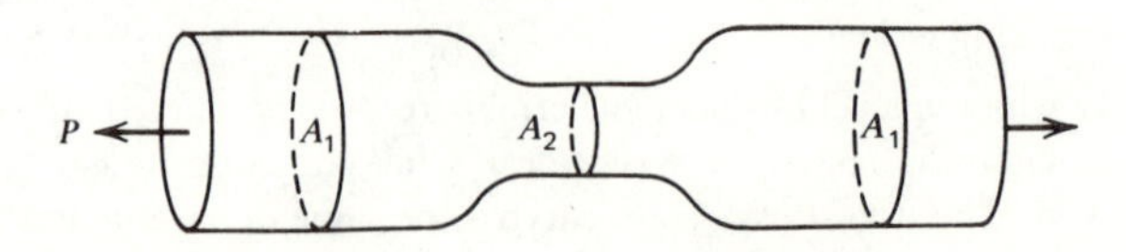

(a) Show that if it behaves in accordance with linear viscous flow (i.e., when subjected to small stresses well above the softening temperature), then the difference between the cross-sectional area of the constricted region and the cross-sectional area outside of the constricted region remains constant as the rod is pulled.

(b) What does the result to part (a) demonstrate about the tendency toward necking?

20.20 The drawing of a semicrystalline polymer, during which a neck forms and lengthens along the specimen, is quite similar phenomenologically to what happens at the lower yield point of a mild steel, where an inhomogeneous deformation band propagates along the specimen. What are the similarities and the differences?

20.21 List what you consider to be the important strengthening mechanisms in martensitic steel.

20.22 Would you expect impurity atoms to segregate to stacking faults in metals? Explain. How would this affect the yield strength?

20.23 (a) Consider a powder metallurgy product made up of 50 % Al and 50 % Cu by volume. Draw the expected stress–strain diagram for this material assuming the stress–strain curve of the aluminum in the mixture is shown in Fig. 19.4 and that of copper is shown in Fig. 20.9*a*.
(b) How would you expect the stress–strain behavior of the mixture to vary as successively finer powder is used in making up the composite?
(c) If this powder were held for a long period of time at 500°C, would the internal structure change?

20.24 Assuming that the yield strength of Ag lamellae is given by an equation of the form $\sigma_{Ag} = \sigma_{Ag_0} + k_{Ag}\lambda_{Ag}^{-1/2}$ where λ_{Ag} is the lamellar thickness of the silver and a similar equation holds for copper lamellae, show that the yield strength of the composite illustrated in Fig. 20.13 should follow an equation of the form, $\sigma_{\alpha-\beta} = (\sigma_{\alpha-\beta})_0 + k_{\alpha-\beta}\lambda^{-1/2}$ (Fig. 20.14). Identify the quantities $(\sigma_{\alpha-\beta})_0$ and $k_{\alpha-\beta}$.

20.25 Compare Fig. 20.15 and Fig. 20.7. Explain why similar equations describe the stress necessary to generate the loop in Fig. 20.7 and to allow the dislocation to pass between dispersed particles, as shown in Fig. 20.15.

21

Mechanical Failure

This chapter describes the various modes in which materials may fail by mechanical means in service.

21.1 TENSILE FRACTURE

The amount of reduction in area, as measured in a tension test, can be used to characterize tensile failure. If failure is accompanied by little or no reduction in area (Fig. 21.1*a*), brittle fracture is said to have occurred. Materials that exhibit brittle fracture include most ceramics and inorganic glasses at room temperature, long-chain polymers below their glass transition temperatures, network polymers, and certain metals and alloys below their ductile-to-brittle transition temperatures. If, as with most metals and alloys, a measurable permanent reduction in area is observed upon failure, the failure is designated as ductile fracture and usually is preceded by necking (Fig. 21.1*b*). A special case of ductile fracture, rupture, is illustrated in Fig. 21.1*c*. Rupture corresponds to essentially a 100% reduction in area at the point of fracture and is observed with some polymers and inorganic glasses above their glass transition temperatures, very malleable single-phase metals, and some metals and alloys at high temperatures.

21.2 DUCTILE FRACTURE OF METALS

The most common type of ductile fracture in tension occurs after appreciable plastic deformation and exhibits three stages: (1) after the onset of necking, cavities form (usually at inclusions or second-phase particles) in the necked region because the geometry has induced a hydrostatic tensile component of stress; (2) the cavities grow and coalesce in the center of the specimen, forming a penny-shaped crack that grows outward in a direction perpendicular to the applied tensile stress; (3) final failure involves rapid crack propagation along directions that tend to be at an angle of 45° with the tensile axis. This final stage results in a shear lip, and the 45° angle between the shear lip and the tensile axis corresponds to the direction of maximum shear stress. Figure 21.2 illustrates the series of events that culminate in a cup-cone fracture. The fracture surface consists of a central fibrous zone and an outer shear lip zone (Fig. 21.2*e*). Certain ductile metals, when tested in tension, also exhibit stages (1) and (2), but final failure involves rupture of the cylindrical rim that surrounds the central crack in the necked region. This gives a double cup fracture, and there is no shear lip.

Although cup-cone, double cup, and pure rupture are observed frequently in tensile specimens of metals and alloys having low strengths, various other ductile fracture surfaces occur with medium and high strength alloys. Certain quenched and tempered steels, for example, exhibit a small, central fibrous zone and a small, outer shear lip, separated by a serrated radial zone that covers much of the cross section. The radial zone, which has star-shaped markings consisting of ridges and valleys that extend in radial directions, is a result of rapid crack propagation.

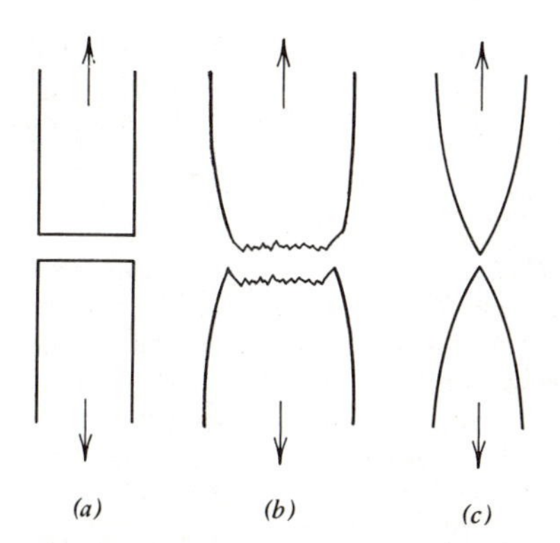

FIGURE 21.1 Types of tensile fracture observed in materials. (*a*) Brittle fracture occurs before measurable plastic deformation. (*b*) Ductile fracture occurs after measurable plastic deformation and a neck forms before final tensile failure. (*c*) Rupture is a special case of ductile fracture in which the reduction in area at fracture is essentially 100%.

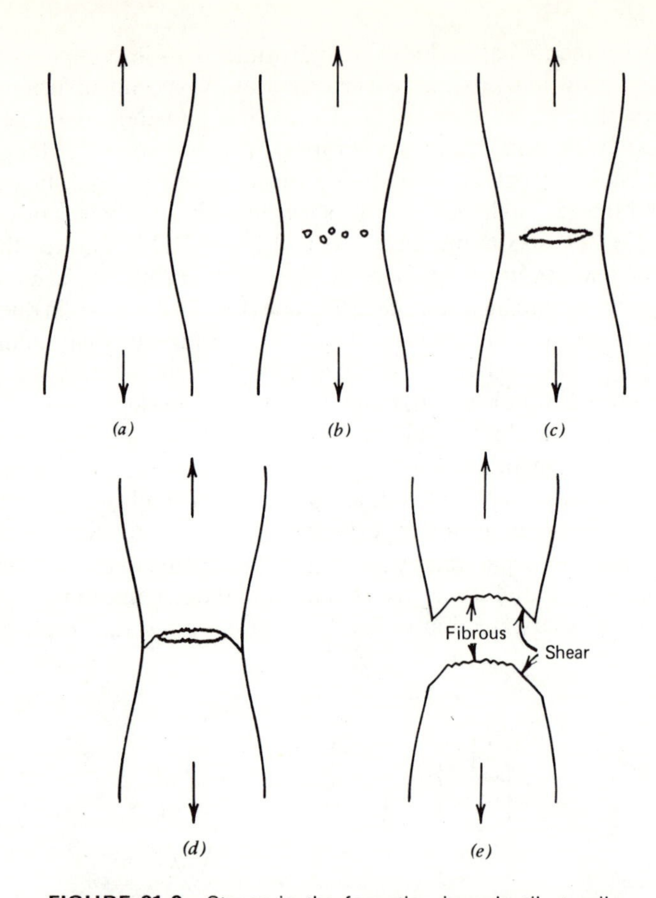

FIGURE 21.2 Stages in the formation in a ductile tensile fracture. Cavities form in the center of the specimen (*b*) subsequent to neck formation (*a*). The cavities coalesce to form a penny-shaped crack (*c*) that then grows outward toward the specimen edge (*d*). The final fracture occurs by shear at a 45° angle to the tensile axis (*d*, *e*). The cup-cone fracture resulting has a dull, fibrous nature over the center of the specimen.

Ductile fracture is not particularly important in terms of mechanical behavior because it usually is associated with good toughness and, under conditions of tensile loading, the onset of necking constitutes failure insofar as tensile load bearing capability is concerned. Certain classes of materials, such as long-chain polymers and body-centered cubic transition metals, may exhibit either ductile or brittle behavior, and the latter is of great engineering importance.

21.3 BRITTLE FRACTURE

Totally brittle fracture is characterized by failure without any prior permanent deformation. Brittle fracture usually occurs at stress levels well below those predicted theoretically from the inherent strength due to atomic or molecular bonds. This situation in some respects is analogous to the discrepancy between the theoretical shear strength of perfect crystals and the much lower observed yield strength values. As with plastic flow, the presence of defects is responsible for the low fracture strengths observed with brittle materials. It was proposed by Griffith in 1920 that microscopic cracks exist both within and on the surface of all real materials. Such cracks are deleterious to the strength of any material that possesses negligible ductility, that is, a small capacity for permanent deformation. In particular, the presence of cracks whose long dimension is perpendicular to the direction of an applied tensile stress gives rise to especially large stress concentrations at the crack tip, where the actual local stress can approach the theoretical strength of the material. If the cracks are assumed to be elliptical in shape as shown in Fig. 21.3, the highest stress (σ_m) at the tip of the crack is given approximately by

$$\sigma_m \simeq \sigma\left(\frac{2c}{\rho}\right)^{1/2} \tag{21.1}$$

where σ is the nominal (i.e., applied) tensile stress, ρ is the radius of curvature at the crack tip, and c is the *length of an interior crack* or *twice the full length of a surface crack*. Since the radius of curvature of the crack tip cannot be less than the order of an atomic diameter, σ_m always must remain finite but can attain the theoretical fracture strength of the material (problem 21.1) even though the bulk of the material is under a fairly low applied stress.

A high local stress at the tip of a crack by itself does not guarantee that a crack will propagate and lead to brittle failure. Rather, a crack will propagate only if there is an accompanying net reduction in energy. If an internal crack of the type shown in Fig. 21.3 propagates, new surface is created and this leads to an increase in total surface energy with increasing crack size. The total surface energy per unit thickness (U_s) of a flat elliptical crack extending through a thin plate is approximately

$$U_s = 2c\gamma \tag{21.2}$$

where the factor of two arises because the crack has two surfaces, each of length c. This surface energy tends to restrict crack growth, but propagation of the crack can lead to a decrease in elastic strain energy. For a thin plate subjected to a fixed

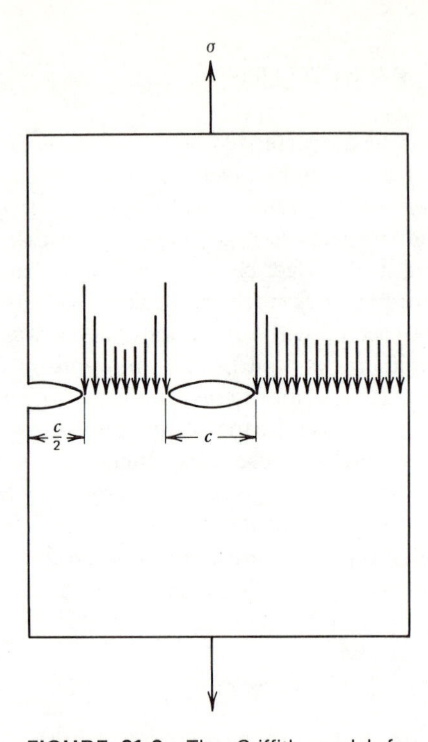

FIGURE 21.3 The Griffith model for microcrack-induced brittle failure. Cracks oriented perpendicular to the tensile axis act as effective stress raisers near the crack tip as noted by the increasing lengths of the arrows signifying the amount of stress.

strain, the total elastic energy per unit thickness (U_e) due to the presence of an elliptical crack extending through the plate is given approximately by

$$U_e = -\frac{\pi c^2 \sigma^2}{4E} \tag{21.3}$$

where σ is the nominal tensile stress and E is the elastic modulus of the plate. That is, Eq. 21.3 gives the difference between the elastic energy for a plate containing a crack and that for a plate without a crack. The total energy (U) associated with the crack is the sum of Eqs. 21.2 and 21.3, which is shown schematically in Fig. 21.4*a*. The form of this plot is similar to the variation of free energy associated with nucleus formation during a phase change (cf. Fig. 17.1), and Fig. 21.4*a* can be interpreted in much the same manner. For a given applied stress σ, cracks having lengths less than the critical value c^* will not propagate because the energy increase due to creation of new surface exceeds the elastic strain energy released. However, cracks with lengths greater than c^* will propagate spontaneously with a resultant decrease in total energy. Although it is theoretically possible for thermal fluctuations to nucleate cracks or to activate subcritical cracks having $c < c^*$, the magnitude of the activation energy U^* is so large that this cannot happen in actuality. In fact, only an applied stress is capable of causing preexisting cracks to propagate. Thus, for the situation represented in Fig. 21.4*b*, a crack of length c_1 is stable at the stress σ_1, but increasing the stress to σ_2 leads to spontaneous propagation of this crack.

The critical stress σ_F required to make a crack of a given size unstable can be obtained by differentiating $U\ (= U_s + U_e)$ with respect to c and setting the derivative equal to zero. The result is

$$\sigma_F = \left(\frac{4\gamma E}{\pi c}\right)^{1/2} \tag{21.4}$$

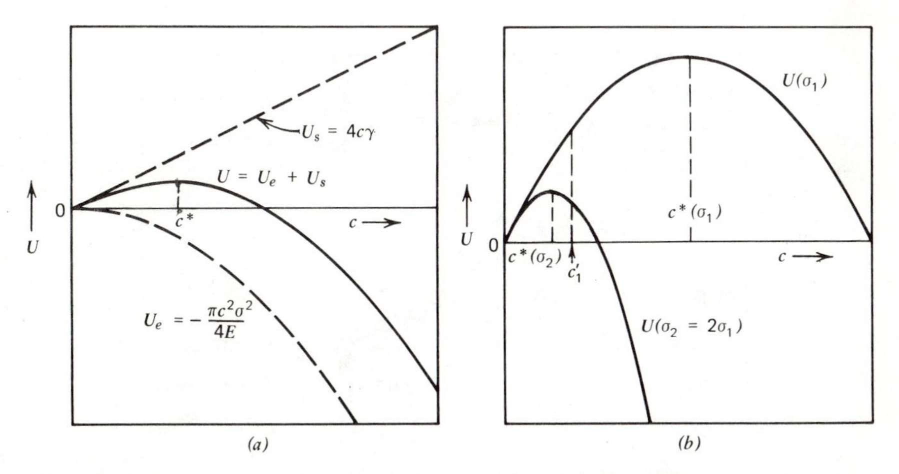

FIGURE 21.4 (*a*) The total energy, U, because of an internal crack, as a function of its length, c, depends on two terms. The surface energy term, U_s, is linear in c and dominates the total energy for small values of c, whereas the elastic strain energy release term, U_e, is quadratic in c and dominates the total energy for large values of c. For the case shown, cracks longer than c^* will propagate spontaneously at the given stress level. Cracks shorter than c^* will not. (*b*) The total energy, U, as a function of crack length for two different stress levels, σ_1 and $\sigma_2 = 2\sigma_1$. Doubling the stress reduces c^* by a factor of four. Note that a crack of length c_1 would be stable at stress σ_1 but is unstable at stress σ_2.

which applies strictly to the case of a flat elliptical crack through a thin plate (Fig. 21.3). For other configurations, the numerical factor changes but is still on the order of unity; thus, the fracture strength is given approximately by

$$\sigma_F \approx \left(\frac{\gamma E}{c}\right)^{1/2} \tag{21.5}$$

Either Eq. 21.4 or Eq. 21.5 leads to the conclusion that if a distribution of crack sizes exists within (or on the surface of) a brittle body, the fracture strength will be determined by the most severe flaw, that is, the longest crack. Some very simple experiments illustrate this principle. Testing an inorganic glass rod in tension will result in failure at a certain stress level. According to the above model, this is due to the propagation of the most severe preexisting surface crack, since only surface flaws are of importance in the case of most inorganic glasses. Upon testing the fragments of the broken rod, higher fracture strengths will be observed because the flaws in these portions must be less severe than in the original rod. That such behavior is observed provides qualitative verification of the correctness of the Griffith theory as applied to intrinsically brittle materials. Quantitative confirmation also has been achieved since γ calculated with Eq. 21.4 compares favorably with the actual surface energy of an inorganic glass.

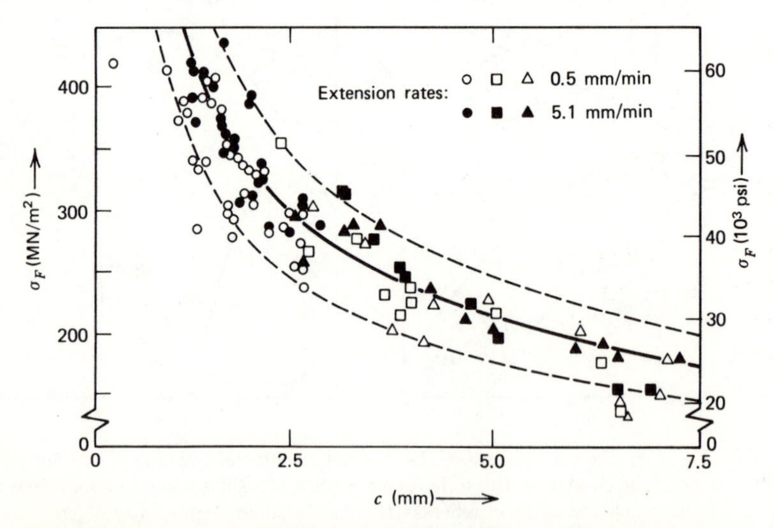

FIGURE 21.5 The fracture strength of polystyrene as a function of crack length. The functional variation of σ_F with c fits the form of the equation for brittle fracture (Eq. 21.4), but γ calculated with this equation (1300 J/m^2) is much greater than the actual surface energy ($\gamma_s \approx 0.5$ J/m^2). [After J. P. Berry, *J. Polymer Sci.*, **50**, 313 (1961).]

The above treatment applies fairly well to rigid network polymers, and we might expect that it should be valid for glassy long-chain polymers, which exhibit macroscopic brittle behavior. Experiments on glassy polymers, having artificially introduced cracks of various sizes, demonstrate that the functional dependence of fracture strength on crack size as given by Eq. 21.4 is correct (Fig. 21.5); however, quantitative agreement is lacking in that γ calculated with Eq. 21.4 is two or three orders of magnitude larger than actual surface energies. This discrepancy arises because energy is dissipated locally as a result of molecular orientation as a crack propagates (Fig. 21.6), and this is in addition to the energy associated with new surface area. To account for this, γ in Eq. 21.4 or Eq. 21.5 is replaced by $\gamma_s + \gamma_p$,

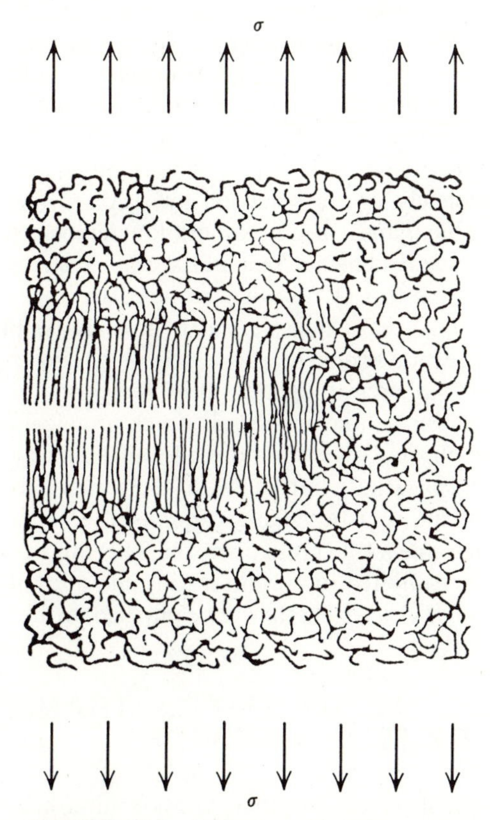

FIGURE 21.6 Molecular orientation in the vicinity of a crack in a glassy polymer. As the crack propagates, viscous flow leading to molecular rearrangement occurs locally in front of the crack tip. (After J. P. Berry and A. M. Bueche, General Electric Co.)

where γ_s is the surface energy and γ_p is a permanent deformation (viscous flow) work term. In effect, the capacity for permanent deformation leads to a *blunting* of the crack, and crack propagation requires a higher applied stress level.

Metals, which always possess some capacity for permanent deformation by plastic flow, behave in a manner similar to glassy polymers, and the Griffith theory must be modified accordingly. That is, γ in Eq. 21.4 or Eq. 21.5 is replaced by $\gamma_s + \gamma_p$, where γ_p is a plastic work term in this case. In a macroscopically brittle metal, a local plastic zone exists in front of the crack tip where the stress essentially equals the yield strength and may be considerably less than the stress calculated with Eq. 21.1. The ratio of γ_p to γ_s can be estimated roughly as one-sixth of the ratio of a material's theoretical fracture strength to its observed yield strength (i.e., $\sim E/60\ \sigma_Y$). For most metals, therefore, γ_p is much greater than γ_s.

With crystalline solids there are three cases where γ_p is important. First, in the case of nonmetallic crystals, which have a low density of dislocation sources and often relatively immobile dislocations, only a limited amount of plastic flow occurs in the vicinity of the crack tip even though the local tensile stress exceeds the yield strength, and γ_p is only slightly larger than γ_s. Consequently, these materials usually exhibit completely brittle behavior by cleavage (i.e., separation along definite crystallographic planes) at normal and low temperatures. Second, in the case of most fcc metals and some hcp metals, the applied stress necessary to propagate a preexisting crack (Eq. 21.5 with γ replaced by $\gamma_s + \gamma_p$) is much greater than the yield strength so that extensive plastic flow always precedes failure and preexisting cracks are not prone to propagate or produce premature failure. Such materials possess inherent toughness at all temperatures. Third, in the case of bcc transition metals and some hcp metals, crack propagation may occur prior to the onset of gross plastic flow even though localized plastic deformation does accompany crack propagation. Indeed, with such materials, preexisting flaws need not be present because plastic yielding can cause the formation of microcracks having lengths comparable to the grain diameter. This third case, where a ductile-to-brittle transition is exhibited by certain metals and alloys that display a large temperature variation of yield strength, will be pursued further in the next section.

21.4 THE DUCTILE TO BRITTLE TRANSITION AND QUASI-BRITTLE BEHAVIOR

Various bcc transition metals and their alloys exhibit brittle behavior at low temperatures, and the change in fracture mode from ductile to brittle is related to the strong temperature dependence of yield strength found with these materials. Above the ductile-to-brittle *transition temperature*, the yield strength is lower than the tensile stress necessary to cause brittle failure. With decreasing temperature, however, the yield strength increases rapidly to the point where it equals the tensile stress for brittle failure, and below this temperature, fracture usually occurs on yielding but without gross plastic deformation. Inhomogeneous plastic flow, such as that accompanying yielding in a metal having a yield point phenomenon, can cause cleavage cracks (microcracks) to form across some grains if the temperature is sufficiently low. Thus, grain size imposes an inherent flaw size (i.e., $c = d$, where d is the mean grain diameter) in certain polycrystalline metals, but the flaws must be created by dislocation motion.

At the transition temperature, the microcracks that form are of critical size for crack propagation (i.e., $c^* = d$) relative to the applied tensile stress, σ_Y (T_2 in Fig. 21.7), and at lower temperatures, the microcracks exceed the critical size (T_3 in Fig. 21.7). Accordingly, at and below the transition temperature, quasi-brittle fracture occurs on yielding. (The designation quasi-brittle is utilized to

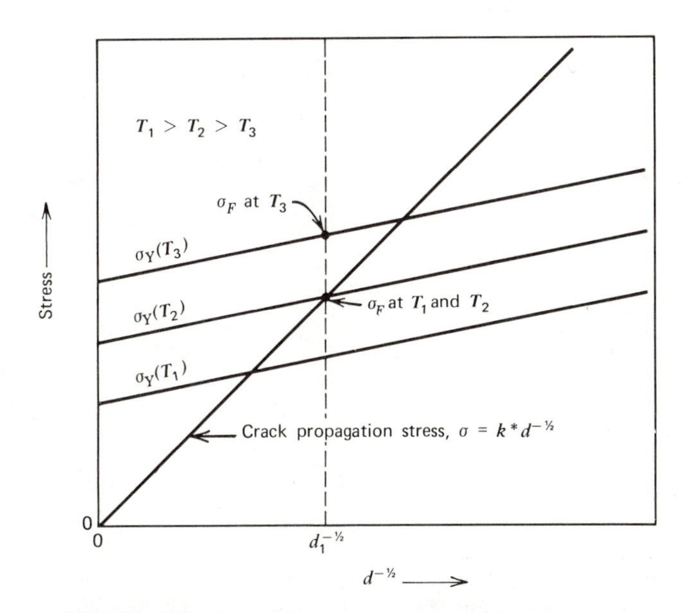

FIGURE 21.7 Dependence of crack nucleation stress and crack propagation stress on grain size at low temperatures for a bcc transition metal subjected to tension. The crack nucleation stress is the same as the yield strength given by Eq. 20.5 (i.e., $\sigma_Y = \sigma_0 + kd^{-1/2}$), and this is the fracture stress for large grain sizes where σ_Y exceeds the crack propagation stress (i.e., $\sigma = k^* d^{-1/2}$, where $k^* \approx \sqrt{\gamma_p E}$). For small grain sizes, the fracture stress corresponds to the crack propagation stress, which is larger than σ_Y. Thus, for a given grain size d_1, strain hardening precedes fracture at T_1, whereas at T_2 ($<T_1$) and at T_3 ($<T_2$), fracture coincides with yielding. (In this example, T_2 is the transition temperature.)

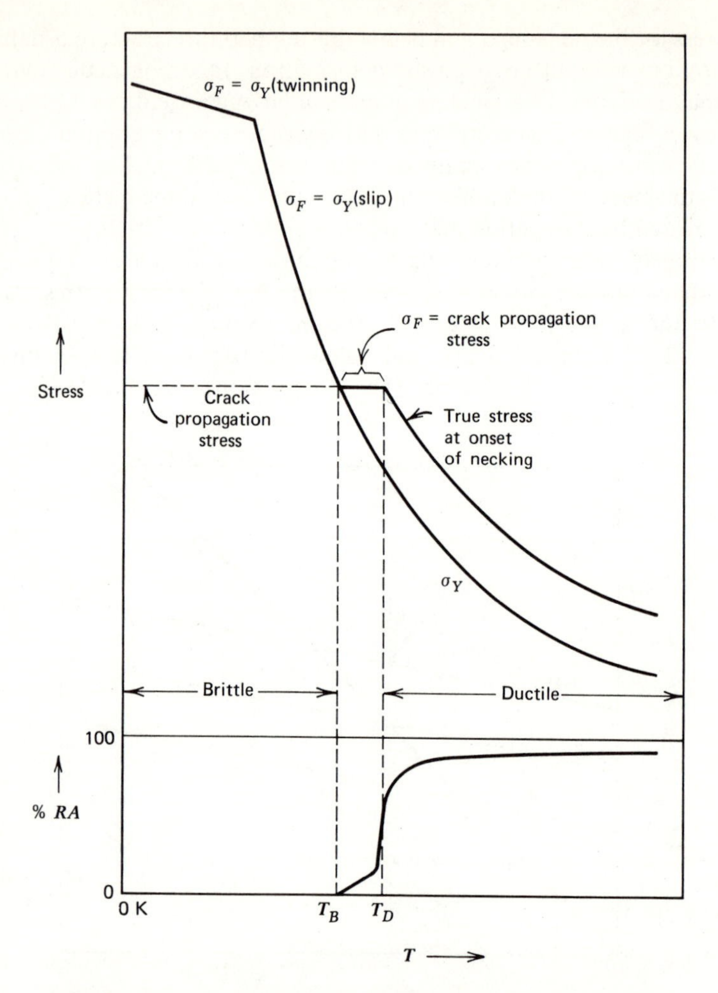

FIGURE 21.8 Idealized tensile behavior as a function of temperature for a metal that undergoes a ductile-to-brittle transition. Below the transition temperature (T_B), fracture occurs on plastic yielding either by slip or by twinning, and the fracture strength equals the yield strength, with no macroscopic plastic flow. Above T_B, strain hardening occurs before fracture, and the percent reduction in area increases with increasing temperature. At the toughness or ductility transition temperature (T_D), there is a very rapid change in % RA, and at higher temperatures, necking and ductile fracture occur.

distinguish macroscopically brittle behavior involving local permanent deformation from completely brittle behavior.) Over a temperature range just above the transition temperature, the microcracks formed are initially subcritical so that further plastic deformation and strain hardening must proceed before the tensile stress level becomes sufficient to cause crack propagation (T_1 in Fig. 21.7). Under these conditions, the metal undergoes uniform deformation without necking prior to failure and the extent of uniform elongation increases with increasing temperature, but fracture is almost completely by cleavage. With further increase in temperature, microcracks no longer form and the fracture mode changes from cleavage to ductile. A large change in toughness and ductility accompanies the change in fracture. The ductility variation with temperature and some of the above features are illustrated schematically in Fig. 21.8. The crack propagation stress (essentially Eq. 21.5 with γ replaced by γ_p and c replaced by d) usually is relatively insensitive to temperature, as shown in Fig. 21.8. Certain polycrystalline

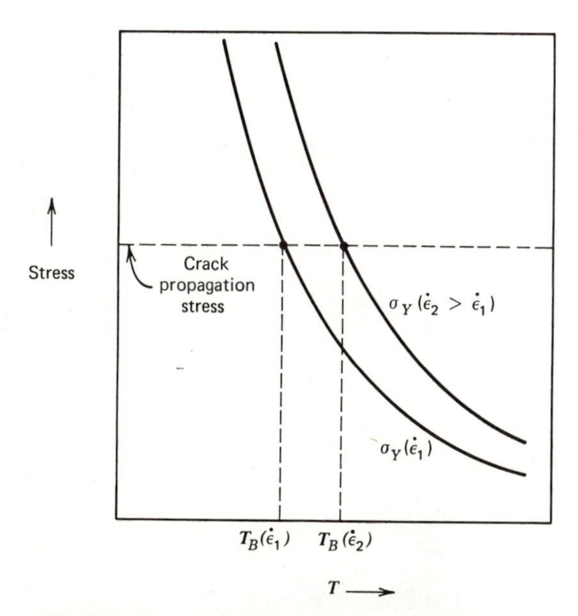

FIGURE 21.9 Schematic effect of strain rate on the transition temperature of a tensile specimen which undergoes a ductile-to-brittle transition. When the yield strength exceeds the crack propagation stress, microcracks that form on yielding propagate to give cleavage fracture. A higher strain rate increases the yield strength but not the crack propagation stress; thus, increasing the strain rate increases the transition temperature (and also the toughness or ductility transition temperature).

metals, such as Mo and W below their transition temperatures, undergo brittle failure at the crack propagation stress rather than at the yield strength because microcracks can form or be initiated at impurity-embrittled grain boundaries prior to general yielding (cf. Fig. 20.17). Even in this case, however, the microcracks have a size characteristic of the grain size.

Strain rate, which increases the yield strength but not the crack propagation stress, increases the ductile-to-brittle transition temperature (Fig. 21.9). In addition

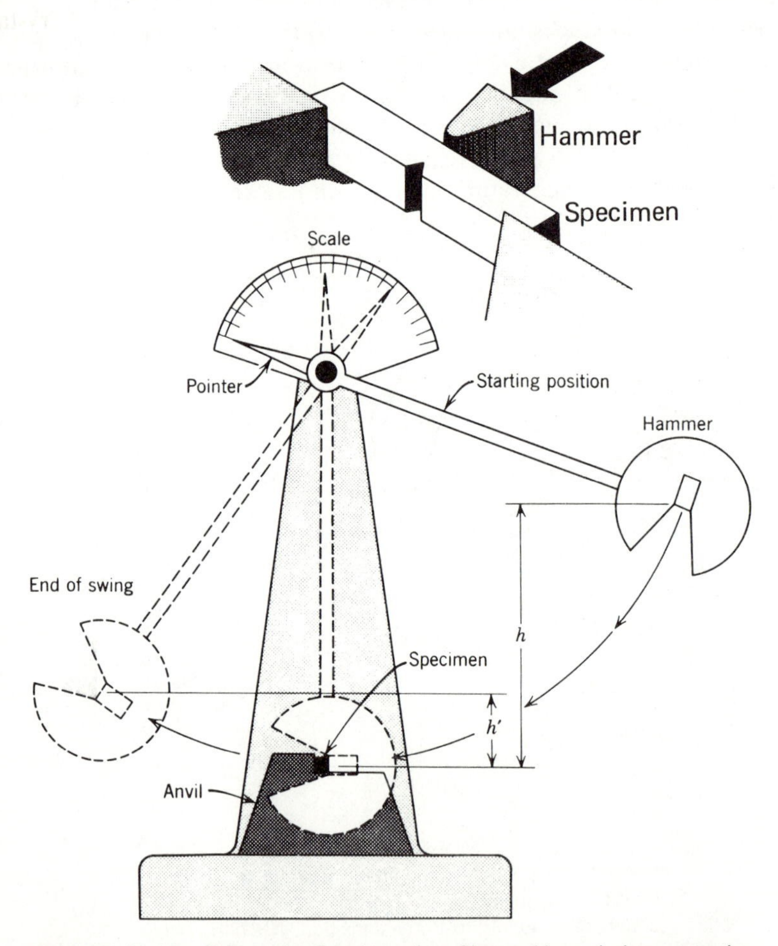

FIGURE 21.10 Schematic drawing of a Charpy V-notch impact-testing apparatus. The hammer is released from a fixed height h and strikes the notched specimen in the manner shown in the insert. The amount of energy expended in breaking the specimen is given by $mg(h - h')$, which is the loss in potential energy of the hammer. (From *The Structure and Properties of Materials*, Vol. III, by H. W. Hayden, W. G. Moffatt, and J. Wulff, Wiley, New York, 1965.)

to low temperature and high strain rate, quasi-brittle behavior is favored by virtually any factor that raises the stress necessary to cause yielding and thus impedes the onset of plastic flow in a susceptible metal. Accordingly, a state of multidirectional (or triaxial) tensile stress promotes quasi-brittle fracture, and preexisting cracks in metals that can exhibit quasi-brittle behavior are of great importance because they raise the transition temperature and also lower the fracture strength and toughness, particularly at low temperatures.

A simple and convenient way to determine a material's propensity for quasi-brittle behavior is afforded by an impact test. In one commonly used impact test, the Charpy V-notch test (Fig. 21.10), a pendulumlike hammer released from a fixed height strikes the specimen at a high velocity and rises to a height lower than the original. The V-notch in the specimen (Fig. 21.10) provides triaxiality of stress, and the high hammer velocity insures a high strain rate. The specimen usually

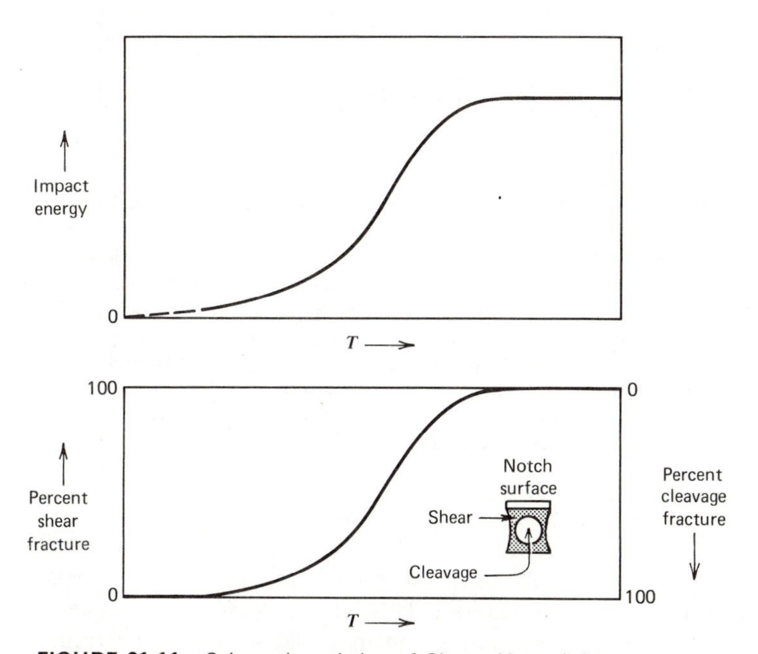

FIGURE 21.11 Schematic variation of Charpy V-notch impact energy and fracture surface appearance as a function of testing temperature. The observed decrease in impact energy correlates well with increasing brittle (i.e., cleavage) fracture character. The transition from ductile-to-brittle behavior in the case of the Charpy V-notch test occurs at much higher temperatures than if an unnotched tensile specimen were being tested. This is a consequence of both stress triaxiality, due to the notch, and the high strain rate. A metal that is not susceptible to a transition to brittle behavior would maintain a high impact energy even at very low temperatures.

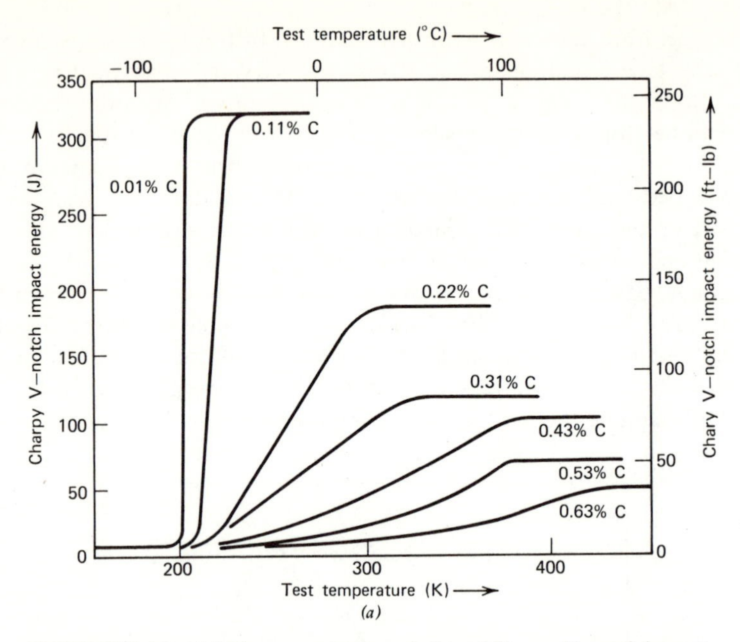

FIGURE 21.12 (*a*) The temperature variation of Charpy V-notch impact energy for steel as a function of carbon content. For iron and steel containing low carbon contents, the impact transition temperature is easily defined and the maximum impact energy is relatively large. Higher carbon contents lead to a broader transition, and the transition temperature must be arbitrarily defined. For example, the temperature corresponding to the average impact energy can be taken as the transition temperature. Note that the maximum impact energy decreases with increasing carbon content (i.e., increasing amount of pearlite). [After J. A. Rinebolt and W. J. Harris, Jr., *Trans. ASM,* **43**, 1175 (1951).]

breaks or tears in this process, and the difference between the initial and final potential energies of the hammer (i.e., the energy expended) is a measure of the fracture resistance of the material. This impact energy can be determined as a function of specimen temperature, and its variation with temperature is related to the ductile-to-brittle transition—for the very special conditions of such an impact test (Fig. 21.11). Because the state of stress is not uniform across the cross section of a Charpy V-notch specimen, the fracture surface often contains regions where fibrous fracture followed macroscopic plastic flow and regions where cleavage fracture occurred with only localized plastic deformation. Crack initiation and propagation by cleavage (i.e., the separation of grains along specific atomic planes) are common features of brittle behavior in bcc transition metals and in other brittle crystalline solids, and both polycrystals and single crystals

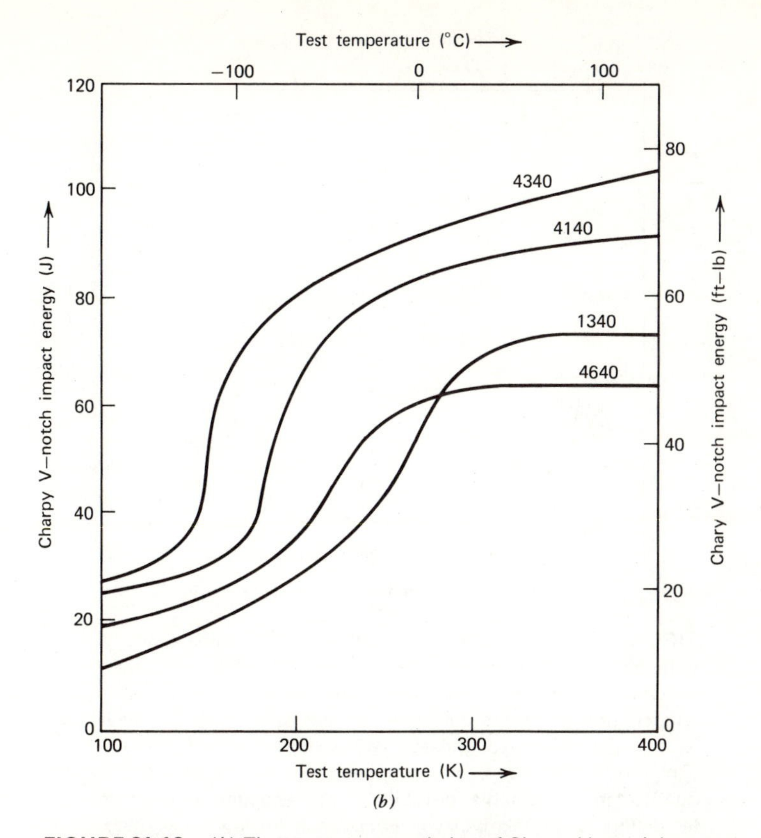

(*b*)

FIGURE 21.12 (*b*) The temperature variation of Charpy V-notch impact energy for some quenched and tempered alloy steels containing 0.40 wt. % C. Each steel was tempered to the same Rockwell hardness (R_C = 35). [After H. J. French, *Trans. AIME,* **206**, 770 (1956).]

can fail by cleavage. In addition, polycrystalline metals may fail in a quasi-brittle manner by grain boundary separation because of weakness resulting from brittle second-phase particles or impurity segregation.

Many steels are among the technologically important materials that exhibit quasi-brittle behavior, and under certain conditions their ductile-to-brittle transition temperatures are close to room temperature. As shown in Fig. 21.12, the transition temperature and the maximum level of impact energy are sensitive to both composition and microstructure. The effect of microstructure and processing treatment is illustrated further in the case of tempered martensite, where the room temperature impact energy is increased significantly by tempering treatments

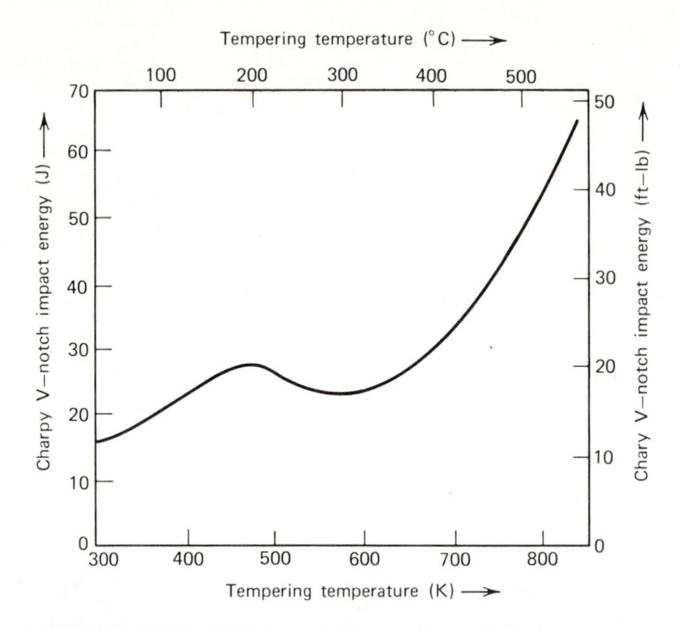

FIGURE 21.13 Variation of Charpy V-notch impact energy with tempering temperature for a quenched 4340 steel. Subsequent to tempering, all specimens were tested at room temperature. In general, a higher tempering temperature leads to a higher notch toughness (i.e., a higher impact energy). (The dip in the curve near a tempering temperature of 300°C is due to an unfavorable distribution of cementite formed on tempering. Since this behavior is common for low-alloy steels, tempering treatments in this range usually are avoided.) [After H. J. French, *Trans. AIME,* **206**, 770 (1956).]

at higher temperatures (Fig. 21.13). It is important to point out, however, that the level of impact energy at room temperature is insufficient to ascertain the relative propensity of two steels for quasi-brittle behavior. For example, a 1340 steel* has a higher impact energy at room temperature than a 4640 steel, but the transition temperature of the 1340 steel also is higher (Fig. 21.12*b*).

A final comment concerning the ductile-to-brittle transition temperature is warranted. First, it does not have a unique value for a given material but depends on flaw size, strain rate, triaxiality of stress, and so forth. Second, there are various conventions used for characterizing it, such as those based on the fracture initiation mode, on the crack propagation mode, or on the appearance of the fracture surface.

* AISI code number for steels: the last two digits indicate carbon content (0.40 wt. % C in this case); the first two digits are a code for the alloying elements in the steel.

From a practical point of view, design utilizing quasi-brittle materials is a most important problem. That is, one wishes to utilize quasi-brittle materials because of their inherently high yield strengths but engineering design must incorporate factors that prevent brittle failure in use. Indeed, a whole new field of engineering science, fracture mechanics—which lies outside the scope of this text—has developed around the engineering utilization of quasi-brittle materials.

21.5 FATIGUE FAILURE

Fatigue failure refers to the fracture of a material as a result of repeated stress (or strain) cycling, typically at levels *below* those necessary for yielding or, in some cases, above the yield strength but below the tensile strength of the material. Time varying stresses are expeienced in aircraft structures, bridges, and a wide variety of machine components. Fatigue failures occur in both metallic and nonmetallic materials and are responsible for a large fraction of identifiable service failures of metals.

Although the detailed stress-time history of a material subject to cyclic loading may be complicated, the intrinsic response of a material to fatigue conditions can be ascertained by fatigue testing. One simple loading cycle used to study fatigue is shown in Fig. 21.14. In this case the material is loaded sinusoidally between tensile and compressive stresses equal in magnitude. The number of cycles necessary to produce failure is determined as a function of the stress amplitude (or, in this case, peak stress level). The higher the stress amplitude (S), the fewer cycles (N) to failure; typical results, in terms of S–N curves, are shown in Fig. 21.15. The fatigue behavior of mild steel and of polymethyl methacrylate are of particular interest, for these materials exhibit fatigue limits. That is, stress amplitudes below

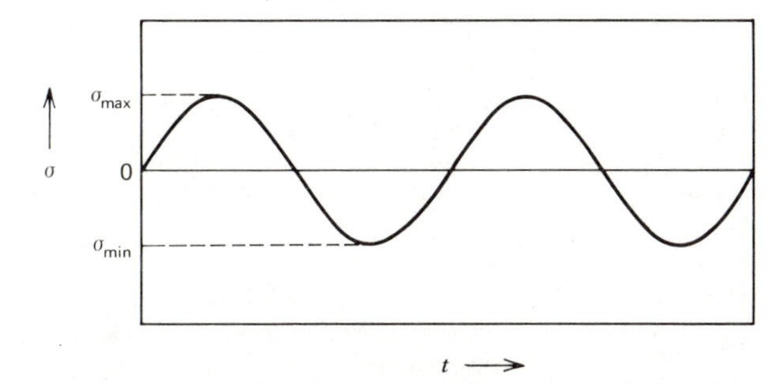

FIGURE 21.14 A time-varying loading cycle that is often used to determine the fatigue properties of a material. In the most common fatigue test, $\sigma_{min} = -\sigma_{max}$ and the mean stress, $\sigma_m = (\sigma_{max} + \sigma_{min})/2$ is therefore zero.

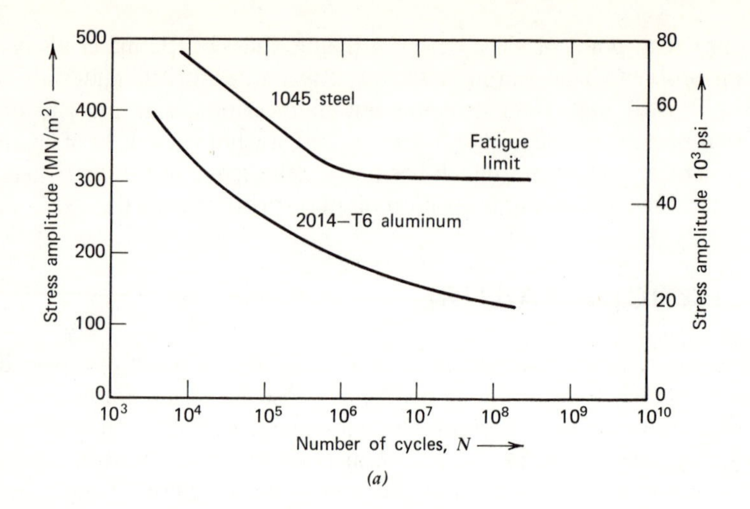

(a)

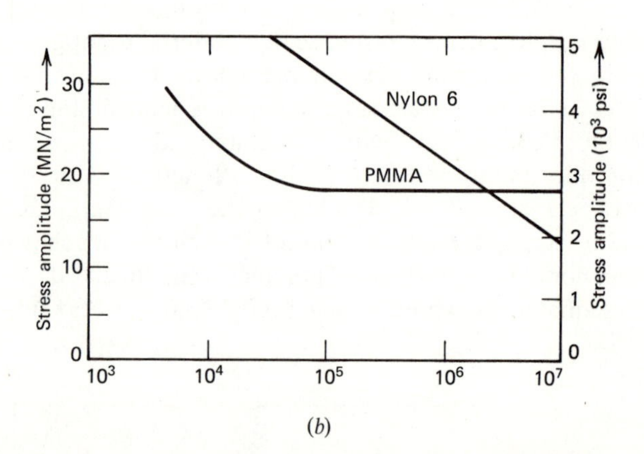

(b)

FIGURE 21.15 (*a*) Experimentally determined *S–N* curves for a plain-carbon steel (0.47% carbon) and an age-hardened aluminum alloy. These curves are for a mean stress of zero. The steel shows a fatigue limit, that is, a stress below which it will not fail regardless of the number of cycles. The aluminum alloy shows no fatigue limit, but the slope of stress vs. log *N* becomes less negative as the stress decreases. (Adapted from *The Structure and Properties of Materials*, Vol. III, by H. W. Hayden, W. G. Moffatt, and J. Wulff, Wiley, New York, 1965.)

(*b*) *S–N* curves for the polymers nylon and PMMA (Lucite). Lucite shows a fatigue limit, but nylon does not. [After M. N. Riddell, G. P. Koo, and J. L. O'Toole, *Polymer Eng. Sci.*, **6**, 363 (1966).]

a certain level will not cause failure regardless of the number of stress cycles. Fatigue limits are exhibited by almost all ferrous alloys, and a general rule of thumb gives the fatigue limit as approximately 0.4 to 0.5 of the steel's tensile strength. In contrast to ferrous materials, most nonferrous alloys and many polymers, such as nylon, do not have a fatigue limit. In principle, this means that these materials will fail eventually when subjected to cyclic loading regardless of the stress amplitude. In practice, however, the number of cycles required for fatigue failure at low stress amplitudes (say, less than about 30% of the tensile strength) is sufficiently large that failure does not occur over the expected lifetime of the part.

With the exception of rotating shafts, most structural members are not subject to symmetrical stress cycles having a mean stress of zero. For a given stress amplitude, a mean tensile stress reduces the fatigue life of a material. The Goodman diagram (Fig. 21.16) provides a convenient way to estimate allowable stress ranges for a material under such conditions, providing that the tensile strength

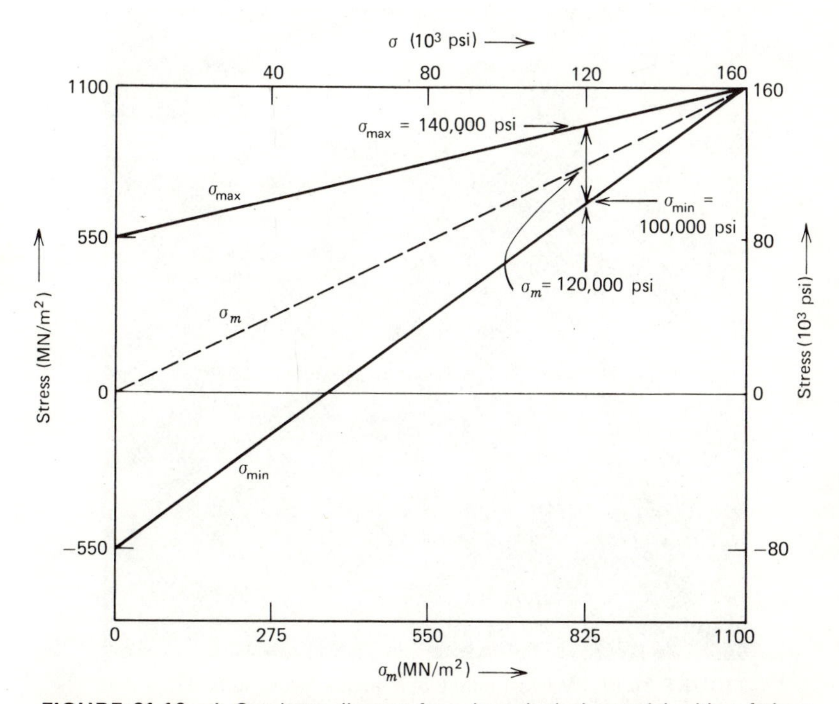

FIGURE 21.16 A Goodman diagram for a hypothetical material with a fatigue limit of 80,000 psi for $\sigma_m = 0$ and a tensile strength of 160,000 psi. The allowable range of stress for $\sigma_m \neq 0$ is estimated by the difference between the upper and lower lines. Thus for $\sigma_m = 120{,}000$ psi, the allowable range of stress is 40,000 psi, that is, $\sigma_{max} = 140{,}000$ psi and $\sigma_{min} = 100{,}000$ psi.

and the fatigue limit (or allowable stress amplitude, with zero mean stress, for a material without an endurance limit) are known for conditions of a symmetrical stress cycle. A Goodman diagram is constructed as follows. The abscissa is chosen to represent the mean stress, which is defined as $\sigma_m = (\sigma_{\max} + \sigma_{\min})/2$, and the ordinate represents the maximum and minimum allowable stresses, $\sigma_{\max}$ and $\sigma_{\min}$. The latter two define the allowable range of stress (i.e., $\sigma_{\max} - \sigma_{\min}$), which is also twice the allowable stress amplitude. At a mean stress of zero, the material can be cycled between tensile and compressive stresses whose magnitudes equal the endurance limit stress. When the mean stress equals the tensile strength, the allowable range of stress must go to zero. It is assumed that both $\sigma_{\max}$ and $\sigma_{\min}$ vary linearly with σ_m between $\sigma_m = 0$ and $\sigma_m = TS$; thus, the allowable range of stress decreases linearly within these limits, going to zero when $\sigma_m = TS$. It should be remarked that the Goodman diagram only provides an estimate of the allowable safe stress limits under cyclic loading conditions.

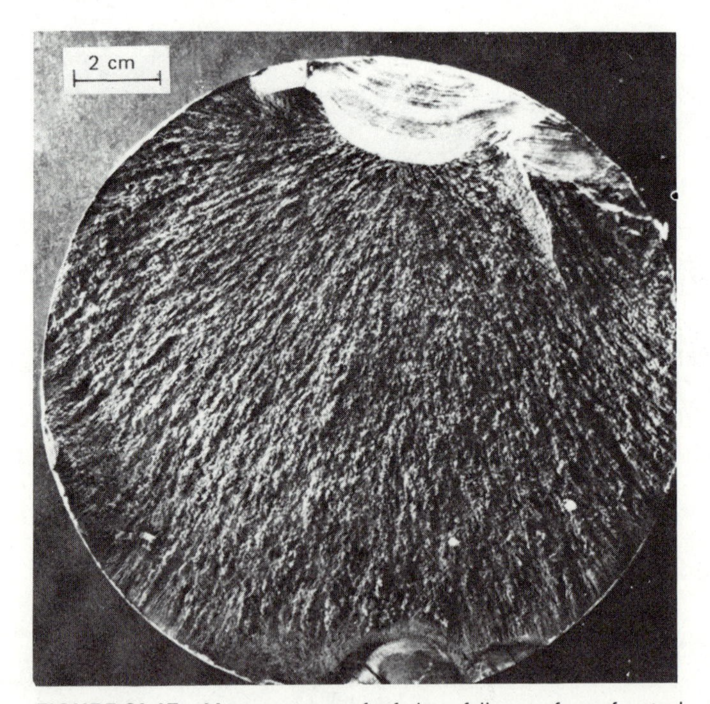

FIGURE 21.17 Macrostructure of a fatigue failure surface of a steel piston rod. Fracture originated at the top edge. The smooth area, with the "clamshell" markings, corresponds to slow fatigue crack growth, and the dull fibrous section is the region of fast fracture. (Reproduced by permission, from *Metals Handbook*, 8th Edition, Vol. 9, American Society for Metals, 1974.)

Fatigue failures usually are found to initiate at a free surface or at internal flaws such as inclusions, where the local stress causes some heterogeneous permanent flow, leading to formation of a small crack. The initiation of a fatigue crack does not lead to immediate failure; rather, the crack propagates slowly and discontinuously across the specimen under the action of cyclic stress. The amount of crack motion per cycle depends on the material and the stress level; high stresses favor larger crack growth increments per cycle. Eventually, the crack propagates to the point where the remaining intact cross section of material no longer can support the applied load, and further crack propagation is rapid, leading to catastrophic failure. The final fracture surface is composed of an area over which there was slow crack propagation and an area where the crack moved rapidly. Frequently, these areas are discernible on a macroscopic scale (Fig. 21.17). If the final fracture surface is sufficiently clean of debris, high magnification electron microscopy can be employed to observe the individual "steps" of crack propagation during the slow growth stage (Fig. 21.18).

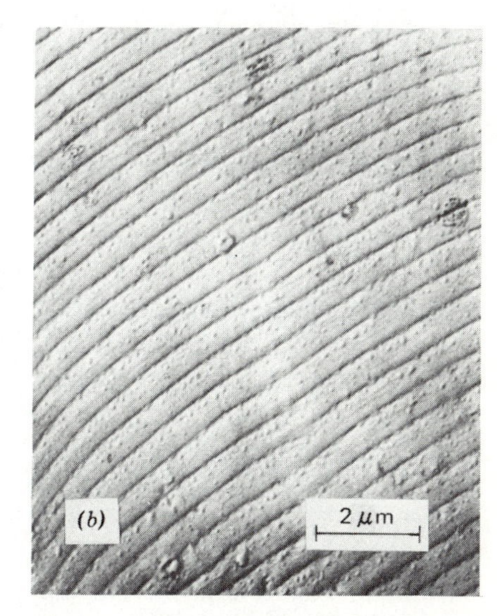

FIGURE 21.18 Striations corresponding to individual increments of fatigue crack growth. The separation between striations corresponds to the amount the crack progressed each stress cycle. (*a*) Fatigue striations in a copper specimen that was subjected to a high stress level; the crack propagated rapidly. (From *Fracture of Structural Materials*, by A. S. Tetelman and A. J. McEvily, Jr., Wiley, New York, 1967.) (*b*) Fatigue striations in an aluminum alloy that was subjected to a low stress level, the crack propagated at a much lower rate than in (*a*). Note that the magnification is much greater than that in (*a*). (Reproduced by permission, from *Metals Handbook*, 8th Edition, Vol. 7, American Society for Metals, 1972.)

The relative proportion of fracture surface created by slow versus fast crack propagation depends on the stress level and the material. At high stress levels, the extent of slow crack growth is small because the cross section capable of withstanding the stress will be a reasonably high fraction of the initial cross-sectional area. At a fixed stress level, a small extent of slow crack growth often is realized with quasi-brittle and brittle materials because the fatigue crack attains a critical size (Sect. 21.4).

Many more fatigue failures start at free surfaces than at internal defects. Surface defects and notches, which serve as stress concentrators, provide especially favorable sites for fatigue crack initiation. Proper design often can minimize such effects and, beyond that, service life can be enhanced through alteration of the surface so as to inhibit crack initiation. Carburizing a steel to produce a hard surface or shot peening a metal to induce residual compressive stresses in a surface are but two of the more common procedures used to improve service life.

21.6 CREEP FAILURE

Creep is time-dependent deformation of a material subjected to a constant stress or load. By this definition, viscoelasticity (Sect. 19.5), anelasticity (Sect. 19.7), and viscous flow (Sect. 20.8) represent aspects of creep. Although dimensional changes resulting from creep may constitute failure for a particular structural member regardless of whether the deformation is elastic or permanent, the ultimate creep failure involves both permanent deformation and fracture. We will focus on creep in terms of time-dependent permanent deformation.

Purely viscous substances and noncrystalline solids in the supercooled liquid state undergo continuing deformation when a constant stress is applied. The amount and rate of straining during creep are a function of the material and conditions of stress and temperature. Creep rate increases with increasing temperature because the diffusive rearrangements responsible for viscous flow occur more readily. Creep rate also increases with increasing applied stress, which biases the direction of diffusive motion. In crystalline solids, creep may proceed by preferential vacancy diffusion (diffusional creep) and exhibit kinetics similar to those for linear viscous flow (i.e., $\dot{\epsilon} \propto \sigma$), but diffusional creep is important only at temperatures close to the melting point. At temperatures greater than about one-half the absolute melting point, various viscouslike processes become operative in addition to plastic flow processes, whereas at temperatures below about 0.5 T_m, viscous processes tend to be unimportant. This division is reflected by typical creep curves for a metal at high and low temperatures (Fig. 21.19).

Besides purely diffusional creep and grain boundary sliding, the most common mechanism of creep (also termed viscoplasticity) in crystalline solids involves thermal recovery processes. For example, at temperatures above about 0.5 T_m

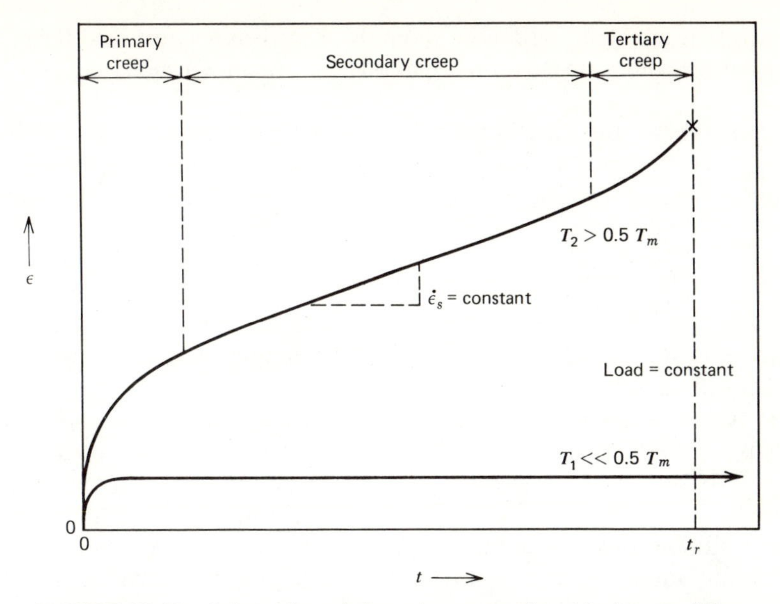

FIGURE 21.19 Schematic variation of creep strain with time at different temperatures. At T_2, the three stages of creep are shown and final failure occurs at a time t_r, the time to rupture. At T_1, the temperature is low enough so that the creep strain is stabilized and fracture does not occur.

where the vacancy concentration becomes sufficiently high and the self-diffusion rate sufficiently rapid, dislocations can climb out of their slip planes and thus avoid obstacles in the process. [Climb involves diffusion of vacancies (or self-interstitials) either to or away from an edge or mixed dislocation.] In essence, then, strain hardening during creep deformation and concurrent recovery processes are competitive, and the relative dominance of one or the other largely determines the creep rate, which increases with increasing temperature and increasing applied stress.

A high-temperature creep curve (Fig. 21.19) can be analyzed as follows. Upon application of a stress at $t = 0$, almost instantaneous elastic and plastic deformation gives rise to a finite strain. This is followed by a stage characterized by a continually decreasing creep rate (primary creep) where strain hardening dominates. Next comes a stage of constant creep rate (secondary or steady state creep) where the rates of strain hardening and thermal recovery exactly balance. Frequently, this stage has the greatest duration. A final stage (tertiary creep), with accelerating creep rate, may follow steady state creep as a result of necking or internal void formation, and fracture always terminates tertiary creep. In each stage, grain boundary sliding may contribute significantly to the creep strain.

In contrast, at temperatures below about 0.5 T_m, creep may proceed through the primary stage, but extensive deformation and fracture usually do not occur (Fig. 21.19).

As noted above, the time duration of steady state creep often is a very large fraction of the total time to rupture (t_r), and t_r can be related inversely to the steady state creep rate $\dot{\varepsilon}_s$. Empirical formulas relating t_r, absolute temperature and stress generally are of the form

$$t_r e^{-Q/RT} = f(\sigma). \tag{21.6}$$

Alternative formulas, similar in form to Eq. 21.5, are utilized frequently for design purposes with materials used at high temperatures.

Most creep resistant alloys are dispersion or precipitation-strengthened materials (see Sects. 18.6, 18.9, and 20.6). If the latter are used, the precipitate not only must be stable in service, but the average size of the precipitate must not increase rapidly with time. That is, the microstructure must be stable under the service conditions of stress and temperature. A number of nickel-base *superalloys* are especially suitable in this regard. Although the precise mechanism is not well understood, dispersion-strengthened alloys can be made remarkably resistant to thermal recovery processes; in fact, high dislocation densities can be maintained in these alloys at temperatures very close to the melting point.

21.7 EMBRITTLEMENT

A number of service failures occur at stress levels that would be perfectly safe in the absence of certain environmental conditions. Strictly speaking, all such effects can be considered to be a form of stress-corrosion cracking. As a matter of convention, however, metallic embrittlement due to hydrogen or liquid metals is usually classified separately.

Lucite (polymethyl methacrylate) sheet, if tested in tension after soaking in ethyl alcohol, exhibits a decreased tensile strength and elongation to fracture. Such behavior is due to the nonuniform dissolution of Lucite by alcohol. The alcohol dissolves into and extends preexisting microcracks on the surface, inducing premature brittle failure. A large number of metals and alloys also are susceptible to similar embrittlement but usually in widely different solutions (Table 21.1). Such embrittlement is similar phenomenologically to that of Lucite in that cracks form and grow under stress in the presence of the *corroding* medium until a crack of sufficient size is produced, subsequent to which fracture occurs. In metals, such attack is usually, although not always, intergranular in nature; that is, the cracks progress between the grains of the polycrystalline material. Fortunately, only a few of the essentially infinite possible combinations of alloys and environments

TABLE 21.1 Some Environments That May Cause Stress-Corrosion of Metals and Alloys

Alloy Type	*Environment*
Aluminum alloys	Sodium chloride–hydrogen peroxide solutions Sodium chloride–water solutions Water vapor
Copper alloys	Ammonia vapors and solutions Amines Water and water vapor
Gold alloys	Ferric chloride solutions Acetic acid–salt solutions
Magnesium alloys	Sodium chloride–potassium chromate solutions
Ferritic steels	Sodium hydroxide solutions Calcium, ammonium and sodium nitrate solutions Sulfuric–nitric acid mixtures Acidic hydrogen sulfide solutions Sea water
Stainless steels	Acidic chloride solutions Sodium chloride–hydrogen peroxide solutions Sea water Hydrogen sulfide Sodium hydroxide–hydrogen sulfide solutions
Titanium alloys	Salt solutions

lead to stress-corrosion cracking; however, the problem can be extremely serious under those conditions in which stress-corrosion cracking is favored.

Hydrogen embrittlement is of particular concern in high-strength ferritic steels in which, during processing such as casting or welding, the solubility limit of hydrogen in the steel at normal temperatures is exceeded. Since this limit can be as low as one part per million by weight, exclusion of hydrogen is not a simple task. Hydrogen embrittlement is manifested by a marked reduction in tensile ductility and a tendency to fail under a static load after a given period of time. While the exact details as to the mechanism of hydrogen embrittlement are not known, it is now evident that hydrogen precipitates as gaseous bubbles and these act to promote crack initiation and propagation. At very low temperatures, hydrogen embrittlement does not occur because the hydrogen diffusivity is so low that precipitation does not occur. Furthermore, hydrogen embrittlement does not occur at high temperatures either, because the diffusivity is sufficiently high so that the hydrogen "bakes out" of the steel. For steels, hydrogen embrittlement

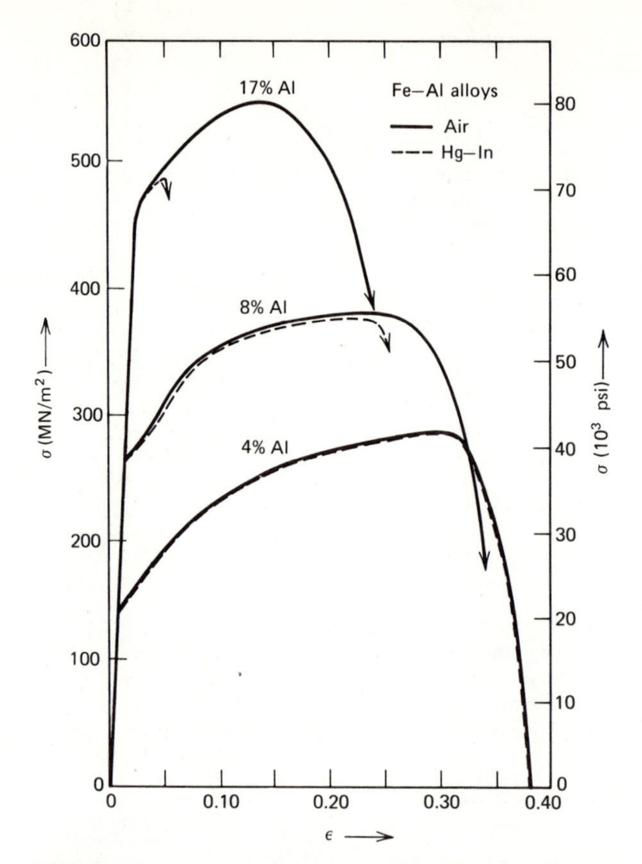

FIGURE 21.20 (*a*) Embrittlement of Fe–Al alloys by liquid Hg–In as evidenced by a loss in tensile ductility. The embrittlement effect becomes more pronounced with increasing aluminum content. (with permission of publisher: after N. S. Stoloff, R. G. Davis, and T. L. Johnston, in *Environment-Sensitive Mechanical Behavior,* Gordon and Breach, New York, 1966.)

thus occurs in the range of approximately $-130°C$ to $+100°C$. It is interesting to note that within this temperature range, hydrogen embrittlement is not favored by increasing strain rate since the time involved in testing decreases with increasing strain rate and, therefore, there is less time available for hydrogen diffusion, which is a necessary condition for embrittlement.

The embrittlement of a stressed material due to the presence of a molten metal is known as liquid-metal embrittlement (*LME*). As with stress-corrosion cracking, *LME* occurs only with selected liquid metal–materials combinations, several of which are noted in Table 21.2. In order for LME to occur, it is necessary for the

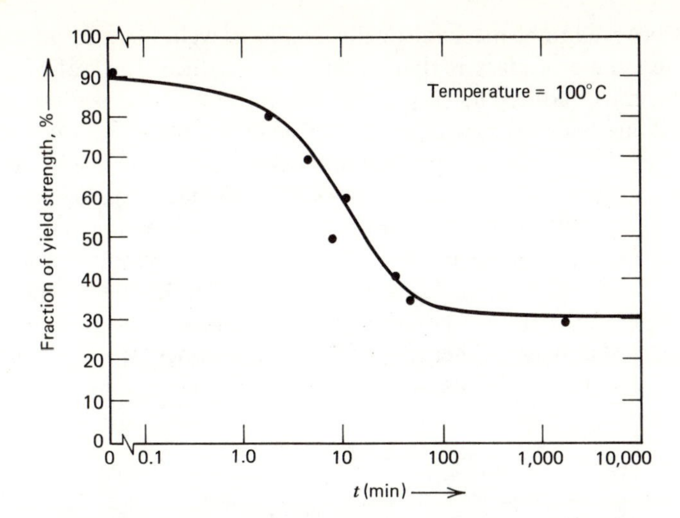

FIGURE 21.20 (*b*) The embrittlement of a commercial aluminum alloy wetted with mercury amalgam. This figure shows, for example, that if the alloy is statically stressed at 50% of its yield strength it will fail in about 10 minutes at this temperature. At stress levels below about 30% of the yield strength, there is no apparent embrittling effect. (From *Embrittlement by Liquid Metals*, by W. Rostoker, J. M. McCaughey, and H. Markus, Reinhold, New York, 1960.)

liquid metal to wet the material. LME usually is associated with intergranular failure and is manifested by a reduction in tensile ductility (Fig. 21.20*a*) or by a time-delayed failure at a stress well below the yield strength of the material (Fig. 21.20*b*).

As mentioned previously, nonmetallic materials also are susceptible to embrittlement arising from environment. When glass is subjected to a stress slightly below the fracture strength, failure occurs after a period of time. This is phenomenologically similar to the effect caused by hydrogen in high strength steels and

TABLE 21.2 Liquid Metals That May Cause Embrittlement of Metals and Alloys

Alloy Type	*Liquid Metal Environment*
Aluminum alloys	Mercury–zinc, gallium, sodium, indium, tin, or zinc
Magnesium alloys	Sodium or zinc
Steels	Indium, lithium, cadmium, or zinc
Titanium alloys	Mercury–zinc

presently is believed to be mechanistically similar as well. The current explanation of this static fatigue of glass is that water vapor reduces the surface energy for fracture and, thus, reduces σ_F (Eq. 21.5). The loosely bound sodium ions in the glass network are believed to play a critical role in this phenomenon by catalyzing the attack on the silicon–oxygen bond by hydrogen ions. Thus, diffusion of sodium plays a role in static fatigue of glass similar to that of hydrogen in hydrogen embrittlement of steel. As evidence for this explanation, experimental studies show that the activation energy for the delay time of glass fracture is the same as that for diffusion of the sodium ion in glass. The absence of static fatigue in glass at liquid nitrogen temperatures also supports this explanation.

The strength of polymeric fibers depends on environment (Fig. 21.21), but in the example illustrated this effect is not always adverse. The strengths of nylon and

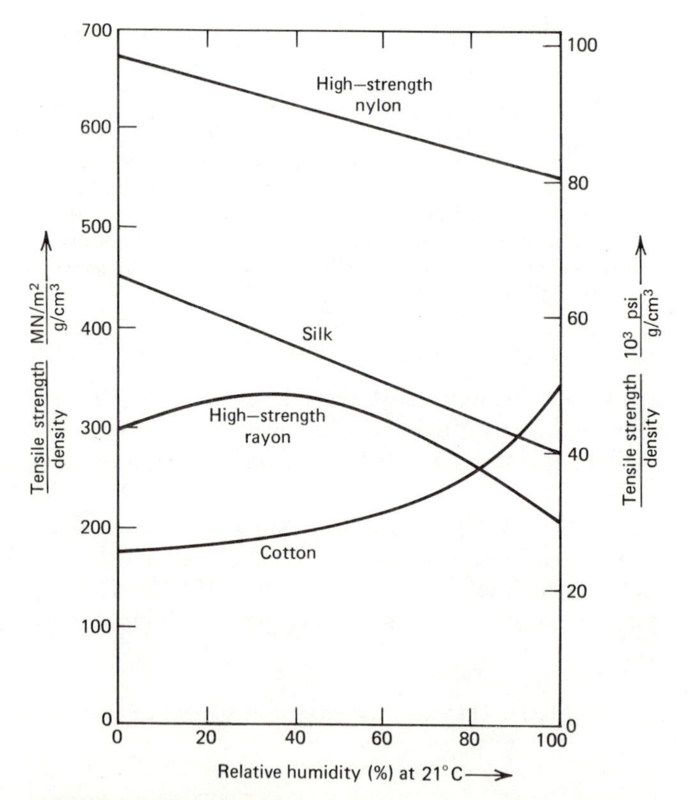

FIGURE 21.21 Strength vs. humidity for several polymeric fibers. Rayon (at high humidity), nylon, and silk are all adversely affected by water vapor. Cotton, by contrast, is not. (The densities of the fibers when dry are between about 1.1 g/cm³ and 1.5 g/cm³.) (Data courtesy of E. I. duPont deNemours & Co.)

silk decrease with relative humidity since in both of these substances absorbed water weakens the interchain bonding by competing for the hydrogen bonds (see Fig. 6.1). The strength of cotton fibers, by contrast, increases with relative humidity, presumably because their cellular structure is able to distort upon absorption of water and allow more molecular chains to support stress along the fiber axis. Fibers of rayon, which is chemically the same as cotton but differs crystallographically, exhibit intermediate behavior.

It is important to emphasize that the nature of mechanical failure can be quite complex as a result of the interplay between a material, its properties, the service conditions, and the type of environment. One result of this complexity is that service failures remain an important problem in the utilization of engineering materials.

REFERENCES

1. Guy, A. G. *Introduction to Materials Science*, McGraw-Hill, New York, 1972.
2. Cottrell, A. H. *The Mechanical Properties of Matter*, Wiley, New York, 1964.
3. Hayden, H. W., Moffatt, W. G., and Wulff, J. *The Structure and Properties of Materials*, Vol. III, Wiley, New York, 1965.

QUESTIONS AND PROBLEMS

21.1 The theoretical stress σ_{max} required to fracture a crystalline solid (theoretical cohesive strength) may be estimated by using a uniaxial interatomic stress curve (see Fig. 19.6, for example). The figure shown below illustrates schematically the theoretical variation of applied stress with interplanar separation.

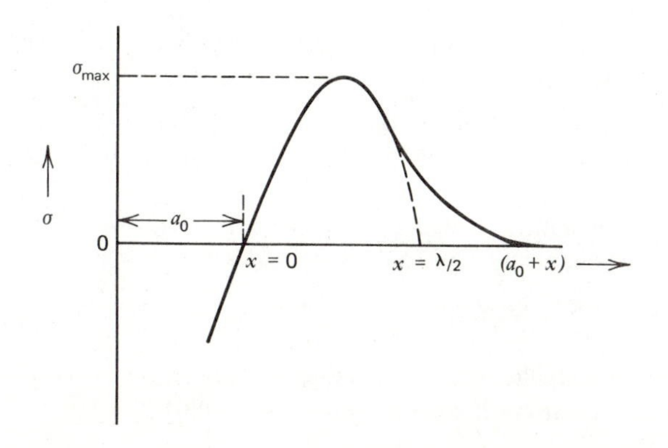

Assume that between $x = 0$ (i.e., at the equilibrium interplanar spacing, a_0) and $x = \lambda/2$ this curve can be approximated by $\sigma = \sigma_{max} \sin(2\pi x/\lambda)$. The work done per unit fracture area to cause fracture is given by the integral of $\sigma\, dx$ between 0 and $\lambda/2$. This work should equal the surface energy per unit fracture area (2γ) created during fracture.

(a) Show that $\sigma_{max} = 2\pi\gamma/\lambda$.

(b) Using Hooke's law for linear elasticity (i.e., $E = \sigma/\varepsilon = d\sigma/d\varepsilon$ at $x = 0$), $\cos x \simeq 1$ for small x, and the result in part (a), show that $\sigma_{max} = \sqrt{E\gamma/a_0}$.

(c) Estimate σ_{max} for copper, silicon, and Al_2O_3. Use data given in Tables 2.5, 11.2, and 19.1.

21.2 Using the results from problem 21.1, estimate the applied macroscopic tensile stress required to reach the theoretical strength of silicon containing an internal crack 1 μm in length. Assume that the radius of curvature at the crack tip is $\rho = 3a_0$. Compare your answer with the stress required to propagate a crack (Eq. 21.4). (*Note*. The important result is that the applied macroscopic stress that is required to cause σ_{max} at the crack tip is of the same order of magnitude as the stress necessary for crack propagation.)

21.3 (a) Derive Eq. 21.4.

(b) Show that U^* is given by $4E\gamma^2/\pi\sigma^2$ (Fig. 21.4*a*).

(c) Calculate U^* for a sheet of silicon 0.5 cm thick subjected to a tensile stress of 35 MN/m^2 (5000 psi). Compare this value with the thermal energy available at room temperature (at 300 K, $kT \simeq 1/40$ eV).

21.4 (a) A glass plate has sharp surface cracks whose sizes range up to 1 μm in length. Calculate its strength when tested in tension. ($E = 6.9 \times 10^{10}$ N/m^2; $\gamma = 0.3$ J/m^2.)

(b) Explain why the strength of glass can be increased by etching off a thin surface layer with hydrofluoric acid.

21.5 A rapid change in external temperature can cause many ceramic objects (crystalline or glassy) to crack or break (thermal shock). Explain.

21.6 An inorganic glass can be strengthened by the following treatment. After the glass is heated above the softening point, its surface is cooled rapidly by blasts of air or an oil bath, while the interior remains at a higher temperature. When the temperature becomes uniform throughout, the resulting *tempered* glass exhibits a greater strength than would be realized with more uniform, slower cooling. Why?

21.7 (a) Assuming the solid line shown in Fig. 21.5 represents the data well, replot this figure as σ_F vs. $c^{-1/2}$. From the slope of this line, determine γ. (For polystyrene, $E = 3.4 \times 10^9$ N/m^2.)

(b) The theoretical value of surface energy for polystyrene is 0.5 J/m^2. Compare this with your answer to part (a) and explain any discrepancy.

21.8 Can the glass transition temperature of a long-chain polymer be considered to be the ductile-to-brittle transition temperature? Explain.

21.9 With a sketch, illustrate how a pile-up of parallel edge dislocations on the same slip plane can cause crack initiation at a grain boundary.

21.10 Assume that the crack propagation stress of a bcc transition metal is independent of temperature.

(a) Demonstrate graphically that a reduction in grain size lowers the ductile-to-brittle transition temperature even though the yield strength is increased.

(b) Demonstrate graphically that solid solution strengthening, precipitation strengthening or moderate strain hardening will raise the ductile-to-brittle transition temperature.

21.11 The yield strength (in psi) of a low-carbon steel having a grain size $d = 0.16$ mm varies with temperature (in K) as

$$\sigma_Y = 113{,}000 - 500T$$

below 175 K. At a given temperature, the yield strength also varies with grain size as

$$\sigma_Y = \sigma_0 + kd^{-1/2}$$

where $k = 13{,}000 \text{ psi} \cdot \text{mm}^{1/2}$.

(a) If the plastic work associated with crack propagation is given by $\gamma_p \approx 200 \text{ J/m}^2$, what is the transition temperature for this steel? (For steel, $E = 20.7 \times 10^{10}$ N/m^2; conversion factor: 145 psi $\simeq$ 1 MN/m^2.)

(b) Assuming the plastic work term (γ_p) is independent of temperature, calculate the transition temperature for grain sizes of 0.1 mm, 0.5 mm, and 1 mm.

21.12 Using Fig. 21.16, determine the allowable range of stress (i.e., $\sigma_{max} - \sigma_{min}$) for mean stresses of 40,000 psi and 80,000 psi.

21.13 What do you expect the effect of surface finish to be on the fatigue life of a material? Explain briefly.

21.14 Using Fig. 21.16, replot the Goodman diagram in terms of allowable stress range as a function of mean stress. Which of the two representations appears most convenient to use?

21.15 (a) Using a ruler, determine how much the fatigue cracks grew per cycle in the copper and the aluminum alloy shown in Fig. 21.18. (Be sure to take magnification into account.)

(b) Assume that the aluminum alloy was originally a 2 in. square bar and that the crack propagated from a point on the surface in a semicircular manner (e.g., Fig. 21.17). If the bar failed when the crack propagated across half of the cross-sectional area, estimate the lifetime of the material assuming there was one stress cycle per second.

21.16 (a) It is often observed that the extent of crack propagation per cycle in a fatigue test increases with time even under a constant nominal stress amplitude. Explain.

(b) In most service applications, the distance between striations of the type illustrated in Fig. 21.18 varies from portion to portion along the fracture surface. Explain.

21.17 The steady state creep rate of a metal at elevated temperatures is proportional to the self-diffusivity of the metal. This is related to the process of dislocation climb in which edge or mixed dislocations move out of their slip plane.

(a) With words and sketches, illustrate how an edge dislocation can climb as a result of self-diffusion.

(b) Explain why dislocation climb will allow continuing deformation at a constant stress level.

21.18 What effect would you expect a fine dispersion of hard, stable second-phase particles to have on creep rate? Explain.

21.19 In the high-temperature creep regime (i.e., $T > 0.5\ T_m$), a fine-grained polycrystalline material exhibits a lower creep rate than a coarse-grained material at temperatures just above $0.5\ T_m$, whereas the opposite is true at temperatures just below T_m. Why?

21.20 How do you expect the high-temperature creep characteristics to differ for a ceramic body formed by vitrification and for the same ceramic formed by sintering?

21.21 Certain precipitation-strengthened nickel-base alloys used in high-temperature service have grain boundary precipitates as well as intragranular precipitates. How do the grain boundary precipitates, which are hard intermetallic compounds, contribute to the creep resistance of these alloys?

22

Electrical Conduction in Solids

The effects of bonding, crystal structure, and imperfections on the electrical conductivity of matter are discussed in this chapter.

22.1 ELECTRICAL CONDUCTION

The flow of electrical charge through a gas, liquid, or solid constitutes an electric current I, which is defined as the rate of charge passage with respect to time:

$$I = \frac{dq}{dt} \tag{22.1}$$

The quantity of charge is measured in coulombs (1 C equals the magnitude of the charge on 6.24×10^{18} electrons), and current is measured in amperes (1 A = 1 C/sec). It is sometimes more convenient to express charge flow as current density J, which is the current per unit area,

$$J = \frac{I}{A} \tag{22.2}$$

where A is the specimen cross-sectional area.

When a power source such as a battery is connected to a conductor as illustrated in Fig. 22.1, a current results. The current through the sample may be determined

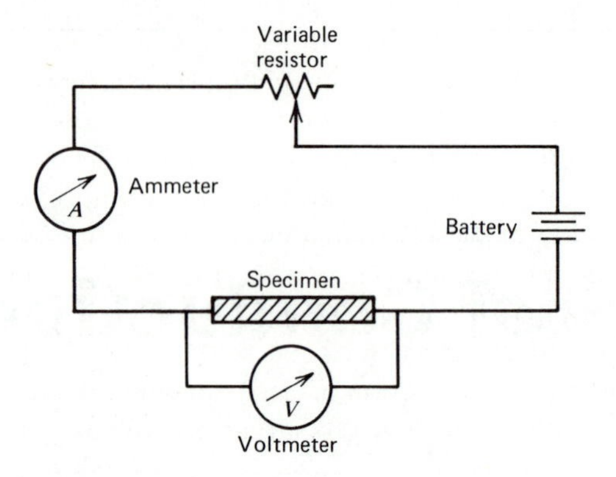

FIGURE 22.1 Schematic circuit for determining the electrical resistance of a conductor. The battery is a source of electrical power that imposes a voltage across the conductor and supplies a current limited by the variable resistor. The voltage drop across the specimen and the current passing through it, measured by a voltmeter and ammeter, respectively, are used in Ohm's law to calculate the specimen resistance.

with an ammeter, and the voltage drop across the sample may be measured with a voltmeter. These quantities are related through Ohm's law

$$V = IR \tag{22.3}$$

where V is the voltage drop in volts and R is the electrical resistance of the conductor in ohms. For a given material, the resistance varies with sample dimensions, increasing linearly with sample length l and inversely with sample area A:

$$R = \rho \frac{l}{A} = \frac{1}{\sigma} \frac{l}{A} \tag{22.4}$$

The quantity ρ, called the electrical resistivity, has units of ohm · m (or ohm · cm) and is a property of the material. Its reciprocal, σ, is called electrical conductivity. Equations 22.3 and 22.4 can be rearranged to give

$$\sigma = \frac{(I/A)}{(V/l)} = \frac{J}{(V/l)} \tag{22.5}$$

The voltage per unit length defines the electrical field strength, E. Thus, Eq. 22.5 becomes

$$\sigma = \frac{J}{E} \tag{22.6}$$

It can be shown that the current density is equal to the product of the number of charge carriers per unit volume of material, n, the quantity of charge per carrier, q, and the average velocity of the charge carriers, $\bar{v}$. Consequently, Eq. 22.6 assumes the form

$$\sigma = nq\frac{\bar{v}}{E} = nq\mu \tag{22.7}$$

where the ratio $(\bar{v}/E)$ is called the carrier mobility μ.

As indicated in Table 22.1, the electrical conductivities of pure metals are nearly 23 orders of magnitude higher than those for electrical insulators such as polyethylene. Silicon, germanium, and similar materials are called semiconductors because their conductivities fall between those of metals and insulators.

TABLE 22.1 Electrical Conductivities of Various Materials at Room Temperature

Metals and Alloys	σ (ohm^{-1}m^{-1})	*Nonmetals*	σ (ohm^{-1}m^{-1})
Silver	6.3×10^7	Graphite	10^5 (average)
Copper, commerical purity	5.85×10^7	SiC	10
Gold	4.25×10^7	Germanium, pure	2.2
Aluminum, commerical purity	3.45×10^7	Silicon, pure	4.3×10^{-4}
Al–1.2% Mn alloy	2.95×10^7	Phenol-formaldehyde	
Sodium	2.1×10^7	(Bakelite)	10^{-7}–10^{-11}
Tungsten, commercial purity	1.77×10^7	Window glass	$<10^{-10}$
Cartridge brass		Alumina (Al_2O_3)	10^{-10}–10^{-12}
(70% Cu–30% Zn)	1.6×10^7	Mica	10^{-11}–10^{-15}
Nickel, commerical purity	1.46×10^7	Polymethyl methacrylate	
Ingot iron, commerical purity	1.03×10^7	(Lucite)	$<10^{-12}$
Titanium, commerical purity	0.24×10^7	Beryllia (BeO)	10^{-12}–10^{-15}
TiC	0.17×10^7	Polyethylene	$<10^{-14}$
Stainless steel, type 301	0.14×10^7	Polystyrene	$<10^{-14}$
Nichrome (80% Ni–20% Cr)	0.093×10^7	Diamond	$<10^{-14}$
		Silica glass	$<10^{-16}$
		Polytetrafluoroethylene	
		(Teflon)	$<10^{-16}$

22.2 IONIC CONDUCTORS

At room temperature, most ionic solids are insulators, and their conductivities increase with temperature (Fig. 22.2*a*). Because of the large band gap in purely ionic materials, no electrons can be excited into the conduction band and electrical conduction corresponds to the passage of charge because of ion diffusion in the presence of an electric field. In many pure ionic materials, diffusion is able to proceed because Schottky defects are present (see Sect. 10.2), and the net passage of charge corresponds to each vacancy in the pair *drifting* in opposite directions. Thus, ionic conduction in crystals, like ionic conduction in fused salts and aqueous electrolytes, involves the transport of matter in the form of ions.

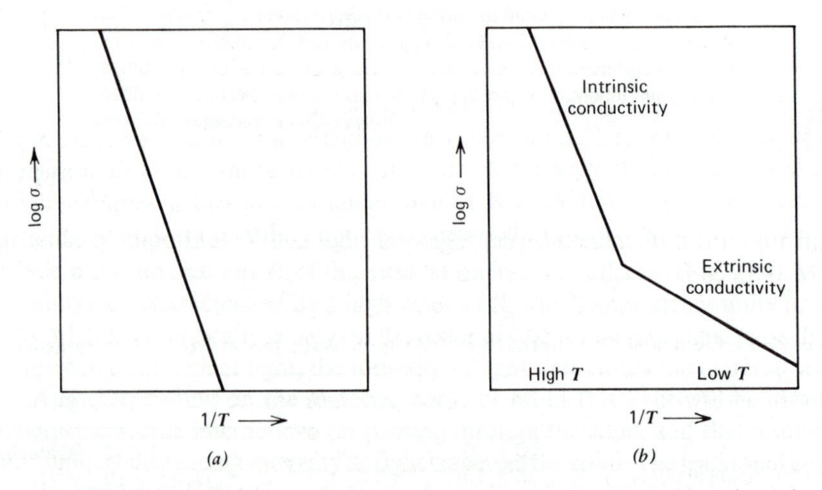

FIGURE 22.2 Schematic variation of the electrical conductivity with reciprocal absolute temperature for an ionic solid that conducts by ion migration. (*a*) The conductivity increases with temperature because the number of defects and the defect mobility both increase as temperature rises. (*b*) Extrinsic conductivity dominates at low temperatures, and intrinsic conductivity dominates at high temperatures (cf. Fig. 12.6).

The electrical conductivity of a solid in which conduction is due solely to Schottky defects can be expressed as

$$\sigma = nq(\mu_c + \mu_A) \tag{22.8}$$

where n is the number of cation–anion vacancy pairs per unit volume, and μ_c and μ_A are the cation and anion mobilities, respectively. Since the number of

Schottky defects varies with temperature according to $n \sim e^{-(\Delta H_s/2RT)}$, where ΔH_s divided by Avogadro's number is the energy to form one cation–anion vacancy pair, and since the mobility is related to the diffusivity which rises exponentially with temperature, the intrinsic conductivity increases rapidly with temperature.

As mentioned in Sect. 10.1, if an ion of different valence forms a substitutional solid solution in an ionic material, then other defects such as vacant sites must be present to maintain charge balance. The number of these vacant sites will not be temperature dependent, although their ability to move (mobility) will be. Thus, the resulting *extrinsic* conductivity in ionic solids is not nearly as sensitive to temperature as the intrinsic conductivity. In impure ionic solids, the total conductivity will be of both the intrinsic and extrinsic types, with the former dominating at high temperatures and the latter, at low temperatures (Fig. 22.2*b*).

22.3 CONDUCTION IN METALS: THE FREE ELECTRON MODEL

In distinction to ionic materials, electrical conduction in metals occurs by the motion of electrons. The older free electron theory of metals pictures the valence electrons as moving freely within a metal very much like the molecules of a gas in a closed container. The force acting on an electron due to an applied electric field is given by

$$F = -eE \tag{22.9}$$

where e is the magnitude of the electronic charge. The acceleration corresponding to this force may be expressed as

$$a = -\frac{eE}{m_e} \tag{22.10}$$

where m_e is the electronic mass. The electron does not accelerate indefinitely, however, because it undergoes collisions with vibrating ion cores and defects in the solid and, as a consequence, loses kinetic energy in the process. Because the mass of the electron is so small, it is reasonable to assume that complete loss of the electron kinetic energy occurs in such a collision and thus the average electron velocity can be related to the average time, 2τ, between collisions. Since, according to Eq. 22.10, the electronic velocity increases linearly with time, the average velocity, $\bar{v}$, is expressed as

$$\bar{v} = \frac{1}{2}a[2\tau] = -\frac{eE}{m_e}\tau \tag{22.11}$$

Thus, in accordance with Eq. 22.7, the conductivity is expressed as

$$\sigma = -ne\frac{\bar{v}}{E} = \frac{ne^2\tau}{m_e} \tag{22.12}$$

and J is given by (cf. Eq. 22.6)

$$J = \frac{ne^2\tau E}{m_e} \tag{22.13}$$

where n is the number of valence electrons per unit volume.

The electrical conductivity of a metal, unlike that of an ionic solid, decreases with increasing temperature. That is, as temperature increases the ion cores vibrate more readily about their equilibrium positions in the crystal structure, giving rise

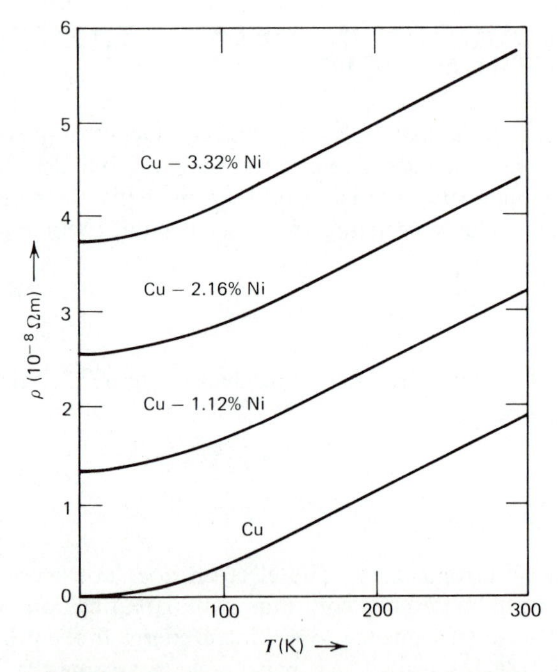

FIGURE 22.3 Electrical resistivity versus temperature for pure and alloyed copper. For a metal, resistivity increases with temperature, and the increase is approximately linear over a wide range of temperatures. Impurities or alloying elements, dissolved in solid solution, raise the entire curve and give rise to finite residual resistivities at absolute zero. [Adapted from J. O. Linde, *Ann. Physik,* **5,** 219 (1932). Reprinted with permission of Verlag Johann Ambrosius Barth, Leipzig.]

to a larger number of thermally excited elastic waves called phonons (see Sect. 24.1) that scatter conduction electrons and thus decrease both the average distance traveled and the time between collisions. In pure metals and solid-solution alloys, the total electrical resistivity may be approximated as a sum of two terms:

$$\rho = \rho_T + \rho_r \tag{22.14}$$

The first term (ρ_T) is temperature dependent and arises from phonon scattering; the second term (ρ_r), called the residual resistivity, is very nearly independent of temperature and arises from scattering caused by solute or impurity atoms and defects such as dislocations. As is shown in Fig. 22.3, resistivity varies with temperature and impurity content. The effect of cold deformation (i.e., an increase in dislocation density) on the resistivity of a pure metal at normal temperatures is not as marked as that of impurity atoms.

The effect of alloying on the resistivity of a metal depends on the crystallographic positions of the solute atoms. This is illustrated in part by the variation of resistivity with temperature for the ordered and disordered forms of Cu_3Au (Fig. 22.4). Above a critical temperature (the order-disorder transformation temperature),

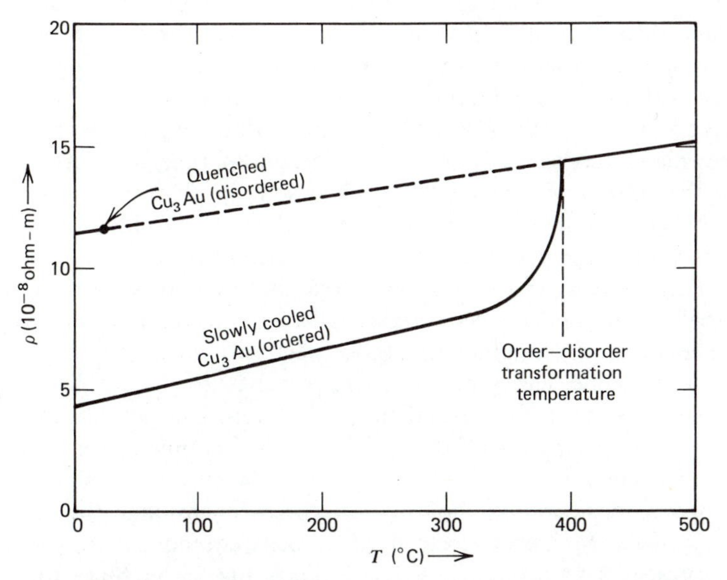

FIGURE 22.4 Temperature variation of the resistivity for 75 at. % Cu–25 at. % Au. The ordered Cu_3Au structure, being inherently more perfect, has a lower resistivity than the disordered structure. [After C. Sykes and H. Evans, *J. Inst. Metals,* **58**, 225 (1936).]

Cu_3Au is a face-centered cubic solid solution with copper and gold atoms occupying atomic sites at random within each unit cell. This disordered structure can be retained at room temperature by quenching from above the order-disorder temperature, and the resistivity of the disordered alloy is significantly higher than the resistivity of the ordered Cu_3Au alloy that can be obtained by slow cooling to room temperature. In the ordered alloy, gold atoms occupy the corner positions and copper atoms occupy the face-centered positions of the unit cell, and such an orderly arrangement results in less electron scattering.

For metals in general, any mechanical or chemical action that alters the crystalline perfection will raise the residual resistivity and, therefore, the total resistivity. Thus, vacancies in metals, in contrast to those in ionic solids, increase the resistivity. The reason for this lies in the inherent differences between conduction mechanisms in these two classes of materials.

22.4 CONDUCTORS, SEMICONDUCTORS, AND INSULATORS: THE BAND MODEL REVISITED

Although we have tacitly assumed that all valence electrons in a metal are free to participate in the conduction process, in actuality only a few percent at most are involved. To adequately explain differences in the number of available charge carriers and their mobilities in various types of solids, the band model of solids must be utilized. As discussed in Sect. 4.5, if N atoms are brought together to form a solid, each atomic energy level beyond the ion core gives rise to N levels in the solid that are discrete, but separated by such small differences in energy that, in effect, they may be considered continuous. Each atomic energy level thus becomes an energy band. In monovalent solids such as sodium, all of the valence electrons are readily accommodated within one band, which also contains a multitude of allowed but unoccupied energy states. Those electrons occupying energy levels near the Fermi energy can be promoted readily to empty states just above the Fermi level by an applied electric field, and these are the electrons involved in electrical conduction. Note that only those electrons lying near the Fermi energy can be so excited since electrons with energies well below E_F are characterized by occupied energy states lying above their energy levels. This point can be amplified further by reconsidering the effect an electric field has on electron velocity. This is most conveniently done by considering a "two-dimensional" model of a crystal, but analogous considerations hold for the three-dimensional case as well. In a *two-dimensional* crystal, each electron has velocity components v_x and v_y, which can be represented by a point in velocity space, Fig. 22.5*a*. Since the energy of an electron is proportional to $v_x^2 + v_y^2$, the magnitude of the velocity vector $v = \sqrt{v_x^2 + v_y^2}$ is a measure of the electronic energy. In addition, because the Fermi energy represents the maximum electronic energy at 0 K, the radius vector

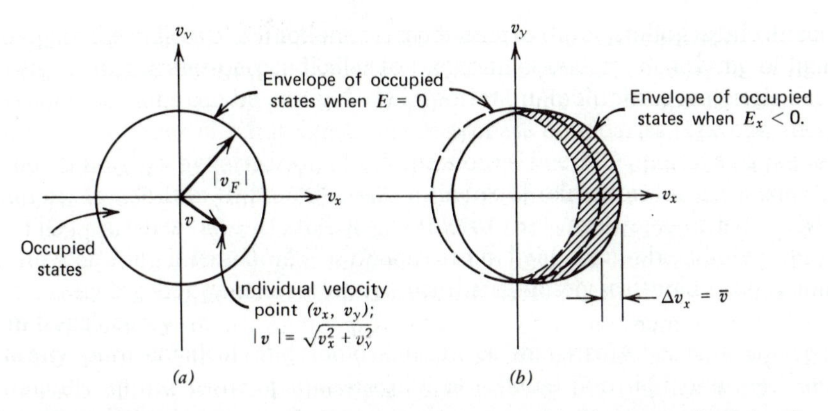

FIGURE 22.5 (*a*) The distribution of electron velocities in a two-dimensional crystal in the absence of an applied field. Each electron velocity is represented by a point having velocity components v_x and v_y. The circle shown represents the Fermi velocity magnitude, v_F, corresponding to the Fermi energy, E_F. At 0 K therefore, all states with $v < v_F$ are occupied and all states with $v > v_F$ are empty. (*b*) Application of an electric field in the $-x$ direction biases the electron velocities in the $+x$ direction by an amount Δv_x ($=\bar{v}$, the drift velocity). The crosshatched area designates the electrons that actually contribute to the conduction process. That is, the motion of these electrons in the $+x$ direction is not balanced by similar motion of electrons in the $-x$ direction because the applied electric field has shifted the occupancy of electronic states.

$v_F = \sqrt{(v_x^2 + v_y^2)_{\max}} = \sqrt{2E_F/m_e}$ defines the Fermi velocity. Velocity points lying outside of this radius represent unoccupied energy states; those lying within this radius represent occupied states. If an electric field is applied in the $-x$ direction, all electrons acquire an incremental velocity component (i.e., a drift velocity) in the $+x$ direction given by Eq. 22.11. This shifts the distribution in velocity space as is noted in Fig. 22.5*b*. Since the increase in velocity due to the electric field is much less than $\sqrt{2E_F/m_e}$ (i.e., $\bar{v} \ll v_F$), electrons, after the application of the electric field, occupy most of the same energy states that were occupied in its absence. Only those electrons whose new velocity magnitudes lie just above and just below $\sqrt{2E_F/m_e}$ will contribute to the conduction process (crosshatched area, Fig. 22.5*b*). The number of these electrons is proportional to the number of electrons having energies $E \approx E_F$ in the absence of the electric field, that is, to the density of states (Sect. 4.7) evaluated at the Fermi energy. In a monovalent metal such as sodium, this quantity is readily obtained (see the equations given in problems 4.13 and 22.14, for example) since band overlap does not occur in the vicinity of E_F (Figs. 4.12 and 4.15). In many metals, such as magnesium, the bands formed from discrete atomic levels (e.g., 3*s* and 3*p* for Mg) overlap and states from both bands are occupied. The resulting density of states curve is a sum of the individual density of states curves for both bands and an analytical evaluation of $N(E_F)$ is complicated. Nevertheless, as with monovalent metals, little energy is required to shift the occupancy of electronic states in a manner that corresponds to a net electric current.

The model of conduction described above can be physically related to the older free electron theory of metals described in Sect. 22.3. Since, in effect, only those electrons having energies near the Fermi level are accelerated by the applied field, these are the ones that are scattered by phonons and defects with a mean time, 2τ, and a mean free path, $l = 2\tau v_F$, between collisions. Upon being scattered, the new electron momentum vector must correspond to an unoccupied point in velocity space which must lie on the circle in Fig. 22.5. On the average, the new velocity vector will be zero since as many scattered electrons will have negative velocities in the x and y directions as vice versa. Once an electron is scattered, its point in velocity space then becomes occupied by an electron which previously had a lesser velocity in the $+x$ direction. Thus, in effect, conduction is a dynamic process in which the electrons with the greatest velocities are scattered, electrons with lesser velocities are then accelerated and these are subsequently scattered, etc. Consequently, the conductivity of a metal can be expressed either by Eq. 22.12, in which $\bar{v}$ is the average drift velocity and n the total number of valence electrons per unit volume, or by

$$\sigma = -\frac{n_c e \bar{v}'}{E} \tag{22.15}$$

where n_c is the effective number of conduction electrons per unit volume (i.e. those with velocities corresponding to the cross-hatched area, Fig. 22.5*b*) and $\bar{v}'$ is the average velocity in the x direction of these electrons. It can be shown that $\bar{v}' = (\frac{2}{3})v_F$ and that $n_c v_F = (\frac{3}{2})n\bar{v}$; thus, $n_c\bar{v}' = n\bar{v}$, and Eqs. 22.15 and 22.12 are equivalent. The older free electron theory of conduction is deficient in many respects; nonetheless, it does offer a good overview of the important parameters affecting electrical conductivity in metals.

The room temperature resistivities of materials like the semiconductor silicon are about 11 orders of magnitude higher than that of a good conductor like copper. This is so because, in the absence of thermal excitation, the valence band in Si or Ge is completely occupied and does not overlap with the conduction band lying at higher energies. The difference in energy between the lowest level in the conduction band and the highest level in the valence band (i.e., the energy gap) represents the width of the forbidden region across which electrons must be promoted in order to be electrical current carriers. Ordinary electric fields are too small to do this and, accordingly, such excitation only can be provided by thermal means. The thermal excitation of electrons to the conduction band leaves holes in the valence band. Each hole, which is simply the lack of an electron in a bonding orbital, acts as a positive charge carrier, whereas an electron in the conduction band acts as a negative charge carrier. Of course, the motion of a hole under the influence of an applied electric field is exactly the opposite of electron motion within the valence band. That is, hole motion is analogous to vacancy motion during diffusion.

In distinction to semiconductors, insulators such as diamond or NaCl have such large energy gaps that even at temperatures comparable to the melting point, thermal excitation is insufficient to promote any electrons into the conduction band. This accounts for the observations that electrical conduction in many ionic solids is solely by cation or anion motion and that diamond is an electrical insulator at all temperatures.

22.5 SUPERCONDUCTIVITY

At extremely low temperatures, the resistivities of some metals like Pt become nearly temperature independent because the thermal contribution to resistivity becomes negligible compared to the residual resistivity (Fig. 22.6). Other metals lose all electrical resistivity abruptly and completely at some low temperature and are called superconductors. The resistivity of a metal below its superconducting transition temperature T_c is vanishingly small. For example, measurements made on lead ($T_c = 7.19$ K) at 4.2 K indicate that its resistivity is at least less than 10^{-25} ohm · m, whereas very pure copper, a nonsuperconductor, has a resistivity of about 10^{-12} ohm · m at 4.2 K. Many elemental metals, solid-solution alloys, and intermetallic compounds exhibit superconductivity. Often, such materials have relatively high electrical resistivities in the normal state (i.e., above T_c) as compared with Ag and Cu, which do not become superconductors. Observed superconducting transition temperatures range from a few millidegrees in absolute temperature to over 21 K for the compound Nb_3Ge.

The transformation from the normal or nonsuperconducting state to the superconducting state is a phase change that effectively involves alterations in the electronic states without a change in crystal structure. This transformation thus can be represented by a unary phase diagram at constant pressure, Fig. 22.7*a*. Although pressure, as well as other mechanical stresses, can alter the superconducting transition temperature, externally imposed electromagnetic forces such as that due to an applied magnetic field have a much greater effect. Thus, a unary temperature-external magnetic field (H) phase diagram can be constructed as shown in Fig. 22.7*b*. As indicated here, superconductivity can be destroyed either by increasing the temperature at constant magnetic field or by increasing the external field at constant temperature. The critical magnetic field (H_c) required to destroy superconductivity is a function of temperature, and the locus of points representing $H_c(T)$ thus divides the phase diagram into normal and superconducting regions.

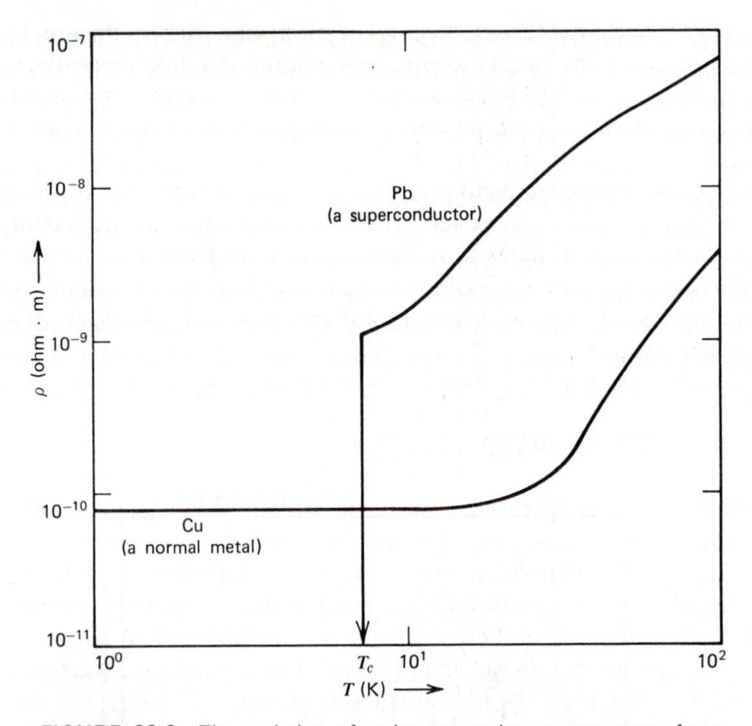

FIGURE 22.6 The variation of resistance at low temperatures for a nonsuperconductor and a superconductor. The resistance of the former approaches a finite value at very low temperatures; that of the latter decreases to essentially zero at the superconducting transition temperature, T_c.

The low values of T_c and the small amount of external magnetic energy required to destroy superconductivity in real materials indicate that the interaction energy associated with the superconducting state is very weak. It has been shown that superconductivity results from an electron–electron interaction and, in the superconducting state, the motion of two electrons becomes correlated so that scattering by impurity atoms and defects is ineffective. When one electron of a transiently bound electron pair is scattered, the other correlated electron responds in such a manner that there is no net effect on the motion of the pair. Because this interaction is so weak, thermal agitation readily disrupts it and superconductivity is observed only at very low temperatures.

Different types of superconductors behave differently in the presence of an external applied magnetic field. Some, designated type I superconductors, completely exclude the external magnetic field from penetrating into the material

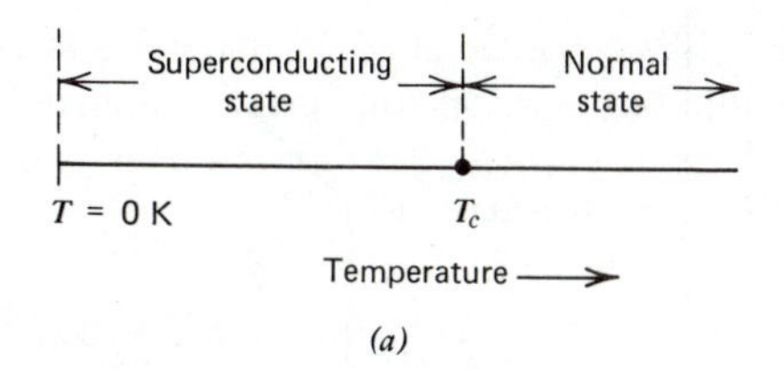

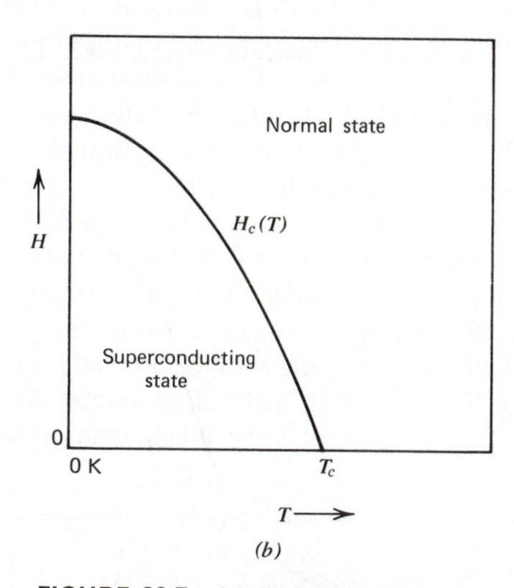

FIGURE 22.7 (*a*) The unary phase diagram showing the superconducting and normal regions as a function of temperature at constant pressure and magnetic field. (*b*) The unary phase diagram showing the superconducting and normal regions as a function of temperature and applied magnetic field. The boundary line between the normal and superconducting phases is the locus of points generated by the superconducting critical magnetic field (H_c) as a function of temperature.

hile in the superconducting state. A plot of internal magnetic field or magnetic nduction B vs. external field H for such a material is as shown in Fig. 22.8. The nternal field is zero below H_c and becomes virtually equal to the external field xcept for an appropriate conversion factor; see Chapter 25) above H_c. Very ure lead and tin are examples of type I superconductors. For type II supercon-uctors, the relationship between magnetic induction and external field is different

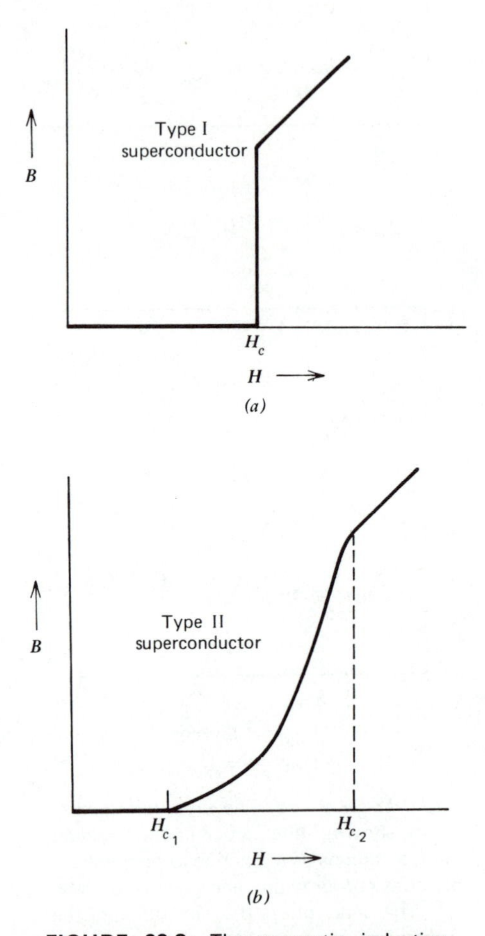

FIGURE 22.8 The magnetic induction (B)–magnetic field (H) relationship for (*a*) type I superconductors and (*b*) type II superconductors. In (*a*) the magnetic field does not penetrate the material until the critical field H_c is reached, where complete magnetic flux penetration takes place. In (*b*) the field begins to penetrate at H_{c1} and the magnetic induction gradually increases until at H_{c2} the material is fully penetrated and normal. Between H_{c1} and H_{c2}, a type II superconductor is in a mixed state.

(Fig. 22.8*b*). In this case, magnetic flux lines begin to penetrate the sample at a lower critical field, H_{c1}, and the penetration continues to increase with increasing field until the upper critical field, H_{c2}, is reached at which point complete field penetration has occurred. Between H_{c1} and H_{c2} the material is said to be in the mixed state and is considered to consist of both normal and superconducting regions (Fig. 22.9). Actually, the magnetic field penetrates the superconductor as discrete magnetic flux lines, each of which contains one quantum of magnetic flux (2×10^{-15} Wb), and a very fine cylindrical filament of normal material having a diameter typically between 10 nm and 100 nm extends along the center of each flux line. With increasing external field, more and more flux lines enter the superconductor until at H_{c2} magnetic field penetration is complete and the normal regions fully occupy the material. Type II superconductors, which include Nb–Ti and Nb–Zr solid-solution alloys as well as Nb_3Ge, Nb_3Sn, and a number of other compounds having the same crystal structure, are technologically the most important class of superconducting materials.

Type II superconductors generally have higher values of T_c than do type I superconductors, and H_{c2} usually is much greater than H_c for type I superconductors. Of particular importance is the amount of externally impressed current or current density that a type II superconductor can carry without destruction of superconductivity because, in addition to temperature and magnetic field, an imposed external current also can cause a transition to normal behavior. The largest current density that a superconductor can carry and still remain resistanceless is called the critical current density J_c and is a function of H and T. From a practical viewpoint, J_c is just as important as H_{c2} and T_c in determining the usefulness of type II superconductors because their primary application is in electromagnets that can generate high magnetic fields. For example, electromagnets will be needed to produce magnetic fields necessary for plasma confinement in most fusion reactors as currently envisioned. In addition to fusion power generation, certain other proposed energy conversion schemes also require magnetic fields that can be supplied practically only by superconducting coils (or solenoids). Conventional solenoids made out of normal conductors are impractical because simply maintaining a magnetic field involves a large power consumption due to the resistance of the conductor, which leads to power dissipation in the form of heat. In contrast, a superconducting solenoid can generate magnetic fields with essentially no consumption of electrical power, providing $T < T_c$, $H < H_{c2}$ and $J < J_c$. Since the strength of a magnetic field generated by a solenoid is proportional to the number of turns of the conductor and to the current passing through the conductor, a material having a high value of J_c at the operating magnetic field and temperature is necessary to minimize the cost of superconducting solenoid production.

While H_{c2} and T_c are determined primarily by the composition and crystal structure of the superconducting material (i.e., these properties are insensitive to

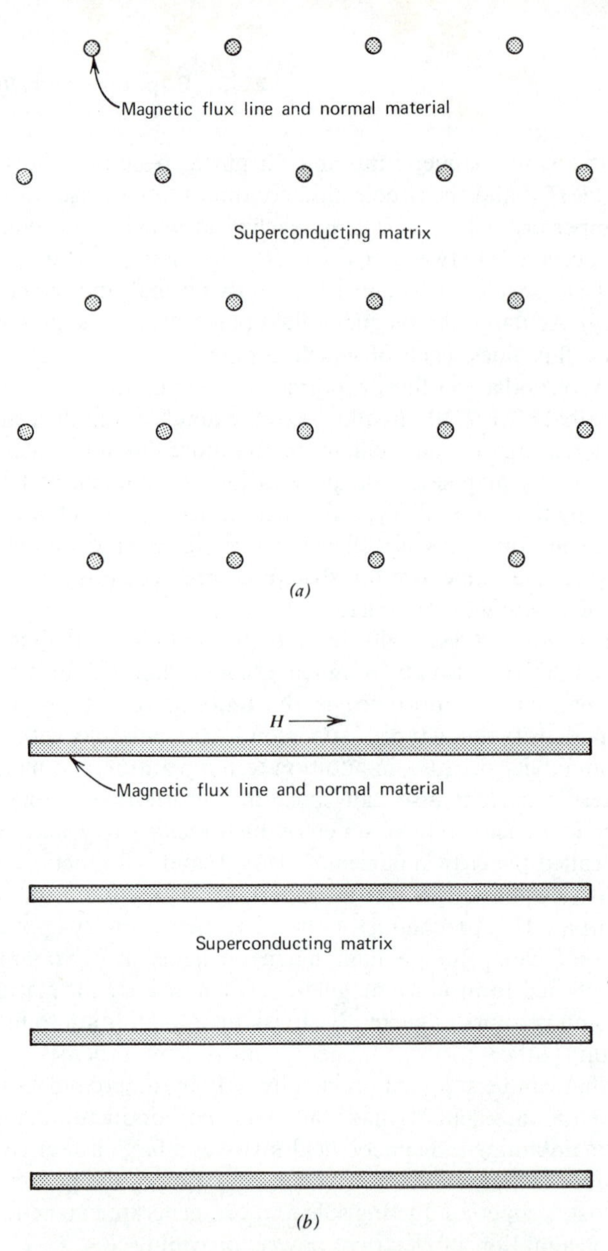

Figure 22.9 Schematic representation of the arrangement of normal and superconducting regions in the mixed state of a type II superconductor as viewed (*a*) parallel to the applied magnetic field and (*b*) perpendicular to the applied magnetic field. In the center of each flux line there is a small filament of normal material; the magnetic flux is concentrated in this region but also extends somewhat into the surrounding superconducting region. The flux lines usually are arranged in a regular geometric array, and an increase in the average internal magnetic induction B corresponds to more flux lines that are spaced more closely.

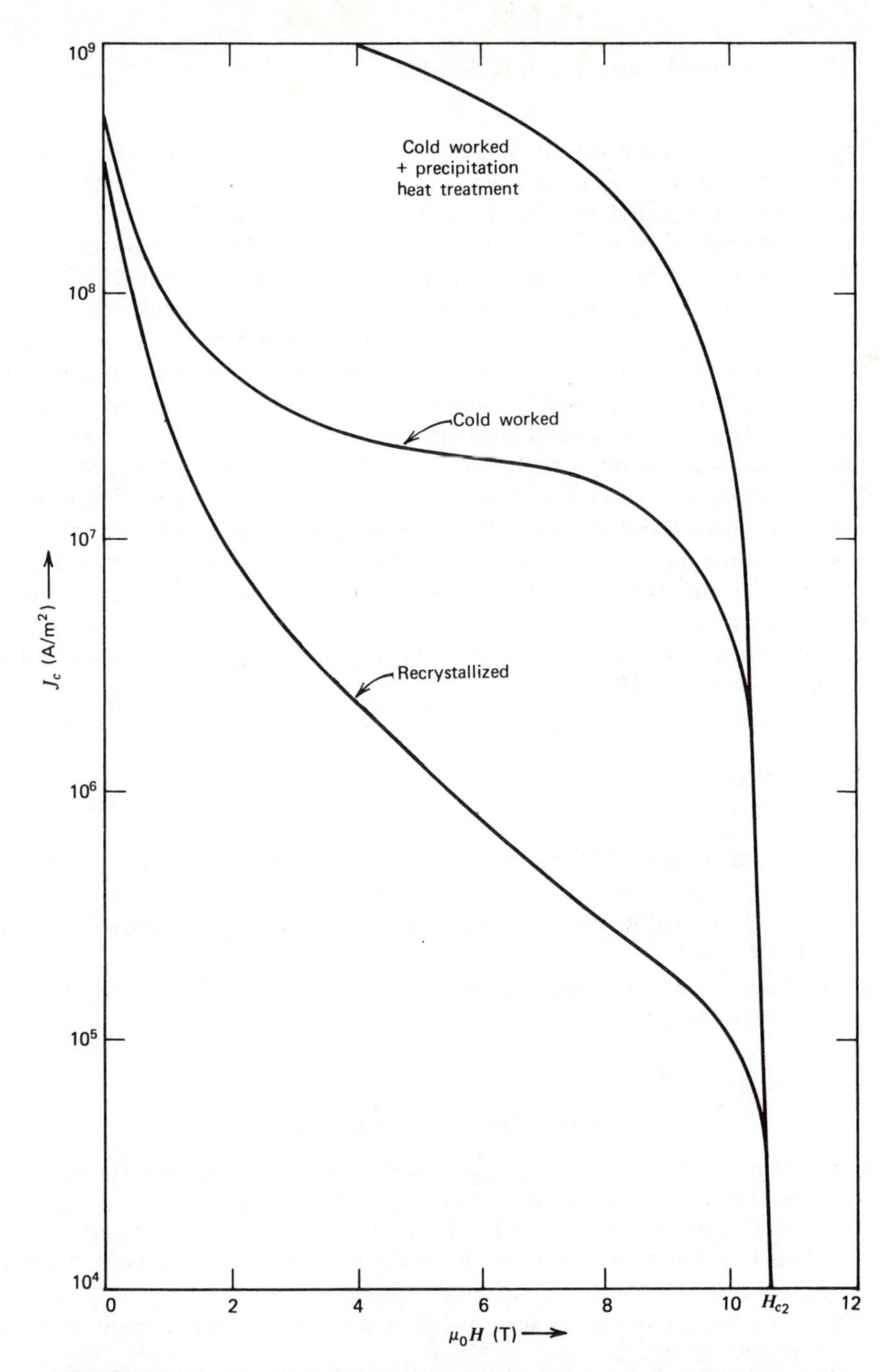

FIGURE 22.10 The effect of thermal and mechanical processing treatments on the critical current density (J_c) versus applied magnetic field (H) relationship for a 60 wt. % Nb–40 wt. % Ti alloy at 4.2 K. Both cold work and precipitation heat treatments increase J_c because precipitate particles, grain boundaries, and dislocations pin flux lines, and a combination of cold work followed by precipitation gives the best properties. Note that J_c falls drastically when the critical field H_{c2} is approached.

microstructure), J_c is probably the most microstructurally sensitive property that has ever been studied. With a given alloy, J_c can be varied by several orders of magnitude as a result of microstructural manipulation through appropriate thermomechanical processing treatments (Fig. 22.10). The dependence of J_c on microstructure is related to the cause of resistance and power dissipation in a type II superconductor carrying current in an applied magnetic field. Basically, the flux lines tend to move in a direction perpendicular to both the current and the field and this, by reason of the laws of electromagnetism, is associated with the appearance of a voltage and power dissipation. If the flux lines are prevented from moving, no voltage results and high supercurrents are possible. Thus, preventing the motion of flux lines in the presence of the current-induced electromagnetic force increases J_c in much the same way as preventing the motion of dislocations in the presence of an applied stress strengthens a material. It is especially interesting that J_c and mechanical hardness are correlated in most type II superconductors; this indicates that the type of defects responsible for increasing J_c are the same as those responsible for increasing hardness; in other words, flux lines in a type II superconductor are pinned by dislocations, grain boundaries, and precipitate particles (see Fig. 22.10).

REFERENCES

1. Rose, R. M., Shepard, L. A., and Wulff, J. *Structure and Properties of Materials*, Vol. IV, Wiley, New York, 1966.
2. Azároff, L. V., and Brophy, J. J. *Electronic Processes in Materials*, McGraw-Hill, New York, 1963.
3. Ehrenreich, H. The Electrical Properties of Materials, *Scientific American*, Vol. 217, No. 3, September, 1967, pp. 194–204.

QUESTIONS AND PROBLEMS

22.1 Aluminum has a room temperature resistivity of 2.65×10^{-8} ohm · m and copper has a room temperature resistivity of 1.67×10^{-8} ohm · m.

(a) Calculate the resistance of an Al rod 1 m long and 5 mm in diameter.

(b) How long must a Cu rod of the same cross-sectional area be in order to have the same resistance?

22.2 To a first approximation, the temperature dependence of the resistivity of a metal may be expressed as

$$\frac{\Delta\rho}{\rho} = \alpha\,\Delta T$$

where $\Delta\rho/\rho$ is the fractional change in resistivity due to a change in temperature ΔT and α is the temperature coefficient of resistivity. For Cu, $\alpha = 4 \times 10^{-3}\ °C^{-1}$ and the resistivity increases by 0.4% for a temperature increase of 1°C near room temperature. At what temperature will the resistance of a Cu conductor be twice that at 0°C?

22.3 A wire 0.2 cm in diameter must carry 25 A; however, the maximum power dissipation allowed is 4 W per meter of wire. (Power $= IV = I^2R$; 1 W = 1 A · V = 1 A^2 · ohm.)
(a) Which of the materials listed in Table 23.1 are suitable for this wire?
(b) Calculate the current density.

22.4 Why does electrical resistivity decrease as temperature increases for an ionic solid, while resistivity increases with temperature for a metal?

22.5 (a) Schematically sketch the logarithm of the number of charge carriers in an impure ionic solid as a function of reciprocal absolute temperature.
(b) Schematically sketch the logarithm of charge carrier mobility versus reciprocal absolute temperature.
(c) Combine your sketches in parts (a) and (b), and sketch the electrical conductivity of the impure ionic solid as a function of reciprocal absolute temperature. Denote the ranges where intrinsic and extrinsic conductivity dominate.

22.6 Would a soda-lime-silica glass be a better electrical conductor than a pure silica glass? Explain.

22.7 Using Eq. 22.10 and assuming that every time a conduction electron is scattered its velocity falls to zero, derive Eq. 22.11 for the average velocity of conduction electrons.

22.8 Show that $J = nq\bar{v}$. Compare this equation to equations that give the rate of viscous flow (Eq. 20.14) and mass flow (Eq. 12.2). In what respects are the equations similar?

22.9 Fine aluminum wires quenched in cold water from 504°C and tested immediately have a room temperature resistivity 0.18% higher than that of slowly cooled Al. When determined 25 minutes later, the resistivity of the quenched wire is the same as that of the slowly cooled Al. Explain.

22.10 Thin metal films deposited onto a cold substrate in a vacuum have resistivities that are considerably higher than bulk samples of the same metal. After being heated to several hundred degrees Celsius and then cooled to room temperature, the resistivity of the films approaches that of the bulk metal. Explain.

22.11 The resistivity ratio of a conductor is defined as $\rho_{300}/\rho_{4.2}$, where ρ_{300} is the resistivity at 300 K and $\rho_{4.2}$, the resistivity at the boiling temperature of liquid helium (4.2 K). The resistivity ratio often is used as a semiquantitative measure of the purity (and perfection) of a metallic material. Explain why this measurement is useful for such a purpose.

22.12 Refer to Fig. 4.12 and assume that the energy difference between the bottom and top of the 3s band is 5.5 eV.
(a) Estimate the spacing between individual energy levels for one mole of solid sodium. (Assume that there is one state per energy level and that the energy levels are spaced equally.)

(b) Compare this spacing with the energy of thermal activation at room temperature (i.e., with $kT \approx 0.025$ eV).

(c) Estimate the maximum energy (Fermi energy) that an electron in the 3*s* band of sodium has with respect to the bottom of the band at 0 K. Compare your estimation with the value given in Table 4.2.

(d) Assuming thermal energy at room temperature can excite electrons only within 1/40 eV of the Fermi energy to unoccupied states above E_F, what fraction of the electrons in the 3*s* band of sodium are thermally excited at room temperature?

(e) For most metals for which the free electron model of conductivity is valid, v_F is approximately 10^6 m/sec. Using Eq. 22.12, compute the drift velocity $\bar{v}$ in sodium at room temperature for an electric field of $E = 100$ V/m. (Na is bcc with $a = 0.429$ nm.) Compare your value of $\bar{v}$ with $\bar{v}_F$, and comment on this comparison relative to what is shown in Fig. 22.5*b*.

22.13 Derive an expression for the Fermi energy E_F at 0 K in terms of the number of free electrons per unit volume (n) by integrating the density of states expression,

$$N(E) = 2\pi V \left(\frac{2m_e}{h^2} \right)^{3/2} E^{1/2},$$

from zero to E_F. Explain why this integration should give the total number of *occupied* electronic states in the solid at 0 K. (*Note.* Two electrons, having opposite spins, can occupy each state.)

22.14 (a) Distinguish between a superconductor and a normal metal.

(b) If a battery is connected to a superconductor, what will happen? Why?

22.15 Calculate the current (at $\mu_0 H = 5$ T) a 60 wt. % Nb–40 wt. % Ti alloy of 0.025 cm diameter could carry assuming that its J_c-H behavior is given by Fig. 22.9. Do the calculation for all three microstructural conditions shown in Fig. 22.9.

22.16 (a) Figure 22.7*b* is actually a representative phase diagram only for type I superconductors. Sketch the temperature-magnetic field phase diagram for type II superconductors.

(b) If the diagram you have sketched is interpreted as a true "equilibrium" phase diagram, is Gibb's phase rule violated? Comment on your answer.

22.17 (a) The resistivities of two-phase alloys generally fall between those predicted by two simple models. In one model, the two phases are assumed to be in series and in the other model, they are assumed to be in parallel. Using the rule for adding resistances in series, show that for the series model

$$\rho_{\text{ser}} = \rho_\alpha V_\alpha + \rho_\beta V_\beta$$

where ρ_α and ρ_β are the resistivities and V_α and V_β are the volume fractions of the alpha and beta phases, respectively.

(b) Using the rule for adding resistances in parallel, show that for the parallel model,

$$\rho_{\text{par}} = \frac{\rho_\alpha \rho_\beta}{V_\alpha \rho_\beta + V_\beta \rho_\alpha}$$

23

Semiconductors

The electrical conductivity of semiconductors, as affected by temperature and impurity content, is discussed in this chapter.

23.1 INTRINSIC SEMICONDUCTORS

The most important elemental semiconductors are Ge and Si from group IVB in the periodic table. Both of these assume the diamond cubic crystal structure (see Fig. 7.9) and have highly directional covalent bonds. Since the sp^3 hybrid bonding orbitals that join each atom to its four nearest neighbors are discrete and occupied by electron pairs, the bonding electrons are unable to move through the crystal and to conduct electricity unless sufficient energy is available to excite some of them from their bonding state. The energy required to do this corresponds to the energy gap between valence and conduction bands. At any finite temperature, thermal energy imparted to the bonding electrons causes a certain small fraction to jump the energy gap. More electrons are excited across the gap at increasingly higher temperatures, and the electrical conductivity of a semiconductor rises rapidly as temperature increases. Conversely, at very low temperatures, completely pure Ge and Si should behave as electrical insulators.

Every electron that reaches the conduction band leaves behind a vacant site or hole in the valence band. Under the influence of an applied electric field, conduction electrons and holes move in opposite directions and, hence, contribute additively to a current. Even though the motion of a hole corresponds identically to the net opposite motion of electrons in the valence band, it is convenient to treat holes as particles of positive charge. The promotion of a single electron into the conduction band thus creates not one, but two charge carriers, although, as might be expected, the mobilities of holes and electrons are distinctly different.

23.2 DENSITY OF CHARGE CARRIERS

In *intrinsic* semiconductors, such as pure Ge and Si, the density (i.e., number per unit volume) of conduction electrons n_e equals the density of holes n_h. Since electrons are thermally activated into the conduction band, n_e and n_h have a temperature dependence that is given by an equation similar to that for other thermally activated processes:

$$n_e = n_h \simeq Ce^{-E_g/2kT} \tag{23.1}$$

or

$$\ln n_e = \ln n_h \simeq \ln C - \frac{E_g}{2kT} \tag{23.2}$$

where k is the Boltzmann constant and E_g, the energy gap between the valence and conduction bands, is the energy necessary to create one conduction electron–hole pair. [The factor of $\frac{1}{2}$ in the exponent of Eq. 23.1 arises because the product of n_e and n_h is proportional to $e^{-E_g/kT}$ (see problem 23.12).] The variation of n_e with temperature for the intrinsic semiconductor Ge is shown in Fig. 23.1. With such a

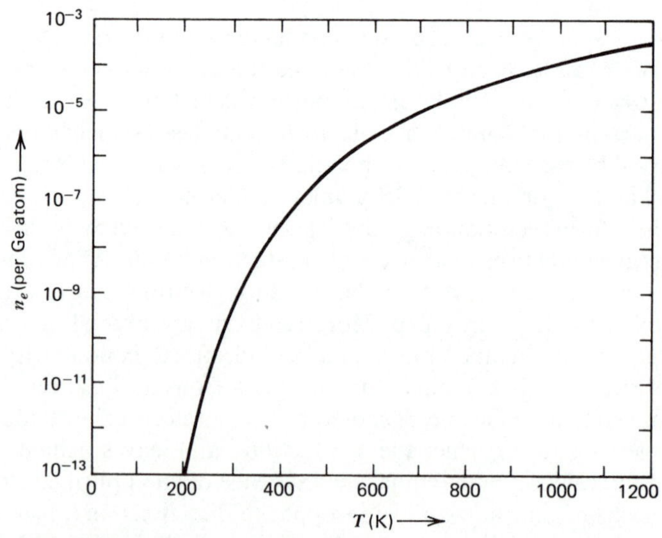

FIGURE 23.1 The variation of the number of conduction electrons with temperature for pure germanium.

strong dependence of the number of charge carriers on temperature, the electrical conductivity of an intrinsic semiconductor is very temperature sensitive.

23.3 CARRIER MOBILITY

The electrical conductivity of a metal, in terms of Eq. 22.7, is

$$\sigma = ne\mu \tag{23.3}$$

where n is the conduction electron density, e is the magnitude of the electron charge, and μ is the electron mobility. The equivalent equation for a semiconductor is

$$\sigma = n_e e\mu_e + n_h e\mu_h \tag{23.4}$$

where μ_e and μ_h are the respective mobilities of a conduction electron and a hole. For intrinsic semiconductors, Eq. 23.4 reduces to

$$\sigma = n_e e(\mu_e + \mu_h) \tag{23.5}$$

Mobilities of holes are always less than those of electrons; in Si and Ge, the ratio μ_e/μ_h is approximately three and two, respectively (Table 23.1). Since the mobilities change only slightly as compared to the change of charge carrier densities with temperature, the temperature variation of conductivity for an intrinsic semiconductor is similar to that of charge carrier density shown in Fig. 23.1, and conductivity data can be used to determine the energy gap:

$$\ln \sigma \simeq \ln C' - \frac{E_g}{2kT} \tag{23.6}$$

TABLE 23.1 Some Room Temperature Properties of Group IV Elements

Element	E_g (eV)	σ ($\text{ohm}^{-1} \cdot \text{m}^{-1}$)	μ_e ($\text{m}^2/\text{V} \cdot \text{sec}$)	μ_h ($\text{m}^2/\text{V} \cdot \text{sec}$)
C (diamond)[a]	~7	10^{-14}	0.18	0.14
Silicon	1.21	4.3×10^{-4}	0.14	0.05
Germanium	0.785	2.2	0.39	0.19
Gray tin[b]	0.09	3×10^{5}	0.25	0.24
White tin[b]	0	10^{7}	—	—
Lead[c]	0	5×10^{6}	—	—

[a] An insulator.

[b] Gray tin has the diamond cubic structure and is an "incipient" semiconductor; white tin, the common allotrope, is tetragonal and metallic.

[c] Lead is metallic.

23.4 EXTRINSIC SEMICONDUCTORS

As mentioned in Sect. 10.2, the presence of an impurity atom in a covalent material such as Si or Ge can substantially alter its electrical properties. The charge carrier density can be increased by impurities of either higher or lower valence. If pentavalent substitutional atoms, such as P, As, or Sb, are present, only four of their five valence electrons are required to participate in covalent bonding. Since the fifth electron remains weakly bound to the impurity or donor atom, it is not entirely free in the crystal. Nevertheless, the binding energy (typically 0.01 eV) is much less than that for an electron in a covalent bonding state (e.g., E_g in Table 23.1), and the extra electron can be detached easily from the impurity or donor atom. The energy state of this electron and the amount of energy necessary to excite it to the conduction band can be represented by the band model as shown in Fig. 23.2. Each electron promoted to the conduction band from a donor state is not associated with a hole in the valence band and, consequently, conduction is mainly by one type of charge carrier. When the majority charge carriers are electrons in the conduction band, the *extrinsic* semiconductor is designated as *n*-type. The analogy with extrinsic conduction in ionic solids should be noted.

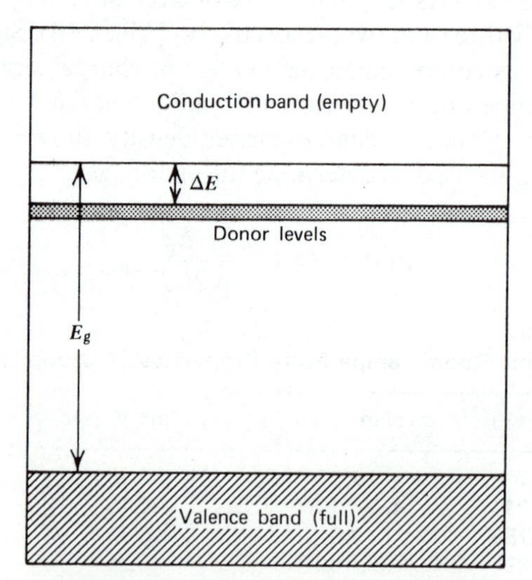

FIGURE 23.2 Schematic energy band diagram showing the energy levels of nonbonding electrons (donor levels) for an *n*-type extrinsic semiconductor. Electrons in donor levels require only a small amount of energy (ΔE) to be excited into the conduction band and, thus, to become conduction electrons.

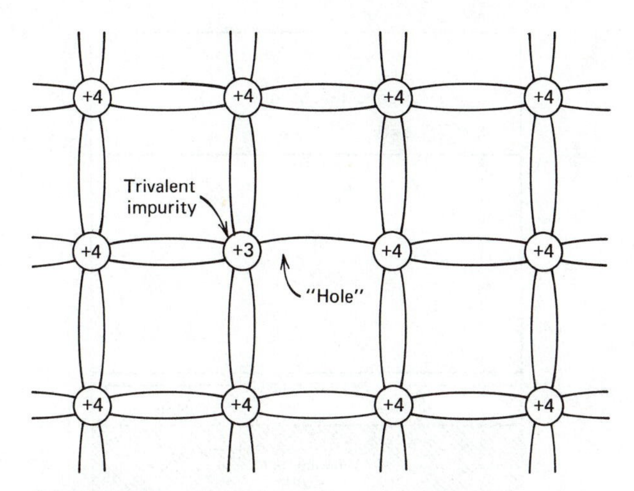

FIGURE 23.3 The addition of a trivalent impurity produces a hole in the bonding structure of a quadrivalent material. The hole is only weakly bound to its parent atom.

Although mechanistically quite different, the addition of an impurity atom (or ion) of different valence results in the donation of a charge carrier to the solid in both cases.

Trivalent substitutional atoms, such as Al, Ga, or In, in a quadrivalent intrinsic semiconductor have an opposite effect. Since these atoms are deficient in bonding electrons, one of their bonding orbitals will contain a hole that is capable of accepting an electron from elsewhere in the crystal. Actually, the hole is weakly bound to the impurity or acceptor atom. As shown in Figs. 23.3 and 23.4, the binding energy is small, and the promotion of an electron from the valence band to the acceptor atom leaves a hole in the valence band that can act as a charge carrier. In this case, when the majority charge carriers are holes in the valence band, the *extrinsic* semiconductor is designated as *p*-type.

Intrinsic semiconductors are prepared routinely with initial total impurity contents less than one part per million (ppm) and harmful impurities (i.e., donor or acceptor atoms) less than one part in 10^7. To such highly purified materials, *controlled* amounts on the order of 1 to 1000 ppm of substitutional donors or acceptors are added intentionally to produce extrinsic semiconductors having specific room temperature conductivities. Since the energy gap with respect to a donor or acceptor level is so small, thermal energy at room temperature gives significant charge carrier concentrations. The type and amount of substitutional element addition is important, however. For example, the addition of transition metals such as Fe, Ni or Co to Si or Ge produces impurity levels far removed from the band edges and increases the room temperature conductivity only slightly.

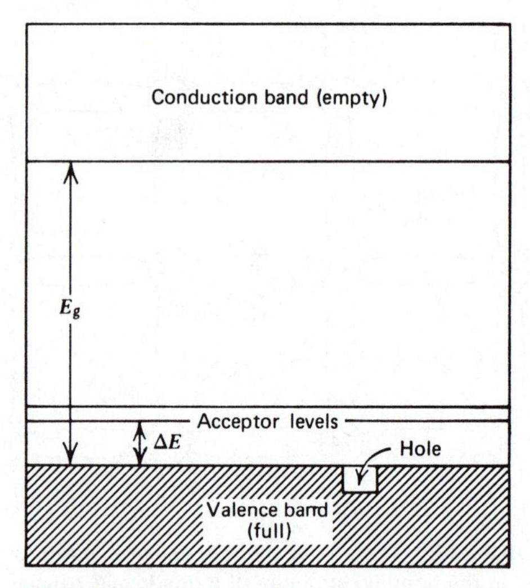

FIGURE 23.4 Schematic energy band diagram showing the energy levels of holes (acceptor levels) for a *p*-type extrinsic semiconductor. The difference in energy between the acceptor level and the valence band (ΔE) is the energy required to excite one bonding electron from a saturated bond to an acceptor level and, thus, to create a hole in the valence band.

23.5 TEMPERATURE VARIATION OF CONDUCTIVITY

The charge carrier concentration for an *n*-type semiconductor varies with temperature as shown in Fig. 23.5. At low temperatures, some electrons are excited from the donor level to the conduction band, but thermal excitation of electrons from the valence to the conduction band is negligible in comparison. At intermediate temperatures, in the so-called exhaustion range, the donor levels tend to become depleted, and the number of charge carriers approaches an almost constant value approximately equal to the number of donor electrons. At higher temperatures, the number of valence electrons excited to the conduction band greatly exceeds the total number of electrons from substitutional donor atoms, and the extrinsic semiconductor behaves like an intrinsic semiconductor. On the basis of charge carrier density alone, the conductivity of an extrinsic semiconductor should vary with temperature as shown in Fig. 23.5, which is similar to the temperature dependence of ionic conductivity in an impure ionic solid (cf. Fig. 22.2*b*). In reality, the behavior of an extrinsic semiconductor differs, as shown in

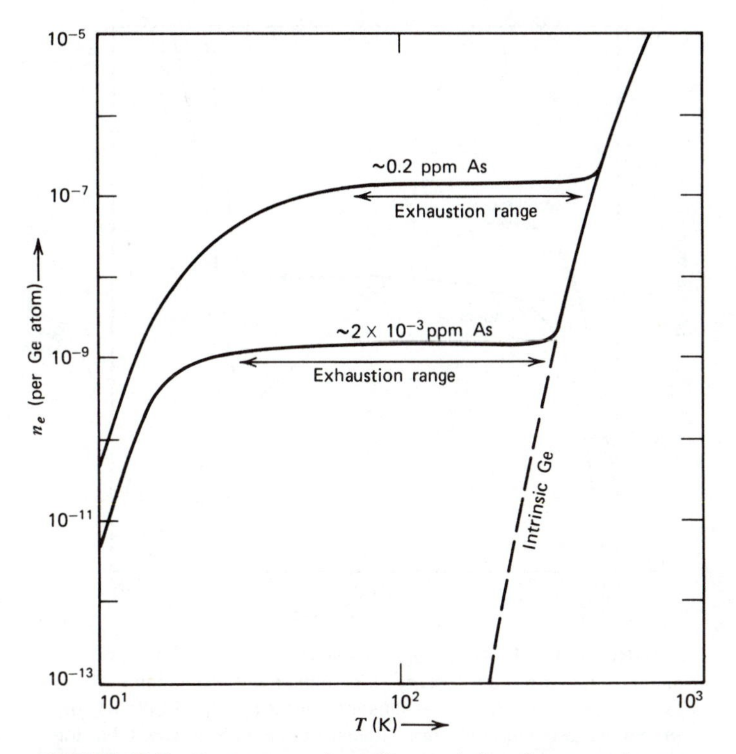

FIGURE 23.5 Temperature dependence of the charge carrier concentration of *n*-type extrinsic germanium with two different impurity atom concentrations. At low temperatures, thermal activation promotes donor electrons to the conduction band. At intermediate temperatures (exhaustion range), most of the donor electrons have been promoted and the number of charge carriers is nearly independent of temperature. At higher temperatures, thermal energy is sufficient to promote measurable numbers of valence (bonding) electrons and to induce intrinsic semiconductivity. [Data from E. M. Conwell, *Proc. IRE*, **40**, 1327 (1952).]

Fig. 23.6, because charge carrier mobility varies with temperature. As mentioned in Sect. 23.3, the temperature dependence of carrier mobility has little effect when intrinsic conductivity predominates. However, when extrinsic conductivity predominates and donor levels are close to the conduction band, the carrier mobility can determine the temperature dependence of conductivity. This is true in the exhaustion range where carrier mobility typically decreases with increasing temperature because of more effective phonon scattering. Thus, in this region, characterized by an approximately constant number of charge carriers and a mobility decreasing with temperature, the temperature dependence of conductivity of an extrinsic semiconductor is similar to that of a metal.

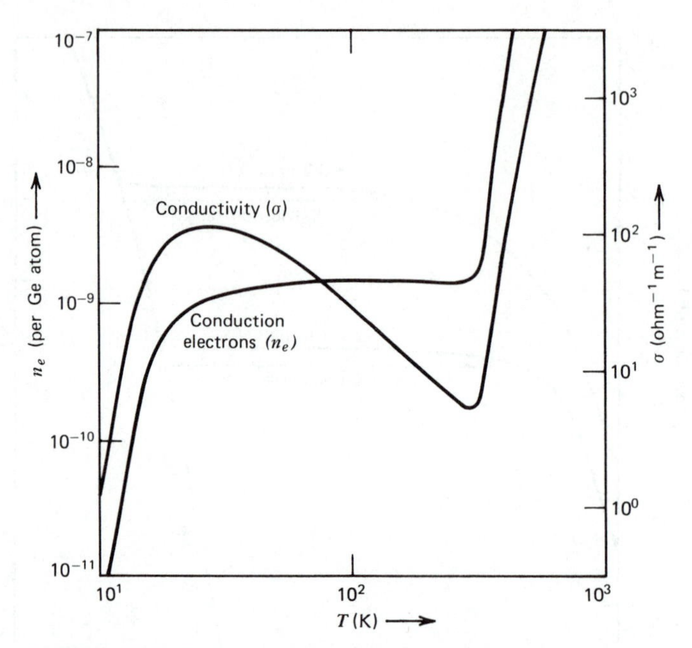

FIGURE 23.6 The temperature variation of the number of charge carriers and the conductivity for germanium containing about 2×10^{-3} ppm As. In the exhaustion range (~30 to 300 K), the number of conduction electrons remains essentially constant, but the conductivity decreases with increasing temperature since the mobility of these electrons decreases with increasing temperature. [Data from E. M. Conwell, *Proc. IRE.*, **40**, 1327 (1952).]

23.6 SEMICONDUCTING COMPOUNDS

A large number of compounds possess structural and electrical characteristics that are similar to those of elemental semiconductors. Many of these are covalent compounds formed between elements of groups IIIB and VB of the periodic table and are termed III–V compounds. In such cases, slight deviations from stoichiometric compositions can lead to extrinsic conduction. An important example is GaAs. The component elements are symmetrically arranged with respect to group IVB and the resulting *average* number of valence electrons is four; the same sp^3 hybrid bonding orbitals occur as in germanium, for example. However, since the elemental substances are from different columns of the periodic table, the bonding is partly ionic in nature and the energy gaps are higher than in the corresponding elemental semiconductors (Table 23.2). These larger energy gaps extend the useful temperature range of exhaustion conduction since intrinsic conduction becomes important only at correspondingly higher temperatures. The general

TABLE 23.2 Energy Gaps (in eV) of Some III–V and II–VI Compounds

	Increasing $Z \rightarrow$				Increasing $Z \rightarrow$		
Group V Element:	P	As	Sb	Group VI Element:	S	Se	Te
Group III Element				Group II Element			
Al	3.0	2.3	1.7	Zn	3.6	2.8	—
Ga	2.5	1.4	0.78	Cd	2.5	1.8	1.6
In	1.4	0.43	0.23	Hg	—	~0.3	0.20

($\leftarrow$ Increasing Z, downward for the rows)

decrease in energy gap with increasing atomic number of group IV semiconductors and III–V compounds is a result of an increasing tendency toward metallic bonding.

Some semiconducting compounds can be of the II–VI type; however, these have much more ionic character than III–V compounds for the reasons mentioned previously. Their energy gaps are thus larger, and in some cases they even may be viewed as electrical insulators (Table 23.2). CdS, CdSe, and CdTe have energy gaps of 2.5, 1.8, and 1.6 eV, respectively, as compared to the lower energy gaps of InP, InAs, and InSb. ZnS is an insulator, whereas ZnSe has an energy gap of 2.8 eV. A wide variety of compounds between transition metals and nonmetals also exhibit semiconductivity, but these materials are used infrequently because impurity control is much more difficult to achieve.

In addition to thermal energy, electromagnetic energy can promote electrons from the valence to the conduction band. Indeed, energy gaps on the order of 1.5 to 3 eV correspond to the optical region of electromagnetic energies, since $E = h\nu = hc/\lambda$ (Eq. 2.3), where ν and λ are the frequency and wavelength, respectively. Thus, bombardment of CdS with visible radiation markedly stimulates photoconductivity and, for this reason, it is used extensively in light-metering applications. Energy gaps of other semiconducting compounds correspond to other frequency ranges in the electromagnetic spectrum; for example, PbS, PbSe, and PbTe, with energy gaps of 0.37, 0.27, and 0.33 eV, respectively, can serve as infrared radiation detectors. Large energy gaps (> 3 eV) result in optical transparency, providing no impurity states exist.

As mentioned, the larger energy gaps of compound semiconductors extend the exhaustion conductivity range exhibited by such materials. In order to develop a better understanding of conduction in these materials, it is worthwhile to examine the mechanism of extrinsic conduction in a few cases. The compound ZnO affords a convenient example. Very pure stoichiometric ZnO, with an energy gap of about 3.3 eV, is almost an insulator in that intrinsic conduction occurs only at very high temperatures. When heated in zinc vapor, however, the conductivity of ZnO is increased markedly. During this process, zinc atoms dissolve interstitially in the relatively open ZnO structure, and this may be written as a chemical reaction,

$$\mathrm{Zn}(g) \longrightarrow \underline{\mathrm{Zn}}_i \tag{23.7}$$

where $\underline{\mathrm{Zn}}_i$ refers to interstitial zinc atoms in the ZnO structure. Within the environment of ZnO, the interstitial zinc atom is readily ionized,

$$\underline{\mathrm{Zn}}_i \longrightarrow \underline{\mathrm{Zn}}_i^+ + e^- \tag{23.8}$$

The ionization energy (or enthalpy) for this process is only 0.05 eV, indicating that zinc atoms readily donate electrons to the conduction band and that ZnO can be considered an n-type semiconductor. Subsequent ionization also can take place,

$$\underline{\mathrm{Zn}}_i^+ \longrightarrow \underline{\mathrm{Zn}}_i^{2+} + e^- \tag{23.9}$$

resulting in two donor electrons per excess zinc atom in solution. Because of the small ionization energies associated with Eqs. 23.8 and 23.9, exhaustion semiconduction is approached in the vicinity of room temperature. It should be noted, however, that appreciably higher temperatures ($\gtrsim 500°\mathrm{C}$) are required to initiate the reaction given in Eq. 23.7. Indeed, if the conductivity of ZnO is measured as a function of temperature in the presence of Zn vapor, the measured activation energy associated with conduction is that for Eq. 23.7, since it is the largest of the activation energies for the series of reactions (Eqs. 23.7 to 23.9). By contrast, if ZnO is saturated with zinc vapor at a high temperature, quenched, and its conductivity then measured over a temperature range where the zinc diffusivity in ZnO is essentially nil, the activation energy for conductivity will be that associated with either Eq. 23.8 or Eq. 23.9, and, as mentioned previously, exhaustion conduction will be achieved at relatively low temperatures. Analogous considerations hold when ZnO is heated in a hydrogen atmosphere. The conductivity of ZnO is increased in this case because gaseous hydrogen dissolves interstitially in the structure and is subsequently ionized by reactions similar to those described by Eqs. 23.7 to 23.8.

The conductivity of magnetite, Fe_3O_4, also illustrates the intimate relationship between atomic and ionic structure and semiconducting properties. Magnetite has the spinel structure in which the oxygen ions form an approximately face-centered cubic array. Trivalent iron ions occupy the tetrahedral interstitial sites in this structure, and both divalent and trivalent ions occupy the octahedral interstitial sites. At room temperature, these latter sites are occupied randomly by the divalent and trivalent ions. This fact allows for relatively large electron mobilities, that is, the position of divalent and trivalent ions can be interchanged by electron transfer between them rather than by direct atomic motion. Since this electron mobility is high, the resistivity of magnetite is low; in fact, at room temperature, the resistivity is about 10^{-4} ohm · m, a value rendering it almost metallic in nature. Below about 100 K, however, the resistivity of magnetite increases markedly. This is due to an order-disorder transformation that occurs within the octahedral ion sublattice; in essence, below this temperature, the occupancy of the octahedral sites is fixed and no longer random with respect to the divalent and trivalent ions. This results in a much lower electron mobility and correspondingly higher resistivity.

As our discussions on ZnO and Fe_3O_4 have indicated, many materials that we would intuitively suspect to be insulators are, in fact, semiconductors. This occurs because in many compounds there are various sources for conduction electrons. Because electron mobility is many orders of magnitude greater than ionic mobility, small numbers of these electrons have a disproportionate influence on conductivity. If a means of introducing such electrons exists, a compound with a relatively large energy gap will therefore exhibit semiconducting properties; if not it will behave as an insulator. Thus, for example, NiO is considered a semiconductor and MgO, an insulator.

23.7 SEMICONDUCTOR JUNCTIONS

The characteristic values and temperature variation of the electrical resistivity of semiconductors render them useful for many applications. Even more important, additional technological uses, particularly in the form of devices, arise from the characteristics of semiconductor junctions. One such device is a rectifier that is composed of a *p–n* semiconductor junction, as illustrated schematically in Fig. 23.7*a*. The current-potential characteristics of this junction are noted in Fig. 23.7*b*; that is, there is a low resistance to current flow in one direction ("forward" bias) and a high resistance to current flow in the opposite direction ("reverse" bias). The form of the $I–V$ curve of a rectifier is similar to that of an electrode in an electrolyte (Fig. 14.10), and the physical basis for rectifying action is similar to that for electrode kinetics, as will be discussed below.

As soon as a *p–n* junction of the type shown in Fig. 23.7*a* is formed, an excess of holes exists on the *p* side and an excess of electrons, on the *n* side of the junction. The resulting concentration of holes and electrons across the junction (Fig. 23.8) leads to hole diffusion from the *p* to the *n* semiconductor and electron diffusion from the *n* to the *p* semiconductor. This flux produces a charge imbalance at the interface (Fig. 23.8*c*) which, in turn, yields an electric field or potential across the interface (Fig. 23.8*d*). At equilibrium the potential difference across the interface is such that no further net hole or electron diffusion occurs, that is, the chemical driving force due to the concentration gradient is balanced by the retarding force due to the electric field. The equilibrium potential difference across the interface

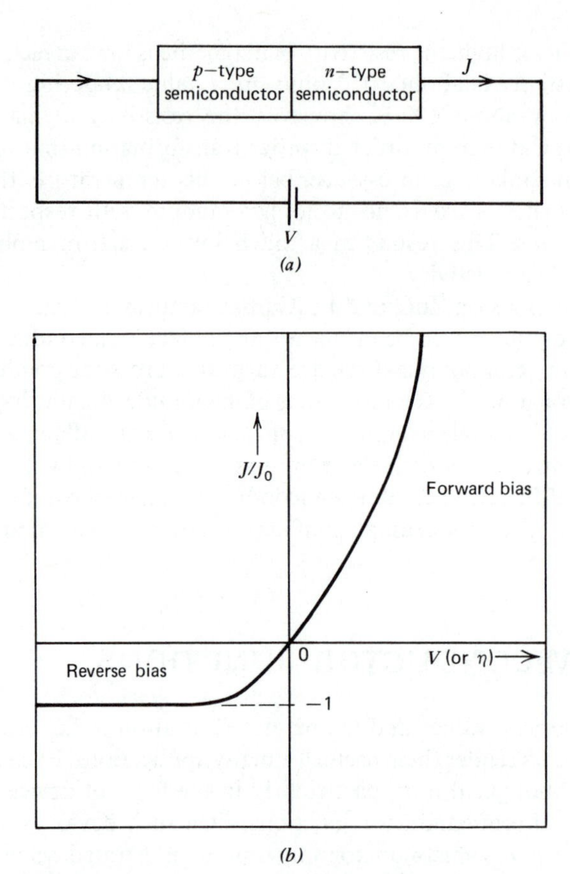

FIGURE 23.7 (*a*) Schematic illustration of a *p–n* semiconductor junction; (*b*) the current–voltage relationship of this junction. The junction is rectifying, that is, it is characterized by a low resistance when forward biased ($V > 0$) and a large resistance when reverse biased ($V < 0$). The limiting reverse current, J_0, depends on the nature and doping levels of the semiconductors.

is denoted as V_{eqm}. In this case, as in others, equilibrium represents a dynamic state in which the flux of holes from the *p* to the *n* side of the junction ($J_{h,\, p\to n}$) is balanced by the flux of holes from the *n* to the *p* side ($J_{h,\, n\to p}$), and there is a similar restriction on electron current densities in each direction. Since the energy of a hole is eV_{eqm} higher on the *n* side of the junction, $J_{h,\, p\to n}$ can be written as (see Chapters 13, 14, and 17)

$$J_{h,\, p\to n} = k_1 n_{h,\, p} e^{(-eV_{\text{eqm}}/kT)} \tag{23.10}$$

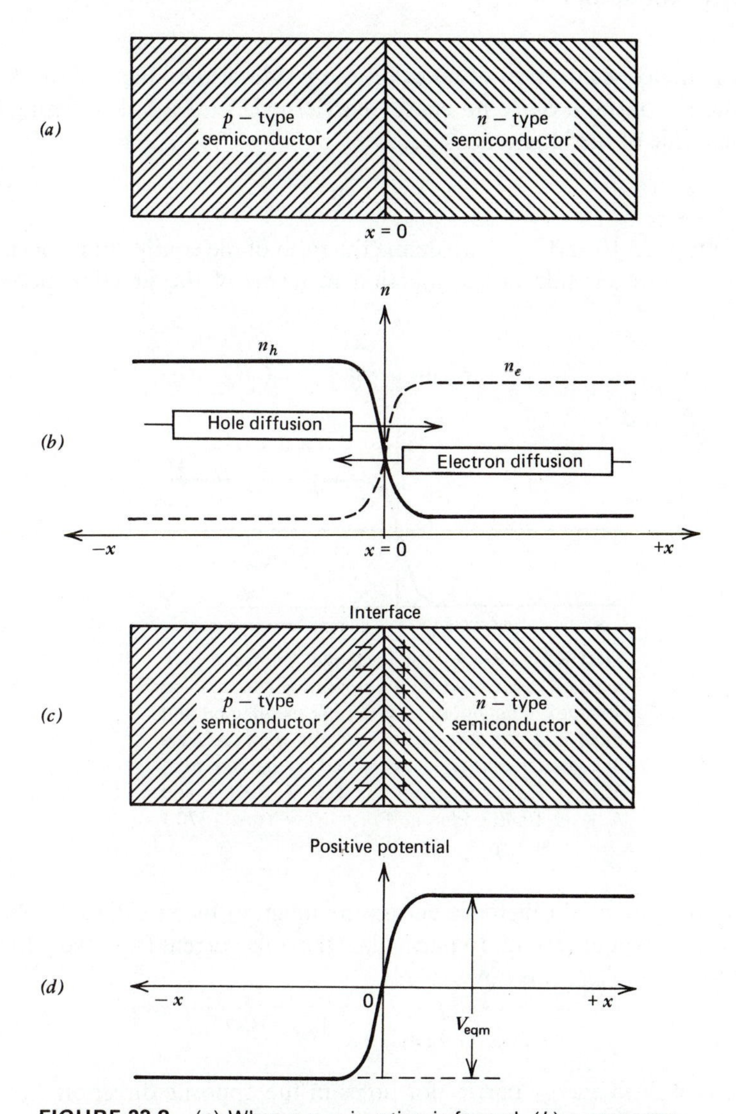

FIGURE 23.8 (*a*) When a *p–n* junction is formed, (*b*) a concentration gradient of holes and electrons exists across the interface that leads to diffusion and (*c*) the formation of a charged interface. (*d*) At equilibrium, a potential difference, V_{eqm}, exists across the interface. (Adapted from Modern Electrochemistry, by J. O'M. Bockris and A. K. N. Reddy, Plenum Press, New York, 1970.)

where $n_{h,p}$ is the concentration of holes on the p side of the junction and k_1 is a proportionality constant. Since no energy barrier exists for a hole crossing from the n to the p side of the junction, $J_{h,n\to p}$ is

$$J_{h,n\to p} = k_1 n_{h,n} \tag{23.11}$$

By equating Eqs. 23.10 and 23.11 we define the ratio of the equilibrium concentrations of holes on each side of the junction in terms of the junction potential, V_{eqm}.

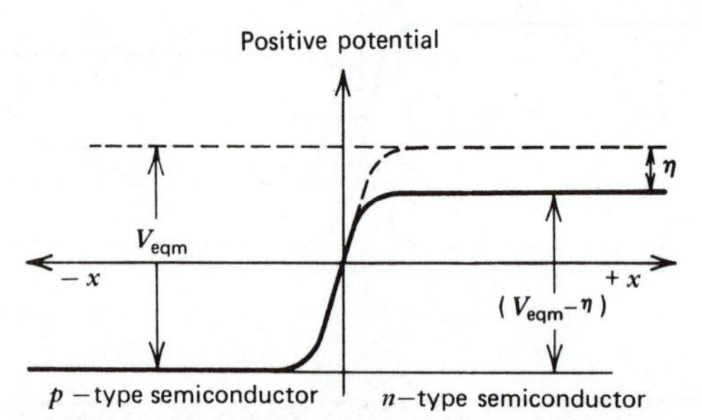

FIGURE 23.9 As the potential across a *p–n* junction is lowered by an amount η (the applied voltage), equilibrium is no longer maintained and there is a net drift of holes to the right because the energy barrier is decreased. (Adapted from *Modern Electrochemistry*, by J. O'M. Bockris and A. K. N. Reddy, Plenum Press, New York, 1970.)

Lowering the potential difference across the junction by an external voltage η (Fig. 23.9) perturbs equilibrium. In particular, the hole current from the p to the n side of the junction, $J'_{h,p\to n}$, is now

$$J'_{h,p\to n} = k_1 n_{h,p} e^{-(eV_{\text{eqm}}-\eta)/kT} \tag{23.12}$$

Since there is still no energy barrier for holes in the opposite direction, $J'_{h,n\to p}$ is still given by Eq. 23.11, and there now exists a net hole current density across the junction:

$$J_{h,\text{net}} = J'_{h,p\to n} - J'_{h,n\to p} = k_1\{n_{h,p} e^{-(eV_{\text{eqm}}-\eta)/kT} - n_{h,n}\} \tag{23.13}$$

By equating Eqs. 23.10 and 23.11, we have $k_1 n_{h,p} e^{-eV_{\text{eqm}}/kT} = k_1 n_{h,n}$ and, thus,

$$J_{h,\text{net}} = J_{0,h}(e^{e\eta/kT} - 1) \tag{23.14}$$

where $J_{0,h} = k_1\, n_{h,n} = k_1 n_{h,p} e^{-eV_{\text{eqm}}/kT}$.

An equation similar to Eq. 23.14 holds for the net electron current density:

$$J_{e,\text{net}} = J_{0,e}(e^{e\eta/kT} - 1) \tag{23.15}$$

and the net total current density across the junction is

$$J_{T,\text{net}} = J_0(e^{e\eta/kT} - 1) \tag{23.16}$$

where $J_0 = J_{0,h} + J_{0,e}$.

The current density-voltage (J–η) relationship (Eq. 23.16) is of the form illustrated in Fig. 23.7*b*. The absolute level of current (as well as the saturation current in reverse bias) depends on J_0, that is, on the doping levels in the p and n semiconductors and on the equilibrium potential, V_{eqm}. Rectifiers not only serve as devices that allow preferential current flow in one direction but can also be used, for example, to convert alternating to direct current (Fig. 23.10).

The current density-voltage relationship defined by Eq. 23.16 does not hold at large negative (i.e., reverse bias) voltages. For negative voltages greater than a characteristic breakdown voltage (Fig. 23.11), electrons and holes are accelerated to sufficient velocities so that upon collision with valence electrons they impart

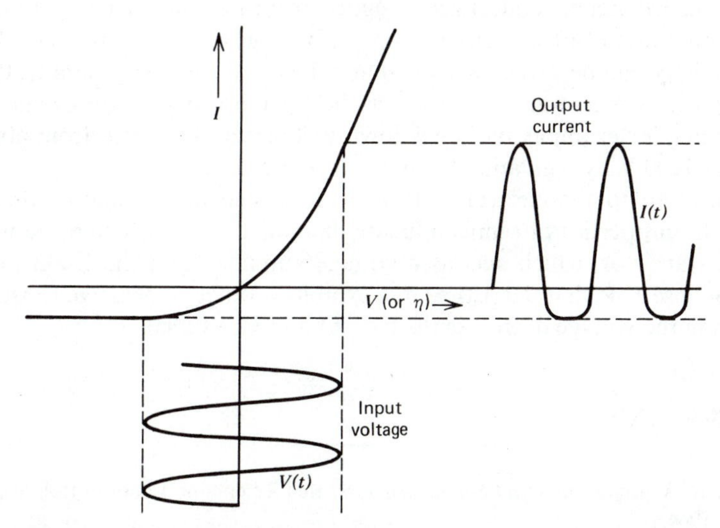

FIGURE 23.10 Use of the rectifying action of a *p–n* junction to convert alternating to direct current (ac to dc). The input voltage is ac; the output current is not strictly dc, but it is predominantly positive. This response can then be supplemented and smoothed out with other electronic devices to produce a smooth dc signal.

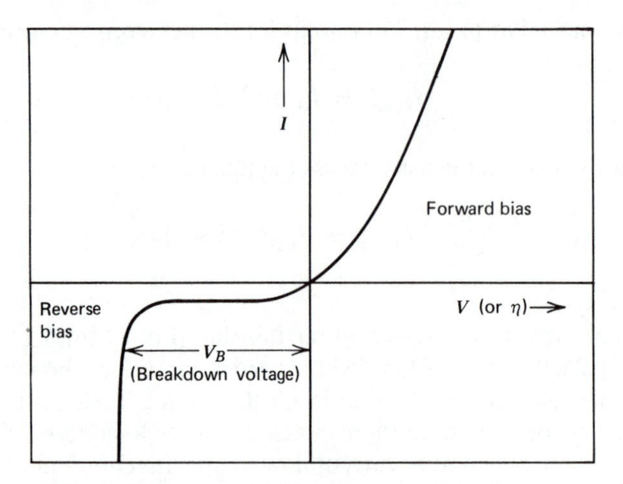

FIGURE 23.11 The current–voltage relationship for a diode showing the breakdown phenomenon. Below a critical negative voltage (or reverse bias), called the breakdown voltage, large currents can be generated.

enough energy to excite them into the conduction band. These liberated electrons and holes, in turn, repeat the process producing an "avalanche" of current. Some devices (called Zener diodes) are designed to utilize this behavior in operating electric circuits and are used, for example, as voltage regulators. In them, the breakdown voltage can be varied by controlling the doping levels; generally, the higher the impurity concentration, the lower the breakdown voltage. In silicon semiconductors, for example, the breakdown voltage can be varied from about 10 V to about 1000 V by regulating the impurity concentration.

In addition to the electronic devices mentioned above, many other devices utilize the properties of semiconductor junctions. The most important of these are the transistor, which acts as a voltage amplifier, and the Esaki (or tunnel) diode, whose I–V characteristic curve exhibits a range of negative resistance (i.e., increasing the voltage decreases the current and vice versa).

REFERENCES

1. Holden, A. *Conductors and Semiconductors*, Bell Telephone Laboratories, Summit Hill N.J., 1964.
2. Adler, R. B., Smith, A. C., and Longini, R. L. *Introduction to Semiconductor Physics* Wiley, New York, 1964.
3. Ehrenreich, H. The Electrical Properties of Materials, *Scientific American*, Vol. 217 No. 3, September, 1967, pp. 194–204.

QUESTIONS AND PROBLEMS

23.1 (a) Draw a simple band diagram for silicon, with the full valence band 1.1 eV below the bottom of the empty conduction band.

(b) Draw a level for acceptor states, with an 0.06 eV binding energy. Suppose 10 atoms of Al per million atoms of silicon are added to one mole of silicon. How many states, per unit volume of silicon, are there in the acceptor level?

(c) Calculate the hole carrier density in silicon because of the aluminum impurity at 300 K, assuming the probability of excitation is proportional to $e^{-\Delta E/kT}$.

(d) Calculate the intrinsic carrier (holes and electrons) density at 300 K. (For Si at room temperature, the coefficient in Eq. 23.1 is approximately $C = 2 \times 10^{26}\,\mathrm{m}^{-3}$.)

(e) Calculate the approximate temperature for onset of intrinsic conduction by assuming this temperature is that at which the intrinsic and extrinsic carrier densities are equal.

23.2 For all of the semiconducting elements and compounds listed in Tables 23.1 and 23.2, the mobility of holes is less than that of electrons. Explain this phenomenon on a physical basis.

23.3 Refer to Fig. 23.5, which shows the charge carrier density as a function of temperature for extrinsic semiconductors. Assume the mobility has a temperature dependence of $T^{-3/2}$. Schematically sketch how the conductivity will vary with temperature and on the same graph compare this with the temperature dependence of the number of charge carriers.

23.4 Compare the energy dependence of the frequency of the electromagnetic spectrum (Chapter 27) with typical values of energy gaps in semiconductors. What type of radiation corresponds to energy gaps of 0.10 eV, 0.50 eV, 1.00 eV, 1.75 eV, and 2.50 eV? Could the energy gap of a semiconductor be measured by the absorption of radiation? Explain.

23.5 Why is a large energy gap between the valence and conduction bands and a small gap between the donor level and the conduction band desirable for useful extrinsic semiconductors?

23.6 If a hybrid semiconductor is made by adding donor and acceptor atoms with twice the concentration of the former than the latter, what can you say about the carrier concentration?

23.7 Discuss the similarities and differences between intrinsic electrical conduction in semiconductors and in an ionic material. Do the same for extrinsic electrical conduction.

23.8 Derive the equation for electron current corresponding to Eq. 23.14 for hole current.

23.9 Consider an ac voltage, $V = V_0 \sin \omega t$ applied to a rectifying device whose I–V characteristics obey Eq. 23.16.

(a) What is the ratio of the maximum (forward) current to the minimum (reverse) current?

(b) Evaluate this ratio for an input voltage of (i) 0.1 V, (ii) 1 V, (iii) −10 V.

(c) In terms of the time duration of the forward and reverse currents, does the rectifying action produce any change?

23.10 Consider hydrogen-doped zinc oxide.

(a) Write out the equations corresponding to Eqs. 23.7 and 23.8 for the donation of electron charge carriers to ZnO.

(b) Assuming that hydrogen atoms are readily ionized upon insertion into ZnO, show that the conductivity of ZnO should vary as $\sigma \sim (p_{H_2})^{1/2}$, where p_{H_2} is the partial pressure of hydrogen gas in the hydrogen atmosphere. (*Hint.* Write the law of mass action.)

23.11 The conductivity of Fe_3O_4 was described in the text as the result of electron mobility. With the aid of sketches including both divalent and trivalent iron ions, show that this type of conduction can formalistically be described also as a result of hole motion. Are the two different descriptions physically compatible?

23.12 In an intrinsic semiconductor, the energy required to form a conduction electron–hole pair equals the energy gap (E_g) between the valence band and the conduction band, and the rate of formation is approximately $k_f e^{-E_g/kT}$. The reverse reaction (termed recombination) occurs simultaneously at a rate proportional to the product of the conduction electron and hole concentrations (i.e., $k_r n_e n_h$).

(a) Using the above information, show that

$$n_e n_h = \frac{k_f}{k_r} e^{-E_g/kT}$$

represents the equilibrium condition.

(b) What is the conduction electron concentration as a function of temperature for an intrinsic semiconductor? The hole concentration? Explain.

(c) If an intrinsic semiconductor is doped with atoms that are electron donors, what will happen to the hole concentration? Why?

24

Thermal Properties of Materials

The thermal properties of materials that depend on energy changes of atoms and free electrons in a solid are discussed in this chapter.

24.1 THERMAL ENERGY AND HEAT CAPACITY

As a substance absorbs heat its temperature rises, and the amount of heat, ΔQ, required to increase the temperature of the substance by ΔT defines the heat capacity of the substance:

$$C = \frac{\Delta Q}{\Delta T} \tag{24.1}$$

If the substance is constrained to constant volume, the heat added exactly equals the increase in its internal energy,

$$(\Delta E)_v = Q \tag{24.2}$$

and the slope of the internal energy versus temperature curve gives the heat capacity at constant volume

$$C_v = \left(\frac{dE}{dT}\right)_v \tag{24.3}$$

Alternatively, if the substance is subjected to constant pressure, the heat addec exactly equals the increase in its enthalpy,

$$(\Delta H)_p = Q \qquad (24.4$$

and the slope of the enthalpy versus temperature curve gives the heat capacity a constant pressure,

$$C_p = \left(\frac{dH}{dT}\right)_p \qquad (24.5$$

Both C_v and C_p are properties of a substance and are usually given on a per mole per unit mass or per unit volume basis. C_v and C_p are related thermodynamically and it can be shown that C_p is always greater than C_v except at the absolute zero o temperature, where $C_p = C_v = 0$. Because C_v can be related more easily to the fundamental nature of a solid, the remaining discussion will concern it. It should be pointed out, however, that C_v is more difficult to measure experimentally than C_p in the case of a condensed phase; however, C_p and C_v are very nearly equal for a condensed phase at and below room temperature.

As mentioned above, the temperature variation of internal energy determines C_v. An ideal monatomic gas provides a simple example. In this case, the molar internal energy, which is totally kinetic in nature, is given by $E = \frac{3}{2}RT$ (see Sect 1.3), and $C_v = \frac{3}{2}R$. A temperature rise and the concurrent increase in therma energy correspond to an increase in translational kinetic energy of the individua particles of a monatomic gas (see Fig. 1.3). For diatomic gases, at successively higher temperatures molecular rotation about the center of mass and atomic vibration along the molecular axis are excited. These motions give rise to additiona kinetic and potential energy terms and result in correspondingly higher values of C_v.

To a first approximation, the atoms in a solid are similar to a vibrating diatomic molecule, and their vibration contributes an energy $3kT$ per atom (or $3RT$ per mole) to the internal energy of the solid. Half of this thermal energy (i.e., $\frac{3}{2}kT$ is the average kinetic energy and half is the average potential energy. According to this simple approach, C_v for a solid should be $3R$ per mole. This relationship called the Dulong–Petit law, is observed to be approximately valid for many nonmetallic solids at and above room temperature but does not hold at lower temperatures. Indeed, it is observed experimentally that the heat capacity of al solids approaches zero as the absolute zero of temperature is approached. The Dulong–Petit law does not apply at low temperatures because the energy o atomic vibrations is quantized and the number of vibrational modes that are excited decreases as the temperature decreases below some value characteristic o each individual material.

In a solid the motion of individual atoms is not only restricted by neighboring atoms but is also coupled to their motion rather than being independent. That part of the internal energy resulting from atomic vibrations may be regarded as a series of superimposed elastic waves whose frequencies depend on the elastic moduli and the density of the solid. As proposed by Debye, it is the energy of these elastic waves (vibrational potential and kinetic energy) that is quantized. One quantum is called a phonon, in analogy with the quantum of electromagnetic energy or photon (see Sect. 2.3). The number of vibrational modes is the same as if the atoms could vibrate independently (i.e., three per atom in the solid), but there is a range of allowed frequencies (ν) and of phonon energies ($h\nu$). The distribution function or number of vibrational modes having frequencies between ν and $\nu + d\nu$ used by Debye is shown in Fig. 24.1. [This is analogous to the distribution of energies and velocities in a monatomic gas (see Fig. 1.3) and to the density of electronic states in a solid (see Fig. 4.15).] The allowed frequencies range from zero to a maximum value (ν_m) determined by the shortest wavelength that can exist in a solid (i.e., the cutoff frequency corresponds to a wavelength equal to twice

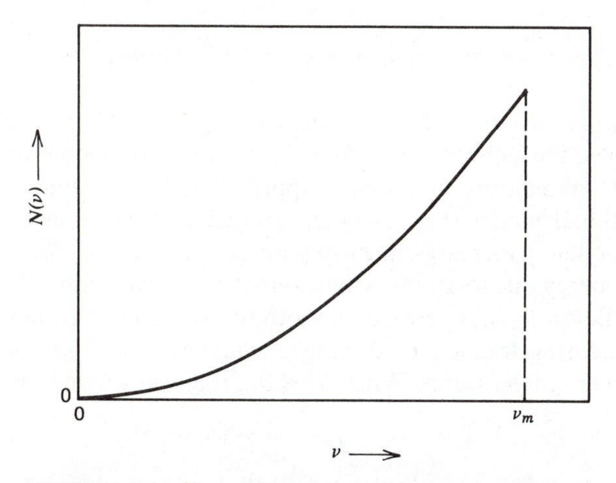

FIGURE 24.1 The distribution function for vibrational modes in a crystal according to the Debye model. The number of vibrational modes having frequencies between ν and $\nu + d\nu$ is $N(\nu)d\nu = (9N\nu^2/\nu_m^3)\,d\nu$, where N is the number of atoms and ν_m, the cutoff frequency, equals $V_s(3N/4\pi V)^{1/3}$. V_s, the velocity of sound in the crystal, is approximately equal to $\sqrt{E/\rho}$ (E = Young's modulus and ρ = density), and $\nu_m \approx V_s/\lambda_{min}$, where λ_{min} is smallest possible wavelength (e.g., twice the interatomic spacing in the close-packed direction of a metal). The total number of vibrational modes (i.e., the area under the distribution function curve) is $3N$.

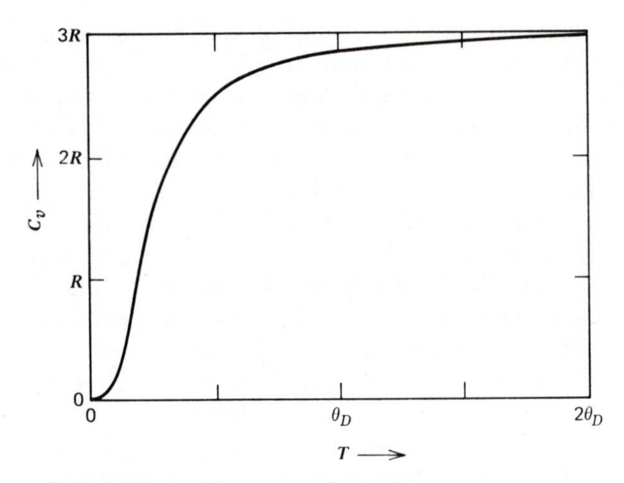

FIGURE 24.2 Calculated heat capacity per mole due to phonons, based on the Debye model, as a function of temperature. Experimental values of C_v, particularly those for nonmetals, often are in good agreement with this curve.

the nearest-neighbor interatomic distance) and by the velocity of sound in the solid.

The temperature variation of C_v is shown in Fig. 24.2. Above a characteristic temperature called the Debye temperature (θ_D), C_v due to phonons becomes almost independent of temperature and equals approximately the Dulong-Petit value. Here all vibrational modes (Fig. 24.1) are excited and the number of phonons having each frequency increases linearly with temperature as does that portion of the internal energy due to atomic vibrations. The Debye temperature is defined by $\theta_D = h\nu_m/k$. Below θ_D, C_v decreases monotonically with decreasing temperature because an increasing fraction of the high-frequency modes are not excited at successively lower temperatures. When $T \ll \theta_D$, the heat capacity is given by

$$C_v = \frac{12\pi^4}{5} R\left(\frac{T}{\theta_D}\right)^3 \tag{24.6}$$

The higher the density of a material, the lower will be the maximum frequency of vibration, and this accounts for the observation that $C_v = 3R$ at room temperature for heavy, but not for light, solid nonmetallic elements.

Since C_v is determined by the temperature variation of the total internal energy, it should include any contribution due to the kinetic energy of electrons in the case of a metal. If all of the valence electrons behaved as free electrons, then each would contribute $\frac{3}{2}k$ to the heat capacity. This is not so because of the distribution of electronic energy states in a metal. In the same way that only electrons having energies near the Fermi energy (E_F) can shift their occupancy of states and participate in electrical conduction when a voltage is imposed across a metal (see Sect. 22.4), only these same electrons can be excited thermally. It can be shown that the electrons that can be excited possess energies within about kT of the Fermi energy, and since these constitute a very small fraction of the total number of valence electrons, the electronic contribution to the heat capacity is relatively small. The number of electrons per unit volume that can be thermally excited is approximately $2N(E_F)kT/V_m$, where $N(E_F)$ is the density of electronic states at the Fermi energy, V_m is the molar volume, and the factor of two has been introduced because each electronic state may be occupied by two electrons. The additional kinetic energy (or thermal energy) of these electrons per unit volume is approximately

$$E_e \approx 2N(E_F)\frac{kT}{V_m}\cdot\frac{3}{2}kT = \frac{3N(E_F)k^2T^2}{V_m} \tag{24.7}$$

and the corresponding electronic contribution to the heat capacity is

$$C_{v_e} \approx \frac{6N(E_F)k^2T}{V_m} \tag{24.8}$$

per unit volume of metal. An exact analysis would yield Eq. 24.8 with a numerical factor of $2\pi^2/3$ rather than six. In the case of free electrons in a single band where the Fermi energy is much lower than any band gap, the density of electronic states at the Fermi energy is $N(E_F) = 3nV_m/4E_F$, where n is the total number of valence electrons per unit volume, and the electronic heat capacity becomes

$$C_{v_e} = \frac{\pi^2 nk^2T}{2E_F} \tag{24.9}$$

using the exact numerical factor. Since E_F is typically 3 to 5 eV and at room temperature $kT \simeq 0.025$ eV, it is apparent that free electrons provide a very minor contribution to the room temperature heat capacity of a metal, although the electronic heat capacity does become more appreciable at high temperatures. At very low temperatures (near absolute zero), where the electronic contribution becomes comparable to and exceeds the vibrational contribution, the total heat capacity can be represented by

$$C_v = \gamma T + AT^3 \tag{24.10}$$

where the first term represents Eq. 24.9 and the second term represents Eq. 24.6. For many metals the electronic heat capacity is much larger than would be expected from the simple free-electron approximation inherent in Eq. 24.9 because the density of electronic states at the Fermi energy is much higher (see Sect. 4.7 and Fig. 4.17). In general, however, we may identify γ with the temperature coefficient in Eq. 24.8 (using the proper numerical factor), and low temperature heat capacity measurements afford one of the most direct experimental techniques for determining the density of electronic states at the Fermi level.

Other thermal excitation processes, such as the destruction of long-range order in certain alloys, the randomization of electron spins in ferromagnetic or ferrimagnetic materials, and changes in the distribution of electrons in a superconductor, also can lead to an increase in the heat capacity. For example, the heat capacity of iron rises very sharply in the vicinity of the ferromagnetic to paramagnetic transition (Fig. 24.3).

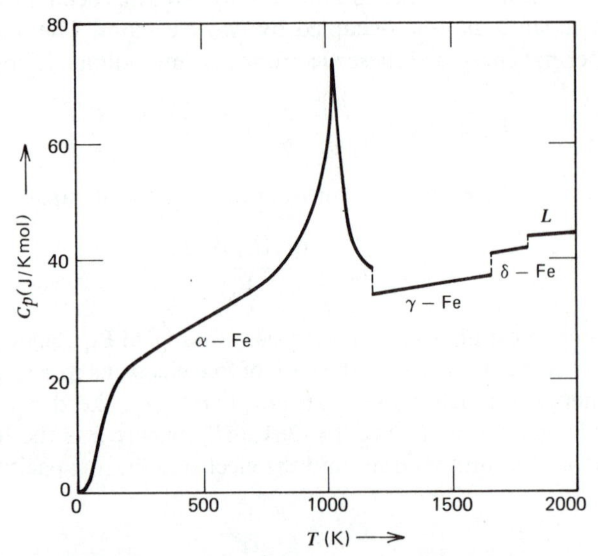

FIGURE 24.3 Experimental heat capacity at constant pressure as a function of temperature for pure iron. Each phase change is accompanied by a discrete change in C_p. The sharp peak in C_p for α-Fe corresponds to the transition from ferromagnetic to paramagnetic behavior with increasing temperature (see Chapter 25). [Reprinted with permission from L. S. Darken and R. P. Smith, *Ind. Eng. Chem.,* **43**, 1815 (1951). Copyright by the American Chemical Society.]

24.2 THERMAL EXPANSION

The volume of most substances increases with increasing temperature. In the case of solids, the volume increase is related directly to an increase in the mean amplitude of atomic (or molecular) vibrations that accompanies a higher thermal energy. Because potential energy curves are asymmetric with respect to changes in interatomic separation or in volume (Figs. 3.4, 19.5, and 19.6), increases in vibrational amplitude with rising temperature result in increases of the average interatomic spacing, and this is reflected on a macroscopic level by an increase in the volume and linear dimensions of a material. The extent of thermal expansion is measured by the fractional volume change per unit increase of temperature, which is called the coefficient of volume expansion,

$$\alpha_v = \frac{1}{V}\frac{dV}{dT} \tag{24.11}$$

An alternative measure of thermal expansion is provided by the coefficient of linear expansion,

$$\alpha_l = \frac{1}{l}\frac{dl}{dT} \tag{24.12}$$

which is the fractional change of a linear dimension per unit increase of temperaure. For anisotropic solids, two or three coefficients of linear expansion are necessary to characterize thermal expansion, but there is only one coefficient of inear expansion for cubic and isotropic solids. In the latter case, $\alpha_v = 3\alpha_l$ holds.
Since thermal expansion is related on the atomic scale to atomic vibrations, the hemical or physical bonding interactions between the atoms or molecules omprising the solid must be important. With increasing binding energy, the otential energy (or restoring attractive or repulsive force) increases more sharply vith atomic displacement from the mean position; this corresponds to a decreasing oefficient of expansion. As mentioned previously, thermal expansion arises ecause of an inherent asymmetry in the potential energy versus interatomic eparation curve. Within a given class of materials, a greater binding energy coresponds to a smaller degree of asymmetry in the potential energy curve and a low oefficient of expansion, whereas a large coefficient of expansion is associated with he greater degree of asymmetry that accompanies a smaller binding energy. t is important to mention that the coefficient of thermal expansion varies with emperature. Typical thermal expansion behavior as a function of temperature is hown in Fig. 24.4. Between 0 K and the Debye temperature the coefficient of hermal expansion increases rapidly, whereas at higher temperatures the increase s more moderate. Vacancies, Schottky defects, or Frenkel defects make a signifiant contribution to thermal expansion only at temperatures approaching the nelting point.
Covalently bonded materials tend to have low expansion coefficients in comparion to metals, and ionically bonded materials tend to have somewhat higher xpansion coefficients than metals. Coefficients of linear thermal expansion range om about $25 \times 10^{-6}\,°C^{-1}$ to over $250 \times 10^{-6}\,°C^{-1}$ for organic polymers at

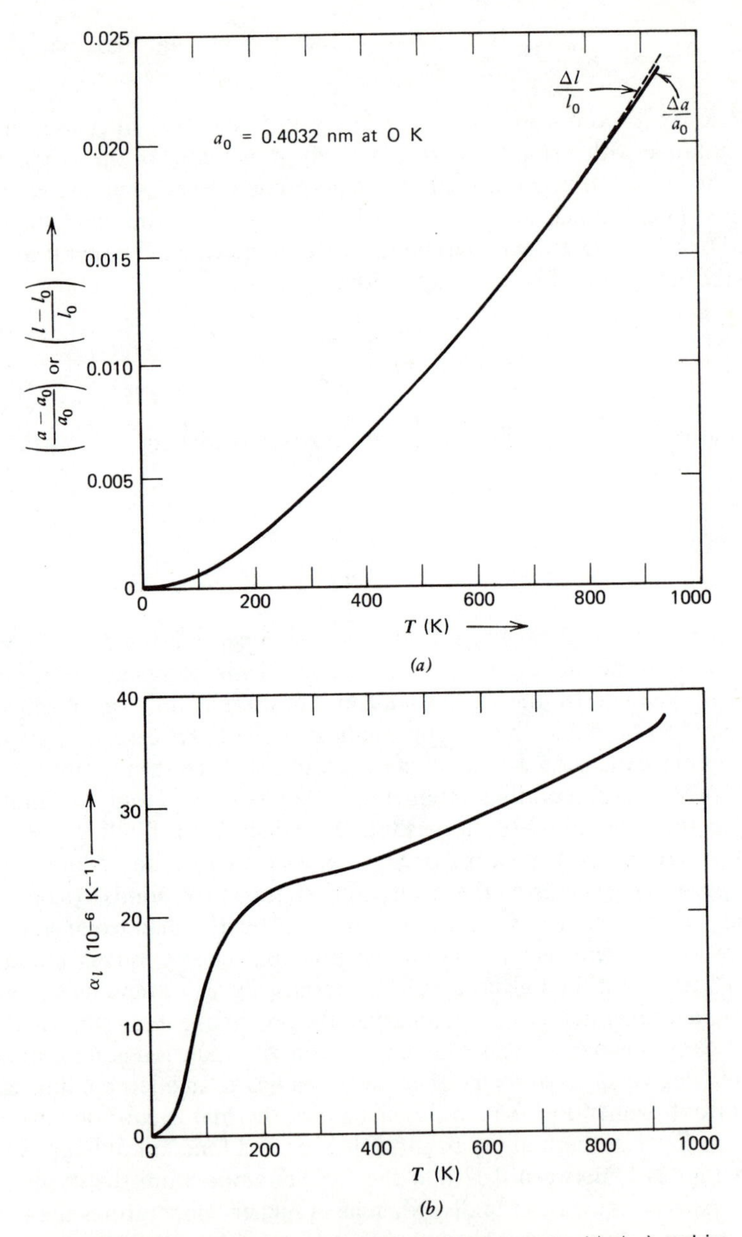

FIGURE 24.4 (*a*) Relative change in lattice parameter ($\Delta a/a_0$) and in linear dimension ($\Delta l/l_0$) as a function of temperature for aluminum. The relative changes in lattice parameter and length are identical until temperatures close to the melting point are reached, where $\Delta l/l_0$ exceeds $\Delta a/a_0$ because of the formation of a significant number of vacancies that contribute to $\Delta l/l_0$ but alter $\Delta a/a_0$ only slightly.

(*b*) Temperature variation of the coefficient of linear thermal expansion for aluminum. The upward turn near the melting point is due to the contribution made by vacancies.

room temperature. Network polymers generally exhibit the lowest values because of their three-dimensional covalent bonding. Higher values are found for long-chain polymers because of weak intermolecular bonding, but the state of a long-chain polymer and the relative amount of crystalline material versus glassy or supercooled liquid material influence the thermal expansion characteristics (see Fig. 6.4). A great amount of variability in the thermal expansion coefficient of any long-chain polymer arises because the precise state of such a polymer depends on its processing history and the rate of cooling from the melt. Elastomers present a particularly interesting case in that they display a negative coefficient of expansion when subjected to an applied stress. This arises because the molecules in an extended elastomer are partially uncoiled, resulting in a decreased entropy, and an increase in temperature favors the more coiled, higher entropy condition such that axial contraction occurs. Among other things, this phenomenon accounts for the increase in elastic modulus with increasing temperature, as discussed in Sect. 19.5.

Thermal expansion is important in many practical situations. For certain types of precise equipment, dimensional changes due to temperature fluctuations must be minimized, and some Fe–Ni and Fe–Co alloys that have low coefficients of thermal expansion ($\alpha_l \approx 1 \times 10^{-6}\,°\mathrm{C}^{-1}$) over a limited temperature range may be used to advantage. In addition, a mechanically constrained body is susceptible to the development of thermal stresses because of a change in temperature. Such stresses are elastic initially but may increase sufficiently so as to cause yielding of a ductile material or fracture of a brittle material. Thermal stresses also may arise because of a difference in cooling rate between the surface and the interior of a body. This is particularly important for ceramic materials, which are prone to thermal shock. It is common experience, for example, that fracture occurs when a piece of heated soda-lime-silica glass ($\alpha_l \approx 9 \times 10^{-6}\,°\mathrm{C}^{-1}$) is plunged into water; that is, thermal contraction of the surface layers induces tensile stresses that cause Griffith cracks to propagate. On the other hand, when cooled at relatively low rates, these same materials develop beneficial residual compressive stresses at the surface. Certain inorganic glasses, such as vitreous silica, possess good thermal shock resistance because they have very low coefficients of thermal expansion (e.g., $\alpha_l = 0.5 \times 10^{-6}\,°\mathrm{C}^{-1}$ for vitreous silica at room temperature).

24.3 THERMAL CONDUCTION

The rate at which heat is transported through a solid depends on the thermal conductivity $\mathscr{K}$ of the solid and the thermal gradient present. Thermal conductivity is defined formally by

$$J = -\mathscr{K}\frac{dT}{dx} \tag{24.13}$$

where J is the heat flux [i.e., the amount of thermal energy crossing a unit area per unit time, $(1/A)(dQ/dt)$, and dT/dx is the temperature gradient]. The minus sign indicates that heat flows down a temperature gradient. Note that this equation has the same form as Ohm's law for electrical conduction (cf. Eqs. 22.5 and 22.6) and Fick's first law for diffusion (cf. Eq. 12.2).

Metals are good thermal conductors because the free electrons in the vicinity of the Fermi level are primarily responsible for heat transport in a metal. Basically, free electrons in a hot region gain kinetic energy and migrate toward a cold region where they transmit their kinetic energy to atoms as a result of collisions with structural imperfections and phonons. A result of such collisions is the generation of more phonons, corresponding to an increase in temperature in the cold region. Concurrently, electrons in cold regions migrate toward hot regions, where they gain kinetic energy as a result of collisions, and the process continues dynamically.

It is reasonable to assume that heat transport by free electrons in a metal is similar to that in an ideal gas. From the kinetic theory of gases, the thermal conductivity is given by

$$\mathscr{K} = \tfrac{1}{3} C_v \bar{v}\, l \tag{24.14}$$

where C_v is the heat capacity at constant volume per unit volume, $\bar{v}$ is the average particle velocity, and l is the average distance traveled or mean free path of the particles between collisions. For free electrons, substitution of Eq. 24.9 into Eq. 24.14 gives

$$\mathscr{K}_e = \frac{\pi^2 n k^2 T}{6E_F} v_F\, l \tag{24.15}$$

Note that the average velocity of the electrons is taken to be the Fermi velocity because only those electrons near the Fermi energy participate in transporting heat. Furthermore, the electronic mean free path l equals $2v_F\tau$, where 2τ is the time between collisions (see Sect. 22.4). Equation 24.15 and that for the electronic conductivity based on the free-electron approach (Eq. 22.12) can be combined in order to make a comparison between electronic thermal and electrical conductivities:

$$\frac{\mathscr{K}_e}{\sigma} = \frac{\pi^2}{3}\left(\frac{k}{e}\right)^2 T = LT \tag{24.16}$$

The relation $E_F = \frac{1}{2} m_e v_F^2$ was used to obtain Eq. 24.16, which is called the Wiedemann–Franz law. The proportionality factor, $L = \mathscr{K}_e/\sigma T$, should be a constant independent of the metal providing the collision processes are the same for electrical and thermal conductivity. Furthermore, if the total thermal conductivity $\mathscr{K}$ of a metal is largely due to heat transport by electrons, then substitution for $\mathscr{K}$ for $\mathscr{K}_e$ should not alter the value of L significantly. Experimental values of L for some metals (Table 24.1) indicate reasonable agreement with Eq. 24.16 in the case of good conductors.

Solid solution alloys like stainless steel (e.g., 74% Fe–18% Cr–8% Ni) are not only poor electrical conductors but also poor thermal conductors because local structural aberrations due to solute atoms serve as scattering sites for electrons. Covalent and ionic electrical insulators are also poor thermal conductors because of the absence of free electrons. Nevertheless, while the thermal conductivities of nonmetals usually are much less than those of metals, important exceptions do exist. Diamond and sapphire, for example, are better thermal conductors than silver in temperature ranges of 30 to 300 K and 25 to 90 K, respectively. Since no free electrons are present in these materials, thermal conduction must occur by an alternative process involving the motion and creation of phonons. That is, elastic wave packets transmit thermal energy from hot to cold regions. As with electrons, phonons also are scattered by other phonons, electrons, and structural imperfections. Phonon heat transport occurs in metals as well as nonmetals but generally provides a minor contribution in comparison to electronic heat transport in most metals.

Since, from a mechanistic point of view, the rate of heat transport by phonons is determined by scattering events similar to those for electrons, Eq. 24.14 may be used to express the phonon thermal conductivity as

$$\mathscr{K}_p = \tfrac{1}{3} C_v \bar{v} l \qquad (24.17)$$

where the various terms are the phonon equivalents of corresponding terms in Eq. 24.14. For pure metals that are good electrical conductors, $\mathscr{K}_p$ is only about 1% of $\mathscr{K}_e$. This situation arises because the electronic mean free path and Fermi velocity are 10 to 100 times those of phonons, which more than compensates for the fact that C_{v_e} is only about 10% of the total heat capacity at normal temperatures. In most semiconductors, by contrast, the density of "free electrons" is so low that heat transport occurs only by a phonon mechanism. As shown in Fig. 24.5, the relationship between electrical and thermal conductivities of a material at a fixed temperature can be compared to the free electron density. For high values of free electron density (i.e., metals), the electrical and thermal conductivities are proportional; for low values (i.e., semiconductors and electrical insulators) the thermal conductivity remains relatively high even though the electrical conductivity becomes quite low.

The temperature dependence as well as the magnitude of thermal conductivity depends on the nature of the material (Fig. 24.6). Pure metals, most pure ceramics, and crystalline polymers exhibit increasing thermal conductivity with decreasing temperature because the electronic and phonon mean free paths increase. At sufficiently low temperatures, however, the mean free paths become constant, being limited by either the specimen size or structural imperfections, and the

TABLE 24.1 Some Thermal Properties of Selected Elements at 25°C

Element	$10^6 \times \alpha_l$ (K^{-1})	C_p (J/K · mol)	$\mathscr{K}$(W/m · K)	$10^{-7} \times \sigma$ ($ohm^{-1}m^{-1}$)	$10^8 \times L\left(\frac{ohm \cdot W}{K^2}\right)$
Al	23.2	24.4	237	3.74	2.13
Ag	19.0	25.4	427	6.21	2.31
Au	14.2	25.4	315	4.22	2.51
Cu	16.5	24.5	398	5.86	2.28
Fe	11.8	25.1	80.3	1.03	2.63
Na	71.0	28.1	134	2.16	2.08
Ni	13.3	26.1	89.9	1.45	2.07
Pb	29.3	26.4	34.6	0.48	2.40
W	4.6	24.8	178	1.77	3.37

$$\left(L_{\text{theoretical}} = \frac{\pi^2 k^2}{3e^2} = 2.44 \times 10^{-8}\ \text{W} \cdot \text{ohm/K}^2\right)$$

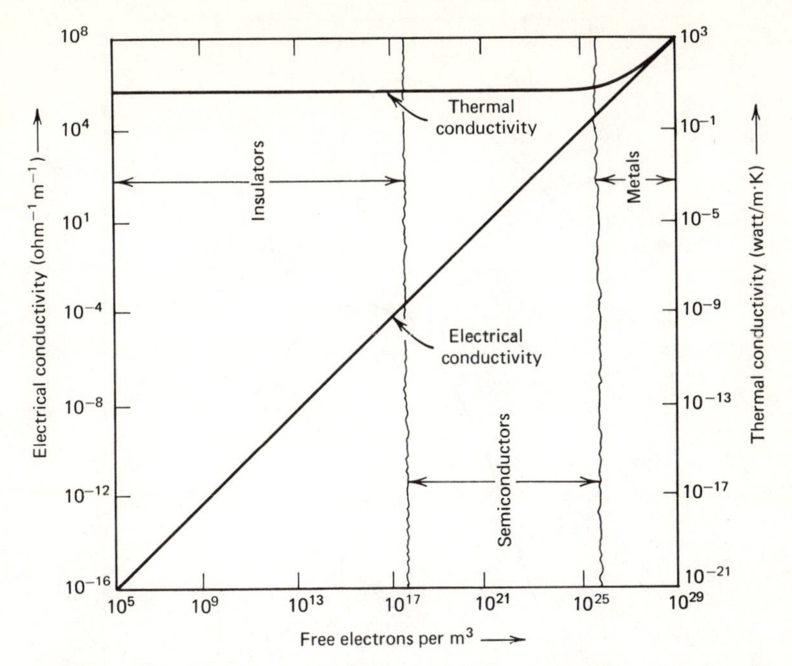

FIGURE 24.5 Schematic variation of electrical conductivity and thermal conductivity with the number of free electrons per unit volume. Although the boundaries separating metals, semiconductors, and electrical insulators are indefinite, there is marked difference in their respective electrical conductivities. Thermal conductivity, affected by the decrease in free electron density in the case of metals, becomes independent of free electron density for semiconductors and electrical insulators as a result of heat transport by phonons. (Adapted from R. L. Sproull, *Scientific American*, Vol. 207, No. 6, p. 92, 1962. Copyright 1962 by Scientific American, Inc. All rights reserved.)

temperature dependence of the heat capacity dominates the variation of thermal conductivity with temperature. Near absolute zero, accordingly, we expect the thermal conductivity to be proportional to T (cf. Eq. 24.8 or Eq. 24.9) for a pure metal and to be proportional to T^3 for a pure ceramic (cf. Eq. 24.6), and such variations are observed. Concentrated solid-solution alloys, inorganic glasses, and glassy polymers, on the other hand, exhibit decreasing thermal conductivity with decreasing temperature. This situation arises because the electronic or phonon mean free path is quite small at all temperatures, and the temperature variation of the heat capacity largely determines the temperature dependence of thermal conductivity.

The low magnitude of $\mathscr{K}$ for common austenitic stainless steel (Fig. 24.6*b*) and the fact that $\mathscr{K}$ decreases continuously as temperature is lowered make it possible to employ stainless steel as a thermal insulator at cryogenic temperatures.

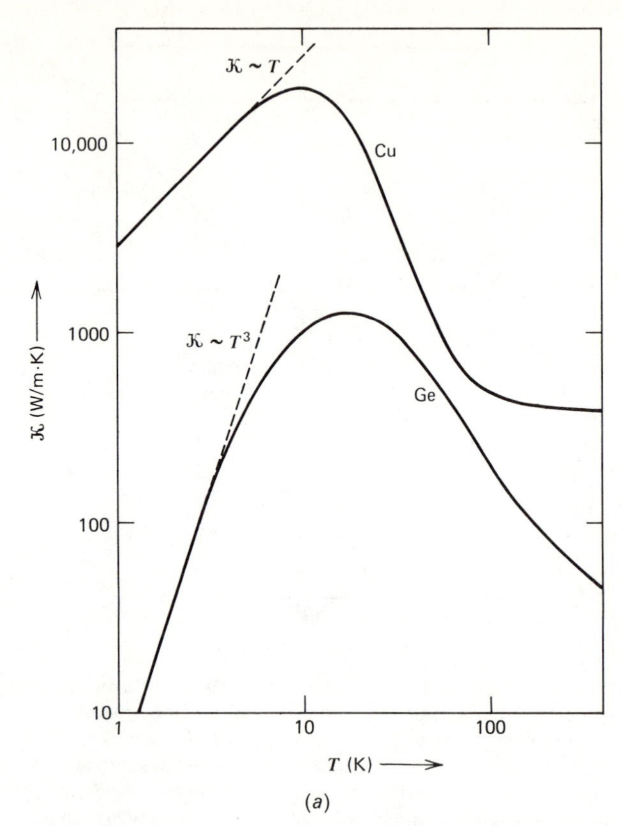

(*a*)

FIGURE 24.6 (*a*) Temperature dependence of thermal conductivity. At very low temperatures, $\mathcal{K}$ increases linearly with T for pure metals and with T^3 for pure semiconductors and pure electrical insulators. This temperature dependence is identical to that for the heat capacity because the mean free path of electrons or phonons is constant at very low temperatures. The decline of $\mathcal{K}$ at higher temperatures is a result of decreasing mean free path.

Inorganic and polymeric glasses also can serve this purpose. With regard to thermal insulation, the best materials are those having compositelike structures in which one of the "phases" is porosity. Gas in closed pores affords a low overall thermal conductivity. Such insulators are particularly effective at cryogenic temperatures; if the gas within the pores condenses and freezes, essentially a vacuum with extremely low conductivity remains. For high temperature thermal insulation, mechanical properties are equally important with thermal conductivity, and many ceramics as well as some network polymers are used widely for this purpose, the former frequently being utilized as refractory brick in steel making and other furnaces.

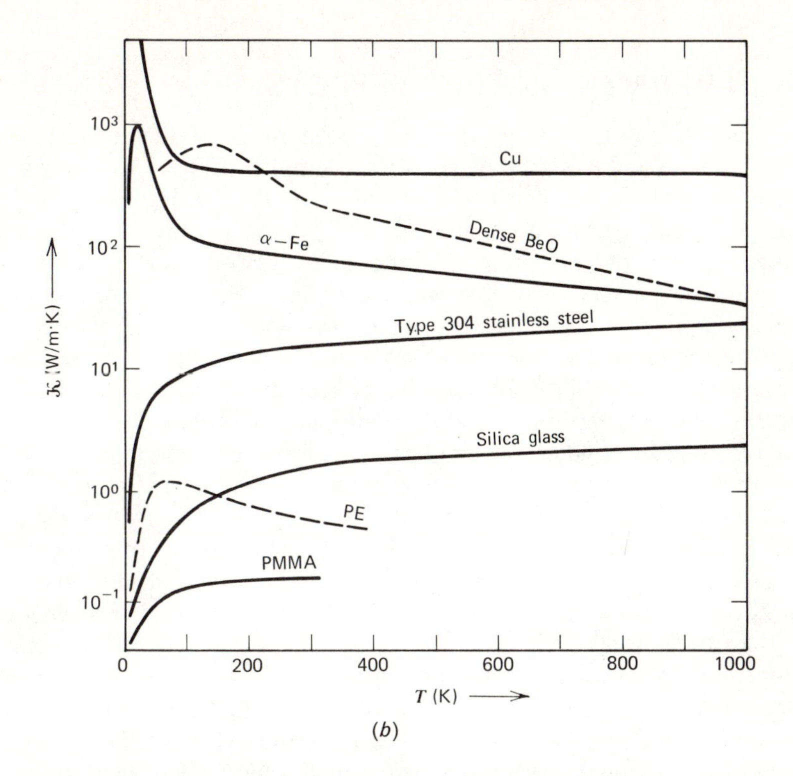

FIGURE 24.6 (*b*) Pure substances exhibit either an almost constant thermal conductivity (e.g., Cu) or a decreasing thermal conductivity [e.g., α-Fe, BeO, and polyethylene (PE)] over a wide range of temperatures because the electronic and phonon mean free paths decrease with increasing temperature. At very high temperatures (not shown), $\mathcal{K}$ may increase as heat capacity (C_p) again becomes dominant.

For noncrystalline solids [e.g., silica glass and polymethyl methacrylate (PMMA)] and concentrated solid-solution alloys (e.g., type 304 stainless steel, essentially 74% Fe–18% Cr–8% Ni), the phonon and electronic mean free paths are small at all temperatures, and the temperature dependence of $\mathcal{K}$ is largely due to the temperature variation of heat capacity.

(Data from: *Thermal Conductivity of Solids at Room Temperature and Below*, NBS Monograph 131, by G. E. Childs, L. J. Ericks, and R. L. Powell, U.S. Government Printing Office, Washington, D.C., 1973; *Thermophysical Properties of High Temperature Solid Materials*, edited by Y. S. Touloukian, Macmillan, New York, 1967.)

24.4 THERMOELECTRIC EFFECTS

Various interesting effects come about as a result of the interrelationship between thermal and electrical behavior. The simplest of these is the variation of electrical resistivity or resistance with temperature. This permits metals and semiconductors to be used as temperature sensors. The strong temperature dependence of electrical resistivity in semiconductors makes them particularly sensitive for the measurement of cryogenic as well as moderately elevated temperatures, and such *thermistors* are in common use. Resistance temperature-measuring devices require calibration, and any structural or chemical change that alters the resistivity will cause irreproducibility. In addition to the above, certain nonequilibrium phenomena also are important, and these are discussed below.

If a temperature gradient is imposed on a metal rod, electrons at the high temperature end, having greater energies than those at the low temperature end, migrate preferentially down the temperature gradient. Physically, this is similar to the behavior of a gas in a closed tube subjected to a temperature gradient, where the number of gas particles per unit volume increases at the cold end and decreases at the hot end until there is an equalization of pressure. For electrons in a metal, there is a transient, net electron flow from the hot to the cold end, but since the electrons are charge carriers, an opposing voltage develops. At steady state, the voltage between the ends of the rod depends only on the temperatures of the ends and not on the distribution of temperature along the rod. This effect is quantified in terms of the thermoelectric power coefficient, S, which is defined as the negative of the temperature-induced voltage increase per unit temperature increase:

$$S = -\frac{dV}{dT} \tag{24.18}$$

Although inappropriate, S often is called simply the thermoelectric power. According to the accepted sign convention, V is positive if the electron density has increased at the cold end and decreased at the hot end, and we would expect S to be negative, although this is not always the case.

The absolute value of the thermoelectric power coefficient is difficult to measure. If, for example, connecting wires and the specimen are of the same material, a thermal voltage is induced in the connecting wires that exactly cancels the voltage induced in the specimen, and the measured voltage is zero. Utilizing a different material as connecting wires, however, does yield a measurable voltage, but this is the difference between the voltages in the specimen and in the connecting wires. If the cold end is maintained at a constant temperature, the slope of the curve for measured voltage versus temperature at the hot end gives the relative thermoelectric power coefficient,

$$S_{12} = S_2 - S_1 = \frac{dV_1}{dT} - \frac{dV_2}{dT} = \frac{dV_{12}}{dT} \tag{24.19}$$

where S_1 and S_2 are the absolute values for the specimen and the connecting wires. In the case of a superconductor, $S = 0$ and absolute thermoelectric power coefficients of other metals can be determined below T_c of the superconductor. Nevertheless, since S is a function of temperature, such studies necessarily provide information only over a limited temperature range.

The thermally induced voltage, V_{12}, of a couple made of two different metals in a temperature gradient is called the Seebeck potential. V_{12} is taken to be positive if an electric current tends to flow from metal 1 to metal 2 at the hotter junction, and $V_{12} = -V_{21}$. The Seebeck potential facilitates measuring temperatures with thermocouples. A thermocouple consists of two different metals joined at two points, with one junction being located in a region where temperature is to be measured and the other placed at a cold or reference temperature (frequently 0°C). The voltage supplied by a thermocouple is purely a function of the temperatures at each junction (not the temperature gradient) and of the metals used, because S_{12} depends on the latter. This voltage is determined with a potentiometer. For a thermocouple to be useful, the metals comprising the couple must possess sufficiently different values of absolute thermoelectric power coefficients so as to produce easily measurable voltages. Table 24.2 lists some common thermocouples, the temperature ranges over which they are useful, and approximate values of their relative thermoelectric power coefficients, and Fig. 24.7 shows the Seebeck potential as a function of temperature for these thermocouples. For temperatures

TABLE 24.2 Some Common Thermocouples

Thermocouple[a]	*Maximum Temperature for Use* (°C)	*Average Sensitivity* (mV/K)	*Range* (°C)
Chromel (90 Ni–10 Cr)–Alumel (94 Ni–2 Al–3 Mn–1 Si)	1250	0.041	0–1250
Iron–Constantan (55 Cu–45 Ni)	850	0.033	−200 to −100
		0.057	0–850
Copper–Constantan	400	0.022	−200 to −100
		0.052	0–400
(Pt–10% Rh)–Pt	1500	0.0096	0–1000
		0.0120	1000–1500
(Pt–13% Rh)–Pt	1500	0.0105	0–1000
		0.0139	1000–1500
Chromel–Constantan	850	0.076	0–850
(W–3% Re)–(W–25% Re)	2500	0.0185	0–1500
		0.0139	1500–2500
(Ir–40% Rh)–Ir	2000	0.005	1400–2000

[a] The first metal or alloy in each thermocouple is positive, and the second is negative.

above 1300°C, either (platinum–rhodium)–platinum or tungsten–(tungsten–rhenium) thermocouples are used in spite of their fairly low sensitivities (i.e., low values of S_{12}). This is because more sensitive thermocouples become too soft, oxidize, or even melt at these high temperatures. The low sensitivity of (Pt–Rh)–Pt thermocouples is tolerated because of their oxidation resistance, but W–(W–Re) thermocouples can be used to higher temperatures in vacuum or an inert atmosphere. One final comment is warranted. Chemical inhomogeneity and structural defects alter the thermoelectric power coefficient of a metal; hence, reproducible Seebeck potentials require controlled processing and service conditions that do not cause structural changes.

A second thermoelectric effect, known as the Peltier effect, is observed when current is passed through a thermocouple junction. Heat is absorbed or evolved depending on the direction of current flow. The Peltier coefficient, π_{12}, is defined as the rate of heat evolution or absorption at the junction per unit current flowing from metal 1 to metal 2,

$$\pi_{12} = \frac{dQ/dt}{I} \tag{24.20}$$

and is related to the relative thermoelectric power coefficient (Eq. 24.19) by

$$\pi_{12} = TS_{12} \tag{24.21}$$

By convention, π_{12} is positive if heat is absorbed; if the direction of the current is reversed, then $\pi_{21} = -\pi_{12}$.

This effect can be explained by the kinetic energy and potential energy of conduction electrons. That is, the Fermi energy is different in different metals, and the conduction electrons therefore have different kinetic energies. If the zero of energy is taken to correspond to an electron at rest just outside of an isolated metal (i.e., the same zero of energy to which the energies of electrons in an atom are referred, see Fig. 2.1), then the bottom of the conduction band equals the potential energy of the electrons in the band (Fig. 24.8*a*). The various occupied electronic energy states from the bottom of the band to the Fermi energy (and somewhat above E_F at any finite temperature) represent increasing kinetic energies. When two different metals are brought into contact, the disparity in total energies at the Fermi level causes transfer of some electrons, but this raises the electronic potential energy of the metal receiving excess electrons and lowers that of the metal losing electrons (Fig. 24.8*b*). (The actual number of electrons transferred is quite small, on the order of 10^{-15} times the valence electron density.) Charge transfer ceases when the Fermi energies of the contacting metals equalize with respect to our chosen zero of energy (Fig. 24.8*c*). For metals and semiconductors under isothermal conditions, the difference between the Fermi energy and the electronic

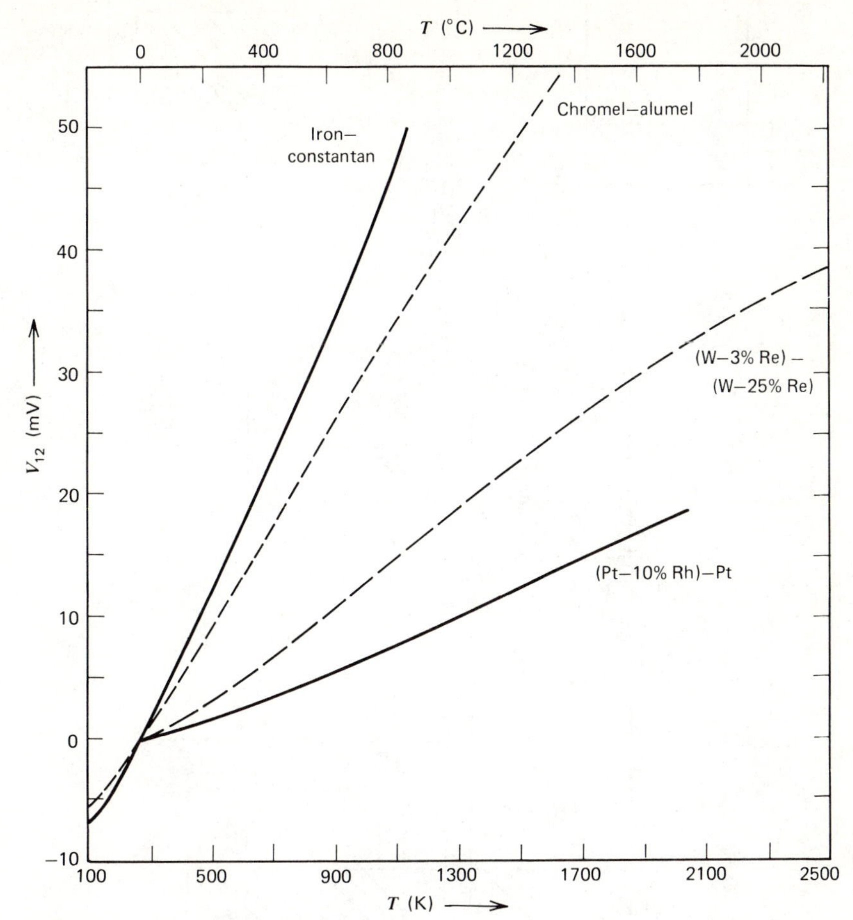

FIGURE 24.7 Seebeck potentials for some common thermocouples as a function of temperature. In each case the reference junction is at 0°C. The sensitivity (i.e., the slope, which is the relative thermoelectric power coefficient, S_{12}) is high for those thermocouples having relatively small temperature ranges of usefulness (e.g., iron–constantan and chromel–alumel) and low for those having usefulness to high temperatures [e.g., (Pt–10% Rh)–Pt and (W–3% Re)–(W–25% Re)]. (Data courtesy of Leeds and Northrup Co. and Engelhard Industries.)

potential energy ($E_F - e\Phi$ in Fig. 24.8*c*) plays the role of electronic free energy per electron, and this is equalized throughout a system at equilibrium. Even when a voltage is imposed across a metal, semiconductor, or various conductors in contact, the electronic free energy remains uniform because both $e\Phi$ and E_F are increased or decreased by the same amount (i.e., eV), and thus their difference is unaltered.

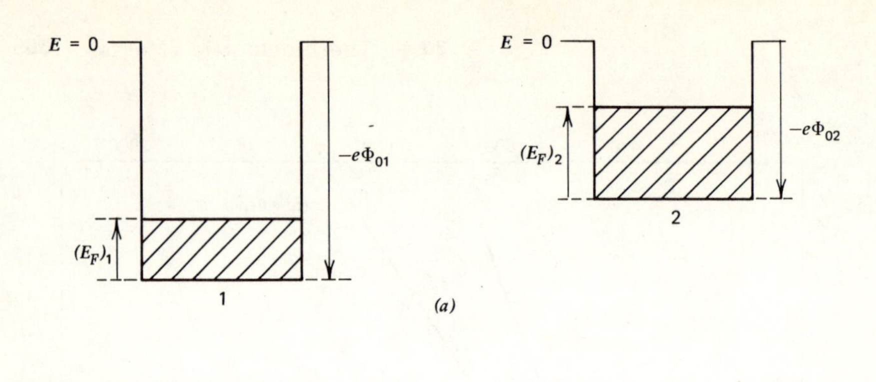

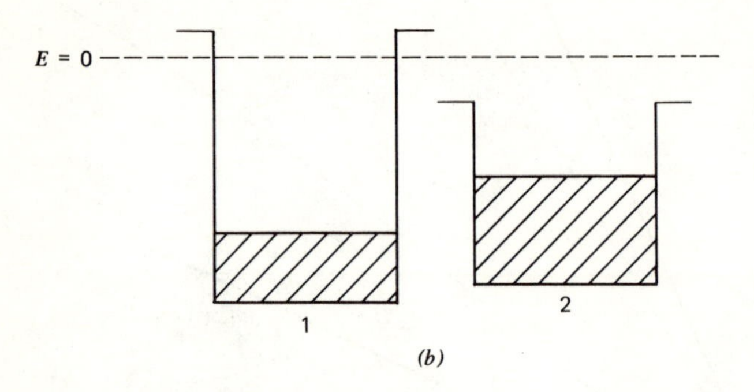

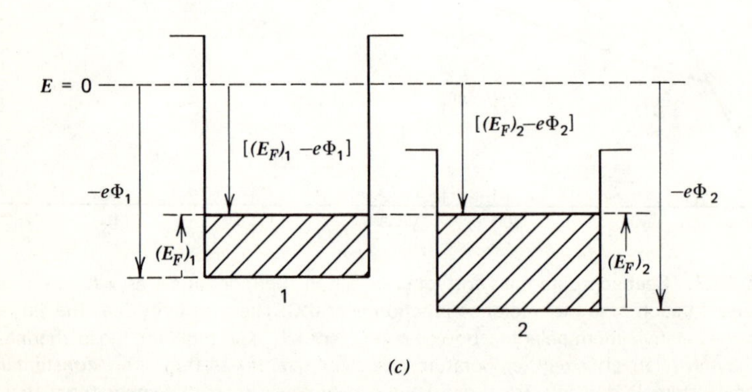

FIGURE 24.8 Schematic representation of the conduction bands for two metals, with the zero of electronic energy taken as a free electron at rest just outside of either metal. (*a*) When the two metals are isolated, the potential energies ($-e\Phi_{01}$ and $-e\Phi_{02}$) and the kinetic energies [or Fermi energies, $(E_F)_1$ and $(E_F)_2$] differ. (*b*) Upon contact, some electrons in metal 2 flow to metal 1, where the conduction electron total energy is lower. The resulting charge imbalance lowers the electronic potential energy in metal 2 and raises it in metal 1. (*c*) Equilibrium is reached after very few electrons have been transferred. At equilibrium, the total energies of conduction electrons at the Fermi levels (actually, the electronic free energies) are equal; that is, $[(E_F)_1 - e\Phi_1]_{eqm} = [(E_F)_2 - e\Phi_2]_{eqm}$. Nevertheless, the internal energies of the conduction electrons remain different (see text).

Now, when an electron goes across a junction from metal 2 to metal 1 (or current flows from metal 1 to metal 2), the electronic free energy is unchanged (Fig. 24.8*c*), but the electronic internal energy suffers a discrete change. In accordance with Eq. 24.3, this internal energy change per electron is given by

$$E_1 - E_2 = \int_0^T \left(\frac{C_{v_e}}{n}\right)_1 dT - \int_0^T \left(\frac{C_{v_e}}{n}\right)_2 dT \tag{24.22}$$

for a junction at temperature T, and it is manifested by heat absorption or heat evolution. The rate of heat absorption or evolution is proportional to the number of electrons crossing the junction per unit time (i.e., I/e),

$$\frac{dQ}{dt} = \frac{I}{e}(E_1 - E_2) \tag{24.23}$$

Combining Eqs. 24.20 and 24.23 and using the heat capacity for free electrons given by Eq. 24.9, we obtain

$$\pi_{12} = \frac{\pi^2 k^2 T^2}{4e}\left[\frac{1}{(E_F)_1} - \frac{1}{(E_F)_2}\right] \tag{24.24}$$

for the Peltier coefficient of the junction at temperature T. When electrons pass from metal 2 to metal 1, π_{12} is positive if $(E_F)_1 < (E_F)_2$ and π_{12} is negative if $(E_F)_2 < (E_F)_1$. It is important to realize that Eq. 24.24 is only an approximation because equilibrium conditions have been utilized to obtain the relationship for a transport process.

If a current is conducted around a loop constructed of two materials, heat will be absorbed at one junction and evolved at the other junction. Moreover, if the junctions are maintained at different temperatures, there will be net heat absorption or generation, considering both junctions. For metals, π_{12} and S_{12} are small, whereas for semiconductors, much larger values are observed frequently. To illustrate this point, it is necessary to consider the definition of Fermi energy in the case of a semiconductor. As defined in Sect. 4.7, the Fermi energy of a metal is taken as the maximum energy (above the bottom of the band) that an electron can have at absolute zero. That is, E_F is the highest filled electronic state at 0 K when there is no applied voltage. For $T > 0$ K, E_F corresponds to the energy state that is half full or occupied by only one electron. (With increasing temperature, electrons are not only excited above E_F, but E_F also decreases slightly for a metal. We have ignored the change in E_F because over normal temperature ranges it is virtually negligible.) For a semiconductor, E_F is defined as the energy at which there would be a 50% probability of occupancy. Thus, the Fermi energy lies halfway between the top of the valence band and the bottom of the conduction band in the case of an intrinsic semiconductor at absolute zero, even though no allowed electronic state exists in this region. Furthermore, the temperature variation of E_F is much larger for extrinsic semiconductors than for metals.

Silent refrigerators that make use of the Peltier effect of semiconductor p–n junctions are currently available. Although the overall efficiency of a Peltier refrigerator is quite low, the compact size and simplicity of construction are advantageous for small units. Thermoelectric generators utilizing semiconductor junctions also have been constructed. In this case, the Seebeck effect is used for direct conversion of thermal energy into electrical energy. At present, thermoelectric generators are inefficient and cannot compete with indirect energy conversion processes; however, they are compact, easily installed and operated, and are used under unique circumstances such as those found with spacecraft.

REFERENCES

1. Rose, R. M., Shepard, L. A., and Wulff, J. *Structure and Property of Materials*, Vol. IV, Wiley, New York, 1966.
2. Ziman, J. The Thermal Properties of Materials, *Scientific American*, Vol. 217, No. 3, September, 1967, pp. 180–188.
3. Cadolf, I. B., and Miller, E. *Thermoelectric Materials and Devices*, Reinhold, New York, 1960.

QUESTIONS AND PROBLEMS

24.1 The thermodynamic relationship between C_p and C_v is given by

$$C_p - C_v = \left[\left(\frac{\partial E}{\partial V}\right)_T + P\right]\left(\frac{\partial V}{\partial T}\right)_P$$

(a) For an ideal gas, $(\partial E/\partial V)_T = 0$. Use the ideal gas law $PV_m = RT$ (where $V_m =$ molar volume) to show that $C_p - C_v = R$.

(b) It can be shown that, for any substance, the right-hand side of the above equation equals $\alpha_v^2 V_m T/\beta$, where α_v is the coefficient of volume thermal expansion and $\beta = -(1/V)(\partial V/\partial P)_T$ is the compressibility. Demonstrate that $\alpha_v^2 V_m T/\beta$ is equal to R for an ideal gas.

24.2 (a) Calculate the velocity of sound ($v_s = \sqrt{E/\rho}$), the Debye cutoff frequency ν_m and the Debye temperature θ_D for Al, Cu, Pb, and diamond, and comment on the trend exhibited by θ_D. (Refer to the caption for Fig. 24.1. See Table 19.1 for values of E and Table 2.5 for atomic radii. $E = 1.6 \times 10^{10}$ N/m^2 for Pb, and $\lambda_{\min} \approx$ 0.4 nm for diamond.)

(b) What is the maximum phonon energy for each material in part (a)?

24.3 The number of phonons having frequency ν varies with temperature as

$$n = \frac{1}{e^{h\nu/kT} - 1}$$

(Unlike electronic energy states, which may be occupied by a maximum of two electrons, a phonon state may have an unlimited number of phonons.)

(a) Using this expression, explain why, at low temperatures, more low-frequency than high-frequency phonons are created.

(b) Plot n as a function of the ratio $kT/h\nu$. Comment on the variation at high temperatures.

(c) The total number of phonons at any temperature is given by

$$n_{\text{total}} = \int_0^{\nu_m} nN(\nu)\, d\nu$$

Qualitatively (i.e., without carrying out the integration), explain why the electrical resistivity of a metal should increase linearly with temperature above θ_D if ρ is directly proportional to n_{total}.

24.4 The internal energy associated with phonons is

$$E = \int_0^{\nu_m} nh\nu N(\nu)\, d\nu$$

where $nh\nu$ is the energy of all phonons having frequency ν (n is given in Problem 24.3). Using this expression, we find that the exact formula for C_v, according to the Debye model, can be obtained:

$$C_v = 9R\left(\frac{T}{\theta_D}\right)^3 \int_0^{\theta_D/T} \frac{x^4 e^x}{(e^x - 1)^2}\, dx$$

Show that, in the high-temperature limit (i.e., $\theta_D/T \ll 1$), C_v reduces to $3R$. (*Hint.* Use the expansion $e^x \approx 1 + x$ for small x; then evaluate the integral.)

24.5 At very low temperatures, C_p and C_v are virtually identical. The following are low-temperature heat capacity data for niobium:

T (K)	C_p (J/K mol)[a]	C_p (J/K mol)[b]
1	0.00766	0.00008
2	0.0161	0.0014
3	0.0261	0.00816
4	0.0385	0.0244
5	0.0540	0.0515
6	0.0736	0.0891
7	0.0979	0.141
8	0.128	0.208
9	0.164	0.297
9.17	0.170	0.315
10	0.207	—

[a] Nb maintained in the normal state (non-superconductor) by means of an applied magnetic field.

[b] Superconducting Nb.

(a) Plot (C_p/T) vs. T^2 for normal Nb, and determine the temperature coefficient of the electronic heat capacity in J/K^2 mol (i.e., γ in Eq. 24.10) and the Debye temperature.

(b) On the same graph used in part (a), plot (C_p/T) vs. T^2 for superconducting Nb. Suggest a reason why the latter increases more sharply with increasing temperature than does that for normal Nb (see Sect. 22.5).

24.6 Given the following low-temperature heat capacity data for white tin (metallic) and gray tin (diamond cubic structure):

T (K)	C_p (J/K mol)[a]	C_p (J/K mol)[b]
1.0	0.00201	0.00021
1.5	0.00350	0.00069
2.0	0.00556	0.00163
2.5	0.00845	0.00320
3.0	0.0129	0.00552
3.72	0.0232	0.0113

[a] White Sn, in the normal (nonsuperconducting) state.
[b] Gray Sn.

(a) Plot (C_p/T) vs. T^2 for white and gray tin. Determine the temperature coefficient of the electronic heat capacity (γ in Eq. 24.10) and the Debye temperature for each.

(b) Why is C_p greater for white Sn than for gray Sn?

24.7 Vanadium and titanium exhibit complete solid solubility at elevated temperatures. Such alloys, containing up to about 80 at. % titanium, can be maintained as single-phase bcc solid solutions at room temperature and below. The temperature coefficient of the electronic heat capacity for the normal state and the superconducting transition temperature have been determined for a number of V–Ti alloys:

	V	V–15 at. % Ti	V–25 at. % Ti	V–50 at. % Ti	V–70 at. % Ti	V–80 at. % Ti
γ (10^{-3} J/K^2 mol):	8.9	10.3	10.6	10.8	10.0	6.9
T_c (K)	4.59	7.02	7.16	7.30	6.14	3.5

(a) The basic theory of superconductivity indicates that T_c is a strong (but nonlinear) function of γ for the normal metal. Do the above data support this?

(b) What leads to the variation of γ with titanium content in V–Ti alloys? (*Hint.* Consider what γ is fundamentally related to.)

24.8 Using the definitions given by Eqs. 24.11 and 24.12, show that $\alpha_v = 3\alpha_l$ for an isotropic material.

24.9 When a linear elastic material is subjected to both an applied stress and a change in temperature, the mechanical and thermal strains are additive. If ΔT is sufficiently small so that α_l and Young's modulus (E) are approximately constant, then the total strain is given by

$$\varepsilon = \frac{\sigma}{E} + \alpha_l \Delta T$$

(a) A bar of steel at 20°C is heated to 50°C. What applied stress at 20°C would produce the same length change? (For steel, $E = 30 \times 10^6$ psi and $\alpha_l = 11.7 \times 10^{-6}$ °C^{-1}.)

(b) If the ends of a steel bar at 20°C are held by rigid supports that keep its length constant, what is the thermally induced elastic stress when the temperature drops to 0°C? Repeat this for a temperature increase to 40°C.

24.10 Explain why a bimetallic strip, consisting of two thin pieces of different metals bonded well along their lengths, can be used in a thermostat.

24.11 (a) Pure silica glass is much more resistant to thermal shock than is soda-lime-silica glass. Why?

(b) Even when temperature remains uniform throughout a crystalline ceramic, cracking and fracture may occur if there is a polymorphic transformation. Explain.

24.12 Fick's first law for diffusion (Eq. 12.2) completely characterizes diffusion only for steady state, although it gives the local flux even in nonsteady state diffusion where Fick's second law (Eq. 12.4) characterizes the diffusion process. Similarly, the heat flux equation (Eq. 24.13) must be augmented by another equation comparable to Fick's second law in order to solve nonsteady state heat flow problems.

In one dimension, the differential equation that gives temperature as a function of position and time can be derived as follows. Consider a volume element of cross-sectional area A and length dx. The heat flux flowing in at coordinate x is J_x, and that flowing out at coordinate $x + dx$ is $J_x + (\partial J/\partial x)_t \, dx$. The difference between the rate of heat flowing in and the rate of heat flowing out equals the "heat accumulation" per unit time in the volume element or $(C_p/V_m)(\partial T/\partial t)_x A \, dx$, where C_p is on a per mole basis and V_m is the molar volume.

(a) Show that the differential equation is

$$\left(\frac{\partial T}{\partial t}\right)_x = \frac{\mathscr{K} V_m}{C_p}\left(\frac{\partial^2 T}{\partial x^2}\right)_t$$

(b) The quantity $\mathscr{K} V_m/C_p$ is called the thermal diffusivity. What units does it have?

(c) Compute the thermal diffusivity at room temperature for Al, Ag, Cu, Fe, Na, and Ni.

24.13 For metals, it is expected that thermal conductivity will be relatively insensitive to temperature if the electrical resistivity increases linearly with temperature. Justify this statement and cite its limitations.

24.14 (a) Although stainless steel possesses a rather low thermal conductivity (see Fig. 24.6*b*), it is commonly used for cooking ware. Why?

(b) Explain why some stainless steel frying pans are clad with a copper bottom whereas aluminum frying pans are not.

24.15 How do you expect the thermal conductivity of a network polymer to vary with increasing temperature? Explain.

24.16 (a) Give some examples of materials that would be good thermal insulators at cryogenic temperatures. Justify your choices and cite limitations of each material.
(b) Explain why a foamed polymer would be a good choice.

24.17 Consider the advantages of using a thermistor as compared to using a thermocouple to measure temperature.
(a) What types of materials would make the best thermistors?
(b) Over what temperature ranges would a thermistor be superior to a thermocouple for measuring temperature? Over what temperature ranges would a thermocouple be superior? Explain briefly.

24.18 The voltage output of a thermoelectric generator, based on the Seebeck effect, decreases with increasing current drawn.
(a) Compare this with the operation of an electrochemical cell.
(b) Why is it preferable to employ a potentiometer rather than a voltmeter when using a thermocouple to measure accurate temperatures?

24.19 In the construction of Peltier refrigerators and thermoelectric generators, direct *p–n* junctions are not made. Instead, the contact is *p*-semiconductor-metal-*n*-semiconductor. Why are direct *p–n* contacts avoided?

24.20 Consider a simple thermocouple, consisting of metals 1 and 2:

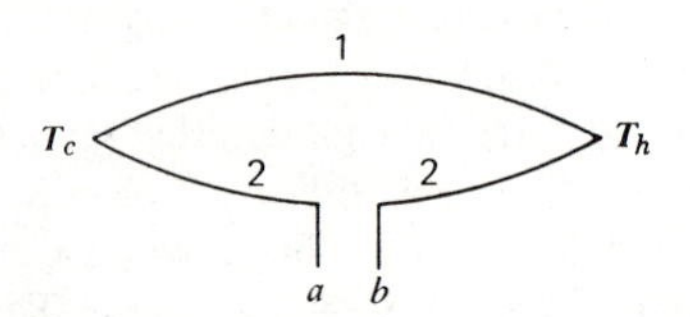

(a) Using Eq. 24.18, show that $V_b - V_a = \int_{T_c}^{T_h} (S_2 - S_1)\, dT$ if $T_a = T_b$. (*Hint*. Set up expressions for voltage differences, for example, $V_a - V_c = -\int_{T_c}^{T_a} S_2\, dT$, and sum these to obtain $V_b - V_a$.)
(b) What is the practical implication of your result in part (a)?

24.21 (a) When a current flows across a junction from a metal to an *n*-type semiconductor, is heat evolved or absorbed? Explain your answer qualitatively in terms of the conduction electron energies. Remember that the Fermi levels equalize.
(b) Repeat part (a) for a current flowing across a junction from a metal to a *p*-type semiconductor.

25

Magnetic Behavior

This chapter introduces the basic principles of magnetism and describes the properties of magnetic materials.

25.1 CONCEPTS OF MAGNETISM

Central to a discussion of magnetic materials is the concept of a magnetic dipole, which is most easily introduced through a brief review of electric dipoles. As mentioned in Sect. 2.11, an electric dipole can be described in terms of two charges, $+q$ and $-q$, separated by a distance a. The dipole moment of this charge distribution, defined as $p_e = qa$, is directed from the negative charge to the positive charge. If an electric field is applied, an electric dipole will tend to align itself in the direction of the field (Fig. 25.1). The torque or moment T acting to align the dipole is given by the resolved force ($eE \sin \phi$) times the separation distance a;

$$T = eEa \sin \phi = p_e E \sin \phi \tag{25.1}$$

where p_e is the electric dipole moment and ϕ is the angle between the electric field vector and the dipole moment vector.

Heteropolar molecules like HF and H_2O possess relatively large permanent dipole moments, and these molecules tend to be aligned by an electric field. Even with nonpolar molecules such as CH_4 and with atoms such as neon, there is interaction with an applied electric field because the field distorts the electronic charge distribution of the outer electrons, effectively separating the centers of

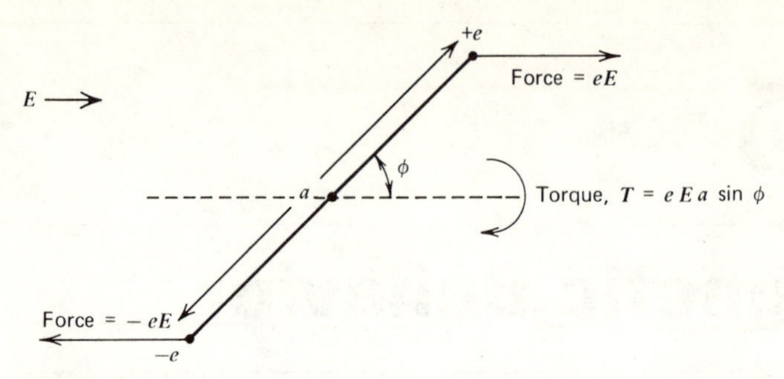

FIGURE 25.1 An electric dipole of strength $p_e = ea$ oriented at an angle ϕ with an applied electric field. The force of the electric field (eE) on the electronic charges produces a torque of magnitude $eEa(\sin\ \phi)$ tending to align the dipole in the direction of the applied field.

electronic and nuclear charge. The van der Waals bonding responsible for cohesion in solid methane or solid neon is a consequence of the interaction between mutually induced fluctuating electric dipoles in these materials (see Sect. 2.10).

Observations of the manner in which atoms and molecules behave in the presence of an applied magnetic field show that such particles possess magnetic dipole moments. These arise from the orbital motion of electrons about the nucleus and from the net spin of unpaired electrons. Furthermore, the nucleus itself possesses a magnetic moment, but it is very small in comparison to the electronic contributions. Magnetic dipole moments interact with magnetic fields in much the same way as electric dipoles interact with electric fields. A simple experiment with a magnetic compass needle illustrates this. Such a needle aligns itself with the earth's magnetic field and, if the needle is rotated by an angle ϕ away from the direction of the earth's field, a torque must be applied to balance the torque experienced by the needle,

$$T = p_m B \sin \phi \tag{25.2}$$

where p_m is the total magnetic dipole moment of the needle, dependent on its size, and B is the magnitude of the earth's magnetic field (or, more precisely, magnetic induction).

According to the laws of electromagnetism, a magnetic dipole may be viewed in terms of a model involving a circulating electric current in a wire loop. As shown in Fig. 25.2, a constant circulating current of magnitude I gives rise to a magnetic dipole moment equal to IA, where A is the area inclosed by the current loop. If the loop is composed of N turns, then the resulting moment will be NIA, and this principle is utilized in the construction of solenoidal electromagnets. The resultant magnetic field of a solenoid lies parallel to its axis and perpendicular

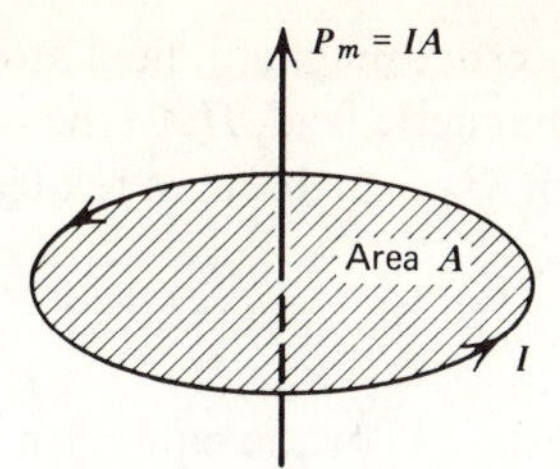

FIGURE 25.2 A circulating current of magnitude I produces a magnetic dipole moment of strength $p_m = IA$ where A is the area enclosed by the current loop. The direction of the moment is normal to the enclosed area.

to the coils and also extends in a circuit through the region outside of the solenoid. In general, a magnetic dipole creates a magnetic field around itself just as an electric dipole has an associated self-field around itself.

The orbital motion of an electron about the nucleus of an atom results in an intrinsic magnetic moment. In the presence of an external magnetic field, the component of magnetic moment that is parallel to the applied field is given by $m_l(eh/4\pi m_e)$, where m_l is the magnetic quantum number (see Sect. 2.4), e and m_e are the magnitude of the electronic charge and mass, respectively, and h is Planck's constant. Any filled shell or subshell contributes nothing to the orbital magnetic moment of an atom, but occupied outer electronic orbitals contribute to a net atomic magnetic moment (parallel to the field) if the summation of magnetic quantum numbers is not equal to zero. The quantity $eh/4\pi m_e$ is a fundamental quantity in magnetism called a Bohr magneton and assigned the symbol μ_B.

Another intrinsic magnetic moment is associated with the spin of an electron. When the spin quantum number is $m_s = +\frac{1}{2}$ (spin up), the magnetic moment is $+\mu_B$ (parallel to the magnetic field); when $m_s = -\frac{1}{2}$ (spin down), it is $-\mu_B$ (opposite to the magnetic field). For free atoms like Ne or Mg, no permanent magnetic moment exists because there are no unpaired electrons and the individual orbital moments sum to zero. Free atoms like Na and molecules like O_2, each of which has unpaired electrons (see Sect. 4.3 concerning O_2). possess a permanent magnetic moment. When both spin and orbital components of magnetic moment exist, they are additive (although not in a simple arithmetic way).

Since the magnetic moments discussed above tend to become aligned by an applied magnetic field and, therefore, reinforce it by virtue of their own magnetic fields, it is useful to measure the interaction. This can be done by comparing the magnetic induction, B, or internal magnetic field strength, within a substance in the presence of an external magnetic field, H, to the magnetic induction produced by the same field in a vacuum, B_0. The latter is related to the applied field by

$$B_0 = \mu_0 H \tag{25.3}$$

B and B_0 are measured, in teslas (T) or the equivalent webers/m^2(Wb/m^2), H is in A/m, and μ_0 is a constant conversion factor equal to $4\pi \times 10^{-7}$ Wb/A · m. A solenoid, consisting of N closely spaced turns of conductor and having a length l, generates a magnetic field of

$$H = \frac{NI}{l} \tag{25.4}$$

and the corresponding induction (flux density) in a vacuum is

$$B_0 = \mu_0 \frac{NI}{l} \tag{25.5}$$

If a substance (gas, liquid, or solid) occupies the core of the solenoid, then the magnetic field remains the same but the induction becomes

$$B = \mu_0 H + \mu_0 M \tag{25.6}$$

where there is an additional term involving the magnetization M of the substance. This equation holds within a substance whether or not the external magnetic field is created by a solenoid. The magnetization represents the net magnetic dipole moment (per unit volume) aligned parallel to the external field. For many substances, a direct proportionality exists between the magnetization and the magnetic field,

$$M = \chi H \tag{25.7}$$

with the proportionality constant χ, called the magnetic susceptibility, being a property of the substance. Even when direct proportionality does not hold, Eq. 25.7 defines the instantaneous magnetic susceptibility. Substitution of Eq. 25.7 into Eq. 25.6 yields

$$B = \mu_0(1 + \chi)H \tag{25.8}$$

and the quantity $(1 + \chi)$ is called the relative permeability of a substance.

25.2 DIAMAGNETISM AND PARAMAGNETISM

All atoms, even those without intrinsic magnetic moments, interact with a magnetic field. When an atom or molecule is subjected to a magnetic field, the motion of its electrons is perturbed in such a way that a very weak magnetic moment opposing the applied field is *induced*. The origin of this opposing moment, which is directly proportional to the magnetic field, is somewhat analogous to the transient shielding experienced by a metal in a changing magnetic field. In the latter case, an electric current is induced, and the magnetic field associated with the induced current opposes the change in applied field. This induced current dies away in a metal because of conduction electron scattering, but the equivalent effect persists in individual atoms and molecules.

Substances consisting of atoms or molecules that do not possess permanent magnetic moments exhibit diamagnetism in a magnetic field. That is, they display a negative, but generally quite small, magnetic susceptibility because of the weak, induced opposing magnetic moment. An induced diamagnetic component of magnetization is inherent in all substances but often is masked by the larger intrinsic moments that reinforce an applied field. Inert gases and solids like Bi, Cu, MgO, and diamond are examples of diamagnetic materials, having typical values of χ on the order of -10^{-5}. The diamagnetic component of susceptibility is independent of temperature and magnetic field.

For the most part, diamagnetic behavior has no engineering importance. A very important exception to this statement is provided by superconductors (see Sect. 22.5). Type I superconductors can be perfectly diamagnetic (i.e., $M = -H$, $\chi = -1$, or $B = 0$) up to the critical magnetic field because induced shielding currents persist indefinitely. Type II superconductors exhibit perfect diamagnetism or partial diamagnetism depending on the applied field (see Fig. 22.8). The strong diamagnetism of superconductors makes it possible to use such materials for magnetic field shielding purposes.

Many solids exhibit susceptibilities greater than zero as a result of net intrinsic magnetic moments associated with electronic orbitals and electron spins. Substances having values of χ in the range from about 10^{-6} to 10^{-2} are called paramagnetic. For paramagnetic materials, the interaction between atomic moments and an applied field can be discussed in terms of thermodynamics. When the individual atomic moments are randomly arrayed, the entropy of the material is high because of the associated disorder. Countering this tendency is the magnetic interaction energy, which is decreased when the moments align with the field. The final degree of moment alignment, therefore, depends on the competing factors of energy and entropy. Thus, the susceptibility, which reflects the degree of alignment, tends to decrease as temperature increases and entropy becomes more dominant. The results of thermodynamic calculations minimizing the free energy as a function of temperature show that the susceptibility decreases with

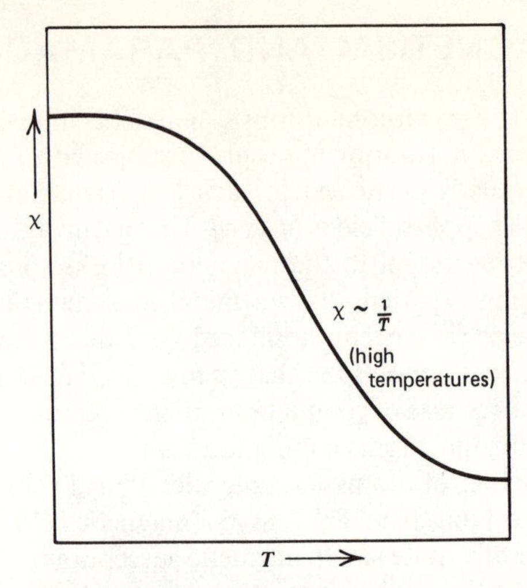

FIGURE 25.3 The variation of magnetic susceptibility with temperature for a paramagnetic material. At ordinary temperatures the susceptibility follows the Curie law, being inversely proportional to temperature. At low temperatures, the susceptibility variation is much less, and the susceptibility tends to saturate at a finite value.

temperature and at intermediate and high temperatures is given approximately by

$$\chi = \frac{\mu_0 N p_m^2}{3kT} = \frac{C}{T} \tag{25.9}$$

where p_m is the magnetic moment per atom and N is the number of atoms per unit volume. This equation holds for most paramagnetic nonmetallic materials above room temperature. The dependence of χ on reciprocal absolute temperature is designated the Curie law, and the constant C is called the Curie constant. At low temperatures, as shown in Fig. 25.3, the Curie law no longer holds, and the susceptibility saturates at a finite value.

In metals, conduction electrons provide both paramagnetic and diamagnetic moments, with a net paramagnetic moment, in addition to the induced diamagnetic moment resulting from core electrons. The paramagnetic component arising from the conduction electrons usually predominates, and most metals have positive susceptibilities. The origin of this paramagnetism can be explained in terms of the conduction electron energy band viewed as two subbands, one containing valence electrons with spin up and the other containing valence electrons with spin down. With no applied magnetic field, the subbands have equal populations of electrons and there is no net spin. Application of a magnetic field lowers the potential energy for the subband having spins parallel to the field and raises the potential energy of the other subband. Since the Fermi energy must be at the same level in both subbands when equilibrium exists (i.e., the electronic free energy is constant; see Sect. 24.4 and Fig. 24.8), a greater number of conduction electrons reinforce the field than oppose it (Fig. 25.4). The conduction electron

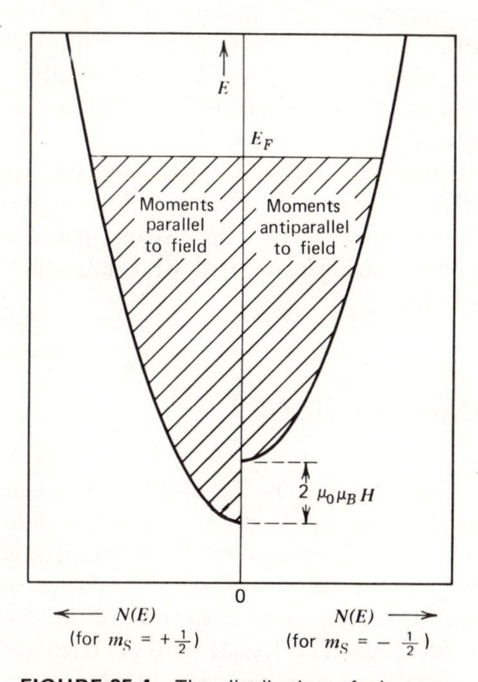

FIGURE 25.4 The distribution of electrons within the conduction band of a metal in the presence of an external magnetic field. Those electrons with moments aligned parallel to the field ($m_s = +\frac{1}{2}$) have a magnetic potential energy $2\mu_0 \mu_B H$ lower than those electrons with moments antiparallel ($m_s = -\frac{1}{2}$). Equalization of the Fermi energy within these subbands leads to a greater population of electrons with moments parallel to the field and a resulting net magnetic moment. ($2\mu_0 \mu_B H$ is generally much smaller than E_F; for clarity the former quantity has been exaggerated in this figure).

net spin susceptibility is proportional to the temperature coefficient of the electronic heat capacity (see Eq. 24.8) and, for free electrons in a single band, having the Fermi energy much lower than any band gap, is given by

$$\chi = \frac{\mu_0 n \mu_B^2}{E_F} \qquad (25.10)$$

where n is the total number of valence electrons per unit volume. Examining Fig. 25.4, we see that only those electrons having energies near the Fermi level are responsible for spin paramagnetism, and this has been taken into account in deriving Eq. 25.10. In general, the spin susceptibility is proportional to the density of states at the Fermi energy, so that transition metals exhibit larger paramagnetic susceptibilities than nontransition metals (Table 25.1) just as they exhibit larger electronic heat capacities.

Some metals are diamagnetic because the conduction electron spin susceptibility is smaller than the induced diamagnetic susceptibility component. On the other hand, various rare earth metals display very strong paramagnetism because of unpaired f electrons that remain associated with individual atoms rather than entering into energy bands.

25.3 FERROMAGNETISM

The elements Fe, Co, Ni, Gd, Tb, Dy, Ho, and Tm, as well as a number of alloys and compounds, exhibit very large positive susceptibilities below a critical temperature θ_c known as the ferromagnetic Curie temperature. Such substances display spontaneous magnetization, the origin of which arises from an internal interaction that causes a permanent offset in the electronic subbands (Fig. 25.5). Only certain metals and other solids having partially filled d subshells or partially filled f subshells are capable of such behavior. In addition to this necessary, but not sufficient, condition for spontaneous permanent magnetization, the interatomic spacing must be such that an internal interaction—termed the exchange interaction—occurs. As shown in Table 25.2, the elements Fe, Co, and Ni in the third period of the periodic table are magnetized spontaneously. That is, the average number of electrons in the d band having spin up is not the same as that having spin down, and the difference between the two gives a net magnetic moment (Table 25.2). As noted in Table 25.2, Ni is ferromagnetic whereas Cu is not. When a series of solid solutions between these elements is prepared, it is observed that the spontaneous magnetization decreases linearly with Cu content until no spontaneous magnetization remains at 60 at. % Cu. At this composition, the 10.6 valence electrons per atom are equally divided between $m_s = +\frac{1}{2}$ and $m_s = -\frac{1}{2}$.

TABLE 25.1 Susceptibilities, χ, ($\times 10^3$) of Common Paramagnetic Metals at 20°C and the Variation of Magnetic Behavior in the Periodic Table for Metallic Substances

IA	IIA	IIIA	IVA	VA	VIA	VIIA		VIII		IB	IIB	IIIB	IVB	VB
Li 0.014	Be Dia-magnetic													
Na 0.008	Mg 0.012	Al 0.021												
K 0.006	Ca 0.019	Sc 0.264	Ti 0.182	V 0.375	Cr 0.313	Mn 0.871	Fe Ferro-magnetic	Co Ferro-magnetic	Ni Ferro-magnetic	Cu Dia-magnetic	Zn Dia-magnetic	Ga Dia-magnetic		
Rb 0.004	Sr 0.034	Y 0.114	Zr 0.109	Nb 0.226	Mo 0.119	Tc 0.395	Ru 0.066	Rh 0.169	Pd 0.802	Ag Dia-magnetic	Cd Dia-magnetic	In Dia-magnetic	Sn 0.002	
Cs 0.005	Ba 0.007	Rare Earths	Hf 0.070	Ta 0.178	W 0.078	Re 0.094	Os 0.015	Ir 0.038	Pt 0.279	Au Dia-magnetic	Hg Dia-magnetic	Tl Dia-magnetic	Pb Dia-magnetic	Bi Dia-magnetic

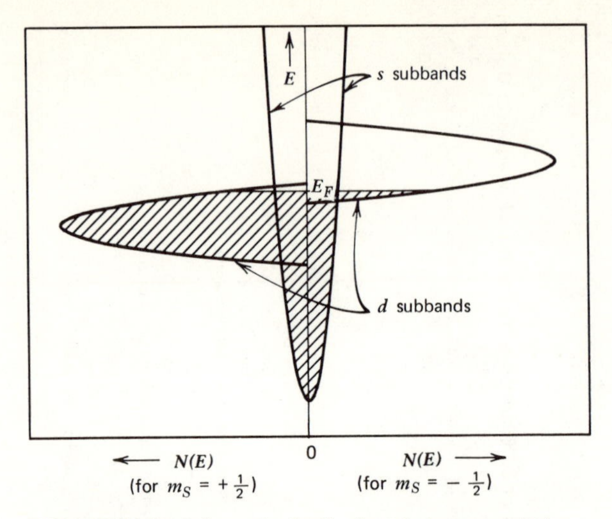

FIGURE 25.5 Schematic distribution of electrons within an s and a d band for a ferromagnetic material in the absence of an external magnetic field. Here the occupation of energy states in the $m_s = \pm\frac{1}{2}$ subbands for the s electrons are identical since $H = 0$. However, a permanent magnetization results because of the exchange energy that arises from a permanent offset of the two d subbands.

TABLE 25.2 Distribution of Electrons and Electronic Magnetic Moments in the Metallic Elements Fe through Cu in the Periodic Table

Element	Fe	Co	Ni	Cu
Total number ($s + d$) valence electrons	8	9	10	11
Average number of electrons in s band	0.2	0.7	0.6	1.0
Average number of electrons in d band	7.8	8.3	9.4	10.0
Number of d electrons with moments up	5.0	5.0	5.0	5.0
Number of d electrons with moments down	2.8	3.3	4.4	5.0
Net moment per atom (units of Bohr magnetons, μ_B)	2.2	1.7	0.6	0

With increasing temperature, the lowering of internal energy because of the exchange interaction is opposed by a greater entropy that results from the disorder of randomized spins. This accounts for the peak in heat capacity observed when θ_c is approached (see Fig. 24.3). Ferromagnetic materials become paramagnetic above θ_c, and their susceptibilities decrease with further increases in temperature (Fig. 25.6*a*). Below θ_c, susceptibilities can be on the order of 250 to 100,000 or higher and continue to increase with decreasing temperature. Typically, susceptibilities at 0.9 θ_c are approximately half those at 0 K.

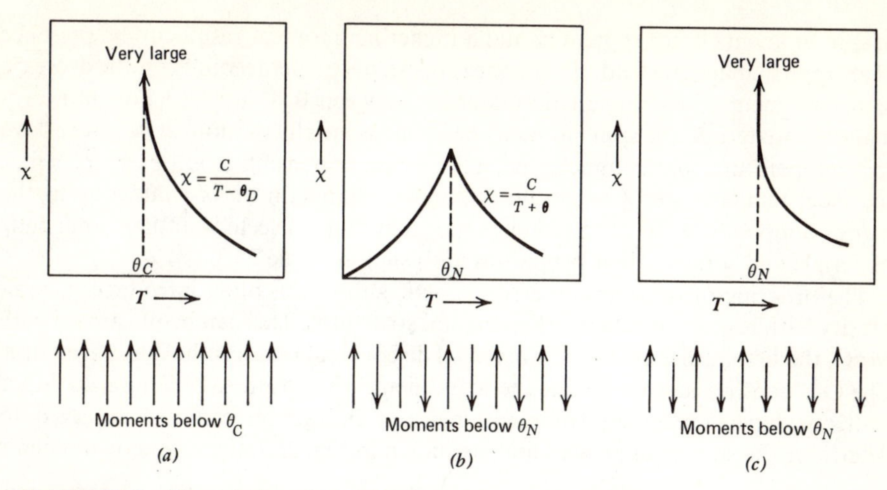

FIGURE 25.6 The variation of susceptibility with temperature for:

(*a*) A ferromagnetic material. Above the Curie temperature the susceptibility varies inversely with temperature and is of the magnitude of a typical strongly paramagnetic material. Below the Curie temperature, the susceptibility becomes very large.

(*b*) An antiferromagnetic material. Above the Néel temperature, the behavior is paramagnetic. (The experimental parameter θ is not the same as the Néel temperature θ_N.) Below the Néel temperature, the susceptibility increases with increasing temperature. The very low value of susceptibility at 0 K corresponds to perfectly aligned moments that are antiparallel.

(*c*) A ferrimagnetic material. The behavior is similar to that for a ferromagnetic material, but the susceptibility below the Néel temperature is usually less than that of a ferromagnet below the Curie temperature, since the net moment often is not as large.

25.4 ANTIFERROMAGNETISM AND FERRIMAGNETISM

In ferromagnetic materials, the exchange interaction gives rise to a parallel alignment of spins; in some metals like Cr and many compounds, this exchange interaction, which is extremely sensitive to interatomic spacing and to atomic positions, results in an antiparallel alignment of spins. When the strength of antiparallel spin magnetic moments are equal, no net spin moment exists and the resulting susceptibilities are quite small. Such materials are called antiferromagnetic. The most noticeable characteristic of an antiferromagnetic material is that the susceptibility attains a maximum at a critical temperature θ_N, the Néel temperature (Fig. 25.6*b*).

The maximum in susceptibility as a function of temperature can be explained using thermodynamic reasoning. At the absolute zero of temperature, the spin moments are aligned as antiparallel as possible, yielding a small susceptibility. With increasing temperature, the tendency for disorder or more random alignment of spin moments become greater, and a higher *net* moment results in the presence of an applied magnetic field. (This same process, of course, accounts for the decrease in net moment of ferromagnetic materials between 0 K and θ_c.) For antiferromagnetic materials, the spin moments become essentially random at θ_N, and above this temperature, paramagnetic behavior with decreasing χ is observed. Thus, the Néel temperature of an antiferromagnetic temperature is analogous to the Curie temperature of a ferromagnetic material. Selected antiferromagnetic materials and some of their properties are listed in Table 25.3.

The ordering of spins in antiferromagnetic substances often introduces a complexity with respect to viewing their crystal structures. This can be illustrated with MnO, the first antiferromagnetic material discovered, that has the NaCl structure. The O^{2-} ions possess no net magnetic moment, whereas the Mn^{2+} ions do. Since neutrons have a magnetic moment, neutron diffraction can be employed to determine the spin arrangement and, as shown in Fig. 25.7, the magnetic moments

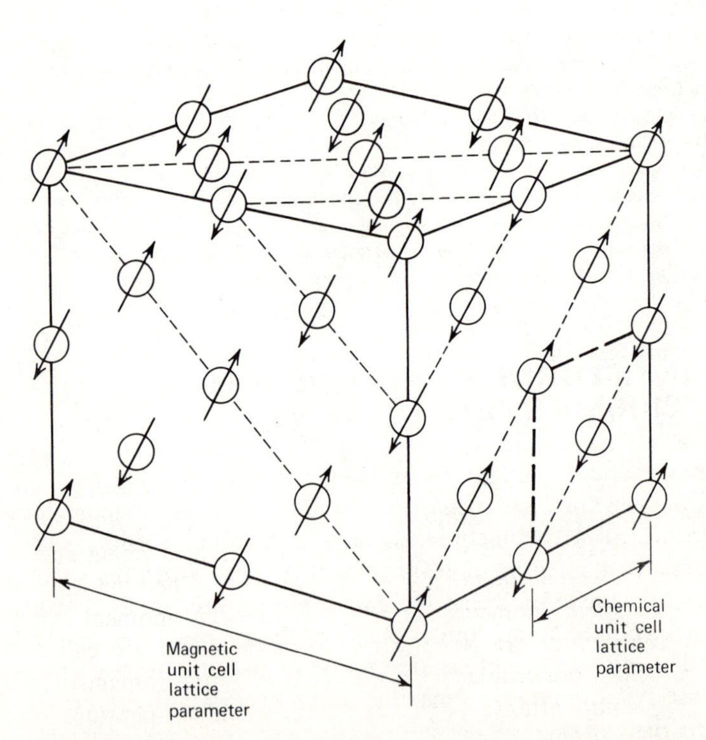

FIGURE 25.7 The crystal structure of MnO. Below the Néel temperature, the moments of Mn^{2+} ions tend to be parallel within each (111) plane, but antiparallel on neighboring (111) planes, giving rise to a very small net moment. Oxygen ions do not possess a permanent moment and are not shown in this figure.

TABLE 25.3 Some Properties of Some Antiferromagnetic Materials

Substance	*Néel Temperature,* θ_N, (K)	χ (0 K)/$\chi(\theta_N)$
MnO	116	0.67
MnS	160	0.82
MnF_2	67	0.76
FeF_2	79	0.72
FeO	198	0.80
CrSb	723	~0.25
$FeCO_3$	35	~0.25

of most Mn^{2+} ions are parallel within a given (111) plane but antiparallel in adjacent (111) planes when MnO is antiferromagnetic. Hence, the net moment is zero. Because of the spin ordering, the *magnetic unit cell* has twice the lattice parameter of the standard unit cell (Fig. 25.7).

Ferrimagnetism is similar to antiferromagnetism in that the spins of different atoms or ions tend to line up antiparallel. In ferrimagnetic materials, however, the spins do not cancel each other out, and a net spin moment exists. Below the Néel temperature, therefore, ferrimagnetic materials behave very much like ferromagnetic materials and are paramagnetic above θ_N (Fig. 25.6*c*). Ferrites, a particularly important class of ferrimagnetic materials, will be discussed in Sect. 25.8.

25.5 MAGNETIC DOMAINS

A macroscopic piece of iron below θ_c can exist either in an unmagnetized condition or in a magnetized condition. That is, it can either produce a magnetic field about itself or not. In the macroscopically unmagnetized condition, permanent spontaneous magnetic moments exist on a local scale within magnetic domains. Each domain possesses a permanent moment but, in unmagnetized iron, the domains are arranged such that the net moment is zero (see Fig. 25.8). The domains are separated by domain boundaries (not grain boundaries), across which the direction of the spin moments gradually changes. This kind of an arrangement represents the lowest free energy in the absence of an externally applied magnetic field.

The average magnetic induction of a ferromagnetic material is related intimately to the domain structure. A typical plot of magnetic induction B as a function of applied field H for an initially unmagnetized specimen of iron is shown as curve a in Fig. 25.9. Prior to application of the field, the individual domain moments are oriented randomly within a polycrystalline specimen. As the external field is

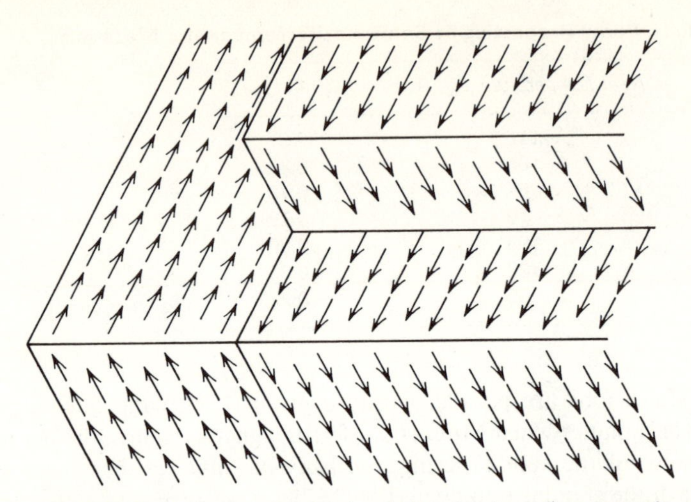

FIGURE 25.8 Schematic sketch of domains in a ferromagnetic material. Within each domain the electronic moments are aligned, giving rise to a net moment per domain. The geometrical arrangement of domains in the unmagnetized state results in a net magnetic moment of zero for the material as a whole.

applied and increased, those domains having moments closely aligned with the field direction grow at the expense of those having moments aligned less favorably. This process continues until only the most favorably oriented domains remain. (The spontaneous moments are aligned naturally with respect to certain crystallographic directions; in the case of iron, they are parallel to $\langle 100 \rangle$ directions.) At sufficiently high magnetic fields, the moments within the remaining domains are reoriented crystallographically so as to become parallel to the applied field. At this point the iron is magnetically saturated (i.e., the magnetization M becomes constant), and the magnetic induction does not increase appreciably with further increase in H. The growth of magnetic domains in an iron single crystal subjected to an applied field is shown in Fig. 25.10.

Once magnetic saturation has been achieved, a decrease in the applied field to zero (curve b in Fig. 25.9) results in a macroscopically permanent or remnant induction denoted by B_r, and the material is in the magnetized condition. With decreasing applied field from the saturated state, spin orientations within domains readily rotate back to their favorable crystallographic orientation, but the original random domain arrangement is not achieved because domain wall motion is limited and, in effect, nucleation of domains having orientations lost during magnetization would have to occur. To bring the induction to zero, a field of magnitude H_c, termed the coercivity, must be applied antiparallel to the original magnetizing field (Fig. 25.9), and complete demagnetization requires cycling the magnetic field with ever decreasing amplitude.

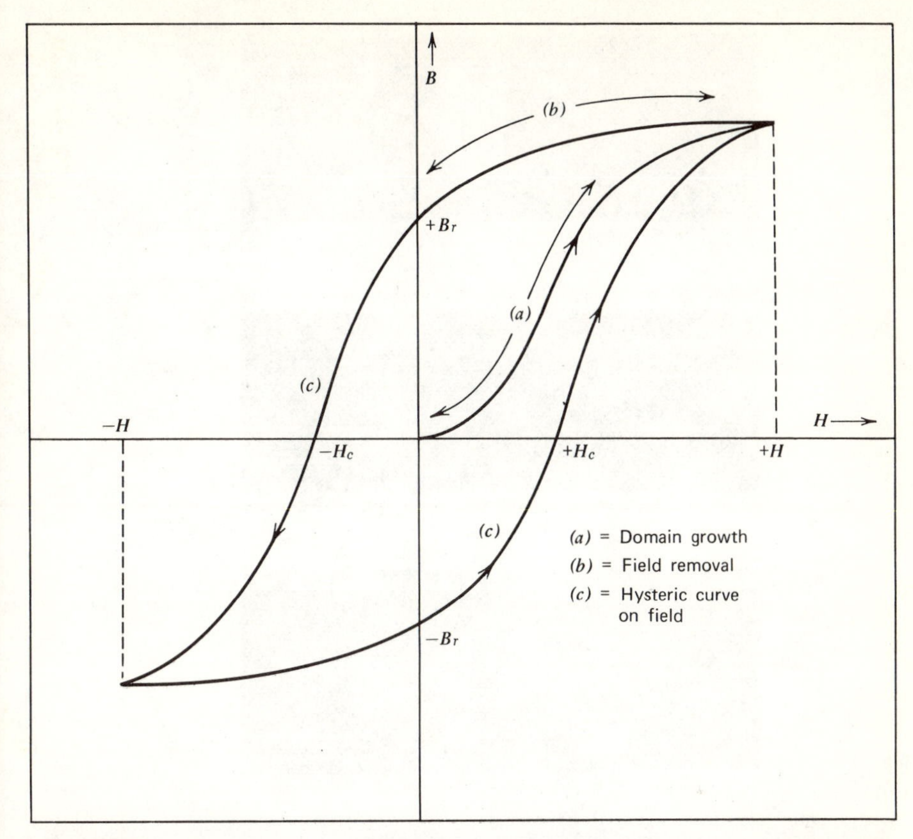

FIGURE 25.9 A magnetic induction versus magnetic field relationship for a ferromagnetic material. Along (*a*), starting from the demagnetized state, the more favorably oriented domains grow at the expense of less favored domains. At high fields, nearing saturation, the induction increases only by domain rotation. Upon removal of the applied field [curve (*b*)], the material maintains a remnant induction, B_r, and can only be demagnetized by imposing a negative field equal in magnitude to the coercivity, H_c. Cyclic applications of fields [such as in curve (*c*)] lead to a hysteresis loop for ferromagnetic materials. The area within the *B–H* loop represents an energy loss per unit volume of material per cycle.

If the ferromagnetic material is magnetized to saturation at applied field H' and then subjected to cyclic fields between H' and $-H'$, a hysteresis loop in the B vs. H relationship is produced (curve *c*, Fig. 25.9). The area within the hysteresis loop represents the energy loss per unit volume of material for one cycle and, unless suitable cooling is provided, the material heats up. Smaller hysteresis loops result for cycling between applied fields that do not cause saturation. Before discussing the causes of hysteresis, it should be noted that such behavior is similar to the stress vs. strain hysteresis in the case of anelasticity and viscoelasticity (see Sects. 19.6 and 19.7).

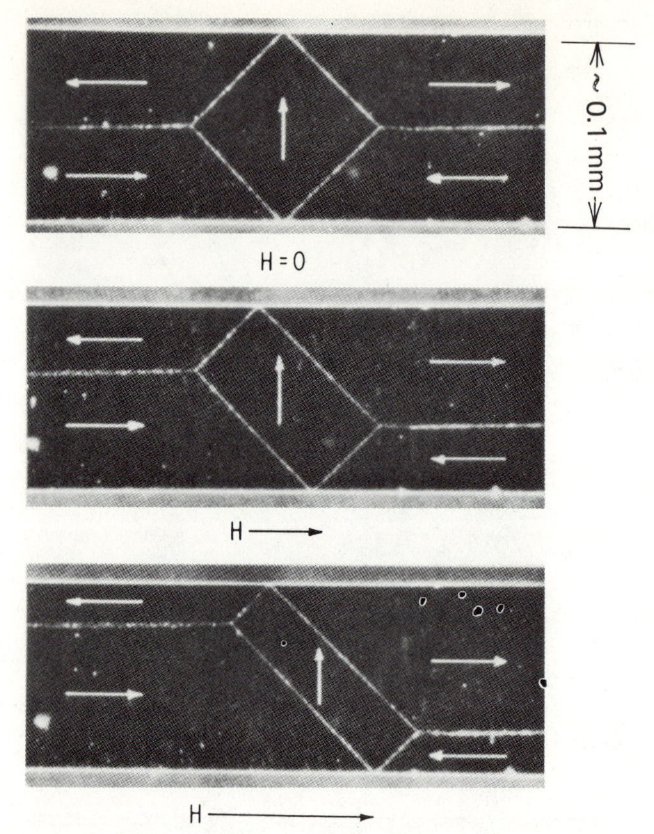

FIGURE 25.10 Photomicrographs of domains and domain growth in an iron single crystal. Note the growth of favorably oriented domains as the magnetic field is increased. (Courtesy of R. W. De Blois and C. D. Graham, General Electric Co.)

25.6 SOFT MAGNETIC MATERIALS

The shape of the hysteresis loop for a ferromagnetic substance is of great practical importance. In devices subjected to alternating fields, such as transformer cores, not only is a high magnetic susceptibility and a high saturation induction desirable, but it is also necessary that the material dissipate as little energy per cycle as possible. That is, a transformer core material must have a hysteresis loop of small area, characterized by a low value of H_c. In ferromagnetic materials, the saturation magnetization is primarily a function of composition, but the susceptibility, coercivity, and shape of the hysteresis curve are very sensitive to structure. A low coercivity coincides with easy motion of domain walls. Domain wall motion is restricted by structural defects such as nonmagnetic inclusions, voids, and precipitates of a nonmagnetic phase. Domain walls are pinned by these defects because the wall energy is lowered. Consequently, the number of such defects must be minimized in a soft magnetic material. Impurities and dislocation structures resulting from cold deformation also lead to higher hysteresis.

In addition to magnetic energy losses, energy losses in soft magnetic materials also arise because of electrical currents (eddy currents) induced in the material by the time-varying magnetic field. Eddy current losses can be reduced by increasing the electrical resistivity of the magnetic material, and this is one reason why solid-solution iron-silicon alloys ($\lesssim 4\%$ Si) are used at power frequencies (60 Hz) and iron-nickel alloys, at audio frequencies. Some magnetically soft ferrites (Sect. 25.8) are very nearly electrical insulators and, thus, immune to eddy current losses. This accounts for their use at high frequencies where the power dissipation problem is particularly acute for electrically conducting magnetic materials. Unfortunately, the susceptibilities of ferrites generally are too small for them to be useful in most generator, motor, and transformer applications. A list giving some of the more commonly used soft magnetic materials, along with selected properties, is given in Table 25.4.

As mentioned in the previous section, the direction of the spin moment within a given magnetic domain is parallel to a crystallographic direction characteristic of the material. For this reason, the *B* vs. *H* behavior of ferromagnetic single crystals depends on the crystallographic orientation of the crystal with respect to the applied field. This is illustrated for Fe and Ni single crystals in Fig. 25.11. The easy

TABLE 25.4 Some Common Soft Magnetic Materials and Some of their Properties

Material	*Initial Susceptibility* (χ *at* $H \sim 0$)	*Hysteresis Loss per Cycle* (J/m^3)	*Saturation Induction* (Wb/m^2)
Commerical iron ingot	250	500	2.16
Fe–4% Si, random	500	50–150	1.95
Fe–3% Si, oriented	15,000	35–140	2.0
45 Permalloy (45% Ni–55% Fe)	2,700	120	1.6
Mu metal (75% Ni–5% Cu–2% Cr–18% Fe)	30,000	20	0.8
Supermalloy (79% Ni–15% Fe–5% Mo–0.5% Mn)	100,000	2	0.79
Ferroxcube A[a] (Mn, Zn)Fe_2O_4	1,200	~40	0.36
Ferroxcube B[a] (Ni, Zn)Fe_2O_4	650	~35	0.29

[a] Ferrites.

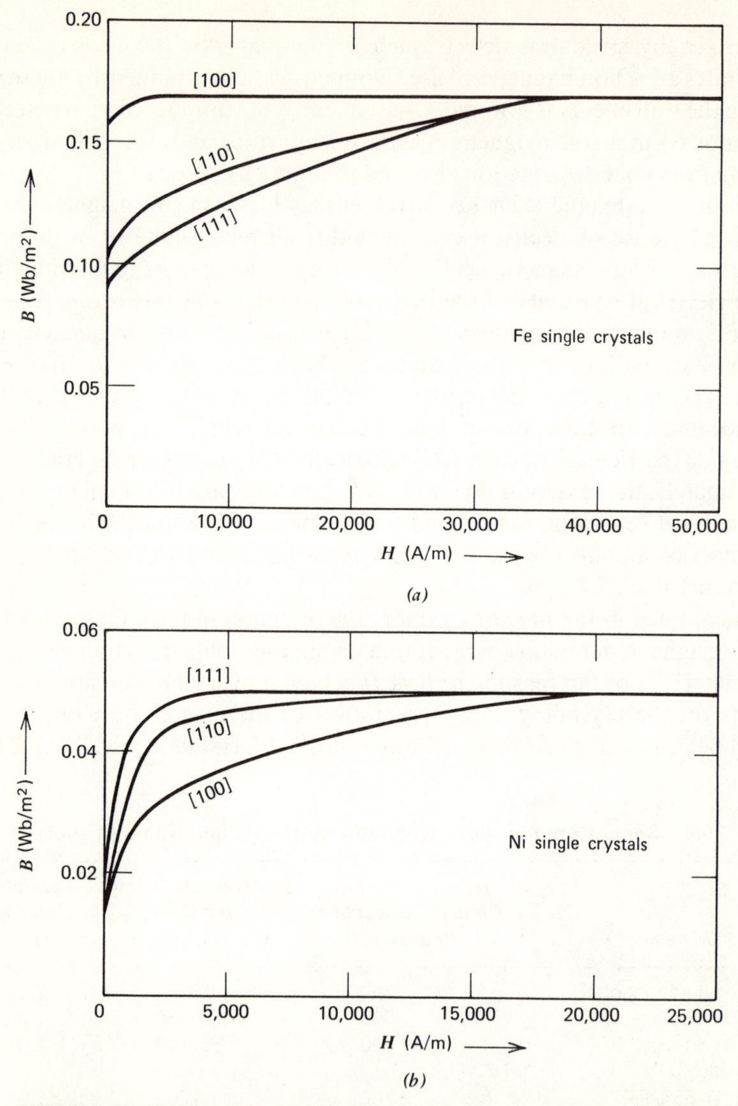

FIGURE 25.11 Induction–field curves for single crystals of (*a*) Fe and (*b*) Ni with different crystal orientations parallel to the applied field. The direction of easy magnetization is [100] in Fe and [111] in Ni. [After K. Honda and S. Kaya, *Sci. Rep. Tôhoku Imp. Univ.*, **15**, 721 (1926); **17**, 639 (1928); **17**, 1157 (1928).]

magnetization directions are ⟨100⟩ for bcc iron and ⟨111⟩ for fcc nickel since these correspond to the natural spin moment alignments. In other words, spin reorientation is not necessary during magnetization, and only domain wall motion occurs. A crystal oriented in the easy magnetization direction is easier to magnetize (*softer*) than one oriented in another direction. Beneficial use of this characteristic is realized in the case of Fe–Si alloys that are formed by rolling into sheet, with appropriate intermediate annealing treatments. When the thermomechanical processing is controlled carefully, a preferred crystallographic orientation develops within the polycrystalline sheet, such that the direction of easy magnetization is parallel to the rolling direction. Such a textured alloy is magnetized readily, and the hysteresis loop is small.

Magnetically soft Fe–Ni alloys can have their properties altered by heat treatment. The ordered phase Ni_3Fe, which has the same structure as Cu_3Au (see Sect. 22.3), undergoes an order-disorder transformation at about 500°C. Since the susceptibility of the ordered phase is only about half that of the disordered phase, a higher susceptibility is realized when the alloy is quenched from 600°C, a process that retains the high-temperature, disordered structure. Heat treatment of Fe–Ni alloys in a magnetic field further enhances their magnetic characteristics (Fig. 25.12), and the square hysteresis loop of 65 Permalloy so

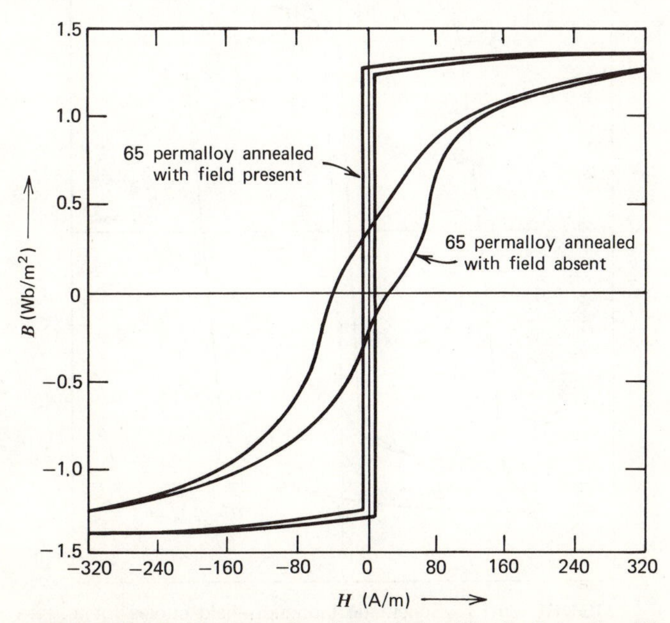

FIGURE 25.12 The induction–field relationships for Permalloy 65 heat treated with and without an applied magnetic field. The hysteresis loss is markedly reduced by heat treating in an applied field. [After J. F. Dillinger and R. M. Bozorth, *Physics*, **6**, 279 (1935).]

processed is desirable in computer, magnetic amplifier and pulse transformer applications. Supermalloy, a sister alloy (Table 25.4), when optimally processed, can have an initial susceptibility of approximately one million.

25.7 HARD MAGNETIC MATERIALS

Permanent magnets require materials with high remnant inductions and high coercivities. As a corollary, they also exhibit large hysteresis losses. Magnetic materials possessing these characteristics are called *hard* magnetic materials. The designations *hard* and *soft* relate to mechanical strength as well, for anything that increases mechanical strength in a ferromagnetic material also tends to produce a harder magnetic material. A comparison of hysteresis loops for hard and soft magnetic materials is shown in Fig. 25.13. A convenient measure of the hardness

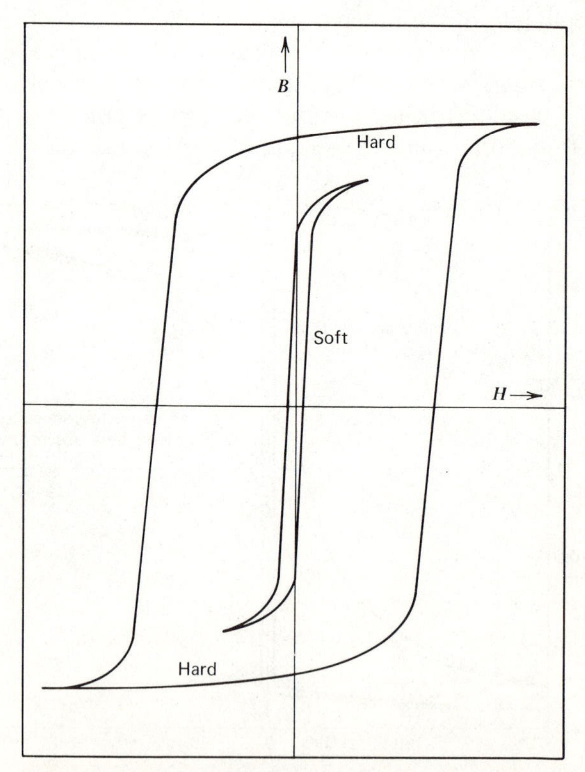

FIGURE 25.13 Schematic induction–field curves for a hard magnetic material and for a soft material. Hard magnetic materials are characterized by high remnant inductions, high coercivities, and large hysteresis losses per cycle.

of magnetic materials is provided by the product $B_r \cdot H_c$, which is roughly twice the energy required to demagnetize a unit volume of material. That is, $(B_r \cdot H_c)/2$ is roughly the area under the B vs. H curve between $H = 0$ and $H = -H_c$. Demagnetization curves for a few useful hard magnetic materials are shown in Fig. 25.14.

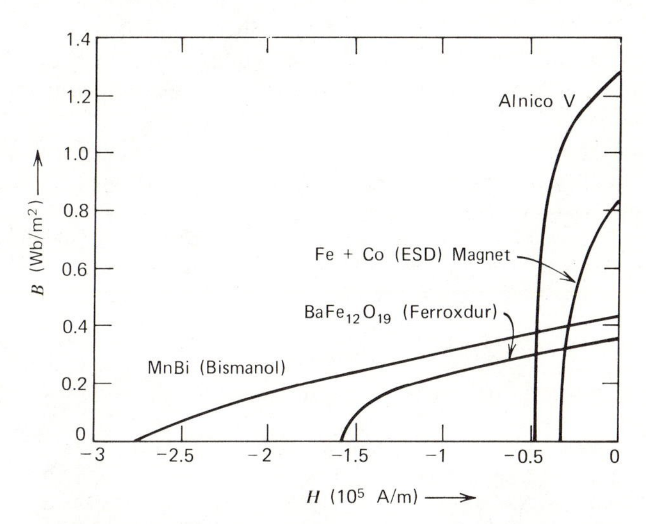

FIGURE 25.14 Demagnetization curves for several hard magnetic materials. (Adapted from R. M. Bozorth, in *The Science of Engineering Materials*, edited by J. E. Goldman, Wiley, New York, 1957.)

Hard magnetic behavior is related intimately to microstructure. Some permanent magnet materials, including high carbon steels and alloy steels containing combinations of Cr, Co, or W, undergo a martensitic transformation on cooling, resulting in a fine-scale microstructure. Others undergo an order-disorder transformation that results in internal strains. For example, Co–50 at. % Pt transforms from high-temperature disordered fcc structure to an ordered fct structure having Co atoms and Pt atoms on alternating (002) planes, and the partially ordered alloy can have H_c as large as 3.7×10^5 A/m, which is one of the highest coercivities known. In addition, there are many permanent magnetic materials having two-phase microstructures produced by solid state precipitation. Among these are Cu–Ni–Fe alloys (Cunife), Cu–Ni–Co alloys (Cunico), and Fe–Al–Ni–Co–Cu alloys (Alnico). Selected hard magnetic materials are listed in Table 25.5.

The Alnico type alloys are the most important hard magnetic materials commercially. At high temperatures these alloys exist as a homogeneous solid solution which, below a miscibility gap, decomposes into a fine two-phase mixture consisting of FeCo-rich particles within a NiAl-rich matrix. Both product phases tend to be ordered with the CsCl structure. A suitable phase decomposition heat treatment yields particles of the strongly magnetic FeCo-rich phase that are sufficiently small so that each particle contains a single magnetic domain. This, in combination with the fact that the NiAl-rich phase is only weakly magnetic, leads to a high coercivity. Indeed, permanent magnets with exceptionally high coercivity can be made by binding together magnetic particles of single domain size with a nonmagnetic binder. In such a case, the magnetization can change only by magnetic moment reorientation since domain growth is impossible. The magnetic properties of

TABLE 25.5 Some Hard Magnetic Materials and their Properties

Material	*Composition* (wt. %)	B_r (Wb/m^2)	H_c (kA/m)	B_rH_c (J/m^3)
Martensitic carbon steel	98.1 Fe–1 Mn–0.9 C	0.95	4.0	3.8
Tungsten steel	92.7 Fe–6 W–0.7 C–0.3 Cr–0.3 Mn	1.05	5.6	5.9
Chromium steel	95.3 Fe–3.5 Cr–0.9 C–0.3 Mn	0.97	5.2	5.0
Cobalt steel	71.8 Fe–17 Co–8 W–2.5 Cr–0.7 C	0.95	11.9	11.3
Cunife	60 Cu–20 Ni–20 Fe	0.54	43.8	23.7
Cunico	50 Cu–29 Co–21 Ni	0.34	52.5	17.9
Alni–a	59.5 Fe–24 Ni–13 Al–3.5 Cu	0.62	38.0	23.6
Alnico-I	63 Fe–20 Ni–12 Al–5 Co	0.72	35.0	25.2
Alnico-V	51 Fe–24 Co–14 Ni–8 Al–3 Cu	1.25	43.8	54.8
Alnico-XII	52 Fe–24.5 Co–13.5 Ni–8 Al–2 Nb	1.20	64.0	76.8
Co–Pt	77 Pt–23 Co	0.52	246.8	128.3
Ferroxdur	60.3 Fe–12.4 Ba–27.3 O ($BaFe_{12}O_{19}$)	0.20	120.0	24.0
Ferroxdur (oriented)		0.39	240.0	93.6

Alnico alloys can be enhanced by various processing techniques. For example, directional solidification leads to a preferred crystallographic orientation, and phase decomposition in the presence of an applied magnetic field causes particles of the strongly magnetic phase to be elongated in the field direction (Fig. 25.15).

Ferroxdur, $BaFe_{12}O_{19}$, a magnetically hard ferrite that has a large anisotropy in magnetic moment, is also processed in a magnetic field. A fine powder of this barium ferrite is pressed in a magnetic field, which aligns the particles, and is then sintered. Additions of minor amounts of other substances often are used to aid in grain alignment and to promote compaction and sintering. These substances also serve to isolate single-domain grains and inhibit grain growth.

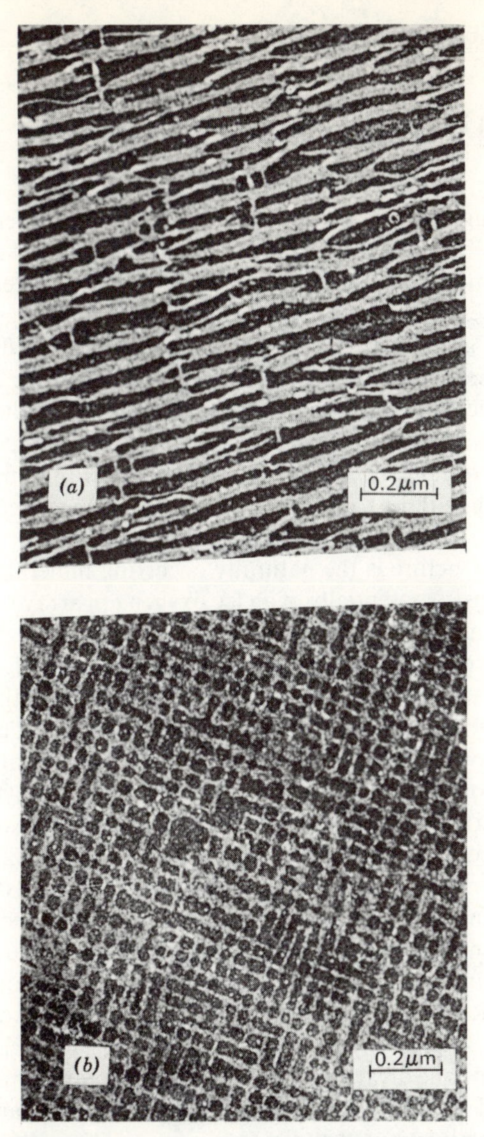

FIGURE 25.15 The microstructure of the alloy Alnico 8 after heat treating in an applied magnetic field. This process produces particles that are elongated in the direction of magnetization as shown in (*a*), which is a longitudinal section. The transverse section shown in (*b*) is normal to the direction of the applied field and shows no anisotropy of particle shape in the two directions normal to the applied field. (Photomicrographs courtesy of K. J. de Vos, Philips Research Lab.)

25.8 FERRITES

A particularly important class of ferrimagnetic materials is known as ferrites or, specifically, ferrimagnetic spinels. In the broadest sense, a ferrite is a magnetic oxide or ceramic. While ferrites generally exhibit lower initial susceptibilities and lower saturation inductions than other soft magnetic materials, they possess electrical resistivities that are many orders of magnitude higher. Consequently, eddy current losses are minimized, and ferrites can be used in devices that operate at very high frequencies, up to and including the microwave range. Most useful ferrites are magnetically soft, although some, such as Ferroxdur or barium ferrite and the structurally equivalent strontium ferrite, are used for permanent magnets.

There are three technically important classes of ceramic magnetic materials: hexagonal ferrites (like $BaFe_{12}O_{19}$), cubic ferrites with the inverse spinel structure, and cubic ferrites with the garnet structure. The prototype of cubic ferrites having the inverse spinel structure is the naturally occurring mineral magnetite, Fe_3O_4, whose structure consists essentially of an fcc oxygen ion array with Fe^{2+} and Fe^{3+} ions in the interstices. The chemical formula of magnetite may be written as $FeO \cdot Fe_2O_3$ in order to emphasize the fact that there are divalent and trivalent metal ions in the ratio of one to two. Ferrite compounds with the inverse spinel structure are similar to magnetite, differing only in that various other metal ions substitute for the iron ions. A portion of the unit cell is shown in Fig. 25.16. As with MnO, the oxygen ions have no permanent magnetic moment. Tetrahedral sites in the fcc oxygen ion array are occupied by half of the trivalent cations, and octahedral sites are occupied equally by divalent cations and the remaining trivalent cations. (For the normal spinel structure, exemplified by $MgAl_2O_4$, all divalent cations are in tetrahedral sites and all trivalent cations are in octahedral sites.)

The permanent moments of the cations in the octahedral and tetrahedral sites are antiparallel. Thus, the moments of the trivalent cations cancel out, and the net moment is due to the divalent cations. If the individual divalent cation moments are known, then it is possible to calculate the net moment per formula unit. In a formula unit of Fe_3O_4 there is one divalent iron ion that has a total moment of $4\mu_B$ resulting from four unpaired electrons, because, as often occurs in many crystals, orbital moments are randomly oriented and average to zero. Thus, the magnetic moment per formula unit of Fe_3O_4 should be $4\mu_B$, which is very close to the observed value. Similarly, in $NiFe_2O_4$ (or $NiO \cdot Fe_2O_3$), the moment per formula unit is $2\mu_B$, identical to that of the divalent nickel ion.

Some ferrimagnetic oxides have the garnet structure, which can accommodate large trivalent rare earth ions with large magnetic moments. The chemical formula for ferrimagnetic garnets is $M_3Fe_5O_{12}$, where M is a rare earth ion (Sm, Eu, Gd, etc.) or an yttrium ion. The garnet $Y_3Fe_5O_{12}$, called yttrium iron garnet (YIG), has a high electrical resistivity and, correspondingly, a very low hysteresis loss even at microwave frequencies.

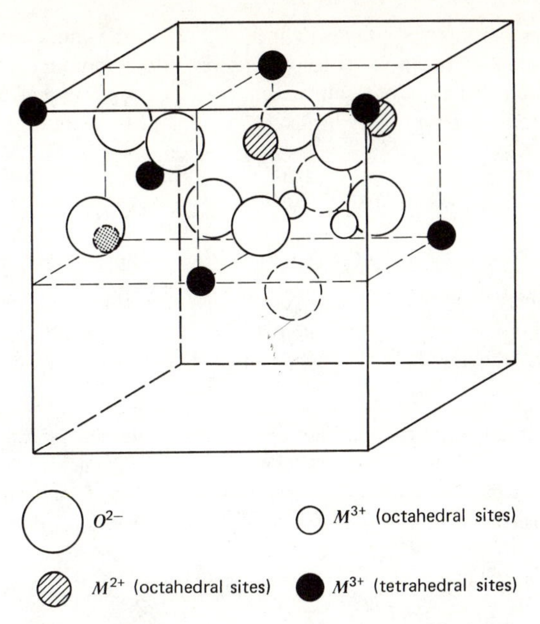

FIGURE 25.16 A portion of the inverse spinel unit cell. Oxygen ions form a face-centered cubic array. Tetrahedral sites are occupied by trivalent metal ions with moments pointing in one direction, and octahedral sites are half occupied with divalent and half with trivalent ions whose moments are antiparallel to those of the ions in tetrahedral sites.

REFERENCES

1. Keffer, F. The Magnetic Properties of Materials, *Scientific American*, Vol. 217, No. 3, September, 1967, pp. 222–234.
2. Bozorth, R. M. Ferromagnetism, *Recent Advances in Science*, New York University Press, New York, 1956.
3. Chikazumi, S. *Physics of Magnetism*, Wiley, New York, 1969.

QUESTIONS AND PROBLEMS

25.1 (a) Explain why oxygen gas (O_2) is paramagnetic whereas fluorine gas (F_2) is diamagnetic. Refer to Sect. 4.3 for the molecular electronic configurations.

(b) Based on electron spins, what do you expect the magnetic moment of an O_2 molecule to be? The observed net magnetic moment is $p_m = 2.84\mu_B$. Suggest a reason why this differs from your prediction.

(c) Calculate the magnetic susceptibility for oxygen gas at 25°C and 1 atm pressure.

25.2 (a) The molar susceptibility is defined as the molar volume times the susceptibility. Explain why the molar susceptibility provides a more meaningful measure of the magnetic behavior of a substance than the susceptibility given by Eq. 25.9.

(b) The molar susceptibility for O_2 gas at various temperatures is

T (K):	250	293	350	400	450	500
$10^8 \times \chi_{\text{molar}}$ (m^3/mol):	5.08	4.33	3.63	3.17	2.82	2.54

Does oxygen gas obey the Curie law? Justify your answer.

25.3 In a magnetic field gradient, a force acts on a paramagnetic material, tending to move it toward the region of highest magnetic field intensity, while the opposite occurs for a diamagnetic material. Briefly justify this.

25.4 A simplified analysis for determining the form of the Curie law (Eq. 25.9) can be done for N magnetic dipoles of strength p_m per unit volume. Assume that the dipoles can either line up parallel or antiparallel with respect to the field. The energy of the former is $-\mu_0 p_m H$ per dipole, and that of the latter is $+\mu_0 p_m H$ per dipole. According to the laws of thermal activation, the fraction of dipoles having spin aligned with the field is

$$f_1 = \frac{e^{\mu_0 p_m H/kT}}{e^{\mu_0 p_m H/kT} + e^{-\mu_0 p_m H/kT}}$$

and the fraction having antiparallel alignment is

$$f_2 = \frac{e^{-\mu_0 p_m H/kT}}{e^{\mu_0 p_m H/kT} + e^{-\mu_0 p_m H/kT}}$$

Consequently, the net moment per unit volume will be $Np_m(f_1 - f_2)$. Derive the approximate Curie law by assuming $\mu_0 p_m H \ll kT$ so that the exponential functions can be expanded (i.e., $e^x \approx 1 + x$ for small x).

25.5 Explain why a paramagnetic transition metal is likely to have a larger magnetic susceptibility than a paramagnetic nontransition metal.

25.6 Calculate the energy shift between the subbands ($2\mu_0 \mu_B H$) for electron spin paramagnetism in a metal (Fig. 25.4) when $\mu_0 H$ equals 0.1 T, 1 T and 50 T. ($\mu_0 H = 50$ T is approximately the most intense magnetic field that can be maintained continuously.) State your answers in both joules ($1 \text{ A} \cdot \text{m}^2 \cdot \text{T} = 1$ J) and electron volts and compare the latter to typical values of E_F.

25.7 Consider the paramagnetic susceptibility of conduction electrons in a metal, with reference to Fig. 25.4. Remember that $2\mu_0 \mu_B H \ll E_F$ in general.

(a) Show that the net number of electrons, per unit volume, having spin moments parallel to the field is given by $2N(E_F)\mu_0 \mu_B H$, where $N(E_F)$ is on a per unit volume basis. (*Hint.* For small Δa, $\int_a^{a+\Delta a} f(x)\,dx \simeq f(a)\,\Delta a$.)

(b) Show that the magnetization is given by $2N(E_F)\mu_0 \mu_B^2 H$ and, therefore, $\chi = 2N(E_F)\mu_0 \mu_B^2$.

(c) Assume we are dealing with a simple free electron energy band so that $N(E) = KE^{1/2}$ (on a per unit volume basis). Recalling that each electronic energy state is occupied by two electrons at 0 K, show that the paramagnetic susceptibility is

$$\chi = \frac{3}{2}\frac{\mu_0 n \mu_B^2}{E_F}$$

where n is the total number of valence electrons per unit volume. (*Note*. The net paramagnetic susceptibility given by Eq. 25.10 includes the above paramagnetic susceptibility as well as the diamagnetic susceptibility, $\chi = -\frac{1}{2}(\mu_0 n \mu_B^2/E_F)$, for conduction electrons.)

25.8 Ferromagnetic cobalt has a saturation value of $\mu_0 M$ equal to 1.76 Wb/m² at room temperature. What is the magnetic moment per atom in units of Bohr magnetons?

25.9 The magnetic induction may be expressed as

$$B = \mu H$$

where μ is called the permeability. The relative permeability $\mu_r = \mu/\mu_0$ is a materials property.

(a) Show that μ_r and the magnetic susceptibility are related by $\mu_r = 1 + \chi$.

(b) Plot schematically the variation of μ_r as a function of H from $H = 0$ to saturation for an initially unmagnetized ferromagnetic material. Use curve a in Fig. 25.9.

25.10 The band structure for transition metals in the first long period (except for Cr and Mn) of the periodic table is represented better by the schematic figure given below than by Fig. 25.5.

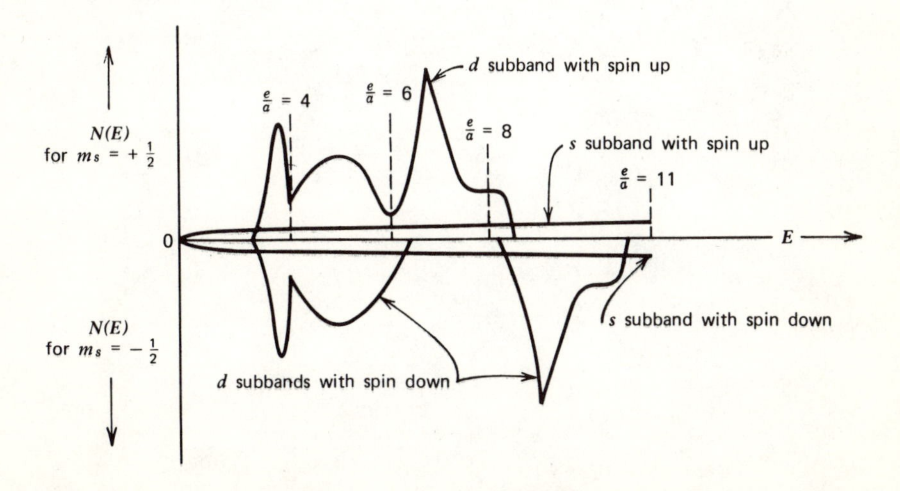

Note that the s subbands and the low-energy portion of the d subbands are symmetric, whereas the d subband portions at higher energies are not. The approximate locations of the Fermi energies are indicated for various values of valence electrons per atom (e/a).

(a) Using sketches of the above diagram and showing occupancy of the subbands, explain qualitatively why Sc, Ti, and V are paramagnetic, while Fe, Co, and Ni are ferromagnetic.

(b) Explain qualitatively why the net magnetic moments per atom are 2.22 μ_B, 1.72 μ_B, and 0.606 μ_B for Fe, Co, and Ni, respectively.

(c) Should Cu be ferromagnetic? Explain.

25.11 Iron alloyed with 30 to 35% Co can be made into a very fine powder with particle size approximately 10 nm. If the powder particles are coated with a thin layer of antimony and then pressed to high density in a magnetic field, the resulting material can be magnetized to produce a permanent magnet of very high coercivity. Explain briefly.

25.12 Laminated iron-silicon alloy sheet is commonly used for transformer cores, in preference to castings or other massive forms. Give at least two reasons for this.

25.13 How do you expect the magnetic hysteresis to vary with grain size for a single-phase ferromagnetic material? Should the coercivity increase or decrease as grain size is decreased? Explain.

25.14 Ferromagnetic materials with high coercivities usually have relatively low initial permeabilities (see problem 25.9). Explain, in terms of domain wall motion, why you would expect this.

25.15 Discuss the similarities and differences between ferromagnetic domain growth and grain growth in metals. What are the respective driving forces? How can growth be stopped or hindered in each case?

26

Dielectric Materials

In this chapter, the characteristics of materials used as electrical insulators, capacitors and as ferroelectrics are discussed.

26.1 FUNDAMENTAL CONCEPTS

At room temperature, many ceramics and polymers have electrical resistivities that are approximately 20 orders of magnitude higher than those of metals, and such materials are used as electrical insulators. Typical inorganic insulators include mica, glass, porcelain, and alumina; polymeric materials such as paper, textiles, rubber, and waxes also find widespread application. The name dielectric has long been associated with insulators in that they all exhibit an electrical dipole structure either because they are composed of polar molecules or because a dipole nature is induced by an electric field or a mechanical strain. As a result of the dipole structure, certain insulators are particularly useful as dielectrics in capacitors.

The interaction of a dielectric material with an electric field can be discussed conveniently in terms of a parallel plate capacitor. When there is a vacuum between the plates and they are subjected to an applied voltage (Fig. 26.1*a*), one plate becomes charged positively and the other becomes charged negatively. The electric field between the plates is directed from the positive to the negative plate and is given by $E = V/l$, where V is the voltage difference and l is the separation distance. [In general, E equals the negative of the potential gradient (i.e., $E = -dV/dl$).] The magnitude of charge per unit area on either plate, called the electric

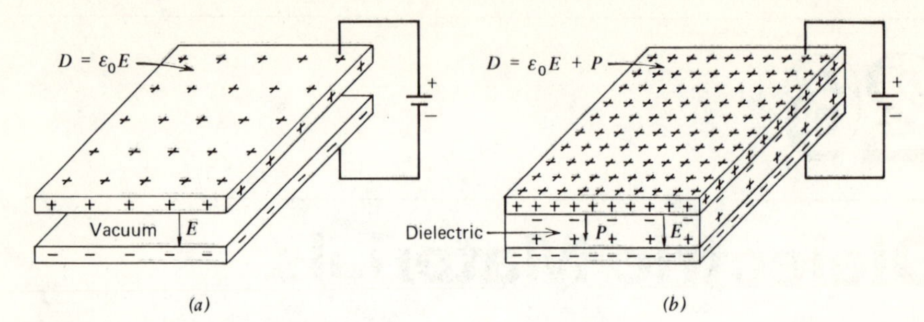

FIGURE 26.1 (*a*) A parallel plate capacitor operating in vacuum. The electric field intensity E ($=V/l$) corresponds to a surface charge density of magnitude $D = \varepsilon_0 E$.
(*b*) A parallel plate capacitor having dielectric material between the plates. If the imposed voltage and plate spacing are the same as in (*a*), D increases to compensate for the induced surface charges of the dielectric. That is, for E ($=V/l$) to remain the same, D must increase by an amount equal to the polarization P of the dielectric.

displacement, is directly proportional to the field,

$$D = \varepsilon_0 E \tag{26.1}$$

where D is measured in C/m^2, E is in V/m, and ε_0 is a constant conversion factor equal to 8.854×10^{-12} C/V · m. Formally, the relationship between D and E is similar to that between B and H, but there are physical differences.

Since D is the surface charge density on a plate, it can be related to the capacitance of a parallel plate capacitor, which is defined as

$$C = \frac{Q}{V} \tag{26.2}$$

where Q is the magnitude of charge on each plate and V is the applied voltage. Taking A as the plate area and l as the plate separation, we have $Q = DA$ and $V = El$, so that

$$C = \frac{DA}{El} = \varepsilon_0 \frac{A}{l} \tag{26.3}$$

when a vacuum exists between the plates. If a polarizable material is inserted between the plates (Fig. 26.1*b*), positive charge will accumulate on its surface toward the negatively charged plate and negative charge will accumulate toward the positively charged plate. This tends to decrease the effective surface charge density on either plate (e.g., the negative charge on one plate and the positive charge on the neighboring surface of the dielectric sum to a value smaller than the charge on the plate). The *expected* decrease in effective charge corresponds to the polarization P of the material, which equals the induced dipole moment per unit volume of polarizable material. However, because the voltage source imposes a fixed voltage between the plates and because charge carriers from the voltage source are mobile, the magnitude of charge per unit area on each plate increases until E in the capacitor reaches the same value as for a vacuum, and the relationship between E, D, and P is

$$E = \frac{1}{\varepsilon_0}(D - P) \tag{26.4}$$

As written, this equation is the electrical equivalent to Eq. 25.6 for magnetism, and E and B, D and H, and P and M play equivalent roles. It should be noted, however, that in most cases E and H are the externally controlled variables.

The surface charge density can be obtained by rearranging Eq. 26.4,

$$D = \varepsilon_0 E + P \tag{26.5}$$

and this is frequently written as

$$D = \varepsilon E \tag{26.6}$$

where ε is called the permittivity. The relative permittivity or dielectric constant, $\varepsilon_r = \varepsilon/\varepsilon_0$, is a property of the dielectric material. Since, from the above discussion, a polarizable material causes an increase in charge per unit area on the plates of a capacitor, the capacitance also increases

$$\frac{C}{C_{\text{vac}}} = \frac{D}{D_{\text{vac}}} = \frac{\varepsilon_0 E + P}{\varepsilon_0 E} = \frac{\varepsilon}{\varepsilon_0} = \varepsilon_r \tag{26.7}$$

Values of dielectric constants for a number of commonly used dielectric materials are given in Table 26.1.

26.2 TYPES OF POLARIZATION

Polarization arises from the interaction between permanent or induced electric dipoles with an applied field and varies linearly with applied field for most dielectric materials. As discussed in Sect. 25.1, an electric field acts to align a permanent electric dipole in the field direction. In addition, an electric field can induce dipoles in a substance that is not permanently polar. The total polarization or dipole

TABLE 26.1 Dielectric Constants ($\varepsilon_r = \varepsilon/\varepsilon_o$) of Some Selected Materials (T = 25°C, $\nu = 10^6$ Hz)

Plastics and Organics		*Glasses*		*Inorganic Crystalline Materials*	
Polytetrafluoroethylene (Teflon)	2.1	Silica glass	3.8	Barium oxide	3.4
Polyisobutylene	2.23	Vycor glasses	3.8–3.9	Mica	3.6
Polyethylene	2.35	Pyrex glasses	4.0–6.0	Potassium chloride	4.75
Polystyrene	2.55	Soda-lime-silica glass	6.9	Potassium bromide	4.9
Butyl rubber	2.56	High-lead glass	19.0	Cordierite ceramics (based on $2MgO \cdot 2Al_2O_3 \cdot 3SiO_2$)	4.5–5.4
Lucite	2.63			Diamond	5.5
Polyvinyl chloride	3.3			Potassium iodide	5.6
Nylon 66	3.33			Forsterite (Mg_2SiO_4)	6.22
Epoxy resins	3.5–3.6			Mullite ($3Al_2O_3 \cdot 2SiO_2$)	6.6
Polyesters	3.1–4.0			Lithium fluoride	9.0
Phenolformaldehyde	4.75			Magnesium oxide	9.65
Neoprene	6.26				
Paper	7.0				

moment associated with atoms, ions, or molecules is due to three different sources: electronic polarization (Fig. 26.2*a*), ionic polarization (Fig. 26.2*b*), and orientation polarization (Fig. 26.2*c*). Electronic polarization arises because the center of the electronic charge cloud around a nucleus is displaced under the action of an applied electric field. The resulting polarization, or dipole moment per unit volume, is

$$P_e = N\alpha_e E_{\text{loc}} \tag{26.8}$$

where N is the number of atoms, ions, or molecules per unit volume, α_e is called the electronic polarizability and E_{loc} is the local electric field at an atom. The local electric field is greater than the average field E because of the polarization of other atoms.

Ionic polarization occurs in ionic materials because an applied field acts to displace cations in the field direction and anions in a direction opposite to the field, giving rise to a net dipole moment per formula unit (Fig. 26.2*b*). For an ionic solid having equal numbers of cations and anions, the dipole moment per ion pair equals the magnitude of ionic charge, q, multiplied by the relative displacement, a, and the polarization is given by

$$P_{\text{ion}} = Nqa = N\alpha_i E_{\text{loc}} \tag{26.9}$$

where N is the number of formula units per unit volume and α_i is the ionic polarizability. The displacement distance a implicity depends on the local electric field

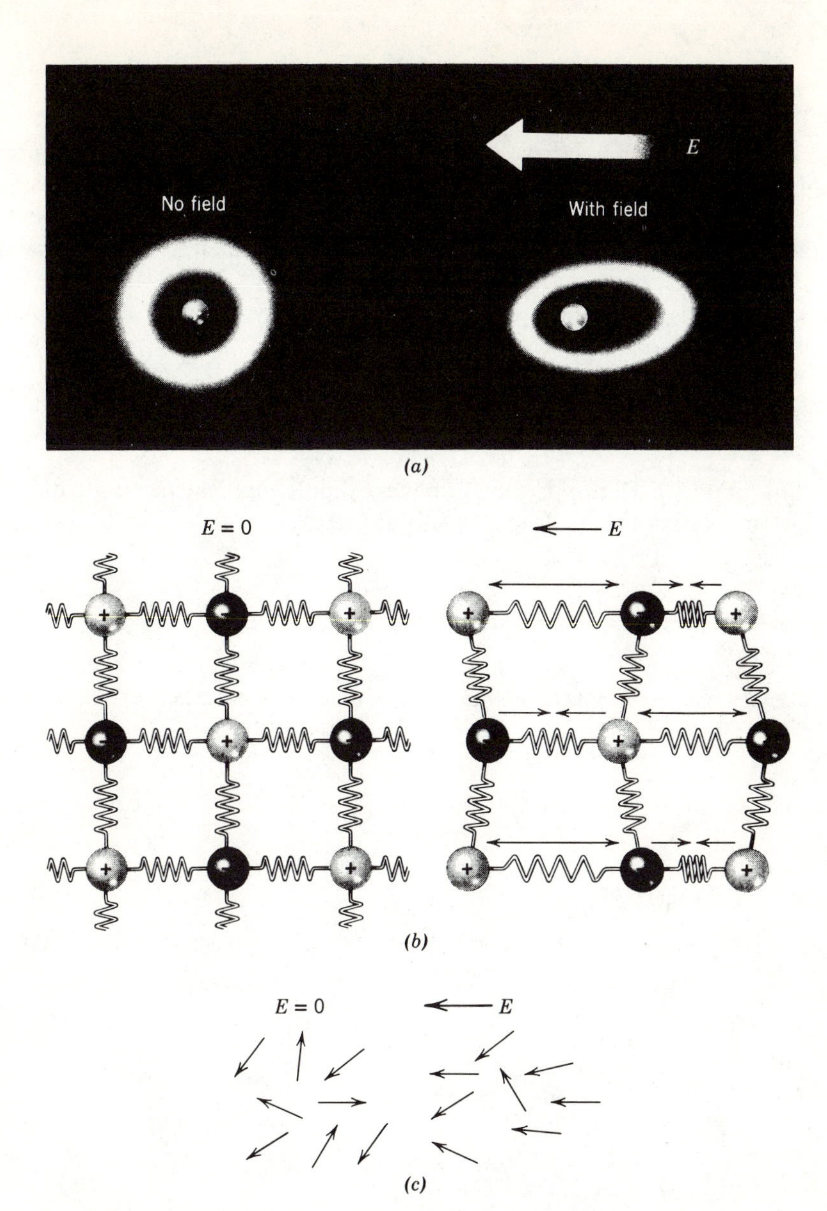

FIGURE 26.2 Types of polarization. (*a*) Electronic polarization results from the displacement of the center of negative charge relative to the center of positive charge within an atom because of the action of the local electric field.

(*b*) Ionic polarization occurs in ionic materials because cations and anions are displaced in opposite directions by the local electric field.

(*c*) Orientation polarization can occur in substances consisting of polar molecules. Under the action of an electric field, the permanent electric dipoles tend to align with the field.

(Adapted from *The Structure and Properties of Materials*, Vol. IV, by R. M. Rose, L. A. Shepard, and J. Wulff, Wiley, New York, 1966.)

at the ions. Orientation polarization (Fig. 26.2*c*) can occur in substances composed of molecules that have permanent electric dipole moments. These tend to become aligned with the applied field, but entropy effects tend to counter this. The physics and thermodynamics of this type of polarization are similar to those describing the interaction between permanent magnetic moments and a magnetic field. Consequently, the orientation polarizability is given by a relationship similar to the Curie law (Eq. 25.9),

$$\alpha_0 = \frac{p^2}{3kT} \tag{26.10}$$

providing $pE \ll kT$. Here p is the permanent dipole moment per molecule. The orientation polarization for N molecules per unit volume is

$$P_0 = \frac{Np^2 E_{\text{loc}}}{3kT} \tag{26.11}$$

TABLE 26.2 Electric Dipole Moments of Some Polar Molecules in the Gaseous Phase

Molecule	10^{30} × *Dipole Moment* (C · m)[a]
AgCl	19.1
LiCl	23.78
NaCl	30.0
KCl	34.26
CsCl	34.76
LiF	21.1
NaF	27.2
KF	28.7
RbF	27.5
CsF	26.3
HF	6.07
HCl	3.60
HBr	2.74
HI	1.47
HNO_3	7.24
H_2O	6.17
NH_3	4.90
CH_3Cl	6.24
CH_2Cl_2	5.34
$CHCl_3$	3.37
C_2H_5OH	5.64
$C_6H_5NO_2$	14.1

[a] To obtain dipole moment in Debyes (1 D = 10^{18} statcoul · cm), multiply value given by 2.9979×10^{29}.

for $pE \ll kT$. Equations 26.10 and 26.11 hold only over temperature intervals in which the molecules are free to rotate, and for a molecular solid a certain polarization (e.g., random polarization) becomes *frozen in* at a temperature below which molecular rotation becomes sluggish. A list of dipole moments for selected polar molecules is given in Table 26.2.

An additional type of polarization warrants mention. This is due to the accumulation of electrical charge at phase boundaries in two-phase and multiphase materials. Such space charge or interface polarization occurs when one of the phases has a much higher resistivity than the other, as is found in a variety of ceramic materials, particularly at elevated temperatures.

26.3 FREQUENCY AND TEMPERATURE DEPENDENCE OF POLARIZATION

The total polarization P and the dielectric constant ε_r of a dielectric material subjected to an alternating electric field depend on the ease with which the permanent or induced electric dipoles can reverse their alignment with each reversal of the applied field. The time required for dipole reversal is called the relaxation time, and its reciprocal is the relaxation frequency. Relaxation times and frequencies depend on the polarization process. When the time duration per cycle of the applied field is much less than the relaxation time of a particular polarization process (or, alternatively, if the field frequency is much greater than the relaxation frequency), the dipoles cannot change their orientation rapidly enough to remain oriented with the applied field (i.e., to remain in phase with the applied field). Thus, the particular polarization process will not contribute to the total polarization.

Because both electronic and ionic polarization involve repositioning of charges over dimensions less than the atomic diameter, these types of polarization persist to high frequencies, that is, up to the ultraviolet frequency range for electronic polarization and up to the infrared frequency range for ionic polarization. By contrast, molecular orientation polarization only persists to frequencies in the microwave frequency range since it involves reorientation on a molecular scale. Space charge polarization exists only through the power frequency range because the net motion of charges is over microscopic distances. Once the frequency of the applied field exceeds the relaxation frequency for a given polarization process, that process becomes ineffective, and measurement of the frequency dependence of total polarization permits determination of the relaxation frequencies for the individual processes (Fig. 26.3*a*). Such measurements also separate the relative contribution that each process makes to the total polarization (Fig. 26.3*a*).

The temperature dependence of polarization strongly depends on the molecular, ionic, or atomic arrangements in a substance as well as the dominant polarization

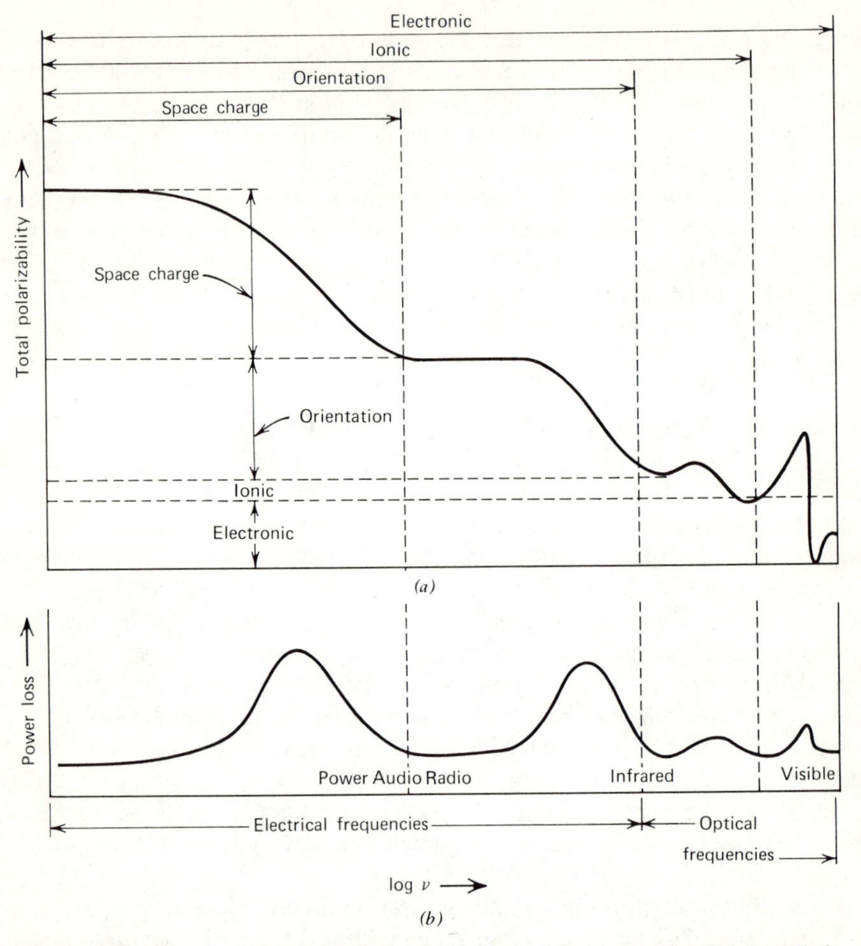

Figure 26.3 (*a*) Total polarizability as a function of the frequency of the applied field. When the frequency of the applied field exceeds the characteristic frequency of a particular polarization process, that process no longer contributes to the total polarization. Depending upon the dielectric material, some polarization processes shown may be absent. Qualitatively, the dielectric constant displays the same frequency dependence as total polarizability.

(*b*) Power loss in a dielectric material as a function of applied field frequency. The loss is at a maximum when the applied frequency equals the relaxation frequency of a particular polarization process. [After E. J. Murphy and S. O. Morgan, *Bell System Tech. J.,* **16**, 493 (1937). Reprinted with permission from *The Bell System Technical Journal*, copyright 1937, the American Telephone and Telegraph Co.]

mechanism. Nitrobenzene ($C_6H_5NO_2$), for example, is a large molecule with a permanent dipole moment that should lead to a large orientation polarization. At temperatures below its melting point (5.7°C), however, molecular motion is hindered by the shape and large size of the molecules, and the permanent dipoles cannot reorient themselves in an alternating field. Thus, the relaxation time for orientation polarization is very large for solid nitrobenzene, and only the temperature-insensitive electronic polarization occurs (Fig. 26.4). When nitrobenzene melts, the individual molecules become much more mobile and the relaxation time decreases so that orientation polarization dominates. As shown in Fig. 26.4, a discontinuous change in the dielectric constant takes place at the melting point. With further increases in temperature above T_m, ε_r decreases (recall Eq. 26.11). HCl, like nitrobenzene, also possesses a permanent dipole moment. Since it is a

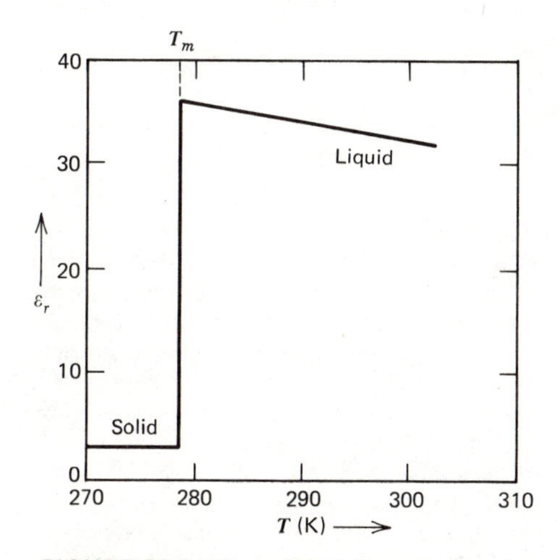

FIGURE 26.4 The dielectric constant for nitrobenzene ($C_6H_5NO_2$) as a function of temperature. Even though nitrobenzene molecules have permanent dipole moments, they are so bulky that there is limited mobility with respect to reorientation in the solid state; thus, orientation polarization is essentially absent below the melting point. Above the melting point, molecular mobility is greatly increased, and orientation polarization provides the dominant contribution to ε_r. With increasing temperature, the degree of orientation polarization decreases according to a relationship similar to the Curie law (see Eq. 25.9). [After C. P. Smyth and C. S. Hitchcock, *J. Am. Chem. Soc.*, **54**, 4631 (1932).]

much smaller molecule, however, it remains rather mobile below its melting temperature, and orientation polarization persists for solid HCl to well below its melting point (Fig. 26.5).

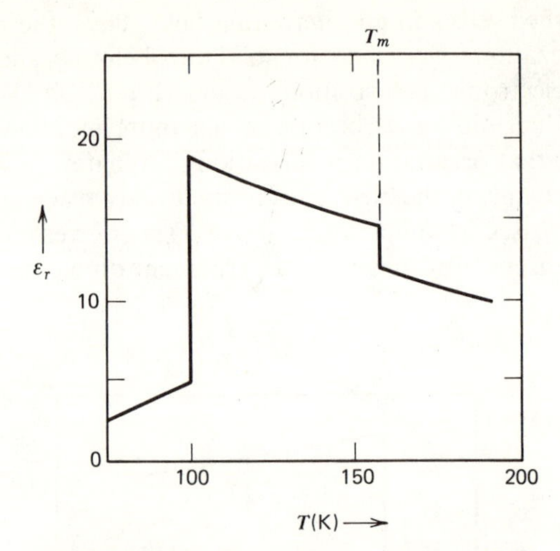

FIGURE 26.5 The dielectric constant of HCl as a function of temperature. An HCl molecule is much smaller and more compact than a nitrobenzene molecule. Orientation polarization persists at temperatures well below the melting point of HCl because molecular mobility with respect to reorientation remains large in the crystalline solid. The discontinuous change in dielectric constant on cooling at 98 K is associated with a polymorphic transformation from the fcc structure (with free molecular rotation) to a more dense structure of lower symmetry, in which molecular rotation is restricted. [After C. P. Smyth and C. S. Hitchcock, *J. Am. Chem. Soc.*, **55**, 1830 (1933).]

In many ionic solids and some having both covalent and ionic natures, polarization increases with temperature, provided that the frequency of the applied field is not too high. This effect is due to the increased ionic mobilities at higher temperatures, which effectively reduce the relaxation time. In the case of soda-lime-silica glass, for example, the motion of Na^+ ions, located in the holes of the silica network, provides the polarization. For crystalline ionic substances, motion of Schottky or Frenkel defects accounts for the increasing dielectric constant with increasing temperature.

26.4 POWER LOSSES, DIELECTRIC STRENGTH, AND USES OF DIELECTRICS

As discussed previously, if the frequency of an applied field is much greater than the relaxation frequency for a given polarization process, then that particular process will be inoperative. Conversely, if the field frequency is much less than the relaxation frequency, the polarization will vary instantaneously with the applied field, remaining in phase with it. As the frequency of the applied field approaches the relaxation frequency, the polarization response increasingly lags behind the applied field. In such cases, a plot of D (or P) against E will exhibit a hysteresis loop representing energy dissipation analogous to that observed in anelasticity. The area enclosed within a D vs. E hysteresis loop represents the energy loss in a unit volume of material per cycle; multiplication of this by the frequency gives the power loss. The dissipated energy serves to heat the material and, if not transported away, can lead to unstable operation. As shown in Fig. 26.3*b*, this type of power loss only becomes appreciable at frequencies in the vicinity of the relaxation frequencies of the individual polarization processes. Dielectric materials may be classified on the basis of power losses as either high-loss or low-loss materials, with the latter being generally more desirable for use in the critical frequency ranges. Typical high-loss materials are polar organic substances and some ceramics with high dielectric constants such as barium titanate.

The power loss depends on the degree to which the polarization lags behind the electric field. This lag can be quantified by measurement of the D–E relationship in a sinusoidally varying field; it is found that if E is given by $E_0 \sin \omega t$, then P is given as $P_0 \sin (\omega t - \delta)$, where δ is the phase angle that the polarization lags behind the field and ω is the angular frequency of the field. Integration of the area enclosed within a hysteresis loop shows that the power loss is directly proportional to the lag angle δ, provided δ is not too great. The propensity for power losses in dielectrics is, therefore, quantified by measurement of the lag angle δ; since δ is frequency dependent, the frequency must be stated when specifying δ.

In addition to power loss, electrical resistivity, and the dielectric constant, another important operating parameter is the maximum voltage gradient that a dielectric can sustain before electric breakdown or failure occurs. At failure, the dielectric no longer performs as a viable insulator. The variation of breakdown voltage or dielectric strength with electrode separation distance for several materials is shown in Fig. 26.6. The approximately linear relationship observed in many cases indicates that the voltage gradient (i.e., the electric field strength) determines the dielectric strength.

Breakdown probably is initiated by the field-induced excitation of a number of electrons from impurity levels to the conduction band. After excitation, the high field accelerates these electrons to kinetic energies that are sufficient to knock other electrons into the conduction band. The result is an avalanche effect in

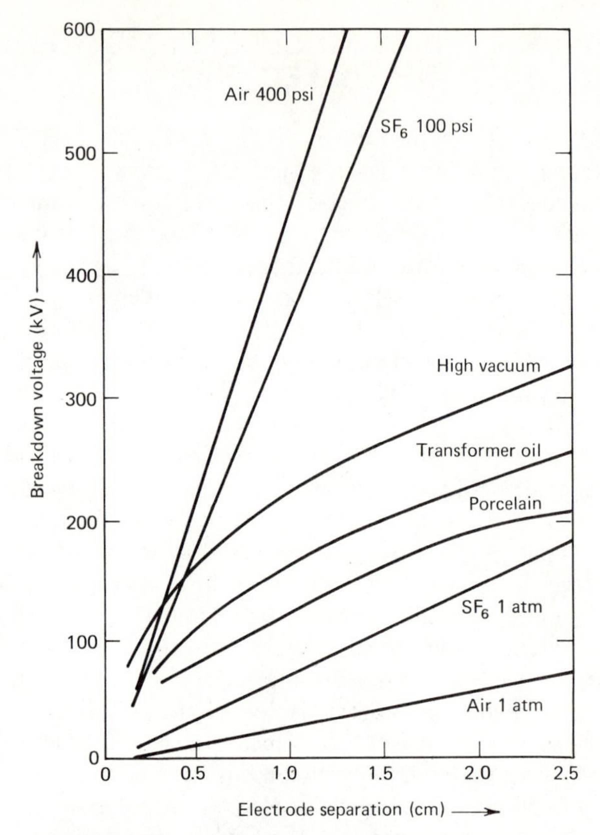

FIGURE 26.6 Static breakdown voltage versus electrode separation (i.e., dielectric thickness) for selected dielectric substances. The ratio of breakdown voltage to electrode separation is approximately constant for a given dielectric substance. (Adapted from J. G. Trump, in *Dielectric Materials and Applications,* edited by A. R. von Hippel, Technology Press and Wiley, New York, 1954. Reprinted by permission of the M.I.T. Press, Cambridge, Mass.)

which current through the dielectric increases greatly and, because of the inherently high electrical resistivity of the material, causes extensive heating. In liquid dielectrics, this heating may lead to localized burning or vaporization; in solid dielectrics, localized melting may occur.

Many dielectrics that have been in service for a period of time may have become contaminated by impurities or mechanically deformed so that localized arcing (i.e., *lightning* on a small scale) occurs because of the high electric field. Since arcing can initiate dielectric breakdown, insulating substances are changed regularly. This is especially true of oil insulation used in high-voltage transformers.

The selection of a dielectric material depends on various factors, including dielectric constant, operating temperature, voltage and frequency, and operating environmental conditions. If the required dielectric constant is less than about 10, insulators such as cotton, silk, paper, and many other polymers and liquids are suitable to temperatures of about 100°C. When higher temperatures and/or breakdown voltages are required, mica, glass, porcelain and similar inorganic materials are useful. Generally, ceramic materials exhibit the best dielectric properties.

26.5 CERAMIC DIELECTRICS

In addition to their value as high-quality dielectric materials, ceramic dielectrics also afford an opportunity to amplify the relationship between structure, processing, and properties. The base components of most ceramic dielectrics are silica (SiO_2), alumina (Al_2O_3), and magnesia (MgO). Some of the common dielectric ceramics and their properties are listed in Table 26.3. Porcelain ceramics additionally contain appreciable amounts of potash feldspar ($K_2O \cdot Al_2O_3 \cdot 6SiO_2$) that is added to improve the ease of manufacture since, as the amount of added feldspar increases, both the amount of liquid formed at the vitrification temperature increases and the temperature at which vitrification is produced decreases. The raw components of insulating porcelain are, in addition to feldspar (typically 35%), dehydrated talc ($3MgO \cdot 4SiO_2$, typically 2%), dehydrated clay ($Al_2O_3 \cdot 2SiO_2$, typically 38%) and raw silica (typically 25%). During the firing, but prior to vitrification, the clay decomposes into mullite and silica according to the reaction,

$$3(Al_2O_3 \cdot 2SiO_2) \longrightarrow 4SiO_2 + 3Al_2O_3 \cdot 2SiO_2\text{(mullite)} \qquad (26.12)$$

and the final microstructure consists of silica particles embedded in a glassy matrix that contains fine crystals of mullite. The ease and economy of manufacture of porcelain ceramics renders them attractive as inexpensive dielectric materials; however, they generally are characterized by relatively high loss factors, a consequence that is related directly to the presence of alkali metal ions in the glassy phase, and a penchant for water absorption. Furthermore, porcelain dielectrics are limited to relatively low temperatures (Table 26.3). A typical use, therefore, is as power line (i.e., low-frequency) insulation. Because of the above considerations, a wide variety of other ceramic dielectrics are prepared that do not contain the feldspar component. Most of these materials are named after prototype compositions; thus, forsterite ceramics are based on the mineral forsterite ($2MgO \cdot SiO_2$), low loss steatites are based on enstatite ($MgO \cdot SiO_2$), commercial steatites are based on dehydrated talc, and cordierite ceramics are based on the mineral cordierite ($2MgO \cdot 2Al_2O_3 \cdot 5SiO_2$). In all cases, however, the actual compositions of these ceramics are nonstoichiometric, that is, they do not correspond to the exact chemical formula, and are not single phase. Steatites are the most economical to produce since they contain a large proportion of talc that is soft and easy to form. Forsterite ceramics are especially useful when dielectric losses must be kept to a minimum (Table 26.3), and cordierite ceramics have exceptionally good thermal shock resistance because of their low coefficients of thermal expansion.

The composition ranges of these ceramics can be illustrated with the aid of a diagram of the type shown in Fig. 26.7. In distinction to a binary alloy, in which the composition can be designated along a line, compositions in ternary alloys are delineated by the relative position within or along the edges of a triangle of the type shown in Fig. 26.7. Here, the corners of the triangle correspond to the pure components, Al_2O_3, SiO_2, and MgO, respectively. An edge of the triangle, for example, between the corner corresponding to Al_2O_3 and that corresponding to SiO_2, corresponds to compositions containing varying proportions of Al_2O_3 and SiO_2 but no MgO. A ternary composition (i.e., one containing all three components) must lie within the triangle. Forsterite compositions, for example, are based on the binary composition, $2MgO \cdot SiO_2$, but generally contain some Al_2O_3 since the shaded region corresponding to forsterite compositions extends into the triangle. Commercial steatites typically contain about 90% talc and 10% clay. Thus, the range of steatite compositions, as shown in Fig. 26.7, lies approximately one-tenth of the distance along a line connecting the talc and clay compositions. The ease with which these ceramic materials can be processed depends on the characteristics of the vitrification process. Successful vitrification requires that some but not too much liquid form during firing, and it is also desirable that the amount of liquid present does not vary greatly with temperature so that precise control over firing temperature is not necessary. The ease of vitrification of the alloys whose compositions correspond to the points *A*, *B*, *C*, and *D* in Fig. 26.7 can be ascertained with reference to Fig. 26.8. Here the relative amount of liquid for ceramics of these different compositions is shown as a function of temperature. For all except the forsterite ceramic, a eutectic occurs at about 1350°C. Cordierites (composition *C*) are difficult to vitrify since the amount of liquid is large at all temperatures above the eutectic and because the amount of liquid increases rapidly over a narrow temperature range. This necessitates precise temperature control during vitrification. In practice this problem is obviated somewhat by small additions of other substances that widen the allowed firing range. Low-loss steatites (composition *B*) and commercial steatites (composition *A*) are somewhat difficult to vitrify for the same reasons. Thus, while commercial steatites are easily cold-formed, they must be fired with care. As with cordierites, additions frequently are made to help circumvent the vitrification problem. The

TABLE 26.3 Some Properties of Typical Ceramic Dielectrics

	A. *Vitrified Products*			
	High-Voltage Porcelain	*Alumina Porcelain*	*Steatites*	*Forsterites*
Typical Application	*Power Line Insulation*	*Spark Plug-Cores, Protection Tubes*	*High Frequency Insulation*	*High Frequency Insulation*
Maximum safe operating temperature (°C)	1000	1350–1500	1000–1100	1000–1100
Tensile strength (MN/m^2)	20–55	55–205	55–70	55–70
Dielectric strength (kV/m)	10,000–16,000	10,000–16,000	8000–14,000	8000–12,000
Dielectric constant	6–7	8–9	5.5–7.5	6.2
δ (at 10^6 Hz)	0.006–0.010	0.001–0.002	0.0008–0.0035	0.0003

	B. *Semivitreous and Refractory Products*			
	Low-Voltage Porcelain	*Cordierite Refractories*	*Alumina and Alumina Silicate Refractories*	*Massive Fired Talc*
Typical Application	*Light Receptacles*	*Arc Chambers*	*High Temperature Insulation*	*High Frequency Insulation*
Maximum safe operating temperature (°C)	900	1250	1300–1700	1200
Tensile strength (MN/m^2)	10–20	7–25	5–20	17
Dielectric strength (kV/m)	2000–4000	2000–4000	2000–4000	3000–4000
Dielectric constant	6–7	4.5–5.5	4.5–6.5	5–6
δ (at 10^6 Hz)	0.010–0.020	0.004–0.010	0.0002–0.010	0.0008–0.010

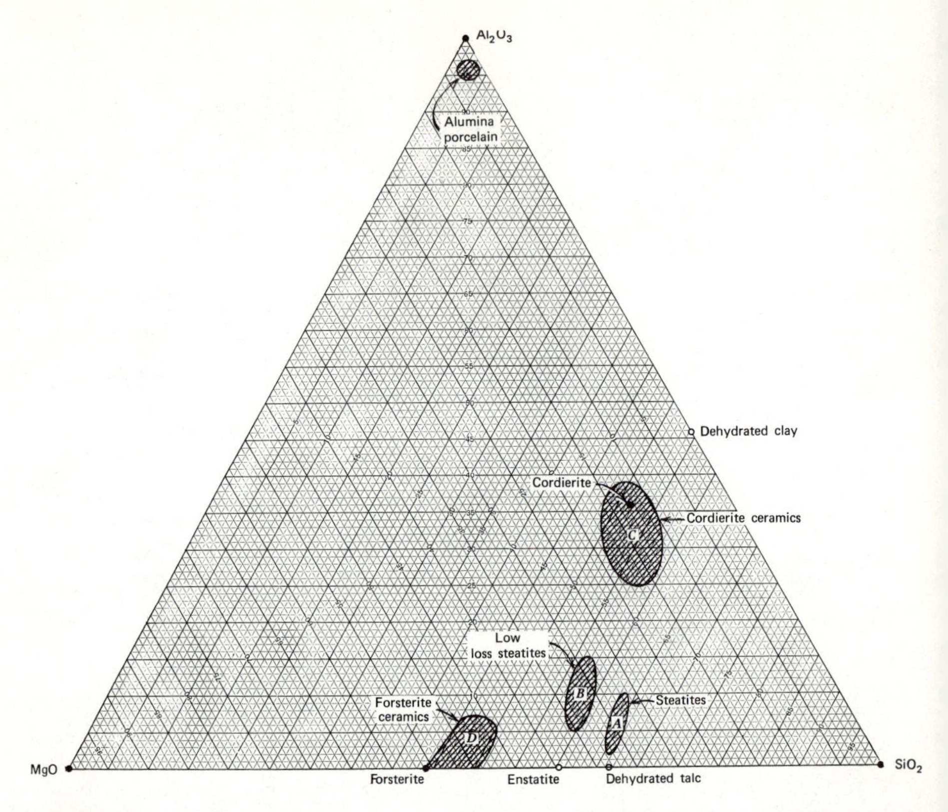

FIGURE 26.7 A ternary diagram showing the composition ranges of commercial dielectric materials based on SiO_2, Al_2O_3, and MgO. The amount of each component in weight percent varies linearly along the respective altitude of the triangle. The locations of selected crystalline phases are denoted by dots (●) and the compositions of dehydrated talc and clay, which are not stable crystalline phases, are denoted by circles (○). Forsterite ceramics consist of forsterite (Mg_2SiO_4) crystals in a glassy matrix; steatities and low-loss steatites consist of enstatite ($MgSiO_3$) and cordierite ($2MgO \cdot 2Al_2O_3 \cdot 5SiO_2$) in a glassy matrix; cordierite ceramics consist of cordierite in a glassy matrix. The raw materials for producing these ceramics include talc, clay, silica and, in some cases, magnesia or alumina. In alumina porcelains, vitreous material binds Al_2O_3 particles together. (Adapted from *Introduction to Ceramics*, by W. D. Kingery, Wiley, New York, 1960.)

forsterites (composition *D*) are vitrified easily because the liquid content at the eutectic is not so great and, more important, because the amount of liquid phase increases less drastically with increasing temperature.

While the above ceramic materials are generally considered to be excellent dielectrics, if high dielectric constants are required or, in particular, if permanent polarization is required, only ferroelectrics are suitable.

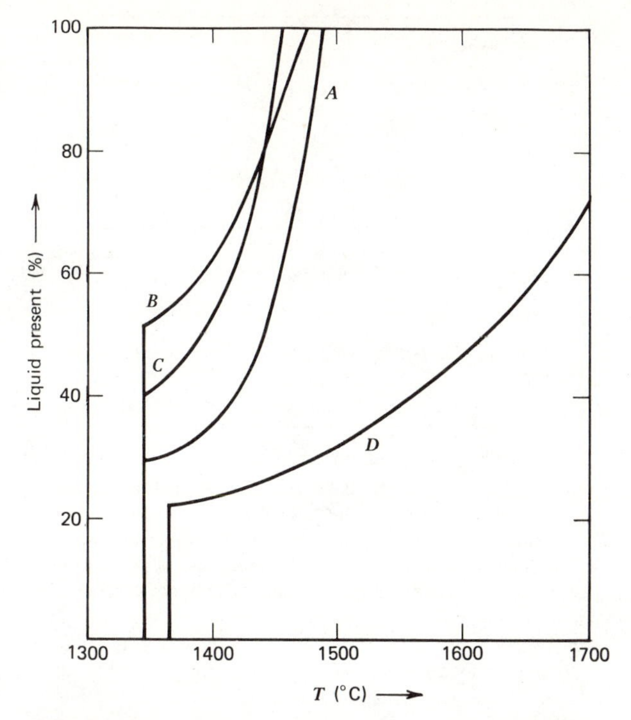

FIGURE 26.8 The fractional amount of liquid present as a function of temperature for compositions *A*, *B*, *C*, and *D* shown in Fig. 26.7. Only forsterite ceramics (composition *D*) can be vitrified easily because the amount of liquid that forms initially at the eutectic temperature is not too large and the amount of liquid with increasing temperature is relatively moderate. Steatites (*A*), low-loss steatites (*B*), and cordierite ceramics, on the other hand, require close temperature control for vitrification. (From *Introduction to Ceramics,* by W. D. Kingery, Wiley, New York, 1960.)

26.6 FERROELECTRICITY AND PIEZOELECTRICITY

Some dielectrics are called ferroelectrics because they exhibit a relationship between electric displacement (or polarization) and electric field (Fig. 26.9) that is similar to the relationship between induction (or magnetization) and magnetic field. The hysteresis characteristics of a ferroelectric material exist down to essentially static frequencies and, thus, have a different origin than the hysteresis discussed in Sect. 26.4. In almost all regards, the former is directly analogous to the hysteresis displayed by a ferromagnetic or ferrimagnetic material. In analogy with

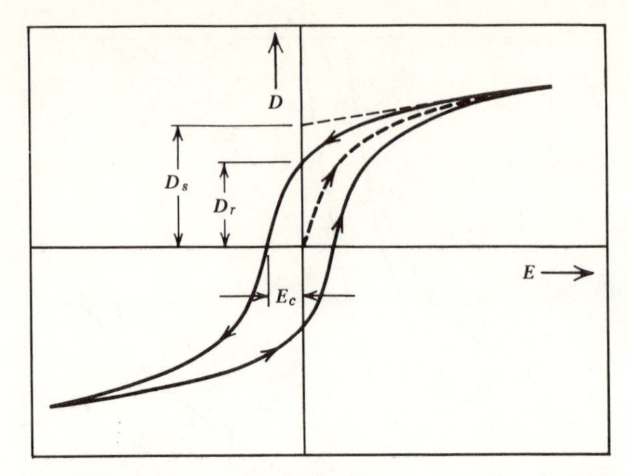

FIGURE 26.9 Electric displacement versus electric field curve for a ferroelectric material. The broken line represents the initial polarization, starting with the material in the unpolarized state (i.e., having no macroscopic polarization). If the field is removed after polarization saturation has been produced, a remnant displacement (or remnant polarization, since $D = P$ when $E = 0$) remains. This can be removed by the application of an opposite field having a magnitude equal to the coercive field E_c. (Compare this figure to Fig. 25.9 for a ferromagnetic material.)

magnetism, the spontaneous polarization remaining upon removal of the applied electric field is called the remnant polarization P_r, and the field required to depolarize a ferroelectric is called the coercive field E_c. A ferroelectric domain structure also is a feature of ferroelectric materials (Fig. 26.10).

One of the most important ferroelectrics is barium titanate ($BaTiO_3$). The spontaneous polarization of $BaTiO_3$ is related directly to the positions of the ions within its tetragonal unit cell (Fig. 26.11*a*). Divalent barium ions occupy each corner position, and the quadrivalent titanium ion is near but just above the center of the cell. The oxygen ions are located just below the centers of (001) planes and just below the centers of (100) and (010) planes. (Above and below are with reference to an [001] direction; see Fig. 26.9*b*.) Consequently, per unit cell, the centers of positive and negative charge do not coincide, and there is a spontaneous polarization (Fig. 26.11*b*). Within a given ferroelectric domain, the unit cells are oriented similarly, but in the unpolarized state the domains are arranged such that the net macroscopic polarization is zero. Application of an electric field leads to polarization in much the way that magnetization develops when a ferromagnetic material is subjected to an applied magnetic field. At the ferroelectric Curie temperature, barium titanate becomes cubic, with Ti^{4+} located at the center of the unit cell and O^{2-} ions occupying face-centered positions, and it loses its spontaneous polarization.

FIGURE 26.10 Ferroelectric domain structure in barium titanate. A large single domain exists over much of the upper and left side of the field of view. Within this domain are fine, needle-shaped domains having differently oriented permanent polarizations than the large domain. When $BaTiO_3$ is in the unpolarized condition, the dipole moments of all domains sum to zero and there is no macroscopic permanent polarization. Application of an electric field favors certain domain orientations, but, unlike the magnetization of a ferromagnetic material where domain walls migrate, more and more new needle-shaped domains nucleate at grain boundaries and grow in their long direction until saturation is reached in a ferroelectric material. [From P. W. Forsbergh, Jr., *Phys. Rev.*, **76**, 1187 (1949).]

Ferroelectric materials can be used for information storage. An applied field of $+E$ or $-E$, sufficient to cause saturation, produces a polarization of $+P_r$ or $-P_r$ that remains upon removal of the field. The remnant polarization can be sensed subsequently by appropriate electronic circuitry. Unfortunately, ferroelectric materials cannot be used for long-time memory storage because, in the absence of an electric field, the material depolarizes at a slow rate.

Ferroelectrics, including barium titanate and Rochelle salt ($KNaC_4H_4O_6 \cdot 4H_2O$), and some other dielectrics like quartz display a spontaneous polarization when subjected to a mechanical stress. These piezoelectric materials can serve as transducers, devices that convert electrical energy into mechanical energy and vice versa. Such devices are used in microphones, phonograph pickups, and equipment for measuring mechanical strain as well as in a variety of other applications. The origin of piezoelectricity can be illustrated with reference to barium titanate. As shown in Figs. 26.12*a* and *b*, application of pressure to $BaTiO_3$

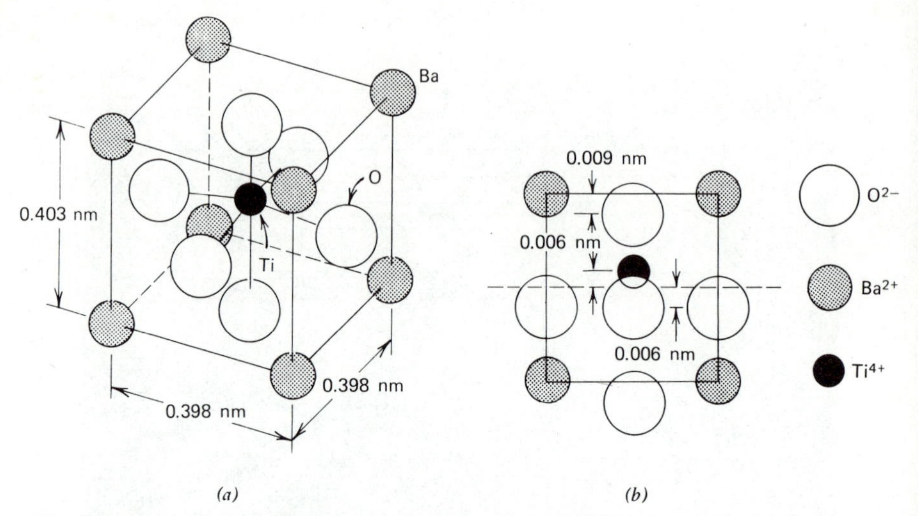

FIGURE 26.11 (*a*) The tetragonal unit cell of $BaTiO_3$. Barium ions occupy the unit cell corners, a titanium ion is *near* the center of the cell, and oxygen ions are *near* the centers of the faces. As shown, the tetragonality is rather small. Above the ferroelectric Curie temperature (about 125°C) the Ba^{2+}, Ti^{4+}, and O^{2-} ions shift to symmetric positions and the unit cell becomes cubic with no spontaneous polarization.

(*b*) A view of ionic positions projected in a [100] direction onto a (100) face of $BaTiO_3$. Because the ions are displaced with respect to symmetric positions, there is an offset between the center of positive charge and the center of negative charge and, consequently, there is a net electric dipole moment per unit cell.

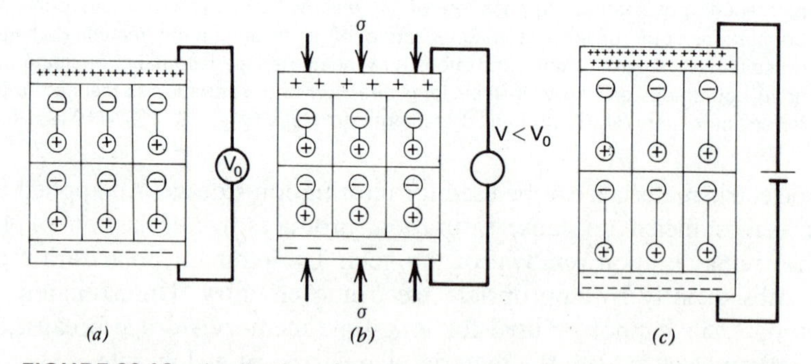

FIGURE 26.12 Application of pressure to a piezoelectric material between two plates of a capacitor changes the net dipole moment [i.e., from what is shown in (*a*) to what is shown in (*b*)], and thus the voltage between the plates changes.

An externally applied voltage (or electric field) alters the net dipole length in a piezoelectric material (*c*), and thus causes a dimensional change. (Adapted from L. H. Van Vlack, *Materials Science for Engineers,* Addison-Wesley, Reading, Mass., 1970.)

changes the net dipole moment per unit cell. Similarly, as shown in Fig. 26.12*c*, application of an electric field produces a net dimensional change in the unit cell because of the increased polarization. In the first case, the change in polarization can be calibrated with respect to the applied pressure, thereby yielding a device for measuring pressure or force. In the second case, the dimensional change is related to the applied field and, after calibration, can be used to measure electric field.

REFERENCES

1. Smyth, C. P. *Dielectric Behavior and Structure*, McGraw-Hill, New York, 1955.
2. Von Hippel, A. R. (editor). *Molecular Science and Molecular Engineering*, Technology Press, M.I.T. and Wiley, New York, 1959.

QUESTIONS AND PROBLEMS

26.1 The electric susceptibility κ can be defined in a way analogous to that for the magnetic susceptibility (Chapter 25); that is, $P = \varepsilon_0 \kappa E$.
(a) Show that $\kappa = \varepsilon_r - 1$.
(b) The local electric field that induces polarization in a gas is essentially the applied field. What is the relationship between κ and the total polarizability per unit volume for a gas consisting of nonpolar molecules? For a gas consisting of polar molecules?

26.2 Given a parallel plate capacitor with plates 10 cm × 25 cm and a dielectric material of dielectric constant equal to 5 filling the 3 cm spacing between plates.
(a) Calculate the capacitance.
(b) Calculate the charge on the plates if the applied voltage is 1000 V.
(c) What is the value of electric displacement D inside the dielectric?

26.3 The index of refraction of light is defined as the ratio of the velocity of light in vacuum, c, to the velocity of light in a substance, v; that is, $n = c/v$. In vacuum, $c = 1/\sqrt{\mu_0 \varepsilon_0}$, where $\mu_0 = B_0/H$ and $\varepsilon_0 = D_0/E$; and $v = 1/\sqrt{\mu\varepsilon}$ in a substance, where $\mu = B/H$ and $\varepsilon = D/E$. Show that, in the case of nonmagnetic materials (i.e., $\mu \simeq \mu_0$), $n^2 = \varepsilon_r$ for optical frequencies.

26.4 In a substance consisting of polar molecules that are free to rotate, the total polarization is related to the sum of the electronic, atomic, and orientation polarizabilities. (Here, the atomic polarizability is comparable to the ionic polarizability in that it results from a small change in relative positions of the nuclei caused by the applied field.) Only the orientation polarizability varies with temperature (see Eq. 26.10).
(a) Schematically plot $P/\varepsilon_0 E$ vs. reciprocal absolute temperature, and identify which features of your curve can be used to obtain the individual polarizabilities of the substance.
(b) If, at some low temperature, the polar molecules become immobile (i.e., incapable of rotation), how will the curve in part (a) be changed? Sketch $P/\varepsilon_0 E$ vs. $1/T$ for this case. (*Note.* The electric field is applied at each temperature and is not applied while temperature changes are made.)

26.5 The dielectric constant for water between 0°C and 100°C is given by the following empirical equation:

$$\varepsilon_r = \frac{4.655 \times 10^{-19}}{kT} - 34.8$$

(a) In terms of the nature of water (see Sect. 3.11), suggest why the observed temperature dependence of ε_r is greater than that based on Eq. 26.11 [i.e., essentially $(Np^2/3\varepsilon_0 kT)$]. Use $p = 6.17 \times 10^{-30}$ C · m for an isolated H_2O molecule.

(b) When water freezes to form ice at 0°C, the dielectric constant decreases discontinuously, whereas when HCl solidifies, ε_r increases discontinuously (see Fig. 26.5). Explain. (*Note.* For both H_2O and HCl, dipole reorientation occurs relatively easily in the crystalline form near the melting point.)

26.6 Neoprene is a copolymer of chloroprene and styrene. The molecular chains of neoprene contain the following basic units:

$-CH_2-CCl{=}CH-CH_2-$; $-CH_2-CH-$ (with a $-CCl-CH_3$ side group on the CH); $-CH_2-CH-$ (with a benzene ring, C_6H_5, on the CH)

Explain why neoprene has such a large dielectric constant (Table 26.1).

26.7 In your own terms, explain why the local electric field in a polarized medium should be different than the applied field.

26.8 The temperature variation of the static dielectric constant for some gases is shown below.

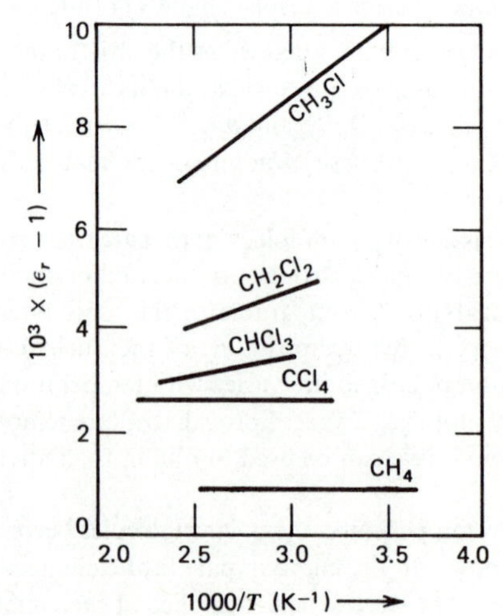

(a) Why do the dielectric constants for CCl_4 and CH_4 not vary with temperature?
(b) What type of polarization occurs in CCl_4 and CH_4?
(c) Why is ε_r greater for CCl_4 than for CH_4?
(d) What type of polarization leads to the inverse temperature dependence of ε_r shown in CH_3Cl, CH_2Cl_2, and $CHCl_3$?
(e) Why is the temperature variation of ε_r greater for CH_3Cl than for $CHCl_3$?

26.9 The variation of electric susceptibility ($\kappa = \varepsilon_r - 1$) with frequency of the applied field can be illustrated in a simple manner by considering the electronic polarization. If we let x be the displacement of a bound electron with respect to the center of positive charge of an atom, the equation of motion of the electron is given approximately as

$$\frac{d^2x}{dt^2} + \omega_0^2 x = \frac{-eE_0}{m_e} \sin \omega t$$

where ω_0 is the natural angular frequency of the electron about the nucleus, $-e$ and m_e are the electronic charge and mass, respectively, and the alternating field, $E = E_0 \sin \omega t$, is applied in the x direction and has an angular frequency ω ($= 2\pi\nu$).

(a) Assume a solution to the above equation of the form $x = x_0 \sin \omega t$. Show that such a solution exists, provided

$$x_0 = -\frac{eE_0}{m_e}\left(\frac{1}{\omega_0^2 - \omega^2}\right)$$

(b) Approximating the induced dipole moment per electron as the product $-(ex)$, show that the polarization P is given by

$$P = \frac{Nze^2}{m_e}\left(\frac{1}{\omega_0^2 - \omega^2}\right)E_0 \sin \omega t$$

where z is the number of electrons per atom and N is the number of atoms per unit volume.

(c) Schematically sketch $\kappa = P/\varepsilon_0 E$ as a function of $\log \omega$. Be sure to cover the frequency range from $\omega \ll \omega_0$ to $\omega \gg \omega_0$. Based on this approximate approach, your result will show that P is in phase with E when $\omega < \omega_0$, and P and E are out of phase when $\omega > \omega_0$. [The singularity you observe at $\omega = \omega_0$ is a consequence of the simplified model used and does not occur in actuality. However, a maximum and a minimum are observed to either side of ω_0 (see Fig. 26.3*a*). This behavior, when ω approaches ω_0, is called resonance and is observed frequently in a wide variety of physical phenomena involving the interaction between a substance and a time varying energy field.]

26.10 In an actual dielectric substance subject to an applied alternating electric field (i.e., $E = E_0 \sin \omega t$, where $\omega = 2\pi\nu$), the polarization and, therefore, the electric displacement may lag behind the applied field. This can be represented mathematically by introducing a lag angle δ such that the time variation of electric displacement is given by $D = \varepsilon_0 \varepsilon_r E_0 \sin(\omega t - \delta)$. When a substance is electrically polarized by application of a field, the work performed per unit volume of dielectric is $\int E\, dD$.

(a) Sketch the D vs. E cycle for δ equal to 0°, 10°, and 20°. In your sketch, indicate the direction of the path with increasing time.

(b) Integrate $E\,dD$ over one cycle (i.e., from $\omega t = 0$ to $\omega t = 2\pi$) and show that the energy loss per cycle associated with a unit volume of dielectric equals $\pi\varepsilon_0\varepsilon_r E_0^2 \sin\delta$.

26.11 When alkali metal ions such as Na^+, K^+, or Rb^+ are added to fused silica, the glass that forms has a higher electrical conductivity and the dielectric energy loss per cycle increases.

(a) Explain the above in terms of structure (see Chapter 7).

(b) A Na^+ ion has a greater mobility than K^+ or Rb^+ ions. How would the loss factor of a silicate glass be affected if Rb^+ ions are substituted for some of the Na^+ ions? Explain briefly.

26.12 Discuss, in your terms, the similarities and differences between dielectric breakdown and the behavior of a Zener diode under a large reverse bias (see Chapter 23).

26.13 The usefulness of a dielectric material as an electrical insulator or in a capacitor very often is limited by electrical conduction on its surface.

(a) Qualitatively explain why surface conductivity might be greater than bulk conductivity.

(b) What environmental conditions will tend to increase the surface conductivity of a dielectric? Why?

26.14 Cubic barium titanate has the perovskite structure, named after the prototype $CaTiO_3$.

(a) How many oxygen anions coordinate a barium cation? How many oxygen anions coordinate a titanium cation?

(b) Are your answers to part (a) consistent with the expected anion coordinations based on radius ratios? Justify your answer.

26.15 Using the ion positions for the tetragonal $BaTiO_3$ unit cell (Fig. 26.11), calculate the electric dipole moment per unit cell and the saturation polarization for $BaTiO_3$. Compare your answer with the observed saturation polarization (i.e., $P_s = 0.26$ $C \cdot m/m^3$).

26.16 Discuss the nature of the boundary between two ferroelectric domains within a single grain of $BaTiO_3$.

26.17 As noted in the caption to Fig. 26.10, favorably oriented ferroelectric domains do not become significantly wider as the applied field is increased. Rather, more and more needle-shaped domains, which are favorably oriented, nucleate at grain boundaries and grow longer into less favorably orientated domains.

(a) Why should D vs. E (or P vs. E) exhibit an inherent hysteresis in the case of a ferroelectric material?

(b) Justify why a long, thin domain, whose tip is within a less favorably oriented domain, can grow longer as the electric field is increased.

(c) With sketches, show schematically how the domain structure within a single grain changes as an applied field is increased from zero to saturation.

26.18 (a) When a single crystal of $BaTiO_3$ is polarized by application of an electric field along various crystallographic directions, are there easy and hard directions as with ferromagnetic materials (see Sect. 25.5)? Explain.

(b) Will the saturation polarization be the same for $BaTiO_3$ single crystals having different crystallographic directions oriented parallel to the applied field? Explain.

26.19 Compare and contrast the polarization process for an initially unpolarized ferroelectric material and the magnetization process for an initially unmagnetized ferromagnetic material in terms of internal changes.

27

Optical Properties

Aspects pertaining to the interaction between electromagnetic radiation and matter are discussed in this chapter.

27.1 ELECTROMAGNETIC RADIATION AND ITS INTERACTION WITH ATOMS

The electromagnetic spectrum, shown in Fig. 27.1*a*, encompasses electromagnetic waves whose wavelengths range from beyond 10^5 m at the low frequency end of the radio region to less than 10^{-13} m at the high frequency end of the gamma ray region. The visible portion of the spectrum covers only a small part of the total wavelength range (Fig. 27.1*b*) and is bounded at shorter wavelengths by the ultraviolet region and at longer wavelengths by the infrared region. Electromagnetic radiation of a given wavelength may be considered to be composed of a group of photons, each of which has energy $E = hc/\lambda = h\nu$ (cf. Eq. 2.3). Using this viewpoint, we can identify the intensity, I, of the radiation with the number of photons impinging on a unit area per unit time. Alternatively, intensity may be expressed as the energy incident on a unit area per unit time, which is the number intensity I multiplied by the energy per photon. Considering electromagnetic radiation in this sense involves the implication of a particulate (i.e., quantized) nature that is distinct from the classical wavelike description of such radiation. As with electrons, electromagnetic radiation has a wave-particle duality. Thus, in describing physical phenomena arising from the interaction of radiation with matter, it is sometimes convenient to use a particulate approach and at other times a wavelike description is more convenient.

Electromagnetic radiation interacts with matter as a result of electronic transitions and polarization effects (see Sect. 26.2). Before considering the resulting properties of materials, it is useful to illustrate the interactions of radiation with a free atom in a dilute gaseous phase. Since electromagnetic radiation may be viewed as a wave consisting of synchronous, alternating electric and magnetic fields having the same frequency and wavelength, it should alter the charge distribution of the electron cloud around the nucleus of an atom. That is, as discussed in Sect. 26.2, electronic polarization (see Fig. 26.2), which depends on the electronic polarizability of the atom, will occur. The total effect, of course, depends on the intensity of the radiation and the number of atoms per unit volume. Electronic polarization is essentially independent of temperature and frequency for frequencies below the ultraviolet frequency range. As a result of the polarization interaction, the velocity of photons (or of the radiation) is lower in the gas than in a vacuum where it is a universal constant given by $c = 1/\sqrt{\mu_0 \varepsilon_0}$. [We see that the product of the

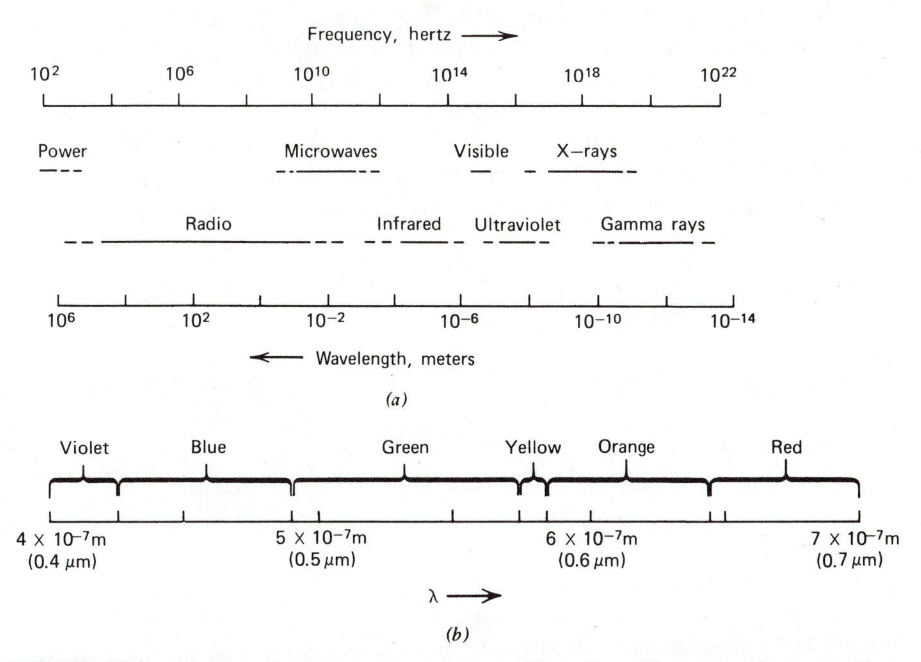

FIGURE 27.1 (*a*) The electromagnetic spectrum. More than 20 orders of magnitude in frequency and wavelength are traversed in going from the power region to the gamma ray region.
(*b*) The visible portion of the electromagnetic spectrum extends from about 0.4 μm (7.49×10^{14} Hz) to about 0.7 μm (4.28×10^{14} Hz). The approximate span of wavelengths corresponding to each primary color is indicated.

conversion factors used to relate electrical quantities (e.g., Eq. 26.5) and magnetic quantities (e.g., Eq. 25.6) has physical significance.] Within a gas, electromagnetic radiation has the velocity $v = 1/\sqrt{\mu\varepsilon}$ and, since its frequency is unaltered, the wavelength changes. The ratio of photon velocity in vacuum to that in a gas (or other substance) defines the index of refraction of the gas,

$$n = \frac{c}{v} = \frac{\mu\varepsilon}{\mu_0\varepsilon_0} = \sqrt{\mu_r\varepsilon_r} \tag{27.1}$$

where μ_r is the relative magnetic permeability and ε_r is the dielectric constant. This relationship applies to any transparent substance (gas, liquid, or solid) in the visible region of the spectrum, where only electronic polarization occurs. Since most substances are but weakly magnetic, the relative permeability is close to unity, and the index of refraction is given by

$$n \approx \sqrt{\varepsilon_r} \tag{27.2}$$

Thus, measurement of n in the visible region can be used to determine the portion of the dielectric constant associated with electronic polarization. In the case of transparent materials, the index of refraction is of practical importance, for it determines the angular change of path when light enters a material (refraction) as well as the fraction of light reflected at the surface.

If the electromagnetic radiation impinging on atoms in a monatomic gas has exactly the right frequency, electrons will be excited from an occupied energy level to a previously unoccupied energy level (Fig. 27.2*a*), and a photon will be absorbed

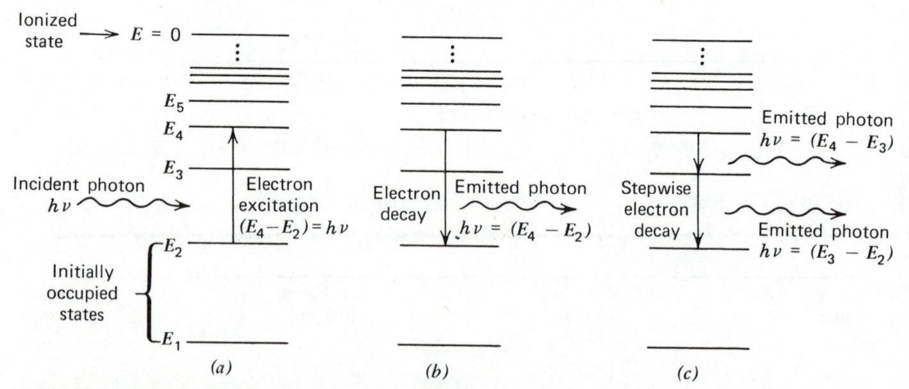

FIGURE 27.2 Schematic energy level diagram for a free atom. (*a*) An incident photon having the precise energy necessary to excite an electron from one energy level to another is absorbed. (*b*) If the excited electron jumps directly from the excited state to its initial state, a photon having the same energy as the absorbed photon is emitted. (*c*) If the excited electron jumps to an excited state of lower energy before returning to its initial state, more than one photon will be emitted, but each photon will have a smaller energy than the absorbed photon.

for each atom excited. This is a quite discrete process because a definite energy difference exists between allowed electronic energy levels in an atom. Consequently, only photons having energy $h\nu = \Delta E$ can be absorbed. Subsequent to excitation, an atom returns to its ground state (i.e., the electron in the high-energy level jumps to a lower energy level), and radiation is emitted. Depending on the exact decay path followed by the excited electron, a photon having energy equal to that of the incident photon may be emitted (Fig. 27.2*b*) or a number of photons having lesser energies may be emitted (Fig. 27.2*c*).

Effects similar to those discussed above occur in solids, and the optical properties of materials depend on the interaction of electromagnetic radiation with electrons, atoms, and defects both on or near the surface as well as within the bulk of the material.

27.2 ELECTRON EMISSION

When a free atom is bombarded with photons having $h\nu$ equal to or greater than the ionization energy (Fig. 27.2*a*), the outermost electron will be ejected from the atom (i.e., ionization occurs) as a photon is absorbed. With increasing photon frequency above the critical or minimum value for ionization, the kinetic energy (i.e., energy of the absorbed photon minus the ionization energy) of the ejected electron is increased. This phenomenon also can occur in a molecule or a solid. As shown in Fig. 27.3 (cf. Fig. 24.8), all electrons in the conduction band of a metal have a potential energy of $-e\Phi$ with respect to the zero of energy (which corresponds to a free electron at rest just outside the metal), and their kinetic energies range from zero at the bottom of the conduction band to E_F at the Fermi energy. In other words, neglecting thermal excitations, electrons at the Fermi level have the highest total energy (i.e., $E_F - e\Phi$), and this equals $-e\phi$ relative to the chosen zero of energy (Fig. 27.3). The quantity $e\phi$, which can be considered as the binding energy of an electron at the Fermi level, is a materials property called the work function and usually is expressed in electron volts.

Incident light quanta can be absorbed by valence electrons in a metal which, consequently, are promoted to higher energy states. If the light quanta have energies greater than or equal to the work function, electrons will escape or be emitted from the metal (the photoelectric effect). At any finite temperature, light quanta having lower energies can give rise to a small number of emitted electrons because there are always a few electrons that are thermally excited to energies considerably in excess of E_F. However, the number of electrons emitted only becomes appreciable when $h\nu \geq e\phi$ since only under these circumstances is there a large population of electrons that can be excited. In this situation, the liberated electrons can result in a measurable photocurrent if the sample emitting the electrons is placed in an appropriate electrical circuit. The photocurrent varies linearly with the number of

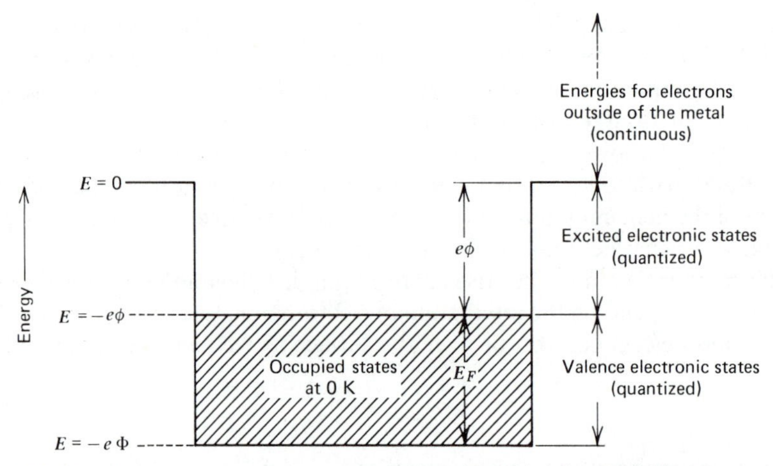

FIGURE 27.3 Energies of electrons within and outside of a metal. The energy of a free electron at rest outside of a metal (i.e., $E = 0$) is higher than the Fermi energy (E_F) by an amount equal to the work function, $e\phi$. This corresponds to the minimum photon energy required to eject a valence electron from the metal at 0 K; larger photon energies can cause emission of electrons with finite kinetic energies (i.e., $E > 0$) or ejection of electrons from lower-energy states within the valence band.

incident light quanta, provided that the applied voltage acts to move the electrons away from the sample. A sufficiently large retarding voltage, on the other hand, will reduce the photocurrent to zero.

The photoelectric yield or efficiency is defined as the number of emitted electrons per incident photon. This depends on the chemical and physical characteristics of the emitter and its surface as well as the energy of the incident radiation. Good electrical conductors ordinarily have low photoelectric yields, which is due in part to their high optical reflectivities (i.e., a very large fraction of incident photons are reflected at the metal surface). Furthermore, while some electrons are excited by incident photons to depths of about 0.1 μm only those electrons within about 1 nm of the surface are able to escape because those excited at larger depths quickly lose their excess energy by scattering processes. Photoelectric yield varies with the incident photon wavelength. Most commercially important photoemitters are composites involving a thin layer of oxide or intermetallic compound on a base metal. These materials frequently are used in devices for monitoring and measuring light.

Electronic transitions induced by radiation are important in many semiconducting devices. For an intrinsic semiconductor, absorption of photons with energy $h\nu \geq E_g$ causes electrons to be excited across the energy gap from the valence band to the conduction band. If the yield of carriers per photon is high enough and the excess carriers exist long enough, the semiconductor will exhibit photoconductivity. Virtually all electrical insulators can be made photoconductive by absorption of photons having suitable energies, but the usefulness of such materials depends on the lifetime of the excess carriers as well as the frequency of radiation absorbed.

In addition to the photoelectric effect, electron emission can result from thermal activation of conduction electrons (thermionic emission). That is, at any finite temperature some electrons are thermally excited to energies greater than $e\phi$ above the Fermi level, and the number of these electrons increases essentially in an exponential manner with reciprocal absolute temperature. In the limit of a high applied voltage, thc cmission current density is given by

$$J = AT^2 e^{-e\phi/kT} \tag{27.3}$$

where A is a materials constant dependent on surface conditions. The most commonly used metallic thermal emitter is tungsten because it can be employed at high temperatures as a result of its high melting point. The work function of pure W (~4.5 eV), however, is too high to produce large emission current densities. Consequently, a tungsten alloy containing 0.5 to 2.0% thoria (ThO_2) and having a work function of only about 2.7 eV is used instead. The low work function of this alloy results from changes in the nature of its surface. This can be achieved by heating the alloy to 2500 or 2800 K for a short time in vacuum, where some of the thoria is reduced to metallic thorium. The metallic Th atoms diffuse to the tungsten surface and result in the significantly lower work function. Obviously, the work function is sensitive to the surface condition, and a vacuum environment is required, in part, so that adsorbed contaminants do not change the characteristics. (Any residual ThO_2 particles and the Th atoms, which are virtually insoluble in tungsten and therefore segregate to grain boundaries, also serve to inhibit excessive grain growth that could lead to brittleness or creep.)

Electrodes coated with thin layers of a BaO–SrO solid solution are used widely as thermal emitters for vacuum tube applications, including cathode ray tubes and television tubes, because they provide a high emission current density at temperatures as low as 1000 K. They are prepared by coating a nickel alloy with $BaCO_3$–$SrCO_3$ paste which, on subsequent heating in vacuum, decomposes to the BaO–SrO layer and CO_2 gas. The resulting work function of the surface coating is only 1.0 eV.

27.3 REFLECTION, ABSORPTION, AND TRANSMISSION

Optical properties of metals, semiconductors, and insulators for infrared, visible, and ultraviolet radiation are of special interest because the corresponding frequency ranges encompass the frequencies necessary for various electronic transitions in solids. Since we perceive many things visually, the visual range is

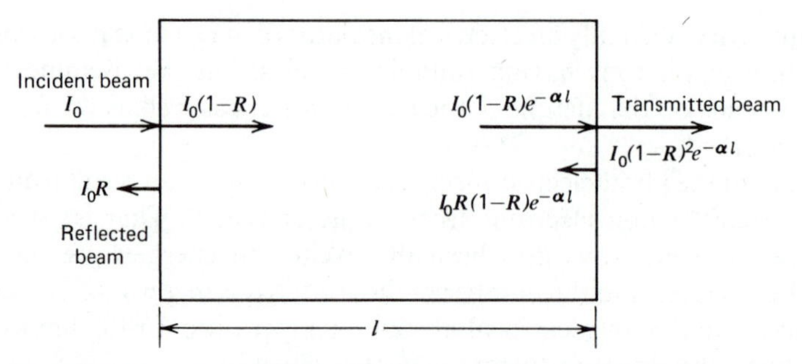

FIGURE 27.4 Reflection, absorption, and transmission by a solid or liquid. A portion ($I_0 R$) of the incident beam intensity is reflected at the front surface (i.e., on the left), and the remainder [$I_0(1 - R)$] enters the medium. The intensity that reaches the back surface (i.e., on the right) is $I_0(1 - R)e^{-\alpha l}$, where α is the absorption coefficient and l is the distance between surfaces. Of this intensity, a fraction is reflected back into the medium at the rear surface, and the remainder is transmitted out of the medium. The parameters α and R depend on the material and are sensitive to frequency or wavelength.

particularly important. When light impinges perpendicular to a solid (or liquid) surface, a certain fraction, R, of the incident intensity is reflected (Fig. 27.4). Metals and alloys are characterized by a high value of R, which approaches unity in some cases, whereas R typically is only on the order of 0.05 for an inorganic glass. If I_0 is the intensity of incident light, the intensity of light that enters the material will be $(1 - R)I_0$. Depending on the material, some or all of this light will be absorbed through electronic interactions on passing through the solid, and this results in a continuously decreasing intensity as light traverses the solid. The fractional change in light intensity dI/I over a distance dx is directly proportional to the linear absorption coefficient α of the material:

$$\frac{dI}{I} = -\alpha\, dx \tag{27.4}$$

Integration of Eq. 27.4 yields

$$I = I_0(1 - R)e^{-\alpha x} \tag{27.5}$$

so that the intensity of light reaching the back surface of a material having thickness l is $I_0(1 - R)e^{-\alpha l}$ (Fig. 27.4). In turn, some of this light is reflected internally at the back surface, and the remainder passes out of the material. Taking these features into account, we find that the fractional amount of transmitted light (i.e., the transmitted intensity divided by the incident intensity) is

$$T = (1 - R)^2 e^{-\alpha l} \tag{27.6}$$

if the same medium is on both sides of the material (Fig. 27.4).

The sum of the intensities of transmitted, reflected, and absorbed light must, of course, equal the intensity of incident light,

$$I_0 = I_T + I_R + I_A \tag{27.7}$$

or, equivalently, the sum of the transmissivity T, reflectivity R, and absorptivity A must equal unity:

$$T + R + A = 1 \tag{27.8}$$

The interrelationships between the various terms in Eq. 27.8 are determined largely by α and R, which are basic materials parameters. For a given material, both α and R depend on the frequency of the incident light as a result of the fundamental interaction of light with matter.

27.4 OPTICAL PROPERTIES OF METALS

The opaqueness and high reflectivity of metals indicates that α and R are large. This arises because many empty electronic states exist just above occupied levels in the conduction band of a metal. In other words, incident radiation, over a wide range of frequencies, will promote or excite electrons to unoccupied states of higher energy and thus will be absorbed. Consequently, light is absorbed within a very short distance of a metal surface, and only very thin films of a metal ($\leq 0.1\ \mu m$) exhibit any transmittance at all. Once the electrons have been excited, they decay back to lower energy levels, and reemission of light from the metal surface occurs. The combination of absorption and reemission in metals gives rise to reflectance. The efficiency of this process depends on the frequency of the incident light, and the color exhibited by a metal exposed to white light arises from the frequency dependence of the reflectivity. The metal silver, for example, is highly reflective over the entire visible range (Fig. 27.5) and, as a consequence, its color is characterized as white metallic and it has a bright luster. The white color of silver indicates that the reemitted photons cover much the same frequency range and are comparable in number to the incident light photons. Furthermore, unlike free atoms where discrete frequencies are absorbed and emitted (Sect. 27.1), photons having a virtually continuous band of frequencies are absorbed and emitted by a metal because of the almost continuous variation of energy levels in the conduction

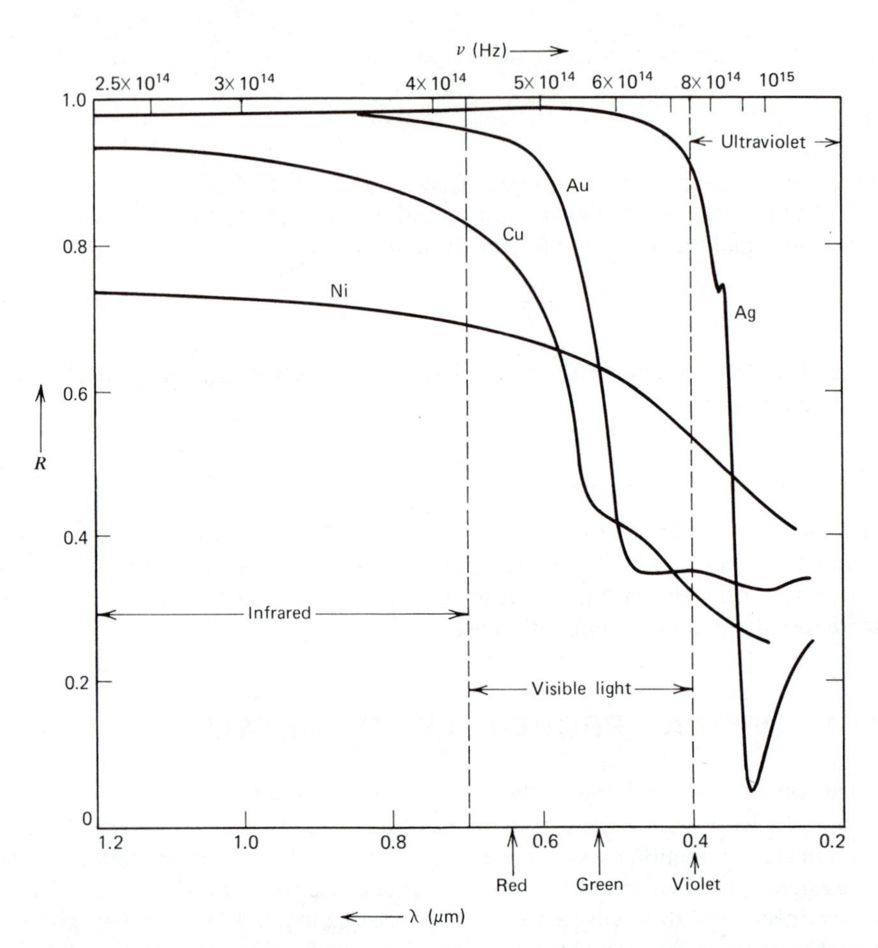

FIGURE 27.5 The wavelength and frequency dependence of reflectivity for copper, silver, gold, and nickel. The color exhibited by a metal depends on the wavelength (or frequency) sensitivity of reflectivity. The white metallic color of Ag arises because of a high reflectivity throughout the visible region. Cu and Au preferentially absorb shorter wavelengths in the visible region, with a correspondingly lower reflectivity for these wavelengths, and they display red-orange and yellow colors, respectively. The grayish color of Ni results from a relatively low reflectivity over the entire visible region.

band. Of course, some light intensity is lost because of phonon generation resulting from collisions suffered by the excited electrons in a metal and because of electronic polarization.

Metals like copper and gold exhibit characteristic red-orange and yellow colors, respectively, because incident photons above a certain frequency cause excitation of electrons from the filled *d* band to empty levels in the *s* band. Since this frequency falls in the visible region for Cu and Au (Fig. 27.5), strong absorption of light with shorter wavelengths occurs, and because the probability that an electron, which has been excited from the *d* band to the *s* band, will jump back to the *d* band is small in comparison to other decay paths, only light of longer wavelengths is strongly reemitted. In the case of Ag, the frequency for interband excitation lies in the ultraviolet region (Fig. 27.5). Many metals, like nickel, iron, and tungsten, have grayish colors because the density of electronic states both for occupied and unoccupied levels exhibits maxima and minima. This, in combination with the fact that the density of electronic states is very large in different parts of the *d* band, leads to relatively strong absorption and, consequently, a relatively low reflectivity throughout the visible region (Fig. 27.5). Most metals become highly reflective for photons having frequencies in the infrared region, whereas many nonmetals are transparent to such radiation.

27.5 OPTICAL PROPERTIES OF NONMETALS

Nonmetals may be opaque or transparent depending on their band structure, and they may be clear or colored as a result of selective absorption, selective reflection, or selective scattering. Most nonmetals absorb infrared radiation to some extent because the photon frequencies are comparable to atomic vibration frequencies, and interactions between photons and electric dipoles induced by atomic vibrations lead to photon absorption through phonon generation. In addition to this, ionic solids exhibit particularly strong absorption of infrared radiation having frequencies that cause ionic polarization (see Sect. 26.2) and, hence, phonon generation. The absorption peak results in a minimum in the transmission curve (Fig. 27.6).

Many semiconductors absorb visible radiation and are opaque because light frequencies are sufficiently large so as to cause excitation of electrons across the energy gap from the valence band to the conduction band. When the energy gap corresponds to a photon frequency in the infrared region (e.g., Si and Ge), all visible light is absorbed and the semiconductor usually is dark gray in color and has a dull metallic luster. Distinctive colors often occur when the energy gap corresponds to a photon frequency in the visible region. Hence, CdS ($E_g = 2.42$ eV) absorbs the blue and violet portions of white light and is yellow-orange in color. Radiation with frequencies less than a critical value determined by the energy gap (and by impurity electronic states) is transmitted.

Electrical insulators tend to be transparent to visible radiation because they possess such large energy gaps that light cannot cause electron excitations. Thus, diverse materials such as diamond, sodium chloride, ice, inorganic glasses, and polymethyl methacrylate readily transmit light. If the incident radiation has a sufficiently high frequency, typically in the ultraviolet range, to excite electrons

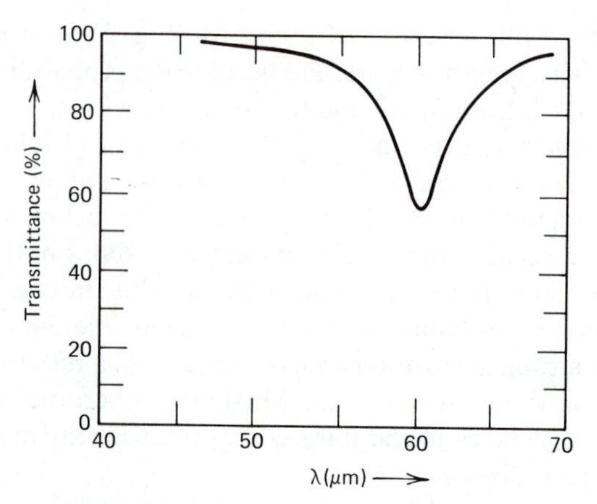

FIGURE 27.6 Transmission of infrared radiation through a thin film (0.17 μm), of sodium chloride, The minimum transmittance and maximum absorption occurs at $\lambda = 61.1$ μm (or $\nu = 4.90 \times 10^{12}$ Hz). Radiation of this frequency interacts strongly with the ions, causing ionic polarization. That is, photon energy is dissipated by induced vibrations of the ions (i.e., phonon generation). [After R. B. Barnes, *Z. Physik,* **75**, 723 (1932). Copyright by Springer-Verlag, Heidelberg, Berlin, New York.]

across the energy gap, strong absorption occurs. For optically transparent materials in a vacuum or in air, which has an index of refraction close to that of vacuum (i.e., $n \approx 1$ for air), the reflectivity of visible radiation incident normal to the surface of the material is given by

$$R = \left(\frac{n-1}{n+1}\right)^2 \tag{27.9}$$

where n is the index of refraction of the material. A perusal of Table 27.1, which gives some index of refraction values, should indicate that R is relatively small for transparent materials. Indeed, such materials are characterized by small values of absorption coefficient as well as small values of reflectivity.

Many inherently transparent materials often appear translucent (i.e., milky white in color) because they transmit light diffusely. Diffuse transmission results from multiple internal reflections. In noncubic polycrystalline materials that have an anisotropic index of refraction, light is reflected or scattered at grain boundaries where the index of refraction changes discontinuously. This effect is essentially the same as the effect that results in reflection at a free surface, and for normal incidence the fraction of light reflected is given by $(n_1 - n_2)^2/(n_1 + n_2)^2$, where n_1

and n_2 are the indices of refraction corresponding to the crystallographic directions in neighboring grains perpendicular to the grain boundary. Scattering of light by internal reflections can be particularly important in multiphase materials because of changes in index of refraction across interphase boundaries. The opacification of inorganic glass by formation of a dispersion of fine TiO_2 particles serves as an example. In addition to the difference in index of refraction for glass and TiO_2, the TiO_2 particles have a size comparable to the wavelength of light, giving a strong scattering interaction. For opacification, light should be diffusely reflected before reaching the opposite surface, rather than diffusely scattered but transmitted as in translucency.

Many pure covalent and ionic substances that are inherently transparent commonly appear to be opaque because of residual porosity resulting from the processing. This dominates in the case of various pure ceramic materials which have been sintered, since only about 1% porosity is needed to produce opaqueness, Conventionally sintered alumina, for example, is opaque, whereas porefree polycrystalline alumina is translucent and even exhibits transparency if not too thick. Indeed, techniques for producing fully dense polycrystalline oxides, which are translucent or transparent, have been developed recently. Such materials maintain their mechanical integrity to much higher temperatures than inorganic glasses and, therefore, are useful for applications such as housings for very high intensity lamps.

TABLE 27.1 Index of Refraction for Various Materials at Room Temperature for Visible Light

Substance	*n (Index of Refraction)*
Air	1.000277
Al_2O_3	1.63–1.68
CaF_2	1.43
Cl_2 (gas)	1.000768
Cl_2 (liquid)	1.385
Diamond	2.417
H_2O (water)	1.33
H_2O (ice)	1.30
Glass (heavy flint)	1.65
Glass (zinc crown)	1.52
KCl	1.49
KF	1.36
NaCl	1.54
Quartz	1.54
Fused silica (silica glass)	1.47
SrO	1.87

27.6 ELECTRONIC DEFECTS AND COLOR

Color variations in ionic and covalent materials can be effected by introducing structural defects that produce energy levels between the valence and conduction bands. This results in a frequency variation of the absorption coefficient such that radiation having certain frequencies is strongly absorbed. Very often, re-emission of these same frequencies does not occur because certain nonradiative electron transition processes are possible when electrons jump back to lower energy levels; furthermore, excitation of electrons by photons of one frequency can result in emission of photons having another frequency.

Inorganic glass, NaCl, alumina, and diamond are examples of transparent materials that can be colored through alteration of their compositions. For inorganic glass, this usually is accomplished by the introduction of transition or rare earth element ions in the molten state. Similarly, the presence of a small amount of Ti^{3+} substituted for Al^{3+} in Al_2O_3 is responsible for the slight bluish color of sapphire, and a small amount of Cr^{3+} substituted for Al^{3+} in Al_2O_3 leads to the characteristic red color of ruby. The fraction of light transmitted through ruby and through clear sapphire is shown as a function of light frequency in Fig. 27.7. Strong absorption of blue-violet light and somewhat weaker absorption of yellow-green light by ruby give rise to its deep red color. The Cr^{3+} ions in ruby are point defects with associated energy states that lie between the valence

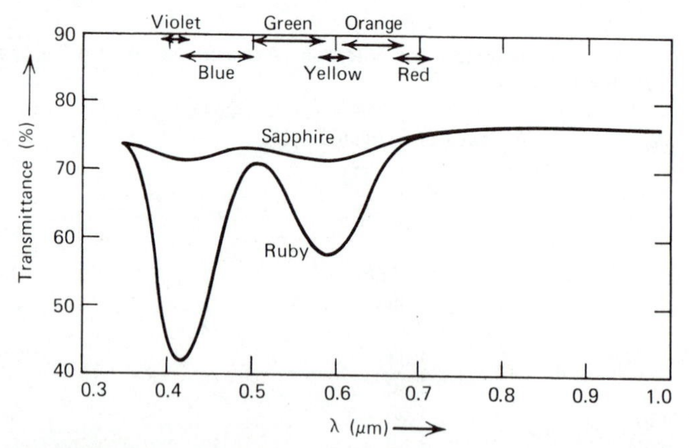

FIGURE 27.7 Transmission of radiation by sapphire (Al_2O_3 containing a small amount of Ti^{3+}) and by ruby (Al_2O_3 containing Cr^{3+}). Whereas sapphire exhibits almost uniform transmission in the visible region and is essentially colorless, ruby has strong absorption for certain wavelengths and is red in color. (After A. Javan, *Scientific American*, Vol. 217, No. 3, p. 238, 1967. Copyright 1967 by Scientific American, Inc. All rights reserved.)

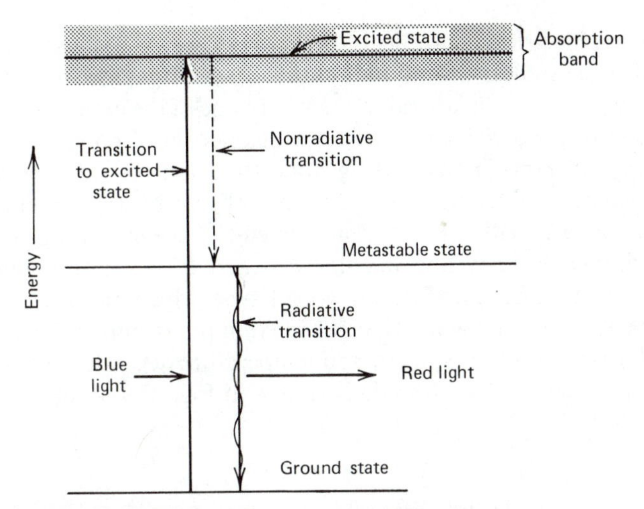

FIGURE 27.8 Schematic energy level diagram for a chromium ion in ruby. The Cr^{3+} ion absorbs a quantum of blue light, causing an electron to be excited to a higher energy level. The excited electron then makes a nonradiative transition to the metastable state from which it can decay spontaneously with the emission of a quantum of red light. (Adapted from A. Javan, *Scientific American*, Vol. 217, No. 3, p. 238, 1967. Copyright 1967 by Scientific American, Inc. All rights reserved.)

and conduction bands of Al_2O_3. These energy levels are highly localized spatially, and, as shown in Fig. 27.8, the separation of levels is such that blue-violet light can be absorbed and red light emitted. A characteristic of such an absorption-emission process, because of the law of conservation of energy, is that the emitted radiation cannot have a frequency higher than the frequency of the absorbed radiation. As indicated in Fig. 27.8, absorption of a blue-violet photon corresponds to the excitation of an electron from the ground state to the excited state, and emission corresponds to the return of the electron to the ground state from an intermediate metastable state. The step from the excited state to the metastable state is nonradiative in that the corresponding energy is not reemitted in the form of a photon but, rather, is absorbed by a phonon. If an impurity ion only interacts weakly with the host atoms or ions, the new set of energy levels will be charateristic of the impurity itself. In cases where the interaction is strong, the situation will be more complex, and prediction of color becomes difficult.

When sodium chloride is heated in sodium vapor and then cooled down, it acquires a yellowish hue. This comes about because sodium atoms are deposited on the NaCl surface, and these ionize quickly and attract chloride anions. That is, at the surface, anion vacancies are produced that can diffuse into the crystal. The resulting sodium chloride is nonstoichiometric with an excess of Na^+ ions and of anion vacancies. Electrical neutrality is preserved, however, because of the electrons liberated when the sodium atoms are ionized. The charge distribution around an anion vacancy makes this a favorable site where one of the excess electrons can be trapped. These trapped electrons possess a distinct set of possible energy levels, and the energy difference between the ground state and excited state corresponds to light frequencies in the blue-green portion of the visible region. Thus, ordinary white light is preferentially absorbed at this end of the spectrum, rendering the Na^+-rich sodium chloride yellow in color. A negative ion site and bound electron is called an *F* center after the German word Farbe, which means color. Behavior similar to the above occurs when KBr is heated in potassium vapor, although the resulting color is blue. The frequency and temperature dependence of absorptivity for K^+-saturated potassium bromide is shown in Fig. 27.9. A situation opposite

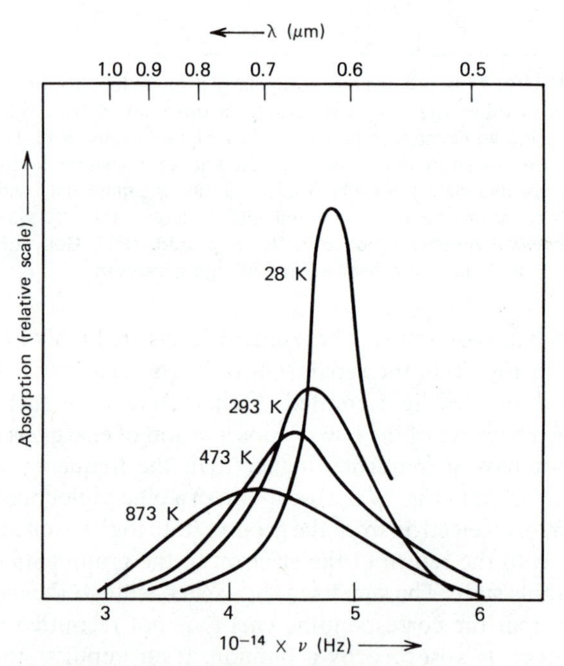

FIGURE 27.9 Absorption of optical and infrared radiation by KBr saturated with potassium as a function of frequency and temperature. Rising temperature decreases the binding of trapped electrons to anion vacancies as a result of ionic thermal vibrations, and the energy levels of the trapped electron are altered as well. These effects lead to a broader absorption band and a shift of maximum absorption toward lower energy photons. [After E. Mollwo, *Z. Physik,* **85**, 56 (1933). Copyright by Springer Verlag, Heidelberg, Berlin, New York.]

to that discussed above arises when NaCl is heated in chlorine gas. The resulting cation vacancies, which can act as traps for holes, also color the salt. This type of defect is called a *V* center.

27.7 LUMINESCENCE

Absorption-emission phenomena occur frequently in a wide variety of materials and are responsible for useful optical characteristics. Zinc sulfide containing small amounts of excess zinc or impurities such as silver, gold, or manganese, for example, emits in the visible region after excitation by higher energy ultraviolet radiation (Fig. 27.10). When absorption of high energy radiation results in emission of visible radiation, the phenomenon is called luminescence. Luminescence itself is subdivided in accordance with the duration of time between absorption and emission. If this delay time is less than about 10^{-8} sec, the phenomenon is called fluorescence; if the delay time is greater, the phenomenon is called phosphorescence. While luminescent materials are used for a variety of purposes, the most familiar example is the fluorescent lamp. The inside of the glass housing of such lamps is coated with special silicates or tungstates that, when excited by ultraviolet light from the mercury glow discharge, emit visible white light.

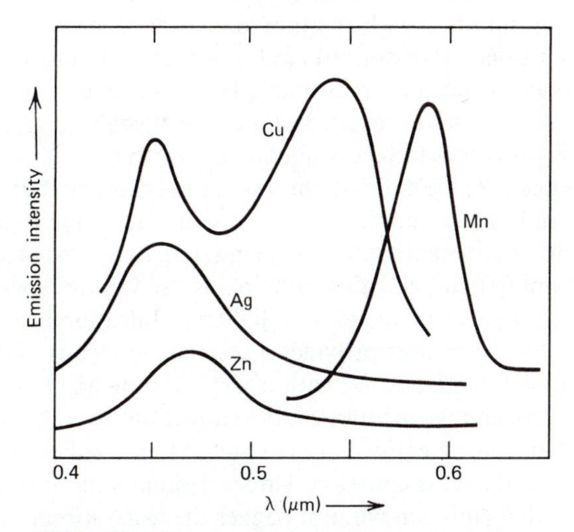

FIGURE 27.10 Luminescence spectra for ZnS activated with Cu, Ag, Mn, and excess Zn. The excitation radiation lies in the ultraviolet region, and the emitted radiation is in the visible region. (Adapted from *An Introduction to Luminescence in Solids,* by H. W. Leverenz, Wiley, New York, 1950.)

Phosphors are useful for detecting X or γ radiation because they emit visible light. Typical phosphors include organic materials such as anthracene as well as inorganic sulfides, silicates, oxides, and tungstates. Zinc sulfide is used routinely as a phosphor for X-ray detection; when inserted into an X-ray beam, it glows with a yellowish light. Such behavior, in combination with a photoconductive substance, can be utilized to quantitatively determine the intensity of the X-ray beam. Light emitted from the phosphor strikes a photoconductive material, and suitable electronics are used to amplify the resulting current through the photoconductor. The final electronic signal will be directly proportional to the intensity of the X-ray beam. Phosphors are also used for more mundane purposes such as luminous numerals and hands on a clock.

27.8 LASERS

The radiative transitions discussed up to now have all been spontaneous; that is, the transition from the excited state to the ground state occurs naturally without any external assistance. In certain cases, by contrast, a radiative transition requires an external stimulus. Stimulated emission occurs in the following way. When a photon of energy $h\nu_s$ interacts with an atom in an excited state at energy $h\nu_s$ above the ground state, it causes the atom to emit a photon that also has energy $h\nu_s$. If the emission does not occur in a reasonable length of time in the absence of the stimulating photon (i.e., if the excited state is metastable, having a relatively long half-life), this process is termed stimulated emission.

In stimulated emission, the emitted photon has the same spatial direction as the incident photon and, furthermore, is in phase with the incident photon, and the photons are said to be coherent. Stimulated emission, under the proper conditions, can be used to amplify light, and this principle is used in the operation of lasers. The term laser is an acronym derived from light amplification by stimulated emission of radiation. The ruby laser provides a convenient means of describing laser action. The energy levels associated with a Cr^{3+} ion in Al_2O_3 have been noted previously (Fig. 27.8), and, as mentioned, the transition from the metastable state to the ground state results in emission of red light. Stimulated emission in ruby can be initiated by a spontaneous emission. Under ordinary circumstances, however, this type of stimulated emission will not trigger the generation of a high intensity light beam since, at any given time, only a very small fraction of the Cr^{3+} ions will have electrons in the metastable state. In order for laser action to occur, therefore, the fraction of electrons in the metastable states must be made relatively large, and this can be accomplished by external means ("optical pumping," as shown schematically in Fig. 27.11). The xenon flash lamp excites Cr^{3+} ions into the upper

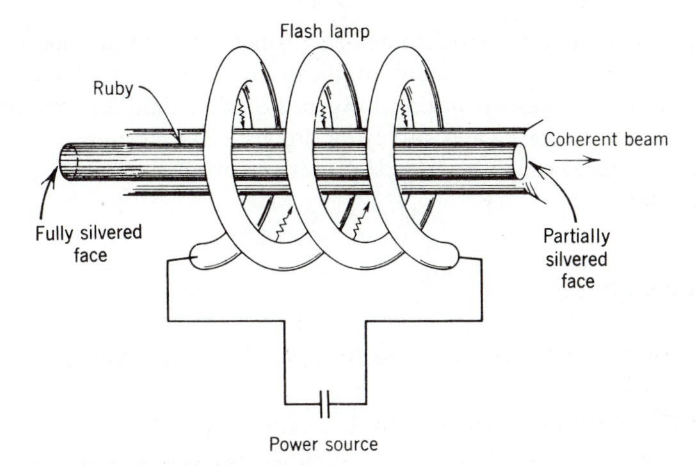

FIGURE 27.11 Schematic diagram of an operating ruby laser. The xenon flash lamp serves as an optical pump, exciting Cr^{3+} ions and insuring a population inversion. A high-intensity, coherent beam of ruby-red light is emitted through the partially silvered end of the crystal. (From *The Structure and Properties of Materials,* Vol. IV, by R. M. Rose, L. A. Shepard, and J. Wulff, Wiley, New York, 1966.)

band, from which they decay rapidly into the metastable state. The lifetime of the metastable state is sufficiently long so that a relatively large number of such states become occupied on pumping. When a spontaneous emission does occur, it triggers stimulated radiation. If this travels parallel to the long crystal axis in Fig. 27.11, it is partially reflected from the partially silvered end and completely reflected from the fully silvered end of the crystal. Thus, photons travel up and down the crystal producing an ever-increasing amount of stimulated emission so that the beam continuously increases in intensity or power. When it is finally emitted through the partially silvered end, it is a high-power coherent light beam. For a ruby laser, the power density is typically around 10^8 W/m^2.

Laser action can occur in a variety of substances including gases. Although gas lasers produce lower intensities or powers because of fewer excited atoms per unit volume, they are more suitable for continuous operation since solid state lasers generate an appreciable amount of heat. Common gas lasers make use of carbon dioxide and a mixture of helium and neon. In the latter, He and Ne are at partial pressures of 1.3×10^{-3} atm and 1.3×10^{-4} atm, respectively, and the gas mixture is contained in a high frequency discharge tube. The neon atoms produce stimulated emission that lies in the infrared region. Other gas mixtures are used to obtain shorter wavelengths. Semiconducting lasers include GaAs and InAs. In GaAs, doping with phosphorus results in laser light in the 0.65 to 0.84 μm range. InAs lasers yield light with longer wavelengths.

The high intensity and monochromatic (i.e., single wavelength) characteristics of laser light make it useful for many applications in photographic and spectroscopic research. The narrow beam divergence of laser light also makes it useful for precision measurement and surveying. Focused laser beams can be used for certain metallurgical and even surgical procedures where highly localized heating on a small scale is necessary.

REFERENCES

1. Javan, A. The Optical Properties of Materials, *Scientific American*, Vol. 217, No. 3, September, 1967, pp. 238–248.
2. Kerr, P. F. *Optical Mineralogy*, McGraw-Hill, New York, 1959.
3. Azároff, L. V. and Brophy, J. J. *Electronic Processes in Materials*, McGraw-Hill, New York, 1963.

QUESTIONS AND PROBLEMS

27.1 Compute the wavelength and frequency of electromagnetic radiation that will ionize atoms in monatomic hydrogen gas. Would radiation having a higher or lower frequency be necessary to ionize H_2 molecules? Explain briefly.

27.2 When a high-energy electron beam impinges on a solid target such as copper, a continuous spectrum of X radiation (called white radiation) is emitted because of the deceleration of the incident electrons. In addition to the white radiation, more intense X rays having discrete frequencies (called characteristic radiation) are emitted if the incident electrons are sufficiently energetic.
(a) Discuss the origin of characteristic X rays.
(b) Why, in an actual X-ray tube, would it be necessary to cool the copper target during operation?

27.3 As shown below, when light goes from one transparent medium to another having a different index of refraction, the path direction is changed.

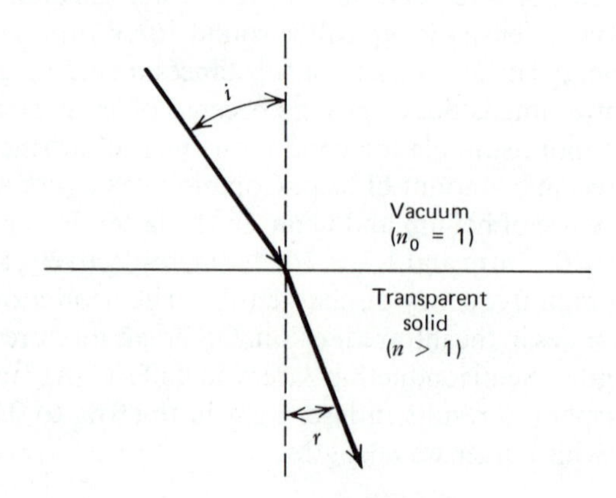

The refraction effect is given by Snell's law.

$$\frac{n}{n_0} = \frac{\sin i}{\sin r}$$

(a) Discuss how the frequency or wavelength variation of n (and, hence, ε_r at optical frequencies) allows prisms made of optically transparent materials such as glass to separate rays of the sun into a colored spectrum.

(b) Light traveling within a transparent solid having $n = 1.5$ can be totally reflected internally on reaching a free surface. What minimum angle must such a light beam make with the free surface for total internal reflection to occur?

27.4 The work functions of copper and sodium are 4.46 eV and 2.28 eV, respectively, at room temperature. What frequencies of electromagnetic radiation will cause the photoelectric effect for these metals?

27.5 The photoelectric threshold is the maximum wavelength of light that will produce photoemission. Photoelectric thresholds for some metals are Al (0.304 μm), Na (0.544 μm), Ni (0.247 μm), and W (0.274 μm). Determine the work function in electron volts for these metals.

27.6 (a) Using a simple energy band diagram, illustrate how a semiconductor can become photoconductive.

(b) Diamond has an energy gap of 5.33 eV. What type of electromagnetic radiation is necessary to make diamond photoconductive? Give the wavelength and frequency.

(c) Why is diamond transparent to optical radiation?

27.7 (a) Rough surfaces scatter light diffusely. Explain why. [*Hint*. Draw a smooth surface and a rough surface in profile and illustrate the path of reflected light for a given path of incident light.]

(b) A well-polished piece of metal is mirror-like, whereas the same metal becomes dull in appearance if etched. Explain.

27.8 Pure Al_2O_3, which is colorless, forms a complete series of solid solutions with Cr_2O_3, which is green. For solid solutions containing less than about 8 mole % Cr_2O_3, a red color is observed (this composition range includes ruby), and the molar volume remains close to the value for Al_2O_3 (25.58 cm^3/mol). At higher Cr_2O_3 contents, the solid solution becomes progressively greener, and the molar volume increases toward the value for Cr_2O_3 (29.09 cm^3/mol).

In the case of ruby, it has been observed experimentally that heating causes it to become green in color, and subsequent cooling restores the red color.

Suggest some qualitative reasons for the behavior cited above.

27.9 (a) Semicrystalline long-chain polymers usually are white and exhibit translucency if not too thick, whereas totally noncrystalline long-chain polymers are transparent if pure. Discuss the reasons for this difference in optical behavior.

(b) When treated thermally or mechanically such that spherulites of very small size result, a semicrystalline polymer can become transparent. Why?

27.10 Certain optically transparent materials can be colored by bombardment with γ rays. Why does this happen? If left unattended for a certain period of time, the color fades. Why?

ADDITIONAL CREDITS

Figure 7.10: Adapted from *The Structure and Properties of Materials,* Vol. I, by W. G. Moffatt, G. W. Pearsall, and J. Wulff, Wiley, New York, 1964.

Figure 8.2: From *The Structure and Properties of Materials,* Vol. 1, by W. G. Moffatt, G. W. Pearsall, and J. Wulff, Wiley, New York, 1964. **Figure 8.4:** From *The Structure and Properties of Materials,* Vol. I, by W. G. Moffatt, G. W. Pearsall, and J. Wulff, Wiley, New York, 1964.

Figure 17.3: Adapted from *Atlas of Isothermal Transformation Diagrams,* United States Steel Corp., Pittsburgh, Pennsylvania, (1951).

INDEX OF SPECIFIC MATERIALS AND SUBSTANCES

INDEX OF SUBJECTS